NUMBERS, INFORMATION AND COMPLEXITY

Numbers, Information and Complexity

Edited by

Ingo Althöfer
Friedrich Schiller-Universität Jena

Ning Cai
National University of Singapore

Gunter Dueck
IBM Germany

Levon Khachatrian
Universität Bielefeld

Mark S. Pinsker
Russian Academy of Sciences

Andras Sárközy
Eötvös Lorand University

Ingo Wegener
Universität Dortmund

and

Zhen Zhang
University of Southern California, Los Angeles

KLUWER ACADEMIC PUBLISHERS
BOSTON / DORDRECHT / LONDON

A C.I.P. Catalogue record for this book is available from the Library of Congress.

ISBN 0-7923-7765-6

Published by Kluwer Academic Publishers,
P.O. Box 17, 3300 AA Dordrecht, The Netherlands.

Sold and distributed in North, Central and South America
by Kluwer Academic Publishers,
101 Philip Drive, Norwell, MA 02061, U.S.A.

In all other countries, sold and distributed
by Kluwer Academic Publishers,
P.O. Box 322, 3300 AH Dordrecht, The Netherlands.

Printed on acid-free paper

Printed in the Netherlands.

Contents

Cryptology

Information Theory and the Related Fields Data Compression, Entropy Theory, Symbolic Dynamics, Probability and Statistics

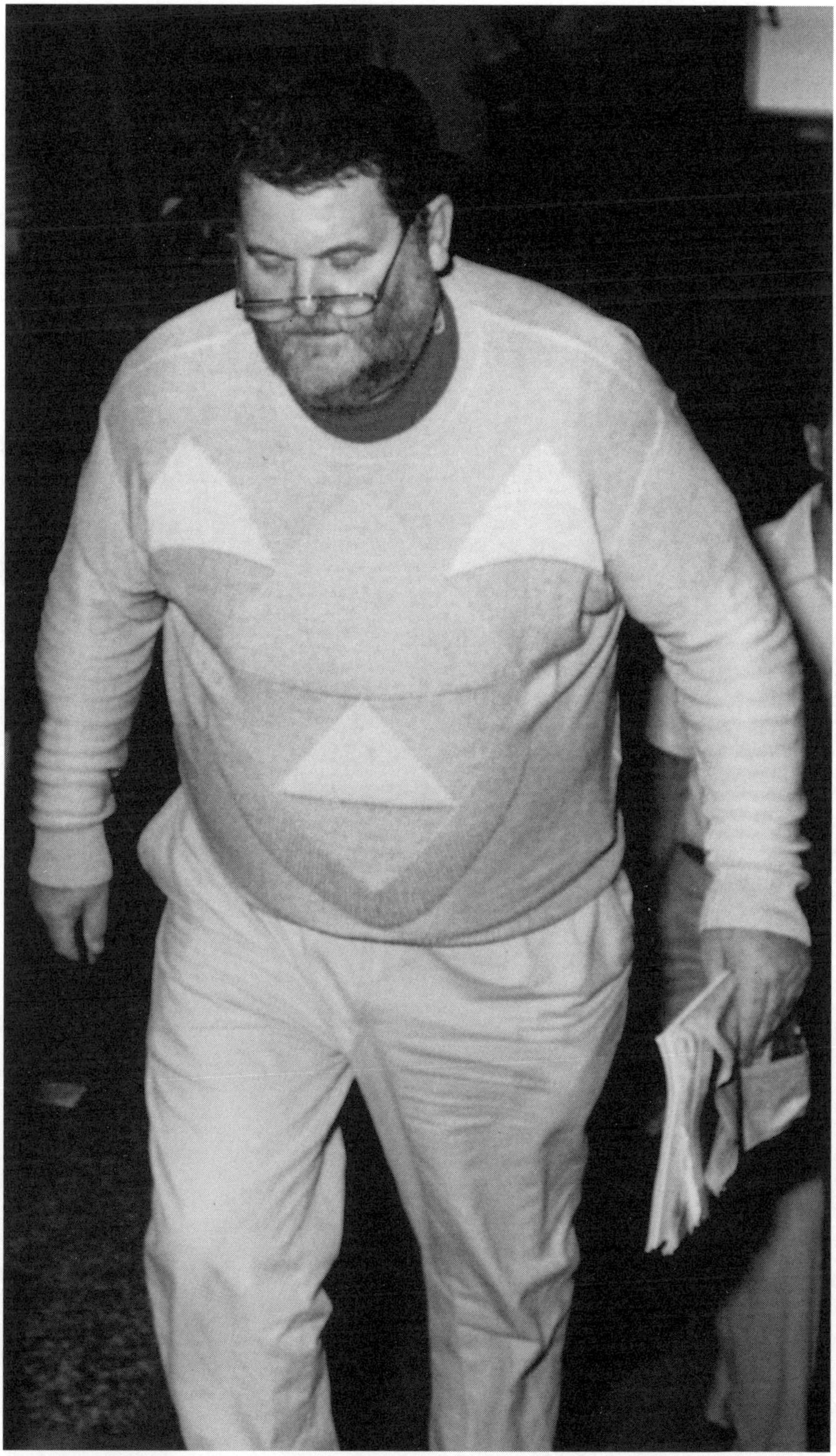

Preface

Numbers, Information and Complexity — these three words stand for research interests of the scientist whose 60–th birthday was celebrated with this volume and a symposium organized at the University of Bielefeld under the same title in October 1998.

Rudolf Ahlswede studied Mathematics, Philosophy, and Physics for one semester in Freiburg and then entirely in Göttingen. He still speaks with excitement about lectures of world–leading mathematicians at that time, Carl Ludwig Siegel and Kurt Reidemeister, and the open–minded atmosphere around his advisor Konrad Jacobs, who, coming from Ergodic Theory, started Information Theory in Germany. He was equally inspired by the theoretical physicist Friedrich Hund, a former assistant to Werner Heisenberg, the philosopher Martin Heidegger (in Freiburg), professors in Philosophy Josef König and Günter Patzig, and in Sociology Plessner and Strelewics.

Ahlswede's path to Information Theory, where he has been world–wide a leader for several decades, is probably unique, because it went without any engineering background through Philosophy: Between knowing and not knowing there are several degrees of knowledge with probability, which can even quantitatively be measured — unheard of in classical Philosophy.

This abstract approach paired with a drive and sense for basic principles enabled him to see new land where the overwhelming majority of information theorists tends to be caught by technical details. Perhaps the most striking example is his creation of the Theory of Identification.

In his doctor thesis he extended Shannon's concept of capacity to that of a capacity function for non–stationary channels. This concept says more about the transmission properties than the familiar supremum of rates capacity concept and is of actual interest in a controversial discussion.

After three years as an Assistant in Göttingen and Erlangen, in 1967 at the beginning of an adventurous life he moved to the US, where at the Ohio State University in Columbus he quickly made his way from Assistant Professor to Full Professor in 1972. Reminiscences about those days from his former PhD student Mike Ulrey can be found at the end of this volume. The time at Ohio

State was interrupted by several visiting professorships in Ithaca, N.Y., Rome, Heidelberg, Urbana and then for almost two years back in Göttingen.
Since then travelling, the discovery of nature, other countries and cultures has become another great passion. By now a great part of the world has been covered — often in risky adventures. Just in the last two years the tours led to Varanasi, San Diego, Galapagos, Peru, Laz Paz, Siberia all the way to lake Baikal, most of Japan, Singapur, Hong Kong, Seoul and South Africa.
The seven years in the US had a lasting influence: above all the constant drive for discoveries and innovations, the inspiring effect of team–work, and the flexibility of administrations. Personally, the influence of the world–renowned statistician Jacob Wolfowitz, the most frequent coauthor of the great Abraham Wald, was very important.
In less than one year of joint work (including one breakthrough for arbitrarily varying channels) Ahlswede had not only learnt Wolfowitz's approach to Information Theory and some of his experiences in mathematical research ("if a conjecture turns out to be false, go for the extreme opposite; let's see what is left after the smoke is gone; let's look at the problem in n–space good enough for my grandfather and therefore also for me") but, perhaps more importantly, he had received a lasting encouragement: "You are like Wald, everything he touched became gold in his fingers".
Probably, Ahlswede's most outstanding result back in those days was the coding theorem for the multiple–access channel — until today this is the only complete characterization of the capacity region for a multi–user channel.
It is largely responsible for the strong interest and progress in Multi–user Information Theory during the seventies. The other impetus came from Tom Cover's work on broadcast channels with the idea of "clouds" of codewords. Ahlswede considers him as the only peer in this subject — at least in craziness. Another lasting contribution was the constructive proof of the coding theorem for discrete memoryless channels with feedback, which led via list codes — independent of Slepian/Wolf and at the same time — to the celebrated idea of binning.
Methodically, it moved beyond Wolfowitz's typical sequences with $\sqrt{n}$ deviation (which he called π–sequences) to exactly typical sequences.
Then Ahlswede left Information Theory. Via the role of the problem of Zarankiewicz in Shannon's two–way channels and the zero–error capacity problem (a special case of the AV–problem) he recognized the importance of Combinatorics, which then became his second major field of research. Since Information Theory was and is not too popular among mathematicians, Ahlswede convinced his colleagues deciding on his last promotion by solving problems in P–adic Analysis (see K. Mahler, "P–adic Numbers and their Functions", sec. ed.). Again and again he solved problems in a variety of fields (he calls this sportsman activities as opposed to far reaching scientific visions).
A first swing back to Information Theory came early in 1974 with a visit of János Körner, who had become interested in multi–user theory. Also Imre Csiszár stopped by for a shorter period. At that time the Hungarian School

was well–prepared by Alfred Renyi in fundamental questions of information measures (Renyi's entropy, f–divergence of Csiszár), but was still lacking a deeper understanding of channel coding theory.

Ahlswede had in Körner, who learnt fast, one of his best students. Many ideas and contributions entered the Csiszár/Körner book "Coding Theorems for Discrete Memoryless Systems". Körner acknowledges this period in "Information Theory: New Trends and Open Problems", G. Longo edited, Springer 1977. The work on sources with side information and broadcast channels was continued together with Peter Gács. The most significant contribution of this period was the "Blowing–up Method".

Later it came to joint work with Csiszár on how to get a bit of information, common randomness in Information Theory and Cryptography, which Ahlswede ever since he heard about it from Martin Hellmann viewed as a kind of dual to Information Theory ("Bad Codes are good Ciphers"), and Hypothesis Testing under Communication Constraints, which gives a novel connection between Information Theory and Statistics. The relation to Hungarian mathematicians continued with work in Combinatorics with G. Katona "Contributions to the Geometry of Hamming Spaces" and others.

This geometrical view on combinatorial extremal problems later was very fruitful. Recently it came to work in Combinatorial Number Theory with András Sarközy, the most frequent coauthor of Paul Erdős. A visit of Te Sun Han for 6 months in Bielefeld in 1980 and of Kingo Kobayashi for two years in the 90's caused spreading of ideas and added to a flourishing school in Information Theory in Japan.

During the last decade Ahlswede had intense contacts with Leonid Bassalygo and Mark Pinsker and thus also learnt a lot about the impressive contributions in the former Soviet Union to unconventional coding problems arising for instance in Memories (Kutznetsov, Tsybakov). In a series of papers presenting several constructions, finally, the optimal rates for nonbinary codes with localized errors were recently found modulo a very small exceptional interval of error frequencies.

In 1975 Ahlswede accepted an offer to Bielefeld, which in those days had a unique profile as a research university. For several years he was devoted to building up the Applied Mathematics Division, which still carries some of his concepts: Inclusion of Theoretical Computer Science, emphasis on stochastical models, algorithmic and combinatorial methods, interdisciplinary activities in the form of Mathematizing the sciences.

About ten years later in 1989 these concepts were essential ingredients for the Sonderforschungsbereich "Diskrete Strukturen in der Mathematik", were for the first time in Germany "pure" and "applied" mathematicians worked together on a large scale on a joint program. Ahlswede has been heading the two projects "Models with Information Exchange" and "Combinatorics on Sequence Spaces".

His book “Suchprobleme” (translated into Russian and English) coauthored by his student Ingo Wegener carries the interdisciplinary flavour and was the first of its kind on this subject.
Over the years his attitude towards Mathematizing has become more critical, if not sceptical, to say the least. Exceptions were the Saturday colloquia with two foreign lecturers from different fields and Reinhard Selten's seminars on coalition games.
Complexity Theory became the main subject in Computer Science. Against all conventions Wolfgang Paul was hired as an Associate Professor at the age of twentyfive and became its prime mover.
Among an impressive group of PHD's we find Ingo Wegener, Friedhelm Meyer auf der Heide and Rüdiger Reischuk, who are now among the leaders in Theoretical Computer Science. Paul and Meyer auf der Heide participated later in two different Leibnitz prizes, the most prestigious monetary award supporting science in Germany. Ingo Wegener is internationally known for his classic on Switching Circuits. Friedhelm Meyer auf der Heide predominently contributed to parallel Computing. Paul and Reischuk made their famous step towards $P \neq NP$. Bridging the connection to Information Theory significant contributions were made to Communication Complexity by Ulrich Tamm, Ning Cai, and Zhen Zhang (see the survey by Tamm). These studies to a large extent are an outgrowth of Ahlswede's “Coloring hypergraphs: A new approach to multi–user source coding I, II”, written at the same time as Yao's pioneering work.
The deep interplay between several disciplines and a broad philosophical view is a thread through Ahlswede's work. For him Information Theory deals with *gaining* information (that is, Statistics), *transfer* of information without and with secrecy constraints (that is Cryptology), and *storing* information (Memories, Data Compression). Applying ideas from one area to another often led to unexpected and beautiful results and even to new theories. Let's give an example involving storage. Motivated by the practical problem of storing data using a new laser technique, code models for reusable memories were introduced in Information Theory. It turned out that the analysis was much more efficient, when stating the question as a combinatorial extremal problem, which led immediately to connections with hypergraph coloring, novel iso–diametrical problems in sequence spaces and finally to the new class of so called “Higher Level Extremal Problems” in Combinatorics.
Ahlswede is rarily frustrated, because the sun is always shining in some part of his universe, that is, one of his over twenty coauthors (some of them over many years) usually has good news when starting the day.
Sometimes it takes a long time for a particular news to come. There is one opening of a research field “Creating order in sequence spaces with simple machines”, coauthored by J. Ye and Z. Zhang, which to his surprise has found only little response. The general aim is to understand how much “order” can be created in a “system” under constraints on our “knowledge about the system” and on the “actions we can perform in the system”. The Maxwell demon

problem falls into this setting. There are amazing results comparing the effects of knowledge of the partial past and future. There is some resemblence of Data Compression, but with the important difference that objects are to be maintained, that is, cannot be mapped to representing symbols.
On the other hand, to keep the balance of justice in the world, the Theory of Identification, in whose development Gunter Dueck significantly participated and subsequently many others joined, again somehow surprising, immediately received worldwide recognition.
The classical transmission problem deals with the question how many possible messages can we transmit over a noisy channel? Transmission means there is an answer to the question "What is the actual message"? In the identification problem we deal with the question how many possible messages the receiver of a noisy channel can identify? Identification means there is an answer to the question "Is the actual message i?" Here i can be any member of the set of possible messages.
Allowing randomized encoding the optimal code size grows double exponentially in the blocklength and somewhat surprisingly the second order capacity equals Shannon's first order transmission capacity.
Striking phenomea are:

— in contrast to the transmission problem feedback increases the capacity for a discrete memoryless channel

— noise increases the identification capacity

— as a key parameter we encounter common randomness.

This new coding theory provides new insight into the old. There are remarkable dualities, problems in one theory often are difficult in the other and vice versa and new areas of study arose: approximation of output statistics via approximation of input distributions, new cryptographic models, and new problems of random number generation.
Since the Theory of Identification cannot be reduced to Shannon's Theory of Transmission, and conversely, Ahlswede presented in "A General Theory of Information Transfer", Preprint 97–118, SFB 343 "Diskrete Strukturen in der Mathematik", a unified model including both these theories as extremal special cases.
On the source coding side it contains a *concept of identification entropy.*
Finally as the perhaps most promising direction it suggests the study of probabilistic algorithms with identification as *concept of solution.* (For example: for any i, is there a root of a polynomial in interval i or not?)
The algorithm should be fast and have small error probabilities. Every algorithmic problem can be thus considered. This goes far beyond Information Theory. Of course, like in general information transfer also here a more general set of questions can be considered. Problems of classification by complexity arise. What rich treasures do we have in the much wider areas of information transfer?!

Lets conclude the contributions to Information Theory with a few remarks. The deepest work was done on AV–channels for several performance criteria. It resulted in methods like the very ingenious Elimination technique, an early, if not the first, case of what is now called Derandomization in Computer Science, several methods to convert coding theorems for sources into those for channels and vice versa, a Robustification technique, Wringing techniques, developed together with Gunter Dueck, leading to the solution of the problem of multiple–descriptions without excess rate within a week — after almost all experts including three Shannon Lecturers, worked in vain (the best known outer bounds for the TW channel are also based on this method), the invention of the maximal probability decoding rule and with Ning Cai the complete solution in case of noiseless feedback in this volume — adding to the Ahlswede *dichotomy*: the random code capacity equals the deterministic capacities for average errors or else the latter equals zero now a *trichotomy* based on code constructions motivated by the Theory of Identification. In a few cases the results have been generalized or completed by others, but in all cases the first breakthroughs were made by Ahlswede.
Also new channels have been introduced. The most interesting seems to be the Matching Channels, whose coding theorems have a remarkable structure involving and enhancing Combinatorial Matching Theory.
Known contributions to Combinatorics are two pearls, the Ahlswede/Daykin – inequality ("4 function theorem"), which is more general and also sharper than known correlation inequalities in Statistical Physics, Probability Theory and Combinatorics (see the survey by Fishburn and Shepp), and the Ahlswede/ Zhang–identity, which improves the LYM–inequality.
A spectacular series of results started with a lecture of Erdős, who raised in 1962 (and repeatedly spoke about) the problem "What is the maximal cardinality of a set of numbers smaller than n with $k+1$ of its members being pairwise relatively prime?"

This stimulated Ahlswede and Khachatrian to make a systematic investigation of this and related number theoretical extremal problems. Its immediate successes are solutions for several well–known conjectures of Erdős and Erdős/Graham. More importantly they gained an understanding for the role of the prime number distribution for such problems, which distinguishes them from combinatorial extremal problems. These investigations had another fruit. The AD–inequality implies a number–theoretical correlation inequality for Dirichlet

densities which implies and is sharper than the classical inequalities by Heilbronn/Rohrbach and Behrend. Number theory came first and AD is a crossroad between pure and applied mathematics.
Finally the analysis led to the discovery of a new "pushing" method with wide applicability.
In particular it led to the solution of well-known combinatorial problems like the famous 4m-conjecture (Erdős/Ko/Rado 1938, one of the oldest problems in combinatorial extremal theory) or the diametric problem in Hamming spaces (optimal anticodes).
Actually, the 4m-conjecture just concerned the first unsolved case of the following much more general problem (see the paper by Bey and Engel):
A system of sets $\mathcal{A} \subset \binom{[n]}{k}$ is called t-intersecting, if $|A_1 \cap A_2| \geq t$ for all $A_1, A_2 \in \mathcal{A}$, and $I(n, k, t)$ denotes the set of all such systems. Determine the function $M(n, k, t) = \max\limits_{\mathcal{A} \in I(n,k,t)} |\mathcal{A}|$ and the structure of maximal systems!
Ahlswede and Khachatrian gave the complete solution for every n, k, t. It has a very clear geometrical interpretation.
There is a lot of writing about methods, combinatorial versus analytical in Information Theory. Ahlswede's position has always been that all languages have their merits and should be used. During the last decade the analytical direction seemed to get the overhand. However, recently Ahlswede, in a few lines, established an Approximation Lemma in the spirit of "Coloring hypergraphs" and thus in support of the combinatorial approach.
When Ahlswede speaks about Number Theory he often goes back in his memories to the time when his grandfather taught him about numbers on the design of the blanket on his table. In the age of seven he then taught the teenagers in a one teacher school. For higher education the next city was often reached hanging at the spare tire at the back of the bus — preparing for later championships in gymnastics. He admired Baron Münchhausen from his home area, who once visited his father from St. Petersburg and when he wanted to leave again at the same day the father said "of course you have been home for at least three hours". Already as a child he was concerned to become a narrow expert on numbers and devoted more time to philosophy and literature. This explains why only in later days he felt free to devote himself to his greatest love: numbers. More recently he left them again, this time for Physics: Quantum Information (see the survey by Freivalds), which has been on his agenda for more than ten years, clearly before the large activity in this area. His acrobatic activities have been replaced by discussions with his son Sasha about literature and law.
Ahlswede's lectures were always among the top rated in the students evaluations and even in the last years, where it has become more difficult to attract students in mathematics his classes still are centers of attraction. (One must spread some life into the "dry mathematics" through humour, anecdotes and jokes!)
He was supervisor of more than 50 Diploma, 29 PhD, and 6 Habilitation theses. The works go in very different directions for example Optimization, Game Theory, Switching Circuits and in one case led through Computer Chess to

Artificial Intelligence: Ingo Althöfer is full of appraisal for this liberal attitude in his book “13 Jahre 3-Hirn – Meine Schach-Experimente mit Mensch-Maschinen- Kombinationen” He introduced several students to do computer supported mathematics. Among them is Bernhard Balkenhol who initiated a group working in data compression and able and willing to perform innovations transfer from the university to industry as for example in time–series analysis for ENEX, concerned about efficient distribution of energy.
Can you imagine Münchhausen to be a member of a singing club? Rudi Ahlswede has turned down invitations to enter organisations. He did, however, organize over a period of almost twenty years meetings in Oberwolfach.

The picture at the right shows him at one such meeting at the bat — a prelude to “Rudi at the board” by James Massey.
In spite of this individualistic life style he has won many prizes, among them are the Best Paper Award of the IEEE Information Theory Society in 1988 and, immediately afterwards, in 1990. However, more important for him than the recognition of contemporaries is his belief that his work may survive some milder storms of history.

ON PREFIX–FREE AND SUFFIX–FREE SEQUENCES OF INTEGERS

Rudolf Ahlswede and Levon H. Khachatrian

Universität Bielefeld, Fakultät für Mathematik,
Postfach 100131, D–33501 Bielefeld, Germany

{ahlswede,lk}@mathematik.uni-bielefeld.de

András Sárközy*

Eötvös University, Department of Algebra and Number Theory,
H-1088 Budapest, Muzeum krt. 6–8,Hungary

sarkozy@cs.elte.hu

INTRODUCTION

The set of the positive integers and positive square–free integers are denoted by $\mathbb{N}$ and $\mathbb{N}^*$, respectively, and we write $\mathbb{N}(n) = \mathbb{N} \cap [1, n], \mathbb{N}^*(n) = \mathbb{N}^* \cap [1, n]$, where $[1, n] = \{1, 2, \ldots, n\}$. The set of primes is denoted by $\mathcal{P}$. The smallest and greatest prime factors of the positive integer n are denoted by $p(n)$ and $P(n)$, respectively. $\omega(n)$ denotes the number of distinct prime factors of n, while $\Omega(n)$ denotes the number of prime factors of n counted with multiplicity:

$$\omega(n) = \sum_{p|n} 1, \quad \Omega(n) = \sum_{p^\alpha \| n} \alpha.$$

$\mu(n)$ denotes the Möbius function.
The counting function of a set $\mathcal{A} \subset \mathbb{N}$, denoted by $A(x)$, is defined by

$$A(x) = |\mathcal{A} \cap [1, x]|.$$

The upper density $\overline{d}(\mathcal{A})$ and the lower density $\underline{d}(\mathcal{A})$ of the infinite set $\mathcal{A} \subset \mathbb{N}$ are defined by

$$\overline{d}(\mathcal{A}) = \limsup_{x \to \infty} \frac{A(x)}{x}$$

and

$$\underline{d}(\mathcal{A}) = \liminf_{x \to \infty} \frac{A(x)}{x},$$

*Research partially supported by the Hungarian National Foundation for Scientific Research, Grant no. T017433. This paper was written while the third author was visiting the Universität Bielefeld.

I. Althöfer et al. (eds.), Numbers, Information and Complexity, 1-16.

respectively, and if $\overline{d}(\mathcal{A}) = \underline{d}(\mathcal{A})$, then the density $d(\mathcal{A})$ of $\mathcal{A}$ is defined as

$$d(\mathcal{A}) = \overline{d}(\mathcal{A}) = \underline{d}(\mathcal{A}).$$

The upper logarithmic density $\overline{\delta}(\mathcal{A})$ of the infinite set $\mathcal{A} \subset \mathbb{N}$ is defined by

$$\overline{\delta}(\mathcal{A}) = \limsup_{x \to \infty} \frac{1}{\log x} \sum_{\substack{a \in \mathcal{A} \\ a \leq x}} \frac{1}{a},$$

and the definitions of the lower logarithmic density $\underline{\delta}(\mathcal{A})$ and logarithmic density $\delta(\mathcal{A})$ are similar.
For $\mathcal{A} \subset \mathbb{N}$, $s > 1$ write

$$f_{\mathcal{A}}(s) = \sum_{a \in \mathcal{A}} a^{-s}.$$

Then the lower and upper Dirichlet densities of $\mathcal{A}$ are defined by

$$\underline{D}(\mathcal{A}) = \liminf_{s \to 1+} (s-1) f_{\mathcal{A}}(s)$$

and

$$\overline{D}(\mathcal{A}) = \limsup_{s \to 1+} (s-1) f_{\mathcal{A}}(s),$$

respectively. If $\overline{D}(\mathcal{A}) = \underline{D}(\mathcal{A})$, then the Dirichlet density $D(\mathcal{A})$ of $\mathcal{A}$ is defined as

$$D(\mathcal{A}) = \overline{D}(\mathcal{A}) = \underline{D}(\mathcal{A}).$$

It is known that for every $\mathcal{A} \subset \mathbb{N}$ we have

$$\overline{\delta}(\mathcal{A}) = \overline{D}(\mathcal{A}), \underline{\delta}(\mathcal{A}) = \underline{D}(\mathcal{A})$$

and

$$0 \leq \underline{d}(\mathcal{A}) \leq \underline{\delta}(\mathcal{A}) \leq \overline{\delta}(\mathcal{A}) \leq \overline{d}(\mathcal{A}) \leq 1.$$

We will study mostly sets of square–free integers. It is well–known that

$$d(\mathbb{N}^*) = \frac{6}{\pi^2}. \tag{1}$$

We will compare the density of a set $\mathcal{A} \subset \mathbb{N}^*$ with the density of $\mathbb{N}^*$, and the density obtained in this way will be denoted by an asterisque. Thus, e.g., for $\mathcal{A} \subset \mathbb{N}^*$ we write

$$\begin{aligned} d^*(\mathcal{A}) &= \tfrac{d(\mathcal{A})}{d(\mathbb{N}^*)} = \tfrac{\pi^2}{6} d(\mathcal{A}), \\ \underline{\delta}^*(\mathcal{A}) &= \tfrac{\underline{\delta}(\mathcal{A})}{\underline{\delta}(\mathbb{N}^*)} = \tfrac{\pi^2}{6} \underline{\delta}(\mathcal{A}), \end{aligned}$$

etc.

A set $\mathcal{A} \subset \mathbb{N}$ is said to be *primitive* if there are no a, a' with $a \in \mathcal{A}$, $a' \in \mathcal{A}$, $a \neq a'$ and $a|a'$. Let $F(n)$ denote the cardinality of the greatest primitive set selected from $\{1, 2, \ldots, n\}$. Then it is easy to see [9] that

$$F(n) = \left\lceil \frac{n}{2} \right\rceil \left(= \left(\frac{1}{2} + o(1) \right) n \right). \tag{2}$$

By the results of Besicovitch [3] and Erdös [6], for all $\varepsilon > 0$

$$\text{there is an infinite primitive set } \mathcal{A} \subset \mathbb{N} \text{ with } \overline{d}(\mathcal{A}) > \frac{1}{2} - \varepsilon. \tag{3}$$

Behrend [4] proved that if $\mathcal{A} \subset \{1, 2, \ldots, N\}$ and $\mathcal{A}$ is primitive then we have

$$\sum_{a \in \mathcal{A}} \frac{1}{a} < c_1 \frac{\log N}{(\log \log N)^{1/2}} \tag{4}$$

(so that an infinite primitive set must have zero logarithmic density) and Erdös [5] proved that if $\mathcal{A} \subset \mathbb{N}$ is a (finite or infinite) primitive set then

$$\sum_{a \in \mathcal{A}} \frac{1}{a \log a} < c_2. \tag{5}$$

These results have been extended in various directions; surveys of this field are given in [2], [8], [9], [10].

Next we will introduce two notions of information theoretical background. If a, b are positive square–free integers with the property that $a|b$ and $p(b/a) > P(a)$, i.e., they are of the form $a = p_1 \ldots p_r$, $b = p_1 \ldots p_r p_{r+1} \ldots p_t$ where $p_1 < \ldots < p_r < p_{r+1} < \ldots < p_t$ are distinct primes (with $t > r$), then we say that a is *prefix* of b and we write $a|_p b$. If $\mathcal{A} \subset \mathbb{N}^*$ is a set such that there are no $a \in \mathcal{A}$, $b \in \mathcal{A}$ with $a|_p b$, then $\mathcal{A}$ is said to be *prefix–free.* Similarly, if $a|b$ and $P(b/a) < p(a)$, then a is called *suffix* of b and we write $a|_s b$. If $\mathcal{A} \subset \mathbb{N}^*$ is a set such that there are no $a \in \mathcal{A}$, $b \in \mathcal{A}$ with $a|_s b$, then $\mathcal{A}$ is said to be *suffix–free.* (Both notions, prefix and suffix, could be extended to the non–squarefree case as well, however, to simplify the discussion here we restrict ourselves to the square–free case.)

A further motivation for introducing and studying these concepts is that there is a close connection between prefix–freeness and primitivity: clearly,

$$\text{if a set } \mathcal{A} \subset \mathbb{N} \text{ is primitive, then it is prefix–free.} \tag{6}$$

Since prefix–freeness appears in connection with primitivity (see the proof of Theorem 3 below), one might like to study how close these concepts are.

Based on these considerations, in this paper our goal is to study density related properties of prefix–free and suffix–free sets.

THE PROBLEMS AND RESULTS

Our first goal is to study the "prefix–free analog" of (2). Let $G(n)$ denote the cardinality of the greatest prefix–free set selected from $\mathbb{N}^*(n)$, and let $P^+(a)$ denote the smallest prime greater than $P(a)$.

Theorem 1. *Write*

$$\mathcal{B}(n) = \{b : b \in \mathbb{N}^*(n), bP^+(b) > n\}. \tag{7}$$

Then $\mathcal{B}(n)$ is prefix–free and

$$G(n) = |\mathcal{B}(n)|.$$

Note that it follows from the prime number theorem that, if $1 > \varepsilon > 0$ and $n > n_1(\varepsilon)$, then for all $b \in \mathbb{N}^*(n)$, $b > (1+\varepsilon)\frac{n}{\log n}$ we have

$$P(b) > \left(1 - \frac{\varepsilon}{2}\right) \log n$$

so that

$$bP^+(b) > bP(b) > (1+\varepsilon)\left(1 - \frac{\varepsilon}{2}\right) \log n > \log n$$

and thus $b \in \mathcal{B}(n)$. It follows that

$$G(n) > \left(1 - \frac{1+\varepsilon}{\log n}\right) N^*(n)$$

so that

$$\lim_{n\to\infty} \frac{G(n)}{N^*(n)} = 1; \tag{8}$$

compare this with (2). A combination of (8) with result of Erdös [6] gives

Corollary 1. *For all $\varepsilon > 0$ there is an infinite prefix–free set $\mathcal{A} \subset \mathbb{N}^*$ with*

$$\overline{d}^*(\mathcal{A}) > 1 - \varepsilon.$$

Since this can be derived trivially from (8) by using ideas of [6], we will not present the details here.

The "prefix–free analog" of Behrend's theorem (4) reflects an interesting difference between primitive sets and prefix–free sets. Indeed, consider now instead of $G(n)$

$$E(n) = \max_{\text{prefix–free } \mathcal{A}\subset\mathbb{N}^*(n)} \sum_{a\in\mathcal{A}} \frac{1}{a}. \tag{9}$$

Theorem 2. *For every $\varepsilon > 0$ and $n > n_2(\varepsilon)$, suitable,*

$$0,2689 - \varepsilon < \frac{E(n)}{\sum\limits_{b\in\mathbb{N}^*(n)} \frac{1}{b}} < 0,7311 + \varepsilon.$$

Actually, we know for every $n \in \mathbb{N}$ the unique optimal prefix–free $\mathcal{A} \subset \mathbb{N}^*(n)$ for which $E(n)$ in (9) is assumed, but the value, and particularly also $\lim_{n\to\infty} E(n)$, which we conjecture to exist, is hard to estimate.
We shall show that the proofs of both, Theorem 1 and Theorem 2, can be given by the same approach via the Basic Lemma 1 in Section 3 involving multiplicative functions. Actually, this lemma seems to be useful also for other cases.
For instance it shades a new light on a well–known conjecture of Erdös concerning (finite or infinite) primitive sets, which says that for every primitive set $\mathcal{A} \subset \mathbb{N}$

$$\sum_{a\in\mathcal{A}} \frac{1}{a\log a} \leq \sum_{p\in\mathcal{P}} \frac{1}{p\log p}.$$

Consider now for any positive, multiplicative function f

$$L_f(\infty) = \max_{\text{prefix–free } \mathcal{A}\subset\mathbb{N}^*} \sum_{a\in\mathcal{A}} f(a) \tag{10}$$

then we have the
Proposition 1. Let f be a multiplicative function such that

$$\sum_{p\geq 3, p\in\mathcal{P}} f(p) < 1,$$

then $L_f(\infty)$ is assumed at the set of primes. In particular, if $f(m) = m^\alpha$, then for every $\alpha \leq \alpha_0$, where $\alpha_0 \in \mathbb{R}$ and $\sum\limits_{p\geq 3} p^{\alpha_0} = 1$, the primes are the optimal set.
Next we will extend Erdös's theorem (5) to prefix–free sets:
Theorem 3. *There is an absolute constant c_3 such that if $\mathcal{A} \subset \mathbb{N}^*$ is a (finite or infinite) prefix–free set, then*

$$\sum_{a\in\mathcal{A}} \frac{1}{a\log a} < c_3.$$

Indeed, in Erdös's proof [5] only the prefix property of primitive sequences is used (that they possess by (6)) so that it also gives the more general result Theorem 3. To see that indeed it is so, for the sake of completeness we will sketch the proof in Section 5 (leaving some technical details to the reader).
It follows easily from Theorem 3 (proving by contradiction and using partial summation) that
Corollary 2. *If $\mathcal{A} \subset \mathbb{N}^*$ is an infinite prefix–free set then we have*

$$A(x) < \frac{x}{\log\log x \, \log\log\log x} \tag{11}$$

for infinitely many x (and, by (5), if $\mathcal{A} \subset \mathbb{N}$ is primitive then (11) also holds infinitely often).

One might like to know how far the upper bound in (11) is from the best possible. This is closely related to one of the favourite problems of Erdös. In [8] this problem is formulated in the following way (and Erdös mentioned it in numerous problem papers as well):
"The following problem seems difficult: Let $b_1 < b_2 \dots$ be an infinite sequence of integers. What is a necessary and sufficient condition that there should exist a primitive sequence $a_1 < a_2 \dots$ satisfying $a_n < cb_n$ for every n?
From (5) ... we obtain that we must have

$$\sum_{i=1}^{\infty} \frac{1}{b_i \log b_i} < \infty \dots \tag{12}$$

We know that (12) is not sufficient — it is not clear whether a simple necessary and sufficient condition exists."
This is followed by a lengthy discussion of the problem how large one can make $\sum_{a \le x} \frac{1}{a}$ uniformly in x for a primitive set $a_1 < \dots$ (see also [7]).
It seems to be a more natural (although more difficult) problem to replace here the sum $\sum_{a \le x} \frac{1}{a}$ by the counting function $A(x)$, i.e. to study the problem how large one can make $A(x)$ uniformly in x for a primitive set $\mathcal{A}$ — and this is the question asked by us also for prefix–free sets. In [1] we gave a quite satisfactory answer by proving that (11) is best possible apart from a factor $(\log\log\log x)^{\varepsilon}$:
Theorem 4. [1] *For all $\varepsilon > 0$ there is an infinite primitive (and therefore also prefix–free) set $\mathcal{A} \subset \mathbb{N}$ such that for $a > x_0(\varepsilon)$ we have*

$$A(x) > \frac{x}{\log\log x (\log\log\log x)^{1+\varepsilon}}.$$

By a standard argument it can be shown that here $\mathcal{A} \subset \mathbb{N}$ can be replaced by $\mathcal{A}^* \subset \mathbb{N}$ (and the same lower bound holds), and by (6), this $\mathcal{A}^*$ also is prefix–free. Thus the behaviour of primitive and prefix–free sets is similar as far as the maximal rate of growth of the counting function is concerned: in both cases the estimates (11) and the one in Theorem 4 can be given.
Problem 1. Is it true that if $\mathcal{A} \subset \mathbb{N}^*$ is an infinite set with

$$\overline{\delta}^*(\mathcal{A}) > 0, \tag{13}$$

then $\mathcal{A}$ contains an infinite "prefix chain", i.e., there is an infinite subset $\{a_{i_1}, a_{i_2}, \dots\}$ of $\mathcal{A}$ with $a_{i_1} |_p a_{i_2} |_p a_{i_3} \dots$?
Note that by a theorem of Davenport and Erdös [11], (13) implies that $\mathcal{A}$ contains an infinite divisibility chain $a_{i_1} | a_{i_2} | a_{i_3} \dots$.
The finite analog of Problem 1 is easier. Indeed, we will prove in Section 6
Theorem 5.

(i) *If $n > n_3$,*

$$\mathcal{A} = \{a_1, \dots, a_t\} \subset \mathbb{N}^*(n) \tag{14}$$

and

$$E(\mathcal{A}) \triangleq \sum_{a\in\mathcal{A}} \frac{1}{a\log a} > c_3 \tag{15}$$

(where c_3 is the constant defined in Theorem 3), then, writing

$$k = \left[\frac{E(\mathcal{A})}{c_3}\right] + 1, \tag{16}$$

$\mathcal{A}$ contains a prefix chain of length k, i.e., there is a subset $\{a_{i_1}, a_{i_2}, ..., a_{i_k}\}$ of $\mathcal{A}$ with $a_{i_1}|_p a_{i_2}|_p \dots a_{i_k}$.

(ii) *There are numbers c_4 and n_4 with the following properties: there is an infinite set $\mathcal{A} \subset \mathbb{N}^*$ such that*

$$d^*(\mathcal{A}) = 1 \tag{17}$$

and, writing

$$E(\mathcal{A}, n) = \sum_{a\in\mathcal{A}, a\leq n} \frac{1}{a\log a},$$

for $n > n_4$ the set $\mathcal{A} \cap \mathbb{N}^*(n)$ does not contain a prefix chain longer than $c_4 E(\mathcal{A}, n)$.

(So that (i) is best possible apart from a constant factor in the length of the maximal chain.)

While the behaviour of prefix–free and primitive sets is similar as far as the maximal rate of growth of the counting function is concerned, the behaviour of the suffix–free sets is very much different and, indeed, they can be much "denser".

We consider now the cardinality and the asymptotic density of suffix–free sets. Let $H(n)$ denote the cardinality of the largest suffix–free set selected from $\mathbb{N}^*(n)$.

Theorem 6. *The set*

$$\mathbb{C}(n) = \{c \in \mathbb{N}^*(n) : 2|c\} \cup \left\{\mathbb{N}^*(n) \cap \left(\frac{n}{2}, n\right]\right\}$$

is suffix–free and $|\mathbb{C}(n)| = H(n)$.

Corollary 3.

$$\lim_{n\to\infty} \frac{H(n)}{|\mathbb{N}^*(n)|} = \frac{2}{3}.$$

Using ideas of Besicovitch [3] and Erdös [5, 6] one can easily get the following result, whose proof is not presented in this paper.

Corollary 4. *For every $\varepsilon > 0$ there exists an infinite suffix–free set $\mathbb{C}$ such that*

$$\overline{d^*}\mathbb{C} > \frac{2}{3} - \varepsilon.$$

Finally we discuss logarithmic densities of sufix–free sets. Let

$$K(n) = \max_{\text{suffix–free } \mathcal{A} \in \mathbb{N}^*} \sum_{a \in \mathcal{A}} \frac{1}{a}.$$

In contrast to the case of prefix–free sets, here Basic Lemma 2 of Section 3 gives a very simple description of the optimal set.

Theorem 7. *Let $\mathcal{B}$ be the set from Basic Lemma 2. We have*
$\mathcal{B} = \mathcal{B}_1 \dot{\cup} \mathcal{B}_2$, *where*
$\mathcal{B}_1 = \{2 \cdot a, 3 \cdot a, 5 \cdot a : a \in \mathbb{N}^* \left(\frac{n}{5}\right) \text{ and } (a, 30) = 1\}$ *and*
$\mathcal{B}_2 = \{a \in \mathbb{N}^* : \frac{n}{5} < a \leq n \text{ and } (a, 30) = 1\}$.
Simple calculations yield

Corollary 5.

$$\lim_{n \to \infty} \frac{K(n)}{\sum\limits_{a \in \mathbb{N}^*(n)} \frac{1}{a}} = \frac{31}{72}.$$

Corollary 6.

(i) *For any infinite suffix–free set $\mathbb{C}$ holds*

$$\overline{D^*} \mathbb{C} = \overline{\delta^*} \mathbb{C} \leq \frac{31}{72}.$$

(ii) *Define*

$$\mathbb{C} = \{2 \cdot a, 3 \cdot a, 5 \cdot a : a \in \mathbb{N}^* \text{and } (a, 30) = 1\}.$$

Then $\mathbb{C}$ is an infinite suffix–free set and

$$d^* \mathbb{C} = \frac{31}{72}.$$

Similarly to $L_f(\infty)$ for infinite prefix–free sets define the quantity $S_f(\infty)$ for infinite suffix–free sets, where f is a positive multiplicative function:

$$S_f(\infty) = \max_{\text{suffix–free } \mathcal{A} \subset \mathbb{N}^*} \sum_{a \in \mathcal{A}} f(a).$$

Proposition 2. *Let f be a multiplicative function such that $\sum_{p \in \mathcal{P}} f(p) < 1$. Then $S_f(\infty)$ is assumed at the set of primes. In particular, if $f(m) = m^\beta$, then for every $\beta \leq \beta_0$, where $\beta_0 \in \mathbb{R}$ and $\sum\limits_{p \in P} p^{\beta_0} = 1$, the primes are the optimal set.*

Remark: We note the difference to Proposition 1, where the summation starts from $p \geq 3$, and hence clearly $\beta_0 < \alpha_0$.

TWO BASIC LEMMAS

For any positive, multiplicative function f define

$$L_f(n) = \max_{\text{prefix-free } \mathcal{A} \in \mathbb{N}^*(n)} \sum_{a \in \mathcal{A}} f(a).$$

Basic Lemma 1. *Write*

$$\mathcal{A} = \left\{ \begin{array}{ll} a \in \mathbb{N}^*(n): & \text{(i)} \sum\limits_{P(a)<p\leq \frac{n}{a}} f(p) < 1 \text{and} \\ & \text{(ii)} \sum\limits_{P(a')<p\leq \frac{n}{a'}} f(p) \geq 1, \text{where } a' = \frac{a}{P(a)}, \end{array} \right\}$$

We assume that (i) *always holds if* $P(a) \geq \frac{n}{a}$ *or* $P(a) < \frac{n}{a}$, *but there is no prime in the interval* $\left(P(a), \frac{n}{a}\right]$. *We also assume that* (ii) *always holds if* $a \in \mathcal{P}$. *Then* $\mathcal{A}$ *is prefix–free and*

$$\sum_{a \in \mathcal{A}} f(a) = L_f(n).$$

Proof: We show that $\mathcal{A}$ is prefix–free. Assume to the opposite that there are $a, b \in \mathcal{A}$ such that $a|_p b$, that is $b = a \cdot c$, $p(c) > P(a)$.
We have from condition (i) for $a \in \mathcal{A}$

$$\sum_{P(a)<p\leq \frac{n}{a}} f(p) < 1 \tag{18}$$

and from condition (ii) for $b' = \frac{b}{P(b)} \geq a$

$$\sum_{P(b')<p\leq \frac{n}{b'}} f(p) \geq 1. \tag{19}$$

Since $P(b') \geq P(a)$, $b' \geq a$ and consequently $\frac{n}{a} \geq \frac{n}{b'}$, (18) and (19) are not compatible. Hence $\mathcal{A}$ is prefix–free.
Now we show that

$$\sum_{a \in \mathcal{A}} f(a) = L_f(n). \tag{20}$$

Let

$$\mathcal{L}_f(n) = \left\{ \mathcal{B} \subset \mathbb{N}^*(n) : \mathcal{B} \text{ is prefix–free and } \sum_{b \in \mathcal{B}} f(b) = L_f(n) \right\}.$$

So, equivalent to (20) is $\mathcal{A} \in \mathcal{L}_f(n)$.
Let $\mathcal{B} \in \mathcal{L}_f(n)$ be a set for which

$$\sum_{b \in \mathcal{B}} b \text{ is maximal among elements of } \mathcal{L}_f(n). \tag{21}$$

We claim that $\mathcal{B} = \mathcal{A}$. For this we have to prove that (i) and (ii) hold for every element $b \in \mathcal{B}$. We show that (i) holds. Assume to the opposite that for an element $b \in \mathcal{B}$ we have

$$\sum_{P(b)<p\leq \frac{n}{b}} f(p) \geq 1. \tag{22}$$

Define

$$\mathcal{B}' = \left(\mathcal{B} \setminus \{b\}\right) \cup \left\{b \cdot p : p > P(b), p \leq \frac{n}{b}\right\}.$$

Since $\mathcal{B}$ is prefix–free necessarily $b \cdot p \notin \mathcal{B}$ for all $p(b) < p \leq \frac{n}{b}$. Clearly $\mathcal{B}' \subset \mathbb{N}^*(n)$.
It is easy to see that $\mathcal{B}'$ is prefix–free and

$$\sum_{b\in\mathcal{B}} b < \sum_{b\in\mathcal{B}'} b. \tag{23}$$

Moreover, since f is a multiplicative function, we have

$$\sum_{P(b)<p\leq \frac{n}{b}} f(p \cdot b) = f(b) \cdot \sum_{P(b)<p\leq \frac{n}{b}} f(p) \geq f(b) \text{ (by assymption (22))}$$

and consequently

$$\sum_{b\in\mathcal{B}} f(b) \leq \sum_{b\in\mathcal{B}'} f(b). \tag{24}$$

Hence $\mathcal{B}' \in \mathcal{L}_f(n)$, which is a contradiction (see (21) and (23)).
Therefore for all $b \in \mathcal{B}$ (i) holds.
Now we show that for all $b \in \mathcal{B}$ (ii) holds. Assume to the opposite that for a $b \in \mathcal{B}$ we have

$$\sum_{P(b')<p\leq \frac{n}{b'}} f(p) < 1, \text{where } b' = \frac{b}{P(b)}. \tag{25}$$

Among such elements $b \in \mathcal{B}$ we choose one which has maximal b'.
Let $\mathcal{B}_1(b') \subset \mathcal{B}$ be the set of all elements of $\mathcal{B}$ for which b' is prefix, that is, $b_1 \in \mathcal{B}_1(b')$ implies $b_1 = b' \cdot c$, $p(c) > P(b')$.
In particular $b \in \mathcal{B}_1(b')$ and $b = b' \cdot P(b)$. We claim that $c \in \mathcal{P}$.
Indeed, assume $b_1 = b' \cdot c$ and $c \notin \mathcal{P}$. Then

$$b_1' = \frac{b_1}{P(b_1)} > b'$$

and (25) also holds for $b_1 \in \mathcal{B}$ and b_1', a contradiction to the maximality of b.
Consider

$$\mathcal{B}_2 = \left(\mathcal{B} \setminus \mathcal{B}_1(b')\right) \cup \{b'\}.$$

We have that $\mathcal{B}_2$ is prefix–free and, since f is multiplicative, that

$$\sum_{b\in\mathcal{B}_1(b')} f(b) \leq f(b') \cdot \sum_{P(b')<p\leq \frac{n}{b'}} f(p) < f(b') \text{ (by assumption (25))}$$

and consequently

$$\sum_{b\in\mathcal{B}_2} f(b) > \sum_{b\in\mathcal{B}} f(b),$$

a contradiction to $\mathcal{B} \in \mathcal{L}_f(n)$.
Hence $\mathcal{B} = \mathcal{A} \in \mathcal{L}_f(n)$.
Define now for any positive, multiplicative function f

$$S_f(n) = \max_{\text{suffix–free } B\subset\mathbb{N}^*(n)} \sum_{b\in B} f(b).$$

Basic Lemma 2. *Write*

$$\mathcal{B} = \left\{ \begin{array}{ll} b \in \mathbb{N}^*(n) : & \text{(i)} \sum\limits_{p<\min\{\frac{n+1}{b},P(b)\}} f(p) < 1 \text{and} \\ & \text{(ii)} \sum\limits_{p<\min\{\frac{n+1}{b'},P(b')\}} f(p) \geq 1, \text{where } b' = \frac{b}{P(b)} \end{array} \right\}.$$

We assume that (i) always holds if $\min\left\{\frac{n+1}{b}, P(b)\right\} \leq 2$ *and that (ii) holds if* $b \in \mathcal{P}$. Then B is suffix–free and $\sum\limits_{b\in B} f(b) = S_f(n)$.
Since the proof is almost identical with the one given for Basic Lemma 1, we do not present it here.

PREFIX–FREE SETS: PROOFS OF THEOREM 1, 2

Proof of Theorem 1: This case concerns maximal cardinalities $G(n)$. Notice that $G(n) = L_f(n)$, if f is the constant function with value 1. Furthermore, we verify that the set $\mathcal{B}$ in Theorem 1 equals the set $\mathcal{A}$ in the Basic Lemma 1, which implies the result.
Proof of Theorem 2: Now we apply the Basic Lemma 1 to the multiplicative function f defined by

$$f(m) = \frac{1}{m} \text{ for } m \in \mathbb{N}^*(n).$$

Then

$$E(n) = L_f(n)$$

and the set $\mathcal{A}$ has the properties claimed.
Moreover, the uniqueness can be seen from the proof of the Basic Lemma 1 by observing that we cannot have equality in (22) and consequently in (24), because $\sum\limits_{p\in\mathcal{P}_1} \frac{1}{p}$ is never an integer for any set $\mathcal{P}_1$ of primes.
To prove the lower bound we consider the set

$$\mathcal{A}' = \left\{ a \in \mathbb{N}^*(n) : P(a) > n^{\frac{1}{1+\varepsilon}+\varepsilon} \text{and } \frac{a}{P(a)} < n^{\frac{1}{1+\varepsilon}-\varepsilon} \right\}.$$

By

$$\sum_{p\leq x} \frac{1}{p} = \log\log x + c_5 + O(1)$$

we have for every $a \in \mathcal{A}'$

$$\sum_{P(a)<p\leq \frac{n}{a}} \frac{1}{p} < \sum_{n^{\frac{1}{1+e}+\varepsilon}<p\leq n^{\frac{e}{1+e}+\varepsilon}} \frac{1}{p} < 1$$

if $n > n_5(\varepsilon)$.
Similarly we have for $a' = \frac{a}{P(a)}$

$$\sum_{P(a')<p\leq \frac{n}{a'}} \frac{1}{p} > 1.$$

Therefore $\mathcal{A}' \subset \mathcal{A}$, where $\mathcal{A}$ is the set defined in the Basic Lemma 1. Hence $\mathcal{A}'$ is a prefix–free set.
We have

$$\sum_{a\in\mathcal{A}'} \frac{1}{a} = \sum_{\substack{p > n^{\frac{1}{1+e}+\varepsilon} \\ b < n^{\frac{1}{1+e}-\varepsilon} \\ p\cdot b \leq n \\ b\in\mathbb{N}^*}} \frac{1}{b\cdot p} = \sum_{p>n^{\frac{1}{1+e}+\varepsilon}} \frac{1}{p} \cdot \left(\sum_{\substack{b < n^{\frac{1}{1+e}-\varepsilon} \\ b < \frac{n}{p} \\ b\in\mathbb{N}^*}} \frac{1}{b} \right)$$

$$\geq \sum_{n^{\frac{e}{1+e}+\varepsilon}>p>n^{\frac{1}{1+e}+\varepsilon}} \frac{1}{p} \cdot \sum_{\substack{b < n^{\frac{1}{1+e}-\varepsilon} \\ b\in\mathbb{N}^*}} \frac{1}{b}$$

$$\sim \frac{6}{\pi^2} \log n^{\frac{1}{1+e}-\varepsilon} \cdot \sum_{n^{\frac{e}{1+e}+\varepsilon}>p>n^{\frac{1}{1+e}+\varepsilon}} \frac{1}{p} \sim \frac{6}{\pi^2} \log n^{\frac{1}{1+e}-\varepsilon}.$$

Hence

$$\frac{\sum\limits_{a\in\mathcal{A}'} \frac{1}{a}}{\sum\limits_{a\in\mathbb{N}^*(n)} \frac{1}{a}} > \frac{1}{1+e} - \varepsilon \sim 0,2689 - \varepsilon$$

and this proves the lower bound.
To show the upper bound we consider the set

$$\mathcal{A}'' = \mathbb{N}^*\left(n^{\frac{1}{1+e}-\varepsilon}\right).$$

For every element $a \in \mathcal{A}''$ we have

$$\sum_{P(a)<p\leq \frac{n}{a}} \frac{1}{p} > \sum_{n^{\frac{1}{1+e}-\varepsilon}<p\leq n^{\frac{e}{1+e}+\varepsilon}} \frac{1}{p} > 1.$$

Therefore $\mathcal{A}'' \cap \mathcal{A} = \varnothing$, where $\mathcal{A}$ is the set defined in Theorem 2.
Since

$$\sum_{a\in\mathcal{A}''} \frac{1}{a} = \sum_{a\leq n^{\frac{1}{1+e}-\varepsilon}} \frac{1}{a} \sim \frac{6}{\pi^2} \log n^{\frac{1}{1+e}-\varepsilon}$$

we get

$$\frac{\sum\limits_{b\in\mathcal{A}} \frac{1}{b}}{\sum\limits_{b\in\mathbb{N}^*(n)} \frac{1}{b}} = \frac{E(n)}{\sum\limits_{b\in\mathbb{N}^*(n)} \frac{1}{b}} < \frac{e}{1+e} + \varepsilon \sim 0,7311 + \varepsilon.$$

Remark: We are sure that by more detailed consideration of the set $\mathcal{A}$ one can get much better estimates. However to tighten the gap between upper and lower bounds to, say 0.1, seems difficult.
A proof of Proposition 1 can be given directly and easily with the Basic Lemma 1.

SKETCH OF THE PROOF OF THEOREM 3

If $\mathcal{A}$ is an infinite prefix–free set and for every finite subset $\mathcal{A}'$ of $\mathcal{A}$ we have

$$\sum_{a} \frac{1}{a\log a} \leq c, \tag{26}$$

where the summation is extended over all $a \in \mathcal{A}'$, then (26) also holds if the summation is extended over all $a \in \mathcal{A}$. Thus we may assume that $\mathcal{A}$ is finite. Let x be large enough in terms of the greatest element of $\mathcal{A}$ (and later we will take $x \to \infty$). Consider the sum

$$S \triangleq \sum_{a\in\mathcal{A}} \sum_{\substack{q\leq x/a\\ p(q)>a}} \frac{1}{aq}. \tag{27}$$

Since $\mathcal{A}$ is prefix–free, thus $aq = a'q'$, $a \in \mathcal{A}$, $a' \in \mathcal{A}$, $q \leq x/q$, $q' \leq x/q'$, $p(q) > a$, $p(q') > a'$ implies that $a = a'$, $q = q'$. In other words, the denominators aq in (27) are distinct, each of them is $\leq x$, so that for $x \to \infty$ we have

$$S \leq \sum_{n\leq x} \frac{1}{n} = \big(1 + o(1)\big) \log x \ (\text{as } \ x \to \infty). \tag{28}$$

On the other hand, we have

$$S = \sum_{a\in\mathcal{A}} \frac{1}{a} \sum_{\substack{q\leq x/a\\ p(q)>a}} \frac{1}{q}. \tag{29}$$

Since by Mertens' theorem we have

$$\prod_{p\leq y} \left(1 - \frac{1}{p}\right) \sim \frac{c_5}{\log y}, \tag{30}$$

thus by using an elementary sieving process, for $x \to \infty$ we may estimate the inner sum in (29) in the following way:

$$\begin{aligned}
\sum_{\substack{q \le x/a \\ p(q) > a}} \tfrac{1}{q} &= \sum_{d|\prod_{p\le a} p} \mu(d) \sum_{\substack{q \le x/a \\ d|q}} \tfrac{1}{q} \\
&= \sum_{d|\prod_{p\le a} p} \mu(d) \sum_{t \le x/ad} \tfrac{1}{dt} = \sum_{d|\prod_{p\le a} p} \tfrac{\mu(d)}{d} \sum_{t\le x/ad} \tfrac{1}{t} \\
&= (1+o(1)) \sum_{d|\prod_{p\le a} p} \tfrac{\mu(d)}{d} \log x \\
&= (1+o(1)) \log x \prod_{p\le a} \left(1 - \tfrac{1}{p}\right) = (1+o(1)) c_5 \tfrac{\log x}{\log a}. \qquad (31)
\end{aligned}$$

By (29) and (31) we have

$$S = (1+o(1)) c_5 \log x \sum_{a\in\mathcal{A}} \frac{1}{a \log a} \quad (\text{as } \ x \to \infty). \qquad (32)$$

Now the desired bound follows from (28) and (32).
Note that we did not use the fact that the a's are square–free so that, extending the notion of prefix to non–squarefree integers, the result could be extended to this more general case as well.

PROOF OF THEOREM 5

(i) Let $\mathcal{A}_1 = \mathcal{A}$, and for $j > 1$ let $\mathcal{A}_j$ denote the set of the integers a such that $a \in \mathcal{A}$ and there is a prefix chain of length j in $\mathcal{A}$ whose last element is a:

$$a_{i_1} |_p a_{i_2} |_p \dots |_p a_{i_{j-1}} |_p a, a_{i_1} \in \mathcal{A}, \dots, a_{i_{j-1}} \in \mathcal{A}.$$

We will show by induction on j that if (14) and (15) hold, and $1 \le j \le k$ (where k is defined by (16)), then

$$E(\mathcal{A}_j) = \sum_{a\in\mathcal{A}_j} \frac{1}{a\log a} \begin{cases} = E(\mathcal{A}) - (j-1)c_3 & \text{for } j = 1 \\ > E(\mathcal{A}) - (j-1)c_3 & \text{for } j > 1. \end{cases} \qquad (33)$$

Indeed, this is trivial for $j = 1$ since then we have $E(\mathcal{A})$ on both sides of (33). Assume now that (33) holds for some j with $1 \le j < k$. Then we have to show that it also holds with $j+1$ in place of j:

$$E(\mathcal{A}_{j+1}) = \sum_{a\in\mathcal{A}_{j+1}} \frac{1}{a\log a} > E(\mathcal{A}) - jc_3. \qquad (34)$$

We will prove this by contradiction: assume that contrary to (34) we have

$$E(\mathcal{A}_{j+1}) = \sum_{a\in\mathcal{A}_{j+1}} \frac{1}{a\log a} \le E(\mathcal{A}) - jc_3. \qquad (35)$$

Write $\mathcal{A}^* = \mathcal{A}_j \setminus \mathcal{A}_{j+1}$. Then by (33) and (35) (and since clearly $\mathcal{A}_{j+1} \subset \mathcal{A}_j$) we have

$$E(\mathcal{A}^*) = \sum_{a\in\mathcal{A}^*} \frac{1}{a\log a} = E(\mathcal{A}_j) - E(\mathcal{A}_{j+1})$$
$$\geq \big(E(\mathcal{A}) - (j-1)c_3\big) - \big(E(\mathcal{A}) - jc_3\big) = c_3.$$

Thus by Theorem 3 there are $a' \in \mathcal{A}^*$, $a'' \in \mathcal{A}^*$ with $a'|_p a''$. Since $a' \in \mathcal{A}^* \subset \mathcal{A}_j$, thus there is a prefix chain of length j in $\mathcal{A}$ whose last element is $a' : a_{i_1}|_p \dots |_p a'$. Then $a_{i_1}|_p \dots |_p a_{i_{j-1}}|_p a'|_p a''$ is a prefix chain of length $j+1$ in $\mathcal{A}$ whose last element is a'', and thus we have $a'' \in \mathcal{A}_{j+1}$. This contradicts $a'' \in \mathcal{A}^* = \mathcal{A}_j \setminus \mathcal{A}_{j+1}$ which proves (34), and this completes the proof of (33) (with $1 \leq j \leq k$).

Using (33) with k in place of j (so that $k \geq 2$ by (15)) we obtain

$$E(\mathcal{A}_k) > E(\mathcal{A}) - (k-1)c_3 = E(\mathcal{A}) - \left\lceil \frac{E(\mathcal{A})}{c_3} \right\rceil c_3 \geq O$$

so that $\mathcal{A}_k$ is non–empty, which completes the proof of (i).

(ii) Let

$$\mathcal{B} = \{b : b \in \mathbb{N}, |\omega(b) - \log\log b| < (\log\log b)^{3/4}\}$$

and $\mathcal{A} = \mathcal{B} \cap \mathbb{N}^*$. Then by a theorem of Hardy and Ramanujan [13] we have $d(\mathcal{B}) = 1$, which implies (17). Moreover, by (1) clearly we have

$$E(\mathcal{A}, n) = \sum_{a\in\mathcal{A}, a\leq n} \frac{1}{a\log a} = \big(1 + o(1)\big) \sum_{a\in\mathbb{N}^*(n)} \frac{1}{a\log a}$$
$$= \big(1 + o(1)\big)\frac{6}{\pi^2}\log\log n. \tag{36}$$

If $a \in \mathcal{A}$, $a \leq n$ and a is the last element of a prefix chain of length k in $\mathcal{A} : a_{i_1}|_p \dots |_p a_{i_{k-1}}|_p a$, then by (36) and the definition of $\mathcal{A}$ we have

$$k \leq \omega(a) < \log\log a + (\log\log a)^{3/4} \leq \log\log n + (\log\log n)^{3/4}$$
$$= \big(1 + o(1)\big)\log\log n = \big(1 + o(1)\big)\tfrac{\pi^2}{6}E(\mathcal{A}, n),$$

which completes the proof of (ii) (with $c_4 = \frac{\pi^2}{6} + \varepsilon$).

PROOF OF THEOREM 6

We apply Basic Lemma 2 with respect to the function $f(m) = 1$, $m \in \mathbb{N}^*$. For this function we have $H(n) = S_f(n)$. It is easy to verify, that $\mathbb{C}(n) \subset \mathcal{B}$, where $\mathbb{C}(n)$ is the set described in Theorem 6 and $\mathcal{B}$ is the set from Basic Lemma 2. Moreover, for every $a \in \mathbb{N}^*(n) \setminus \mathbb{C}(n)$ we have $2 \nmid a$, $a \leq \frac{n}{2}$ and hence

$$\min\left\{\frac{n+1}{a}, P(a)\right\} > 2.$$

Consequently, the condition (i) in Basic Lemma 2 does not hold. Therefore $\mathbb{C}(n) = \mathcal{B}$ and $\mathbb{C}(n)$ is the optimal set.

PROOF OF THEOREM 7

In Basic Lemma 2 consider the set $\mathcal{B}$ with respect to the multiplicative function $f(m) = \frac{1}{m}$, $m \in \mathbb{N}^*$. Using the inequalities $\frac{1}{2} + \frac{1}{3} = \frac{5}{6} < 1$ and $\frac{1}{2} + \frac{1}{3} + \frac{1}{5} = \frac{31}{30} > 1$, it is easy to verify that $(\mathcal{B}_1 \dot{\cup} \mathcal{B}_2) \subset \mathcal{B}$, where $\mathcal{B}_1, \mathcal{B}_2$ are defined in the Theorem. Moreover, using the mentioned inequalities one easily gets that every $b \in \mathbb{N}^*(n) \setminus \{\mathcal{B}_1 \cup \mathcal{B}_2)$ violates one of the conditions (i), (ii) in Basic Lemma 2. Hence $\mathcal{B} = \mathcal{B}_1 \dot{\cup} \mathcal{B}_2$, proving the Theorem.
Corollary 6 and 7 directly follow from Theorem 7 and from the construction. Finally, Proposition 2 is an immediate consequence of Basic Lemma 2.

References

[1] R. Ahlswede, L. Khachatrian and A. Sárközy, "On the counting function of primitive sets of integers", *Preprint 98–077, SFB 343 "Diskrete Strukturen in der Mathematik"*, Universität Bielefeld, submitted to J. Number Theory.

[2] R. Ahlswede and L.H. Khachatrian, "Classical results on primitive and recent results on cross–primitive sequences", in: *The Mathematics of Paul Erdös*, vol. I, eds.R.L. Graham and J. Nesetril, Algorithms and Combinatorics 13, Springer–Verlag, 1997; 104–116.

[3] A.S. Besicovitch, "On the density of certain sequences", *Math. Ann. 110*, 1934, 336–341.

[4] F. Behrend, "On sequences of numbers not divisible by one another", *J. London Math. Soc.*, 10, 1935, 42–44.

[5] P. Erdös, "Note on sequences of integers no one of which is divisible by any other", *J. London Math. Soc.*, 10, 1935, 126–128.

[6] P. Erdös, "A generalization of a theorem of Besicovitch", *J. London Math. Soc.*, 11, 1935, 92–98.

[7] P. Erdös, A. Sárközy and E. Szemerédi, "On a theorem of Behrend", *J. Australian Math. Soc.*, 7, 1967, 9–16.

[8] P. Erdös, A. Sárközy and E. Szemerédi, "On divisibility properties of sequences of integers", *Coll. Math. Soc. J. Bolyai*, 2, 1970, 35–49.

[9] H. Halberstam and K.F. Roth, "Sequences", Springer–Verlag, Berlin–Heidelberg–New York, 1983.

[10] A. Sárközy, "On divisibility properties of sequences of integers", in: *The Mathematics of Paul Erdös*, eds. R.L. Graham and J. Nesetril, Algorithms and Combinatorics 13, Springer–Verlag, 1997, 241–250.

[11] A. Selberg, "Note on a paper by L.G. Sathe", *J. Indian Math. Soc.*, 18, 1954, 83–87.

[12] H. Davenport and P. Erdös, "On sequences of positive integers", *Acta Arith.*, 2, 1936, 147–151.

[13] G.H. Hardy and S. Ramanujan, "The normal number of prime factors of a number n", *Quarterly J. Math.*, 48, 1920, 76–92.

ALMOST ARITHMETIC PROGRESSIONS

Egbert Harzheim

Mathematisches Institut, Heinrich Heine Universitẗ Düsseldorf,
Universitätsstr. 1, 40225 Düsseldorf, Germany

Abstract: We investigate almost arithmetic progressions $x_1, x_2, \ldots, x_L$ of real numbers, that means sequences for which there exist non-overlapping intervals $A_i = [a_i, b_i]$ of equal length, where the a_i constitute an arithmetic progression, and which satisfy $x_i \in A_i$ for $i = 1, ..., L$.

Several papers study the existence of arithmetic progressions in sequences of integers, where the gaps between consecutive elements are below a given bound, e.g. [7], [6], [1], [2], [3]. In [8], [4],[5] sequences were considered which can be well approximated by arithmetic progressions. So e.g. in [8] it was proved - roughly spoken - that a sequence of positive density contains long "almost arithmetic" progressions. We now precise our concepts:

Definition 1. An *arithmetic progression* of length L is a set $\{x_1, ..., x_L\}$ of real numbers, where L is an integer ≥ 2, such that all differences $x_{i+1} - x_i$, $i = 1, ..., L-1$, are equal, say $= s > 0$. Then s is called the *step length* of $\{x_1, ..., x_L\}$.

Let $\mathbf{N}$ (resp. $\mathbf{N}_0$) denote the set of positive (resp. nonnegative) integers.

Definition 2. An *arithmetic interval sequence* is a finite set of closed intervals $A_\nu = [a_\nu, b_\nu]$, $\nu = 1, ..., L$,where L is an integer ≥ 2,which has the following two properties:

1) All intervals A_ν have the same length $b_\nu - a_\nu = w$, and their open kernels are pairwise disjoint.

2) The initial elements a_ν, $\nu = 1, ..., L$, form an arithmetic progression with $a_1 < .... < a_L$. Because of 1) the same then also holds for the final elements b_ν, $\nu = 1, ..., n$.

Again we call the step-length of the arithmetic progression $\{a_1, ..., a_L\}$ also the *step-length* of the arithmetic interval sequence.

We call the number $\lambda := \frac{w}{s}$ the *shrink factor* of $\{A_1, ..., A_L\}$.

If $\lambda = 1$, we have $w = s$, and then we call $\{A_1, ..., A_L\}$ a *sequence of consecutive intervals of equal length.* Every arithmetic interval sequence arises from a sequence of consecutive intervals of equal length by shrinking the intervals by the factor λ to the left endpoint, - this explains the choice of the naming.

I. Althöfer et al. (eds.), Numbers, Information and Complexity, 17-20.

If $\lambda = 0$, we have $w = 0$, and then the arithmetic interval sequence $\{A_1, ..., A_L\}$ is identical in character with an arithmetic progression $\{a_1, ..., a_L\}$. Generalizing a notion of (4) we define:

Definition 3. A set of real numbers $x_1, ..., x_L$ with $x_1 < .... < x_L$, where L is an integer ≥ 2, is said to be an *almost arithmetic progression* of *length* L and with a *shrink factor* $\lambda \in [0, 1]$, shortly an AAP(L,λ), if there exists an arithmetic interval sequence $A_\nu = [a_\nu, b_\nu]$, $\nu = 1, ..., L$, with shrink factor λ which satisfies $x_\nu \in A_\nu$ for $\nu = 1, ..., L$. Of course, λ is not uniquely determined by the $x_1, ..., x_L$. (By the way, then the family $(x_\nu)_{\nu=1,...,L}$ is a system of distinct representatives of $(A_\nu)_{\nu=1,...,L}$.)

The number λ can be considered as a measure, how close to an arithmetic progression the sequence $(x_\nu)_{\nu=1,...,L}$ is. In the papers [4] and [5] the case $\lambda = 1$ was treated in detail.

The question arises how many elements a set $A \subset \{0, ..., n\}$ can have without containing an AAP(L, λ) for given numbers L (≥ 2) $\in \mathbf{N}$ and $\lambda \in [0, 1]$. In this context in [5] the following was proved for the case $\lambda = 1$:

Proposition 1. *Let L,n be integers with* $5 \leq L < n$. *Then there exists a subset* $M \subset [0, n) \cap \mathbf{N}_0$ *with* $|M| > n^{1 - \frac{2}{(L-4)\cdot \log 2}} \cdot f(L)$, *where*

$$f(L) := (L-1)^{-(1/d)} \cdot \lfloor \tfrac{L-1}{2} - \tfrac{L-1}{L-2} \rfloor \text{ with } d = 1 + \tfrac{2}{(L-4)\cdot \log 2} \text{ ,}$$

such that M does not contain an $AAP(L, 1)$.

It can easily be verified that $f(L)$ tends to $\frac{1}{2}$ for $L \to \infty$.

In this context one of the reviewers of [5] proved the following

Proposition 2. *Suppose* $L \geq 6$, $r := 6/\log 2 = 8{,}656...$. *Then for each positive integer* $n > L$ *there is a subset* $M \subset [0, n) \cap \mathbf{N}_0$ *with* $|M| > \frac{1}{20} \cdot n^{1-\frac{r}{L}}$ *which contains no* $AAP(L, 1)$.

Before we come to the general case we present the following

Lemma. *Let* n, $L \in \mathbf{N}$, $\lambda \in [0, 1]$, $L > 4$ *resp.* $L \geq 4$ *if* $\lambda < 1$. *We define two subintervals of* $[0, n]$, *namely*

$I_0 := [0, \frac{n}{2} - \frac{n(1+\lambda)}{2\cdot(L-1-\lambda)})$ *and* $I_1 := [\frac{n}{2} + \frac{n(1+\lambda)}{2(L-1-\lambda)}, n)$.

(They arise by deleting from $[0, n)$ *a middle segment, left closed, right open, of length* $\frac{n(1+\lambda)}{L-1-\lambda}$.)

Let now M be an $AAP(L, \lambda)$ *which is* $\subset I_0 \cup I_1$. *Then we have already* $M \subset I_0$ *or* $M \subset I_1$.

Proof. Let M be $= \{a_1, ..., a_L\}$ and $\mathcal{A} := (A_\nu)$, $\nu = 1, ..., L$, an arithmetic interval sequence with shrink factor λ, which satisfies $a_\nu \in A_\nu$ for $\nu = 1, ..., L$. If M would intersect I_0 and I_1, there would exist a last element a of M in I_0 and a first element b of M in I_1. Then we have

$$b - a > \tfrac{n}{L-1-\lambda} \cdot (1 + \lambda \qquad (1)$$

Let s be the step length of $\mathcal{A}$. Then there holds

$$(L-1-\lambda) \cdot s \leq n. \qquad (2)$$

Indeed, we have $n \geq a_L - a_1 \geq (L-2) \cdot s + (1-\lambda) \cdot s = (L-1-\lambda) \cdot s$.

On the other hand we have

The distance of two consecutive elements of M *is* $\leq s \cdot (\lambda + 1)$. (3)

Finally we have $\frac{n}{L-1-\lambda} \cdot (1+\lambda) < b - a \leq s \cdot (\lambda+1)$, which leads to $\frac{n}{L-1-\lambda} < s$, and this contradicts (2).

Definition 4. For the following we abbreviate $c := \frac{1}{2} - \frac{1+\lambda}{2(L-1-\lambda)}$ and $d := \frac{1+\lambda}{L-1-\lambda}$. The length of an interval I shall be denoted by $l(I)$.
In the previous lemma then I_0 and I_1 have the length $n \cdot c$. And the eliminated middle segment has the length $n \cdot d$. We have $c > 0$ because of $\frac{1+\lambda}{L-1-\lambda} < 1$. Now we formulate the main theorem:

Theorem 1. *Let n, L be natural numbers with $n \geq L > 4$, $\lambda \in [0,1]$. Then there exists a set $A \subset [0,n)$ of integers with $|A| \geq \lfloor n \cdot c^k \rfloor \cdot 2^k$ elements, where $c = \frac{1}{2} - \frac{1+\lambda}{2(L-1-\lambda)}$ and $k := \lceil \frac{\log \frac{L-1}{n}}{\log c} \rceil$, which has no $AAP(L,\lambda)$.*

Proof. We put $I := [0,n)$.We define the intervals I_0, I_1 of length $n \cdot c$ in the same way as in the lemma. Then we repeat the construction which lead from I to I_0 and I_1 : Starting with I_ν $(\nu = 0, 1)$ instead of I we construct two subintervals $I_{\nu 0}$ and $I_{\nu 1}$(left closed, right open) of I_ν of equal length $l(I_\nu) \cdot c$ by deleting from I_ν a middle segment (left closed, right open) of length $l(I_\nu) \cdot d$. Then $I_{\nu 0}$ and $I_{\nu 1}$ have the length $n \cdot c^2$. Again we delete from the four intervals $I_{\alpha\beta}$,where $\alpha, \beta \in \{0,1\}$ a middle segment of length $l(I_{\alpha\beta}) \cdot d = l(I_\alpha) \cdot d^2$ and obtain eight intervals $I_{\alpha\beta\gamma}$ of length $n \cdot c^3$. We continue this process so long until we obtain a set of intervals $I_{\alpha_1,\ldots,\alpha_k}$ $(\alpha_1, \ldots, \alpha_k \in \{0,1\})$, where k *is* the first natural number, so that the length of $I_{\alpha_1,\ldots,\alpha_k}$, which is $= n \cdot c^k$, becomes $\leq L - 1$.
The least integer k for which $l(I_{\alpha_1,\ldots\alpha_k}) \leq L - 1$ holds, satisfies $n \cdot c^k \leq L - 1$, that means $k \cdot \log c \leq \log \frac{L-1}{n}$. (Here and in the following log always denotes the natural logarithm.) This is equivalent with $k \geq \frac{\log \frac{L-1}{n}}{\log c}$, because log c is negative. So the least k is $k = \lceil \frac{\log \frac{L-1}{n}}{\log c} \rceil$.
The union A of the 2^k sets $I_{\alpha_1,\ldots,\alpha_k} \cap \mathbf{N}_0$ now has no AAP(L,λ), because according to our lemma, such a set had to be a subset of an $I_{\alpha_1}$with $\alpha_1 \in [0,1]$, and further a subset of an I_{α_1,α_2} and so on. Finally it had to be a subset of an interval $I_{\alpha_1,\ldots,\alpha_k}$, which is impossible since this half-open interval has a length $\leq L-1$, and insofar at most $L-1$ integers. (A half-open interval, whose length is an integer g, has at most g integers.) A half-open interval of length l has at least $\lfloor l \rfloor$ integers, and so the half-open interval $I_{\alpha_1 \cdots \alpha_k}$ has at least $\lfloor n \cdot c^k \rfloor$ integers. Then A has $|A| \geq \lfloor n \cdot c^k \rfloor \cdot 2^k$ elements and thus satisfies our assertion.

In the above formulas we now put in the values for c and k to obtain a better oversight.

Theorem 1'. *Let the assumptions of theorem 1 be satisfied. Then there exists a set $A \subset [0,n)$ of integers, which has no $AAP(L,\lambda)$, with a cardinality $|A| \geq n^{1-\frac{1+\lambda}{(L-2(1+\lambda))\log 2}} \cdot f(L)$, where $f(L) = (L-1)^{-(1+\frac{1+\lambda}{(L-2(1+\lambda))\log 2})^{-1}} \cdot \lfloor \frac{L-1}{2} - \frac{(L-1)(1+\lambda)}{2(L-1-\lambda)} \rfloor$. The function $f(L)$ tends to $\frac{1}{2}$ for $L \to \infty$.*

Proof. By definition of k we have $n \cdot c^{k-1} > L-1$ and thus $n \cdot c^k > c \cdot (L-1)$. This yields

$$|A| \geq \lfloor c \cdot (L-1) \rfloor \cdot 2^k. \tag{4}$$

We have

$$2^k \geq (2^{\log \frac{L-1}{n}})^{(\log c)^{-1}} = (e^{(\log 2) \cdot \log \frac{L-1}{n}})^{(\log c)^{-1}} = (e^{\log \frac{L-1}{n}})^{\frac{\log 2}{\log c}} = (\frac{L-1}{n})^{\frac{\log 2}{\log c}}. \tag{5}$$

Concerning $\frac{\log 2}{\log c}$ we obtain from the mean value theorem

$\frac{\log \frac{1}{2} - \log c}{\frac{1}{2} - c} = \log' \zeta = \frac{1}{\zeta}$ for a number $\zeta \in (c, \frac{1}{2})$. Then $\zeta = \frac{1}{2} - \delta \cdot (\frac{1}{2} - c) = \frac{1}{2} - \delta \cdot \frac{1+\lambda}{2(L-1-\lambda)}$ for some $\delta \in (0,1)$. (6)

From (6) we obtain $\zeta = \frac{\frac{1+\lambda}{2(L-1-\lambda)}}{-\log 2 - \log c} = \frac{1}{2} - \delta \cdot \frac{1+\lambda}{2(L-1-\lambda)}$ and from this $\frac{1+\lambda}{-\log 2 - \log c} = L - 1 - \lambda - \delta \cdot (1+\lambda)$. Then $-\log 2 - \log c = \frac{1+\lambda}{L-1-\lambda-\delta \cdot (1+\lambda)}$.

$$\log c = -\log 2 - \frac{1+\lambda}{L-(1+\delta)(1+\lambda)} > -\log 2 - \frac{1+\lambda}{L-2(1+\lambda)}. \tag{7}$$

Because of $\log c < 0$ we obtain from (7)

$\frac{\log 2}{\log c} < \frac{\log 2}{-\log 2 - \frac{1+\lambda}{L-2(1+\lambda)}} = -(1 + \frac{1+\lambda}{(L-2(1+\lambda)) \log 2})^{-1} = -d^{-1}$, where $d := 1 + \frac{1+\lambda}{(L-2(1+\lambda)) \log 2}$. (8)

The definition of d yields

$$d^{-1} > 1 - \frac{1+\lambda}{(L-2(1+\lambda)) \log 2}. \tag{9}$$

From (4),(5) and (8) and because of $\frac{L-1}{n} < 1$ we obtain finally

$|A| \geq \lfloor \frac{L-1}{2} - \frac{(L-1)(1+\lambda)}{2(L-1-\lambda)} \rfloor \cdot (\frac{L-1}{n})^{-(d^{-1})} = n^{(d^{-1})} \cdot (L-1)^{-(d^{-1})} \cdot \lfloor \frac{L-1}{2} - \frac{(L-1)(1+\lambda)}{2(L-1-\lambda)} \rfloor$. Because of (9) this is $\geq n^{1 - \frac{1+\lambda}{(L-2(1+\lambda)) \log 2}} \cdot f(L)$. The rest is easily verified.

References

[1] T.C. Brown, P. Erdös, A.R. Freedman, "Quasi-progressions and descending waves", *J. Combin. Theory* Ser. A 53, 1990, 81-95.

[2] T.C. Brown and D.R. Hare, "Arithmetic progressions in sequences with bounded gaps", *J. Combin. Theory* Ser. A 77, 1997, 222-227.

[3] P. Ding, A.R. Freedman, "Semi-progressions", *J.Combin.Theory* Ser. A 76, 1996, 99-107.

[4] E. Harzheim, "Weakly arithmetic progressions in sets of natural numbers", *Discrete Math.* 89, 1991, 105-107.

[5] E. Harzheim, "On weakly arithmetic progressions", *Discrete Math.* 138, 1995, 255-260.

[6] M.B. Nathanson, "Arithmetic progressions contained in sequences with bounded gaps", *Canad. Math. Bull* 23, 1980, 491-493.

[7] J.R. Rabung, "On applications of van der Waerden's theorem", *Math. Mag.* 48, 1975, 142-148.

[8] A. Sarközy, "Some metric problems in the additive number theory I", *Annales Univ. Sci. Budapest*, Eötvös 19, 1976, 107-127.

A METHOD TO ESTIMATE PARTIAL-PERIOD CORRELATIONS

Aimo Tietäväinen *

Department of Mathematics and TUCS
University of Turku
FIN-20014 Turku, Finland

Abstract: Many applications require large families of sequences with good correlation properties. Some of the best families can be constructed by means of cyclic codes. The full-period correlation of such a family is closely connected with a complete sum of additive characters. In several important special cases it can be easily estimated. On the other hand, the partial period correlations, which are connected with certain incomplete sums of additive characters, are not easy to estimate. A device for estimating is the finite Fourier transform. This approach, which in fact is a modification of an old number theoretic method due to Vinogradov, needs bounds for hybrid sums of additive and multiplicative characters. In this survey we apply this approach in three cases: the m-sequence, the set of dual-BCH sequences, and the small Kasami set.

CORRELATION

Assume that there are $K(>1)$ sender-receiver pairs (called users), all of whom simultaneously want to communicate over the same channel. To allow each receiver to distinguish its signal from that of the other users, each user U_i uses its own code word $\mathbf{x}_i = (x_i(t))_{t=0}^{n-1}$. Consider in this talk the binary case which is most often used in practice. Then $\mathbf{x}_i \in \mathbf{F}_2^n$. Let $\xi_i = (\xi_i(t))_{t=0}^{n-1}$ where

$$\xi_i(t) = \begin{cases} 1 \text{ if } x_i(t) = 0, \\ -1 \text{ if } x_i(t) = 1. \end{cases}$$

When the user U_i wants to transmit the symbol $a \in \{0, 1\}$, it in fact sends ξ_i if $a = 0$ and $-\xi_i$ if $a = 1$.

*The work was supported by the Academy of Finland under Grant 37358

I. Althöfer et al. (eds.), Numbers, Information and Complexity, 21-27.

The transmissions of the K users are not necessarily synchronized in time and usually only a small fraction of the K users will be transmitting at any given time. Thus the received signal σ will be a sum of certain shifts of the vectors $\delta_j\xi_j$ ($j = 1, 2, \dots, K$) where δ_j is 1 or -1 if S_j has been active andδ_j is 0 otherwise. Consider now σ from the point of view of the ith receiver. If the transmissions would be synchronized and if we know that the ith sender is active, we could calculate the dot product

$$\xi_i \cdot \sigma = \xi_i \cdot \sum_{j=1}^{K} \delta_j \xi_j = \delta_i n + \sum_{\substack{j=1 \\ j \neq i}}^{K} \delta_j \xi_i \cdot \xi_j$$

and decode

$$\sigma \to \begin{cases} 0 \text{ if } \xi_i \cdot \sigma > 0, \\ 1 \text{ if } \xi_i \cdot \sigma < 0. \end{cases}$$

(If it is not known whether the ith sender is active or not, we could decode in the following way:

$$\sigma \to \begin{cases} 0 \text{ if } \xi_i \cdot \sigma > n/2, \\ 1 \text{ if } \xi_i \cdot \sigma < -n/2, \\ \text{"not active" otherwise.)} \end{cases}$$

Let $d(\mathbf{x}, \mathbf{y})$ be the Hamming distance between the vectors $\mathbf{x}$ and $\mathbf{y}$. In order to get good results in the decoding above we should demand that for $i \neq j$ the moduli of the correlations

$$\vartheta(\mathbf{x}_i, \mathbf{x}_j) := \xi_i \cdot \xi_j = \sum_{t=0}^{n-1} (-1)^{x_i(t)+x_j(t)} = n - 2d(\mathbf{x}_i, \mathbf{x}_j)$$

are small. Since the transmissions of the users are not necessarily synchronized and since both the words ξ_j and $-\xi_j$ can be sent, we have to demand much more. For any binary sequence $\mathbf{x} = (x(0), x(1), \dots, x(n-1))$, let $S\mathbf{x} = (x(n-1), x(0), \dots, x(n-2))$ and $T\mathbf{x} = (1 + x(n-1), x(0), \dots, x(n-2))$. If $X = \{\mathbf{x}_1, \dots, \mathbf{x}_K\}$, the numbers $\vartheta(\mathbf{x}_i, S^k\mathbf{x}_j)$ are called the even correlations and the numbers $\vartheta(\mathbf{x}_i, T^k\mathbf{x}_j)$ are called the odd correlations of X. In the trivial case $i = j$, $k = 0$ these numbers are equal to n. In order to get good results in the decoding above it is natural to demand that the following two numbers are small:

$$\begin{aligned} \vartheta(X) &= \max_{i,j,k} |\vartheta(\mathbf{x}_i, S^k\mathbf{x}_j)|, \\ \hat{\vartheta}(X) &= \max_{i,j,k} |\vartheta(\mathbf{x}_i, T^k\mathbf{x}_j)|, \end{aligned}$$

where the maxima are taken over all values $1 \leq i \leq K$, $1 \leq j \leq K$, $0 \leq k \leq n-1$; $k \neq 0$ if $i = j$. If the ith sender is active, the ith receiver now achieves synchronization by computing $S^k\xi_i \cdot \sigma$ and varying k until the modulus of this dot product reaches a peak. Having achieved synchronization the decoding rule given above is used.

The (maximal nontrivial) even (or full-period) correlation $\vartheta(X)$ can be quite precisely estimated for several important sets of sequences. The (maximal nontrivial) odd correlation $\hat{\vartheta}(X)$ is closely connected with partial-period correlations in the following way.

Let B be a subset of consecutive elements of the set N, the set of residues modulo n. Define

$$\vartheta_B(\mathbf{x}, \mathbf{y}) = \sum_{t \in B} (-1)^{x(t)+y(t)}$$

and

$$\vartheta_B(X) = \max |\vartheta_B(\mathbf{x}_i, S^k \mathbf{x}_j)|$$

where the maximum is taken over the same values i, j, k as above. Then

$$\vartheta(\mathbf{x}_i, T^k \mathbf{x}_j) = -\vartheta_{B_k}(\mathbf{x}_i, S^k \mathbf{x}_j) + \vartheta_{\bar{B}_k}(\mathbf{x}_i, S^k \mathbf{x}_j),$$

where $B_k = \{0, 1, \ldots, k-1\}$ and $\bar{B}_k = \{k, k+1, \ldots, n-1\}$, and so

$$\hat{\vartheta}(X) \leq \max\{\vartheta_{B_k}(X) + \vartheta_{\bar{B}_k}(X) | 1 \leq k \leq n-1\}.$$

Thus

$$\hat{\vartheta}(X) \leq 2\bar{\vartheta}(X)$$

where the (maximal nontrivial) *partial-period correlation* $\bar{\vartheta}(X)$ is defined by the equation

$$\bar{\vartheta}(X) = \max_B \vartheta_B(X)$$

when the maximum is taken over all subsets B of consecutive elements of the set N.

SETS CONSTRUCTED BY MEANS OF CYCLIC CODES

Assume that $q = 2^m$, $n = q - 1$ and γ is a primitive element of $\mathbf{F}_q$. Define the trace function Tr by

$$Tr(a) = a + a^2 + a^4 + \ldots + a^{2^{m-1}} \text{ for all } a \in \mathbf{F}_q.$$

Since

$$Tr(a + b) = Tr(a) + Tr(b) \text{ for all } a, b \in \mathbf{F}_q,$$

the function e defined by

$$e(x) = (-1)^{Tr(x)} \text{ for all } x \in \mathbf{F}_q$$

is an additive character of $\mathbf{F}_q$.Let P be an additive subgroup of $\mathbf{F}_q[x]$. Define

$$C = C(P) = \{\mathbf{c} = \mathbf{c}(f) | f \in P\}$$

where

$$\mathbf{c} = \mathbf{c}(f) = (Tr(f(1)),\ Tr(f(\gamma)),\ Tr(f(\gamma^2)), \ldots,\ Tr(f(\gamma^{n-1}))).$$

Then

$$\begin{aligned}\sum_{i=0}^{n-1} e(f(\gamma^i)) &= |\{i|e(f(\gamma^i)) = 1\}| - |\{i|e(f(\gamma^i)) = -1\}| \\ &= |\{i|Tr(f(\gamma^i)) = 0\}| - |\{i|Tr(f(\gamma^i)) = 1\}| \\ &= (n - wt(\mathbf{c})) - wt(\mathbf{c}) \\ &= n - 2wt(\mathbf{c}),\end{aligned}$$

where $wt(\mathbf{c})$ is the Hamming weight of the vector $\mathbf{c}$, and therefore

$$wt(\mathbf{c}) = \frac{1}{2}(n - \sum_{i=0}^{n-1} e(f(\gamma^i))).$$

Any subspace C of $\mathbf{F}_2^n$ with at least two elements may be called a linear binary code of length n. A linear code C is called cyclic if for all vectors $\mathbf{c}$ in C the cyclic shift $S\mathbf{c}$ is also in C. It is known (see, e.g., [2, Theorem 4.2]) that if C is cyclic and of length n then there is a polynomial set P of the form $\{\sum_{i=1}^{u} \alpha_i x^{s_i} | \alpha_i \in \mathbf{F}_q\}$ such that $C = C(P)$.

Assume that C is a binary cyclic code of length n. We say that two code words are conjugate if one is a cyclic shift of the other. Assume that $X = \{\mathbf{x}_1, \ldots, \mathbf{x}_K\}$ is a set of representatives of all the conjugacy classes that have n code words. If $k \neq 0$ or $i \neq j$ then $\mathbf{x}_i + S^k\mathbf{x}_j$ is a nonzero code word, say $\mathbf{c}$, in C. Thus

$$\vartheta(\mathbf{x}_i, S^k\mathbf{x}_j) = n - 2wt(\mathbf{c}) = \sum_{t=0}^{n-1} e(f(\gamma^t))$$

and so

$$\vartheta(X) \leq \max_{f\in P, f\neq 0} |\sum_{t\in N} e(f(\gamma^t))| \leq 1 + \max_{f\in P, f\neq 0} |\sum_{x\in \mathbf{F}_q} e(f(x))|.$$

Similarly,

$$\vartheta_B(X) \leq \max_{f\in P, f\neq 0} |\sum_{t\in B} e(f(\gamma^t))|,$$

and, as we defined above,

$$\bar{\vartheta}(X) = \max_B \vartheta_B(X).$$

Thus $\vartheta(X)$ and $\bar{\vartheta}(X)$ can be estimated by means of complete (full-period) and incomplete (partial-period) character sums, respectively. In this talk we concentrate on partial-period correlations and so on incomplete sums.

INCOMPLETE SUMS

Let n, N and B be defined as above and let φ be a mapping from N to the field of complex numbers. Now we use finite Fourier transforms in order to consider the incomplete sum $\sum_{t\in B}\varphi(t)$. Let N' be the dual group of the additive group N and let ξ_0 be the trivial character in N'. Then it is very well known that

$$\sum_{\xi\in N'}\xi(x-t)=\begin{cases} n \text{ if } x=t,\\ 0 \text{ otherwise.}\end{cases}$$

If we define

$$\Phi(\xi)=\sum_{x\in\mathbf{N}}\varphi(x)\xi(x),$$

we thus have

$$\sum_{\xi\in N'}\Phi(\xi)\xi(-t)=\sum_{x\in N}\varphi(x)\sum_{\xi\in N'}\xi(x-t)=n\varphi(t).$$

Therefore

$$\begin{aligned} n\sum_{t\in B}\varphi(t) &= \sum_{\xi\in N'}\Phi(\xi)\sum_{t\in B}\xi(-t)\\ &= |B|\Phi(\xi_0)+\sum_{\xi\neq\xi_0}\Phi(\xi)\sum_{t\in B}\xi(-t)\\ &= |B|\sum_{x\in N}\varphi(x)+\theta\Delta(\varphi)E(B)\end{aligned}$$

where θ is a complex number with modulus at most 1,

$$\Delta(\varphi)=\max\{|\sum_{x\in N}\varphi(x)\xi(x)|:\xi\in N',\ \xi\neq\xi_0\}$$

and

$$E(B)=\sum_{\xi\neq\xi_0}|\sum_{t\in B}\xi(-t)|.$$

It is well known [11, Problem III.11.c] that

$$E(B)<n\ln n.$$

Thus

$$|\sum_{t\in B}\varphi(t)|<|\sum_{x\in N}\varphi(x)|+\Delta(\varphi)\ln n. \tag{1}$$

MAIN RESULTS

Let $\varphi(t)=e(f(\gamma^t))$ where γ is a primitive element of $\mathbf{F}_q$. Thus, by inequality (1) and page 24,

$$\bar{\vartheta}(X)<\vartheta(X)+\Delta(\varphi)\ln n. \tag{2}$$

Then in order to estimate $\bar{\vartheta}(X)$ we should estimate the function

$$\Delta(\varphi) = \max_{\chi \neq \chi_0} |\sum_{t=0}^{n-1} e(f(\gamma^t))\chi(\gamma^t)| = \max_{\chi \neq \chi_0} |\sum_{x \in \mathbf{F}_q} e(f(x))\chi(x)|$$

where $\chi(\gamma^t) = \xi(t)$ and so χ runs over all nontrivial multiplicative characters of $\mathbf{F}_q$ (by definition $\chi(0) = 0$ if $\chi \neq \chi_0$). We may use the following two lemmas (see [12] and [5]).

Lemma 1. *If $f \in \mathbf{F}_q[x]$, $\deg f = d$, $\gcd(d, q) = 1$ and χ is a nontrivial multiplicative character of $\mathbf{F}_q$ then*

$$|\sum_{x \in \mathbf{F}_q} e(f(x))\chi(x)| \leq d\sqrt{q}.$$

It is well known (see, e.g., [6, p. 193]) that the special case of Lemma 1 where $f(x) = x$ can be proved very easily.

Lemma 2. *Assume that $q = 2^{2r}$, $\alpha x + \beta x^{2^r+1} \in \mathbf{F}_q[x]$, $\alpha \neq 0$ and χ is a nontrivial multiplicative character of $\mathbf{F}_q$. Then*

$$|\sum_{x \in \mathbf{F}_q} e(\alpha x + \beta x^{2^r+1})\chi(x)| \leq 2\sqrt{q}.$$

Using these lemmas we get the following results.

I. m-sequences ([8], [9]). Now $P = \{\alpha x | \alpha \in \mathbf{F}_q\}$. It is well known [1, Section 3.1.1] that $\vartheta(X) = 1$ and thus, by (2) and Lemma 1,

$$\bar{\vartheta}(X) < 1 + \sqrt{q}\ln(q-1).$$

II. The set of dual-BCH sequences [7, Proposition 3]. Now $P = \{\sum_{i=1}^{u} \alpha_i x^{2i-1} | \forall i : \alpha_i \in \mathbf{F}_q\}$. Then [2, Theorem 4.10] $\vartheta(X) \leq (2u-2)\sqrt{q} + 1$ and so, by (2) and Lemma 1,

$$\bar{\vartheta}(x) < (2u-1)\sqrt{q}(\ln(q-1)+1).$$

III. The small Kasami set [4]. In this case $q = 2^{2r}$ and $P = \{\alpha x + \beta x^{2^r+1} | \alpha \in \mathbf{F}_q,\ \beta \in \varepsilon \mathbf{F}_{2^r}\}$ where ε is a fixed element of the set $\mathbf{F}_q - \mathbf{F}_{2^r}$. Now [1, Example 6.4] $\vartheta(X) = \sqrt{q} + 1$ and so, by (2) and Lemma 2,

$$\bar{\vartheta}(X) < 2\sqrt{q}(\ln(q-1)+1).$$

Shanbhag, Kumar and Helleseth [10] applied this approach to the Galois ring sequences and Koponen and Lahtonen [3] to binary Shanbhag-Kumar-Helleseth sequences.

PROBLEMS

Above we have used the finite Fourier transform approach to get upper bounds for the maximal nontrivial partial-period correlation $\bar{\vartheta}(X)$ in three classical cases: the m-sequence, the set of dual-BCH sequences, and the small Kasami set. We have two important open problems:

1) Can similar results be found in the other classical cases; i.e., for the Gold set and for the large and very large Kasami set?

2) Using the approach above we get for $\bar{\vartheta}(X)$ upper bounds of the form $O(\sqrt{q}\ln q)$. Is it possible to replace $\ln q$ by an essentially smaller function; e.g., by $\ln\ln q$?

References

[1] T. Helleseth and P.V. Kumar, "Sequences with low correlation", In: *Handbook of Coding Theory* (eds. V.S. Pless, R.A. Brualdi and W. C. Huffman), to appear.

[2] I. Honkala and A. Tietäväinen, "Codes and number theory", In: *Handbook of Coding Theory* (eds. V.S. Pless, R.A. Brualdi and W. C. Huffman), to appear.

[3] S. Koponen and J. Lahtonen, "On the aperiodic and odd correlations of the binary Shanbhag-Kumar-Helleseth sequences", *IEEE Trans. Information Theory,* 43, 1997, 1593–1596.

[4] J. Lahtonen, "On the odd and the aperiodic correlation properties of the Kasami sequences", *IEEE Trans. Information Theory,* 41, 1995, 1506–1508.

[5] J. Lahtonen, "Examples of small hybrid sums", *London Mathematical Society, Lecture Notes,* Series 233, 1996, 155–161.

[6] R. Lidl and H. Niederreiter, *Finite Fields,* Addison-Wesley, 1983.

[7] S. Litsyn and A. Tietäväinen, "Character sum constructions of constrained error-correcting codes", *Appl. Algebra in Engineering, Communication and Computing,* 5, 1994, 45–51.

[8] R.J. McEliece, "Correlation properties of sets of sequences derived from irreducible cyclic codes", *Information and Control,* 45, 1980, 18–25.

[9] D.V. Sarwate, "An upper bound on the aperiodic autocorrelation function for a maximal-length sequence", *IEEE Trans. Information Theory,* 30, 1984, 685–687.

[10] A.G. Shanbhag, P.V. Kumar and T. Helleseth, "An upper bound for the aperiodic correlation of weighted-degree CDMA sequences", *Proc. of the 1995 IEEE International Symposium on Information Theory,* 1995, 92.

[11] I.M. Vinogradov, *Elements of Number Theory,* Dover, 1954.

[12] A. Weil, "Sur les courbes algébriques et les veriétés qui s'en déduisent", *Actualités Sci. Ind.,* no. 1041, Hermann, Paris, 1948.

SPLITTING PROPERTIES IN PARTIALLY ORDERED SETS AND SET SYSTEMS

Rudolf Ahlswede and Levon H. Khachatrian

Universität Bielefeld, Fakultät für Mathematik
Postfach 100131, 33501 Bielefeld, Germany

Abstract: It was shown in [1] that in any "dense" finite poset $\mathcal{P} = (P, <)$ (e.g. in the Boolean lattice) every maximal antichain S may be partitioned into disjoint subsets S_1 and S_2, such that the union of the upset of S_1 with the downset of S_2 yields the entire poset: $U(S_1) \cup D(S_2) = P$.
Under suitable denseness assumptions we establish splitting properties in great generality for infinite posets, directed graphs and set systems. We show also that for countable posets the conjecture (4.4) of [1] is not true. The poset of squarefree integers serves as an example.
It seems also to be of interest that already for the finite Boolean lattice there are antichains which splitt cardinalitywise only in an extremely unbalanced way.
Finally we introduce new notions of splitting, called Y–splitting, λ–splitting and X–splitting. For instance in a Y–splitting $\{S_1, S_2\}$ in addition to the property above we have also that $U(S_1) \cup D(S_1) \cup S_2 = P$. We establish first results in a challenging new area.

BASIC DEFINITIONS FOR POSETS

Downsets, upsets, generators, antichains

Let $\mathcal{P} = (P, <)$ be a partially ordered set (poset) and let H be a subset of P. The *downset* $D(H)$ of the subset H is

$$D(H) = \{x \in P : \exists s \in H (x \leq s)\}. \tag{1.1}$$

The upset $U(H)$ of H is

$$U(H) = \{x \in P : \exists s \in H (s \leq x)\}. \tag{1.2}$$

I. Althöfer et al. (eds.), Numbers, Information and Complexity, 29-44.

We introduce also the sets

$$D^*(H) = \{x \in P : \exists s \in H(x < s)\} \tag{1.3}$$

and

$$U^*(H) = \{x \in P : \exists s \in H(s < x)\}. \tag{1.4}$$

A subset $G \subset P$ is called a *generator* of $\mathcal{P}$, if

$$U(G) \cup D(G) = P. \tag{1.5}$$

A generator G of $\mathcal{P}$ is called *minimal*, if no proper subset of G is a generator of $\mathcal{P}$.
A subset $S \subset P$ is called *antichain or Sperner system*, if no two elements of S are comparable. An antichain S is maximal (or saturated) if for every antichain $S' \subset P$, $S \subset S'$ implies $S = S'$. It is easy to see that an antichain S is maximal iff it is a generator of $\mathcal{P}$. We also remark that a minimal generator of $\mathcal{P}$ is not necessary an antichain.

A splitting property and notions of denseness

We say that $H \subset P$ has the *splitting property*, if there exists an $H_1 \subset H$ with

$$U(H_1) \cup D(H \setminus H_1) = P. \tag{1.6}$$

Of course, for H to have the splitting property it is necessary that H is a generator of $\mathcal{P}$. We say that $\mathcal{P}$ has the splitting property, if every maximal antichain has the splitting property.
Now we introduce notions of denseness in $\mathcal{P}$ for $H \subset P$.
If for every *open interval* $< x, y >= \{z \in P : x < z < y\}$ with endpoints $x, y \in P \setminus H$:
(d_1) $\quad \langle x, y\rangle \cap H \neq \phi \Rightarrow |\langle x, y\rangle \cap P| \geq 2$,
then we call H d_1*–dense* in $\mathcal{P}$,
(d_2) $\quad \langle x, y\rangle \cap H \neq \phi \Rightarrow |\langle x, y\rangle \cap H| \geq 2$,
then we call H d_2*–dense* in $\mathcal{P}$.
Furthermore, if for every open interval $\langle x, y\rangle$ with endpoints $x, y \in P$:
(d_2^*) $\quad \langle x, y\rangle \cap H \neq \phi \Rightarrow |\langle x, y\rangle \cap H| \geq 2$,
then we call H d_2^**–dense* in $\mathcal{P}$.
Clearly, a d_2^*–dense set is also d_2–dense and a d_2–dense set is also d_1–dense.
Remarks:

- In the special case $H = P$ in [1] for d_2^*–denseness the term "$\mathcal{P}$ is weakly dense" is used. Also, $\mathcal{P}$ is strongly dense, if for any non–empty interval $\langle x, y\rangle$ and any $z \in \langle x, y\rangle$ there is a $z' \in \langle x, y\rangle$ *incomparable* with z. For finite P the notions coincide. Then $\mathcal{P}$ is said to be dense.

- If H is an antichain, then d_2–dense coinsides with d_2^*–dee and they are the same as "the antichain H is dense in $\mathcal{P}$".

Finally it is convenient to have the following notation:
For $H, G \subset P$ we write $H >\!\!|\!\!< G$ iff for all $h \in H$ and all $g \in G$ elements h and g are incomparable. For $s, s' \in P$ and $G \subset P$ we also write $s >\!\!|\!\!< s'$ instead of $\{s\} >\!\!|\!\!< \{s'\}$ and $s >\!\!|\!\!< G$ instead of $\{s\} >\!\!|\!\!< G$.
Similarly, we write

$$U(s) = U(\{s\}), U^*(s) = U^*(\{s\}), D(s) = D(\{s\}), D^*(s) = D^*(\{s\}). \quad (1.7)$$

REDUCTION OF GENERATORS TO ANTICHAINS

We begin with an auxiliary result.
Lemma 1 For any poset $\mathcal{P}$ let $C \subset P$ be a set such that every element $c \in C$ is comparable with at least one other element c' of C. Then

(i) there exists a $C_1 \subset C$ such that for $C_2 = C \setminus C$, we have the properties: $\forall a \in C_1\ \exists b \in C_2$ such that $a > b$, $\forall b \in C_2\ \exists a \in C_1$ such that $b < a$.

(ii) there exists a $C_1 \subset C$ with $D(C) \cup U(C) = D(C_1) \cup U(C_2)$.

Proof: (i) Let $A \subset C$ be a maximal antichain in C. Its existence is guaranteed by Zorn's Lemma. By the maximality of the antichain A

$$C \subset D^*(A) \cup U^*(A) \cup A.$$

We write A in the form

$$A = A_{\max} \cup A_{\min} \cup A_0,$$

where

$$A_{\max} = \{a \in A : \nexists c \in C \text{ with } c > a\}, A_{\min} = \{a \in A : \nexists c \in C \text{ with } c < a\},$$
$$A_0 = A \setminus (A_{\max} \cup A_{\min}).$$

By our assumption on C $A_{\max} \cap A_{\min} = \phi$ and also one of the sets $D^*(A)$ and $U^*(A)$ is not empty. W.l.o.g. we can assume that $D^*(A) \neq \phi$ and consider the sets

$$C_1 = \big(A_{\max} \cup U^*(A) \cup A_0\big) \cap C, \quad (2.1)$$

$$C_2 = \big(A_{\min} \cup D^*(A)\big) \cap C, \quad (2.2)$$

which clearly satisfy $C_2 = C \setminus C_1$.
One also readily verifies that they can serve as sets whose existence is claimed in (i) and (ii).
Let now $G \subset P$ be a generator of $\mathcal{P}$. Partition it into $G = G_1 \dot\cup G_2$, where

$$G_1 = \{g \in G : \exists g' \in G, g' \neq g \text{ and } g >\!\!|\!\!< g'\}, \quad (2.3)$$

and $G_2 = G \setminus G_1$. Obviously G_2 is an antichain in $\mathcal{P}$.
We consider the poset $\mathcal{P}' = (P', <)$, where

$$P' = P \setminus (D(G_1) \cup U(G_1)). \tag{2.4}$$

Since G is a generator of $\mathcal{P}$, G_2 is a generator and hence maximal antichain in $\mathcal{P}'$. This and Lemma 1 yield the following result on reduction.
Proposition 1. Let $G \subset P$ be a generator of $\mathcal{P}$ and let G_1, G_2 be defined as above, $G_1 \dot{\cup} G_2 = G$. Now G has the splitting property in $\mathcal{P}$ iff the maximal antichain G_2 in $\mathcal{P}'$ has the splitting property in $\mathcal{P}'$.
The next and last result on reduction is readily verified.
Proposition 2. Let G be any d_1–dense (resp. d_2-dense) subset of P (not necessarily a generator) and define G_1, G_2 and $\mathcal{P}'$ as in (2.3) and (2.4). Then G_2 is d_1–dense (resp. d_2–dense) in the poset $\mathcal{P}'$.

SPLITTING OF D_1–DENSE ANTICHAINS

Under the weakest of our density assumptions and further regularity conditions we present next a splitting result for not necessarily finite posets.

Theorem 1 Let $\mathcal{P}$ be a poset and let $S \subset P$ be a maximal antichain, which is d_1–dense.
Additionally, we assume that

(i) in $D^*(S)$ exists an antichain $\underline{S}$ with $D(\underline{S}) = D^*(S)$

(ii) in $U^*(S)$ exists an antichain $\overline{S}$ with $U(\overline{S}) = U^*(S)$

(iii) S carries a well–ordering μ with the property: for all $u \in \overline{S}$ the set $A(u) = \{s \in S : s < u\}$ has a maximal element according to μ.

Then S has the splitting property.

Proof: For every $d \in \underline{S}$ we consider the set

$$B(d) = \{s \in S : d < s\}. \tag{3.1}$$

Let $f(d)$ be its minimal element according to μ. We consider $S_1 = \bigcup_{d \in \underline{S}} \{f(d)\}$ and prove that it gives the desired splitting. Since S is a maximal antichain, of course

$$D(S) \cup U(S) = P.$$

From condition (i) and the construction of S_1 we get

$$D(S_1) = D(S) \setminus (S \setminus S_1).$$

It remains to prove that

$$U(S \setminus S_1) = U(S) \setminus S_1.$$

By condition (ii) for this it suffices to show that

$$\overline{S} \subset U(S \setminus S_1).$$

Suppose then, to the opposite, that for some $u \in \overline{S}$ we have $u \notin U(S \setminus S_1)$. We consider the set

$$A(u) = \{s \in S : s < u\}. \tag{3.2}$$

Since $u \notin U(S \setminus S_1)$, necessarily $A(u) \subset S_1$. Let $s_0 \in A(u)$ be according to μ the maximal element of $A(u)$, which exists by (iii). From the construction of S_1 it follows that $s_0 = f(d_0)$ for some $d_0 \in \underline{S}$.
We consider now the open interval $\langle d_0, u \rangle$, which contains $s_0 \in S$. Since S is d_1–dense there is a $t \in P$ with $t \neq s_0$ and $d_0 < t < u$.
Furthermore, since $d_0 \in \underline{S}$ and by (i) $\underline{S}$ is antichain with $D(\underline{S}) = D^*(S)$, we know that $t \notin D^*(S)$. Symmetrically, by (ii), $t \notin U^*(S)$, and hence $t \in S$. Now we have $t \in A(u)$, since $t < u$, and $t \in B(d_0)$, since $d_0 < t$. However, s_0 is the maximal element of $A(u)$ in the well–ordering μ. Hence, s_0 is not the minimal element in $B(d_0)$ according to μ. Therefore, $s_0 \neq f(d_0)$, which is a contradiction.

Corollary 1 Let S be a maximal antichain in a finite poset $\mathcal{P}$. If S is d_1–dense in $\mathcal{P}$, then S has the splitting property.
Remark 3: Theorem 2.1 of [1] is a special case of this Corollary and also Theorem 3.1 of [1] easily follows. Actually in case of finite posets the proof above closely resembles the second proof of [1].
An instructive infinite poset is $\mathcal{Z} = (Z, <)$, where Z is the set of 0–1–sequences and for two sequences $a = (a_1, a_2, \dots), b = (b_1, b_2, \dots) \in Z$ $a \leq b$ exactly if $a_i \leq b_i$ for all $i = 1, 2, \dots$. Clearly, any subset $H \subset Z$ is d_1–dense.

Corollary 2 Let $S \subset Z$ be a maximal antichain, whose members have at most k ones. Then S has the splitting property.

Proof: The maximal elements in $D^*(S)$ form an antichain $\underline{S}$ and the minimal elements in $U^*(S)$ form an antichain $\overline{S}$. They guarantee (i) and (ii). Since for $u \in \overline{S}$ $A(u)$ is finite, also (iii) holds.

THE LATTICE OF SQUARE–FREE NUMBERS DOES NOT HAVE THE SPLITTING PROPERTY

Let $Z^* \subset Z = \{0,1\}^\infty$ be the set of all 0–1–sequences with finitely many ones. Those sequences can be identified with the sequences of exponents in the prime number representation of square–free numbers $\mathbb{N}^*$. The order relation in Z, and thus in Z^* says in terms of $\mathbb{N}^*$: for $a, b \in \mathbb{N}^*$ $a \leq b$ iff $a \mid b$ (a divides b). According to this relation the upset of $H \subset \mathbb{N}^*$ is the set of multiples of H

$$M(H) = \{n \in \mathbb{N}^* : \ell | n \text{ for some } \ell \in H\} \tag{4.1}$$

and the downset is the set of divisors of H

$$D(H) = \{n \in I\!N^* : n|\ell \text{ for some } \ell \in H\}. \tag{4.2}$$

Theorem 2 The poset of square–free numbers does not have the splitting property.

Remark 4: $I\!N^*$ is a countable and strongly dense poset. Therefore Theorem 2 refutes Conjecture 4.4 of [1].

Proof of Theorem 2: We construct a maximal antichain S without the splitting property as follows:
We choose an arbitrary $T_1 \in I\!N$ and consider the set

$$A_1 = \{n \in I\!N^* : T_1 < n \le 2\,T_1\}.$$

Next we choosse any $T_2, T_2 > 8\,T_1^2$, and define the set

$$A_2 = \{n \in I\!N^* : n \in (T_2, 2\,T_2] \setminus M(A_1)\}.$$

Inductively, for every $k > 1$ we choose $T_k, T_k > 8\,T_{k-1}^2$, and define the set

$$A_k = \left\{n \in I\!N^* : n \in (T_k, 2T_k] \setminus M\left(\bigcup_{i=1}^{k-1} A_i\right)\right\}.$$

Finally we define

$$S = \bigcup_{i=1}^{\infty} A_i. \tag{4.3}$$

- Clearly, numbers in A_i are incomparable and $a \in A_i$, $b \in A_j$ $(i < j)$ are incomparable, because we have excluded the multiples of A_i in the definition of A_j and $b > a$. Thus S is an antichain (also called primitive sequence in Number Theory).

- We show next that S is maximal, that is, $I\!N^* = M(S) \cup D(S)$. If this is not the case, then an $\alpha \in I\!N^*$ with $\alpha \notin M(S) \cup D(S)$ and, particularly, $\alpha \notin (T_i, 2\,T_i]$ for $i = 1, 2, \ldots$ exists. Hence $2\,T_k < \alpha \le T_{k+1}$ for some $k \in I\!N$ or $2 \le \alpha \le T_1$. It follows from Bertrand's postulate that there exists a prime $p \in \mathbb{P}$ (the set of all primes) such that

$$\frac{T_{k+2}}{\alpha} < p \le 2\frac{T_{k+2}}{\alpha} \text{ or, equivalently, } T_{k+2} < \alpha \cdot p \le 2 \cdot T_{k+2}.$$

Since $T_{k+2} > 8\,T_{k+1}^2$ and $2\,T_k < \alpha \le T_{k+1}$, we conclude that

$$2\,T_{k+1} < p < T_{k+2}.$$

Hence $p > \alpha$ and $\alpha \cdot p \in I\!N^*$.

Now, if $\alpha \cdot p \in M(S)$ or (equivalently) $\alpha'|\alpha \cdot p$ for some $\alpha' \in S$ ($\alpha' \leq 2\ T_{k+1}$), then, since $p \in \mathbb{P}$ and $p > 2\ T_{k+1}$ we have $\alpha'|\alpha$ and hence $\alpha \in M(S)$, a contradiction.

On the other hand, if $\alpha\ p \notin M(S) \setminus S$ then the conditions $T_{k+2} < \alpha\ p \leq 2\ T_{k+2}$, $\alpha\ p \in I\!N^*$ yield $\alpha\ p \in S$. But then $\alpha \in D(S)$, again a contradiction.

- Finally we show that the maximal antichain S does not have the splitting property.

  Let us assume to the opposite that for some $S_1 \subset S$

$$D(S_1) \cup M(S \setminus S_1) = I\!N^*.$$

Necessarily $S_1 \neq \phi$, because for example all squarefree integers from $[1, T_1)$ and all primes from $(2\ T_k, T_{k+1}]$, $k \in I\!N$, are not in $M(S)$.

Let then $\beta \in S_1$ and $T_k < \beta \leq 2\ T_k$ for some $k \in \mathbb{N}$. From Bertrand's postulate we know that there is a prime q with $2\ T_k < q \leq 4\ T_k$. Consider the integer $\beta \cdot q$. Obviously $\beta \cdot q \in I\!N^*$ and since $T_{k+1} > 8\ T_k^2$ we have

$$2\ T_k < \beta \cdot q < T_{k+1}.$$

Clearly, $\beta \cdot q \notin D(S)$, because S is an antichain and $\beta \in S$.

On the other hand $\beta{\cdot}q \in M(S{\setminus}S_1)$ would imply $\beta'|\beta q$ for some $\beta' \in S{\setminus}S_1$ and then $\beta' \leq 2\ T_k$, because $\beta \cdot q < T_{k+1}$, and hence $\beta'|\beta$, because $2\ T_k < q$. But then β', β are in the antichain S and at the same time comparable. This contradiction implies that for the integer $\beta \cdot q \in \mathbb{N}^*$

$$\beta q \notin D(S_1) \cup M(S \setminus S_1).$$

ON THE SPLITTING RATIO OF MAXIMAL ANTICHAINS IN THE BOOLEAN POSET $\mathcal{L}^N = \{0,1\}^N$

To fix ideas, let us consider the maximal antichain $S = \binom{[n]}{\ell}$ in $\mathcal{L}^n$. For a splitting $S = S_1 \dot{\cup} S_2$ necessarily $D(S_1) \supset \binom{[n]}{\ell-1}$ and $U(S_2) \supset \binom{[n]}{\ell+1}$, $1 \leq \ell \leq n-1$, and therefore

$$\begin{aligned} \frac{\ell}{\ell+1}\binom{n}{\ell} &\geq |S_1| \geq \tfrac{1}{\ell}\tbinom{n}{\ell-1} = \tfrac{1}{n-\ell+1}\tbinom{n}{\ell}, \\ \frac{n-\ell}{n-\ell+1}\binom{n}{\ell} &\geq |S_2| \geq \tfrac{1}{n-\ell}\tbinom{n}{\ell+1} = \tfrac{1}{\ell+1}\tbinom{n}{\ell}. \end{aligned}$$

Thus

$$\ell \geq \frac{|S_1|}{|S_2|} \geq \frac{1}{n-\ell} \tag{5.1}$$

or $\max\left(\frac{|S_1|}{|S_2|}, \frac{|S_2|}{|S_1|}\right) \leq \max(\ell, n-\ell) \leq n$.
So the ratio of the cardinalities is at most linear in n. However, we construct antichains whose splitting ratios $\rho(n) = \min\left\{\frac{|S_2|}{|S_1|} : \{S_1, S_2\} \text{ is a splitting of } \mathcal{L}^n\right\}$ satisfy for large n

$$\rho(n) \geq 2^{\varepsilon n} \text{ for some constant } \varepsilon. \tag{5.2}$$

Construction: For a $k \in I\!N$, $2|k$, let $L = L_k \subset \binom{[k]}{\frac{k}{2}}$ be a code with minimal Hamming distance ≥ 4 and with a maximal number of codewords. We consider the poset $\mathcal{P}_k = \{0,1\}^k \smallsetminus U(L)$ and define $E = E_k$ as the set of all maximal elements in $\mathcal{P}_k$. Every element of E has at least $\frac{k}{2}$ ones.
For $n = k \cdot r \in I\!N$ partition $[n]$ into r blocks $R_1, R_2, \ldots, R_r$ each of cardinality k.
We denote by I_t, $1 \leq t \leq r$, the 0–1–sequence of length n, which has ones exactly in the positions from block R_t. For any $\ell \in L$, $e \in E$ and $t, 1 \leq t \leq r$, we denote by ℓ_t and q_t the 0–1–sequences of length n, which have zeros in the blocks R_i, $i \neq t$, and ℓ resp. e in the block R_t.
Define $L_t^* = \{\ell_t : \ell \in L\}$ and $E_t^* = \{e_t : e \in E\}$. We consider now $S = A \cup B \subset \{0,1\}^n$, where

$A = \{a \in \{0,1\}^n : a \wedge I_t \in L_t^* \text{ for all } 1 \leq t \leq r\}$

and

$B = \{b \in \{0,1\}^n : \exists t \in \{1, \ldots, r\} \text{ with } b \wedge I_t \in E_t^* \text{ and } b \wedge T_{t'} = I_{t'} \text{ for } t' \neq t\}$.

One can verify that S is a maximal antichain and by Corollary 2 possesses the splitting property.
We observe that $A \subset \binom{[n]}{\frac{n}{2}}$ and consider the set

$$X = U(A) \cap \binom{[n]}{\frac{n}{2}+1}.$$

It satisfies $X \cap D(B) = \phi$, because S is antichain and for any $x \in X$ there exists exactly one $a \in A$ with $a < x$, since $a_1, a_2 \in A$ implies $d_H(a_1, a_2) \geq 4$. Hence, for every splitting $S = S_1 \dot{\cup} S_2$, $D(S_1) \cup U(S_2) = \{0,1\}^n$ we always have $A \subset S_2$.
Therefore, using a familiar lower bound on $|L|$,

$$|S_2| \geq |A| = |L|^{\frac{n}{k}} > \left(\frac{c\binom{k}{\frac{k}{2}}}{k^2}\right)^{\frac{n}{k}}$$

and

$$|S_1| \leq |B| = \frac{n}{k} \cdot |E| < \frac{n}{k} \cdot 2^k.$$

Now $\frac{|S_2|}{|S_1|} \geq 2^{\varepsilon(c)n}$ for large n, if we choose $k \sim \sqrt{n}$.

THE SET–THEORETICAL FORMULATION OF THE SPLITTING PROPERTY, D_2–DENSENESS

Let $\mathcal{P}$ be a poset and let $S \subset P$ be a maximal antichain in $\mathcal{P}$. Consider the families of sets $\mathcal{A}, \mathcal{B} \subset 2^S$ defined by

$$\mathcal{A} = \{A(u) : u \in \mathcal{U}^*(S)\},\ \mathcal{B} = \{B(d) : d \in D^*(S)\}. \tag{6.1}$$

Here we use again the definitions (3.1) and (3.2) for $A(u)$ and $B(d)$.
The splitting property of S can equivalently be written in the set–theoretic formulation: There exists a partition of S; $S = S_1 \dot\cup S_2$; such that

$$S_1 \cap A \neq \phi \text{ for all } A \in \mathcal{A} \text{ and } S_2 \cap B \neq \phi \text{ for all } B \in \mathcal{B}. \tag{6.2}$$

We can forget now how $\mathcal{A}, \mathcal{B}$ originated in (6.1) from $(\mathcal{P}, S)$ and can consider abstractly any set S and two families $\mathcal{A}, \mathcal{B}$ of subsets of S and ask whether they have the splitting property (6.2).
Of course any abstract system $(S, \mathcal{A}, \mathcal{B})$ can be viewed as coming via (6.1) from a suitable poset. The new language creates new associations. For instance in [2] for any set system $\mathcal{M} \subset 2^S$ a so called *B–property* was introduced, which means that S has a partition $S = S_1 \dot\cup S_2$ with

$$H \cap S_1 \neq \phi \text{ and } H \cap S_2 \neq \phi \text{ for all } H \in \mathcal{M}. \tag{6.3}$$

Obviously, if $\mathcal{M} = \mathcal{A} \cup \mathcal{B}$ has the B–property, then S possesses the splitting property with respect to $\mathcal{A}, \mathcal{B}$, but the converse is not always true.
In the following special situation it is easy to establish the B–property.

Proposition 3. Let S be an infinite set and let $\mathcal{M} \subset 2^S$ be countable, $\mathcal{M} = \{H_1, H_2, \dots,\}$; and let every $H_i \in \mathcal{M}$ be infinite. Then $\mathcal{M}$ has the B–property.

Proof: Since $|H_i| = \infty$ for $i = 1, 2, \dots$, we can sequentially choose two different elements $h_i, g_i \in H_i$ for $i = 1, 2, \dots$ such that $h_i \neq h_j$, $h_i \neq g_j$, $g_i \neq g_j$ $(i \neq j)$. Now we define

$$S_1 = \{h_1, h_2, \dots\} \text{ and } S_2 = S \setminus S_1.$$

Here we consider for the first time the property d_2–dense for a maximal antichain $S \subset \mathcal{P}$. We study it right away in the new setting. The set S is d_2–dense for the set systems $\mathcal{A}, \mathcal{B} \subset 2^S$, if for all $A \in \mathcal{A}$ and all $B \in \mathcal{B}$ necessarily

$$|A \cap B| \neq 1. \tag{6.4}$$

We also say that $\mathcal{A}, \mathcal{B}$ have property d_2.

Theorem 3 Let $\mathcal{A}, \mathcal{B} \subset 2^S$ have property d_2, let $\phi \notin \mathcal{A} \cup \mathcal{B}$ and let both, $\mathcal{A}$ and $\mathcal{B}$, be countable. Then S has the splitting property for $(\mathcal{A}, \mathcal{B})$.

Proof: First note that this theorem is not a consequence of Proposition 3, where we require all members of $\mathcal{A}$ and $\mathcal{B}$ to be infinite.

Let now $\mathcal{A} = \{A_1, A_2, \ldots\}$, $\mathcal{B} = \{B_1, B_2, \ldots\}$ and by property d_2 $|A_i \cap B_j| \neq 1$ for all $A_i \in \mathcal{A}$, $B_j \in \mathcal{B}$. Then we can choose $a_1 \in A_1$ and $b_1 \in B_1$; $a_1 \neq b_1$. We remove all sets from $\mathcal{A}$ which contain a_1 and all sets from $\mathcal{B}$, which contain b_1. We remove also the element a_1 from every set in $\mathcal{B}$ and the element b_1 from every set in $\mathcal{A}$. We denote the remaining sets by $\mathcal{A}^1$ and $\mathcal{B}^1$. Now verify that $\phi \notin \mathcal{A}^1 \cup \mathcal{B}^1$ and $\mathcal{A}^1, \mathcal{B}^1$ have again property d_2!
We note also that the set system $\mathcal{A}^1$ (as well as $\mathcal{B}^1$) is ordered according to the ordering of $\mathcal{A}$, i.e. $\mathcal{A}^1 = \{A_1^1, A_2^1, \ldots\}$ $A_k^1 = A_m \setminus \{a_1\}$ is followed by $A_t^1 = A_\ell \setminus \{a_1\}$ for $k < t$ iff $m < \ell$.
Now we choose $a_2 \in A_1^1$, $b_2 \in B_1^1$, $a_2 \neq b_2$ and construct set systems $\mathcal{A}^2, \mathcal{B}^2$, etc. Continuation of this procedure leads to the subsets of $S : S_1 = \{a_1, a_2, \ldots\}$ and $S_2 = \{b_1, b_2, \ldots,\}$. They splitt $\mathcal{A}, \mathcal{B}$.
Next we show how important it is that in Theorem 3 both, $\mathcal{A}$ and $\mathcal{B}$, are countable.

Example 2: *(S countable, $\mathcal{A}, \mathcal{B} \subset 2^S$, $\phi \notin \mathcal{A} \cup \mathcal{B}$, $\mathcal{A}, \mathcal{B}$ have property d_2 (and even a stronger property), $\mathcal{A}$ is countable, $\mathcal{B}$ is non–countable, but S does not have the splitting property.)*
$S = \mathbb{N}$, $\mathcal{A} = \{A \subset \mathbb{N} : |A^c| < \infty\}$, where A^c is the complement of A, $\mathcal{B} = \{B \subset \mathbb{N} : |B| = \infty\}$. Clearly for every $A \in \mathcal{A}$ and $B \in \mathcal{B}$

$$|A \cap B| = \infty \text{ (stronger than } d_2).$$

Suppose that $S = S_1 \dot{\cup} S_2$ and that

$$S_1 \cap A \neq \phi \quad \forall\, A \in \mathcal{A} \text{ and } S_2 \cap B \neq \phi \quad \forall\, B \in \mathcal{B}. \tag{6.5}$$

In case $|S_1| < \infty$ we have $S_1^c \in \mathcal{A}$ and hence $S_1 \cap S_1^c = \phi$ violates the first relation in (6.5). In case $|S_1| = \infty$ we have $S_1 \in \mathcal{B}$ and hence $S_2 \cap S_1 = \phi$ violates the second relation.

SPLITTING OF SETS WITH PROPERTY D_2, MINIMAL REPRESENTATIVE SETS AND MINIMAL COVERINGS

The results of the last Section gave the motivation for introducing a further concept.
Let S be a set and $\mathcal{M} \subset 2^S$. The set $R \subset S$ is a *representative set* for $\mathcal{M}$, if

$$R \cap H \neq \phi \text{ for all } H \in \mathcal{M}. \tag{7.1}$$

A representative set for $\mathcal{M}$ $R \subset S$ is *minimal*, if no proper subset $R' \subset R$ is representative set for $\mathcal{M}$.

Theorem 4 For a set S and $\mathcal{A}, \mathcal{B} \subset 2^S$ with property d_2 and $\phi \notin \mathcal{A} \cup \mathcal{B}$ let also $\mathcal{A}$ (or $\mathcal{B}$) have a minimal representative set.
Then S has the splitting property.

Proof: We show that we can choose as S_1 in the partition of S the minimal representative set $R \subset S$ of $\mathcal{A}$.

Since by definition $R \cap A \neq \phi$ for all $A \in \mathcal{A}$ and it remains to be seen that there does not exist a $B_0 \in \mathcal{B}$ with $(S \setminus R) \cap B_0 = \phi$, or equivalently $B_0 \subset R$. Assume the opposite.

We choose an arbitrary $b \in B_0$ and consider the set $R' = R \setminus \{b\}$. Since R' is not representative for $\mathcal{A}$ there is an $A \in \mathcal{A}$ with $A \cap R \neq \phi$ and $A \cap R' = \phi$. Therefore $A \cap R = \{b\}$ and since $b \in B_0$, $B_0 \subset R$ we have $|A \cap B_0| = 1$. This contradicts d_2.

Remark 5: The existence of minimal representatives is not necessary for the splitting property.

Example 3: Let $S = \{s_1, s_2, s_3, \ldots\}$ be any infinite countable set and $\mathcal{A} = \mathcal{B} = \{S, S \setminus \{s_1\}, S \setminus \{s_1, s_2\}, \ldots\}$.

Since $|A \cap B| = \infty$ for $A \in \mathcal{A}$ and $B \in \mathcal{B}$, we have property d_2. Neither $\mathcal{A}$ (nor $\mathcal{B}$) has a minimal representative. However, for every infinite $S_1 \subset S$, for which $S \setminus S_1$ is also infinite, we have a splitting of $\mathcal{A}$ and $\mathcal{B}$. Moreover, in this case the existence of a splitting follows from Proposition 3.

Minimal representative sets are related to minimal coverings:

The set $\mathcal{M} \subset 2^X$ is a *covering* of the set X, if $\bigcup_{H \in \mathcal{M}} = X$, and it is a *minimal covering* if no proper subset is a covering of X.

Now, let $S \subset \mathcal{P}$ be a maximal antichain in the poset $\mathcal{P}$. Recall the definitions of $\mathcal{U}^*(s)$ and $D^*(s)$ for $s \in S$ in Section 1 and consider the systems of sets

$$\mathcal{U} = \{U^*(s) : s \in S\}, \mathcal{D} = \{D^*(s) : s \in S\}.$$

Since $\bigcup_{s \in S} U^*(s) = \mathcal{U}^*(S)$ and $\bigcup_{s \in S} D^*(s) = D^*(S)$, the systems $\mathcal{U}$ and $\mathcal{D}$ are coverings of $U^*(S)$ and $D^*(S)$ resp.

The following statement is immediately proved by inspection.

Proposition 4. Let $S \subset \mathcal{P}$ be a maximal antichain in the poset $\mathcal{P}$ and let $\mathcal{A}$, $\mathcal{B}$, $\mathcal{U}$, and $\mathcal{D}$ be the associated set systems. Thus $\mathcal{A}$ (resp. $\mathcal{B}$) has a minimal representative set iff $\mathcal{U}$ (resp. $\mathcal{D}$) contains a minimal covering of $U^*(S)$ (resp. $D^*(S)$).

From here we get an equivalent formulation of Theorem 4.

Theorem 4' Let $S \subset \mathcal{P}$ be a maximal antichain in the poset $\mathcal{P}$ with property d_2 and let the associated set system $\mathcal{U}$ (resp. $\mathcal{D}$) have a minimal covering of $\mathcal{U}^*(S)$ (resp. $\mathcal{D}^*(S)$). Then S possesses the splitting property.

Klimo [2] has studied minimal coverings and proved the following result.

Theorem [2] Let $\mathcal{M} \subset 2^X$ be a covering of X.

(i) Suppose that there is a well–ordering μ of $\mathcal{M}$ with the property: for all $x \in X$ the sets $\{H \in \mathcal{M} : x \in H\}$ have a maximal element according to μ. Then $\mathcal{M}$ contains a minimal covering of X.

(ii) Suppose that for all $H \in \mathcal{M}$ $|H| \leq k$ for some $k \in \mathbb{N}$, then $\mathcal{M}$ contains a minimal covering of X.

Remark 6: As explained in [2], this Theorem implies that a point–finite covering $\mathcal{M}$ of X (i.e. $\forall\, x \in X$ $|\{H \in \mathcal{M} : x \in H\}| < \infty$) contains a minimal covering of X.

From Theorems 4, 4', [2] and Proposition 4 we obtain

Corollary 3 Let S be a set, $\mathcal{A}, \mathcal{B} \subset 2^S$, $\phi \notin \mathcal{A} \cup \mathcal{B}$ and $\mathcal{A}, \mathcal{B}$ have property d_2.

(i) Let μ be a well–ordering of S such that every $A \in \mathcal{A}$ has a maximal element according to μ. Then S has the splitting property.

(ii) Suppose that for some $k \in I\!N$ every element of S is contained in at most k sets from $\mathcal{A}$, then S has the splitting property.

Remark 7: An immediate consequence of this Corollary is, that for $\mathcal{A}, \mathcal{B}$ with property d_2 and all $A \in \mathcal{A}$ finite S has the splitting property.

NEW AND STRONGER SPLITTING PROPERTIES

We say that S, a maximal antichain in the poset $\mathcal{P}$, has a *Y–splitting*, if for some partition $S = S_1 \dot{\cup} S_2$

$$U^*(S_1) \cup D^*(S_1) = U^*(S) \cup D^*(S) \tag{8.1}$$

and

$$U^*(S_2) = U^*(S). \tag{8.2}$$

Symmetrically, we say that S has a *λ–splitting*, if for some partition $S = S_1 \dot{\cup} S_2$

$$D^*(S_2) = D^*(S) \tag{8.3}$$

and (8.1) holds.

Finally, S has an *X–splitting*, if for some partition $S = S_1 \dot{\cup} S_2$

$$U^*(S_1) \cup D^*(S_1) = U^*(S_2) \cup D^*(S_2) = U^*(S) \cup D^*(S). \tag{8.4}$$

Clearly, all these properties imply the familiar splitting property.

We begin their exploration with one of the basic posets, namely $\mathcal{Z} = \{0,1\}^\infty$. At first we analyse d_2–dense antichains S for this poset. For this we look for $b \in S$ at intervalls $\langle c, a \rangle$ with $b \in S \cap \langle c, a \rangle$ and

$$\begin{aligned} a &= b_1 b_2 \dots b_{i-1}\ 1\ b_{i+1} \dots b_{j-1}\ 1\ b_{j+1} \dots \\ b &= b_1 b_2 \dots b_{i-1}\ 1\ b_{i+1} \dots b_{j-1}\ 0\ b_{j+1} \dots \\ c &= b_1 b_2 \dots b_{i-1}\ 0\ b_{i+1} \dots b_{j-1}\ 0\ b_{j+1} \dots . \end{aligned}$$

Clearly $c \in D^*(S)$, $a \in U^*(S)$ and $c < b < a$. Since S is b_2–dense, we must have

$$b' = b_1 b_2 \dots b_{i-1}\ 0\ b_{i+1} \dots b_{j-1}\ 1\ b_{j+1} \dots \in S.$$

Thus property d_2 implies the

Exchange property: S is closed under exchanging any two positions in its elements.

So, if S contains an element $s = (s_1, s_2, \ldots)$ with finitely many, say k, ones, then necessarily

$$S = \binom{\mathbb{N}}{k}. \tag{8.5}$$

We know from Remark 7 that this S has the splitting property. Actually we can choose $S_1 = \{s = (s_1, s_2, \ldots) \in S : s_1 = 1\}$ and $S_2 = S \setminus S_1$.
Next we consider $\mathcal{Z}^* \subset \mathcal{Z}$, the poset of all 0–1–sequences with finitely many ones, $\mathcal{O}^* \subset \mathcal{Z}$, the poset of all 0–1–sequences with finitely many zeros, and

$$\mathcal{P}_\infty = \mathcal{Z} \setminus (\mathcal{Z}^* \cup \mathcal{O}^*) \tag{8.6}$$

the poset of all 0–1–sequences with infinitely many ones and infinitely many zeros.

Proposition 5. Every maximal antichain in $\mathcal{P}_\infty$ is uncountable.

Proof: Cantor's diagonal argument shows that countability is contradictory.

Theorem 5

(i) In the poset $\mathcal{Z}^*$ every maximal d_2–dense and non–trivial $\left(S \neq \binom{\mathbb{N}}{0}\right)$ antichain S has a λ–splitting.

(ii) In the poset $\mathcal{P}_\infty$ every maximal d_2–dense antichain S has an X–splitting.

Proof:
(i) We have already demonstrated that for some k $S = \binom{\mathbb{N}}{k}$.

Case k even:
We choose $S_1 = \left\{a = (a_1, a_2, \ldots) \in \binom{\mathbb{N}}{k} : \sum_{i=1}^{\infty} i\ a_i \equiv 0 \mathbf{mod} 2\right\}$. and $S_2 = S \setminus S_1$. Verification of the λ–splitting:
For $b = (b_1, b_2, \ldots) \in \binom{\mathbb{N}}{k+1}$ either $\sum_{i=1}^{\infty} i\ b_i \equiv 1 \mathbf{mod} 2$ and then $b \in U^*(S_1)$, because for some odd i_0 $b_{i_0} = 1$ and its replacement by 0 produces an $a \in S_1$, or $\sum_{i=1}^{\infty} i\ b_i \equiv 0 \mathbf{mod} 2$ and then $b \in U^*(S_1)$, because $k+1$ being odd enforces $b_{i_0} = 1$ for some even i_0 and its replacement by 0 produces an $a \in S_1$. Similarly we show that $D^*(S_1) = D^*(S_2) = D^*(S)$.

Case k odd:
Define $\mathbb{N}_1 = \{n \in \mathbb{N} : 2 \nmid n\}$, $T = \binom{\mathbb{N}}{k}$ and let $T = T_1 \dot\cup T_2$ be a splitting (guaranteed by Corollary 2) of $\mathcal{Z}_1^*$, the poset of all 0–1–sequences with finitely many ones in the positions $\mathbb{N}_1$ and zeros in the positions $\mathbb{N} \setminus \mathbb{N}_1$.
Now we take

$$L_1 = S_1 \cup T_1 \text{ and } L_2 = \binom{\mathbb{N}}{k} \setminus L_1$$

and again verify the λ–splitting.

(ii) Let $S \subset \mathcal{P}_\infty$ be a maximal and d_2–dense antichain. We have to show that there is a partition $S = S^1 \dot\cup S^2$ with

$$U^*(S) \cup D^*(S) = U^*(S^1) \cup D^*(S^1) = U^*(S^2) \cup D^*(S^2). \tag{8.7}$$

By the *exchange property* S is uniquely partitioned into equivalence classes $\{S_i\}_{i\in I}$ such that every class $S_i (i \in I)$ consists of those elements of S which can be obtained from each other by finitely many exchanges.
Clearly, $S_i (i \in I)$ is countable and hence by Proposition 5 the set of indices I must be uncountable.
Now we consider the sets

$$\overline{S}_i = \{a = (a_1, a_2, \ldots) \in \mathcal{P}_\infty : \exists\, s = (s_1, s_2, \ldots) \in S_i \text{ with } s_\ell = 0, a_\ell = 1 \text{ for some } \ell \in I\!N \text{ and } a_j = s_j \text{ for } j \neq \ell\}$$

and

$$\underline{S}_i = \{a = (a_1, a_2, \ldots) \in \mathcal{P}_\infty : \exists\, s = (s_1, s_2, \ldots) \in S_i \text{ with } s_\ell = 1, a_\ell = 0 \text{ for some } \ell \in I\!N \text{ and } a_j = s_j \text{ for } j \neq \ell\}.$$

Let $\overline{S}$ and $\underline{S}$ be the "parallel levels" of S, that is, $\overline{S} = \bigcup_{i\in I} \overline{S}_i$ and $\underline{S} = \bigcup_{i\in I} \underline{S}_i$.
It is clear that a partition $S = S^1 \dot\cup S^2$ satisfies (8.7) exactly if

$$\overline{S} \cup \underline{S} \subset U(S^1) \cup D(S^1) \text{ and } \overline{S} \cup \underline{S} \subset U(S^2) \cup D(S^2). \tag{8.8}$$

We observe that $\overline{S}$ and $\underline{S}$ are maximal antichains in $\mathcal{P}_\infty$ and their equivalence classes are $\{\overline{S}_i\}_{i\in I}$ and $\{\underline{S}_i\}_{i\in I}$ resp.
Moreover, for $u \in \overline{S}_i$ and $d \in \underline{S}_i$ the sets $A(u) = \{s \in S : s < u\}$ and $B(d) = \{s \in S : s > d\}$ are contained in S_i. For every $i \in I$ we consider now the systems of sets

$$\mathcal{A}_i = \{A(u) : u \in \overline{S}_i\}, \mathcal{B}_i = \{B(d) : d \in \underline{S}_i\}, \text{ and } \mathcal{M}_i = \mathcal{A}_i \cup \mathcal{B}_i.$$

We observe that $\mathcal{M}_i \subset 2^{S_i}$, $\mathcal{M}_i$ is countable and every subset of $\mathcal{M}_i$ is infinite. By Proposition 3 $\mathcal{M}_i$ has property B. This is equivalent to the following: there exists a partition $S_i = S_i^1 \cup S_i^2$ such that $\overline{S}_i \cup \underline{S}_i \subset U^*(S_i^1) \cup D^*(S_i^1)$ and $\overline{S}_i \cup \underline{S}_i \subset U^*(S_i^2) \cup D^*(S_i^2)$. Finally we choose

$$S^1 = \bigcup_{i\in I} S_i^1 \text{ and } S^2 = \bigcup_{i\in I} S_i^2.$$

In conclusion we return to our best friend, the Boolean poset $\{0,1\}^n$. Under an exchange property its maximal antichains are of the form $S = \binom{[n]}{k}$.

Theorem 6 If there exists a partition $S = S_1 \dot\cup S_2$ for $S = \binom{[n]}{k} \subset \{0,1\}^n$ such that

$$U^*(S_1) = U^*(S_2) = U^*(S),$$

then S has a Y–splitting.
Proof: We consider the set of partitions

$$\mathcal{V}(S) = \{(S_1, S_2) : S_1 \dot{\cup} S_2 = S, U(S_1^*) = U(S_2^*) = U^*(S)\}.$$

Let $(S_1', S_2') \in \mathcal{V}(S)$ be extremal in the sense that $S_1' \subset S_1$, $S_1' \neq S_1$ implies $(S_1, S \setminus S_1) \notin \mathcal{V}(S)$. It suffices to show that $D^*(S_1') = D^*(S)$.
Suppose, in the opposite, that there exists an $\alpha \in \binom{[n]}{k-1}$ with $\alpha \notin D^*(S_1')$.
Hence, the elements $\beta_1, \beta_2, \dots, \beta_{n-k+1} \in \binom{[n]}{k}$ with $\beta_i > \alpha$ are from the set S_2'.
But then $(S_1' \cup \{\beta_1\}, S_2' \setminus \{\beta_1\}) \in \mathcal{V}(S)$, because $\gamma > \beta_1$ implies also $\gamma > \beta_i$ for some $i \geq 2$.

SPLITTING PROPERTIES FOR DIRECTED GRAPHS

We consider directed graphs $\mathcal{G} = (V, \mathcal{E})$ with multiple edges, that is, both edges, (v_1, v_2) and (v_2, v_1) can be in $\mathcal{E}$.
They can be viewed as generalizations of posets, because with every poset $\mathcal{P} = (P, <_p)$ we can associate a graph $G(\mathcal{P}) = (P, \mathcal{E}(<_p))$ as follows:

$$\text{For } v_1, v_2 \in P \qquad (v_1, v_2) \in \mathcal{E}(<_p) \Leftrightarrow v_1 <_p v_2. \tag{9.1}$$

In such a graph there are no directed cycles, so the class of directed graphs is wider than the class of posets.
If S is an antichain in $\mathcal{P}$, then for $s_1, s_2 \in S$

(a) there is no edge in $G(\mathcal{P})$ between s_1 and s_2

(b) there is no directed path in $G(\mathcal{P})$ from s_1 to s_2.

For $G(\mathcal{P})$ properties (a) and (b) are the same. However, for general graphs they are different. If for a set $S \subset V$ (a) holds, then we call S an antichain, and if (the stronger) (b) holds, we call S a pathwise or (shortly) p–antichain.
We extend now the notion of a *dense poset* in the sense of [1], discussed in Section 1, to graphs. We use abbreviations like $a \rightsquigarrow b$ (resp. $a \not\rightsquigarrow b$), if there is (resp. is not) a directed path from a to b.
We say that $G = (V, \mathcal{E})$ *is* p–dense, if for every directed path $[a_1, a_2, \dots, a_t]$ of length $t - 1 \geq 2$ and every a_i $(2 \leq i \leq t-1)$ there exists a directed path $a_t \rightsquigarrow a_i$, a directed path $a_i \rightsquigarrow a_1$ or there exists a b_i on a directed path from a_1 to a_t and p–independent of a_i.
All notions of splitting in the previous Section 8 can be extended. However, we consider here only the original concept of [1].
Let S be a maximal p–antichain, then S possesses a *p–splitting of* $\mathcal{G}$, if there is a partition $S = S_1 \dot{\cup} S_2$ with

$$U(S_1) \cup D(S_2) = V,$$

where

$$U(S_1) = \{v \in V : \exists\ s \rightsquigarrow v \text{ for some } \ s \in S\},$$

$$D(S_2) = \{v \in V : \exists\ v \rightsquigarrow s \text{ for some } \ s \in S\}.$$

Here is our generalization of the main result in [1].

Theorem 7 Let $\mathcal{G}$ be a finite p–dense, directed graph, then every maximal p–antichain S in $\mathcal{G}$ possesses a splitting of $\mathcal{G}$.

Sketch of proof:

We follow the idea of the first proof of Theorem 3.1 in [1], which is by induction on $|V|$.

If $s \in S$ is needed for "up" to u and for "down" to d, then for the chain $d \rightsquigarrow s \rightsquigarrow u$ by p–denseness either we find a chain $u \rightsquigarrow d$ and we have a contradiction, because d can be attained in $U(S)$ (does not use full strength of (c)!), or by (d) there is a v with $d \rightsquigarrow v \rightsquigarrow u$ and $s \not\rightsquigarrow v$, $v \not\rightsquigarrow s$.

In this case independence of s from S would contradict maximality of S, so we have either for some $s_1 \in S$ $s_1 \rightsquigarrow v$ or for some $s_2 \in S$ $v \rightsquigarrow s_2$.

Therefore either $s_1 \rightsquigarrow u$ or $d \rightsquigarrow s_2$ and in any case a contradiction to the definition of s.

It remains to discuss the case where some $U(s)$ (or $D(s)$) is removed from the graph. As in [1] we show by inspection that the induced graph on $V \setminus U(s)$ is p–dense.

Remark 8: It is interesting to analyse number–theoretic examples such as $G = (V, \mathcal{E})$, where $V \subset \mathbb{N}$ and for $m, n \in V$ $(m, n) \in \mathcal{E}$ iff g.c.d $\{m, n\} = 1$ and $m < n$.

We thank Peter Erdös for proposing the study of splitting properties in infinite posets.

References

[1] R. Ahlswede, P.L. Erdös, and N. Graham, "A splitting property of maximal antichains", *Combinatorica* 15 (4), 1995, 475-480.

[2] J. Klimó, "On the minimal covering of infinite sets", *Discrete Applied Mathematics* 45, 1993, 161–168.

OLD AND NEW RESULTS FOR THE WEIGHTED T–INTERSECTION PROBLEM VIA AK–METHODS

Christian Bey and Konrad Engel

Universität Rostock, FB Mathematik
18051 Rostock, Germany

Dedicated to Professor Rudolf Ahlswede on occasion of his 60th birthday.

Abstract: Let $[n] := \{1, \ldots, n\}$, $2^{[n]}$ be the power set of $[n]$ and $s \in [n]$. A family $\mathcal{F} \subseteq 2^{[n]}$ is called *t-intersecting in* $[s]$ if

$$|X_1 \cap X_2 \cap [s]| \geq t \text{ for all } X_1, X_2 \in \mathcal{F}.$$

Let $\omega : 2^{[n]} \to \mathbb{R}_+$ be a given weight function and

$$M_s(n, t; \omega) := \max\{\omega(\mathcal{F}) : \mathcal{F} \text{ is } t\text{-intersecting in } [s]\}.$$

For several weight functions, the numbers $M_n(n, t; \omega)$ can be determined using three important methods of Ahlswede and Khachatrian: Generating Sets [2], Comparison Lemma [4], and Pushing–Pulling [3]. We survey these methods.

Also, sufficient conditions on ω for the equality

$$M_s(n, t; \omega) = M_n(n, t; \omega)$$

are presented which simplify the method of Generating Sets. In addition, analogous conditions are given for the case that $|\cap_{X \in \mathcal{F}} X| < t$ is required (nontrivial t-intersection).

Applications of these methods include new intersection theorems for chain– and star products.

INTRODUCTION AND NOTATION

In this paper we give a survey and discuss some new results for and insights into the problem of determining the maximum weight of t-intersecting families

I. Althöfer et al. (eds.), Numbers, Information and Complexity, 45-74.

of subsets of a finite set. The ingenious, relatively elementary methods were elaborated by Ahlswede and Khachatrian in several papers [2, 1, 4, 3]. Our aim is to provide a unifying approach such that most of the results are covered. Since Erdős, Ko and Rado [11] have initiated the study of such problems in the thirties many results were obtained by several authors. Here we cite only the recent papers which are related to the new AK-methods. More on the history of the results can be found in the corresponding papers. Moreover, in order to avoid too much technical details we describe only one and not all optimal families though in most cases Ahlswede and Khachatrian also proved the uniqueness of the optimal family up to permutation of the elements.

Let $\mathbb{N}$ be the set of natural numbers, $[n] := \{1,\dots,n\}$ and for $i,j \in \mathbb{N}, i<j$, let $[i,j] := \{i, i+1, \dots, j\}$. Let $2^{[n]}$ (resp. $\binom{[n]}{k}$) be the family of all (resp. all k-element) subsets of $[n]$. Each subfamily of $\binom{[n]}{k}$ is said to be *k-uniform*. A family $\mathcal{F} \subseteq 2^{[n]}$ is called *t-intersecting* if $|X_1 \cap X_2| \geq t$ for all $X_1, X_2 \in \mathcal{F}$ (1-intersecting is abbreviated by *intersecting*). We will suppose throughout that $1 \leq t \leq n-1$. Let $I(n,t)$ be the class of all t-intersecting families of subsets of $[n]$.

Suppose that we are given a *weight function* $\omega : 2^{[n]} \to \mathbb{R}_+$ (the set of all nonnegative reals). For $\mathcal{F} \subseteq 2^{[n]}$ let

$$\omega(\mathcal{F}) := \sum_{X \in \mathcal{F}} \omega(X).$$

The *weighted t-intersection problem* is the problem of determining

$$M(n,t;\omega) := \max\{\omega(\mathcal{F}) : \mathcal{F} \in I(n,t)\}.$$

In several applications the weight function depends only on the size of the subsets, i.e. we have $\omega(X_1) = \omega(X_2)$ if $|X_1| = |X_2|$. In this case we call ω *size–dependent* and we set $\omega_i := \omega(X)$ for $|X| = i$. Each family $\mathcal{F} \subseteq 2^{[n]}$ may be partitioned into (possibly empty) subfamilies $\mathcal{F}_i := \{X \in \mathcal{F} : |X| = i\}$. We put $f_i := f_i(\mathcal{F}) := |\mathcal{F}_i|$. The vector $(f_0, \dots, f_n)$ is called the *profile* of $\mathcal{F}$. A special case of the weighted t-intersection problem is the *size–dependent weighted t-intersection problem*: For $\boldsymbol{\omega} = (\omega_0, \dots, \omega_n) \in \mathbb{R}_+^{n+1}$ determine

$$M(n,t;\boldsymbol{\omega}) := \max\left\{\sum_{i=0}^{n} \omega_i f_i : \mathcal{F} \in I(n,t)\right\}.$$

Candidates for the solution are the families

$$\mathcal{S}_r^n := \{X \subseteq [n] : |X \cap [t+2r]| \geq t+r\}, r = 0, \dots, \lfloor \tfrac{n-t}{2} \rfloor$$

which are easily seen to be t-intersecting. Some further candidates are

$$\begin{aligned} \mathcal{S}_{r,i}^n := &\{X \subseteq [n] : |X \cap [t+2r]| \geq t+r \text{ and } i \leq |X| < n+t-i\} \\ &\cup \{X \subseteq [n] : |X| \geq n+t-i\}, r = 0, \dots, \lfloor \tfrac{n-t}{2} \rfloor, i = t+r, \dots, \lfloor \tfrac{n+t}{2} \rfloor. \end{aligned}$$

In the following we will omit the upper index n if the basic set $[n]$ is clear from the context. Note that $\mathcal{S}_r = \mathcal{S}_{r,t+r}$.

For instance, if $n = 8$, $t = 2$, $\boldsymbol{\omega} = (0,0,0,0,1,0,1,0,0)$ we have $\omega(\mathcal{S}_{1,4}) = 45$, whereas $\omega(\mathcal{S}_0) = 30$, $\omega(\mathcal{S}_1) = 39$, $\omega(\mathcal{S}_2) = 43$, $\omega(\mathcal{S}_3) = 28$, so these further candidates should not be forgotten.

In particular, for $t = 1$, Erdős, Frankl and Katona [12] proved:

Theorem 1. *The optimum in the size–dependent weighted* 1*-intersection problem is attained at one of the families* $\mathcal{S}_0, \mathcal{S}_{0,2}, \ldots, \mathcal{S}_{0,\lfloor\frac{n+1}{2}\rfloor}$, *and each of these families is optimal for some weight function.* □

Unfortunately, we do not have such a general theorem for $t > 1$. The so far strongest result is given by the celebrated complete intersection theorem of Ahlswede, Khachatrian [2] which solves the case $\boldsymbol{\omega} = \boldsymbol{e}_k$ where $\boldsymbol{e}_k$ is the $n+1$-dimensional unit vector with 1 at coordinate k, $k = 0, \ldots, n$.

Theorem 2 (Complete Intersection Theorem) *Let* $\boldsymbol{\omega} = \boldsymbol{e}_k$, $k \geq t$. *The optimum in the size–dependent weighted t-intersection problem is attained at* $\mathcal{S}_r$, $r = 0, \ldots, \lfloor\frac{n-t}{2}\rfloor$, *if*

$$(k-t+1)\left(2+\frac{t-1}{r+1}\right) \leq n \leq (k-t+1)\left(2+\frac{t-1}{r}\right) \tag{1}$$

(with the definition $\infty := \frac{t-1}{0}$ *for all* $t \geq 1$*) and at* $\mathcal{S}_{\lfloor\frac{n-t}{2}\rfloor}$ *if*

$$n < (k-t+1)\left(2+\frac{t-1}{\lfloor\frac{n-t}{2}\rfloor+1}\right). \tag{2}$$

□

We note that (2) is equivalent to $n \leq 2k-t$ and that in this case $\mathcal{S}_{\lfloor\frac{n-t}{2}\rfloor} \supseteq \binom{[n]}{k}$.

Define

$$k_r^n := \begin{cases} t & \text{if } r = -1 \\ \frac{(r+1)n}{t+2r+1} + t - 1 & \text{if } r = 0, \ldots, \lfloor\frac{n-t}{2}\rfloor. \end{cases}$$

We omit again the upper index n if the basic set $[n]$ is clear from the context. Note that $k_{-1} \leq k_0 \leq \ldots \leq k_{\lfloor\frac{n-t}{2}\rfloor}$. It is easy to see that an equivalent formulation of Theorem 2 is the following:

Theorem 2a. *The maximum size of a k-uniform t-intersecting family in* $2^{[n]}$ *is attained at* $\mathcal{S}_r \cap \binom{[n]}{k}$, *if* $k_{r-1} \leq k \leq k_r$, $r = 0, \ldots, \lfloor\frac{n-t}{2}\rfloor$, *and at* $\binom{[n]}{k}$ *if* $k > k_{\lfloor\frac{n-t}{2}\rfloor}$. □

As a direct consequence we obtain:

Corollary 3. *Suppose that for some* $r \in \{0, \ldots, \lfloor\frac{n-t}{2}\rfloor\}$

$$\omega_i = 0 \textit{ unless } k_{r-1} \leq i \leq k_r. \tag{3}$$

Then the optimum in the size–dependent weighted t-intersection problem is attained at $\mathcal{S}_r$. □

Remark. *If we replace condition (3) by*

$$\omega_i = 0 \text{ unless } k_{r-1} \leq \ell \leq i \leq k_r \text{ or } n + t - \ell \leq i$$

then the optimum is attained at $\mathcal{S}_{r,\ell}$ since $\mathcal{S}_{r,\ell}$ contains the maximum possible number of members from $\binom{[n]}{i}$ if i belongs to the first interval and all members from $\binom{[n]}{i}$ if $n + t - \ell \leq i$.

Corollary 3 can be sharpened to the following theorem which will be proved in Section 6 (the essential steps are given in Example 4 and Lemma 19 which also provide an independent proof of Theorem 2).

Theorem 3a. *Suppose that for some $r \in \left\{0, \dots, \left\lfloor \frac{n-t}{2} \right\rfloor - 1\right\}$*

$$\omega_i = 0 \text{ unless } k_{r-1} \leq i \leq k_{r+1}.$$

Then $M(n, t, \boldsymbol{\omega})$ is attained at $\mathcal{S}_r$ or $\mathcal{S}_{r+1}$. □

Let $s \in [n]$. It is easy to see that

$$\mathcal{S}_r^n = \{X \cup Y : X \in \mathcal{S}_r^s \text{ and } Y \subseteq [s+1, n]\}, r = 0, \dots, \left\lfloor \tfrac{s-t}{2} \right\rfloor,$$

and that for each i-element subset X of $[s]$ there exist $\binom{n-s}{k-i}$ k-element subsets Z of $[n]$ such that $Z \cap [s] = X$. Since the t-intersection in $\mathcal{S}_r^n$ is already realized in $[s]$ if $r \leq \left\lfloor \frac{s-t}{2} \right\rfloor$ we obtain a further consequence of Theorem 2a.

Corollary 3b. *Let $s \in [n]$ and let $\boldsymbol{\omega}$ be defined by $\omega_i = \binom{n-s}{k-i}$. Then $M(s, t, \boldsymbol{\omega})$ is attained at $\mathcal{S}_r^s$ if $k_{r-1}^n \leq k \leq k_r^n$, $r = 0, \dots, \left\lfloor \frac{s-t}{2} \right\rfloor$.* □

We mention that we cannot conclude automatically that $M(s, t, \boldsymbol{\omega})$ is attained at $\mathcal{S}^s_{\lfloor \frac{s-t}{2} \rfloor}$ if $k > k^n_{\lfloor \frac{s-t}{2} \rfloor}$.

For a succeeding application we change the notation a little bit. We put $m := n$, $s := n$, and $\boldsymbol{\nu} := \boldsymbol{\omega}$. Then a special case of Corollary 3b reads:

Corollary 3c. *The optimum $M(n, t, \boldsymbol{\nu})$ with $\nu_i = \binom{m-n}{k-i}$ is attained at $\mathcal{S}_{\lfloor \frac{n-t}{2} \rfloor}$ if $k^m_{\lfloor \frac{n-t}{2} \rfloor - 1} \leq k \leq k^m_{\lfloor \frac{n-t}{2} \rfloor}$, i.e. if*

$$(k - t + 1)\left(2 + \frac{t-1}{\left\lfloor \frac{n-t}{2} \right\rfloor + 1}\right) \leq m \leq (k - t + 1)\left(2 + \frac{t-1}{\left\lfloor \frac{n-t}{2} \right\rfloor}\right). \tag{4}$$

□

OPTIMALITY OF THE LAST CANDIDATE FAMILY

Corollary 3c gives a first example of a non–trivial weight function for which $\mathcal{S}_{\lfloor \frac{n-t}{2} \rfloor}$ is optimal. Here we look for other weight functions with the same property.

The following theorem is due to Katona [15].

Theorem 4. *Let $t \le i \le \lfloor \frac{n+t-1}{2} \rfloor$. If $\omega_i = \omega_{n+t-i-1} = 1$ and $\omega_j = 0$ for $j \notin \{i, n+t-i-1\}$ then $M(n,t;\boldsymbol{\omega})$ is attained at $\mathcal{S}_{\lfloor \frac{n-t}{2} \rfloor}$, i.e.*

$$M(n,t;\boldsymbol{\omega}) = \begin{cases} \binom{n}{n+t-i-1} & \text{if } i < \frac{n+t-1}{2} \\ \binom{n-1}{\frac{n+t-1}{2}} & \text{if } i = \frac{n+t-1}{2}. \end{cases} \tag{5}$$

□

As an easy consequence of Theorem 4 Engel, Frankl [10] obtained the following:

Theorem 5. *If $\omega_i \le \omega_{n+t-i-1}$, $i = t, \ldots, \lfloor \frac{n+t-1}{2} \rfloor$, then $M(n,t;\boldsymbol{\omega})$ is attained at $\mathcal{S}_{\lfloor \frac{n-t}{2} \rfloor}$.* □

From this theorem we may easily derive that Corollary 3c remains true for all integers m with

$$n \le m \le 2k - t + 1 \tag{6}$$

since for these m the inequalities

$$\binom{m-n}{k-i} \le \binom{m-n}{k-n-t+i+1}, i = 0, \ldots, \lfloor \tfrac{n+t-1}{2} \rfloor,$$

are true.

Theorem 5 contains a fundamental result of Katona [15] as a special case ($\omega_i = 1$ for all i):

Theorem 6. *Among all t-intersecting families in $2^{[n]}$ the family $\mathcal{S}_{\lfloor \frac{n-t}{2} \rfloor}$ has maximum size.* □

In order to apply the strong Corollary 3c, Ahlswede and Khachatrian developed a method which is given in the next theorem.

Theorem 7 (Comparison Lemma) *Let $\mathcal{P}$ be a set of points in $\mathbb{R}_+^{n+1-t}$ whose coordinates are indexed by $t, t+1, \ldots, n$. Let $\boldsymbol{\nu} \in \mathbb{R}_+^{n+1-t}$ be a given positive weight vector. Suppose that there is some $\boldsymbol{f}^* \in \mathcal{P}$ such that*

$$\boldsymbol{\nu} \cdot \boldsymbol{f}^* = \max\{\boldsymbol{\nu} \cdot \boldsymbol{f} : \boldsymbol{f} \in \mathcal{P}\},$$

and for some $p \in [t,n]$

$$f_i^* = 0 \quad \text{if } t \le i < p \tag{7}$$

$$f_i \le f_i^* \quad \text{if } p \le i \le n \text{ and } \boldsymbol{f} \in \mathcal{P}. \tag{8}$$

Let $\boldsymbol{\omega} \in \mathbb{R}_+^{n+1-t}$ be another positive weight vector with the property

$$\frac{\nu_i}{\nu_{i+1}} \ge \frac{\omega_i}{\omega_{i+1}}, i = t, \ldots, n-1. \tag{9}$$

Then also

$$\boldsymbol{\omega} \cdot \boldsymbol{f}^* = \max\{\boldsymbol{\omega} \cdot \boldsymbol{f} : \boldsymbol{f} \in \mathcal{P}\}.$$

Probably this theorem is not as well–known as it should be. Thus we reprove it here:

Proof. First we consider the special case

$$\frac{\nu_i}{\nu_{i+1}} = \frac{\omega_i}{\omega_{i+1}} \text{ for all } i = p, \ldots, n-1.$$

Then

$$\omega_i \begin{cases} = \frac{\omega_p}{\nu_p}\nu_i & \text{if } i = p, \ldots, n \\ \leq \frac{\omega_p}{\nu_p}\nu_i & \text{if } i = t, \ldots, p-1. \end{cases}$$

Consequently, for all $\boldsymbol{f} \in \mathcal{P}$ (using (7))

$$\begin{aligned} \boldsymbol{\omega} \cdot (\boldsymbol{f}^* - \boldsymbol{f}) &= \sum_{i=t}^{p-1} \omega_i(f_i^* - f_i) + \sum_{i=p}^{n} \omega_i(f_i^* - f_i) \\ &\geq \sum_{i=t}^{p-1} \frac{\omega_p}{\nu_p}\nu_i(f_i^* - f_i) + \sum_{i=p}^{n} \frac{\omega_p}{\nu_p}\nu_i(f_i^* - f_i) \\ &= \frac{\omega_p}{\nu_p}\boldsymbol{\nu} \cdot (\boldsymbol{f}^* - \boldsymbol{f}) \geq 0. \end{aligned}$$

Now we prove the general case by induction on the smallest number $s(\boldsymbol{\omega})$ such that

$$\frac{\nu_i}{\nu_{i+1}} = \frac{\omega_i}{\omega_{i+1}} \text{ for all } i = s(\boldsymbol{\omega}), \ldots, n-1.$$

Just before we treated the case $s(\boldsymbol{\omega}) = p$. Let us look at the induction step. Suppose that

$$\frac{\nu_q}{\nu_{q+1}} > \frac{\omega_q}{\omega_{q+1}} \text{ but } \frac{\nu_i}{\nu_{i+1}} = \frac{\omega_i}{\omega_{i+1}} \text{ for all } i = q+1, \ldots, n-1,$$

i.e. $s(\boldsymbol{\omega}) = q + 1$. Let

$$\alpha := \frac{\nu_q}{\nu_{q+1}} \frac{\omega_{q+1}}{\omega_q}$$

and let $\boldsymbol{\omega}'$ be defined by

$$\omega_i' := \begin{cases} \omega_i & \text{if } i = q+1, \ldots, n \\ \alpha\omega_i & \text{if } i = t, \ldots, q. \end{cases}$$

Then $s(\boldsymbol{\omega}') = q$ and $\boldsymbol{\omega}'$ satisfies (9) with $\boldsymbol{\omega}$ replaced by $\boldsymbol{\omega}'$. By the induction hypothesis and (8), for all $\boldsymbol{f} \in \mathcal{P}$

$$\boldsymbol{\omega} \cdot (\boldsymbol{f}^* - \boldsymbol{f}) = \frac{1}{\alpha}\boldsymbol{\omega}' \cdot (\boldsymbol{f}^* - \boldsymbol{f}) + \left(1 - \frac{1}{\alpha}\right) \sum_{i=q+1}^{n} \omega_i(f_i^* - f_i) \geq 0.$$

□

In order to apply Theorem 7 via Corollary 3c to some other size–dependent weighted t-intersection problems we delete from the profiles of the t-intersecting families the coordinates $0, \ldots, t-1$ (they are obviously zero), we take $\boldsymbol{f}^*$ as the (reduced) profile of $\mathcal{S}_{\lfloor \frac{n-t}{2} \rfloor}$, and put $p := \lfloor \frac{n+t}{2} \rfloor$. Then (7) and (8) are satisfied (note (5) in the case $2 \nmid n+t$). Note that $\nu_i = \binom{m-n}{k-i} = 0$ if $k < i$ or $m - n < k - i$ and that

$$\frac{\nu_i}{\nu_{i+1}} = \frac{m-n+1}{k-i} - 1 \text{ if } k - m + n \leq i \leq k - 1.$$

Thus we have:

Corollary 8. *The optimum in the size–dependent t-intersection problem with weight vector ω being positive at coordinates $t, \ldots, n$ is attained at $\mathcal{S}_{\lfloor \frac{n-t}{2} \rfloor}$ if there are integers m and k such that*

$$n \leq \min\{m, k, m + t - k\}, \tag{10}$$

(4) (resp. (6)) is satisfied, and

$$\frac{\omega_i}{\omega_{i+1}} \leq \frac{m-n+1}{k-i} - 1 \tag{11}$$

for all $i = t, \ldots, n-1$. □

In general, it is not easy to find such integers m and k satisfying (4), (10), and (11). One idea is to look for "large" numbers m and k. In order to avoid long and tedious computations we use the following easy lemma.

Lemma 9. *Let $a_j, b_j, c_j, d_j \in \mathbb{R}_+, j = 1, \ldots, p$, and n be a fixed number. If*

$$\max\{a_1, \ldots, a_p\} < \min\{c_1, \ldots, c_p\}$$

then there are positive integers m and k not less than n such that

$$a_j k + b_j \leq m \leq c_j k + d_j, j = 1, \ldots, p.$$

□

Corollary 10. *The optimum in the size–dependent t-intersection problem with weight vector ω being positive at coordinates $t, \ldots, n$ is attained at $\mathcal{S}_{\lfloor \frac{n-t}{2} \rfloor}$ if*

$$\max\left\{\frac{\omega_i}{\omega_{i+1}}, i = t, \ldots, n-1\right\} < 1 + \frac{t-1}{\lfloor \frac{n-t}{2} \rfloor}.$$

Proof. The inequalities (4), (10), and (11) can be written in the form

$$a_j k + b_j \leq m \leq c_j k + d_j, j = 1, 2, 3, \ldots, n - t + 2$$

where

$$\begin{aligned}
a_1 &= 2 + \frac{t-1}{\lfloor \frac{n-t}{2} \rfloor + 1}, \quad c_1 = 2 + \frac{t-1}{\lfloor \frac{n-t}{2} \rfloor},\\
a_2 &= 1, \quad c_2 = \infty,\\
a_j &= \frac{\omega_i}{\omega_{i+1}} + 1, \quad c_j = \infty, \quad j = i - t + 3 = 3, \dots, n - t + 2.
\end{aligned}$$

Corollary 8 and Lemma 9 yield the result. □

We mention that the case $\omega_i = \alpha^i$, i.e. $\frac{\omega_i}{\omega_{i+1}} = \frac{1}{\alpha}$ = constant was considered by Ahlswede, Khachatrian in [4], see also Example 5.

The idea of looking for large numbers m and k does not always work. Sometimes one has luck since there exist "small" numbers m and k.

Corollary 11. *Let* $\omega_i = \binom{\ell-1}{\ell-i}, i = 0, \dots, n$, *and let* $\ell \geq n$. *Then* $M(n, t; \boldsymbol{\omega})$ *is attained at* $\mathcal{S}_{\lfloor \frac{n-t}{2} \rfloor}$.

Proof. We apply Corollary 8 with $k := n$, $m := 2n$. Then (10) is satisfied, (4) is equivalent to

$$\begin{aligned}
2n - \frac{2(t-1)}{n-t+2} \;&\leq 2n \leq\; 2n + \frac{2(t-1)}{n-t} \quad \text{if } 2 \mid n+t,\\
2n \;&\leq 2n \leq\; 2n + \frac{2(t-1)}{n-t-1} \quad \text{if } 2 \nmid n+t,
\end{aligned}$$

hence (4) is satisfied, and finally (11) is equivalent to

$$\frac{\ell}{\ell - i} \leq \frac{n+1}{n-i} \text{ for all } i = t, \dots, n-1$$

which is obviously satisfied since $\ell \geq n$. □

A NEW APPLICATION – PRODUCTS OF INFINITE CHAINS

Let $N_\ell(n,\infty) := \{\boldsymbol{a} = (a_1, \dots, a_n) : a_i \in \mathbb{N}, i = 1, \dots, n, \sum_{i=1}^n a_i = \ell\}$. A family $\mathcal{F} \subseteq N_\ell(n,\infty)$ is called (statically) *t-intersecting* if for all $\boldsymbol{a}, \boldsymbol{b} \in \mathcal{F}$ there exist t coordinates $i_1, \dots, i_t$ such that $a_{i_j}, b_{i_j} \geq 1$ holds for $j = 1, \dots, t$. Let

$$M_\ell(n, \infty, t) := \max\{|\mathcal{F}| : \mathcal{F} \subseteq N_\ell(n,\infty),\ \mathcal{F} \text{ is } t\text{-intersecting}\}.$$

The set $N_\ell(n,\infty)$ can be viewed as the ℓ-th level in the direct product of n chains $0 \lessdot 1 \lessdot \cdots$ or as the family of all ℓ-element multisets over the basic set $[n]$. The property of being t-intersecting means in the case of multisets that any two members of the family have at least t different elements from the basic set in common.

Define for $\boldsymbol{a} \in N_\ell(n,\infty)$ (resp. $\mathcal{F} \subseteq N_\ell(n,\infty)$) the *support* of $\boldsymbol{a}$ (resp. of $\mathcal{F}$) by $\operatorname{supp}(\boldsymbol{a}) := \{i : a_i > 0\}$ (resp. $\operatorname{supp}(\mathcal{F}) := \{\operatorname{supp}(\boldsymbol{a}) : \boldsymbol{a} \in \mathcal{F}\}$). Obviously $\mathcal{F} \subseteq N_\ell(n,\infty)$ is t-intersecting iff $\operatorname{supp}(\mathcal{F}) \subseteq 2^{[n]}$ is t-intersecting.

A fundamental combinatorial formula (combinations with repetitions) says that for each fixed i-element subset X of $[n]$

$$|\{\boldsymbol{a} \in N_\ell(n,\infty) : \text{supp}\,(\boldsymbol{a}) \subseteq X\}| = \binom{i+\ell-1}{\ell}.$$

Deleting from each coordinate of supp $(\boldsymbol{a})$ a one leads to a bijection between the sets $\{\boldsymbol{a} \in N_\ell(n,\infty) : \text{supp}\,(\boldsymbol{a}) = X\}$ and $\{\boldsymbol{a} \in N_{\ell-i}(n,\infty) : \text{supp}\,(\boldsymbol{a}) \subseteq X\}$. Hence, for each fixed i-element subset X of $[n]$

$$|\{\boldsymbol{a} \in N_\ell(n,\infty) : \text{supp}\,(\boldsymbol{a}) = X\}| = \binom{i+\ell-i-1}{\ell-i} = \binom{\ell-1}{\ell-i}. \tag{12}$$

We define the weight vector $\boldsymbol{\omega}$ by $\omega_i = \binom{\ell-1}{\ell-i}$ and obtain easily

$$M_\ell(n,\infty,t) = M(n,t;\boldsymbol{\omega}). \tag{13}$$

Let $\mathcal{F}_r := \{\boldsymbol{a} \in N_\ell(n,\infty) : \text{supp}\,(\boldsymbol{a}) \in \mathcal{S}_r\}$. Clearly, $\mathcal{F}_r$ is t-intersecting. Define

$$\ell_r := \begin{cases} t & \text{if } r=-1 \\ \frac{r+1}{t+r}(n+t-2)+t-1 & \text{if } r=0,\dots,\lfloor\frac{n-t}{2}\rfloor. \end{cases}$$

It is not difficult to verify that $\ell_{-1} \le \ell_0 \le \dots \le \ell_{\lfloor\frac{n-t}{2}\rfloor}$.

Theorem 12. [1] *We have*

$$M_\ell(n,\infty,t) = \begin{cases} |\mathcal{F}_r| & \text{if } \ell_{r-1} \le \ell \le \ell_r \\ & \text{for some } r \in \{0,\dots,\lfloor\frac{n-t}{2}\rfloor\} \\ |\mathcal{F}_{\lfloor\frac{n-t}{2}\rfloor}| & \text{if } \ell > \ell_{\lfloor\frac{n-t}{2}\rfloor}. \end{cases}$$

Proof. In view of (12) and (13) we only have to show that $M(n,t;\boldsymbol{\omega})$ is attained at $\mathcal{S}_r$ if $\ell_{r-1} \le \ell \le \ell_r$, $r \in \{0,\dots,\lfloor\frac{n-t}{2}\rfloor\}$ and at $\mathcal{S}_{\lfloor\frac{n-t}{2}\rfloor}$ if $\ell > \ell_{\lfloor\frac{n-t}{2}\rfloor}$.

First let for some $r \in \{0,\dots,\lfloor\frac{n-t}{2}\rfloor\}$

$$\ell_{r-1} \le \ell \le \ell_r. \tag{14}$$

We put in Corollary 3b $k := \ell$, $s := n$, $n := n+\ell-1$. Then we obtain our weight function $\omega_i = \binom{n-s}{k-i} = \binom{\ell-1}{\ell-i}$, and by Corollary 3b, $M(n,t;\boldsymbol{\omega})$ is attained at $\mathcal{S}_r$ if

$$k_{r-1}^{n+\ell-1} \le \ell \le k_r^{n+\ell-1}. \tag{15}$$

Using the definition of k_r^n it is easy to prove the equivalence of (14) and (15).

Now let $\ell > \ell_{\lfloor\frac{n-t}{2}\rfloor}$. A simple computation shows that this inequality is equivalent to

$$\ell > \begin{cases} n-1+4\frac{t-1}{n+t} & \text{if } 2 \mid n+t \\ n-1+2\frac{t-1}{n+t-1} & \text{if } 2 \nmid n+t, \end{cases}$$

[1]We thank U. Leck for stimulating the study of $M_\ell(n,\infty,t)$.

hence to $\ell \geq n$. The assertion follows directly from Corollary 11. □

Instead of $N_\ell(n,\infty)$ one can consider $N_\ell(n,k) := \{\boldsymbol{a} = (a_1, \dots a_n) : a_i \in \{0,1,\dots,k\}, i = 1,\dots,n, \sum_{i=1}^n a_i = \ell\}$ – the ℓ-th level in the direct product of n chains $0 \lessdot 1 \lessdot \cdots \lessdot k$. We take the same t-intersection property as before and define

$$M_\ell(n,k,t) := \max\{|\mathcal{F}| : \mathcal{F} \subseteq N_\ell(n,k),\ \mathcal{F} \text{ is } t\text{-intersecting}\}.$$

It seems very difficult to determine this number in general. The following asymptotic result of the authors generalizes previous work from [10].

Theorem 13. *For every $t > 1$ (resp. for $t = 1$) there exist real numbers $0 = \lambda_{t,-1} < \lambda_{t,0} < \lambda_{t,1} < \cdots$ (resp. $0 = \lambda_{1,-1} < \lambda_{1,0}$) with $\lim_{r\to\infty} \lambda_{t,r} = \lambda_{1,0}$ such that the following holds. If k, t and λ are fixed and n tends to infinity then*

a) $M_{\lfloor \lambda n \rfloor}(n,k,t) \sim |\mathcal{F}_r|$ if $\lambda_{t,r-1} < \lambda \leq \lambda_{t,r}$,

b) $M_{\lfloor \lambda n \rfloor}(n,k,t) \sim \frac{1}{2}|N_{\lfloor \lambda n \rfloor}(n,k)|$ if $\lambda = \lambda_{1,0}$,

c) $M_{\lfloor \lambda n \rfloor}(n,k,t) \sim |N_{\lfloor \lambda n \rfloor}(n,k)|$ if $\lambda > \lambda_{1,0}$.

Here, of course, $\mathcal{F}_r := \{\boldsymbol{a} \in N_{\lfloor \lambda n \rfloor}(n,k) : \mathrm{supp}\,(\boldsymbol{a}) \in \mathcal{S}_r\}$. □

The proof uses Corollary 3 and (in the case $\lambda = \lambda_{t,r}$) Theorem 3a. See [8] for details.

THE METHOD OF RESTRICTED INTERSECTION

In this section we present a method which can be considered as one key for the proof of many intersection theorems, in particular also of Theorem 2. It is based of but simplifies the original method of generating sets by Ahlswede and Khachatrian [2].

Let $s \in [n]$ and $\mathcal{F} \in 2^{[n]}$. We call $\mathcal{F}$ *t-intersecting in* $[s]$ (briefly *s-t-intersecting*) if

$$|X_1 \cap X_2 \cap [s]| \geq t \text{ for all } X_1, X_2 \in \mathcal{F}.$$

Let $I_s(n,t)$ be the class of all s-t-intersecting families in $2^{[n]}$. Given a weight function $\omega : 2^{[n]} \to \mathbb{R}_+$, the *weighted s-t-intersection problem* is the problem of determining

$$M_s(n,t;\omega) := \max\{\omega(\mathcal{F}) : \mathcal{F} \in I_s(n,t)\}.$$

We define a new weight function $\omega_{n\to s} : 2^{[s]} \to \mathbb{R}_+$ by

$$\omega_{n\to s}(X) := \omega(\{Z \subseteq [n] : Z \cap [s] = X\}).$$

Note that $\omega_{n\to s}$ is size–dependent if so is ω. We then put for $|X| = i$, $i = 0,\dots,s$,

$$\omega_{n\to s}(i) := \omega_{n\to s}(X).$$

Obviously,

$$M(s,t;\omega_{n\to s}) = M_s(n,t;\omega) \le M(n,t;\omega). \tag{16}$$

Moreover, for $s_1 < s_2 < s_3$ and $X \subseteq [s_1]$

$$\omega_{s_3\to s_1}(X) = (\omega_{s_3\to s_2})_{s_2\to s_1}(X). \tag{17}$$

Using (16) and (17) one can derive that

$$\begin{aligned} M(s,t;\omega_{n\to s}) &= M_s(s+1,t;\omega_{n\to s+1}) \\ &\le M(s+1,t;\omega_{n\to s+1}) = M_{s+1}(s+2,t;\omega_{n\to s+2}) \le \dots \\ &\le M(n-1,t;\omega_{n\to n-1}) = M_{n-1}(n,t;\omega) \le M(n,t;\omega). \end{aligned} \tag{18}$$

In the following we will study the question when the inequality in (16) does hold as an equality. Because of (18) it is enough to look for sufficient conditions for the equality

$$M_{n-1}(n,t;\omega) = M(n,t;\omega). \tag{19}$$

First recall the shifting–operation $s_{i,j} : 2^{2^{[n]}} \to 2^{2^{[n]}}$ defined for $i,j \in [n]$ by

$$s_{i,j}(\mathcal{F}) := \{s_{i,j}(X) : X \in \mathcal{F}\} \cup \{X : X \in \mathcal{F}, s_{i,j}(X) \in \mathcal{F}\},$$

where (with the same notation) $s_{i,j} : 2^{[n]} \to 2^{[n]}$ is given by

$$s_{i,j}(X) := \begin{cases} X \setminus \{j\} \cup \{i\}, & \text{if } j \in X \text{ and } i \notin X \\ X & \text{otherwise.} \end{cases}$$

It is well–known (cf. [13]) and can be easily checked that

$$s_{i,j}(\mathcal{F}) \text{ is } t\text{-intersecting iff } \mathcal{F} \text{ is } t\text{-intersecting}. \tag{20}$$

When studying (19) we will apply only $s_{i,n}, i \in [n-1]$. Obviously, $s_{i,n}(\mathcal{F}) = \mathcal{F}$ or $s_{i,n}(\mathcal{F})$ contains less members having n as an element than $\mathcal{F}$. Iterated application of $s_{i,n}$ (with all possible i's) yields a family $\mathcal{F}'$ with the property

$$s_{i,n}(\mathcal{F}') = \mathcal{F}' \text{ for all } i \in [n-1].$$

We call families with this property *n-shifted.* Let $I^*(n,t)$ be the class of all n-shifted t-intersecting families, and

$$M^*(n,t;\omega) := \max\{\omega(\mathcal{F}) : \mathcal{F} \in I^*(n,t)\}.$$

Supposition 1. *For all $i \in [n-1]$ and $A \subseteq [n]$*

$$\omega(A) \le \omega(s_{i,n}(A)).$$

It is easy to see that under this Supposition $M^*(n,t;\omega) = M(n,t;\omega)$. In the following we require the weight function ω to satisfy Supposition 1. Note that this is always true if ω is size–dependent.

Now assume that

$$M(n,t;\omega) > M_{n-1}(n,t;\omega). \tag{21}$$

We will look for further suppositions such that a contradiction can be obtained.

Choose among all optimal t-intersecting families, i.e. $\omega(\mathcal{F}) = M(n,t;\omega)$, one for which the set

$$\mathcal{R} := \mathcal{R}(\mathcal{F}) := \{X \in \mathcal{F} : n \in X, X \setminus \{n\} \notin \mathcal{F}\} \tag{22}$$

has minimum cardinality (note that $\mathcal{R} \neq \emptyset$ since otherwise $\mathcal{F}$ would be already t-intersecting in $[n-1]$ in contradiction to (21)). We may assume that $\mathcal{F}$ has the following property:

$$n \notin X \in \mathcal{F}, X \subseteq Y \text{ implies } Y \in \mathcal{F}. \tag{23}$$

Then, by Supposition 1, the choice of $\mathcal{F}$, and (23), $\mathcal{F}$ is n-shifted. Let

$$\mathcal{R}_i := \{X \in \mathcal{R} : |X| = i\} \text{ and } \mathcal{R}'_i := \{X \setminus \{n\} : X \in \mathcal{R}_i\}. \tag{24}$$

It is not too difficult to verify that $|X \cap Y| \geq t$ for all $X \in \mathcal{R}'_i$ and $Y \in \mathcal{F} \setminus \mathcal{R}_{n+t-i}$ (use that $\mathcal{F}$ is n-shifted t-intersecting). Hence, for any $i \in [t, \frac{n+t}{2})$ the two families

$$\begin{aligned} \mathcal{F}_{1,i} &:= (\mathcal{F} \setminus \mathcal{R}_{n+t-i}) \cup \mathcal{R}'_i, \\ \mathcal{F}_{2,i} &:= (\mathcal{F} \setminus \mathcal{R}_i) \cup \mathcal{R}'_{n+t-i} \end{aligned} \tag{25}$$

are t-intersecting and we have

$$\begin{aligned} \omega(\mathcal{F}_{1,i}) &\geq \omega(\mathcal{F}) \quad \text{iff } \omega(\mathcal{R}'_i) \geq \omega(\mathcal{R}_{n+t-i}), && (26) \\ |\mathcal{R}(\mathcal{F}_{1,i})| &< |\mathcal{R}(\mathcal{F})| \quad \text{iff } \mathcal{R}_i \neq \emptyset \text{ or } \mathcal{R}_{n+t-i} \neq \emptyset, && (27) \\ \omega(\mathcal{F}_{2,i}) &\geq \omega(\mathcal{F}) \quad \text{iff } \omega(\mathcal{R}'_{n+t-i}) \geq \omega(\mathcal{R}_i), && (28) \\ |\mathcal{R}(\mathcal{F}_{2,i})| &< |\mathcal{R}(\mathcal{F})| \quad \text{iff } \mathcal{R}_i \neq \emptyset \text{ or } \mathcal{R}_{n+t-i} \neq \emptyset. && (29) \end{aligned}$$

Hence we obtain a contradiction if (26) and (27) or if (28) and (29) hold since otherwise $\mathcal{F}_{1,i}$ or $\mathcal{F}_{2,i}$ would be "better" than $\mathcal{F}$.

This leads us to the following second supposition (with the definition $\mathcal{A} \cup \{a\} := \{A \cup \{a\} : A \in \mathcal{A}\}$, $\mathcal{A} \subseteq 2^{[n]}$).

Supposition 2. *For all $j \in [t, \frac{n+t}{2})$, $\mathcal{A} \subseteq \binom{[n-1]}{j-1}$, $\mathcal{B} \subseteq \binom{[n-1]}{n+t-j-1}$ not both inequalities are valid:*

$$\begin{aligned} \omega(\mathcal{A}) &< \omega(\mathcal{B} \cup \{n\}), \\ \omega(\mathcal{B}) &< \omega(\mathcal{A} \cup \{n\}). \end{aligned}$$

Under Supposition 2 we obtain $\mathcal{R}_i = \emptyset$ for all $i \in [t,n] \setminus \{\frac{n+t}{2}\}$ which yields the contradiction $\mathcal{R} = \emptyset$ in the case $2 \nmid n+t$. Thus we need a further supposition such that $\mathcal{R}_{\frac{n+t}{2}} \neq \emptyset$ leads to a contradiction.

Case $t = 1$:

Let $A \in \mathcal{R}_{\frac{n+t}{2}}$ and $A' := A \setminus \{n\}$. Let $B' := [n-1] \setminus A'$, $B := B' \cup \{n\}$. If $B \notin \mathcal{R}_{\frac{n+t}{2}}$ (which implies $B \notin \mathcal{F}$) then $\mathcal{F}' := \mathcal{F} \cup \{A'\}$ is also t-intersecting, but $\omega(\mathcal{F}') \geq \omega(\mathcal{F})$ and $|\mathcal{R}(\mathcal{F}')| < |\mathcal{R}(\mathcal{F})|$, a contradiction. Thus $B \in \mathcal{R}_{\frac{n+t}{2}}$. Let

$$\begin{aligned} \mathcal{F}_1 &:= (\mathcal{F} \setminus \{B\}) \cup \{A'\}, \\ \mathcal{F}_2 &:= (\mathcal{F} \setminus \{A\}) \cup \{B'\}. \end{aligned} \tag{30}$$

Obviously $\mathcal{F}_1$ and $\mathcal{F}_2$ are t-intersecting and

$$\begin{aligned} \omega(\mathcal{F}_1) &\geq \omega(\mathcal{F}) \quad \text{iff } \omega(A') \geq \omega(B), \\ \omega(\mathcal{F}_2) &\geq \omega(\mathcal{F}) \quad \text{iff } \omega(B') \geq \omega(A), \\ |\mathcal{R}(\mathcal{F}_i)| &< |\mathcal{R}(\mathcal{F})|, \quad i = 1, 2. \end{aligned}$$

This leads us to the following supposition which yields in the case $t = 1$ the desired contradiction.

Supposition 3.1. *For all* $A \in \binom{[n-1]}{\frac{n-1}{2}}$ *not both inequalities are valid:*

$$\begin{aligned} \omega(A) &< \omega([n] \setminus A), \\ \omega([n-1] \setminus A) &< \omega(A \cup \{n\}). \end{aligned}$$

Supposition 3.1 is true if $\omega(A) \geq \omega(A \cup \{n\})$ for all $A \subseteq \binom{[n-1]}{\frac{n-1}{2}}$.

Case $t \geq 1$ and the weight function ω is size dependent, i.e. there is some $\boldsymbol{\omega}$ such that $\omega(X) = \omega_j$ for all X with $|X| = j$.
We have by double counting

$$\sum_{j=1}^{n-1} \sum_{X \in \mathcal{R}_{\frac{n+t}{2}} : j \notin X} \omega(X) = \sum_{X \in \mathcal{R}_{\frac{n+t}{2}}} \sum_{j \in [n-1] : j \notin X} \omega(X) = \frac{n-t}{2} \omega(\mathcal{R}_{\frac{n+t}{2}}).$$

Hence there is some $j \in [n-1]$ such that

$$\sum_{X \in \mathcal{R}_{\frac{n+t}{2}} : j \notin X} \omega(X) \geq \frac{n-t}{2(n-1)} \omega(\mathcal{R}_{\frac{n+t}{2}}). \tag{31}$$

Let $\mathcal{T} := \{X \in \mathcal{R}_{\frac{n+t}{2}} : j \notin X\}$ and $\mathcal{T}' := \{X \setminus \{n\} : X \in \mathcal{T}\}$. By the size–dependence of ω, (31) is equivalent to

$$|\mathcal{T}| \, \omega_{\frac{n+t}{2}} \geq \frac{n-t}{2(n-1)} \left| \mathcal{R}_{\frac{n+t}{2}} \right| \omega_{\frac{n+t}{2}}. \tag{32}$$

It is easy to see that

$$\mathcal{F}_1 := \left(\mathcal{F} \setminus \mathcal{R}_{\frac{n+t}{2}} \right) \cup \mathcal{T} \cup \mathcal{T}' \tag{33}$$

is t-intersecting,

$$|\mathcal{R}(\mathcal{F}_1)| < |\mathcal{R}(\mathcal{F})| \text{ if } \mathcal{R}_{\frac{n+t}{2}} \neq \emptyset,$$

and that $\omega(\mathcal{F}_1) \geq \omega(\mathcal{F})$ is equivalent to the following inequalities

$$\begin{aligned} \omega(\mathcal{T}) + \omega(\mathcal{T}') &\geq \omega(\mathcal{R}_{\frac{n+t}{2}}), \\ |\mathcal{T}| \left(\omega_{\frac{n+t}{2}} + \omega_{\frac{n+t}{2}-1}\right) &\geq |\mathcal{R}_{\frac{n+t}{2}}| \omega_{\frac{n+t}{2}}. \end{aligned} \tag{34}$$

Thus we obtain the desired contradiction in the case $\mathcal{R}_{\frac{n+t}{2}} \neq \emptyset$ if (34) holds. Finally we claim that the following supposition for our candidate families is sufficient for (34):

Supposition 3.2. *We have*

$$\omega(\mathcal{S}_{\frac{n-t}{2}-1}) \geq \omega(\mathcal{S}_{\frac{n-t}{2}}). \tag{35}$$

Indeed, we have

$$\begin{aligned} \mathcal{S}_{\frac{n-t}{2}-1} \setminus \mathcal{S}_{\frac{n-t}{2}} &= \left\{X \subseteq [n-2] : |X| = \frac{n+t}{2} - 1\right\}, \\ \mathcal{S}_{\frac{n-t}{2}} \setminus \mathcal{S}_{\frac{n-t}{2}-1} &= \left\{X \cup \{n-1, n\} : X \subseteq [n-2], |X| = \frac{n+t}{2} - 2\right\}. \end{aligned}$$

Hence (35) is equivalent to the following inequalities:

$$\begin{aligned} \binom{n-2}{\frac{n+t}{2}-1} \omega_{\frac{n+t}{2}-1} &\geq \binom{n-2}{\frac{n+t}{2}-2} \omega_{\frac{n+t}{2}}, \\ \omega_{\frac{n+t}{2}-1} &\geq \frac{n+t-2}{n-t} \omega_{\frac{n+t}{2}}. \end{aligned} \tag{36}$$

From (32) and (36) we obtain (34).

Herewith we proved the following theorem:

Theorem 14. *We have $M_{n-1}(n,t;\omega) = M(n,t;\omega)$ if Suppositions 1, 2, and 3.1 (if t=1) or Suppositions 2 and 3.2 (if ω is size dependent) are true. In the case $n+t$ odd already Suppositions 1 and 2 (resp. only Supposition 2) are sufficient.* □

Note for all $\ell \in [t+2r+1, n]$ the following two facts:

1) If $\omega(Z) \leq \omega(s_{i,\ell}(Z))$ for all $i \in [\ell-1]$, $Z \subseteq [n]$ then also $\omega_{n\to\ell}(X) \leq \omega_{n\to\ell}(s_{i,\ell}(X))$ for all $i \in [\ell-1]$, $X \subseteq [\ell]$.

2) $\omega(\mathcal{S}_\varrho^n) = \omega_{n\to\ell}(\mathcal{S}_\varrho^\ell)$ for all $\varrho \in \left[0, \left\lfloor \frac{\ell-t}{2} \right\rfloor\right]$.

Hence, iterated application of Theorem 14 together with (18) yields:

Theorem 15. *Let $r \in \{0, \dots, \lfloor \frac{n-t-1}{2} \rfloor\}$. We have*

$$M_{t+2r}(n, t; \omega) = M(n, t; \omega)$$

if $t = 1$ and conditions (i), (ii), and (iii.1) are satisfied, or if $t \geq 1$, ω is size–dependent and conditions (ii) and (iii.2) are satisfied, where

(i) For all $\ell \in [t + 2r + 1, n]$, $i \in [\ell - 1]$, $A \subseteq [n]$

$$\omega(A) \leq \omega(s_{i,\ell}(A)).$$

(ii) For all $\ell \in [t + 2r + 1, n]$, $i \in [t, \frac{\ell+t}{2})$, $\mathcal{A} \subseteq \binom{[\ell-1]}{i-1}$, $\mathcal{B} \subseteq \binom{[\ell-1]}{\ell+t-i-1}$ not both inequalities are valid:

$$\begin{aligned} w_{n\to\ell}(\mathcal{A}) &< \omega_{n\to\ell}(\mathcal{B} \cup \{\ell\}), \\ w_{n\to\ell}(\mathcal{B}) &< \omega_{n\to\ell}(\mathcal{A} \cup \{\ell\}). \end{aligned}$$

(iii.1) For all $\varrho \in [r + 1, \lfloor \frac{n-1}{2} \rfloor]$, $A \in \binom{[2\varrho]}{\varrho}$ not both inequalities are valid:

$$\begin{aligned} \omega_{n\to 2\varrho+1}(A) &< \omega_{n\to 2\varrho+1}([2\varrho + 1] \setminus A), \\ \omega_{n\to 2\varrho+1}([2\varrho] \setminus A) &< \omega_{n\to 2\varrho+1}(A \cup \{2\varrho + 1\}). \end{aligned}$$

(iii.2)

$$\omega(\mathcal{S}_r) \geq \omega(\mathcal{S}_{r+1}) \geq \dots \geq \omega(\mathcal{S}_{\lfloor \frac{n-t}{2} \rfloor}).$$

□

Remark.

a) The following condition (iv) is sufficient for (ii) and (iii.1).

(iv) For all $\ell \in [t + 2r + 1, n]$, $A \in 2^{[n]}$

$$\omega(A) \geq \omega(A \cup \{\ell\}).$$

b) In the case of size–dependence the following conditions (ii') and (iii.2') are sufficient for (ii) and (iii.2), respectively (recall (35), (36)).

(ii') For all $\ell \in [t + 2r + 1, n]$, $i \in [t, \frac{\ell+t}{2})$

$$\omega_{n\to\ell}(i - 1)\, \omega_{n\to\ell}(\ell + t - i - 1) \geq \omega_{n\to\ell}(i)\, \omega_{n\to\ell}(\ell + t - i).$$

(iii.2') For all $\varrho \in [r + 1, \lfloor \frac{n-t}{2} \rfloor]$

$$\omega_{n\to t+2\varrho}(t + \varrho - 1) \geq \left(1 + \frac{t-1}{\varrho}\right) \omega_{n\to t+2\varrho}(t + \varrho).$$

Example 1. Let $\boldsymbol{\omega} = \boldsymbol{e}_k$, $k \geq t$. Then $\omega_{n\to\ell}(i) = \binom{n-\ell}{k-i}$. A simple computation shows that *(ii')* is satisfied if $n > 2k - t$ and *(iii.2')* is satisfied if $k \leq k_r^n$.

Example 2. [2] Let $\boldsymbol{\omega} = \boldsymbol{e}_k + \boldsymbol{e}_{k+1}$, $k \geq t$. Then $\omega_{n\to\ell}(i) = \binom{n-\ell}{k-i} + \binom{n-\ell}{k+1-i} = \binom{n-\ell+1}{k+1-i}$. As in the previous example, *(ii')* is satisfied if $n + 1 > 2(k+1) - t$ (i.e. $k + 1 \leq k^{n+1}_{\lfloor \frac{n+1-t}{2} \rfloor}$) and *(iii.2')* is satisfied if $k + 1 \leq k_r^{n+1}$. Together with Corollary 3c we obtain that $\mathcal{S}_r$ is optimal if $k_{r-1}^{n+1} \leq k + 1 \leq k_r^{n+1}$, $r = 0, \ldots, \lfloor \frac{n-t}{2} \rfloor$. If $k + 1 > k^{n+1}_{\lfloor \frac{n-t}{2} \rfloor}$ then $k \geq k^n_{\lfloor \frac{n-t}{2} \rfloor - 1}$, hence $\mathcal{S}_{\lfloor \frac{n-t}{2} \rfloor}$ is optimal.

Example 3. Let $\omega_i = \binom{\ell-1}{\ell-i}$, $\ell \geq t$ (compare with (12)). Then $\omega_{n\to s}(i) = \binom{n-s+\ell-1}{\ell-i}$. A simple computation shows that *(ii')* is satisfied if $n > \ell - t + 1$ and *(iii.2')* is satisfied if $\ell \leq \ell_r$.

Example 4. Let $\omega_k = 0$ unless $k \leq k_r$. Note that then $\omega_k \neq 0$ implies $n > 2k - t$. We have $\omega_{n\to\ell}(i) = \sum_{k=0}^{k_r} \omega_k \binom{n-\ell}{k-i}$. Then *(ii')* reads:

$$\sum_{j,k=0}^{k_r} \omega_j \omega_k \binom{n-\ell}{j-i+1}\binom{n-\ell}{k-\ell-t+i+1} \geq \sum_{j,k=0}^{k_r} \omega_j \omega_k \binom{n-\ell}{j-i}\binom{n-\ell}{k-\ell-t+i}.$$

Using $j + k - t < n$ it is not difficult to verify that

$$\binom{n-\ell}{j-i+1}\binom{n-\ell}{k-\ell-t+i+1} \geq \binom{n-\ell}{j-i}\binom{n-\ell}{k-\ell-t+i}$$

for all $0 \leq j, k \leq k_r$. Consequently, *(ii')* is satisfied. Using Example 1 it is easy to show that also *(iii.2')* is satisfied.

Example 5. Let α be a positive real number and $\omega_i = \alpha^{-i}$. Then $\omega_{n\to\ell}(i) = \alpha^{-i}(1 + \alpha^{-1})^{n-\ell}$ and *(ii')* is satisfied if $\alpha \geq 1$. Further, *(iii.2')* is satisfied if $\alpha \geq 1 + \frac{t-1}{r+1}$. Together with Corollary 10 we obtain that $\mathcal{S}_r$ is optimal if $1 + \frac{t-1}{r} > \alpha \geq 1 + \frac{t-1}{r+1}$.

This example has the following application: For $\alpha, n \in \mathbb{N}$ consider the set $H_\alpha^n := \{\boldsymbol{a} = (a_1, \ldots, a_n) : a_i \in \{1, \ldots, \alpha\}\}$. On H_α^n one has the Hamming metric d_H which for two tuples $\boldsymbol{a}$, $\boldsymbol{b}$ counts the number of different coordinates: $d_H(\boldsymbol{a}, \boldsymbol{b}) = |\{i : a_i \neq b_i\}|$. As usual, for a subset $\mathcal{F}$ of H_α^n the diameter $d(\mathcal{F})$ is the maximum possible distance between two elements of $\mathcal{F}$. Let $d \in \mathbb{N}$. We are interested in the following diametric problem: Determine the maximum cardinality of a set $\mathcal{F} \subseteq H_\alpha^n$ with diameter d or less. The complete solution was given by Ahlswede and Khachatrian [4]. Independently, Frankl and Tokushige [14] proved the following t-intersection version.

[2] This result was communicated to us by L. Khachatrian.

Call a set $\mathcal{F} \subseteq H_\alpha^n$ *t-intersecting* if any two tuples of $\mathcal{F}$ agree in at least t coordinates. Obviously, subsets of H_α^n with diameter at most d are $(n-d)$-intersecting and vice versa. Thus the diametric problem is equivalent to:

$$\text{Determine } M(n,\alpha,t) := \max\{|\mathcal{F}| : \mathcal{F} \subseteq H_\alpha^n,\ \mathcal{F} \text{ is } t\text{-intersecting}\}. \tag{37}$$

Define for $i,j \in [\alpha]$, $c \in [n]$ the operation $s_{i,j,c} : 2^{H_\alpha^n} \to 2^{H_\alpha^n}$ by

$$s_{i,j,c}(\mathcal{F}) := \{s_{i,j,c}(\boldsymbol{a}) : \boldsymbol{a} \in \mathcal{F}\} \cup \{\boldsymbol{a} \in \mathcal{F} : s_{i,j,c}(\boldsymbol{a}) \in \mathcal{F}\}, \tag{38}$$

where (with the same notation) $s_{i,j,c} : H_\alpha^n \to H_\alpha^n$ is given by

$$s_{i,j,c}(\boldsymbol{a}) := \begin{cases} (a_1,\ldots,a_{c-1},i,a_{c+1},\ldots,a_n), & \text{if } a_c = j \\ \boldsymbol{a} & \text{otherwise.} \end{cases} \tag{39}$$

It is easy to verify that this operation respects the t-intersection property. Furthermore, if $s_{i,j,c}(\mathcal{F}) = \mathcal{F}$ for all i,j,c, $i<j$, then any two tuples of $\mathcal{F}$ have entry 1 in at least t common coordinates. It follows that the determination of $M(n,t;\omega)$ with $\omega_i := (\alpha-1)^{n-i}$ suffices for (37). Thus, Example 5 shows that one of the candidates $\mathcal{S}_r$ is optimal. We refer to [4] for more details and background.

Let us generalize the previous application. For $\boldsymbol{\alpha} = (\alpha_1,\ldots,\alpha_n) \in \mathbb{N}^n$ consider the set $F_{\boldsymbol{\alpha}} := \{\boldsymbol{a} = (a_1,\ldots,a_n) : a_i \in \{0,\ldots,\alpha_i\}\}$. We define an order relation on $F_{\boldsymbol{\alpha}}$ by $\boldsymbol{a} \le \boldsymbol{b}$ iff $a_i = 0$ or $a_i = b_i$ for all $i=1,\ldots,n$. Then $F_{\boldsymbol{\alpha}}$ is a ranked partially ordered set, isomorphic to the direct product of n stars $0 < \alpha_1, \alpha_2, \ldots, \alpha_i$, $i=1,\ldots,n$. Let $N_k(\boldsymbol{\alpha})$ be the k-th level of $F_{\boldsymbol{\alpha}}$, i.e. $N_k(\boldsymbol{\alpha}) = \{\boldsymbol{a} \in F_{\boldsymbol{\alpha}} : |\{i : a_i > 0\}| = k\}$ and define $W_k(\boldsymbol{\alpha}) := |N_k(\boldsymbol{\alpha})|$ (note that if $\alpha := \alpha_1 = \alpha_2 = \cdots = \alpha_n$ then $N_n(\boldsymbol{\alpha}) = H_\alpha^n$). A family $\mathcal{F} \subseteq F_{\boldsymbol{\alpha}}$ is called t-intersecting if for all $\boldsymbol{a},\boldsymbol{b} \in \mathcal{F}$ there exist t coordinats $i_1,\ldots,i_t$ such that $a_{i_j} = b_{i_j} > 0$ holds for $j=1,\ldots,t$ (i.e. the infimum of $\boldsymbol{a}$ and $\boldsymbol{b}$ in $F_{\boldsymbol{\alpha}}$ has rank at least t). Define

$$M_k(n,\boldsymbol{\alpha},t) = \max\{|\mathcal{F}| : \mathcal{F} \subseteq N_k(\boldsymbol{\alpha}),\ \mathcal{F} \text{ is } t\text{-intersecting}\}.$$

For $K \subseteq [n]$ let $\pi_K : \mathbb{N}^n \to \mathbb{N}^{n-|K|}$ be the projection map onto the coordinates that are not contained in K. Define $\omega : 2^{[n]} \to \mathbb{R}_+$ by

$$\omega(\{i_1,\ldots,i_m\}) = W_{k-m}\left(\pi_{\{i_1,\ldots,i_m\}}(\alpha_1-1,\ldots,\alpha_n-1)\right). \tag{40}$$

Using the operation $s_{i,j,c} : 2^{F_{\boldsymbol{\alpha}}} \to 2^{F_{\boldsymbol{\alpha}}}$ defined by (38) and (39) one can derive that

$$M_k(n,\boldsymbol{\alpha},t) = M(n,t;\omega). \tag{41}$$

Example 6. Let $\alpha := \alpha_1 = \ldots = \alpha_n \ge 2$, $n > k$, and ω as in (40). Then ω is size–dependent. Let us use the abbreviation

$$N_k\left((\alpha-1)^a, \alpha^b\right) := N_k(\underbrace{\alpha-1,\ldots,\alpha-1}_{a}, \underbrace{\alpha,\ldots,\alpha}_{b}),$$

and similar for $W_k\left((\alpha-1)^a,\alpha^b\right)$. Then

$$\omega_{n\to\ell}(i) = W_{k-i}\left((\alpha-1)^{\ell-i},\alpha^{n-\ell}\right).$$

It is easy to see that $\omega_{n\to\ell}(i-1) \geq \omega_{n\to\ell}(i)$ holds for all i, hence (ii') is satisfied. Furthermore, purely numerical considerations show that $\omega(\mathcal{S}_0) \geq \omega(\mathcal{S}_1)$ holds iff $(iii.2)$ holds for $r=0$ iff

$$n \geq \left\lfloor \frac{(k-t+\alpha)(t+1)}{\alpha} \right\rfloor. \tag{42}$$

It follows that in this case the family $\mathcal{S}_0$ is optimal for (41). See [6] for details.

In the general case we do not know whether always one of the families $\mathcal{S}_r$ is optimal. However, one can prove the following [5]: If α, t are constant and n is sufficiently large then it holds (for all k)

$$M_k(n,\boldsymbol{\alpha},t) = \max_r \omega(\mathcal{S}_r).$$

Example 7. Let $t=1$, $\alpha_1 \leq \alpha_2 \leq \ldots \leq \alpha_n$, and ω as in (40). Then (i) is clearly satisfied. It is easy to see that (iv) is satisfied if $\alpha_{t+2r+1} \geq 2$. It follows that $\mathcal{S}_0$ is optimal for (41) (with $t=1$) if $\alpha_2 \geq 2$. Using Theorem 1 one can also deal the general case $1=\alpha_1=\ldots=\alpha_m<\alpha_{m+1}\leq\ldots\leq\alpha_n$, see [7] for details.

NONTRIVIAL T–INTERSECTION

A family $\mathcal{F} \subseteq 2^{[n]}$ is called *nontrivial t-intersecting* (resp. *nontrivial t-intersecting in [s]*, briefly *nontrivial s-t-intersecting*, $s \in [n]$) if it is t-intersecting (resp. s-t-intersecting), and if

$$\left|\bigcap_{X\in\mathcal{F}} X\right| < t. \tag{43}$$

Let $\tilde{I}(n,t)$ (resp. $\tilde{I}_s(n,t)$) denote the class of all such families.

Suppose we are given a t-intersecting family $\mathcal{F}$ such that $|X|>t$ for all $X \in \mathcal{F}$ (e.g. a k-uniform t-intersecting family with $k>t$). Then the family $\mathcal{F} \cup \{[n]\setminus\{i\} : i \in n\}$ is nontrivial t-intersecting. Thus, dealing with optimal nontrivial t-intersecting families with respect to some weight function ω, the intersection in (43) should include only sets $X \in \mathcal{F}$ with $\omega(X)>0$.

We require the weight function $\omega : 2^{[n]} \to \mathbb{R}_+$ to satisfy the following supposition:

Supposition 4. *$\omega(X)>0$ implies $\omega(Y)>0$ for all $Y \in 2^{[n]}$ with $|Y|=|X|$.*

Let $\Omega := \Omega(\omega) := \{i : \omega([i])>0\}$ and for $\mathcal{F} \in 2^{[n]}$ let

$$\mathcal{F}_\Omega := \bigcup_{i\in\Omega} \mathcal{F}_i = \{X \in \mathcal{F} : \omega(X)>0\}.$$

Note that for $s \in [n-1]$ the new weight function $\omega_{n\to s}$ satisfies Supposition 4 if so does ω. Moreover, we have

$$i \in \Omega(\omega_{n\to s}) \text{ iff } [i, i+n-s] \cap \Omega(\omega) \neq \emptyset. \tag{44}$$

Finally, we define

$$\begin{aligned} \tilde{M}(n,t;\omega) &:= \max\{\omega(\mathcal{F}) : \mathcal{F}_\Omega \in \tilde{I}(n,t)\}, \\ \tilde{M}_s(n,t;\omega) &:= \max\{\omega(\mathcal{F}) : \mathcal{F}_\Omega \in \tilde{I}_s(n,t)\}. \end{aligned}$$

Note that

$$\begin{aligned} \tilde{M}(s,t;\omega_{n\to s}) &= \tilde{M}_s(s+1,t;\omega_{n\to s+1}) \\ &\leq \tilde{M}(s+1,t;\omega_{n\to s+1}) = \tilde{M}_{s+1}(s+2,t;\omega_{n\to s+2}) \leq \dots \\ &\leq \tilde{M}(n-1,t;\omega_{n\to n-1}) = \tilde{M}_{n-1}(n,t;\omega) \leq \tilde{M}(n,t;\omega). \end{aligned} \tag{45}$$

In this section we shall study the problem of the determination of these numbers. We will see that the method of restricted intersection works as well.

We may always suppose that $n \geq t+2$ since obviously $\tilde{I}(n,n-1) = \tilde{I}(n,n) = \emptyset$.

Let us look at candidates for optimal families. Clearly

$$(\mathcal{S}_1)_\Omega, \dots, \left(\mathcal{S}_{\lfloor \frac{n-t}{2} \rfloor}\right)_\Omega$$

are nontrivial t-intersecting if $\Omega \neq \{n\}$. Furthermore, for $T \subseteq [n]$ let

$$\mathcal{G}'_T := \{X \subseteq [n] : T \subseteq X\} \cup \{[n] \setminus \{j\} : j \in T\}$$

and

$$\mathcal{G}_T := \begin{cases} \mathcal{G}'_T & \text{if } |T| > t, \\ \mathcal{G}'_T \setminus \{T\} & \text{if } |T| = t. \end{cases}$$

It is easy to see that

$$\mathcal{G}_T \supseteq \mathcal{G}_{T'} \text{ if } T \subsetneq T',\ |T| \geq t, \tag{46}$$

where equality holds iff $|T| = t$ and $n = t+2$.

Note that $(\mathcal{G}_T)_\Omega \in \tilde{I}(n,t)$ if $n-1 \in \Omega$.

Theorem 16. *We have for* $n \geq t+2$

$$\tilde{M}(n,t;\omega) = \begin{cases} \tilde{M}_{n-1}(n,t;\omega) & \text{if } n-1 \notin \Omega \\ \max\left\{\tilde{M}_{n-1}(n,t;\omega),\ \omega(\mathcal{G}_T) : T \in \binom{[n-1]}{t}\right\} & \text{if } n-1 \in \Omega \end{cases}$$

if Supposition 4 and the suppositions from Theorem 14 are satisfied.

Proof. We proceed as in the previous section. Assume that

$$\tilde{M}(n,t;\omega) > \tilde{M}_{n-1}(n,t;\omega).$$

Choose among all t-intersecting families $\mathcal{F}$ with $\mathcal{F}_\Omega \in \tilde{I}(n,t)$ and $\omega(\mathcal{F}) = \tilde{M}(n,t;\omega)$ one for which

$$\mathcal{R} := \mathcal{R}(\mathcal{F}) := \{X \in \mathcal{F} : n \in X, X \setminus \{n\} \notin \mathcal{F}\}$$

has minimum cardinality. Note that $\mathcal{R} \neq \emptyset$. We may asume that $\mathcal{F}$ has property (23).
Claim: $\mathcal{F}$ is n-shifted.
Assume the contrary. Then the set

$$I_\Omega := \{i \in [n] : s_{i,n}(\mathcal{F}_\Omega) \neq \mathcal{F}_\Omega\}$$

is not empty since otherwise every family $s_{i,n}(\mathcal{F})$ with $s_{i,n}(\mathcal{F}) \neq \mathcal{F}$ would be "better" than $\mathcal{F}$. Also, we have for all $i \in I_\Omega$

$$\left| \bigcap_{X \in s_{i,n}(\mathcal{F}_\Omega)} X \right| = t. \tag{47}$$

Let $S := \bigcap_{X \in \mathcal{F}_\Omega} X$. Then (47) implies (for all $i \in I_\Omega$) $i, n \notin S$, $|S| = t-1$ and

$$\bigcap_{X \in s_{i,n}(\mathcal{F}_\Omega)} X = S \cup \{i\}. \tag{48}$$

It follows that for all $i \in I_\Omega$ there exist sets $X_i, Y_i \in \mathcal{F}_\Omega$ such that $X_i \cap \{i,n\} = \{n\}$, $Y_i \cap \{i,n\} = \{i\}$. Further, for all $i \in I_\Omega$ and $Z \in \mathcal{F}_\Omega$ we have $Z \cap \{i,n\} \neq \emptyset$. Since $\mathcal{F}$ has maximum weight we must have (for all $i \in I_\Omega$)

$$Z \in \mathcal{F}_\Omega \text{ if } S \cup \{i,n\} \subseteq Z,\ |Z| \in \Omega. \tag{49}$$

Note that for all $i \in I_\Omega$ we have $|X_i| \leq n-2$ or $|Y_i| \leq n-2$ since otherwise $[n-1], [n] \setminus \{i\} \in \mathcal{F}_\Omega$ in contradiction to (48).

Clearly $I_\Omega \subseteq [n-1] \setminus S$. Note that $|[n-1] \setminus S| \geq 2$ since $n \geq t+2$.
Case $|I_\omega| = 1$ **:**
Let $I_\Omega = \{i\}$. We have $j \in X_i$ for all $j \in [n-1] \setminus (S \cup \{i\})$ since otherwise $i, n \notin s_{j,n}(X_i) \in \mathcal{F}_\Omega$ which contradicts (48). Thus $X_i = [n] \setminus \{i\}$, in particular $n-1 \in \Omega$. By property (23) (applied to $Y_i \in \mathcal{F}_\Omega$) we have also $[n-1] \in \mathcal{F}_\Omega$, a contradiction to (48).
Case $I_\Omega = [n-1] \setminus S$ **:**
Let $i \in I_\Omega$. We have $Y_i = [n-1]$ since otherwise $j, n \notin Y_i \in \mathcal{F}_\Omega$ for some $j \in I_\Omega$, a contradiction to (48). In particular, $|Y_i| = n-1 \in \Omega$. By (49) we have also $[n] \setminus \{i\} \in \mathcal{F}_\Omega$, again a contradiction to (48).
Case $|I_\Omega| \geq 2$ **and** $I_\Omega \subsetneq [n-1] \setminus S$ **:**
Let $i, j \in I_\Omega$, $i \neq j$, and $k \in [n-1] \setminus (S \cup I_\Omega)$. Then (49) implies that there exists a set $Z \in \mathcal{F}_\Omega$ such that $S \cup \{i,n\} \subseteq Z$ and $j, k \notin Z$. But then $j, n \notin s_{k,n}(Z) \in \mathcal{F}_\Omega$, a contradiction to (48).

Hence in all three cases a contradiction is obtained, this proves the claim.

Let now

$$T := \bigcap_{X \in (\mathcal{F}\setminus\mathcal{R})_\Omega} X, \ \tau := |T|.$$

Case $\tau < t$:
In (25), (30), and (33) we constructed families by deleting some members of $\mathcal{R}$ from $\mathcal{F}$ and adding some new members such that the new families are still t-intersecting. Since we did not change $\mathcal{F}\setminus\mathcal{R}$ the new families are even nontrivial t-intersecting and we may argue exactly as in the proof of Theorem 14.
Case $\tau \geq t$:
For all $X \in \mathcal{R}_\Omega$ we have

$$\text{either } T \subseteq X \text{ or } X = [n] \setminus \{j\} \text{ for some } j \in T. \tag{50}$$

Indeed, otherwise there would exist two elements $i \in [n]$, $j \in T$ such that $i, j \notin X$. Since $\mathcal{F}$ is n-shifted $s_{i,n}(X) \in (\mathcal{F}\setminus\mathcal{R})_\Omega$. Clearly $j \notin s_{i,n}(X)$ which is a contradiction to $j \in T \subseteq s_{i,n}(X)$. Note that in the case $|T| = t$ the set T is not an element of $\mathcal{R}$ since otherwise $T \subseteq X$ for all $X \in \mathcal{F}$ which is impossible since $\mathcal{F}_\Omega$ is nontrivial t-intersecting. By the definition of T and (50) we have $\mathcal{F}_\Omega \subseteq \mathcal{G}_T$. But then, recalling (46) and the optimality of $\mathcal{F}$, $\omega(\mathcal{F}) = \omega(\mathcal{G}_{T'})$ for some $T' \in \binom{n-1}{t}$. Note that in this case necessarily $n-1 \in \Omega$ since otherwise $\mathcal{F}_\Omega$ would not be nontrivial t-intersecting. □

Note that for $i \notin T, \ell \in T$ the relation

$$\mathcal{G}_{s_{i,\ell}(T)} = s_{i,\ell}(\mathcal{G}_T)$$

holds. Thus, under Supposition *(i)* from Theorem 15, it is enough to consider sets T from $\binom{[t+2r]}{t}$.

In addition to the candidates $\mathcal{G}_T$ we define for $T \in \binom{[t+2r]}{t}$, $\ell \in [t+2r, n]$, $r \geq 1$

$$\begin{aligned}\mathcal{G}_{T,\ell} := \{X \subseteq [n] : T \subseteq X,\ ([\ell]\setminus T) \cap X \neq \emptyset\} \\ \cup \{X \subseteq [n] : [\ell] \setminus T \subseteq X,\ |X \cap T| = t-1\}.\end{aligned}$$

We define further $\mathcal{G}_{T,n+1} := \mathcal{G}_{T,n}$.

Let $\omega_{\max} := \max\{i : i \in \Omega\}$. Iterated application of Theorem 16 together with (44) and (45) yields:

Theorem 17. *Let* $r \in \left\{1, \ldots, \left\lfloor \frac{n-t-2}{2} \right\rfloor\right\}$. *We have*

$$\tilde{M}(n,t;\omega) = \begin{cases} \tilde{M}_{t+2r}(n,t;\omega) & \text{if } \omega_{\max} < t+2r \\ \max\left\{\tilde{M}_{t+2r}(n,t;\omega),\ \omega(\mathcal{G}_{T,\ell}) : T \in \binom{[t+2r]}{t},\ \ell \in [t+2r+1, \omega_{\max}+1]\right\} & \\ & \text{if } \omega_{\max} \geq t+2r \end{cases}$$

if Supposition 4 and the suppositions from Theorem 15 are satisfied. □

For applications, the most important case is if $r = 1$. Since $\tilde{M}_{t+2}(n,t;\omega) = \omega(\mathcal{S}_1)$ and $\mathcal{S}_1 = \mathcal{G}_{[t],t+2}$ the determination of $\tilde{M}(n,t;\omega)$ reduces then to a purely numerical problem: Find the maximum of all numbers $\omega(\mathcal{G}_{T,\ell})$, $T \in \binom{t+2}{t}$, $\ell \in [t+2, \omega_{\max}+1]$. Note that if ω is size–dependent then $\omega(\mathcal{G}_{T,\ell}) = \omega_{n\to t}(t) - \omega_{n\to\ell}(t) + t\,\omega_{n\to\ell}(\ell-1)$.

In general, one has often $M(n,t;\omega) = \omega(\mathcal{S}_r)$ for the smallest r for which the conditions of Theorem 15 are satisfied. If $r \geq 1$ then also $\tilde{M}(n,t;\omega) = \omega(\mathcal{S}_r)$ since $\mathcal{S}_r$ is nontrivial t-intersecting.

Example 8. Let $\boldsymbol{\omega} = \boldsymbol{e}_k$, $t \leq k \leq k_1$. Then one can take $r = 1$ in Theorem 17. We have

$$\omega(\mathcal{G}_{[t],\ell}) = \binom{n-t}{k-t} - \binom{n-\ell}{k-t} + t\binom{n-\ell}{k-\ell+1}.$$

A unimodality–argument (see [1]) yields

$$\tilde{M}(n,t;\omega) = \max\left\{\omega(\mathcal{G}_{[t],t+2}),\, \omega(\mathcal{G}_{[t],k+1})\right\}.$$

Example 9. Let $\omega_i = \binom{\ell-1}{\ell-i}$, $t \leq \ell \leq \ell_1$. Then one can take $r = 1$ and we have

$$\omega(\mathcal{G}_{[t],s}) = \binom{n-t+\ell-1}{\ell-t} - \binom{n-s+\ell-1}{\ell-t} + t\binom{n-s+\ell-1}{\ell-s+1}.$$

As in Example 8 one can show that

$$\tilde{M}(n,t;\omega) = \max\left\{\omega(\mathcal{G}_{[t],t+2}),\, \omega(\mathcal{G}_{[t],\ell+1})\right\}.$$

Example 10. Let $\omega_i = \alpha^{-i}$, $\alpha \geq \frac{t+1}{2}$. Again, one can take $r = 1$. We have

$$\omega(\mathcal{G}_{[t],\ell}) = \alpha^{-t}(1+\alpha^{-1})^{n-t} - \alpha^{-t}(1+\alpha^{-1})^{n-\ell} + t\alpha^{-(\ell-1)}(1+\alpha^{-1})^{n-\ell}.$$

It is not difficult to verify that

$$\tilde{M}(n,t;\omega) = \max\left\{\omega(\mathcal{G}_{[t],t+2}),\, \omega(\mathcal{G}_{[t],n})\right\}.$$

Example 11. Let $\alpha := \alpha_1 = \ldots = \alpha_n \geq 2$, $n > k$, ω as in (40), and let the equivalent conditions of Example 6 be satisfied. Then one can take $r = 1$. It holds

$$\omega(\mathcal{G}_{[t],\ell}) = W_{k-t}\left(\alpha^{n-t}\right) - W_{k-t}\left((\alpha-1)^{\ell-t}, \alpha^{n-\ell}\right) + t\,W_{k-\ell+1}\left(\alpha-1, \alpha^{n-\ell}\right).$$

We have also in this example

$$\tilde{M}(n,t;\omega) = \max\left\{\omega(\mathcal{G}_{[t],t+2}),\, \omega(\mathcal{G}_{[t],k+1})\right\}.$$

Indeed, let us show that $\omega(\mathcal{G}_{[t],\ell}) < \omega(\mathcal{G}_{[t],\ell+1})$ implies $\omega(\mathcal{G}_{[t],\ell+1}) < \omega(\mathcal{G}_{[t],\ell+2})$. Using

$$W_j\left((\alpha-1)^a, \alpha^b\right) + W_{j+1}\left((\alpha-1)^{a+1}, \alpha^b\right) = W_{j+1}\left((\alpha-1)^a, \alpha^{b+1}\right) \tag{51}$$

our claim reads:

$$t\,W_{k-\ell+1}\left((\alpha-1)^2,\alpha^{n-\ell-1}\right) < W_{k-t-1}\left((\alpha-1)^{\ell-t},\alpha^{n-\ell-1}\right)$$

implies

$$t\,W_{k-\ell}\left((\alpha-1)^2,\alpha^{n-\ell-2}\right) < W_{k-t-1}\left((\alpha-1)^{\ell+1-t},\alpha^{n-\ell-2}\right).$$

But this is true since the map

$$\tau : N_{k-t-1}\left((\alpha-1)^{\ell-t},\alpha^{n-\ell-1}\right) \times N_{k-\ell}\left((\alpha-1)^2,\alpha^{n-\ell-2}\right) \longrightarrow$$

$$N_{k-t-1}\left((\alpha-1)^{\ell+1-t},\alpha^{n-\ell-2}\right) \times N_{k-\ell+1}\left((\alpha-1)^2,\alpha^{n-\ell-1}\right)$$

defined for $\boldsymbol{a} \in N_{k-t-1}\left((\alpha-1)^{\ell-t},\alpha^{n-\ell-1}\right)$, $\boldsymbol{b} \in N_{k-\ell}\left((\alpha-1)^2,\alpha^{n-\ell-2}\right)$ by

$$\tau(\boldsymbol{a},\boldsymbol{b}) := \begin{cases} (a_1,\dots,a_{\ell-t},a_{\ell-t+1},\dots,a_{n-t-1},\boldsymbol{b},1) & \text{if } a_{\ell-t+1} < \alpha \\ (a_1,\dots,a_{\ell-t},1,a_{\ell-t+2},\dots,a_{n-t-1},\boldsymbol{b},2) & \text{if } a_{\ell-t+1} = \alpha \end{cases}$$

is injective.

The two previous examples have the following application. Recall the partially ordered set $F_{\boldsymbol{\alpha}}$ defined in the previous section (before Example 6). Here we deal only the case $\alpha := \alpha_1 = \cdots = \alpha_n (\geq 2)$. A family $\mathcal{F} \subseteq F_{\boldsymbol{\alpha}}$ is called *nontrivial t-intersecting* if it is t-intersecting and if the infimum of *all* members of $\mathcal{F}$ has rank less than t. This means that the set

$$\{i : a_i = b_i > 0 \text{ for all } \boldsymbol{a},\boldsymbol{b} \in \mathcal{F}\}$$

has cardinality at most $t-1$. Define

$$\tilde{M}_k(n,\boldsymbol{\alpha},t) = \max\{|\mathcal{F}| : \mathcal{F} \subseteq N_k(\boldsymbol{\alpha}),\ \mathcal{F} \text{ is nontrivial } t\text{-intersecting}\}.$$

We suppose that $k \geq t+2$ since otherwise there are no such families. For a tuple $\boldsymbol{a} \in F_{\boldsymbol{\alpha}}$ let the *support of* $\boldsymbol{a}$ be given by

$$\operatorname{supp}(\boldsymbol{a}) := \{i : a_i = 1\}.$$

Then we have the following nontrivial t-intersecting candidate families:

$$\mathcal{F}_r := \{\boldsymbol{a} \in N_k(\boldsymbol{\alpha}) : \operatorname{supp}(\boldsymbol{a}) \in \mathcal{S}_r\},\ r \geq 1.$$

Recall that

$$M_k(n,\boldsymbol{\alpha},t) = M(n,t;\boldsymbol{\omega})$$

with $\omega_i = W_{k-i}((\alpha-1)^{n-i})$.

Now Example 10 and Example 11 solve the uniform nontrivial t-intersection problem in $\mathcal{F}_{\boldsymbol{\alpha}}$. This is clear with the next Lemma.

Lemma 18. *We have* $\tilde{M}_k(n,\boldsymbol{\alpha},t) = \tilde{M}(n,t;\boldsymbol{\omega})$.

Proof. Let $\mathcal{F}$ be a maximum k-uniform nontrivial t-intersecting family in $F_{\boldsymbol{\alpha}}$. It suffices to show that

$$|\mathcal{F}| \leq \tilde{M}(n, t; \boldsymbol{\omega}). \tag{52}$$

Recall the operation $s_{i,j,c} : 2^{F_{\boldsymbol{\alpha}}} \to 2^{F_{\boldsymbol{\alpha}}}$ defined by (38) and (39). We know that $|s_{i,j,c}(\mathcal{F})| = |\mathcal{F}|$ and that $s_{i,j,c}(\mathcal{F})$ is t-intersecting for all i, j, k. Let

$$I := I(\mathcal{F}) := \{(i, j, c) : 1 \leq i < j \leq \alpha, c \in [n], s_{i,j,c}(\mathcal{F}) \neq \mathcal{F}\}.$$

If $I = \emptyset$ we are done since then the family

$$\{\text{supp}\,(\boldsymbol{a}) : \boldsymbol{a} \in \mathcal{F}\} \subseteq 2^{[n]}$$

is easily seen to be nontrivial t-intersecting.

Thus let $I \neq \emptyset$. We may assume that $s_{i,j,c}(\mathcal{F})$ is not nontrivial t-intersecting for all $(i, j, c) \in I$ (otherwise keep applying the corresponding operations $s_{i,j,c}$). Then the set

$$T := \{i : a_i = b_i > 0 \text{ for all } \boldsymbol{a}, \boldsymbol{b} \in \mathcal{F}\}$$

has cardinality $t - 1$. Let w.l.o.g. $T = [t - 1]$ and $a_i = 1$ for all $\boldsymbol{a} \in \mathcal{F}$, $i \in T$. Moreover, for all $(i, j, c) \in I$ and all $\boldsymbol{a} \in \mathcal{F}$ we have $a_c \in \{i, j\}$, and there are $\boldsymbol{a}, \boldsymbol{b} \in \mathcal{F}$ with $a_c = i$ and $b_c = j$. We have $i = 1$ for all $(i, j, c) \in I$ since otherwise also $(1, i, c) \in I$ and $s_{1,i,c}(\mathcal{F})$ would be nontrivial t-intersecting. Analogously, we have $j = 2$ for all $(i, j, c) \in I$. Define

$$C := \{c : (1, 2, c) \in I\}.$$

W.l.o.g. let $C = [t, t + q]$, $q = 0, \ldots, n - t$.
Case $|C| > 1$, i.e. $q > 0$.
We will show that $|\mathcal{F}| \leq |\mathcal{F}_1|$ which implies (52).

Note that there are not $\boldsymbol{a}, \boldsymbol{b} \in \mathcal{F}$ with $a_t = 1, b_t = 2$ and $a_i = b_i$ for all $i > t$ since otherwise $\boldsymbol{b} \in s_{1,2,t}(\mathcal{F})$ and hence $s_{1,2,t}(\mathcal{F})$ would be nontrivial t-intersecting. Consequently, recalling that for all $\boldsymbol{a} \in \mathcal{F}$ we have $a_1 = \cdots = a_{t-1} = 1$ and $a_t, \ldots, a_{t+q} \in \{1, 2\}$,

$$|\mathcal{F}| \leq 2^q W_{k-t-q}(\alpha^{n-t-q}) \leq 2^q \alpha^{-(q-1)} W_{k-t-1}(\alpha^{n-t-1}) \leq 2W_{k-t-1}(\alpha^{n-t-1}).$$

Note that

$$|\mathcal{F}_1| = (t + 2)W_{k-t-1}(\alpha - 1, \alpha^{n-t-2}) + W_{k-t-2}(\alpha^{n-t-2}).$$

Hence, using (51), $|\mathcal{F}| \leq |\mathcal{F}_1|$ follows from

$$\begin{aligned} 2W_{k-t-1}(\alpha^{n-t-1}) &= 2\left(W_{k-t-1}(\alpha - 1, \alpha^{n-t-2}) + W_{k-t-2}(\alpha^{n-t-2})\right) \\ &\leq (t + 2)W_{k-t-1}(\alpha - 1, \alpha^{n-t-2}) + W_{k-t-2}(\alpha^{n-t-2}). \end{aligned}$$

Case $|C| = 1$, i.e. $q = 0$.
Define for $\ell = 1, 2$ the (nonempty) families

$$\mathcal{H}_\ell := \{\text{supp}\,(\boldsymbol{a}) \setminus [t] : \boldsymbol{a} \in \mathcal{F},\ a_t = \ell\}.$$

Since $(1,j,c) \notin I$ for all $c > t$, $j \in [\alpha]$, $\mathcal{H}_1$ and $\mathcal{H}_2$ are *cross–intersecting*, i.e.

$$H_1 \cap H_2 \neq \emptyset \text{ for all } H_1 \in \mathcal{H}_1, H_2 \in \mathcal{H}_2. \tag{53}$$

Also, since $\mathcal{F}$ is nontrivial t-intersecting,

$$\bigcap_{H \in \mathcal{H}_1 \cup \mathcal{H}_2} H = \emptyset. \tag{54}$$

Since $|\mathcal{F}|$ is maximum, we necessarily have

$$\mathcal{F} = \bigcup_{\ell \in \{1,2\}} \{\boldsymbol{a} \in N_k(\boldsymbol{\alpha}) : [t-1] \subseteq \operatorname{supp}(\boldsymbol{a}),\, a_t = \ell,\, \operatorname{supp}(\boldsymbol{a}) \cap [t+1,n] \in \mathcal{H}_\ell\}.$$

We apply the shift–operation $s_{i,j}$, $t < i < j \leq n$, simultaneously to $\mathcal{H}_1$ and $\mathcal{H}_2$. It is easy to see that $s_{i,j}(\mathcal{H}_1)$ and $s_{i,j}(\mathcal{H}_2)$ still satisfy (53). Let $\mathcal{F}_{i,j}$ be the family which corresponds to the pair $s_{i,j}(\mathcal{H}_1)$, $s_{i,j}(\mathcal{H}_2)$, i.e.

$$\mathcal{F}_{i,j} := \bigcup_{\ell \in \{1,2\}} \{\boldsymbol{a} \in N_k(\boldsymbol{\alpha}) : [t-1] \subseteq \operatorname{supp}(\boldsymbol{a}),\, a_t = \ell,$$
$$\operatorname{supp}(\boldsymbol{a}) \cap [t+1,n] \in s_{i,j}(\mathcal{H}_\ell)\}.$$

Note that $|\mathcal{F}_{i,j}| = |\mathcal{F}|$. If

$$\bigcap_{H \in s_{i,j}(\mathcal{H}_1) \cup s_{i,j}(\mathcal{H}_2)} H \neq \emptyset$$

then we have $a_i = 1$ for all $\boldsymbol{a} \in \mathcal{F}_{i,j}$. Consequently,

$$|\mathcal{F}| = |\mathcal{F}_{i,j}| \leq 2W_{k-t-1}(\alpha^{n-t-1}) \leq |\mathcal{F}_1|$$

which again implies (52). Hence we may assume that $s_{i,j}(\mathcal{H}_1)$ and $s_{i,j}(\mathcal{H}_2)$ also satisfy (54). We now continue the shifting until we obtain a family (also named $\mathcal{F}$) for which the corresponding families $\mathcal{H}_1$ and $\mathcal{H}_2$ are *left–shifted in* $[t+1,n]$, i.e. $s_{i,j}(\mathcal{H}_\ell) = \mathcal{H}_\ell$ for all $t+1 \leq i < j \leq n$, $\ell = 1,2$. But then there are obviously $\boldsymbol{a}, \boldsymbol{b} \in \mathcal{F}$ with $a_t = 1$, $b_t = 2$, $a_{t+1} = b_{t+1} = \cdots = a_k = b_k$, $a_{k+1} = b_{k+1} = \cdots = a_n = b_n = 0$. Now (52) follows since $s_{1,2,t}(\mathcal{F})$ is nontrivial t-intersecting and $I(s_{1,2,t}(\mathcal{F})) = \emptyset$. □

PUSHING–PULLING

Beside the method of generating sets [2, 4] Ahlswede and Khachatrian developed another proof method, called pushing–pulling, which was used in [3] to give a (new) proof of their Theorem 2, and a proof of Katona's Theorem 6. Since it seems difficult to find general suppositions on ω under which the pushing–pulling method works, we shall stay quite closely to the original arguments in [3]. This section finishes the proof of Theorem 3a. We will also deduce Theorem 5.

Recall that a family $\mathcal{F} \in 2^{[n]}$ is called left–shifted if

$$s_{i,j}(\mathcal{F}) = \mathcal{F} \text{ for all } i, j \in [n], i < j.$$

Let $\ell \in [n]$. A family $\mathcal{F} \in 2^{[n]}$ is called *invariant in* $[\ell]$ if

$$s_{i,j}(\mathcal{F}) = \mathcal{F} \text{ for all } i, j \in [\ell].$$

Lemma 19. *Let $t \geq 2$. Suppose that for some $r \in \{0, \ldots, \lfloor \frac{n-t}{2} \rfloor\}$*

$$\omega_i = 0 \textit{ unless } k_{r-1} \leq i.$$

Then there is a left–shifted optimal family which is invariant in $[t + 2r]$.

Proof. First we deal with an arbitrary (nonnegative) weight vector $\boldsymbol{\omega}$. Among all left–shifted optimal families $\mathcal{F}$ choose one for which

$$\ell := \ell(\mathcal{F}) := \max\{i : \mathcal{F} \text{ is invariant in } [i]\}$$

is maximum. We may assume that $\omega_i = 0$ implies $f_i = 0$.

Now we assume that $\ell < t + 2r$ and look for a contradiction. Let

$$\begin{aligned}
\mathcal{L} &:= \mathcal{L}(\mathcal{F}) := \{X \in \mathcal{F} : s_{\ell+1,i}(X) \notin \mathcal{F} \text{ for some } 1 \leq i \leq \ell\}, \\
\mathcal{L}_i &:= \{X \in \mathcal{L} : |X \cap [\ell]| = i\}, \\
\mathcal{L}' &:= \{X : \ell + 1 \in X \text{ and } s_{i,\ell+1}(X) \in \mathcal{L} \text{ for some } i \in [\ell]\}, \\
\mathcal{L}'_i &:= \{X \in \mathcal{L}' : |X \cap [\ell]| = i - 1\}
\end{aligned}$$

The set $\mathcal{L}$ (and hence also $\mathcal{L}'$) is not empty and invariant in $[\ell]$. Hence,

$$\omega(\mathcal{L}_i) = \binom{\ell}{i} \sum_{X \in \mathcal{L}_i^*} \omega_{|X|+i} \text{ and } \omega(\mathcal{L}'_i) = \binom{\ell}{i-1} \sum_{X \in \mathcal{L}_i^*} \omega_{|X|+i} \tag{55}$$

where

$$\mathcal{L}_i^* := \{X \cap [\ell + 2, n] : X \in \mathcal{L}_i\}.$$

It is not too difficult to verify that $|X \cap Y| \geq t$ for all $X \in \mathcal{L}'_i$ and $Y \in \mathcal{F} \setminus \mathcal{L}_{\ell+t-i}$ (use that $\mathcal{F}$ is left–shifted, invariant in $[\ell]$, and t-intersecting). Hence, for any $i \in [t, \frac{\ell+t}{2})$ the two families

$$\begin{aligned}
\mathcal{F}_{1,i} &:= (\mathcal{F} \setminus \mathcal{L}_{\ell+t-i}) \cup \mathcal{L}'_i, \\
\mathcal{F}_{2,i} &:= (\mathcal{F} \setminus \mathcal{L}_i) \cup \mathcal{L}'_{\ell+t-i}
\end{aligned}$$

are t-intersecting and since $\mathcal{F}$ is optimal we have

$$\omega(\mathcal{L}'_i) \leq \omega(\mathcal{L}_{\ell+t-i}) \text{ and } \omega(\mathcal{L}'_{\ell+t-i}) \leq \omega(\mathcal{L}_i). \tag{56}$$

It follows that $\mathcal{L}_i = \emptyset$ for all $i \in [\ell] \setminus \{\frac{\ell+t}{2}\}$ because otherwise (56) together with (55) yields

$$i(\ell + t - i) \leq (\ell - i + 1)(i + 1 - t),$$

which is easily seen to be false since $t \geq 2$.

It follows $2 \mid \ell + t$ since otherwise we have $\mathcal{L} = \emptyset$. The assumption $\ell < t + 2r$ then implies

$$\ell \leq t + 2r - 2. \tag{57}$$

Suppose that we find an *intersecting* subfamily $\mathcal{T}^*$ of $\mathcal{L}^*_{\frac{\ell+t}{2}}$ which satisfies

$$\frac{\sum_{X \in \mathcal{T}^*} \omega_{|X| + \frac{\ell+t}{2}}}{\sum_{X \in \mathcal{L}^*_{\frac{\ell+t}{2}}} \omega_{|X| + \frac{\ell+t}{2}}} > \frac{\ell - t + 2}{2(\ell + 1)} = \frac{\binom{\ell}{(\ell+t)/2}}{\binom{\ell+1}{(\ell+t)/2}}. \tag{58}$$

Let

$$\begin{aligned} \mathcal{T} &:= \{X \in \mathcal{L}_{\frac{\ell+t}{2}} : X \cap [\ell + 2, n] \in \mathcal{T}^*\}, \\ \mathcal{T}' &:= \{X \in \mathcal{L}'_{\frac{\ell+t}{2}} : X \cap [\ell + 2, n] \in \mathcal{T}^*\}. \end{aligned}$$

Then, as in (55),

$$\omega(\mathcal{T}) = \binom{\ell}{(\ell + t)/2} \sum_{X \in \mathcal{T}^*} \omega_{|X| + \frac{\ell+t}{2}}, \tag{59}$$

$$\omega(\mathcal{T}') = \binom{\ell}{(\ell + t)/2 - 1} \sum_{X \in \mathcal{T}^*} \omega_{|X| + \frac{\ell+t}{2}}. \tag{60}$$

It is easy to see that

$$\mathcal{F}_1 := \left(\mathcal{F} \setminus \mathcal{L}_{\frac{\ell+t}{2}}\right) \cup \mathcal{T} \cup \mathcal{T}'$$

is t-intersecting. But (58) together with (55), (59), and (60) yields

$$\omega(\mathcal{T}) + \omega(\mathcal{T}') > \omega(\mathcal{L}_{\frac{\ell+t}{2}}),$$

hence $\omega(\mathcal{F}_1) > \omega(\mathcal{F})$, a contradiction.

Now let $\boldsymbol{\omega}$ satisfy the hypothesis of Lemma 19. We have by double counting

$$\sum_{i \in [\ell+2,n]} \sum_{X \in \mathcal{L}^*_{\frac{\ell+t}{2}} : i \in X} \omega_{|X| + \frac{\ell+t}{2}} = \sum_{X \in \mathcal{L}^*_{\frac{\ell+t}{2}}} \left(\omega_{|X| + \frac{\ell+t}{2}}\right) |X|.$$

Hence there is some $i \in [\ell + 2, n]$ such that

$$\begin{aligned} \sum_{X \in \mathcal{L}^*_{\frac{\ell+t}{2}} : i \in X} \omega_{|X| + \frac{\ell+t}{2}} &\geq \frac{1}{n - \ell - 1} \sum_{X \in \mathcal{L}^*_{\frac{\ell+t}{2}}} \left(\omega_{|X| + \frac{\ell+t}{2}}\right) |X| \\ &\geq \frac{k_{r-1} - \frac{\ell+t}{2}}{n - \ell - 1} \sum_{X \in \mathcal{L}^*_{\frac{\ell+t}{2}}} \left(\omega_{|X| + \frac{\ell+t}{2}}\right), \end{aligned} \tag{61}$$

where the last inequality follows from

$$|X| < k_{r-1} - (\ell + t)/2 \text{ implies } \omega_{|X| + \frac{\ell+t}{2}} = 0.$$

Using the definition of k_r it is easy to show that

$$\frac{k_{r-1} - (\ell + t)/2}{n - \ell - 1} \geq \frac{\ell - t + 2}{2(\ell + 1)} \text{ if } \ell \leq t + 2r - 2.$$

Hence, recalling (57), strict inequality in (61) gives an intersecting family $\mathcal{T}^* \subseteq \mathcal{L}^*_{\frac{\ell+t}{2}}$ satisfying (58). If we have equality in (61) for all $i \in [\ell + 2, n]$ then take $i := \ell + 2$. This gives a $\mathcal{T}^*$ for which the corresponing family $\mathcal{F}_1$ is left–shifted and (obviously) invariant in $[\ell + 1]$, a contradiction to our choice of $\mathcal{F}$. □

Note that if in Lemma 19

$$\omega_i = 0 \text{ unless } k_{r-1} < i,$$

then the above proof yields that all left–shifted optimal families are invariant in $[t + 2r]$.

Now we are ready to prove Theorem 3a.

Proof of Theorem 3a. The case $t = 1$ is trivial. Let $t > 1$. By Example 4 we know $M(n, t; \boldsymbol{\omega}) = M_{t+2r+2}(n, t; \boldsymbol{\omega})$. As in Lemma 19, choose among all left–shifted optimal families $\mathcal{F} \in I_{t+2r+2}(n, t)$ one for which $\ell(\mathcal{F})$ is maximum. Then the proof of Lemma 19 shows that also this family $\mathcal{F}$ is invariant in $[t + 2r]$. (Note that if we take $i := \ell + 2$ in (61) then the corresponding family $\mathcal{F}_1$ is still in $I_{t+2r+2}(n, t)$.) Let $\mathcal{F}'_i := \{X \in \mathcal{F} : |X \cap [t + 2r]| = i\}$. Then the following facts are easy consequences of the $(t + 2r + 2)$-t-intersection and the $[t + 2r]$-invariance property of $\mathcal{F}$:

1) $\mathcal{F}'_i = \emptyset$ for all $i < t + r - 1$,

2) $\{t + 2r + 1, t + 2r + 2\} \in X$ for all $X \in \mathcal{F}'_{t+r-1}$,

3) if $\mathcal{F}'_{t+r-1} \neq \emptyset$ then $|\{t + 2r + 1, t + 2r + 2\} \cap X| \geq 1$ for all $X \in \mathcal{F}'_{t+2r}$.

It follows that $\mathcal{F} = \mathcal{S}_r$ or $\mathcal{F} = \mathcal{S}_{r+1}$. □

Let $\mathcal{F}$ be t-intersecting. Note that if $2 \mid n + t$ and $\mathcal{F}$ is invariant in $[n]$ or if $2 \nmid n + t$ and $\mathcal{F}$ is invariant in $[n - 1]$ then $\mathcal{F} \subseteq \mathcal{S}_{\lfloor \frac{n-t}{2} \rfloor}$. Hence the pushing–pulling method can be used to prove the optimality of the last candidate family.

Proof of Theorem 5. Again, the case $t = 1$ is trivial. Let $t > 1$. It suffices to show the existence of an optimal family which is invariant in $[n]$ resp. $[n - 1]$ if $2 \mid n + t$ resp. if $2 \nmid n + t$. We proceed as in the proof of Lemma 19. Hence we assume $\ell < n$ if $2 \mid n + t$ and $\ell < n - 1$ if $2 \nmid n + t$. Then (57) becomes

$$\ell \leq n - 2 \text{ if } 2 \mid n + t,\ \ell \leq n - 3 \text{ if } 2 \mid n + t. \tag{62}$$

Now we claim that $\mathcal{L}^*_{\frac{\ell+t}{2}}$ is *self–complementary* (in $2^{[\ell+2,n]}$), i.e. $X \in \mathcal{L}^*_{\frac{\ell+t}{2}}$ implies $[\ell + 2, n] \setminus X \in \mathcal{L}^*_{\frac{\ell+t}{2}}$. Indeed, for every set $X \in \mathcal{L}^*_{\frac{\ell+t}{2}}$, $X \neq \emptyset$, there is a set $Y \in \mathcal{L}^*_{\frac{\ell+t}{2}}$ with $X \cap Y = \emptyset$. Otherwise one could add any set $s_{\ell+1,i}(Z)$, with $Z \in \mathcal{L}_{\frac{\ell+t}{2}}$, $Z \cap [\ell + 2, n] = X$, $i \in Z \cap [\ell]$, to the family $\mathcal{F}$ without violating the

t-intersection property, but this contradicts the optimality of $\mathcal{F}$ since we have assumed that $\omega_{|X|} > 0$ for all $X \in \mathcal{F}$. Using

$$0 < \omega_{|X|+\frac{\ell+t}{2}} \leq \omega_{|[\ell+2,n]\setminus X|+\frac{\ell+t}{2}} \text{ for } X \in \mathcal{L}^*_{\frac{\ell+t}{2}},\ |X| + (\ell+t)/2 \leq (n+t-1)/2$$

we deduce that

$$Z \in \mathcal{L}_{\frac{\ell+t}{2}}, |Z \cap [\ell+2,n]| \leq \frac{n-\ell-1}{2} \text{ implies } (Z \cap [\ell]) \cup ([\ell+2,n] \setminus Z) \in \mathcal{F},$$

and hence implies (using the $[\ell]$-invariance of $\mathcal{F}$)

$$(Z \cap [\ell]) \cup ([\ell+2,n] \setminus Z) \in \mathcal{L}_{\frac{\ell+t}{2}}.$$

This establishes that $\mathcal{L}^*_{\frac{\ell+t}{2}}$ is self–complementary.

Now let $\mathcal{T}^*$ be the intersecting family of all sets $X \in \mathcal{L}^*_{\frac{\ell+t}{2}}$ with $|X| > \frac{n-\ell-1}{2}$ and (in the case $2 \mid n-\ell-1$) all sets $X \in \mathcal{L}^*_{\frac{\ell+t}{2}}$ with $|X| = \frac{n-\ell-1}{2}$ and $n \notin X$. Then, using the hypothesis on $\boldsymbol{\omega}$ and the fact that $\mathcal{L}^*_{\frac{\ell+t}{2}}$ is self–complementary, it is easy to deduce that this family $\mathcal{T}^*$ satisfies (58):

$$\frac{\sum_{X \in \mathcal{T}^*} \omega_{|X|+\frac{\ell+t}{2}}}{\sum_{X \in \mathcal{L}^*_{\frac{\ell+t}{2}}} \omega_{|X|+\frac{\ell+t}{2}}} \geq \frac{1}{2} > \frac{\ell-t+2}{2(\ell+1)}.$$

This finishes the proof. □

References

[1] R. Ahlswede and L.H. Khachatrian. "The complete nontrivial–intersection theorem for systems of finite sets". *J. Combin. Theory Ser. A*, **76**:121–138, (1996).

[2] R. Ahlswede and L.H. Khachatrian. "The complete intersection theorem for systems of finite sets". *European J. Combin.*, **18**:125–136, (1997).

[3] R. Ahlswede and L.H. Khachatrian. "A pushing–pulling method: New proofs of intersection theorems". *Combinatorica*, **19**:1–15, (1999).

[4] R. Ahlswede and L.H. Khachatrian. "The diametric theorem in Hamming space – optimal anticodes". *Adv. in Appl. Math.*, **20**:429–449, (1998).

[5] C. Bey. "Durchschnittsprobleme im Booleschen Verband". Ph. D. Thesis. Universität Rostock, (1999).

[6] C. Bey. "The Erdős–Ko–Rado bound for the function lattice". *Discrete Appl. Math.*, **95**:115–125, (1999).

[7] C. Bey. "An intersection theorem for weighted sets". *Discrete Math.*, to appear.

[8] C. Bey and K. Engel. "An asymptotic complete intersection theorem for chain products". *European J. Combin.*, **20**:321–327, (1999).

[9] K. Engel. *Sperner Theory.* Cambridge University Press, Cambridge, (1997).

[10] K. Engel and P. Frankl. "An Erdős–Ko–Rado theorem for integer sequences of given rank". *European J. Combin.*, **7**:215–220, (1986).

[11] P. Erdős, C. Ko, and R. Rado. "Intersection theorems for systems of finite sets". *Quart. J. Math. Oxford Ser.*, **12**:313–320, (1961).

[12] P.L. Erdős, P. Frankl, and G.O.H. Katona. "Extremal hypergraph problems and convex hulls". *Combinatorica*, **5**:11–26, (1985).

[13] P. Frankl. "The shifting technique in extremal set theory". In C. Whitehead, editor, *Surveys in Combinatorics*, volume **123** of *Lond. Math. Soc. Lect. Note Ser.*, pages 81–110, Cambridge, (1987). Cambridge University Press.

[14] P. Frankl and N. Tokushige. "The Erdős–Ko–Rado theorem for integer sequences". *Combinatorica*, **19**:55–63, (1999).

[15] G.O.H. Katona. "Intersection theorems for systems of finite sets". *Acta Math. Acad. Sci. Hung.*, **15**:329–337, (1964).

SOME NEW RESULTS ON MACAULAY POSETS

Sergei L. Bezrukov

Department of Mathematics and Computer Science,
University of Wisconsin - Superior, USA

Uwe Leck

Department of Mathematics, University of Rostock, Germany

Dedicated to Rudolf Ahlswede on his 60th birthday

Abstract: Macaulay posets are posets for which there is an analogue of the classical Kruskal-Katona theorem for finite sets. These posets are of great importance in many branches of combinatorics and have numerous applications. We survey mostly new and also some old results on Macaulay posets. Emphasis is also put on construction of extremal ideals in Macaulay posets.

INTRODUCTION

Macaulay posets are, informally speaking, posets for which an analogue of the classical Kruskal-Katona theorem for finite sets holds. They are related to many other combinatorial problems like isoperimetric problems on graphs [9] (see also section 6) and problems arising in polyhedral combinatorics. Several optimization problems can be solved within the class of Macaulay posets, or at least for Macaulay posets with additional properties (cf. section 6). Therefore, Macaulay posets are very useful and interesting objects.

I. Althöfer et al. (eds.), Numbers, Information and Complexity, 75-94.

A few years ago, the classical Macaulay posets listed in section 6 were the only known essential examples, and, consequently, the theory of Macaulay posets was more or less the theory of these examples. In his book [30, chapter 8], Engel made a first attempt for unification the theory of Macaulay posets. Although the book appeared quite recently, a number of new examples, relations and applications have been found meantime. In this paper, our objective is to give a survey on Macaulay posets that includes these new results and updates [30].

We start with some basic facts and definitions in section 6 and the classical examples in section 6. For all definitions not included here we refer to Engel's book [30]. In section 6 we proceed with constructions for Macaulay posets and relations to isoperimetric problems. New examples of Macaulay posets are presented in section 6. Section 6 is devoted to optimization problems on Macaulay posets.

Some basic definitions

Let P be a partially ordered set (briefly, poset) with the associated partial order $\leq$. For $x, y \in P$, we say that y *covers* x, denoted by $x \lessdot y$, if $x \leq y$ and there is no $z \in P$ such that $z \neq x, y$ and $x \leq z \leq y$. An *antichain* is defined as a subset $X \subseteq P$ such that the conditions $x, y \in X$ and $x \leq y$ imply $x = y$.

A subset $X \subseteq P$ is an *ideal* (or *downset*) if the conditions $x \in X$ and $y \leq x$ imply $y \in X$. If X is an antichain, then the set $I(X) := \{y \in P \mid y \leq x \text{ for some } x \in X\}$ is an ideal, which is called *ideal generated by* X. Conversely, if I is an ideal, then the set $\max(I) := \{x \in I \mid x \not\leq y \text{ for any } y \in I,\ y \neq x\}$ is an antichain, which is called the *set of maximal elements* of I.

A *rank function* on P is a function $r : P \mapsto I\!N$ such that $r(x) = 0$ for some minimal element x of P and $r(y) = r(z) - 1$ whenever $y \lessdot z$. The poset P is called *ranked*, if a rank function on P exists. The *rank* of P is defined by $r(P) := \max\{r(x) \mid x \in P\}$, where $r(P) = \infty$ is allowed. A ranked poset P is called *graded* if all minimal elements have rank 0, and all maximal elements have rank $r(P)$.

The *dual* P^* of P is the poset on the same set of elements with the partial order defined by: $x \leq^* y$ iff $y \leq x$. If P is ranked with $r(P) < \infty$, then P^* is ranked. If P is ranked with $r(P) = \infty$, then P^* is not ranked in the usual sense. In this case $r^*(x) := -r(x)$ will considered to be the rank function for P^*.

If P is ranked, then the set $\{x \in P \mid r(x) = i\}$ is called the i-th *level* of P and is denoted by $N_i(P)$ or P_i. The *(lower) shadow* of an element $x \in P_i$ is the set $\Delta(x) := \{y \in P \mid y \lessdot x\}$, and its *upper shadow* is $\nabla(x) := \{y \in P \mid x \lessdot y\}$. The lower shadow $\Delta(X)$ (resp. upper shadow $\nabla(X)$) of a subset $X \subseteq P_i$ is defined as the union of the lower (resp. upper) shadows of its elements. For given integers i and m with $1 \leq i \leq r(P)$ and $1 \leq m \leq |P_i|$, the *shadow minimization problem* (SMP) consists in finding an m-element subset $X \subseteq P_i$ such that $|\Delta(X)| \leq |\Delta(Y)|$ for all $Y \subseteq P_i$ with $|Y| = m$. We say that a subset $X \subseteq P_i$ is *optimal* if it has minimum shadow among all subsets of P_i of the same size. Obviously, the SMP is at least NP-hard, since it implies a solution to the Minimum Cover Problem.

The *(cartesian) product* $P \times Q$ of two posets P and Q is the set of all pairs (x, y) with $x \in P$, $y \in Q$, where the partial order is given by: $(x, y) \leq_{P\times Q} (x', y')$ iff $x \leq_P x'$, $y \leq_Q y'$. If P and Q are ranked, then the poset $P \times Q$ is ranked too, and the rank function for $P\times Q$ is given by: $r(x, y) := r_P(x)+r_Q(y)$. The n-th *(cartesian) power* of a poset P is the poset $P^n := P \times P \times \cdots \times P$ (n times).

Macaulay posets

Let P be a ranked poset and consider some total order $\preceq$ of its elements. Note that we do not claim the order $\preceq$ to be a *linear extension* of P. For a subset $X \subseteq P$ and a natural number $m \leq |X|$ we will use the notation $C(m, X)$ (resp. $L(m, X)$) for the set of the first (resp. last) m elements of X w.r.t. $\preceq$. In particular, for $X \subseteq P_i$ we abbreviate $C(|X|, P_i)$ and $L(|X|, P_i)$ by $C(X)$ and $L(X)$, respectively. The operation of replacing $X \subseteq P_i$ with $C(X)$ is called *compression*, and we say that X is *compressed* if $X = C(X)$. Compressed subsets will also be called *initial segments* (IS), whereas a *final segment* of P_i is a subset $X \subseteq P_i$ with $X = L(X)$. A *segment* of P_i simply is a set of elements of P_i which are consecutive w.r.t. $\preceq$ (restricted to P_i). For an element $x \in P_i$, the initial segment of P_i whose last element w.r.t. $\preceq$ is x is denoted by $\mathcal{F}_i(x)$.

The poset P is said to be a *Macaulay poset* if there exists a total order $\preceq$ of its elements (called *Macaulay order*) such that

$$\Delta(C(X)) \subseteq C(\Delta(X)) \text{ for all } X \subseteq P_i \text{ and for all } i = 1, \ldots, r(P). \qquad (1)$$

If (1) is satisfied for a ranked poset P with a partial order $\leq$ and for a total order $\preceq$ of the elements of P, then the triple $(P, \leq, \preceq)$ is called *Macaulay structure*.

It is easy to verify (cf, [30] for details) that (1) holds iff the conditions $\mathbf{N}_1$ and $\mathbf{N}_2$ given below are satisfied for all $X \subseteq P_i$ and for all $i = 1, \ldots, r(P)$:

$$\mathbf{N}_1\text{:} \quad |\Delta(C(X))| \leq |\Delta(X)|,$$

$$\mathbf{N}_2\text{:} \quad C(\Delta(C(X))) = \Delta(C(X)).$$

According to $\mathbf{N}_1$, compressed subsets are optimal for the Macaulay poset P. Therefore, $\mathbf{N}_1$ is called the condition of *nestedness* (of the optimal subsets). By $\mathbf{N}_2$, the shadow of a compressed set is a compressed set again. That is why $\mathbf{N}_2$ is said to be the condition of *continuity*.

For a total order $\preceq$ of the elements of P denote by $\preceq^*$ its inverse.

Proposition 1. (Bezrukov [8]). *$(P, \leq, \preceq)$ is a Macaulay structure iff so is $(P^*, \leq^*, \preceq^*)$.*

For many applications it turns out to be natural and useful to choose a Macaulay order rank greedily. We say that a total order $\preceq$ is *rank greedy* (on P), if it is a linear extension of the partial order $\leq$ (i.e. if $x \leq y$ implies $x \preceq y$), and if, in addition, $r(x) = r(y) + 1$ implies $x \preceq y$ whenever the last element of $\Delta(x)$ w.r.t. $\preceq$ precedes y in the order $\preceq$. It can be easily shown (see e.g. [30]) that for every Macaulay poset there exists a rank greedy Macaulay order of its elements. The proof for this and the next assertion can be found in [30].

Proposition 2. *If a total order $\preceq$ is rank greedy for a Macaulay poset P, then $\preceq^*$ is rank greedy for P^*.*

If we associate a rank greedy total order with some Macaulay poset P, then we also say that P is rank greedy. Note that all Macaulay orders presented in sections 6 and 6 are rank greedy.

The shadow function

Let P be a Macaulay poset. The *shadow function* sf_i assigns with each subset $X \subseteq P_i$ the number $sf_i(X) = |\Delta(C(X))|$. We briefly discuss some properties of the shadow function.

The lower and upper *new shadows* of an element $x \in P$ are defined by:

$$\begin{aligned} \Delta_{new}(x) &:= \{y \in P \mid y \lessdot x \text{ and there is no } z \in P \text{ with } z \preceq x,\ z \neq x,\ y \lessdot z\}, \\ \nabla_{new}(x) &:= \{y \in P \mid x \lessdot y \text{ and there is no } z \in P \text{ with } x \preceq z,\ z \neq x,\ z \lessdot y\}, \end{aligned}$$

respectively. Note that the upper new shadow of x in P is exactly the lower new shadow of x in P^*. The lower new shadow $\Delta_{new}(X)$ (resp. upper new shadow $\nabla_{new}(X)$) of a subset $X \subseteq P$ is the union of the lower (resp. upper) new shadows of its elements. The shadow function sf_i is called *additive* if the inequality

$$|\Delta_{new}(X)| \geq |\Delta_{new}(Y)| \geq |\Delta_{new}(Z)|$$

is satisfied for all segments $X, Y, Z \subseteq P_i$ with X being initial, Z being final, and $|X| = |Y| = |Z|$. We say that P is additive if sf_i is additive for all $i = 0, \ldots, r(P)$.

Proposition 3. (Engel [30]). *Let P be a Macaulay poset. P is graded and additive iff its dual P^* is graded and additive.*

The Macaulay poset P is called *shadow increasing* if for all $i = 0, \ldots, r(P)-1$ and for any initial segments $X \subseteq P_i$ and $Y \subseteq P_{i+1}$ with $|X| = |Y|$ the inequality $|\Delta(X)| \leq |\Delta(Y)|$ holds. We say that P is *final shadow increasing* if we have $|\Delta_{new}(X)| \leq |\Delta_{new}(Y)|$ for all $i = 0, \ldots, r(P) - 1$ and for any final segments $X \subseteq P_i$ and $Y \subseteq P_{i+1}$ with $|X| = |Y|$. Finally, P is said to be *weakly shadow increasing* if $|\Delta_{new}(X)| \leq |\Delta_{new}(Y)|$ holds for any segments $X \subseteq P_i$ and initial segments $Y \subseteq P_j$ such that $i \leq j$, $|X| = |Y|$ and $X \cup Y$ is an antichain.

Proposition 4. (Engel, Leck [31]). *Let P be a Macaulay poset.*

a. If P is final shadow increasing, then P^* is shadow increasing.

b. Let P be graded, additive, and shadow increasing. If P^* is shadow increasing, then P is final shadow increasing.

c. If P is a graded, additive and shadow increasing, then P is weakly shadow increasing.

SOME KNOWN MACAULAY POSETS

Boolean lattices

Boolean lattices are certainly the most popular examples of Macaulay posets. For a natural number n the Boolean lattice B^n is defined as the collection of all subsets of $[n] := \{1, 2, \ldots, n\}$ partially ordered by inclusion, i.e. $X \leq Y$ for $X, Y \subseteq [n]$ iff $X \subseteq Y$. The unique rank-function on B^n maps a set $X \subseteq [n]$ to $|X|$. Representing the subsets of $[n]$ by their characteristic vectors, it is obvious that B^n is isomorphic to the n-th cartesian power of the chain $0 < 1$ of length one.

The *lexicographic order* of the elements of B^n is defined by $X \preceq_{lex} Y$ iff $\max(X \setminus Y) \leq \max(Y \setminus X)$, where $\max(\emptyset) := 0$. The following theorem, which meantime became a classical one, was proved by Kruskal [39] and Katona [37].

Theorem 5. (Kruskal-Katona theorem). *$(B^n, \subseteq, \preceq_{lex})$ is a Macaulay structure.*

The solution to the SMP provided by Kruskal-Katona theorem is not unique, in general. However, for at least 2^{n-1} cardinalities m the IS of the lexicographic order of size m is essentially a unique optimal subset, as it is shown in the next theorem. Denote $\mathcal{T}(m, k) = |\mathcal{T}(C(m, B^n_k))|$.

Theorem 6. (Füredi, Griggs [32]). *If $\mathcal{T}(m + 1, k) > \mathcal{T}(m, k)$ for some $k \geq 1$, then the set $C(m, B^n_k)$ is a unique optimal subset of size m (up to isomorphism).*

This result, however, is a corollary of more general results [7, 8] which concern the VIP. Without going into details, for which readers are referred to a survey [8], we mention another corollary of results on VIP.

Theorem 7. (Bezrukov [7]). *If $A \subseteq B^n_k$ is optimal for some $k \geq 0$, then so is for $\Delta(A)$.*

Presently it is not known if this property is valid for other Macaulay posets.

Chain products

Cartesian product of chains, called also *lattice of multichains*, is a well-studied generalization of Boolean lattices. For positive integers n and $k_1 \leq k_2 \leq \cdots \leq k_n$ the chain product $S(k_1, k_2, \ldots, k_n)$ consists of all vectors $\mathbf{x} = (x_1, x_2, \ldots, x_n)$ such that $x_i \in \{0, 1, \ldots, k_i\}$ for $i = 1, 2, \ldots, n$. The partial order is a coordinatewise one: $\mathbf{x} \leq \mathbf{y}$ iff $x_i \leq y_i$ for $i = 1, 2, \ldots, n$. Again we have a uniquely determined rank-function, namely $r(\mathbf{x}) = \sum_{i=1}^{n} x_i$. Obviously, $S(k_1, k_2, \ldots, k_n)$ is the cartesian product of the chains $0 \lessdot 1 \lessdot \cdots \lessdot k_i$, $i = 1, 2, \ldots, n$.

A natural extension of the lexicographic order to chain products is established by: $\mathbf{x} \preceq_{lex} \mathbf{y}$ iff $\mathbf{x} = \mathbf{y}$ or $x_j < y_j$, where j is the smallest index with $x_j \neq y_j$.

Theorem 8. (Clements-Lindström theorem). *$(S(k_1, \ldots, k_n), \leq, \preceq_{lex})$ is a Macaulay structure.*

A short proof of this theorem is based on shifting technique and is published in [41]. A principally different approach used in [17] for the MWI problem (cf. Section 5.2) implies a short proof too. The properties of chain products given in the following theorem are important for many applications (see section 6 for instance).

Theorem 9. (Clements [18]). *Chain products are additive and shadow increasing.*

The star posets

Another natural way to generalize Boolean lattices is to consider the chain $0 \lessdot 1$ as a star with just two vertices. This leads to cartesian products of stars. For positive integers n and $k_1 \leq k_2 \leq \cdots \leq k_n$ the star poset $T(k_1, k_2, \ldots, k_n)$ consists of all vectors $\mathbf{x} = (x_1, x_2, \ldots, x_n)$ such that $x_i \in \{k_n - k_i, k_n - k_i + 1, \ldots, k_n\}$ for $i = 1, 2, \ldots, n$, where the partial order is given by: $\mathbf{x} \leq \mathbf{y}$ iff $x_i = y_i$ or $y_i = k_n$ for $i = 1, 2, \ldots, n$. The unique rank-function on $T(k_1, k_2, \ldots, k_n)$ is given by $r(\mathbf{x}) = |\{i \mid x_i = k_n\}|$.

To introduce a Macaulay order $\preceq$ on $T(k_1, k_2, \ldots, k_2)$, define $\mathbf{x}(j) := \{i \in [n] \mid x_i = j\}$ for $\mathbf{x} \in T(k_1, k_2, \ldots, k_n)$ and $j = 0, 1, \ldots, k_n$. Now $\preceq$ is defined as follows: $\mathbf{x} \preceq \mathbf{y}$ iff $\mathbf{x} = \mathbf{y}$ or $\mathbf{y}(h) \prec_{lex} \mathbf{x}(h)$, where h is the smallest number with $\mathbf{x}(h) \neq \mathbf{y}(h)$.

Theorem 10. *$(T(k_1, k_2, \ldots, k_n), \leq, \preceq)$ is a Macaulay structure.*

This theorem is found by Lindström [48] for the case $k_1 = \cdots = k_n = 2$ (his proof, however, contains a gap), and is proved by Leeb [47] and Bezrukov [6] in the case $k_1 = \cdots = k_n$. Actually, both mentioned proofs can be extended for the case $k_1 \neq k_n$. Explicit proofs for this general case are given in [30, 42].

Theorem 11. *Star products are additive and shadow increasing.*

The additivity part of this theorem is due to Clements [20] (see [30] for simplification), the shadow increase property was shown by Leck [43] by using an idea of Kleitman.

Colored complexes

Obviously, for $k_n \geq 2$ the star product $T(k_1, k_2, \ldots, k_n)$ is not isomorphic to its dual. Engel [30] observed that the duals of star products are isomorphic to colored complexes which were introduced by Frankl, Füredi and Kalai [34] in the case $k_n - k_1 \leq 1$.

To define colored complexes in general, for positive integers n and $k_1 \leq k_2 \leq \cdots k_n$, and for $i = 1, 2, \ldots, n$, let the i-th *color class* be the set

$$A_i := \{i, n+i, 2n+i, \ldots, (k_i - 1)n + i\}.$$

Now the colored complex $Col(k_1, k_2, \ldots, k_n)$ consists of all subsets $X \subseteq A := \bigcup_{i=1}^{n} A_i$ such that $|X \cap A_i| \leq 1$ for $i = 1, 2, \ldots, n$, i.e. of all subsets of A which meet every color class at most once. The corresponding partial order is the usual set inclusion.

Due to the isomorphism mentioned above, Proposition 1 and Theorem 10, and, respectively, Proposition 3, yield the following corollaries.

Corollary 12. **(Colored Kruskal-Katona theorem** [34]). *$(Col(k_1, k_2, \ldots, k_n), \subseteq, \preceq_{lex})$ is a Macaulay structure.*

Corollary 13. *The colored complexes are additive.*

The following theorem is the result of yet another application of the Kleitman's idea mentioned above.

Theorem 14. (Leck [43]). *Colored complexes are shadow increasing.*

CONSTRUCTION OF MACAULAY POSETS

Posets with a given shadow function

Here we show that for any shadow function sf_i there exists a Macaulay poset with this shadow function. Obviously, it suffices to construct Macaulay posets with two levels only. Let P be a ranked poset with $r(P) = 1$ and consider the SMP on its top level P_1. Denote by $\mathcal{T}(m)$ the minimal size of the shadow of a set consisting of m elements of P_1. Obviously, the sequence $\{\mathcal{T}(m)\}$ is nondecreasing.

Proposition 15. *For any nondecreasing sequence $\{\mathcal{T}(1), ..., \mathcal{T}(p)\}$ there exists corresponding Macaulay poset P with $r(P) = 1$.*

To construct such a poset, denote $P_1 = \{a_1, \ldots, a_p\}$ and $P_0 = \{b_1, \ldots, b_{\mathcal{T}(p)}\}$. We define a partial order $\leq$ on $P = P_0 \cup P_1$ as follows. For any $i = 1, \ldots, p$ set $a_i > b_j$ for $j = 1, \ldots, \mathcal{T}(i)$. Obviously, the constructed poset is Macaulay and the labelings of a_i's and b_i's provide Macaulay orders on P_1 and P_0 respectively.

Similarly Macaulay posets with more levels can be constructed. This construction is, in a sense, invertible. Given a Macaulay poset $(P, \leq, \preceq)$, construct another poset $Q = (P, \sqsubseteq)$ as follows. Take an element $a \in P_i$ for some $i > 1$ and consider $\mathcal{F}_i(a)$. Then $\mathcal{T}(\mathcal{F}_i(a)) = \mathcal{F}_{i-1}(b)$ for some $b \in P_{i-1}$. Let $c \in \mathcal{F}_{i-1}(b)$ and assume $c \not\leq a$. Now we extend the partial order $\leq$ by setting $c \leq a$.

Proposition 16. (Bezrukov, Portas, Serra [16]). *The poset Q is Macaulay.*

Posets related to isoperimetric problems on graphs

Let $G = (V_G, E_G)$ be a graph. For $A \subseteq V_G$ denote

$$\begin{aligned} E(A) &= \{(u,v) \in E_G \mid u \in A,\ v \notin A\}, \\ E(m) &= \max_{|A|=m} |E(A)|. \end{aligned}$$

Consider an edge-isoperimetric problem (EIP): for any $m \leq |V_G|$ find $A \subseteq V_G$ such that $|A| = m$ and $|E(A)| = E(m)$. We say that the edge-isoperimetric problem has ***nested solutions*** if there exists a numbering of V such that each IS is an optimal set. For more information on edge-isoperimetric problems on graphs readers are referred to the survey [12].

Assume that the EIP has nested solutions for the graph G. We construct a Macaulay poset $(P, \leq)$ with $|P| = |V_G|$ by induction on $|V_G|$ (cf. [11]). If $|V_G| = 1$, then the poset is trivial. For $|V_G| > 1$ let $V_G = \{1, \dots, |V_G|\}$ and assume that for each $m = 1, \dots, |V_G|$ the subset $\{v_1, \dots, v_m\} \subseteq V_G$ is optimal. Note that for $m < |V_G|$ this subset is also optimal for the subgraph G' which is induced by the vertex set $\{1, \dots, |V_G| - 1\}$. Construct the representing poset $(P', \leq')$ for G' by induction. Now extend P' by adding a new element v at level $i = E(|V_G|) - E(|V_G| - 1)$ and extend the partial order $\leq'$ by setting v to be greater than any element of P' at level $i - 1$. This procedure results in the poset $(P, \leq)$.

Proposition 17. (cf. [11]). *A poset obtained according to the EIP-construction is Macaulay.*

What is interesting that if a poset P represents a graph G, and if P^n is Macaulay, then the EIP on G^n has nested solutions [9, 10]. The inverse proposition is, however, not correct, in general. However, the posets P^n are good candidates for being Macaulay (cf. the discussion in section 5.3).

Now we turn to a *vertex-isoperimetric problem* on $G = (V_G, E_G)$. For $A \subseteq V_G$ denote

$$\begin{aligned} \Gamma(A) &= \{v \in V_G \setminus A \mid (v,u) \in E_G,\ u \in A\}, \\ \Gamma(m) &= \min_{|A|=m} |\Gamma(A)|. \end{aligned}$$

The vertex-isoperimetric problem (VIP) consists in finding for a given $m \leq |V_G|$ a set $A \subseteq V_G$ such that $|A| = m$ and $|\Gamma(A)| = \Gamma(m)$. Such problems often arise in combinatorics. For a survey we refer to [8].

We additionally assume that for any IS $A \subseteq V_G$ the set $A \cup \Gamma(A)$ is an IS, too. This property corresponds to the continuity in the definition of Macaulay poset and holds for many graph families.

Let $V_G = \{1, \dots, |V_G|\}$, where any IS represents an optimal set. We construct a poset $(P, \leq)$ with $r(P) = 1$ and $|P| = 2|V_G|$ as follows. Let $P_0 = \{b_1, \dots, b_{|V_G|}\}$ and $P_1 = \{a_1, \dots, a_{|V_G|}\}$. We set $b_i < a_i$ for $i = 1, \dots, |V_G|$. Furthermore, if $(i,j) \in E_G$, then set $b_i < a_j$ and $b_j < a_i$.

Proposition 18. *The poset obtained according to the VIP-construction from a graph G is Macaulay iff G satisfies the nestedness and continuity properties with respect to the VIP.*

Product theorems

Counterexamples show that if P and Q are Macaulay posets, then $P \times Q$ is not necessarily Macaulay. For example, if P is a poset whose Hasse diagram is isomorphic to $K_{p,p}$ for $p \geq 2$ (i.e. we have a special case of a so-called *complete* poset [28]) then $P \times P$ is not Macaulay in contradistinction to a conjecture in [28]. Indeed, if $m \leq p$, then a set of m elements of P_1^2 has minimal shadow iff these elements agree in some entry whose rank in P is 0. However, the shadow of any element of P_2^2 consists of $2p$ elements of P_1^2, which do not contain p elements of the form above.

Thus, a condition on P and Q is needed for a product theorem. The situation is, however, simple if Q is a trivial poset with $r(Q) = 0$. In this case a necessary and sufficient condition for P is found by Clements:

Theorem 19. (Clements [21]). *If $r(Q) = 0$, then $P \times Q$ is additive and Macaulay iff so is P.*

Probably, the next case in this hierarchy are posets of the form $P \times C_q$ with C_q being a chain with q elements. Counterexamples show that a condition on P is required for $P \times C_q$ to be Macaulay. However, this is not the case for $r(P) = 1$, as our result shows.

Theorem 20. *Let P be a poset with $r(P) = 1$ and let $q \geq 1$. Then $P \times C_q$ is a Macaulay poset iff P is Macaulay.*

A local-global principle

Consider the SMP on a cartesian power P^n of a Macaulay poset P. There exists a powerful technique for establishing the Macaulayness of such posets, which, in particular, involves induction on the number n of posets in the product. However, the general arguments within this technique work for $n \geq 3$ only. The case $n = 2$ is a special one and must be considered separately.

A similar situation also occurs in the *edge isoperimetric* problem on graphs (see section 3.3). Ahlswede and Cai proved in [1] that if the *lexicographic order* (see section 2) provides nestedness in EIP, then it is so for any $n \geq 3$. It turns out that the last result, which is called the *local-global principle* in [1], is valid for the edge-isoperimetric problem also with respect to some other total orders [12].

In what concerns the SMP, the above approach can not be directly applied because of the necessity to maintain the level structure of a poset. It turns out, however, that for the validity of such a principle with respect to the lexicographic order it is important that the poset satisfies some additional conditions, which have no analogies for graphs yet.

We call a Macaulay poset P *strongly Macaulay* if it is additive, shadow increasing and final shadow increasing. Note that Theorems 19 and 20 are valid with respect to strongly Macaulay posets too. Denote by $\mathcal{M}$ the class of ranked posets having only one maximum and only one minimum element.

Proposition 21. *A poset $P \in \mathcal{M}$ is strongly Macaulay iff so is its dual P^*.*

Theorem 22. (Bezrukov, Portas, Serra [16]). *Let $(P, \leq, \preceq) \in \mathcal{M}$ be strongly Macaulay and rank-greedy. Let the lexicographic order $\preceq^2$ be Macaulay for P^2. Then for any $n \geq 2$ the lexicographic order $\preceq^n$ is a Macaulay order for P^n.*

The assumptions concerning the poset P in Theorem 22 are essential, as the following result shows.

Theorem 23. (Bezrukov, Portas, Serra [16]). *Let $(P, \leq, \preceq)$ be a Macaulay poset. Furthermore, let $r(P) \geq 3$ and assume the orders $\preceq^2$ and $\preceq^3$ are Macaulay for P^2 and P^3, respectively. Then for any $n \geq 1$ one has: $P^n \in \mathcal{M}$, P^n is rank greedy, and P^n is strongly Macaulay.*

As an application of the local-global principle consider the following poset $(T(k), \leq) \in \mathcal{M}$ of rank k. For $1 \leq i \leq k-1$ the ith level of $T(k)$ consists of two elements a_i and b_i. Denote by b_0 and a_k the elements of T_0 and T_k, respectively. The partial order is defined as follows: $x < y$ iff $r(x) < r(y)$. We define the total order $\preceq$ on $T(k)$ by setting $b_{i-1} \prec a_i$ for $i = 1, \ldots, k$ and $a_i \prec b_i$ for $i = 1, \ldots, k-1$. Obviously, the order $\preceq$ is Macaulay on $(T(k), \leq)$.

Theorem 24. (Bezrukov, Portas, Serra [16]). *For any $k \geq 1$ and any $n \geq 1$ the poset $(T^n(k), \leq_\times, \preceq^n)$ is Macaulay.*

Further posets for which the local-global principle is applicable can be constructed using Proposition 16. Let P satisfy the assumptions of Theorem 22, and construct the poset $Q = (P, \sqsubseteq)$ as in section 3.3. Then Theorem 22 is applicable to Q. Indeed, the poset Q is Macaulay by Proposition 16. Now consider P^2. Since

$$\mathcal{T}_{P^2}(\mathcal{F}_i((x,y))) = \{(x,\xi) \mid \xi \in \mathcal{T}_P(\mathcal{F}_{i-r_P(x)}(y))\} \cup \{(\xi, y) \mid \xi \in \mathcal{T}_P(\mathcal{F}_{i-r_P(y)}(x))\},$$

then $\mathcal{T}_{P^2}(\mathcal{F}_i((x,y)) = \mathcal{T}_{Q^2}(\mathcal{F}_i((x,y)))$. Therefore, if P satisfies the assumptions of Theorem 22, then so does Q. On the other hand, since the lexicographic order is Macaulay for P^2, then so it is for P^4, for example. Extending P^2 as shown in section 3.1 results in a new poset, for which Theorem 22 is applicable.

NEW MACAULAY POSETS

In this section we present some further new families of Macaulay posets. We start with posets which are factorable by using the cartesian product operation in subsections 1 - 3 and proceed with two posets which do not appear to be cartesian products.

The products of trees and spider poset

Evidently, the classical Macaulay posets mentioned in Section 2 (we mean the Boolean lattice, the chain products, and the star poset) have something in common. Namely, the Hasse diagrams of the underlying posets in the product are trees. These posets are also *upper-semilattices*. For $a, b \in P$ denote by $\sup_P(a, b)$ an element $c \in P$ (if it exists) such that $a \prec c$, $b \prec c$ and $c \prec d$ if $a \prec d$ and $b \prec d$. The poset P is an *upper-semilattice* if for any $a, b \in P$, $\sup_P(a, b)$ exists and is unique.

Denote by $\mathcal{P}$ the class of upper semilattices P whose Hasse diagrams are trees. For which posets $P \in \mathcal{P}$ any their cartesian posers P^n are Macaulay ? Denote by $Q(k, l) \in \mathcal{P}$ the poset with the element set $\{0, 1, \ldots, (k+1)l\}$, and the partial order $\leq$ being defined as follows: $\alpha \leq \beta$ iff (i) $\alpha = \beta \pmod{k+1}$ and $\alpha \leq \beta$, or (ii) $\beta = (k+1)l$. The Hasse diagram of $Q(k, l)$ is a regular *spider* with k legs consisting of l vertices each.

Theorem 25. (Bezrukov [10]). *Suppose for some poset $P \in \mathcal{P}$ that P^n is Macaulay for some integer $n \geq r(P) + 3$. Then P is isomorphic to $Q(k, l)$ for some $k \geq 1$ and $l \geq 1$.*

It turns out that the inverse theorem is also valid.

Theorem 26. (Bezrukov, Elsässer [15]). *The poset $Q^n(k, l)$ is Macaulay for all integers n, k and l.*

The Macaulay order for $Q^n(k, l)$ is quite complicated and involves, in particular, the star poset order. We refer readers to [15] for exact definitions.

Looking back at Theorem 10 for star posets it is natural to ask if all cartesian products of the form $Q(k_1, l) \times Q(k_2, l) \times \cdots \times Q(k_n, l)$ are Macaulay. We conjecture an affirmative answer. On the other hand, it is easily seen that products of the form $Q(k, l_1) \times Q(k, l_2) \times \cdots \times Q(k, l_n)$ are not Macaulay in general.

Generalized submatrix orders

Let n and $k_1 \leq k_2 \leq \ldots k_m$ be positive integers such that $k_0 := n - \sum_{i=1}^{m} k_i \geq 0$. Furthermore, let $A_0, A_1, \ldots, A_m$ be the sets defined by

$$\begin{aligned} A_0 &:= \{1, 2, \ldots, k_0\}, \\ A_i &:= \left\{ \sum_{j=0}^{i-1} k_j + 1, \sum_{j=0}^{i-1} k_j + 2, \ldots, \sum_{j=0}^{i} k_j \right\} \text{ for } i = 1, 2, \ldots, m. \end{aligned}$$

Clearly, the sets A_i $(i = 0, 1, \ldots, m)$ form a partition of $[n] = \{1, 2, \ldots, n\}$.

The *generalized submatrix order* $\mathcal{S} := SM(n; k_1, \ldots, k_m)$ consists of all subsets X of $[n]$ such that $A_i \not\subseteq X$ for all $i = 1, 2, \ldots, m$. The corresponding partial order is given by: $X \leq Y$ iff $X \subseteq Y$. According to this definition, $\mathcal{S}$ is isomorphic to the cartesian product $B^{k_0} \times \tilde{B}^{k_1} \times \cdots \times \tilde{B}^{k_m}$, where $\tilde{B}^s$ denotes the Boolean lattice B^s without its maximal element.

The name generalized submatrix order refers to the work of Sali [51, 53] who actually considered the dual of $\mathcal{S}$ in the case $m = 2$, $k_0 = 0$. Sali proved for this poset several analogies to classical theorems on finite sets (Sperner, Erdös-Ko-Rado). For this poset, he also solved the problem of minimizing the number of atoms which are covered by an m-element subset of the i-th level for given i, m and conjectured Theorem 27 below in an equivalent form.

Theorem 27. (Leck [45, 46]). *$(\mathcal{S}, \subseteq)$ is a Macaulay poset.*

Before the above theorem was established, the closely related problem of finding ideals of maximum rank (cf. section 5.3) was solved by Vasta [54] for $\mathcal{S}^*$ with $k_0 = 0$. Using Theorem 27, a more general statement is now implied by Theorem 39.

In the proof of Theorem 27, again the case $m = 2$ required some special treatment, a modification of the well-known shifting operator for finite sets was used to settle this case. The following theorem is commonly used in the proof for $m > 2$, which is done by induction.

Theorem 28. (Leck [46]). *Generalized submatrix orders are additive.*

Another interesting poset which is related to the generalized submatrix orders is the poset M^n of square submatrices of a square matrix of order n ordered by inclusion. This poset also was studied by Sali [50, 52] with respect to Sperner and intersecting properties. For $n \leq 3$ the poset M^n is Macaulay, but not for $n \geq 4$ in contradistinction to a conjecture in [28].

The torus poset

Denote by T_k the poset whose Hasse diagram can be obtained from two disjoint chains of length k each by identifying their top and bottom vertices. Obviously, the Hasse diagram of T_k is a cycle of length $2k$.

Let $T^n_{k_1,\ldots,k_n} = T_{k_1} \times \cdots \times T_{k_n}$. The solution to the SMP for this poset follows from a solution to a more general problem: the VIP (cf. Section 3.2). In order to show the relation, let us consider a bipartite graph G. Fix a vertex $v_0 \in V_G$ and denote by G_i the set of all vertices of G at distance i from v_0. This leads to a ranked poset P with $P_i = G_i$ whose Hasse diagram is isomorphic to G. Assume that a solution to VIP on G satisfies the nestedness and continuity properties. Moreover, we assume that the total order $\mathcal{O}$ which provides a solution to the VIP orders the vertices of G_i in sequence. In other words, if A is an IS of $\mathcal{O}$ and $\sum_{i=0}^{r} |G_i| \leq |A| \leq \sum_{i=0}^{r+1} |G_i|$, then A contains a ball of radius r centered in v_0 and is contained in the ball of radius $r+1$ with the same center.

Obviously, a solution to the SMP with respect to the minimization of $\nabla(\cdot)$ for the subsets of P_r follows. Moreover, each IS of the order $\mathcal{O}$ restricted to P_r provides an optimal set. This problem is equivalent to the SMP with respect to the minimization of $\mathcal{T}(\cdot)$ for the dual of P. Thus, both P^* and P are Macaulay.

The Macaulay order for $T^n_{k_1,\dots,k_n}$, thus, can be obtained from the VIP-order $\mathcal{T}$ for the torus. This order is first established in [36], mentioned in the survey [8] and recently rediscovered in [49] and the readers are referred to these papers for exact definitions.

Theorem 29. (Karachanjan [36], Riordan [49]). *Any IS of the $\mathcal{T}$-oder provides a solution to the VIP. Moreover, the $\mathcal{T}$-oder satisfies the continuity property.*

Subword orders

Let us now turn to a first example of a Macaulay poset which is not representable as a cartesian product of nontrivial factors.

Let $n \geq 2$ be an integer, and let Ω denote the set $\{0, 1, \dots, n-1\}$. In the sequel, we call Ω the *alphabet*. The *subword order* $SO(n)$ consists of all strings (called *words*) that contain symbols (called *letters*) from Ω only. The partial order on $SO(n)$ is the subword relation, i.e. we have $x_1x_2\dots x_k \leq y_1y_2\dots y_l$ iff there is a set $\{i_1, i_2, \dots, i_k\} \subseteq \{1, 2, \dots, l\}$ of indices such that $i_1 < i_2 < \dots < i_k$ and $x_j = y_{i_j}$ for $j = 1, 2, \dots, k$. In other words, $\mathbf{x} \leq \mathbf{y}$ holds iff the word $\mathbf{x}$ can be obtained from the word $\mathbf{y}$ by successively deleting letters. By this definition, the rank of an element of $SO(n)$ equals its length, that means $r(x_1x_2\dots x_i) = i$. The only element of $N_0(SO(n))$ is the *empty word* ε.

Consider the case $n = 2$. Clearly, the level $N_i(SO(2))$ consists of all 0-1-words of length i and, therefore, in an obvious way its elements can be considered as the elements of the Boolean lattice B^i. It was shown by Harper [35] that, among all subsets $X \subseteq B^i$ of fixed cardinality, the IS in the *VIP-order* minimizes $|\Gamma_B(X)|$ (the size of the vertex-boundary of X in the Boolean lattice B^i). This order induces a total order of the elements for each level of $SO(2)$. For convenience, we define $w(x_1x_2\dots x_i) := |\{j \mid x_j = 1, 1 \leq j \leq i\}|$. Now the rank greedy extension of the VIP-order to the whole poset $SO(2)$ is given by the following conditions:

(1) $\mathbf{x} \preceq_{vip} \mathbf{y}$ if $w(\mathbf{x}) < w(\mathbf{y})$,

(2) $\mathbf{x} \preceq_{vip} \mathbf{y}$ if $w(\mathbf{x}) = w(\mathbf{y})$ and there is some $j \leq \min\{r(\mathbf{x}), r(\mathbf{y})\}$ such that $x_j > y_j$ and $x_h = y_h$ for $h = 1, 2, \dots, j-1$,

(3) $\mathbf{x} \preceq_{vip} \mathbf{y}$ if $w(\mathbf{x}) = w(\mathbf{y})$, $r(\mathbf{x}) \leq r(\mathbf{y})$ and $x_j = y_j$ for $j = 1, 2, \dots, r(\mathbf{x})$.

The next theorem reflects the importance of the VIP-order.

Theorem 30. (Ahlswede, Cai [2], Daykin, Danh [24, 25], Bezrukov [9]). *$(SO(2), \leq, \preceq_{vip})$ is a Macaulay structure.*

Let us remark that there are also several other Macaulay orders for $SO(2)$ which are specified by Daykin [29].

Based on the numerical approach of Ahlswede and Cai in [2], Engel and Leck [31] provided a relatively simple proof of Theorem 30. One of the main observations relates the SMP for $SO(2)$ to the VIP for Boolean lattices: If

$X \subseteq N_i(SO(2))$ is a final segment, then $|\nabla(X)| = |\Gamma_B(X)| + 2|X|$ holds. Another interesting observation is that $C(X)$ and $L(X)$ are isomorphic for any $X \subseteq N_i(SO(2))$. Clearly, this implies $|\Delta(C(X))| = |\Delta(L(X))|$ for all $X \subseteq N_i(SO(2))$ and all i. Macaulay posets satisfying this equality are called *shadow symmetric*.

Theorem 31. (Engel, Leck [31]). *Let P be a Macaulay poset. If P is shadow symmetric, then P additive.*

According to the above theorem, $SO(2)$ and its dual are additive.

Theorem 32. (Engel, Leck [31]). *The subword order $SO(2)$ is shadow increasing and weakly shadow increasing.*

Unfortunately, the dual of $SO(2)$ is obviously not shadow increasing In fact, this poset is even *shadow decreasing* (see [31] for a proof). However, for some applications (see section 6) the weak shadow increase property can serve as a substitute.

Let us now briefly discuss the case of larger alphabets. In [14] a Kruskal-Katona type theorem for $SO(n)$ with $n \geq 2$ was presented but there is a mistake in the proof, as pointed out by Danh and Daykin [26]. They also provided an example showing that the statement itself is not true at all for $n > 2$.

Daykin [28] introduced the *V-order*, an extension of the VIP-order for $SO(n)$ with $n \geq 2$. He conjectured that this order is a Macaulay order for $SO(n)$. For $n \geq 3$, a counterexample to this conjecture is given in [44]. Even worse, this example and a tedious case study yield the following result.

Theorem 33. (Leck [44]). *If $n > 2$, then the subword order $SO(n)$ is not a Macaulay poset.*

The linear lattice

The linear lattice L^n is another example of a poset which is not representable as a cartesian product of other posets. This poset is defined to be the collection of all proper nonempty subspaces of $\mathrm{PG}(n, 2)$ ordered by inclusion.

Note that $2^{n+1} - 1$ points of $\mathrm{PG}(n, 2)$ are just $(n+1)$-dimensional non-zero binary vectors $(\beta_1, \ldots, \beta_{n+1})$. Using the lexicographic ordering of the points, let us represent each subspace $a \in L^n$ by its characteristic vector, i.e. by the $(2^{n+1} - 1)$-dimensional binary vector $(\alpha_{2^{n+1}-1}, \ldots, \alpha_1)$, where α_i corresponds to the ith point of $\mathrm{PG}(n, 2)$.

For two subspaces $a, b \in L^n$, we say that a is greater than b in the order $\mathcal{O}$ if the characteristic vector of a is greater than the one of b in the lexicographic order. Now for $t > 0$ and $A \subseteq L_t^n$ denote

$$\hat{\mathcal{T}}(A) = \{x \in L_0^n \mid x \leq y,\ y \in A\}$$

and consider the SMP for the levels L_t^n and L_0^n.

Theorem 34. (Bezrukov, Blokhuis [13]). *Let $n \geq 1$ and $t > 0$. Then any IS of the order $\mathcal{O}_t$ has minimal shadow $\hat{\mathcal{T}}(\cdot)$. The shadow $\hat{\mathcal{T}}(\cdot)$ of any IS is an IS itself.*

However, as it is shown in [13], this poset is not Macaulay for $n \geq 3$.

EXTREMAL IDEALS IN MACAULAY POSETS

In this section we will be concerned with some optimization problems for which solutions are known for a rich class of Macaulay posets.

Let P be a poset, and let $\mathbb{R}^+$ denote the set of nonnegative real numbers. Furthermore, let there be a *weight function* $w : P \mapsto \mathbb{R}^+$ on P. If $w(x) = w(y)$ whenever $r(x) = r(y)$, the function $w(\cdot)$ is called *rank-symmetric*. If $w(\cdot)$ is a rank-symmetric weight function and $w(x) \leq w(y)$ whenever $r(x) < r(y)$, then $w(\cdot)$ is called *monotone*. Now define the weight of a subset $X \subseteq P$ as $w(X) = \sum_{x \in X} w(x)$.

Generated ideals of minimum weight

Consider the problem of constructing an antichain $X \subseteq P$ of given cardinality $m \leq d(P)$ such that the ideal generated by X has minimum weight for some monotone weight function.

This problem was considered by Frankl [33] for the Boolean lattice. For chain products, the problem was solved by Clements [19] who generalized preliminary results of Kleitman [38] and Daykin [27]. A further generalization is due to Engel [30] who provided a solution for the class of Macaulay posets P such that P and P^* are graded, additive, and shadow increasing. Unfortunately, the subword order $SO(2)$ is not included in this class since its dual is not shadow increasing (see section 6). Therefore, Engel and Leck [31] gave the following strengthening which applies to the classical Macaulay posets as well as to $SO(2)$.

Theorem 35. (Engel, Leck [31]). *Let P be a Macaulay poset such that P and P^* are weakly shadow increasing. Furthermore, let $m \leq d(P)$ be a positive integer, and put $i := \min\{j \mid m \leq |P_j|\}$ and $a := \min\{b \mid b+|P_{i-1}|-|\Delta(C(b, P_i))| = m\}$. Then the set*

$$X := C(a, P_i) \cup (P_{i-1} \setminus \Delta(C(a, P_i)))$$

is an antichain of size m. Moreover, $w(I(X)) \leq w(I(Y))$ holds for all antichains $Y \subseteq P$ with $|Y| = m$ with respect to any monotone weight function.

This theorem provides a sufficient condition for a poset to be Sperner (cf. [31] for details).

Corollary 36. *Let P be a Macaulay poset such that P is not an antichain. If P and P^* are weakly shadow increasing, then P is graded and has the Sperner property, i.e. the size of maximum antichain of P is equal to $\max_i |P_i|$.*

Ideals with maximum number of maximal elements

Now consider a dual to the last problem. Namely, we are looking now for an ideal of a given size, which has maximum number of maximal elements. In order to present a solution to this problem, we first introduce quasispheres. A *quasisphere* of size m in a ranked poset P is a set of the form

$$P_0 \cup P_1 \cup \cdots \cup P_i \cup C(a, P_{i+1}),$$

where the numbers a and i are (uniquely) defined by $m = \sum_{j=0}^{i} |P_j| + a$, $0 \leq a < |P_{i+1}|$. Obviously, any quasisphere is an ideal.

Theorem 37. (Engel, Leck [31]). *Let P be a Macaulay poset such that P and P^* are weakly shadow increasing. Then a quasisphere of size m has the maximum number of maximal elements in the class of all ideals of size m in P.*

Clearly, the set of maximal elements of some ideal is an antichain. For Boolean lattices, a related problem was considered by Labahn [40]. He determined the maximum size of an antichains X such that the ideal generated by X contains exactly m elements of P_i.

Maximum weight ideals

Now consider a problem of finding an ideal $I^* \subseteq P$ such that $w(I) \geq w(I)$ for any other ideal $I \subseteq P$ with $|I| = |I^*|$. We call this problem the Maximum Weight Ideal problem (MWI for brevity). Denote $w_i = w(x)$ for any $x \in P_i$.

The MWI problem is closely related to the edge-isoperimetric problems (cf. Section 3.2 and [8, 11] for more details) and was first considered by Bernstein and Steiglitz in [5] for the Boolean lattice and applied to a problem in coding theory.

Theorem 38. (Bernstein, Steiglitz [5]). *If $\preceq$ is a lexicographic order, then for any $m = 0, \ldots, 2^n$ the set $C(m, B^n)$ is a solution to the MWI problem for B^n with respect to any monotone weight function.*

Clements and Lindström in [23] extended Theorem 38 to the chain products in the case $w_i = i$ for all i, where a similar solution with respect to the lexicographic order was obtained by using Theorem 8. It turns out that the MWI problem is a direct consequence of the shadow minimization problem, as presented in the following theorem (see [6, 30]).

Theorem 39. *Let $(P, \leq, \preceq)$ be a rank-greedy Macaulay structure with a monotone weight function. Then the set $C(m, P)$ is a solution to the MWI problem for P.*

What if the weight function is not monotone? It is easily seen that if $w_0 \geq w_1 \geq \cdots \geq w_n$ then a solution to the MWI problem is attained on a quasisphere for any ranked poset P. For some less trivial nonmonotone weight functions a solution to the MWI is known for the Boolean lattice.

Theorem 40. (Ahlswede, Katona [4]). *Consider the Boolean lattice and let* $\preceq$ *be the lexicographic order.*

a. *If* $w_0 \leq w_1 \leq \cdots \leq w_{i-1} \geq w_i \geq \cdots \geq w_n$*, then a solution to the MWI problem is attained on an intersection of* $C(m', B^n)$ *with a quasisphere for some* $m' \leq m$.

b. *If* $w_0 \geq w_1 \geq \cdots \geq w_{i-1} \leq w_i \leq \cdots \leq w_n$*, then a solution to the MWI problem is attained on an union of* $C(m', B^n)$ *with a quasisphere for some* $m' \leq m$.

Bezrukov and Voronin in [17] proposed a new approach to this problem which significantly explores the Macaulayness property. They showed that similar result holds for the chain products. Note that the methods of neither [4] nor [17] provide exact values of m'. The corresponding results describe the situation just qualitatively and only ensure that such m' does exist. We guess that the approach of [17] can be extended to qualitatively describe maximum weight ideals for any rank-symmetric weight function, at least for the Boolean lattice and the products of chains.

Let us return back to Theorem 39. Evidently, the MWI and the SMP are closely related. The principal question is what should we claim on the solutions to the MWI problem in order to deduce the Macaulayness of the corresponding poset? Counterexamples show that the nestedness in the MRI problem on a poset P does not imply the Macaulayness of P in general. Thus, the SM problem is, in a sense, a more difficult problem than MWI.

References

[1] R. Ahlswede, N. Cai, "General edge-isoperimetric inequalities, Part II: A local-global principle for lexicographic solution", *Europ. J. Combin.*, **18** (1997), 479–489.

[2] R. Ahlswede, N. Cai, "Shadows and isoperimetry under the sequence-subsequence relation", *Combinatorica*, 17 (1997), 11–29.

[3] R. Ahlswede, N. Cai, "Isoperimetric theorems in the binary sequences of finite length", *SFB 343 Diskrete Strukturen in der Mathematik*, preprint 97-047, Universität Bielefeld (1997).

[4] R. Ahlswede, G.O.H. Katona, "Contributions to the geometry of Hamming spaces", *Discr. Math.*, 17 (1977), No. 1, 1–22.

[5] A.J. Bernstein, K. Steiglitz, "Optimal binary coding of ordered numbers", *J. SIAM*, 13 (1965), 441-443.

[6] S.L. Bezrukov, "Minimization of the shadows in the partial mappings semilattice", (in Russian), *Discretny Analiz*, 47 (1988), 3–18.

[7] S.L. Bezrukov, "On the construction of solutions of a discrete isoperimetric problems in Hamming space", *Math. USSR Sbornik*, 63 (1989), No. 1, 81–96.

[8] S.L. Bezrukov, "Isoperimetric problems in discrete spaces", in: *Extremal Problems for Finite Sets*, Bolyai Soc. Math. Stud. 3, P. Frankl, Z. Füredi, G. Katona, D. Miklos eds., Budapest 1994, 59–91.

[9] S.L. Bezrukov, *Discrete extremal problems on graphs and posets*, Habilitationsschrift, Universität–GH Paderborn (1995).

[10] S.L. Bezrukov, "On Posets whose products are Macaulay", *J. Comb. Theory*, A-84 (1998), 157–170.

[11] S.L. Bezrukov, "On an equivalence in discrete extremal problems", *Discr. Math.*, 203 (1999), 9–22.

[12] S.L. Bezrukov, "Edge-isoperimetric problems of graphs", in *Graph Theory and Combinatorial Biology*, Bolyai Soc. Math. Stud. 7, L. Lovász, A. Gyarfas, G.O.H. Katona, A. Recski, L. Székely eds., Budapest, 1999, 157–197.

[13] S.L. Bezrukov, A. Blokhuis, "A Kruskal-Katona theorem for the linear lattice", *Europ. J. Combin.*, 20 (1999), 123–130.

[14] S.L. Bezrukov, H.-D.O.F. Gronau, "A Kruskal-Katona type theorem", *Rostock Math. Kolloq.*, 46 (1992), 71–80.

[15] S.L. Bezrukov, R. Elsässer, "The spider poset is Macaulay", to appear in J. Comb. Theory.

[16] S.L. Bezrukov, X. Portas, O. Serra, "A local-global principle for Macaulay posets", to appear in *Order*.

[17] S.L. Bezrukov, V.P. Voronin, "Extremal ideals of the lattice of multisets with respect to symmetric functionals", (in Russian), *Discretnaya Matematika*, 2 (1990), No. 1, 50–58.

[18] G.F. Clements, "More on the generalized Macaulay theorem II", *Discr. Math.*, 18 (1977), 253–264.

[19] G.F. Clements, "The minimal number of basic elements in a multiset antichain", *J. Comb. Theory A*, 25 (1978), 153–162.

[20] G.F. Clements, "The cubical poset is additive", *Discr. Math.* 169 (1997), 17–28.

[21] G.F. Clements, "Characterizing profiles of k-families in additive Macaulay posets", *J. Comb. Theory*, A-80 (1997), 309–319.

[22] G.F. Clements, "Additive Macaulay Posets", *Order*, 4 (1997), 39–46.

[23] G.F. Clements, B. Lindström, "A generalization of a combinatorial theorem of Macaulay", *J. Comb. Th.* 7 (1969), No. 2, 230–238.

[24] T.N. Danh, D.E. Daykin, "Ordering integer vectors for coordinate deletion", *J. London Math. Soc.*, 55 (1997), 417–426.

[25] T.N. Danh, D.E. Daykin, "Sets of 0,1 vectors with minimal sets of subvectors", *Rostock. Math. Kolloq.*, 50, to appear.

[26] T.N. Danh, D.E. Daykin, "Bezrukov-Gronau order is not optimal", *Rostock. Math. Kolloq.*, 50, to appear.

[27] D.E. Daykin, "Antichains in the lattice of subsets of a finite set", *Nanta Math.*, 8 (1975), 84–94.

[28] D.E. Daykin, "Ordered ranked posets, representations of integers and inequalities from extremal poset problems", *Graphs and Order*, I. Rival ed., Proc. Conf. Banff Alta., 1984, NATO Adv. Sci. Inst. Ser. C: Math. Phys. Sci., 147, 1985, 395–412.

[29] D.E. Daykin, To find all "suitable" orders of 0,1-vectors, *Congressus Numerantium*, special volume in honour of C. Nash-Williams, to appear.

[30] K. Engel, *Sperner theory*, Cambridge University Press, 1997.

[31] K. Engel, U. Leck, "Optimal antichains and ideals in Macaulay posets", *Preprint* 96/21, University of Rostock, to appear in *Graph Theory and Combinatorial Biology*, Bolyai Soc. Math. Stud. 7, L. Lovász, A. Gyarfas, G.O.H. Katona, A. Recski, L. Székely eds., Budapest.

[32] Z. Füredi, J.R. Griggs, "Families of finite sets with minimum shadows", *Combinatorica*, 6 (1986), No. 4, 355–363.

[33] P. Frankl, "A lower bound on the size of a complex generated by an antichain", *Discr. Math.*, 76 (1989), 51–56.

[34] P. Frankl, Z. Füredi, G. Kalai, "Shadows of colored complexes", *Math. Scand.*, 63 (1988), 169–178.

[35] L.H. Harper, "Optimal numberings and isoperimetric problems on graphs", *J. Comb. Theory*, 1 (1966), 385–393.

[36] V.M. Karachanjan, "A discrete isoperimetric problem on multidimensional torus", (in Russian), *Doklady AN Arm. SSR*, vol. LXXIV (1982), No. 2, 61–65.

[37] G.O.H. Katona, "A theorem of finite sets", in: *Theory of graphs*, Academia Kiado, Budapest, 1968, 187–207.

[38] D.J. Kleitman, "On subsets contained in a family of non-commensurable subsets of a finite set", *J. Comb. Theory*, 7 (1969), 181–183.

[39] J.B. Kruskal, "The optimal number of simplices in a complex", in: *Math. Optimization Tech.*, Univ. of Calif. Press, Berkeley, California, 1963, 251–268.

[40] R. Labahn, "Maximizing antichains in the cube with fixed size of a shadow", *Order*, 9 (1992), 349–355.

[41] U. Leck, *Shifting for chain products and a new proof of the Clements-Lindström theorem*, Freie Universität Berlin, FB Mathematik, preprint A 16-94 (1994).

[42] U. Leck, *Extremalprobleme für den Schatten in Posets*, Ph. D. Thesis, FU Berlin, 1995; Shaker-Verlag Aachen, 1995.

[43] U. Leck, "A property of colored complexes and their duals", *Proc. Minisemester on Discr. Math., Warsaw 1996*, special issue of Discrete Math., to appear.

[44] U. Leck, *Nonexistence of a Kruskal-Katona type theorem for subword orders*, Universität Rostock, FB Mathematik, preprint 98/6 (1998), submitted.

[45] U. Leck, *Optimal shadows and ideals in submatrix orders*, Universität Rostock, FB Mathematik, preprint 98/15 (1998), submitted.

[46] U. Leck, *Another generalization of Lindström's theorem on subcubes of a cube*, Universität Rostock, FB Mathematik, forthcoming preprint.

[47] K. Leeb, "Salami-Taktik beim Quader-Packen", *Arbeitsberichte des Instituts für Mathematische Maschinen und Datenverarbeitung*, Universität Erlangen, 11 (1978), No. 5, 1–15.

[48] B. Lindström, "The optimal number of faces in cubical complexes", *Ark. Mat.*, 8 (1971), 245–257.

[49] O. Riordan, "An ordering on the discrete even torus", *SIAM J. Discr. Math.*, 11 (1998), No. 1, 110–127.

[50] A. Sali, "Constructions of ranked posets", *Discr Math.* 70 (1988), 77–83.

[51] A. Sali, "Extremal theorems for submatrices of a matrix", in *Proc. Int. Conf. on Combinatorics* (Eger, 1987), 439–446, Colloq. Math. Soc. Janos Bolyai 52, North-Holland, Amsterdam, 1988.

[52] A. Sali, "Extremal theorems for finite partially ordered sets and matrices", *Ph. D. Thesis*, Math. Inst. Hungar. Acad. Sci, Budapest 1991.

[53] A. Sali, "Some intersection theorems", *Combinatorica*, 12 (1992), 351–362.

[54] J.C. Vasta, *The maximum rank ideal problem on the orthogonal product of simplices*, PhD thesis, Univ. of California, Riverside (1998).

MINIMIZING THE ABSOLUTE UPPER SHADOW

Béla Bollobás

Department of Mathematics, University of Memphis
Memphis, TN 38152, U.S.A.

Imre Leader

Department of Mathematics, University College London
London WC1E 6BT, England

Abstract: The absolute upper shadow of a family $\mathcal{A}$ of r-sets on $\{1,\ldots,n\}$ is $\overline{\partial}\mathcal{A} = \{A \cup \{i\} : A \in \mathcal{A},\ i \notin A,\ i \in \cup\mathcal{A}\}$. Given $|\mathcal{A}|$, how small can $\overline{\partial}\mathcal{A}$ be? Our aim in this note is to give an exact solution to this question. Curiously, the extremal sets turn out not to form a nested nestedfamily.

Our main tool is an inequality concerning the colex ordering that may be of independent interest.

INTRODUCTION

The Kruskal-Katona theorem [7],[5] states that the minimum lower shadow of a set system of given size is attained for initial segments of the colex order . More precisely, let $\mathcal{A}$ be a family of r-sets from an n-element ground set: $\mathcal{A} \subset [n]^{(r)} = \{1, 2, \ldots, n\}^{(r)}$. The *shadow* or *lower shadow* of $\mathcal{A}$ is

$$\partial\mathcal{A} = \partial^{-}\mathcal{A} = \{A - \{i\} : A \in \mathcal{A},\ i \in A\}.$$

The *colexicographic* or *colex* order on $[n]^{(r)}$ is defined as follows. Given distinct $A, B \in [n]^{(r)}$, write $A = \{a_1, \ldots, a_r\}$, $B = \{b_1, \ldots, b_r\}$, where $a_1 < \ldots < a_r$

I. Althöfer et al. (eds.), Numbers, Information and Complexity, 95-100.

and $b_1 < \ldots < b_r$. Then we set $A < B$ in the colexicographic order if $a_s < b_s$, where $s = \max\{t : a_t \neq b_t\}$. Equivalently, we have $A < B$ if and only if $\max(A \triangle B) \in B$, where as usual $\triangle$ denotes symmetric difference. For example, for every k, the set $[k]^{(r)}$ is an initial segment of colex. Then the Kruskal-Katona theorem states that if $\mathcal{A} \subset [n]^{(r)}$, and $\mathcal{C}$ is the set of the first $|\mathcal{A}|$ elements of $[n]^{(r)}$ in the colex order, then $|\partial\mathcal{A}| \geq |\partial\mathcal{C}|$.

By taking complements, one immediately obtains a similar result for upper shadows, where the *upper shadow* of $\mathcal{A}$ is

$$\partial^+\mathcal{A} = \{A \cup \{i\} : A \in \mathcal{A},\ i \notin A\}.$$

To formulate this result, we define the *lexicographic* or *lex* ordering on $[n]^{(r)}$ by setting $A < B$ if $\min(A \triangle B) \in A$. For example, the set $\{A \in [n]^{(r)} : 1 \in A\}$ is an inital segment of lex. Then the Kruskal-Katona theorem may be rephrased as: if $\mathcal{A} \subset [n]^{(r)}$, and $\mathcal{B}$ is the set of the first $|\mathcal{A}|$ elements of $[n]^{(r)}$ in the lex ordering, then $|\partial^+\mathcal{A}| \geq |\partial^+\mathcal{B}|$.

However, there is a difference between the upper and lower shadows. The lower shadow is *absolute*, meaning that if $A \subset [n]^{(r)}$, and we subsequently regard $\mathcal{A}$ as a subset of $[m]^{(r)}$, for some $m > n$, then the lower shadow of $\mathcal{A}$ is unchanged. Whereas to determine the upper shadow $\partial^+\mathcal{A}$ of $\mathcal{A}$ we need to know the 'ground set' of $\mathcal{A}$ – for example, if $\mathcal{A} = \{12, 13\} \subset [3]^{(2)}$ then $\partial^+\mathcal{A} = \{123\}$, but if $\mathcal{A} = \{12, 13\} \subset [4]^{(2)}$ then $\partial^+\mathcal{A} = \{123, 124, 134\}$.

What would the 'absolute' notion corresponding to the upper shadow be? The natural choice is to allow the addition of only those members of the ground set that have been 'mentioned' in the system – in other words, that belong to at least one set from the family. So, for $\mathcal{A} \subset [n]^{(r)}$, we define the *absolute upper shadow* of $\mathcal{A}$ to be

$$\overline{\partial}\mathcal{A} = \{A \cup \{i\} : A \in \mathcal{A},\ i \notin A,\ i \in \cup\mathcal{A}\}.$$

Then the analogue of the question answered by the Kruskal-Katona theorem is the following: given $|\mathcal{A}|$, how should we choose $\mathcal{A} \subset [n]^{(r)}$ to minimize $|\overline{\partial}\mathcal{A}|$?

For this problem, lex and colex seem to pull in opposite directions. Of course, if we know $|\cup\mathcal{A}|$, we should choose $\mathcal{A}$ as an initial segment of lex, by the Kruskal-Katona theorem for upper shadows. But $|\cup\mathcal{A}|$ itself is minimized by initial segments of colex.

It turns out that there is no single ordering on $[n]^{(r)}$ whose initial segments are extremal. A little experiment suggests that we should keep $\cup\mathcal{A}$ as small as possible, and then use lex inside that. In other words, if we are to choose $\mathcal{A}$ with $|\mathcal{A}| = m$, having minimum absolute upper shadow, then we should choose the minimal k with $m \leq \binom{k}{r}$, and take $\mathcal{A}$ to be the first m elements of $[k]^{(r)}$ in lex. Our main result, Theorem 3, states that this is indeed the case.

It is clear that, as m varies, the extremal sets mentioned above do *not* form a nested family. For example, suppose that $r > n/2$, so that $\binom{n-1}{r-1} > \binom{n-1}{r}$. Then for $|\mathcal{A}| = \binom{n-1}{r}$ the extremal system is $[n-1]^{(r)}$, while for $|\mathcal{A}| = \binom{n-1}{r}$ the extremal system is $\{A \in [n]^{(r)} : 1 \in \mathcal{A}\}$. This means that the direct compression methods usually used on isoperimetric questions (see eg. [1],[2],[4],[6]) cannot be applied.

Our main lemma, which is almost equivalent to Theorem 3, is a result about the colex ordering. It states that the first m elements of the colex ordering on $[n]^{(r)}$ have lower shadow at most as large as the lower shadow of the first m elements of the colex ordering on $[n]^{(r+1)}$. Such a simple result is very believable, as larger sets ought to be worse for the lower shadow, but it seems to be rather elusive. Indeed, remarkably, it seems that the simplest proof makes use of the Kruskal-Katona theorem itself.

We prove this lemma, and our main result, in the next section. In the following section we place the absolute upper shadow in a more general framework, and give some related problems and conjectures.

Finally, we note that there is a superficial resemblance between our problem and the problem of minimizing the lower shadow over all set systems $\mathcal{A} \subset [n]^{(r)}$ (with $|\mathcal{A}|$ given) satisfying $\cup\mathcal{A} = [n]$. This problem was solved by Mörs [8], but the two problems do not seem to be related.

THE MINIMUM ABSOLUTE UPPER SHADOW

We need a small amount of notation. Write $[2,n]$ for $\{2,\ldots,n\}$. For $\mathcal{A} \subset [n]^{(r)}$, the *sections* of $\mathcal{A}$ are the systems $\mathcal{A}_+ \subset [2,n]^{(r-1)}$ and $\mathcal{A}_- \subset [2,n]^{(r)}$ given by

$$\mathcal{A}_+ = \left\{A \in [2,n]^{(r-1)} : A \cup \{1\} \in \mathcal{A}\right\}$$

and

$$\mathcal{A}_- = \left\{A \in [2,n]^{(r)} : A \in \mathcal{A}\right\}.$$

Thus $|\mathcal{A}| = |\mathcal{A}_+| + |\mathcal{A}_-|$. Note that the lower shadow $\partial\mathcal{A}$ of $\mathcal{A}$ has sections given by:

$$(\partial\mathcal{A})_+ = \partial(\mathcal{A}_+)$$

and

$$(\partial\mathcal{A})_- = \partial(\mathcal{A}_-) \cup \mathcal{A}_+.$$

Let us also point out that the sections of an initial segment of colex on $[n]^{(r)}$ are themselves initial segents of colex on $[2,n]^{(r-1)}$ and $[2,n]^{(r)}$ (where of course the colex order on say $[2,n]^{(r)}$ is that induced from the colex order on $[n]^{(r)}$ – i.e. $A < B$ if $\max(A \,\triangle\, B) \in B$).

Lemma 1. *Let $1 \le r \le n-1$, and let $\mathcal{A} \subset [n]^{(r+1)}$ and $\mathcal{B} \subset [n]^{(r)}$ be initial segments of colex with $|\mathcal{A}| = |\mathcal{B}|$. Then $|\partial\mathcal{A}| \ge |\partial\mathcal{B}|$.*

Proof. We proceed by induction on n: the result is trivial for $n = 2$ (or $n = 1$), so we turn to the induction step. Given $\mathcal{A} \subset [n]^{(r+1)}$ and $\mathcal{B} \subset [n]^{(r)}$, initial segments of colex with $|\mathcal{A}| = |\mathcal{B}|$, let us suppose first that we have $|\mathcal{A}_+| \le \binom{n-1}{r-1}$ and $|\mathcal{A}_-| \le \binom{n-1}{r}$. In that case, we may define a set system $\mathcal{C} \subset [n]^{(r)}$ by giving its sections: we let $\mathcal{C}_+ \subset [n-1]^{(r-1)}$ and $\mathcal{C}_- \subset [n-1]^{(r)}$ be the initial segments of colex of sizes $|\mathcal{A}_+|$ and $|\mathcal{A}_-|$ respectively.

We claim that $|\partial\mathcal{C}| \leq |\partial\mathcal{A}|$. Indeed, we have

$$|\partial\mathcal{A}| = |\partial(\mathcal{A}_+)| + |\partial(\mathcal{A}_-) \cup \mathcal{A}_+|$$

and

$$|\partial\mathcal{C}| = |\partial(\mathcal{C}_+)| + |\partial(\mathcal{C}_-) \cup \mathcal{C}_+|.$$

Actually, since $\mathcal{A}$ is an initial segment of colex, a moment's thought shows that $\partial(\mathcal{A}_-) \subset \mathcal{A}_+$ – we shall need this fact a little later.

Now, by induction we have $|\partial(\mathcal{C}_+)| \leq |\partial(\mathcal{A}_+)|$. Similarly, we have $|\partial(\mathcal{C}_-)| \leq |\partial(\mathcal{A}_-)|$. Also, $|\mathcal{C}_+| = |\mathcal{A}_+|$. However, the sets $\partial(\mathcal{C}_-)$ and $\mathcal{C}_+$ are nested, as each is an initial segment of colex on $[2,n]^{(r-1)}$. It follows that $|\partial(\mathcal{C}_-) \cup \mathcal{C}_+| \leq |\partial(\mathcal{A}_-) \cup \mathcal{A}_+|$, and hence $|\partial\mathcal{C}| \leq |\partial\mathcal{A}|$, as claimed.

Since $\mathcal{C} \subset [n]^{(r)}$, and $|\mathcal{C}| = |\mathcal{B}|$, the Kruskal-Katona theorem tells us that $|\partial\mathcal{B}| \leq |\partial\mathcal{C}|$. Thus $|\partial\mathcal{B}| \leq |\partial\mathcal{A}|$, as required. We now turn to the case when

$|\mathcal{A}_+| > \binom{n-1}{r-1}$ or $|\mathcal{A}_-| > \binom{n-1}{r}$. If $|\mathcal{A}_-| > \binom{n-1}{r}$ then, by applying the induction hypothesis to $\mathcal{A}_-$, we see that $|\partial(\mathcal{A}_-)| \geq \binom{n-1}{r-1}$, whence $|\mathcal{A}_+| \geq \binom{n-1}{r-1}$.

So we may assume that $|\mathcal{A}_+| \geq \binom{n-1}{r-1}$. The induction hypothesis tells us that $|\partial(\mathcal{A}_+)| \geq \binom{n-1}{r-2}$. Thus

$$|\partial(\mathcal{A}_+)| + |\mathcal{A}_+| \geq \binom{n-1}{r-2} + \binom{n-1}{r-1} = \binom{n}{r-1},$$

so that certainly $|\partial\mathcal{A}| \geq |\partial\mathcal{B}|$. □

We remark that there are other ways to prove Lemma 1. Indeed, after we had publicised Lemma 1, we received alternative proofs from David Daykin [3] and Mark Ryten [9], based on cascade-type arguments. What makes the above proof simpler seems to be the fact that, by using Kruskal-Katona, one just needs to exhibit *some* system of r-sets with shadow no larger than that of $\mathcal{A}$, as opposed to having to deal with $\mathcal{B}$ itself.

It is natural to ask how much larger $\partial\mathcal{A}$ must be than $\partial\mathcal{B}$ – in other words, how small $|\partial\mathcal{A}|/|\partial\mathcal{B}|$ can be. We do not know the answer to this question. It seems very plausible that the minimum value of $|\partial\mathcal{A}|/|\partial\mathcal{B}|$ occurs when $|\mathcal{A}| = |\mathcal{B}| = r+2$. Indeed, the size $r+2$, besides being very small, is good for $\partial\mathcal{A}$ (as $\mathcal{A}$ is exactly of the form $[k]^{(r+1)}$) and bad for $\partial\mathcal{B}$ (as $\mathcal{B}$ is a set of the form $[k]^{(r)}$, together with one more set). In this case we have $|\partial\mathcal{A}| = \binom{r+1}{2}$ and $|\partial\mathcal{B}| = \binom{r}{2} + r - 1$.

Conjecture 2. *Let $1 \leq r \leq n-1$, and let $\mathcal{A} \subset [n]^{(r+1)}$ and $\mathcal{B} \subset [n]^{(r)}$ be initial segments of colex with $|\mathcal{A}| = |\mathcal{B}|$.*
Then $|\partial\mathcal{A}| \geq \left(1 + 4/(r^2+3r-2)\right) |\partial\mathcal{B}|$.

Armed with Lemma 1, we are ready for our main result.

Theorem 3. *For $\mathcal{A} \subset [n]^{(r)}$, choose k with $\binom{k-1}{r} < |\mathcal{A}| \leq \binom{k}{r}$, and let $\mathcal{B}$ consist of the first $|\mathcal{A}|$ elements in the lex order on $[k]^{(r)}$. Then $|\overline{\partial}\mathcal{A}| \geq |\overline{\partial}\mathcal{B}|$. In particular, if $|\mathcal{A}| = \binom{k}{r}$, for some k, then $|\overline{\partial}\mathcal{A}| \geq \binom{k}{r+1}$.*

Proof. If $|\mathcal{A}| > \binom{n-1}{r}$ then certainly $\cup\mathcal{A} = [n]$, so that $\overline{\partial}\mathcal{A} = \partial^+\mathcal{A}$, and our assertion reduces to the Kruskal-Katona theorem. So we may assume that $|\mathcal{A}| \leq \binom{n-1}{r}$.

Our aim is to show that there is a set system $\mathcal{C} \subset [n-1]^{(r)}$ with $|\mathcal{C}| = |\mathcal{A}|$ and $|\overline{\partial}\mathcal{C}| \leq |\overline{\partial}\mathcal{A}|$ – we will then be done, as induction on n gives $|\overline{\partial}\mathcal{C}| \geq |\overline{\partial}\mathcal{B})|$.

If $|\cup\mathcal{A}| \leq n-1$ then we have nothing to prove, as we may take $\mathcal{C} = \mathcal{A}$ (up to a permutation of the ground set). So we may assume that $\cup\mathcal{A} = [n]$, so that $\overline{\partial}\mathcal{A} = \partial^+\mathcal{A}$. Let $\mathcal{C}$ consist of the first $|\mathcal{A}|$ elements of $[n-1]^{(r)}$ in lex. We are done if we can show that $|\partial^+\mathcal{A}| \geq |\partial^+\mathcal{C}|$ (where, for the upper shadow of $\mathcal{C}$, we regard the ground set of $\mathcal{C}$ as $[n-1]$).

Taking complements, this is equivalent to the following assertion: if $\mathcal{A}'$ is an initial segment of colex on $[n]^{(n-r)}$, and $\mathcal{C}'$ is an initial segment of colex on $[n-1]^{(n-r-1)}$, with $|\mathcal{C}'| = |\mathcal{A}'|$, then $|\partial\mathcal{C}'| \leq |\partial\mathcal{A}'|$. However, because $[n-1]^{(n-r-1)}$ is an initial segment of the colex order on $[n]^{n-r-1}$, this assertion follows immediately from Lemma 1. □

SOME RELATED QUESTIONS

The absolute upper shadow is actually just one of a family of related notions, as we now describe. For a set system $\mathcal{A} \subset [n]^{(r)}$, and any $t = 1, \ldots, r$, we define the *t-shadow* of $\mathcal{A}$ to be

$$\mathcal{A}_t = \left\{B \in [n]^{(t)} : \ B \subset A \text{ for some } A \in \mathcal{A}\right\}.$$

In other words, $\mathcal{A}_t$ is the $(r-t)$-fold iterated lower shadow of $\mathcal{A}$. So for example we have $\mathcal{A}_r = \mathcal{A}$, $\mathcal{A}_{r-1} = \partial\mathcal{A}$, and $\mathcal{A}_1 = \cup\mathcal{A}$. For $1 \leq s,t \leq r$ we define the *(s,t)-shadow* of $\mathcal{A}$ to be

$$\mathcal{A}_{s,t} = \{A \cup B : \ A \in \mathcal{A}_s, \ B \in \mathcal{A}_t, \ A \cap B = \emptyset\}.$$

So $\mathcal{A}_{s,t}$ consists of those $(s+t)$-sets that may be partitioned into an s-set and a t-set, each contained in members of $\mathcal{A}$. Thus for example the absolute upper shadow $\overline{\partial}\mathcal{A}$ is precisely $\mathcal{A}_{r,1}$.

Given s and t, how should we choose $\mathcal{A} \subset [n]^{(r)}$ to minimize $\mathcal{A}_{s,t}$? For which s and t do we have a similar 'globally colex' situation, in that all sets of the form $[k]^{(r)}$ are extremal?

It is easy to see that this is the case if $s+t \leq r$. Indeed, if $s+t \leq r$ then we certainly have $\mathcal{A}_{s,t} \supset \mathcal{A}_{s+t}$. However, if $\mathcal{A} = [k]^{(r)}$ then $\mathcal{A}$ not only minimizes $|\mathcal{A}_{s+t}|$ (among systems of size $\binom{k}{r}$), but also has $\mathcal{A}_{s,t} = \mathcal{A}_{s+t}$. Hence $[k]^{(r)}$ is extremal for the problem of minimizing $|\mathcal{A}_{s,t}|$.

It is also easy to see that this is *not* the case if $s+t \geq r+2$. Indeed, if $s+t \geq r+2$ then any system $\mathcal{A}$ all of whose members contain some fixed $(r-1)$-set clearly has $\mathcal{A}_{s,t} = \emptyset$. So for example the system $[s+t]^{(r)}$ is not extremal (for $n \geq \binom{s+t}{r} + r - 1$).

This leaves only the case when $s+t = r+1$. We believe that sets of the form $[k]^{(r)}$ *are* still extremal.

Conjecture 4. *Let $\mathcal{A} \subset [n]^{(r)}$ with $|\mathcal{A}| = \binom{k}{r}$, and let $1 \leq s,t \leq r$ with $s+t = r+1$. Then $|\mathcal{A}_{s,t}| \geq \binom{k}{r+1}$.* □

In view of the fact that the case $s = r$ and $t = 1$ of Conjecture 4 is precisely Theorem 3, perhaps the most appealing special case of Conjecture 4 is the symmetric case $s = t$.

Conjecture 5. *Let $\mathcal{A} \subset [n]^{(2r-1)}$ with $|\mathcal{A}| = \binom{k}{2r-1}$. Then $|\mathcal{A}_{r,r}| \geq \binom{k}{2r}$.* □

Finally, of course, it would be desirable to know the exact extremal sets for the problem of minimizing $|\mathcal{A}_{s,t}|$. In other words, for $\mathcal{A} \subset [n]^{(r)}$, with $|\mathcal{A}|$ given, and $1 \leq s,t \leq r$, how small can $|\mathcal{A}_{s,t}|$ be?

References

[1] B. Bollobás, *Combinatorics*, Cambridge University Press, 1986, xii + 177 pp.

[2] B. Bollobás and I. Leader, "Compressions and isoperimetric inequalities", *J. Combinatorial Theory* (A) **56** (1991), 47-62.

[3] D. Daykin, personal communication.

[4] P. Frankl, "The shifting technique in extremal set theory", *Surveys in Combinatorics 1987* (Whitehead, C., ed.), Cambridge University Press, 1987, 81-110.

[5] G.O.H. Katona, "A theorem on finite sets", *Theory of Graphs* (Erdős, P. and Katona, G.O.H., eds.), Akadémiai Kiadó, Budapest, 1968, 187-207.

[6] D.J. Kleitman, "Extremal hypergraph problems", *Surveys in Combinatorics* (Bollobás, B., ed.), Cambridge University Press, 1979, 44-65.

[7] J.B. Kruskal, "The number of simplices in a complex", *Mathematical Optimization Techniques*, Univ. California Press, Berkeley, 1963, 251-278.

[8] M. Mörs, "A generalization of a theorem of Kruskal", *Graphs Combin.* **1** 1985, 167-183.

[9] M. Ryten, personal communication.

CONVEX BOUNDS FOR THE 0,1 CO–ORDINATE DELETIONS FUNCTION

David E. Daykin

Mathematics Department, University of Reading, England RG6 2AX

INTRODUCTION

Let $\mathcal{V}(n)$ be the set of 0,1 co–ordinate vectors of dimension n. For $\mathcal{A} \subseteq \mathcal{V}(n)$ let $\triangle\mathcal{A}$ be the set of vectors in $\mathcal{V}(n-1)$ obtained by deleting a co–ordinate from a vector of $\mathcal{A}$ in all ways. The 0,1 co–ordinate deletions function $\delta(k,n)$ is $\min|\triangle\mathcal{A}|$ over all $\mathcal{A} \subseteq \mathcal{V}(n)$ with $|\mathcal{A}| = k$.

If $\underset{\sim}{a} = a_1, a_2, \ldots, a_n$ then $\omega\underset{\sim}{a} = a_1 + \ldots + a_n$. We order $\mathcal{V}(n)$ by $\underset{\sim}{a} < \underset{\sim}{b}$ if either (i) $\omega\underset{\sim}{a} < \omega\underset{\sim}{b}$ or (ii) $\omega\underset{\sim}{a} = \omega\underset{\sim}{b}$ and the least j with $a_j \neq b_j$ has $1 = a_j > b_j = 0$.

Theorem 1. (Danh–Daykin [2–6]). *If I is the first k vectors of $\mathcal{V}(n)$ then $\delta(k,n)$ is the number of $\underset{\sim}{a} \in I$ with $a_n = 0$.*

In Part 2 we give new lower bounds for δ, and in Part 3 we show that the slopes of the convex hull of δ form the Farey sequence.

CONVEX LOWER BOUNDS FOR $\delta(k,n)$

We put $\mu\underset{\sim}{a} = k$ if $\underset{\sim}{a}$ is the k–th vector, with $\mu\phi = 0$, and allow $\delta(\underset{\sim}{a}) = \delta(k,n) = \delta(k)$. On the real (x,y) plane we plot $\big(x, \delta(x)\big)$ for $x = 0, 1, \ldots$. If S is a section of $\mathcal{V}(n)$ then αS (resp. $\beta S, \gamma S$) is the vector just before S (resp. first in S, last in S). If $\underset{\sim}{a} = \alpha S$, $\underset{\sim}{c} = \gamma S$ the line through $\big(\mu\underset{\sim}{a}, \delta(\underset{\sim}{a})\big)$ and $\big(\mu\underset{\sim}{c}, \delta(\underset{\sim}{c})\big)$ we call the *line of* S, and its slope is *slope* S.

Given $r + s = n$, $\underset{\sim}{a} \in \mathcal{V}(r)$, $0 \leq h \leq s$ we put

$$T = T(\underset{\sim}{a}, h, s) = \{\underset{\sim}{a}\,\underset{\sim}{b} : \underset{\sim}{b} \in \mathcal{V}(s), \omega\underset{\sim}{b} = h\} \subseteq \mathcal{V}(n),$$

and call T a *tunnel*. By Theorem 1 $\operatorname{slope} T = (s-h)/s$. If $|T| \geq 2$ then T is $T(\underset{\sim}{a}\,1, h-1, s-1)$ followed by $T(\underset{\sim}{a}\,0, h, s-1)$, and $(s-h)/(s-1) \geq (s-h)/s \geq (s-1-h)/(s-1)$. Induction on $|T|$ gives the tunnel lemma.

I. Althöfer et al. (eds.), Numbers, Information and Complexity, 101-104.

Lemma 1. *In any tunnel the plot of δ is above ($\geq$) the tunnel line.*

The lines of the tunnels with $s = 0$ from a convex lower bound for δ. So too do those with $s = 1$. These facts form

Theorem 2. *If $0 \leq k \leq 2^n$ we get two representations*

$$k = \binom{n}{n} + \binom{n}{n-1} + \ldots + \binom{n}{n-g+1} + G \text{ with } 0 \leq G \leq \binom{n}{n-g}, \quad (1)$$

$$k = 2\binom{n-1}{n-1} + 2\binom{n-1}{n-2} + \ldots + 2\binom{n-1}{n-h+1} + H \text{ with } 0 \leq H \leq 2\binom{n}{n-h}. \quad (2)$$

Then $$\binom{n-1}{n-1} + \binom{n-1}{n-2} + \ldots + \binom{n-1}{n-g} + (\tfrac{n-g}{n})G \leq$$

$$\leq 2\binom{n-2}{n-2} + 2\binom{n-2}{n-3} + \ldots + 2\binom{n-2}{n-h} + (\tfrac{n-h}{n})H \leq \delta(k,n).$$

THE CONVEX HULL $CH(n+1)$ OF $\delta(k, n+1)$

Let S be a section of $\mathcal{V}(n+1)$. To get S' replace each $\underset{\sim}{a} \in \mathcal{S}$ by $\underset{\sim}{a}'$ its succesor. We call S an *h–sec* if $\underset{\sim}{b}, \underset{\sim}{b}' \in S$ where $\underset{\sim}{b}$ is the last $\underset{\sim}{a}$ with $\omega\underset{\sim}{a} = h$, so $\omega\underset{\sim}{b}' = h+1$. We say S is *Nice* if $\alpha S = 0\underset{\sim}{c}$ and $\gamma S = 1\underset{\sim}{c}$ for some $\underset{\sim}{c} \in \mathcal{V}(n)$. Considering $S, S'(S')', \ldots$ we see that an h–Sec S is Nice iff $|S| = \binom{n}{h}$. Such an S has $\operatorname{slope} S = (n-h)/n$.

There is a sequence $\phi = \underset{\sim}{e}_0 < \underset{\sim}{e}_1 < \ldots < \underset{\sim}{e}_u = 11\ldots1$ such that $(\mu\underset{\sim}{e}_j, \delta(\underset{\sim}{e}_j))$ are the extreme points of $CH(n+1)$. We call $\mathcal{E}_j = \{\underset{\sim}{a} : \underset{\sim}{e}_{j-1} < \underset{\sim}{a} \leq \underset{\sim}{e}_j\}$ an *Exsec*, and the lines of these Exsec form $CH(n+1)$.

The Farey sequence $F(n)$ consists of all fractions $1 \geq (q-p)/q \geq 0$ in descending order, where $1 \leq q \leq n$ and p, q are coprime [7].

Theorem 3. *In the above notation* $\operatorname{slope}\mathcal{E}_1$, $\operatorname{slope}\mathcal{E}_2, \ldots,$ $\operatorname{slope}\mathcal{E}_u$ *is* $F(n)$.

Proof: As part of the induction hypothesis we need

$$\begin{cases} \text{Let } 0 < h < n \text{ and } \underset{\sim}{b} \text{ be the last } \underset{\sim}{a} \text{ with } \omega\underset{\sim}{a} = h. \\ \text{Then } \underset{\sim}{b} \in \mathcal{E} \text{ iff } \underset{\sim}{b}' \in \mathcal{E} \text{ iff } \operatorname{slope}\mathcal{E} = (n-h)/n. \end{cases} \quad (3)$$

Clearly $|\mathcal{E}_1| = |\mathcal{E}_u| = n$ with slopes 1 and 0. Let $\mathcal{D}_1, \mathcal{D}_2, \ldots, \mathcal{D}_t$ be the Exsec for $CH(n)$. Put $m = n-1$ and recall $F(m) \subseteq F(n)$. Let $\mathcal{D}$ be any $\mathcal{D}_i$ with $1 < i < t$. Using $\mathcal{D}$ we now describe the $\mathcal{E}$ with $\operatorname{slope}\mathcal{E} = \operatorname{slope}\mathcal{D} = \eta$ say, and (3) will hold. We keep $0 < h < n$.

Case 1. $(m-h+1)/m > \eta > (n-h)/n$. Here $\mathcal{E}$ is $0\mathcal{D}$.

Case 2. $(n-h)/n > \eta > (m-h)/m$. Here $\mathcal{E}$ is $1\mathcal{D}$.

Case 3. $\eta = (m-h)/m$. Here (3) shows $\mathcal{D}$ is an h–Sec. We map $\underset{\sim}{a} \in \mathcal{D}$ to $1\underset{\sim}{a}$ (resp. $0\underset{\sim}{a}$) if $\omega\underset{\sim}{a}$ is h (resp. $h+1$). Then $\mathcal{E}$ is the image of $\mathcal{D}$.

Case 4. $\eta = (n-h)/n$. Here $\mathcal{E}$ is $0\mathcal{D}$ and $1\mathcal{D}$ and the set $\mathcal{B}$ of vectors between them. Let $\mathcal{A}$ be $\mathcal{B}$ and $1\mathcal{D}$. Trivially $\mathcal{A}$ is Nice. Because we are not in Case 3, we know $\mathcal{D}$ is not an h–Sec, so $\mathcal{A}$ is an h–Sec. Hence $0\mathcal{D}$, $\mathcal{B}$, $1\mathcal{D}$, $\mathcal{A}$, $\mathcal{E}$ all have $\operatorname{slope}\eta$, and in fact the same line. If $0\underset{\sim}{d} \in \mathcal{B}$ then all vectors between $0\mathcal{D}$ and $0\underset{\sim}{d}$ start 0. Since $\mathcal{D}$ is an Exsec, and $\underset{\sim}{d}$ follows $\mathcal{D}$, we have $\delta(\underset{\sim}{d})$ below ($\leq$) the line of $\mathcal{D}$. So $\delta(0\underset{\sim}{d})$ is below the line of $0\mathcal{D}$. Similarly if $1\underset{\sim}{d} \in \mathcal{B}$ then $\delta(1\underset{\sim}{d})$ is below the line of $1\mathcal{D}$. Thus δ is below the line of $\mathcal{E}$ in $\mathcal{E}$. Note that $|\mathcal{E}| = |\mathcal{D}| + |\mathcal{A}| = |\mathcal{D}| + \binom{n}{h}$.

Finally we want an $\mathcal{E}$ for any $\operatorname{slope}\theta = (n-h)/n \notin F(m)$.

There is an i with $(m-h+1)/m \geq \operatorname{slope}\mathcal{D}_i > \theta > \operatorname{slope}\mathcal{D}_{i+1} \geq (m-h)/m$.

Let $\underset{\sim}{c} = \gamma\mathcal{D}_i$, $\underset{\sim}{b} = \beta\mathcal{D}_{i+1}$, so $\underset{\sim}{c}' = \underset{\sim}{b}$. Since (3) applies to both $\mathcal{D}$ we have $\omega\underset{\sim}{c} = \omega\underset{\sim}{b} = h$. Some cases above sent $\underset{\sim}{c}$ to $0\underset{\sim}{c}$ and $\underset{\sim}{b}$ to $1\underset{\sim}{b}$. We take all vectors between $0\underset{\sim}{c}$ and $1\underset{\sim}{b}$ for $\mathcal{E}$. Trivially $\mathcal{E}$ is a Nice h–Sec with $\operatorname{slope}\theta$. If $0\underset{\sim}{d} \in \mathcal{E}$ then all vectors of $\mathcal{E}$ before $0\underset{\sim}{d}$ start 0. If we delete the 0 they start in $\mathcal{D}_{i+1}$. So $\delta(\underset{\sim}{d})$ is below the line of $\mathcal{D}_{i+1}$. Hence $\delta(0\underset{\sim}{d})$ is below the line of $\operatorname{slope}\mathcal{D}_{i+1}$ through $(\mu 0\underset{\sim}{c}, \delta(0\underset{\sim}{c}))$. Similarly if $1\underset{\sim}{d} \in \mathcal{E}$ then $\delta(1\underset{\sim}{d})$ is below the line of $\operatorname{slope}\mathcal{D}_i$ through $(\mu 1\underset{\sim}{b}, \delta(1\underset{\sim}{b}))$. It follows that δ is below the line of $\mathcal{E}$ in $\mathcal{E}$. Note that $|\mathcal{E}| = \binom{n}{h}$.

All vectors and slopes have been accounted for.

Remark. Let $\mathcal{E}$ be an Exsec for $CH(n+1)$. (A) If $\operatorname{slope}\mathcal{E} = (q-p)/p$ with p, q comprime then $|\mathcal{E}| = \sum\{1 \leq r \leq n/q\}\binom{rq}{rp}$. (B) If $\operatorname{slope}\mathcal{E} = 1/2$ then δ lies between the line $2y = x + f(n)$ of $\mathcal{E}$ and the line $2y = x$. Also $\gamma\mathcal{E}$ is the end of $\ldots 101010110$, so valleys [2] yield

$$f(n) = \binom{n}{\lfloor n/2 \rfloor} + \sum\{0 \leq r \leq (n-1)/2\}\binom{2r}{r}/(r+1).$$

Footnote 1. In [2] is not only Theorem 1, but also an evaluation of $\delta(k,n)$ using shifts of valleys/ cascades . The work was continued in [1]. where $\delta(k,n)$ is in (1.11) on page 13 as $\nabla\{G(n,k)\}$. The referee asked for a derivation of Theorem 2 from [1], but D.E.D. could not give one.

Footnote 2. The author D.E.D. had geometry lectures from Prof. E.H. Neville at Reading in 1953/54. This was the first university year for D.E.D., and the last year before retirement for E.H.N., who had written [7].

References

[1] R. Ahlswede and N. Cai, "Shadows and isoperimetry under the sequence subsequence relation", Combinatorica (1) 17, 1997, 11–29.

[2] T-N. Danh and D.E. Daykin, "Ordering integer vectors for co–ordinate deletions", *J. London Math. Soc.* (2) 55, 1997, 417–426.

[3] T-N. Danh and D.E. Daykin, "Sets of 0,1 vectors with minimal sets of subvectors", *Rostock Math. Kolloq.* 50, 1997, 47–52.

[4] D.E. Daykin, "To find all "suitable" orders of 0,1 vectors", *Congr. Numer* 113, 1996, 55–60.

[5] D.E. Daykin, "A cascade proof of a finite vectors theorem", *Southeast Asian Bull. Math.* 21, 1997, 167–172.

[6] D.E. Daykin, "On deleting co–ordinates from integer vectors", submitted.

[7] E.H. Neville, *The Farey Series of order 1025. Displaying solutions of the Diophantine equation* $bx - ay = 1$. (University Press: Cambridge, 1950).

THE EXTREME POINTS OF THE PROBABILISTIC CAPACITIES CONE PROBLEM

David E. Daykin

Mathematics Department, University of Reading, England RG6 2AX

THE PROBLEM

Let φ be the empty set and R be the reals. Let $N = \{1, 2, \ldots, n\}$ and S be the set of subsets of N. Let C be the set of maps $p : S \to R$ satisfying

$$0 \leq p(\varphi) \leq p(X) \leq p(N) \text{ for all } X \subseteq N, \tag{1}$$

$$0 \leq f(X,Y) \equiv p(X) + p(Y) - p(X \cap Y) - p(X \cup Y) \text{ for all } X, Y \subseteq N. \tag{2}$$

We call such a p a **cap**, and show below that (1), (2) imply $p(X) \leq p(Y)$ for $X \subseteq Y$. Let D be the set of all caps p with $0 = p(\varphi)$ and $p(N) = 1$. These are well known as **probabilistic capacities** .

Recall that $B \subseteq C$ is **convex** if $p, r \in B$ and $0 < \alpha < 1$ imply $\alpha p + (1-\alpha)r \in B$. Suppose B is convex, and let T be the set of all $t \in B$ for which there is no such $\alpha p + (1-\alpha)r = t$, with p, r distinct. Then T is the set of **extreme points** of B. Moreover each $p \in B$ is a finite sum $p = \Sigma \alpha_i t_i$ with $0 < \alpha_i < 1$ and $t_i \in T$. Clearly C, D are convex, and the open problem is to find the extreme points of D. We give a partial solution in Theorem 1 below.

THE CONE

If we restrict $p \in D$ to a subset of S which is closed under unions and intersections, then we get a cap. For this reason we study C. Let z, u have $z(X) = 0, u(X) = 1$ for all $X \in S$. So z, u are the zero, unit caps. Note $z, u \notin D$. If $0 \leq \alpha$ and $p \in C$ then $\alpha p \in C$, so C is a **cone**. We call C the **probabilistic capacities cone**. The **unit ray** is the set $\{\alpha u : 0 < \alpha\}$ of non-zero constant caps. Define a map π on non-constant caps p by $\pi p = \big(p - p(\varphi)u\big) / \big(p(N) - p(\varphi)\big)$. Clearly $\pi p \in D$, and the set of all p with the same πp form a **ray**. Any member of a ray represents the ray. Thus the extreme points of D represent ("are the same as") the extreme rays of C, except for z, u.

I. Althöfer et al. (eds.), Numbers, Information and Complexity, 105-107.

INTEGER CAPS

Let any non–zero cap p be given. We will construct a map p from S to the rationals Q. We need

$$q(X) = 0 \Leftarrow p(X) = 0, \tag{3}$$

and

$$g(X,Y) \equiv q(X) + q(Y) - q(X \cap Y) - q(X \cup Y) = 0 \Leftarrow f(X,Y) = 0. \tag{4}$$

The general solution to the simultaneous equations (3), (4) has the matrix form

$$\text{dependent variables } = A^{-1}B \text{ (independent variables)},$$

where A, B are over Q. For each independent variable $q(X)$ we give $q(X)$ a value in Q close $p(X)$. Then for all X, Y we have $q(X), g(X,Y)$ in Q and close to $p(X), f(X,Y)$ respectively. So $q(X) > 0, g(X,Y) > 0$ outside (3), (4) respectively, and q is a rational cap.

For $0 < \varepsilon \in R$ sufficiently small $p - \varepsilon q$ is a cap. By increasing ε we will get $p - \varepsilon q$ a cap with more zeros in (1) or (2) than we had for p. Repetition gives us a finite sum $p = \Sigma \varepsilon_i q_i$ with $0 < \varepsilon_i$ and q_i over Q. We conclude that the extreme rays of C are integer valued caps.

Let us define a partial order for integer caps by $p_1 \le p_2$ if firstly, for all X, Y we have both $p_1(X) \le p_2(X)$ and $f_1(X,Y) \le f_2(X,Y)$, and secondly, for all $X \subseteq Y$ we have $p_1(Y) - p_1(X) \le p_2(Y) - p_2(X)$. Clearly if $p_1 < p_2$ are in different rays, then $p_2 - p_1$ is an integer cap, and p_2 is not extreme. Next we apply Lemma 1 below, which is easy to prove by induction on m, and our Theorem 1 below is established.

Lemma 1. *Let V be a set of integer vectors $\mathbf{a} = (a_1, a_2, \ldots, a_m)$ with the $a_i \ge 0$. Suppose there are no distinct $\mathbf{a}, \mathbf{b} \in V$ with $a_i \le b_i$ for each i. Then V is finite*

Theorem 1. *The cone C of caps (resp. D) has only a finite number of extreme rays (resp. points), they may be represented by integer (resp. rational) caps. The integer caps are minimal in the above partial order.*

ELEMENTARY FACTS

Some notation will be helpful. For $i = 0, 1, 2, 3$ a set of 2^i subsets of N of the form $\{K : J \subseteq K \subseteq L\}$ with $|L \setminus J| = i$, we call a **dot, edge, face, cube** respectively. An edge of the form $\{K, K \cup k\}$ we call a **k–rung**.

Given any $X, Y \subseteq N$ put $I = X \cap Y$ and $U = X \cup Y$. Then one can plot on a plane, as a rectangular lattice, all the dots $\{W : I \subseteq W \subseteq U\}$. Have I at the bottom and U at the top. Any k–rung in the diagram has $k \in U \setminus I$. A face has two pairs of rungs.

Now suppose we have a cap p, and let us write numbers on the diagram. At each dot we write the value of p. On a rung $\{K, K \cup k\}$ we write $e \equiv p(K \cup k) - p(K)$. On a face $\{K, K \cup j, K \cup k, K \cup j \cup k\}$ we write $d = p(K \cup$

$j) + p(K \cup k) - p(K) - P(K \cup j \cup k)$ from (2). We call e, d the **edge, face functions**. For $k \in U \setminus I$, as we move down the diagram the value of k–rungs is ≥ 0 and increasing, so p decreases. The d values addup to $f(X, Y)$.

Writing 123 for $\{1, 2, 3\}$ and so on, for the cube $\{W : \varphi \subseteq W \subseteq 123\}$ we get

$$f(12, 13) - f(2, 3) = f(12, 23) - f(1, 3) = f(13, 23) - f(1, 2) = c \text{ say.5} \quad (1)$$

In the obvious notation (5) holds for any cube, and so we have defined the **cube function** c. By the **cube equations** we mean all of the equations of the form (5).

RESULTS

The cases $n = 1, 2, 3$ are not hard. When $n = 4$, it takes some effort on the cube equations to show that they have 20 extreme points, all with $0, 1$ edges. It seems that it would take a computer to complete the case $n = 4$.

EXAMPLES OF EXTREME CAPS

These are $z, u, w_i, r_i, s, p^{\#}, p^{\#\#}$. First we have z, u. Using u shows that all other examples have $p(\varphi) = 0$. Next for each $i \in N$ define w_i by $w_i(X)$ is 1 if $i \in X$ but is 0 otherwise. These show that all further examples have $e(\text{edge}) = 0$ for edges on N. For $1 \leq i < n$ define r_i by $r_i(X) = \min\{i, |X|\}$.

Let $N^{\#} = \{1, \ldots, n+1\}$ and $N^{\#\#} = \{1, \ldots, n+2\}$. Assume that p is an extreme cap on N. We can extend p to an extreme cap s on $N^{\#}$ by making $s(n+1\text{–rungs}) = 0$. Alternatively we can extend p to $p^{\#}$ by putting $p^{\#}(X) = p(N)$ if $n+1 \in X$. Repeating we get $p^{\#\#}$ on $N^{\#\#}$. Let us start this process with the example r_i above for p. Then $p^{\#\#}$ is an extreme cap whose set of dot values, set of edge values, and set of face values are all equal to $\{1, 2, \ldots, i\}$.

The author was unable to describe all $0, 1$ edge valued extreme caps. These alone appear to have interesting structure worthy of study.

CHANGING CAPS

Let p be any cap, and $0 < \varepsilon$. The cap $t(X) = \min\{\varepsilon, p(X)\}$ is the **ε–trim** of p. Next we define the **invert** v of p. Let μ be the maximum of the edge values of p. The v edge value of $\{X, Y\}$ is $\mu - \lambda$ where λ is the edge value of $\{N \setminus Y, N \setminus X\}$ from p, and $v(\varphi) = 0$. If we invert, trim, invert p we get the **flood** of p. **Direct sums** of caps are caps. There are self–inverse $0, 1$ edge extreme caps.

ACKNOWLEDGEMENT

It was on 15th November 1973 that J.D. Maitland–Wright told D.E.D. that Dominic Welsh had proposed the problem of finding the extreme points.

ON SHIFTS OF CASCADES

David E. Daykin *

Mathematics Department, University of Reading, England RG6 2AX

Abstract: For $k, n \geq 1$ the cascade

$$\mathcal{C}_k(n) = \binom{c_k}{k} + \binom{c_{k-1}}{k-1} + \ldots + \binom{c_t}{t}$$

has

$$c_k > c_{k-1} > \ldots > c_t \geq t \geq 1.$$

The $(i, j) = \Delta$ shift of binco $\binom{r}{s}$ is $\binom{r+i+j}{s+j}$. We show when $\Delta\mathcal{C}_k(n) + \Delta\mathcal{C}_k(p) \geq \Delta\mathcal{C}_k(n+p)$, and when $\leq$. We compare $\Delta\mathcal{C}_k(n)$ and $\Delta\mathcal{C}_{k+1}(n)$. If $n = \binom{x}{k}$ with x real, we show when $\Delta\mathcal{C}_k(n) \geq \Delta\binom{x}{k}$, and when $\leq$. This generalises results for $(1, -1)$ known from Kruskal–Katona and Lovász. For $(1, -1)$, if $\binom{x}{k} = \binom{y}{k+1}$ and $F(x) = \binom{y}{k}/\binom{x}{k-1}$ then F increases with x. Most results are best possible.

INTRODUCTION

We study shifts of cascades of *bincos* (binomial coefficients) over PT (Pascal's Triangle). Our bincos $\beta(r, s)$ cover the plane and

$$0 \leq \beta(r, s) = \binom{r}{s} \equiv \binom{r-1}{s} + \binom{r-1}{s-1} \text{ for } \beta \neq bomb = \binom{0}{0}. \tag{1}$$

Given h, the set of $\beta(r, s)$ with $s = h$ (resp. $r - s = h$) we call *col h* (resp. *row h*). If $g < h$ then col g is *right* of col h and row g is *above* row h.

By a *k–shade* we mean an integer n represented as

$$n = \mathcal{C} = \mathcal{C}_k(n) = \binom{c_k}{k} + \binom{c_{k-1}}{k-1} + \ldots + \binom{c_t}{t}, \tag{2}$$

satisfying

$$c_k > c_{k-1} > \ldots > c_t. \tag{3}$$

*Address for all correspondence: Sunnydene, Tuppenny Lane, Emsworth, Hants, England PO10 8HG.

I. Althöfer et al. (eds.), Numbers, Information and Complexity, 109-116.

This shade is a *cade* iff it lies in PT iff $c_t \geq t \geq 0$. We get a cade from a shade by deleting zero bincos. The cades are partitioned into *cascades* with $t \geq 1$ and *imcades* (improper cascades) [8] with $t = 0$, also we allow the empty shade $\varnothing$ to be both. If $k, n \geq 1$ it is well known that n has a unique k–cascade. Given $k, n \geq 1$ there may be no imcade, but if (2) is one it is unique to within c_0 where $c_1 > c_0 \geq 0$.

The (i,j) *shift* Δ of a binco β is defined by

$$\Delta\beta = (i,j)\beta = \binom{r+i+j}{s+j} \text{ where } \beta = \beta(r,s) = \binom{r}{s}. \tag{4}$$

If $\mathcal{A}$ is any sum of bincos then $\Delta\mathcal{A}$ is the sum of the shifts of the non–zero bincos of $\mathcal{A}$ with $\Delta 0 = \Delta\varnothing = \varnothing = 0$. Thus the shift of a cade is a shade. Also if (2) is an imcade then $\Delta\mathcal{C}$ increases ($\leq$) as c_0 moves down col 0, because every col, row is monotone.

Theorem 1. *Let $\Delta = (i,j)$ be a shift, and $\mathcal{C}, \mathcal{D}$ be k–cades for n, p respectively. Then $\Delta\mathcal{C} \geq \Delta\mathcal{D}$ if $n = p$ and $\mathcal{C}$ is a cascade, or if $n > p$. Also $\Delta\mathcal{C} = \Delta\mathcal{D}$ if $n = p$ and $k \geq 1$ and $i \geq 0 \geq j$.*

COMPARING SHIFTS OF CADES

Theorem 2. *Let $\Delta = (i,j)$ be a shift, and $\mathcal{C}, \mathcal{D}$ be k–cades for $n, p \geq 1$ respectively. Also let $\mathcal{E}$ be any k–cade for $n + p$. Then*

$$\Delta\mathcal{C} + \Delta\mathcal{D} \geq \Delta\mathcal{E} \quad \text{if } i \geq 0 \geq j, \tag{5}$$

$$\Delta\mathcal{C} + \Delta\mathcal{D} \leq \Delta\mathcal{E} \quad \text{if } i \leq 0 \leq j. \tag{6}$$

The case $(i,j) = (1,-1)$ of (5) is well known, and gives a proof of the Kruskal–Katona Theorem [1,2]. The Danh–Daykin Theorem [6,7,10] in Part 5, and its generalisation by Ahlswede–Cai [1], use the shift $(0,-1)$. We do not get (5) or (6) for any more shifts by Examples 1.

Let $rod(r,s,u)$ be the sum of the $u \geq 1$ bincos starting at $\beta(r,s)$ and moving right along the row. Iterating (1) gives

$$\beta(r,s) \equiv rod(r-1, s, s+1) \text{ if } r > s \geq 0, \tag{7}$$

$$\beta(r,s) \equiv rod(r-1,s,u) + \beta(r-u, s-u) \text{ if } u \geq 1 \text{ and } bomb \notin rod(r,s,u). \tag{8}$$

Observe that (7) remains $\equiv$ under shifts $(1,0)$ or $(0,-1)$ but becomes an inequality $\geq$ under shifts $(-1,0)$ or $(0,1)$. Clearly $(i,j) = (i,0)(0,j)$ and $(2,0) = (1,0)(1,0)$ and so on. So (i,j) keeps $\equiv$ in (7) if $i \geq 0 \geq j$, but gives $\geq$ otherwise.

Proof of Theorem 1. We assume $k \geq 1$ because $k = 0$ is trivial. Let $n = p$ and $\mathcal{D}$ be an imcade. First move the last binco of $\mathcal{D}$ as far as possible down col 0. If $i \geq 0 \geq j$ the increase in $\Delta\mathcal{D}$ is zero. Second take the right side of (7) from $\mathcal{D}$ and add the left to get a cascade. By the above remarks the theorem holds for $n = p$. Next let $\mathcal{D}$ be a cascade for $p \geq 1$. To get a cade for $p+1$ we add a binco in row 0 to $\mathcal{D}$ increasing $\Delta\mathcal{D}$.

Remark. Given $k, n \geq 1$, the above proof shows that there is a k–imcade for n iff the last binco in the k–cascade for n is below row 0.

Proof of Theorem 2. We use Daykin's algorithm for k–cades [8,9,10]. Each *Job* starts with two cades $\mathcal{A}, \mathcal{B}$ and produces two more $\mathcal{A}', \mathcal{B}'$ which replace $\mathcal{A}, \mathcal{B}$. Moreover $\mathcal{A} + \mathcal{B} = \mathcal{A}' + \mathcal{B}'$ and $\mathcal{A}' \leq \mathcal{B}'$. Initially $\mathcal{A} = \mathcal{C}$ and $\mathcal{B} = \mathcal{D}$.

Programme. Start, J1,J2,J1,J2 (after which $\mathcal{A}, \mathcal{B}$ will be cascades with $\mathcal{A} \leq \mathcal{B}$), then if $\mathcal{A}$ lies below row 0 do J3 and start again, else do J4, then if $\mathcal{A} = \varnothing$ stop, else start again.

Job J1. (Bigger and smaller cades.) In every col: – If both $\mathcal{A}$ and $\mathcal{B}$ have a binco give the bigger to $\mathcal{B}'$ and the smaller to $\mathcal{A}'$. If only one of $\mathcal{A}$ and $\mathcal{B}$ have a binco give it to $\mathcal{B}'$. **Note.** $\Delta\mathcal{A} + \Delta\mathcal{B} \equiv \Delta\mathcal{A}' + \Delta\mathcal{B}'$.

Job J2. Let $\mathcal{A}' = \mathcal{A}$ and make $\mathcal{B}$ into a cascade $\mathcal{B}'$. **Note.** Theorem 1 gives $\Delta\mathcal{B} = \Delta\mathcal{B}'$ for (5) but $\Delta\mathcal{B} \leq \Delta\mathcal{B}'$ for (6).

Job J3. Let $\mathcal{B}' = \mathcal{B}$ and use (7) to make $\mathcal{A}$ into an imcade $\mathcal{A}'$. **Note.** Theorem 1 gives $\Delta\mathcal{A} = \Delta\mathcal{A}'$ for (5) but $\Delta\mathcal{A} \geq \Delta\mathcal{A}'$ for (6).

Job J4. (Single binco transfer.) Here $\mathcal{A}' = \mathcal{A} - \theta$ and $\mathcal{B}' = \mathcal{B} + \eta$, where θ, η are bincos in row 0, and θ is the last one in $\mathcal{A}$, while η is the first one that can be added to $\mathcal{B}$. **Note.** $\theta = \eta = 1$ and θ is left of η. Also $\Delta\mathcal{A} + \Delta\mathcal{B} \geq \Delta\mathcal{A}' + \Delta\mathcal{B}'$, with equality for (6), because there $\Delta\theta$ is 0 or 1 and $\Delta\theta = \Delta\eta$.

Case $i \geq 0 \geq j$. Here (5) holds because every Job helps.

Case $i \leq 0 \leq j$. For (6) it is sufficient to prove our claim that the sequence J3,J1,J2,J1 produces zero change in $\Delta\mathcal{A} + \Delta\mathcal{B}$. So suppose we are about to do J3 with $\mathcal{A} = \alpha_k + \ldots + \alpha_s$ and $\mathcal{B} = \beta_k + \ldots + \beta_t$ as the cades. The programme will have just finished J1,J2,J1,J2 so $s \geq t \geq 1$ and $\alpha_q \leq \beta_q$ for $k \geq q \geq s$. The bincos α_q, β_q with $q > s$ play no part, so we may assume $k = s$ and $\mathcal{A} = \alpha_s = \beta(r, s)$ say. Now J3 uses (7) with α_s on the left to get $\alpha_s = \delta_s + \ldots + \delta_0 = \mathcal{A}'$ say. Let

$$\mathcal{U} = \{q : s \geq q \geq 1 \text{ and } \beta_q > \delta_q\} \text{ so } s \in \mathcal{U} \text{ and } u = |\mathcal{U}| \geq 1 \text{ and } s - u + 1 \geq t.$$

Case $s - u + 1 = t$. A routine check shows that the effect of J3,J1,J2 is to replace $\mathcal{A}$ by rod $(r-1, s, u)$ and to add $\beta(r-u, s-u)$ to $\mathcal{B}$. Then J1 makes no change. Thus J3,J1,J2,J1 deletes the left of (8) from $\mathcal{A} + \mathcal{B}$ and adds the right. Clearly this case of (8) is preserved under $(0,1)$ and $(-1,0)$, and hence for $i \leq 0 \leq j$. Our claim holds for this case.

Case $s - u \geq t$. This time, in addition to $\mathcal{U}$, we must consider the interval where $\beta_q = \delta_q$ and the one where $\beta_q < \delta_q$. Each interval could be $\varnothing$. This time the effect of J3,J1,J2,J1,J2 is the same as before with the addition that β_{s-u} goes from $\mathcal{B}$ to $\mathcal{A}'$. The rest of this case is the same as the last one. Our claim is proved.

The programme stops when $\mathcal{A} = \varnothing$, and its last Job was J4. Thus either $\mathcal{B}$ ends at $\eta = bomb$ with $\Delta\mathcal{B} \leq \Delta\mathcal{E}$, or there is no imcade for $n + p$ and $\mathcal{B} = \mathcal{E}$ uniquely.

Examples 1. Let $\Delta = (i,j)$ and $g > 0$. We show we can have

$$g + \Delta\mathcal{C} + \Delta\mathcal{D} \leq \quad \Delta\mathcal{E} \text{ if } i < 0 \text{ or } j > 0, \tag{9}$$

$$\Delta\mathcal{C} + \Delta\mathcal{D} \geq g + \Delta\mathcal{E} \text{ if } i > 0 \text{ or } j < 0, \tag{10}$$

for k–cades $\mathcal{C}, \mathcal{D}, \mathcal{E}$ with $\mathcal{C} + \mathcal{D} = \mathcal{E}$. In each example we only choose an h–cade $\mathcal{B}$ and a k–cade $\mathcal{C}$ with $k > h \geq 1$ and $\mathcal{B} = \mathcal{C}$. Then $\mathcal{D}$ is any k–cade such that $\mathcal{D}$ followed by $\mathcal{B}$ is a k–cade $\mathcal{E}$. Thus $\mathcal{E} = \mathcal{D} + \mathcal{B} = \mathcal{C} + \mathcal{D}$ and $\Delta\mathcal{E} = \Delta\mathcal{D} + \Delta\mathcal{B}$. So we only compare $\Delta\mathcal{B}$ with $\Delta\mathcal{C}$.

Case $i < 0$. Here $\mathcal{B}$ is a binco θ with $\Delta\theta \geq g$, while every binco of $\mathcal{C}$ is in row 0, so $\Delta\mathcal{C} = 0$.

Case $j < 0$. Here $\mathcal{C}$ is a binco θ with $\Delta\theta \geq g + 1$ and $\mathcal{B}$ is a binco in col 1 so $\Delta\mathcal{B}$ is 0 or 1.

Case $i > 0$. Every binco of $\mathcal{B}, \mathcal{C}$ is in row 0, so for $k - h$ large $\Delta\mathcal{C} \geq g + \Delta\mathcal{B}$.

Case $j > 0$. Let $\mathcal{C} = \beta(r,k)$ so $\Delta\mathcal{C}$ is a polynomial in r of degree $k + j$. Let $n = \mathcal{C}$ and $\mathcal{B} = \beta(n,1)$ so $\Delta\mathcal{C}$ is a polynomial in r of degree $k(j+1)$. So we can have $\Delta\mathcal{B} \geq g + \Delta\mathcal{C}$.

Bollobás–Leader proved [4] the case $(i,j) = (1,-1)$ of

Theorem 3. *Let $\Delta = (i,j)$ be a shift. Let $\mathcal{C}$ be a k–cade and $\mathcal{D}$ be a $(k+1)$–cade with $\mathcal{C} = \mathcal{D} = n \geq 1$. Then*

$$\Delta\mathcal{D} \geq \beta(k+i+j, k+1+j) + \Delta\mathcal{C} \quad \textit{if} \quad i \geq 0 \geq j. \tag{11}$$

Proof. For $k = 0$ or $k + 1 + j \leq 0$ the result is easy, so assume otherwise. Let $\theta = \beta(q+1, k+1)$ be the first binco of $\mathcal{D}$, and $\mathcal{E} = \mathcal{D} - \theta$. For $q + i + j \geq s \geq k + j$ put $f(s) = \beta(s, k+j)$, so like (7) we have $\Delta\theta = \Sigma f$. Moreover $f(k+i+j-1) + \ldots + f(k+j)$ is the β in (11). Next for $q \geq r \geq k$ put $e(r) = \beta(r,k)$ so $\Delta e(r) = f(r+i+j)$ and $\theta = \Sigma e$. We trivially get (11) for $n = 1$, with equality unless $\Delta\mathcal{D} = \textit{bomb}$. Now let $n \geq 2$. Observe firstly $\mathcal{E} + \Sigma e = \mathcal{E} + \theta = \mathcal{D} = n = \mathcal{C}$, secondly that $\mathcal{E}$ and each e is a k–cade, and thirdly there are at least two cades not $\varnothing$ among them. So

$$\Delta\mathcal{D} = \Delta\mathcal{E} + \Delta\theta = \Delta\mathcal{E} + \Sigma f = \beta + \Delta\mathcal{E} + \Sigma\Delta e \geq \beta + \Delta\mathcal{C},$$

by iterating Theorem 2.

Theorem 4. *Let $\Delta = (i,j)$ be a shift with $i \leq 0 \leq j$. Put*

$$\lambda = \beta(k-i+1, k+1), \;\; \mu = \beta(k-i, k+1), \;\; \nu = \beta(k-i, k),$$

so $\lambda = \mu + \nu$ is (1). Let $k \geq 1$ so $\lambda, \nu \geq 1$. Suppose $p > \lambda$ and put $n = p - \mu$ so $n > \nu$. Let $\mathcal{C}, \mathcal{D}$ be $k, (k+1)$ cades for n, p respectively. Then

$$\Delta\mathcal{C} \geq \Delta\mathcal{D} \quad \textit{if} \quad i \leq 0 \leq j. \tag{12}$$

Proof. Let $\theta, q, \mathcal{E}, f, e$ be as in the last proof. Under Δ bincos above row $-i$ go to 0. So $\Delta e(r) = 0$ iff $r \notin \mathcal{S} = \{\amalg, \amalg - \infty, \ldots, \| - \rangle\}$. Let Σ' denote summation over $\mathcal{S}$. Then

$$\Delta\mathcal{D} = \Delta\mathcal{E} + \Delta\theta = \Delta\mathcal{E} + \Sigma f = \Delta\mathcal{E} + \Sigma'\Delta e(r), \text{ and}$$

$$n = p - \mu = \mathcal{E} + \theta - \mu = \mathcal{E} + \Sigma' e(r).$$

Moreover in the last sum there appear at least two not $\varnothing$ cades because if $q = k - i$ then $\mathcal{E} \neq \varnothing$. So iterating Theorem 2 yields $\Delta\mathcal{E} + \Sigma'\Delta e(r) \leq \Delta\mathcal{C}$ as required. With $p = \lambda, n = \nu$ both $\Delta\mathcal{C}$ and $\Delta\mathcal{D}$ can be 0 or 1, so (12) does not hold, but it is best possible. If $p < \lambda$ then $\Delta\mathcal{D} = 0$. So (12) holds for all p if n or p get $\varnothing$ when ≤ 0 and $\mathcal{C}$ is a cascade when $n = \nu$.

APPROXIMATING SHIFTS OF CADES

We call $\gamma(x, k)$ a *genco* (generalised binco), where

$$\gamma(x,k) = \begin{cases} x(x-1)\ldots(x-k+1)/(k!) & \text{if } 1 \leq k \text{ and } k-1 \leq x, \\ 1 & \text{if } 0 = k \text{ and } -1 < x, \\ 0 & \text{otherwise.} \end{cases}$$

Thus γ increases with real x, and like (1) we have

$$0 \leq \gamma(x,k) \equiv \gamma(x-1,k) + \gamma(x-1,k-1) \text{ if } 1 \leq k \leq x \text{ or } 0 = k < x. \quad (13)$$

Now we extend a classical (1979) result of Lovász [1, p. 123].

Theorem 5. *Let $\Delta = (i, j)$ be a shift. Let $k, n \geq 1$ and $\mathcal{C}$ be a k-cade for n and let $\gamma(x, k) = n$. Then*

$$\Delta\mathcal{C} \geq \Delta\gamma \quad \textit{if} \quad i \geq 0 \geq j, \quad (14)$$

$$\Delta\mathcal{C} \leq \Delta\gamma \quad \textit{if} \quad i \leq 0 \leq j. \quad (15)$$

Proof. Case $(1, -1)$. Due to Lovász. The footnote [1]

[1]Let $\mathcal{S} = \{\infty, \in, \ldots, \backslash\}$ and $\mathcal{D}$ be a set of subsets of $\mathcal{S}$. Suppose $\mathcal{D}$ is a *down-set*, which means $A \subseteq B \in \mathcal{D}$ implies $A \in \mathcal{D}$. We give a new proof of **Lemma 1.** (Bollobás–Thomason) [5]. *If $0 \leq j \leq k \leq n$ and $J, K \subseteq \mathcal{S}$ with $j = |J|$, $k = |K|$ then*

$$(\textit{Probability } K \in \mathcal{D})^j \leq (\textit{Probability } J \in \mathcal{D})^k.$$

Proof. We may assume $j + 1 = k \geq 1$. For $p = k - 1, k$ let d_p be the number of $A \in \mathcal{D}$ with $|A| = p$. Take real $x \geq k - 1$ with

$$d_k = \binom{x}{k} \text{ then } \left(\frac{d_k}{\binom{n}{k}}\right)^{k-1} = \left(\frac{\binom{x}{k}}{\binom{n}{k}}\right)^{k-1} \leq \left(\frac{\binom{x}{k-1}}{\binom{n}{k-1}}\right)^{k} \leq \left(\frac{d_{k-1}}{\binom{n}{k-1}}\right)^{k},$$

as required, where the second $\leq$ uses Lovász result, and we get the first $\leq$ by direct expansion, because $x \leq n$.

shows its power.

Case $(1,0)$. Here (13) gives $(1,0)\gamma(x,k) \equiv \gamma(x,k) + (1,-1)\gamma(x,k)$. Summing (1) we get $(1,0)\mathcal{C} \equiv \mathcal{C} + (1,-1)\mathcal{C}$. The $(1,-1)$ case says $(1,-1)\mathcal{C} \geq (1,-1)\gamma(x,k)$ so $\Delta\mathcal{C} \geq \Delta\gamma$.

Case $(g,0)$ **with** $g \geq 2$. We do the $(2,0)$ case, but omit the induction because it is similar. Let $(1,0)\mathcal{C} = \gamma(y,k)$. The $(1,0)$ case says $\gamma(y,k) \geq \gamma(x+1,k)$ so $y \geq x+1$. It also gives $(2,0)\mathcal{C} = (1,0)(1,0)\mathcal{C} \geq \gamma(y+1,k)$. So $\Delta\mathcal{C} \geq \Delta\gamma$ since $\gamma(y+1,k) \geq \gamma(x+2,k)$.

Case $(0,-1)$. Let $\mathcal{A} = (-1,0)\mathcal{C}$ and $\mathcal{B} = (0,-1)\mathcal{C}$ so $\mathcal{A} \geq 0$, $\mathcal{B} \geq 1$ and $\mathcal{B} \geq (1,-1)\mathcal{A}$. We must show $\mathcal{B} \geq \gamma(x-1,k-1) = z$ say. If $\mathcal{A} = 0$ then $\mathcal{C} = 1$ and $x = k$ and $1 = \mathcal{B} = z$, so assume $\mathcal{A} \geq 1$. Note that $\mathcal{A}+\mathcal{B} = \mathcal{C} = \gamma(x-1,k)+z$. If $\mathcal{A} \leq \gamma(x-1,k)$ we are finished. If $\gamma(y,k) = \mathcal{A} > \gamma(x-1,k)$ then $y > x-1$ and, using the $(1,-1)$ case, $\mathcal{B} \geq (1,-1)\mathcal{A} \geq \gamma(y,k-1) > z$.

Case $(0,-g)$ **with** $g \geq 2$. Induction as in $(g,0)$ case.

Case $i \geq 0 \geq j$. Use $(i,j) = (i,0)(0,j)$.

Case $i \leq 0 \leq j$. Suppose $1 \leq (i,j)\mathcal{C} = \mathcal{B} = \gamma(y+i,k+j)$. Then $\gamma(x,k) = \mathcal{C} \geq (-i,-j)\mathcal{B} \geq \gamma(y,k)$ so $x \geq y$ and $\mathcal{B} \leq \gamma(x+i,k+j)$.

Notice that (14), (15) are sharp because we get equality each time $\mathcal{C}$ is a binco. Computer results suggest we cannot add more shifts to (14), (15). Bounding (14), (15) gives the approximations

$$\Delta\gamma(p+1,k) \geq \Delta\mathcal{C} \geq \Delta\gamma \geq \Delta\gamma(p,k) \quad \text{if } i \geq 0 \geq j, \tag{16}$$

$$\Delta\gamma(q,k) \leq \Delta\mathcal{C} \leq \Delta\gamma \leq \Delta\gamma(q+1,k) \quad \text{if } i \leq 0 \leq j, \tag{17}$$

where p, q are the obvious integers.

THE RATIO OF TWO SHIFTED BINCOS

Let $\Delta = (i,j)$ and $k \geq 2$ and $\mathcal{C}, \mathcal{D}$ be $k, k+1$–cades with $1 \leq \mathcal{C} = \mathcal{D}$. So study $\Delta\mathcal{D}/\Delta\mathcal{C}$ one can use (16), (17), but now we approximate it more closely by a function F. For $x \geq k-1$ the unique $y \geq k$ with $\gamma(x,k) = \gamma(y,k+1)$ has $x+1 < y$ for $k-1 < x < k$, but $y < x+1$ for $k < x$, by $\gamma(x+1,k+1) = (x+1)\gamma(x,k)/(k+1)$. With this y we put

$$F(x) = \Delta\gamma(y,k+1)/\Delta\gamma(x,k) = \binom{y+i+j}{k+1+j} \Big/ \binom{x+i+j}{k+j}.$$

Of course the denominator must not be zero. We have $F \to 0, 1, \infty$ as $x \to \infty$ according as j is $> 0, = 0, < 0$. We wonder if F is always monotone or unimodal, and so it is interesting that $F(k-1), F(k), F(2k+1)$ are $(k+i+j)/D$, $(k+1+i+j)/D$, $(k+1+i)/D$ with $D = k+1+j$, for $i \geq 1$, $j \geq -k$.

From now on $\Delta = (1,-1)$. We conjecture that $F(x_0) < F(x)$ for $x_0 < x$. The more precise statement, and our results, are in Theorem 6.

Theorem 6. *If* $k \geq 3$ *and* $\gamma(x_0,k) = \gamma(y_0,k+1) < \gamma(x,k) = \gamma(y,k+1)$ *then*

$$\binom{x}{k-1}\binom{y_0}{k} < \binom{x_0}{k-1}\binom{y}{k} \quad \text{for } k-1 \leq x_0 \leq 2k+1. \tag{18}$$

In particular $K\gamma(x, k-1) < k\gamma(y, k)$ *with* $K = k, k+1, k+2$ *when* $x_0 = k-1, k, 2k+1$ *respectively.*

Proof. Think of x_0 as fixed and x as a variable.

Case $k - 1 < x_0$. Here $k < y_0$. Using $\gamma(z,s) = (s+1)\gamma(z, s+1)/(z-s)$ four times, and then the given equations, changes (18) into

$$(x_0 - k + 1)(y - k) < (x - k + 1)(y_0 - k) \text{ for } k - 1 < x_0.$$

Let $y_1 = y_1(x)$ be that value of y which makes this an equality. We need $y < y_1$ and it is sufficient to show $\gamma(y, k+1) = \gamma(x, k) < \gamma(y_1, k+1)$. So, multiplying by $(k+1)!$, we want $0 < \pi(x) = \lambda(x) - \rho(x)$ for $x_0 < x$ where

$$\lambda(x) = y_1(y_1 - 1)\dots(y_1 - k) \text{ and } \rho(x) = (k+1)x(x-1)\dots(x-k+1).$$

As we would expect $\pi(x_0) = 0$, because $y_1 = y_0$ when $x = x_0$ and $\lambda(x_0) = (k+1)!\gamma(y_0, k+1) = (k+1)!\gamma(x_0, k) = \rho(x_0)$. Also $\pi(k-1) = 0$ because $y_1 - k = 0$ when $x = k-1$. Clearly $k-1$ is the biggest root of both λ and ρ. The roots of λ (resp. ρ) are equally spaced distance σ where $\sigma > 0$ is $(x_0 - k + 1)/(y_0 - k)$ (resp. 1). By the remark on $x + 1 \lessgtr y$ we have $\sigma < 1$ for $k - 1 < x_0 < k$, but $1 < \sigma$ for $k < x_0$. Also $\sigma = 1$ for $x_0 = k$.

Next we calculate $\pi'(k-1)$, with dash meaning d/dx. For $\rho'(k-1)$ we write out $\rho'(x)$ and look for the factor $(x - k + 1)$, to see $\rho'(k-1) = (k-1)!(k+1)$. Next we note that $(y_1)' = 1/\sigma$ for all x. We find λ' as we found ρ' to get $\pi'(k-1) = (k-1)!\{(k/\sigma) - (k+1)\}$. All we will use is that $\pi'(k-1) < 0$ if $k/(k+1) < \sigma$.

Case $k = x_0$. Here $y_0 = k+1$ and (18) is the triviality $y - k < x - k + 1$.

Case $k < x_0$ **and** $1 < \sigma < (k-1)/(k-2)$. The condition on σ ensures that there is exactly one root of λ between each of the roots $0, 1, \dots, k-2$ of ρ. Hence the sequence $\pi(k-2), \pi(k-3), \dots, \pi(0)$ goes $-ve, +ve, -ve, +ve, \dots$. This gives $k-2$ roots of π, and we already have $k-1, x_0$ as roots. The final root lies between $k-2$ and $k-1$ because $\pi'(k-1) < 0$. For this case we have shown $0 < \pi(x)$ for $x_0 < x$, and (18) is proved.

We have in fact shown that the only $x > k-1$ with $\lambda(x) = \rho(x)$ is $x = x_0$. Now ρ depends only on k. Given x_0 we get in turn y_0, λ, σ. We only use beautiful polynomials, so σ is a smooth function of x_0. We have $\sigma = 1$ when $x_0 = k$. As x_0 increases from k, the value of σ must change. It cannot revert to an earlier value, so σ increases from 1.

Case $x_0 = 2k+1$. Here $y_0 = x_0$ and $1 < \sigma = (k+2)/(k+1) < (k-1)/(k-2)$. So this is a special case of the last one, and (18) holds for $k \le x_0 \le 2k+1$, because σ increases with x_0.

Case $x_0 = k-1$. Doing this case in effect takes the limit $x_0 \to k-1$ of our earlier work. Here $y_0 = k$ and (18) is $\gamma(x, k-1) < \gamma(y, k)$. By our previous method this simplifies to $ky < (k+1)x + 1$, so $y_1 = \big((k+1)x + 1\big)/k$. With the same λ, ρ, π we want $0 < \pi(x)$ for $k - 1 < x$. The spacing σ is now $k/(k+1)$. We would expect it, and easily check, that this time $k-1$ is a double root of π.

The sequence $\pi(k-2), \pi(k-3), \ldots, \pi(0), \pi(-\infty)$ goes $+ve, -ve, +ve, -ve, \ldots$ so all roots of π are located and $k-1$ is the biggest root. Again $0 < \pi$ where required.

Case $k-1 < x_0 < k$ and $(k-1)/k < \sigma < 1$. Our earlier arguments carry over with the double root now become two roots $k-1, x_0$.

DELETION OF COORDINATES FROM 0,1 VECTORS

For $1 \leq N \leq 2^d$ there is a *valley* $\mathcal{V}$ representation

$$N = \mathcal{V}_k(N) = \binom{d}{d} + \binom{d}{d-1} + \ldots + \binom{d}{k+1} + \mathcal{C}_k(n) \text{ with } 0 \leq \mathcal{C}_k(n) < \binom{d}{k}. \tag{19}$$

Suppose $\mathcal{I}$ is a set of 0,1 vectors of dimension d. Let $\mathcal{W}$ be the set of dimension $d-1$ vectors obtainable by deleting a coordinate from a vector in $\mathcal{I}$. The Danh–Daykin theorem [6,7,10] says, if $N = |\mathcal{I}|$ then $|\mathcal{W}| \geq (0,-1)\mathcal{V}$. In $(0,-1)\mathcal{V}$ is $(0,-1)\mathcal{C}$, and it was trying to prove (14) for $(0,-1)\mathcal{C}$ which started this paper.

References

[1] R. Ahlswede and N. Cai, "Shadows and isoperimetry under the sequence–subsequence relation", *Combinatorica* 17 (1), 1997, 11–29.

[2] I. Anderson, *Combinatorics of finite sets*, Clarendon Press, Oxford, 1987.

[3] B. Bollobás, *Combinatorics*, Cambridge University Press, 1986.

[4] B. Bollobás and I. Leader, Lecture at Reading University, 26 January 1998.

[5] B. Bollobás and A. Thomason, "Threshold functions", *Combinatorica* 7 (1), 1986, 35–38.

[6] T.–N. Danh and D.E. Daykin, "Ordering integer vectors for coordinate deletions", *J. London Math. Soc.(2)* 55, 1997, 417–426.

[7] T.–N. Danh and D.E. Daykin, "Sets of 0,1 vectors with minimal sets of subvectors", *Rostock, Math. Kolloq.* 50, 1997, 47–52.

[8] D.E. Daykin, "An algorithm for cascades giving Katona–type inequalities", *Nanta Math.* 8, 1975, 78–83.

[9] D.E. Daykin, "Ordered ranked posets, representations of integers, and inequalities from extremal poset problems", Graphs and order, *Proc. Conf.*, Banff, Canada, Ed. I. Rival, 395–412, 1984.

[10] D.E. Daykin, "A cascade proof of a finite vectors theorem", *Southeast Asian Bull. Math.* 21, 1997, 167–172.

ERDŐS-KO-RADO THEOREMS OF HIGHER ORDER

Péter L. Erdős and László A. Székely

Abstract: We survey conjectured and proven Ahlswede-type higher-order generalizations of the Erdős-Ko-Rado theorem.

This paper is dedicated to the 60th birthday of Professor Rudolf Ahlswede.

INTRODUCTION

Rudolf Ahlswede's seminal work in extremal combinatorics includes:
• the Ahlswede-Daykin (or *Four Function*) inequality [4, 5] which provides for a common generalization of many correlation inequalities;
• the Ahlswede-Zhang identity, which unexpectedly turns the familiar LYM inequality into an identity [13];
• the complete solution (in joint work with L. Khachatrian [6, 7]) for maximizing the number of t-intersecting k-element sets—a problem dating back to the 30's [20];
• breakthrough results in Erdős type number theory (using the shifting technique in joint works [9, 10, 11] with L. Khachatrian) on problems like what is the maximum number of positive integers up to n such that no k of them are relatively primes, and related results.

The present survey paper focuses on *higher order extremal problems* in the sense of Ahlswede [3, 14]. The traditional questions about set systems sound like "how many sets can one have under certain restrictions" while the new higher order questions ask "how many *families of sets* can one have under certain restrictions". R. Ahlswede *et al.* have started this research, with strong motivation from information theory [3, 14]. They propose that any problem about set systems may give rise to four higher-order problems. For illustration, the classic Erdős-Ko-Rado theorem [20] sets an upper bound, on how many pairwise intersecting k-element subsets of an n-element set can one find. The four higher-order problems each ask how many *pairwise disjoint* families of k-element subsets of an n-element set can have such that for any two families:

(1) there exists an element of the first family which intersects all elements of the second family;

I. Althöfer et al. (eds.), Numbers, Information and Complexity, 117-124.

(2) there exists an element of the first family and an element of the second family that intersect;
(3) for all elements of the first family there exists an element of the second family, which intersects it;
(4) all elements of the first family intersect all elements of the second family.

One may not expect, of course, that all new problems generated in this way make sense and are interesting. But some of them yield elegant generalizations of known results. Ahlswede conjectured a bound $\binom{n-1}{k-1}$ for the problem (1), which would have given a higher-level generalization of the classic Erdős-Ko-Rado theorem. (For an intersecting family of k-sets $\{A_i : i \in I\}$ one makes the family of *singleton* families $\{\{A_i\} : i \in I\}$. If an upper bound holds for the second family, then it holds for the first family.) However, it was shown in [1] that although the conjecture holds for $k = 2, 3$, it is false for $k \geq 8$. The proof of the counterexample uses the *probabilistic method.* In this paper we restrict our interest to higher order generalizations of the Erdős-Ko-Rado theorem. The higher order generalizations of Sperner's theorem [8, 14, 15] will not be considered here.

In this paper we do not take narrowly the definition of Ahlswede-type higher-order extremal problems, since we rather do not insist on the pairwise disjointness of the families, but require that the sets in the same family have a certain additional structural property (make classes of a partition or be comparable for inclusion, etc.).

It is instructive to compare the concept of higher order generalization to other generalizing principles in combinatorics. Gian-Carlo Rota taught us to look for analogues of theorems valid on the power set lattice on the subspace lattice and the partition lattice. In the setting of Erdős-Ko-Rado theorems, Miklós Simonovits and Vera Sós initiated the study of "structured intersection theorems" [32, 33]: they look for the largest number of "structures" (graphs, arithmetic progressions, etc.) that pairwise intersect in a required type of "substructre".

If we understand higher order generalization in a broader sense, where we want to bound the number of families instead of the number of sets, it turns out that these three directions for generalization frequently overlap.

Excellent references on Erdős-Ko-Rado type theorems for set systems are [18, 26, 28].

INTERSECTING CHAINS IN POSETS

This section reviews results on intersecting chains in posets. A *k-chain* in a poset is a set of k distinct poset elements, such that any two elements are comparable in the poset. We say that two chains in a poset *intersect*, if they share at least one poset element. P. L. Erdős, Faigle, and Kern [22] pointed out that certain frequently studied problems well belong to this line. For example, let $M_1, M_2, \ldots, M_n$ be n pairwise disjoint sets of the same cardinality q. The associated *generalized Boolean algebra* (or *sequence space*) consists of the family

$$\mathcal{B}(n,q) = \{C \subseteq M_1 \cup \ldots \cup M_n : |C \cap M_i| \leq 1,\ i = 1, \ldots, n\}$$

ordered by inclusion. Observe that $\mathcal{B}(n,q)$ may be viewed as the collection of chains of an order $P = P(n,q)$ on $M_1 \cup \ldots \cup M_n$ with order relation

$$x < y \qquad if \qquad i < j$$

for all $x \in M_i, y \in M_j$. Frankl and Füredi [27] and Deza and Frankl [18] proved that, for $q \geq 2$ and $k = 1, \ldots, n$, there are at most $\binom{n-1}{k-1} q^{k-1}$ pairwise intersecting k-chains. Their method did not apply for the case $q = 1$. This is, however, the "classical" power set case, and therefore the original Erdős-Ko-Rado theorem also fits into this framework by solving the case $q = 1$.

It is worth pointing out, that these results can be strengthened to Bollobás type inequalities (see [22] or Engel [19]).

P. L. Erdős, Faigle, and Kern [22], among other results on intersecting chains, posed the problem of finding the largest number of pairwise intersecting k-chains in B_n^c, where B_n^c denotes the poset of sets $\{X \subseteq \{1, 2, ..., n\} : c \leq |X| \leq n - c\}$ for inclusion. Füredi solved this problem first, using the kernel method, for $c = 0, 1$ and $n > 6k \log k$ (personal communication). Ahlswede and Cai [2] solved the problem for $c = 0$. For an arbitrary value of c it was solved by Ákos Seress and the authors ([23, 24]). More precisely:

Definition. For $c \leq m \leq n - c$, let $\mathcal{T}_{n,k}^c(m)$ denote the set of those k-chains in B_n^c, which contain as element the initial segment $\{1, \ldots, m\}$. Clearly $|\mathcal{T}_{n,k}^c(m)|$ is also the cardinality of the set of those k-chains in $B_{n,k}^c$ which contain a specified (but otherwise arbitrary) subchain of length 1 with specified size m.

Theorem 1 ([24]) *Let $c \geq 1$ and let $\mathcal{F}$ be a family of intersecting k-chains in B_n^c. Then $|\mathcal{F}| \leq |\mathcal{T}_{n,k}^c(m)|$, and there is an injection $\phi : \mathcal{F} \to \mathcal{T}_{n,k}^c(c)$ such that every chain $\mathcal{L} = (L_1, L_2, \ldots, L_k) \in \mathcal{F}$ and its image $\phi(\mathcal{L}) = \mathcal{H} = (H_1, H_2, \ldots, H_k) \in \mathcal{T}_{n,k}^c(c)$ satisfy*

$$|L_k| \geq |H_k|.$$

The proof is based on a version of the shifting technique and uses mathematical induction. It is interesting to remark, that the same technique could apply for t-intersecting k-chains, if we had an easy base case for the induction, which we do not have. In lack of a good base case, the corresponding result in [24] uses the *kernel method*, and therefore does not give all n's for which the theorem holds. Finding the exact threshold for t-intersecting problem seems to be a very challenging problem.

The following problem fits the scheme of "structured intersection theorems" [32, 33] of Simonovits and Vera Sós: given a graph G, what is the maximum number of pairwise intersecting complete k-subgraphs? The maximum number of pairwise intersecting k-chains in a poset is exactly this problem if G is the comparability graph of the poset.

Whenever the poset elements are sets, the maximum number of pairwise intersecting k-chains in a poset fits the description of higher-order problems. Rota type analogues also came into play. Czabarka [16] obtained a q-analogue of the shifting proof of the theorem of Seress and the authors on intersecting

k-chains in B_n^c to intersecting k-chains for subspaces in an n-dimensional linear space over $GF(q)$, although for $c = 0, 1$ only. Note that the classic Erdős-Ko-Rado theorem also has a q-analogue, found by Hsieh [31].

Here we cite two other, general theorems of Seress and the authors [24] on intersecting chains in posets. These are the basis to prove result on t-intersecting chains in B_n^c. The first is an Erdős-Ko-Rado type result, the second is a Hilton-Milner type result.

Let us be given a fixed k and a sequence of posets P_n. For a given t-chain $\mathcal{L}$, let $\mathcal{T}_{n,k}(\mathcal{L})$ denote the set of k-chains in P_n which contain $\mathcal{L}$ as a subset. Define $T_{n,k}(\mathcal{L}) = |\mathcal{T}_{n,k}(\mathcal{L})|$. Also define $r_t(n) = \max T_{n,k}(\mathcal{L})$, where the maximum is taken for t-chains $\mathcal{L}$ in P_n.

Theorem 2. *For fixed $1 \le t < k$, and a sequence of posets P_n, let us be given a family $\mathcal{F}_n$ of t-intersecting k-chains in P_n. Assume that*

$$\lim_{n\to\infty} r_{t+1}(n)/r_t(n) = 0.$$

Then, for n sufficiently large, $|\mathcal{F}_n| \le r_t(n)$, and equality implies that the elements of $\mathcal{F}_n$ share a t-subchain.

For a t-chain $\mathcal{X} \subset P_n$ and $y \notin \mathcal{X}$, let $T(\mathcal{X}, y)$ denote the number of k-chains which contain X and y. For a t-chain $\mathcal{X}$ and a k-chain $\mathcal{L}$ in P_n, such that $|\mathcal{X} \cup \mathcal{L}| = k+1$, let $y_{\mathcal{L}}^* \in \mathcal{L} \setminus \mathcal{X}$ such that $T(\mathcal{X}, y_{\mathcal{L}}^*)$ minimize $T(\mathcal{X}, y_{\mathcal{L}})$ for the elements $y \in \mathcal{L} \setminus \mathcal{X}$, and set

$$\tau(\mathcal{X}, \mathcal{L}) = \sum_{y \in \mathcal{L} \setminus \mathcal{X},\ y \neq y_{\mathcal{L}}^*} T(\mathcal{X}, y).$$

Also define

$$M_\tau(n) = \max_{\mathcal{X}, \mathcal{L}} \tau(\mathcal{X}, \mathcal{L}),$$

and

$$M_\tau^*(n) = \max_{\substack{\mathcal{X}, \mathcal{L}: \\ \tau(\mathcal{X}, \mathcal{L}) = M_\tau(n)}} T(\mathcal{X}, y_{\mathcal{L}}^*).$$

Now the following Hilton-Milner type theorem [30] holds:

Theorem 3. *For fixed $1 \le t < k$, and a sequence of posets P_n, let us be given a maximum sized family $\mathcal{F}_n$ of non-trivially t-intersecting k-chains in P_n. Assume further that*

$$\lim_{n\to\infty} r_{t+2}(n)/M_\tau^*(n) = 0.$$

then, for n sufficiently large, $\mathcal{F}_n$ has one of the following two descriptions:

(i) there exists a t-chain $\mathcal{X}$ and a $(k+1-t)$-chain $\mathcal{Y}$, such that $\mathcal{X} \cap \mathcal{Y} = \emptyset$; and $\mathcal{F}_n$ is the following set of k-chains:

$$\{\mathcal{L}:\ \mathcal{X} \subseteq \mathcal{L} \text{ and } \mathcal{L} \cap \mathcal{Y} \neq \emptyset\} \cup \{\mathcal{L}:\ \mathcal{Y} \subseteq \mathcal{L} \text{ and } |\mathcal{L} \cap \mathcal{X}| = t-1\},$$

where the second set of chains is non-empty;

(ii) there exists a $(t+2)$-chain $\mathcal{Z}$, and $\mathcal{F}_n$ is the following set of k-chains:

$$\{\mathcal{L} : |\mathcal{L} \cap \mathcal{Z}| \geq t+1\},$$

and $|\bigcap_{\mathcal{L}\in\mathcal{F}_n} \mathcal{L} \cap \mathcal{Z}| \leq t-1$.

These theorems provide for a common generalization of the classic Erdős-Ko-Rado theorem and the theorem on intersecting chains in B_n^c. The proofs depend on the kernel method and may allow for generalization to other hereditary families than chains.

INTERSECTING PARTITIONS

This section poses some new problems on intersecting set partitions. A *partition* is a collection of disjoint non-empty sets whose union is the universe. We are going to consider different definitions for intersecting partitions. All of them are related to the type (2) higher-order problem. First, we say that two partitions of n elements *intersect in a class* if the two partitions share a class. It is natural to conjecture, that the largest number of k-partitions of an n-set that pairwise intersect in a class can be obtained by taking a fixed singleton and all $(k-1)$-partitions of the remaining $n-1$ elements.

Second, we say that that two partitions of an n-element set *intersect in a pair* if there exist respective classes C_1, C_2 of the two partitions such that $|C_1 \cap C_2| \geq 2$. This is the Rota type analogue of the intersection property to the partition lattice: two partitions intersect if their meet is above an atom. (We think about the partition lattice such that 0 is the finest partition and 1 is the coarsest partition.) This problem fits well the scheme of Simonovits and Vera Sós: consider those graphs on n vertices, which are vertex-disjoint unions of cliques. Give the largest number of those graphs which pairwise share at least one edge.

Conjecture 4. *If $n \leq 2k-1$, then the largest number of k-partitions of an n-set that pairwise intersect in a pair is $S(n-1,k)$. This bound can be attained by taking a fixed pair and all k-partitions of the n elements that have this pair in one class.*

Note that if $n = 2k$, then we can freely add to the above construction any partition which has a single class of size $k+1$ and $k-1$ singletons. Therefore, for $n = 2k$, the construction in the conjecture is no longer optimal.

Third, we say that that two partitions of n elements *intersect in a co-pair* if there exist a two-partition $\{C_1, C_2\}$ of $\{1, 2, ..., n\}$ such that both partitions refine $\{C_1, C_2\}$. This is also a Rota type analogue of the intersection property on the partition lattice: two partitions intersect if their join is under a co-atom.

Conjecture 5. *If $n \geq 2k-1$, then the largest number of k-partitions of an n-set that pairwise intersect in a co-pair is $S(n-1,k-1)$. This bound can be attained by taking a fixed singleton and all $(k-1)$-partitions of the remaining $n-1$ elements.*

Note that if $n = 2k-2$, then we can freely add to the above construction any partition which has a single class of size $k-1$ and $k-1$ singletons. Such k-partitions intersect in a co-pair every other k-partitions, otherwise the other partition would have a class whose size exceeds k, which is impossible. Therefore, for $n = 2k - 2$, the construction in the conjecture is no longer optimal. The threshold in this conjecture is somewhat bold, the conjecture might require a larger value of n.

Theorem 6. *For fixed $k > t \geq 1$ and $n > n_0(k)$, the largest number of k-partitions of an n-set that pairwise intersect in at least t classes is $S(n-t, k-t)$. This bound can be attained by taking t singletons fixed and all $(k-t)$-partitions of the remaining $n - t$ elements.*

For the proof of the theorem we review facts about sunflowers that we use in the kernel method. A set system $\{A_1, A_2, \dots, A_m\}$ is called a *sunflower* or *delta-system*, if $A_i \cap A_j = \bigcap_{l=1}^{m} A_l$ for all $1 \leq i < j \leq m$. The sets A_i are called the *petals* and $\bigcap_{l=1}^{m} A_l$ is called the *kernel* of the sunflower.

We say that a set system is of *rank* k, if $|H| \leq k$ for all $H \in \mathcal{H}$; and $\mathcal{H}$ is *t-intersecting*, if $|H_1 \cap H_2| \geq t$ for all $H_1, H_2 \in \mathcal{H}$. For $t \geq 1$, we say that $\mathcal{H}$ is *non-trivially* t-intersecting, if it is t-intersecting, and $|\bigcap \mathcal{H}| < t$. We say that $\mathcal{H}$ is *critically* t-intersecting, if it is t-intersecting, and deleting any $x \in H$ from any $H \in \mathcal{H}$, the resulting set system $\mathcal{H} \setminus \{H\} \cup \{H \setminus \{x\}\}$ is not t-intersecting.

Estimates in the kernel method are usually based on the following simple observation.

Lemma 7. *Let $\mathcal{H}$ be a critically t-intersecting system $(t \geq 1)$ of rank k. Then $\mathcal{H}$ does not contain a sunflower with $k + 1$ petals.*

Proof. Indeed, if $\{H_1, H_2, \dots, H_{k+1}\}$ is a sunflower i n $\mathcal{H}$, then any $H \in \mathcal{H}$ must intersect the kernel K of the sunflower in at least t elements, since a $\leq k$-element set cannot intersect each of the $k+1$ disjoint sets $H_1 \setminus K$, $H_2 \setminus K$,..., $H_{k+1} \setminus K$. Hence the deletion of $H_1 \setminus K$ from H_1 (if $H_1 \neq K$) results a t-intersecting set system, contradicting the minimality of $\mathcal{H}$. □

We will also need the Erdős-Rado theorem [21]:

Lemma 8. *For every i and l, there exists a number $f(i, l)$, such that any family of $f(i, l)$ sets of size i each, contains a sunflower with l petals.* □

Now we return to the proof of the theorem. Identify a partition $\mathcal{P}$ with the k-element set of its classes. Throw out classes of partitions until we obtain a critically intersecting family $\mathcal{H}$. Let $\mathcal{H}_i$ denote the set of i-element collections in $\mathcal{H}$. If $\mathcal{H}_t \neq \emptyset$, then we have t identical classes present in all partitions, and the theorem follows by the monotonicity of $S(n, k)$, the Stirling number of the second kind, in n.

If $\mathcal{H}_t = \emptyset$, then from the Lemmas we have $|\mathcal{H}_i| \leq f(i, k + 1)$. Any element of $\mathcal{H}_i$ can be extended in at most $S(n - i, k - i)$ ways toward a partition $\mathcal{P}$. Hence the total number of partitions in this case is at most

$$\sum_{i=t+1}^{k} f(i, k + 1) S(n - i, k - i).$$

Using the fact that for fixed k the asymptotic formula

$$S(n, k) \sim \frac{k^n}{k!}$$

holds ([17] p. 293), it follows that the number of partitions is $o(S(n-t, k-t))$.

Acknowledgment. The authors are indebted to Éva Czabarka for Conjecture 4. The research of the first author was supported in part by the Hungarian NSF contract T 016 358. The research of the second author was supported in part by the Hungarian NSF contract T 016 358, and by the NSF contract DMS 970 1211.

References

[1] R. Ahlswede, N. Alon, P. L. Erdős, M. Ruszinkó and L. A. Székely, "Intersecting systems", *Combinatorics, Probability, and Computing* 6, 1997, 127–137.

[2] R. Ahlswede and N. Cai, "Incomparability and intersection properties of Boolean interval lattices and chain posets", *Europ. J. Combinatorics*, 17, 1996, 667–687.

[3] R. Ahlswede, N. Cai, and Z. Zhang, "A new direction in extremal theory", *J. Combinatorics, Information & System Sciences* 19, 1994, 269–280.

[4] R. Ahlswede and D. E. Daykin, "An inequality for the weights of two families of sets, their unions and intersections", *Z. Wahrsch. Verw. Gebiete*, 43, 1978, 183–185.

[5] R. Ahlswede and D. E. Daykin, "Inequalities for a pair of maps $S \times S \to S$ with S a finite set", *Math. Z.* 165, 1979, 267–289.

[6] R. Ahlswede and L. H. Khachatrian, "The complete nontrivial-intersection theorem for systems of finite sets", *J. Combin. Theory Ser. A* 76, 1996, 121–138.

[7] R. Ahlswede and L. H. Khachatrian, "The complete intersection theorem for systems of finite sets", *European J. Combin.* 18, 1997, 125–136.

[8] R. Ahlswede and L. H. Khachatrian, "The maximal length of cloud-antichains", *Discrete Math.* 131, 1994, 9–15.

[9] R. Ahlswede and L. H. Khachatrian, "Maximal sets of numbers not containing $k+1$ pairwise coprime integers", *Acta Arith.* 72, 1995, 77–100.

[10] R. Ahlswede and L. H. Khachatrian, "Sets of integers and quasi-integers with pairwise common divisor", *Acta Arith.* 74, 1996, 141–153.

[11] R. Ahlswede and L. H. Khachatrian, "Sets of integers and quasi-integers with pairwise common divisor and a factor from a specified set of primes", *Acta Arith.* 75, 1996, 259–276.

[12] R. Ahlswede and L. H. Khachatrian, "Optimal pairs of incomparable clouds in multisets", *Graphs Combin.* 12, 1996, 97–137

[13] R. Ahlswede and Z. Zhang, "An identity in combinatorial extremal theory", *Adv. Math.* 80, 1990, 137–151.

[14] R. Ahlswede and Z. Zhang, "On cloud-antichains and related configurations", *Discrete Math.* 85, 1990, 225–245.

[15] N. Alon and B. Sudakov, "Disjoint systems", *Random Structures and Algorithms*, 6, 1995, 13–20.

[16] Éva Czabarka, "Structure of intersecting chains of subspaces in finite vector spaces", *Combinatorics, Probability, and Computing*, to appear.

[17] L. Comtet, *Advanced Combinatorics*, Reidel, Boston, Ma., 1974.

[18] M. Deza and P. Frankl, "Erdős-Ko-Rado theorem – 22 years later", *SIAM J. Alg. Disc. Methods* 4, 1983, 419–431.

[19] K. Engel, "An Erdős-Ko-Rado theorem for the subcubes of a cube", *Combinatorica* 4, 1984, 133–140.

[20] P. Erdős, C. Ko and R. Rado, "Intersection theorems for systems of finite sets", *Quart. J. Math. Oxford Ser. 2* 12, 1961, 313–318.

[21] P. Erdős, R. Rado, "A combinatorial theorem", *J. London. Math. Soc.* 25, 1950, 249–255.

[22] P. L. Erdős, U. Faigle and W. Kern, "A group-theoretic setting for some intersecting Sperner families", *Combinatorics, Probability and Computing* 1, 1992, 323–334.

[23] P. L. Erdős, Á. Seress and L. A. Székely, "On intersecting chains in Boolean algebras", *Combinatorics, Probability, and Computing* 3, (1994), 57–62. Reprinted in *Combinatorics, Geometry, and Probability. A tribute to Paul Erdős. Papers from the Conference in Honor of Erdős' 80th Birthday held at Trinity College, Cambridge, March 1993. Eds. B. Bollobás and A. Thomason, Partial reprinting of Combinatorics, Probability and Computing*, Cambridge University Press, Cambridge, 1997, 299–304.

[24] P. L. Erdős, Á. Seress and L. A. Székely, "Erdős-Ko-Rado and Hilton-Milner type theorems for intersecting chains in posets", submitted.

[25] P. Frankl, "On intersecting families of finite sets", *J. Combin. Theory*, Ser. A, 24, 1978, 146–161.

[26] P. Frankl, "The shifting technique in extremal set theory", *Combinatorial Surveys* (C. Whitehead, Ed.), Cambridge Univ. Press, London/New York, 1987, 81–110.

[27] P. Frankl and Z. Füredi, "The Erdős-Ko-Rado theorem for integer sequences", *SIAM J. Alg. Disc. Methods*, 1, 1980, 376–381.

[28] Z. Füredi, "Turán type problems", *London Math. Soc. Lecture Note Series*, Cambridge Univ. Press, 166, 1991, 253–300.

[29] A. Hajnal and B. Rothschild, "A generalization of the Erdős-Ko-Rado theorem on finite set systems", *J. Combin. Theory Ser. A* 15, 1973, 359–362.

[30] A. J. W. Hilton and C. Milner, "Some intersection theorems for systems of finite sets", *Quart. J. Math. Oxford*, 2, 18, 1967, 369–384.

[31] W. N. Hsieh, Systems of finite vector spaces, *Discrete Mathematics*, 2, 1975, 1–16.

[32] M. Simonovits and Vera T. Sós, "Intersection theorems on structures", *Ann. Discrete Math.*, 6, 1980, 301–313.

[33] M. Simonovits and Vera T. Sós, "Intersection properties of subsets of integers", *European J. Combin.*, 2, No. 4, 1981, 363–372.

ON THE PRAGUE DIMENSION OF KNESER GRAPHS

Zoltán Füredi

Department of Mathematics, University of Illinois, Urbana, IL 61801
Mathematical Institute of Hungarian Academy, POB 127, Budapest 1364, Hungary

z-furedi@math.uiuc.edu,furedi@math-inst.hu

Abstract: In this note we point out another connection between the Prague dimension of graphs and the dimension theory of partially ordered sets by giving a very short proof of a theorem of Poljak, Pultr and Rödl [10]. We show that the dimension of the Kneser graph is bounded as $\dim_{\mathrm{P}}(K(n,k)) < C_k \log\log n$, where C_k is depending only on k.

DIMENSION OF GRAPHS

The Kneser graph $K(n,k)$ is the graph whose vertices are the k-subsets of the n-element set $[n] := \{1, 2, ..., n\}$, with vertices being adjacent when the corresponding k-sets are disjoint.

The *product* of the graphs (V_1, E_1) and (V_2, E_2) is a graph with vertex set $V_1 \times V_2$; two vertices (v_1, v_2) and (w_1, w_2) are adjacent in the product graph if (v_1, w_1) is adjacent in G_1 and (v_2, w_2) is adjacent in G_2. In particular, v_i and w_i must be distinct. The *Prague dimension* (or product dimension) of the graph G, $\dim_{\mathrm{P}}(G)$, is the minimum number d such that G is an induced subgraph of the product of d complete graphs. In other words, it is the minimum d such that the vertices x of G can be represented by vectors $\mathbf{v}(x) = (v_1(x), \ldots, v_d(x))$ such that (x, y) forms an edge if and only if $v_i(x) \neq v_i(y)$ for all $1 \leq i \leq d$. Again, another form, it is the minimum number of good colorings of the vertices of G, $\varphi_1, \ldots, \varphi_d$, (not necessarily with minimum number of colors), such that for every non-edge (a, b) one has at least one i with $\varphi_i(a) = \varphi_i(b)$.

The Prague dimension was introduced and investigated in a series of papers by Nešetřil, Pultr [9], and other Czech mathematicians. Poljak, Pultr and Rödl [10] proved that

$$\log_2 \log_2 (n/(k-1)) \leq \dim_{\mathrm{P}}(K(n,k)) \leq C_k \lceil \log_2 \lceil \log_2 n \rceil \rceil \ , \qquad (1)$$

I. Althöfer et al. (eds.), Numbers, Information and Complexity, 125-128.

with $C_k \leq (k-1)k^2$. Later (for n sufficiently large) they [11] improved this to $C_k \leq (81/64)k^2/(\ln k)$. Very recently Körner [4] showed $C_k \leq (k/2) + o(1)$ (again for $n \to \infty$), which is conjectured to be tight in [7]. The case $n = 2k$ was discussed by Lovász, Nešetřil and Pultr [8], they proved that the dimension of the product of d (nontrivial) complete graphs is d. This implies $\dim_{\mathrm{P}}(K(2k,k)) = \lceil \log_2 \binom{2k}{k} \rceil = 2k - O(\log k)$.

The aim of this note is to point out another connection between the Prague dimension of graphs and the dimension theory of partially ordered sets by giving a very short proof of the upper bound in (1).

SCRAMBLING PERMUTATIONS AND DIMENSION OF POSETS

The dimension of a partially ordered set P is the minimum d such that P can be embedded into $\mathbf{R}^d$ in an order preserving way. In other words, it is the minimum number of linear extensions $\pi_1, \dots, \pi_d$ such that for all $x, y \in P$ there exists a π_i with $x <_i y$ (x precedes y in π_i) except, of course, if $y <_P x$. In the latter case y precedes x in all linear extensions. Additional background material on dimension theory can be found in the monograph [13].

Let 2^S denote the collection of subsets of S, and let $\mathbf{B}_n = (2^{[n]}, \subseteq)$ denote the *Boolean lattice*, the subsets of $[n]$ ordered by inclusion. For a set S, let $\binom{S}{k}$ denote the collection of k-element subsets of S. For $0 \leq s < t \leq n$ let $\mathbf{B}_n(s,t)$ denote the restriction of $\mathbf{B}_n$ to $\binom{[n]}{s} \cup \binom{[n]}{t}$. Finally, let $\dim(n;s,t)$ denote the (order) dimension of $\mathbf{B}_n(s,t)$. The function $\dim(n;s,t)$ was first studied by Dushnik [1] in 1950, he determined the exact value for $\dim(n;1,t)$ when $2\sqrt{n} - 2 \leq t < n-1$.

Call the set of permutations of $[n]$, Π, *t-scrambling* if for every (now unordered) t-subset $\{p_1, \dots, p_t\} \subset [n]$ and for every distinguished element of the set, say p_j, there is a permutation $\pi \in \Pi$ such that $\pi(p_j)$ precedes all the other $(t-1)$ $\pi(p_i)$'s. The cardinality of the smallest t-scrambling family is denoted by $N(n,t)$. It is easy to see that determination of $N(n,t)$ is equivalent to the question of the dimension of the partially ordered set formed by the $(t-1)$ and 1-element subsets of $[n]$ and ordered by inclusion, i.e., $N(n,t) = \dim(n;1,t-1)$. For t fixed and $n \to \infty$ an argument due to Hajnal and Spencer [12] gives that

$$\log_2 \log_2 n \leq N(n,t) \leq \frac{t}{\log_2(2^t/(2^t-1))} \log_2 \log_2 n. \tag{2}$$

In [3] the asymptotic $N(n,3) = \log_2 \log_2 n + (\frac{1}{2} + o(1)) \log_2 \log_2 \log_2 n$ was proved.

Theorem 1. $\dim_{\mathrm{P}}(K(n,k)) \leq N(n, 2k-1)$.

Proof: Let $\pi_1 \dots, \pi_d$ be a $(2k-1)$-scrambling set of permutations of $[n]$. We define $\varphi_1, \dots, \varphi_d$ good colorings of the Kneser graph $K(n,k)$, $\varphi_i : \binom{[n]}{k} \to [n]$, as follows. Let $\varphi_i(K) = x$ where $x \in K$ is the smallest element of K in the linear order π_i.

As $\varphi_i(K) \in K$, for disjoint k-sets, $K, L \in \binom{[n]}{k}$, we have that $\varphi_i(K) \neq \varphi_i(L)$ for all i. However, for a non-edge, i.e, for an intersecting pair (K, L), for $x \in K \cap L$, one can find a permutation π_i which puts x to the first place among the elements in $K \cup L$.

Remark 2.2. The constructions in [10, 11, 4] use qualitatively independent partitions and k-independent families of sets. Let us note that the upper bound in (2) also uses k-independent families of sets so it cannot give a better bound for C_k as 2^k. However, together with the upper bound from [3] for $N(n,3)$, it gives the asymptotics for the case $k = 2$, which was also shown in [10]. Finally, Theorem 1 also gives a number of new upper bounds for $\dim_P(K(n,k))$, when n is "not too large" with respect to k, e.g., $k \sim \log n$, where Kierstead's bound [5] gives $O(\log^3 n / \log\log n)$.

Remark 2.3. One can easily see, that, similarly to the examples in [10, 11, 4], our construction is *faithful*, i.e., $\varphi(K) \cap \varphi(L) = K \cap L$ holds for every two k-sets, where $\varphi(K) := \{\varphi_i(K) : 1 \leq i \leq d\}$.

Remark 2.4. (*Binary intersection representations.*) Körner and Monti [6] defined the *Bohemian representation* of the Kneser graph $K(n,k)$ as a set of colorings of its vertex set, $\varphi_1, \ldots, \varphi_t$, where now $\varphi_i : \binom{[n]}{k} \to N$ is not necessarily a good coloring of the graph, and a function $\varphi : 2^{[t]} \to 2^{[n]}$ with the following property. For a pair of distinct sets $A, B \in \binom{[n]}{k}$ let $\delta(A,B)$ denote a sequence from $\{0,1\}^t$ with $\delta_i = 1$ for $\varphi_i(A) = \varphi_i(B)$ and 0 otherwise. In a Bohemian representation $(\varphi_1, \ldots, \varphi_t, \varphi)$ we want to be able to read out the intersection structure of the complete hypergraph knowing only the binary vectors, $\delta(A,B)$, i.e., we have $\varphi(\delta(A,B)) = A \cap B$. The minimum of such t is called the *Bohemian dimension*, and is denoted by $T(n,k)$. Körner and Monti [6] proved that

$$k - 1 \leq \liminf_{n\to\infty} \frac{T(n,k)}{\log_2 n} \leq \limsup_{n\to\infty} \frac{T(n,k)}{\log_2 n} \leq k(k-1).$$

Using a different kind of set of scrambling permutations, one can see that $T(n,k) = O(\log n)$ as k is fixed and $n \to \infty$ as follows. Call a family of permutations $\pi_1, \ldots, \pi_t$ of $[n]$ *completely k-scrambling* if for every ordered k-subset $\{p_1, \ldots, p_k\}$ of k distinct elements of $[n]$ there is a permutation π_i with $\pi_i(p_1) < \ldots < \pi_i(p_k)$. This means that all k-subsets appear in all $k!$ possible orderings. The cardinality of the smallest completely k-scrambling family is denoted by $N^\star(n,k)$. It is known (for $k \geq 3$) that $\frac{1}{2}(k-1)!\log_2 n < N^\star(n,k) \leq (1+o(1))\frac{k}{\log_2(k!/(k!-1))}\log_2 n$. Here the lower bound is from [2] and the upper bound is due to Spencer [12].

Now, one can easily see, that a completely $(4k-2)$-scrambling set of permutations in the same way as in Theorem 2.1 provides a Bohemian representation of $K(n,k)$ thus proving $T(n,k) \leq N^\star(n, 4k-2)$. Even more, again, the obtained φ_i's are proper colorings of the Kneser graph.

Further problems and connections between permutations and order dimensions can be found in [2].

ACKNOWLEDGEMENTS

This research was supported in part by the Hungarian National Science Foundation grant OTKA 016389, and by a National Security Agency grant MDA904-98-I-0022.

References

[1] B. Dushnik, "Concerning a certain set of arrangements", *Proc. Amer. Math. Soc.* 1, 1950, 788–796.

[2] Z. Füredi, "Scrambling permutations and entropy of hypergraphs", *Random Structures and Algorithms* 8, 1996, 97–104.

[3] Z. Füredi, P. Hajnal, V. Rödl, and W. T. Trotter, "Interval orders and shift graphs", *Sets, Graphs and Numbers*, A. Hajnal and V. T. Sós, Eds., *Proc. Colloq. Math. Soc. János Bolyai* 60, 297–313, (Budapest, Hungary, 1991), North-Holland, Amsterdam 1992, 297–313.

[4] L. Gargano, J. Körner and U. Vaccaro, "Capacity and dimension", Lecture by J. Körner, *Symposium Numbers, Information and Complexity in honor of R. Ahlswede*, Bielefeld, Germany, October 1998.

[5] H. A. Kierstead, "On the order dimension of 1-sets versus k-sets", *J. Combin. Theory Ser. A* 73, 1996, 219–228.

[6] J. Körner and A. Monti, "Compact representations of the intersection structure of families of finite sets", manuscript, November 1998.

[7] J. Körner and A. Orlitzky, "Zero-error information theory", *IEEE Trans. Information Theory*, 50'th anniversary volume, to appear.

[8] L. Lovász, J. Nešetřil and A. Pultr, "On the product dimension of graphs", *J. Combin. Theory Ser. B* 29, 1980, 47–67.

[9] J. Nešetřil and A. Pultr, "A Dushnik-Miller type dimension of graphs and its complexity", *Fundamentals of Computation Theory*, Proc. Conf. Poznań–Kórnik, 1977, *Springer Lect. Notes in Comp. Sci.* 56, 1977, 482–493.

[10] S. Poljak, A. Pultr and V. Rödl, "On the dimension of Kneser graphs", *Algebraic Methods in Graph Theory*, Proc. Colloq. in Szeged, Hungary, 1978, L. Lovász and V. T. Sós, Eds., *Proc. Colloq. Math. Soc. J. Bolyai* 25, 1981, 631–646.

[11] S. Poljak, A. Pultr and V. Rödl, "On qualitatively independent partitions and related problems", *Discrete Applied Math.* 6, 1983, 193–205.

[12] J. Spencer, "Minimal scrambling sets of simple orders", *Acta. Math. Hungar.* 22, 1972, 349–353.

[13] W.T. Trotter, *Combinatorics and Partially Ordered Sets: Dimension Theory*, John Hopkins University Press, Baltimore, Maryland, 1991. Also: "Progress and new directions in dimension theory for finite partially ordered sets", *Extremal Problems for Finite Sets*, Proc. Colloq., Visegrád, Hungary, 1991, P. Frankl et al., Eds., *Bolyai Soc. Math. Studies* 3, 1994, 457–477.

THE CYCLE METHOD AND ITS LIMITS

Gyula O.H. Katona*

Alfréd Rényi Institute of Mathematics, Hungarian Academy of Sciences, Budapest, P.O.B. 127, H-1364, HUNGARY

ohkatona@renyi-inst.hu

Abstract: A powerful tool of extremal set theory, the cycle method is surveyed in the paper. It works, however only when the non-emptyness of the pairwise intersections of the members of the family is assumed. If these intersections have to be at least 2, the method fails: the celebrated Complete Intersection Theorem by Ahlswede and Khachatrian cannot be proved by this method. We show the reasons and some attempts to overcome the difficulties.

THE BEGINNING

Let $X = \{1, 2, \ldots, n\}$ be a finite set of n elements, we will consider families $\mathcal{F}$ of its subsets: $\mathcal{F} \subset 2^X$. The family of all k-element subsets of X will be denoted by $\binom{X}{k}$. A family $\mathcal{F}$ of distinct subsets is called *intersecting* if $F, G \in \mathcal{F}$ implies $F \cap G \neq \emptyset$. One of the fundamental theorems of the theory of *extremal families* is the Erdös-Ko-Rado theorem ([8]). It answers the question, what is the maximum size of an intersecting family of subsets of an n-element set. If $k > \frac{n}{2}$ then the question is uninteresting, one can choose all k-element subsets, this family will be intersecting. This is not true when $k \leq \frac{n}{2}$. In this case one can choose all k-element subsets containing the element $1 \in X$. The theorem states that this is the best we can do .

Theorem 1 (Erdős-Ko-Rado) *Let $|X| = n, k \leq \frac{n}{2}$, and suppose that $\mathcal{F} \subset \binom{X}{k}$ is an intersecting family. Then*

$$|\mathcal{F}| \leq \binom{n-1}{k-1}. \tag{1}$$

The cycle method will be illustrated by the proof of this theorem.

*The work was supported by the Hungarian National Foundation for Scientific Research grant number T029255

I. Althöfer et al. (eds.), Numbers, Information and Complexity, 129-141.

Theorem 1 (Erdős-Ko-Rado) *Let* $|X| = n, k \leq \frac{n}{2}$, *and suppose that* $\mathcal{F} \subset \binom{X}{k}$ *is an intersecting family. Then*

$$|\mathcal{F}| \leq \binom{n-1}{k-1}. \tag{1}$$

The cycle method will be illustrated by the proof of this theorem.

Proof (**[20]**) Place the elements of X listed along a cycle and consider the *intervals* along this cycle, that is, the sets of form $\{i, i+1, \ldots, i+l\}$ where these numbers are taken mod n. Solve the question of Erdős-Ko-Rado for intervals of length k, first. The number of intervals of length k containing the element 1 is obviously k and this family of intervals is intersecting. We will see that this is the best.

Lemma 2. *If* $A_1, A_2, \ldots, A_s$ *is a family of intersecting* k*-element intervals in* X *then*

$$s \leq k. \tag{2}$$

Proof of the lemma Suppose that one of the A's, say, $A_1 = \{1, 2, \ldots, k\}$. The intersection property implies that every other A has either its first or last element in A_1. However, i cannot be the last element of an A when $i+1$ is the first element of another A, since $2k < n$, the two intervals cannot meet at the "other end". Therefore there is at most one further A for each pair $i, i+1$ $(1 \leq i < k)$. The total number of As is at most $1 + k - 1 = k$, proving the lemma. □

The rest of the proof is based on *double counting.* Let $\mathcal{F}$ be the family in the theorem. Count the number of pairs $(\mathcal{C}, F)$ where $\mathcal{C}$ is a cyclic permutation of X, $F \in \mathcal{F}$ is an interval in the permuted X. First fix F. The number of permutations of X where F is an interval is $k!(n-k)!$ since the elements of F and the other elements can be permuted independently. Therefore the number of pairs is $|\mathcal{F}|k!(n-k)!$.

Now fix the permutation $\mathcal{C}$. The lemma can be applied for any permuted version of X therefore, by (2), there are at most k members $F \in \mathcal{F}$ which are intervals in this permutation. Since the number of cyclic permutations is $(n-1)!$, the number of pairs is at most $(n-1)!k$. Comparing the two countings:

$$|\mathcal{F}|k!(n-k)! \leq (n-1)!k.$$

This is equivalent to (1). □

Observe that the "miracle" works because we found a subfamily (intervals) of $\binom{X}{k}$ in which the intersecting property ensures proportionally the same bound as in the original "big" case. Namely, as the lemma states we can have at most k sets out of the n intervals. The proportion is $\frac{k}{n}$. This proportional bound is sufficient for the original problem, since

$$\frac{\binom{n-1}{k-1}}{\binom{n}{k}} = \frac{k}{n}.$$

UNICITY IN THE SPERNER THEOREM

The very first theorem of the theory of extremal families was the theorem of Sperner ([28]). A family $\mathcal{F}$ of distinct subsets is called *inclusion-free* if $F, G \in \mathcal{F}$ implies $F \not\subset G$. It is obvious that the family of all k-element subsets is inclusion-free. The largest one of the numbers $\binom{n}{k}$ is $\binom{n}{\lfloor n/2 \rfloor}$, therefore we have an inclusion-free family of this size. Sperner's theorem states that this is the best.

Theorem 3 (Sperner) *Let $\mathcal{F} \subset 2^X$ be an inclusion-free family, then*

$$|\mathcal{F}| \leq \binom{n}{\lfloor \frac{n}{2} \rfloor} \tag{3}$$

with equality only when

$$\mathcal{F} = \binom{X}{\lfloor \frac{n}{2} \rfloor} \quad or \quad \binom{X}{\lceil \frac{n}{2} \rceil}. \tag{4}$$

The simplest proof of (3) is due to Lubell [24]. His proof is somewhat simpler than the cycle method. The application of this latter method, however, gives an easy proof for the second part of the theorem, too.

Proof (Füredi [14]) The following lemma solves the analogous question for the cycle.

Lemma 4. *If $A_1, A_2, \ldots, A_s$ is a family of inclusion-free intervals in X then*

$$s \leq n \tag{5}$$

with equality only when the family consists of all possible intervals of a fixed length.

Proof of the lemma Since the family of intervals is inclusion-free, at most one of them can start with i $(1 \leq i \leq n)$. This proves (5). In the case of equality $s = n$, suppose that the interval starting with i is denoted by A_i. It is easy to see that $|A_i| \leq |A_{i+1}|$ holds, otherwise $A_{i+1} \subset A_i$ contradicts our assumption. Finally $|A_1| \leq |A_2| \leq \cdots \leq |A_n| \leq |A_1|$ proves the statement. □

Count the number of pairs $(\mathcal{C}, F)$ where $\mathcal{C}$ is a cyclic permutation of X, $F \in \mathcal{F}$ is an interval in $\mathcal{C}$. For any fixed F the number of cyclic permutations in which F is an interval is $|F|!(n - |F|)!$, therefore the number of pairs is $|\mathcal{F}||F|!(n - |F|)!$. On the other hand, for any fixed $\mathcal{C}$ there are at most n intervals from $\mathcal{F}$. The number of pairs is at most $(n-1)!n$. Hence we obtained the inequality

$$|\mathcal{F}||F|!(n - |F|)! \leq n! \tag{6}$$

which is equivalent to (3).

Suppose that there is an equality in (6). Then there are exactly n intervals from $\mathcal{F}$ along each cycle. Using the second part of the lemma all intervals along a given cycle must have the same length. Let $F, G \in \mathcal{F}$. It is easy to see that there is a cycle in which both F and G are intervals. (It can be formed from the intervals $F-G, F\cap G, G-F, X-F-G$.) This proves $|F| = |G|$ for any two members. Hence

$$\mathcal{F} = \binom{X}{k}$$

for some k. The latter expression is maximum only when $k = \lfloor \frac{n}{2} \rfloor$ or $k = \lceil \frac{n}{2} \rceil$. □

DOUBLE COUNTING WITH WEIGHT

Combine the above conditions and find the largest intersecting, inclusion-free family. It is easy to see that

$$\binom{X}{\lceil \frac{n+1}{2} \rceil}$$

satisfies these conditions. The following theorem states that this is the best one.

Theorem 5 (Milner [25]) *Let $\mathcal{F} \subset 2^X$ be an intersecting, inclusion-free family, then*

$$|\mathcal{F}| \leq \binom{n}{\lceil \frac{n+1}{2} \rceil}. \tag{7}$$

Proof ([22]) We will use double counting with a weight function. This is why the lemma does not simply upperbound the number of intervals in question.

Lemma 6. *If $A_1, A_2, \ldots, A_s$ is a family of intersecting, inclusion-free intervals in X then*

$$\sum_{i=1}^{s} \binom{n}{|A_i|} \leq n \binom{n}{\lceil \frac{n+1}{2} \rceil}. \tag{8}$$

Without **proof**, see [22]. □

The number of pairs $(\mathcal{C}, F)$ where $\mathcal{C}$ is a cyclic permutation of X, $F \in \mathcal{F}$ is an interval in $\mathcal{C}$ will be counted with the weight $\binom{n}{|F|}$, that is, the sum

$$\sum_{(\mathcal{C},F)} \binom{n}{|F|} \tag{9}$$

will be considered. On one hand it is equal to

$$\sum_{F\in\mathcal{F}} \sum_{\{\mathcal{C}:F \text{ is an interval in } \mathcal{C}\}} \binom{n}{|F|} = \sum_{F\in\mathcal{F}} |F|!(n-|F|)!\binom{n}{|F|} = |\mathcal{F}|n!. \tag{10}$$

On the other hand, (9) can also be written in the form

$$\sum_{\mathcal{C}} \sum_{\{F\in\mathcal{F}:\text{ is an interval in }\mathcal{C}\}} \binom{n}{|F|}$$

that is, by the lemma

$$\sum_{\mathcal{C}} n\binom{n}{\lceil\frac{n+1}{2}\rceil} = n!\binom{n}{\lceil\frac{n+1}{2}\rceil} \tag{11}$$

is an upper bound on (9). Comparing (10) and (11) the statement of the theorem is obtained. □

INEQUALITIES FOR INTERSECTING, INCLUSION-FREE FAMILIES

One can prove more complicated inequalities rather than just an upper bound on the number of members of $\mathcal{F}$.

Theorem 7 (Bollobás [2]) *If $\mathcal{F}$ is an intersecting, inclusion-free family of subsets of X then*

$$\sum_{\substack{F\in\mathcal{F}\\ |F|\le n/2}} \frac{1}{\binom{n-1}{|F|-1}} \le 1. \tag{12}$$

Proof Again, the analogous inequality for intervals is needed for the proof of the theorem.

Lemma 8. *If $\mathcal{A}$ is a family of intersecting, inclusion-free intervals in X then*

$$\sum_{\substack{A\in\mathcal{A}\\ |A|\le n/2}} \frac{1}{|A|} \tag{13}$$

holds.

Without **proof**, see [2] . □

The obvious weight function will be used in the double counting, the sum

$$\sum_{(\mathcal{C},F)} \frac{1}{|F|} \tag{14}$$

will be considered. On one hand, it is equal to

$$\sum_{\substack{F\in\mathcal{F}\\ |F|\le n/2}} \sum_{\{\mathcal{C}:F\text{ is an interval in }\mathcal{C}\}} \frac{1}{|F|} = \sum_{\substack{F\in\mathcal{F}\\ |F|\le n/2}} |F|!(n-|F|)!\frac{1}{|F|}. \tag{15}$$

On the other hand, by the lemma we have

$$\sum_{C} 1 = (n-1)! \tag{16}$$

as an upper bound for (14). The comparison of (15) and (16) proves (14). □

The above theorem does not say anything about the large members of the family. The following theorem tries to improve this situation.

Theorem 9 (**[18]**) *If $\mathcal{F}$ is an intersecting, inclusion-free family of subsets of X then*

$$\sum_{\substack{F \in \mathcal{F} \\ |F| \le n/2}} \frac{1}{\binom{n}{|F|-1}} + \sum_{\substack{F \in \mathcal{F} \\ |F| > n/2}} \frac{1}{\binom{n}{|F|}} \le 1. \tag{17}$$

Proof Here the small and large members need different kinds of weights.

Lemma 10. *If $\mathcal{A}$ is a family of intersecting, inclusion-free intervals in X then*

$$\sum_{A \in \mathcal{A}, |A| \le n/2} \frac{n-|A|+1}{|A|} + \sum_{A \in \mathcal{A}, |A| \le n/2} \frac{1}{n} \tag{18}$$

holds.

Without **proof**, see [18] . □

The rest of the proof is the same as in the case of Bollobás's theorem. □

CONVEX HULLS

Introduce the notation $p_i(\mathcal{F}) = |\{F \in \mathcal{F} : |F| = i\}|$ $(1 \le i \le n)$. Furthermore, the vector $p(\mathcal{F}) = (p_0, p_1, \ldots, p_n) \in R^{n+1}$ is called the *profile vector* of $\mathcal{F}$. Then, e.g. the Bollobás inequality (12) can be written in the form

$$\sum_{i=1}^{\lfloor \frac{n}{2} \rfloor} \frac{f_i}{\binom{n-1}{i-1}} \le 1.$$

Observe that this is a linear inequality which has to be satisfied for the profile vector of an intersecting inclusion-free family. The coefficients are

$$c_3(n,i) = \begin{cases} \frac{1}{\binom{n-1}{i-1}} & \text{if} \quad 1 \le i \le \frac{n}{2}, \\ 0 & \text{if} \quad \frac{n}{2} < i. \end{cases}$$

Our other statements can also be written in a form of a linear inequality for the profile vector:

$$\sum_{i=1}^{n} f_i c(n,i) \le 1. \tag{19}$$

Supposing $k \le \lfloor \frac{n}{2} \rfloor$ and choosing

$$c_1(n,i) = \begin{cases} \frac{1}{\binom{n-1}{k-1}} & \text{if} \quad i = k, \\ 0 & \text{if} \quad i \ne k. \end{cases}$$

(19) becomes the Erdős-Ko-Rado theorem.

$$c_2(n,i) = \frac{1}{\binom{n}{\lceil \frac{n+1}{2} \rceil}}$$

makes the Milner theorem from (19). Finally, if

$$c_4(n,i) = \begin{cases} \frac{1}{\binom{n}{i-1}} & \text{if} \quad 1 \le i \le \frac{n}{2}, \\ \frac{1}{\binom{n}{i}} & \text{if} \quad \frac{n}{2} < i. \end{cases}$$

then Theorem 9 is obtained from (19). One can determine all linear inequalities of type (19) which are satisfied for the profile vectors of intersecting, inclusion-free families (see [23]). These inequalities (hyperplanes) determine the *convex hull* of the profile vectors of intersecting, inclusion-free families. This convex hull can be easier described by its *extreme points* (= vertices).

Theorem 11 (([6])) *The extreme points of the convex hull of the profile vectors of intersecting, inclusion-free families on an n-element set are*

$$(0, \dots, 0),$$

$$\left(0, \dots, \binom{n-1}{i-1}, \dots, 0\right) \quad \left(0 \le i \le \binom{n}{2}\right),$$

$$\left(0, \dots, \binom{n}{j}, \dots, 0\right) \quad \left(\binom{n}{2} < j\right),$$

$$\left(0, \dots, \binom{n-1}{i-1}, \dots, \binom{n-1}{j}, \dots, 0\right) \quad \left(0 \le i \le \binom{n}{2}, n < i + j\right)$$

where the non-zero components are the ith and jth ones, resp.

Proof It is easy to see that there are intersecting, iclusion-free families with the above profiles. We only have to prove that any profile can be expressed as a convex linear combination of the given extreme points. This can be proved with the cycle method again. First we have to see the analogous problem for the intervals.

Lemma 12. *The extreme points of the convex hull of the profile vectors of intersecting, inclusion-free families of intervals on a cyclically ordered n-element set are*

$$(0, \dots, 0),$$

$$(0,\ldots,i,\ldots,0) \quad \left(0 \le i \le \binom{n}{2}\right),$$

$$(0,\ldots,n,\ldots,0) \quad \left(\binom{n}{2} < j\right),$$

$$(0,\ldots,i,\ldots,n-j,\ldots,0) \quad \left(0 \le i \le \binom{n}{2}, n < i+j\right),$$

where the non-zero components are the i*th and* j*th ones, resp., and the non-zero* 0*th and* n*th components are replaced by 1.*

Without **proof**, see [6]. □

Proof of the theorem It is easy to see that there are intersecting, inclusion-free families with profile vectors listed in the theorem. It remained to prove that the profile vector of any such family is in the convex linear combination of these given vectors. The proof of this statement will use the cycle method with a vector-valued weight function:

$$w(F) = \left(0,\ldots,\frac{1}{(n-1)!},\ldots,0\right)$$

where the non-zero component is the $|F|$th one. As before, the double sum of this weight will be calculated for the pairs $(\mathcal{C}, F)$ where $\mathcal{C}$ is a cyclic permutation of X, $F \in \mathcal{F}$ and F is an interval along $\mathcal{C}$. Let $\mathcal{F}(\mathcal{C})$ denote the family of those members $F \in \mathcal{F}$ which are intervals along $\mathcal{C}$.

For a fixed $\mathcal{C}$ we obtain

$$\sum_{F\in\mathcal{F}(\mathcal{C})} w(F) = \frac{1}{(n-1)!} p(\mathcal{F}(\mathcal{C})).$$

Denote the extreme points in the lemma by $e_1,\ldots,e_N$. The lemma implies that $p(\mathcal{F}(\mathcal{C}))$ is a convex linear combination of these vectors, that is,

$$p(\mathcal{F}(\mathcal{C})) = \sum_{i=1}^{N} \lambda_i(\mathcal{C}) e_i$$

where the λ's are non-negative and their sum is 1.

Hence

$$\sum_{\mathcal{C},F} w(F) = \sum_{\mathcal{C}}\sum_{F} w(F) = \sum_{\mathcal{C}} \frac{1}{(n-1)!} \sum_{i=1}^{N} \lambda_i(\mathcal{C}) e_i$$

$$= \sum_{i=1}^{N} \frac{1}{(n-1)!} \left(\sum_{\mathcal{C}} \lambda_i(\mathcal{C})\right) e_i$$

follows where $\sum_{i=1}^{N} \frac{1}{(n-1)!} \sum_{\mathcal{C}} \lambda_i(\mathcal{C}) = 1$. We have proved that $\sum_{\mathcal{C},F} w(F)$ is a convex linear combination of the e_i's.

Summing in the reverse order we obtain

$$\sum_{C,F} w(F) = \sum_F \sum_C w(F) = \sum_F{}^* \left(0, \dots, \frac{|F|!(n-|F|)!}{(n-1)!}, \dots, 0\right)$$

$$= \left(p_0, p_1 \frac{n}{\binom{n}{1}}, \dots, p_i \frac{n}{\binom{n}{i}}, \dots, p_{n-1} \frac{n}{\binom{n}{n-1}}, p_n\right),$$

where $\sum^*$ denotes that $(1, 0, \dots, 0)$ and $(0, \dots, 0, 1)$ are taken for $F = \emptyset$ and $F = X$, resp., as the number of cyclic permutations along which F is an interval is $|F|!(n-|F|)!$ for $0 < |F| < n$ but it is $(n-1)!$ for $|F| = 0, n$. It follows that the last vector is a convex linear combination of $e_1, \dots, e_N$, therefore $(p_0, \dots, p_n)$ is convex linear combination of the vectors listed in the theorem, since they can be obtained from the e_i's by multiplication with $\binom{n}{i}/n$ $(0 < i < n)$. □

OTHER RESULTS

There are many other applications of the method, see e.g. [4], [10], [12], [15], [16], [17] and [27]. Most of these are contained in the excellent book of Engel ([3]). In [7] the convex hull of several other classes of families are determined using the cycle method. [5] extends the method for more general structures. The most sophisticated application of the method is due to Pyber ([26]). He proved a special case of the following conjecture.

Conjecture 13 (Frankl-Füredi-Pyber) *Let $\mathcal{F}$ be an inclusion-free family of subsets of an n-element set, $2 \le k \le n$ be a fixed integer and suppose that any two members $F, G \in \mathcal{F}$ satisfy the conditions*

$$|F| \le n - k,$$

$$1 \le |F \cap G| \le k - 1.$$

Then

$$|\mathcal{F}| \le \binom{n-1}{k-1}$$

holds.

This would be an extension of the Erdős-Ko-Rado theorem. One can easily modify the method of [11] to prove the conjecture for the case

$$\frac{100k^2}{\log k} \le n.$$

Pyber proved it for the case

$$6k \le n \le \frac{r^2}{5}.$$

In all other applications of the method, an analogous problem is solved for the cycle and then double counting makes it valid for the original problem. Here Pyber considers mutual relationship between cycles. He uses statements, that if something happens in a cycle, then it strongly influences cycles which are not "far" from this cycle.

LARGER INTERSECTIONS

The most important recent theorem in extremal set theory is the following theorem what will be formulated here in a somewhat weaker form. We say that a family $\mathcal{F}$ is *t-intersecting* if $t \leq |F \cap G|$ holds for any pair of members $F, G \in \mathcal{F}$.

Theorem 14 (Ahlswede-Khachatrian [1]) *Let $1 \leq t \leq k \leq n, X = \{1, 2, \ldots, n\}$ and suppose that $\mathcal{F} \subset \binom{X}{k}$ is t-intersecting. Then $|\mathcal{F}|$ cannot exceed the size of the largest one of the following families*

$$\mathcal{A}_r = \left\{ A \in \binom{X}{k} : |A \cap \{1, 2, \ldots, t + 2r\}| \geq t + r \right\} \quad \left(0 \leq r \leq \frac{n-t}{2}\right). \tag{20}$$

The problem has a long history. It was posed in the original paper of Erdős, Ko and Rado ([8]). They proved that the family in (20) with $r = 0$ is the best when n is large enough, and posed the statement of Theorem 14 as a conjecture for the case when n is divisible by 4, $k = \frac{n}{2}$, $t = 2$ and $r = \frac{n-2}{2}$. Frankl has generalized this conjecture in the above form in [9]. He also determined in [9] the exact threshold in n when $15 \leq t$: the conjecture is true when $(k-t+1)(t+1) < n$ with $r = 0$, otherwise the construction with $r = 1$ gives a larger family. The cases $t = 2, \ldots, 14$ were solved by Wilson ([30]). Therefore the following theorem is a special case of Theorem 14, we formulate it separately because it will be used later.

Theorem 15 (Frankl-Wilson) *The largest t-intersecting family $\mathcal{F} \subset \binom{X}{k}$ has $\binom{n-t}{k-t}$ members if $(k - t + 1)(t + 1) \leq n$, otherwise it has more members.*

Frankl and Füredi ([13]) proved Frankl's conjecture (that is, the Ahlswede-Khachatrian theorem) for $c\sqrt{t/\log(t+1)}(k - t + 1) < n$.

Summarizing, a longstanding effort, for many decades was needed to solve the problem. Why does the cycle method which proved to be very effective in many cases fail when one of the conditions is the t-intersecting property with $2 \leq t$? Try the trivial generalization: determine the maximum number of t-intersecting intervals of length k. It is easy to see that the answer is $k - t + 1$ when $k \leq \frac{n+t-1}{2}$. The ratio selected/total number of intervals is much more than in the case of all sets: $\binom{n-t}{k-t} / \binom{n}{k}$.

One has to find a "more dense" substructure rather than the intervals along a cycle. A candidate is a *Steiner system* $\mathcal{S}(n, k, t)$, which is such a subfamily of $\binom{X}{k}$ that every t-element subset of X is contained in exactly one member. Observe that

$$|\mathcal{S}(n, k, t)| = \frac{\binom{n}{t}}{\binom{k}{t}}. \tag{21}$$

It is obvious that if $\mathcal{F}$ is a t-intersecting family of k-element subsets of X then $\mathcal{F}$ and $\mathcal{S}(n, k, t)$ have at most one common member. This is true for the family obtained from $\mathcal{S}(n, k, t)$ by permuting X. Consider the pairs $(\mathcal{P}, F)$ where $\mathcal{P}$ is a permutation of X, $F \in \mathcal{F}$ and $\mathcal{P}$ brings F to a member of $\mathcal{S}(n,k,t)$. There are $k!(n-k)!$ permutations bringing a given F to a given $S \in \mathcal{S}(n,k,t)$. Using (21) we obtain that the number of pairs in question is

$$|\mathcal{F}|\frac{\binom{n}{t}}{\binom{k}{t}}k!(n-k)!. \tag{22}$$

On the other hand, if $\mathcal{P}$ is fixed, there is at most one F by the above remark. Therefore the number of pairs is at most $n!$, consequently (22) is $\leq n!$. This inequality implies $\mathcal{F} \leq \binom{n-t}{k-t}$.

Theorem 16 (Frankl-Katona) *If there is a Steiner system $\mathcal{S}(n,k,t)$ for the given integers $2 \leq t < k < n$ and $\mathcal{F} \subset \binom{X}{k}$ is a t-intersecting family, then*

$$|\mathcal{F}| \leq \binom{n-t}{k-t}.$$

As the existence of Steiner systems is a difficult question, this result did not seem to be very effective. This is why it was not published before except for a short remark ($k=3, t=2$) in [21] (page 221). However, if it is combined with Theorem 15 then we obtain a new proof of an old theorem of Tits ([29]):

Theorem 17 (Tits) *In any non-trivial Steiner system $\mathcal{S}(n,k,t)$*

$$(k-t+1)(t+1) \leq n$$

holds.

Another attempt to generalize the cyclic method for more-intersecting families can be found in [19]. For sake of simplicity we show the case $t=2$, only. Consider the group S_n of all permutations of X. A subgroup Γ of S_n is called 2-*transitive* if any ordered pair (x_1,y_1) of different elements can be mapped into any other pair (x_2,y_2) (of different elements) by one of $\phi \in \Gamma$. It is called *sharply* 2-*transitive* if there is exactly one such ϕ. If n is prime power then the function $ax+b$ ($a \neq 0$) is a permutation on $GF(n)$ for any $a,b \in GF(n)$. It is easy to see that the group of these functions (for composition) is a sharply 2-transitive subgroup of S_n. The number of elements of this subgroup is $n(n-1)$. Obviously, this must hold for any sharply 2-transitive subgroup Γ. (Note that the subgroup of cyclic shifts $\phi_j(i) = i+j \bmod n$ form a sharply 1-transitive subgroup.) Consider the sets obtained from a given k-element $A \subset X$ by applying the permutations $\phi \in \Gamma$ where Γ is a sharply 2-transitive subgroup: $\phi_1(A), \ldots, \phi_{n(n-1)}(A)$. If we can prove that a 2-intersecting subfamily of this family is of size at most $k(k-1)$ then the ratio of the selected subsets over the total number of subsets is the same as in the family of all sets. Let us formulate it as a theorem.

Theorem 18 (Howard-Károlyi-Székely) *A sharply 2-transitive group Γ acting on X is given. Let $A \subset X, |A| = k$. Suppose that any 2-intersecting* *subfamily of* $\{\phi(A) : \phi \in \Gamma\}$ *has at most* $k(k-1)$ *members. Then any 2-intersecting family* $\mathcal{F} \in \binom{X}{k}$ *satisfies* $|\mathcal{F}| \leq \binom{n-2}{k-2}$.

In [19] the authors find an infinite class of integers n and k for which they are able to use the above theorem to prove Theorem 14 in case of $t = 2$.

References

[1] R. Ahlswede, L. Khachatrian, "The Complete Intersection Theorem for Systems of Finite Sets", *Europ. J. Combinatorics*, 18, 1997, 125-136.

[2] B. Bollobás, "Sperner systems consisting of pairs of complementary subsets", *J. Combinatorial Th. A*, 15, 1973, 363-366.

[3] K. Engel, *Sperner Theory*, Encyclopedia of Mathematics and its Applications, Cambridge University Press, Cambridge, 1997.

[4] K. Engel, Péter L. Erdős, "Sperner families satisfying additional conditions and their convex hulls", *Graphs and Combinatorics*,5, 1988, 50-59.

[5] Péter L. Erdős, U. Faigle, W. Kern, "A group theoretical setting for some intersecting Sperner families", *Combinatorics, Probability and Computing*, 1, 1992, 323-334.

[6] Péter L. Erdős, P. Frankl, G.O.H. Katona, "Intersecting Sperner families and their convex hulls", *Combinatorica*, 4, 1984, 21-34.

[7] Péter L. Erdős, P. Frankl, G.O.H. Katona, "Extremal hypergraph problems and convex hulls", *Combinatorica*, 5, 1985, 11-26.

[8] P. Erdős, Chao Ko, R. Rado, "Intersection theorems for systems of finite sets", *Quart. J. Math. Oxford, (2)* , 12, 1961, 313-318.

[9] P. Frankl, "The Erdős-Ko-Rado theorem is true for $n = ckt$", *Coll. Soc. Math. J. Bolyai*, 18, 1978, 365-375.

[10] P. Frankl, Z. Füredi, "The Erdős-Ko-Rado Theorem for integer sequences", *SIMA J. on Algebraic Discrete Methods*, 1, 1980, 376-381.

[11] P. Frankl, Z. Füredi, "Families of finite sets with a missing intersection", *Finite and Infinite Sets (Proc. 6th Hungar. Colloq. on Combinatorics, Eger, 1981)*, Eds. A. Hajnal, L. Lovász and V.T. Sós, vol. **37**, North Holland, Amsterdam, 1984, 305-318.

[12] P. Frankl, Z. Füredi, "Extremal problems concerning Kneser graphs", *J. Combin. Theory B*, 40, 1986, 270-284.

[13] P. Frankl, Z. Füredi, "Beyond the Erdős-Ko-Rado theorem", *J. Combinatorial Th. A*, 56, 1991, 182-194.

[14] Z. Füredi, personal communication.

[15] Z. Füredi, "The maximum number of balancing sets", *Graphs and Combin.*, 3, 1987, 251-254.

[16] Z. Füredi, "Cross-intersecting families of finite sets", *J. Combinatorial Th. A*, 72, 1995, 332-339.

[17] Z. Füredi, D. Kleitman, "The minimal number of zero sums", in *Combinatorics, Paul Erdős is eighty*, Vol. I, pp; 159-172, Keszthely, Hungary, 1993, D. Miklós et al., Eds., *Bolyai Society Mathematical Studies* **1**(1993), Budapest, Hungary.

[18] C. Greene, G.O.H.Katona, D.J. Kleitman, "Extensions of the Erdős-Ko-Rado theorem", *SIAM*, 55, 1976, 1-8.

[19] R. Howard, Gy. Károlyi, L.A. Székely, "Towards a Katona type proof for the 2-intersecting Erdős-Ko-Rado theorem", preprint.

[20] G.O.H. Katona, "A simple proof of the Erdős-Chao Ko-Rado theorem", *J. Combinatorial Th. A*, 13, 1972, 183-184.

[21] G.O.H. Katona, "Extremal problems for hypergraphs", *Combinatorics*, Ed. by M. Hall, Jr., J.H. van Lint, D. Reidel, Dordrecht/Boston, 1975, 215-244.

[22] G.O.H. Katona, "A simple proof of a theorem of Milner", *J. Combinatorial Th. A*, 83, 1998, 138-140.

[23] G.O.H. Katona, G. Schild, "Linear inequalities describing the class of Sperner families of subsets I", *Topics in Combinatorics and Graph Theory (Essays in Honour of Gerhard Ringel)*, Ed. R. Bodendiek and R. Henn, Physica-Verlag, Heidelberg, 1990, 413-420.

[24] D. Lubell, "A short proof of Sperner's lemma", *J. Combinatorial Th.* , 1, 1966, 299.

[25] E.C. Milner, "A combinatorial theorem on systems of sets", *J. London Math. Soc.*, 43, 1968, 204-206.

[26] L. Pyber, "An extension of a Frankl-Füredi theorem", *Discrete Math.*, 52, 1984, 253-268.

[27] L. Pyber, "A new generalization of the Erdős-Ko-Rado theorem", *J. Combinatorial Th. A*, 43, 1986, 85-90.

[28] E. Sperner, "Ein Satz über Untermengen einer endlichen Menge", *Math. Z.*, 27, 1928, 544-548.

[29] J. Tits, "Sur les systèmes de Steiner associés aux trois 'grands' groupes de Mathieu", *Rend. Math. e Appl.* (5), 23, 1964, 166-184.

[30] R. M. Wilson, "The exact bound on the Erdős-Ko-Rado theorem", *Combinatorica*, 4, 1984, 247-257.

EXTREMAL PROBLEMS ON Δ-SYSTEMS*

Alexandr V. Kostochka

Institute of Mathematics, Siberian Branch
Russian Academy of Sciences

Abstract: A family of sets is called a *Δ-system* (respectively, a *weak Δ-system*) if the intersection of any two sets is the same (respectively, the cardinality of the intersection of any two sets is the same). In 1960, P.Erdős and R.Rado started studying the maximum size of a k-uniform hypergraph not containing a Δ-system of a given size. The aim of the present article is to survey the progress and state of art in this and related problems.

INTRODUCTION

In connection with some problems in Number Theory, P.Erdős and R.Rado [12] introduced the notion of a Δ-system . They called a family $\mathcal{H}$ of sets a *Δ-system* if every two members of $\mathcal{H}$ have the same intersection. Define $f(k,r)$ to be the least cardinal so that any k-uniform family of more than $f(k,r)$ sets contains a Δ-system consisting of r sets. Erdős and Rado [12, 13] completely determined $f(k,r)$ in case at least one of k and r is infinite and found some upper and lower bounds for the case that both k and r are finite.

In 1974, Erdős, E.Milner and Rado [11] introduced the related notion of a weak Δ-system. A *weak Δ-system* is a family of sets where all pairs of sets have the same intersection size. Let $g(k,r)$ be the least cardinal so that every k-uniform family of more than $g(k,r)$ sets contains a weak Δ-system consisting of r sets. Erdős, Milner and Rado [11] found the values of $g(k,r)$ in case of infinite k and r assuming the generalized continuum hypothesis.

*This work was partly supported by the grant RM1-181 of the Cooperative Grant Program of the Civilian Research and Development Foundation and by the grant 96-01-01614 of the Russian Foundation for Fundamental Research.

I. Althöfer et al. (eds.), Numbers, Information and Complexity, 143-150.

Similar problems for families having a fixed cardinality of the ground set were introduced in 1978 by Erdős and E. Szemerédi [14]. They defined $F(n,r)$ to be the largest integer so that there exists a family $\mathcal{F}$ of subsets of an n-element set which does not contain a Δ-system of r sets and $G(n,r)$ to be the largest integer so that there exists a family $\mathcal{F}$ of subsets of an n-element set which does not contain a weak Δ-system of r sets.

The problems of estimating $f(k,r)$, $g(k,r)$, $F(n,r)$ and $G(n,r)$ have been attracting attention of many Mathematicians and were among favorite problems of Erdős for decades.

In this article, we survey the progress in studying these four functions, each of the subsequent sections devoted to a function. We focus the attention more on constructions than on proofs.

THE ORIGINAL PROBLEM

The first and most famous problem is about $f(k,r)$. Erdős and Rado [12] proved that

$$(r-1)^k \le f(k,r) \le (r-1)^k k! \left\{ 1 - \sum_{t=1}^{k-1} \frac{t}{(t+1)!(r-1)^t} \right\}. \qquad (1)$$

The construction providing the lower bound is as follows.

Construction 1. Let $X_1, \ldots, X_k$ be disjoint sets of cardinality $r-1$ each. Let $\mathcal{F} = \{(x_1, \ldots, x_k) \mid x_i \in X_i, i = 1, \ldots, k\}$. Clearly, $|\mathcal{F}| = (r-1)^k$. Suppose that some members $A_1, \ldots, A_r$ of $\mathcal{F}$ form a Δ-system. Since these sets are distinct, there is an element x which belongs to exactly one of $A_1, \ldots, A_r$. We may assume that $x \in A_1 \cap X_1$. Then all the r sets $A_i \cap X_1$, $i = 1, \ldots, r$, (each consisting of a single element) must be disjoint. Since $|X_1| = r-1$, this is impossible.

Erdős and Rado [12] also conjectured that for each r, there exists a constant C_r so that $f(k,r) < C_r^k$. Erdős (see [9]) has offered 1000 dollars for the proof or disproof of this for $r = 3$.

The next remarkable paper in this direction was that of H.L. Abbott, D. Hanson, and N. Sauer [5]. They completely solved the case $k = 2$ (namely, they showed that $f(2,r) = r(r-1)$ for odd r and $f(2,r) = r(r-1.5)$ for even r), improved the upper bound in (1) to $(k+1)!\left(\frac{r-1+\sqrt{r^2+6r-7}}{4}\right)^k$ and the lower bound for $f(k,3)$ to $2 \cdot 10^{k/2 - c\log k}$. This is still the best known lower bound. It is derived from their construction for every positive integer t of an intersecting 3^t-uniform family $\mathcal{F}_t$ of cardinality $10^{(3^t-1)/2}$ not containing a Δ-system of size 3. A description of the construction is as follows.

Construction 2. We use induction on t. It is a routine to check that the family $\mathcal{F}_1 = \{\{1,2,3\}, \{1,2,4\}, \{1,3,5\}, \{1,4,6\}, \{1,5,6\}, \{2,3,6\}, \{2,4,5\}, \{2,5,6\}, \{3,4,5\}, \{3,4,6\}\}$ with the ground set $\{1, \ldots, 6\}$ is what we need for $t = 1$. Suppose that we have constructed an intersecting 3^{t-1}-uniform family $\mathcal{F}_{t-1}$ with a ground set V of cardinality $10^{(3^{t-1}-1)/2}$ not containing a Δ-system of size 3. Let $\mathcal{F}_t$ have the ground set $V_1 \cup \ldots \cup V_6$, where every V_i is a copy of V; the members of $\mathcal{F}_t$ are the sets of the kind $E_\alpha \cup E_\beta \cup E_\gamma$, where $\{\alpha, \beta, \gamma\}$ is an edge in $\mathcal{F}_1$ and E_α, E_β and E_γ are arbitrary members of copies of $\mathcal{F}_{t-1}$ on the sets V_α, V_β and V_γ, respectively. Then

$$|\mathcal{F}_t| = |\mathcal{F}_1| \cdot |\mathcal{F}_{t-1}|^3 = 10 \cdot 10^{3(3^{t-1}-1)/2} = 10^{(3^t-1)/2}.$$

Since $\mathcal{F}_{t-1}$ and $\mathcal{F}_1$ both are intersecting families, $\mathcal{F}_t$ also is an intersecting family. To see that $\mathcal{F}_t$ does not contain a Δ-system of size 3, consider three arbitrary members A, B and C of $\mathcal{F}_t$.
CASE 1. The set $A \cup B \cup C$ meets at least four sets V_i. Then, due to construction of $\mathcal{F}_1$, some V_j (say, V_1) meets exactly two of A, B and C, say, A and B. Since $\mathcal{F}_{t-1}$ is an intersecting family, there exists some $v \in A \cap B \cap V_1$. This v witnesses that A, B and C do not form a Δ-system.
CASE 2. Every of A, B and C meets the same three sets V_i, say, V_1, V_2 and V_3. Since A, B and C are distinct sets, we may suppose that they do not coincide on V_1. Then, due to the properties of $\mathcal{F}_{t-1}$, some element w of V_1 belongs to exactly two of A, B and C. This w witnesses that A, B and C do not form a Δ-system.

It would be very interesting to improve the construction even just a bit. But maybe it is optimal.

The next upper bounds on $f(k,r)$ are due to J. H. Spencer [20]. He proved that for every fixed r and any $\epsilon > 0$,

$$f(k,r) < C(1+\epsilon)^k k!$$

and that

$$f(k,3) < e^{ck^{3/4}} k!\ .$$

Z. Fűredi and J. Kahn (see [10]) proved that $f(k,3) < e^{c\sqrt{k}} k!$. Currently best upper bound on $f(k,r)$ for small r is the following [16]:

For each integers $r > 2$ and $\alpha > 1$, there exists $D(r,\alpha)$ such that for all k,

$$f(k,r) \leq D(r,\alpha) k! \left(\frac{(\log\log\log k)^2}{\alpha \log\log k} \right)^k. \tag{2}$$

This bound is less than $k!$ but not much less and the gap between lower and upper bounds is still drastic.

A better situation takes place for large r and small k. As was mentioned above, Abbott, Hanson, and Sauer [5] completely solved the case $k = 2$. Then Abbott and Hanson [3] proved that $f(3,r) \leq 1.8r^3 + O(r^2)$. Recently, V. Rödl, L. Talysheva and I [18] proved that for every fixed k, Construction 1 by Erdős and Rado is asymptotically (in r) best possible:

Let k be fixed and r be sufficiently large. Then

$$f(k,r) = r^k + o(r^k). \tag{3}$$

We don't know how small is $o(r^k)$ in (3). It seems it is the only known asymptotically exact bound concerning Δ-systems.

Abbott and B. Gardner [2] proved in 1969 that $f(3,3) = 20$, and since then no other exact value of $f(k,r)$ for $k \geq 3$ and $r \geq 3$ became known. Abbott and G. Exoo [1] obtained the lower bounds $f(k,4) \geq C \cdot 38^{n/3}$ and $f(k,6) \geq C \cdot 146^{n/3}$.

WEAK Δ-SYSTEMS

Erdős, Milner and Rado [11] gave the lower bounds $g(k,r) \geq r^k$ and $g(k,2) \geq \frac{5}{4}2^k$ for $k \geq 2$ and showed that for every positive integer k and $r > 1 + k\binom{n}{n/2}$, any k-uniform weak Δ-system is a strong Δ-system. The last result was sharpened by M. Deza [8]: he proved that for every $r > k^2 - k + 1$, any k-uniform weak Δ-system is a strong Δ-system, implying that $g(k,r) = f(k,r)$ for every $r > k^2 - k + 1$.

The lower bound on $g(k,r)$ by Erdős, Milner and Rado was obtained due to the following construction.

Construction 3. Given a $(k-1)$-uniform family $\mathcal{F}$ without weak Δ-systems of size r, a k-uniform family $\mathcal{F}'$ without weak Δ-systems of size r can be constructed from $\mathcal{F}$ by replacing every member A by the members $A_1 = A \cup \{a_1(A)\}$, $A_2 = A \cup \{a_2(A)\}, \ldots, A_{r-1} = A \cup \{a_{r-1}(A)\}$, where all the elements $a_i(A)$ are distinct for all A and i. This gives

$$g(k,r) \geq (r-1)g(k-1,r) \tag{4}$$

and the bound (for $r \geq 4$) follows. The direct construction implied by this argument is as follows. Consider the complete $(r-1)$-nary tree $T_k(r)$ of height k. For every of $(r-1)^k$ pendant vertices v, let A_v be the set of the vertices of the path from v to the root w of $T_k(r)$ excluding w. The family of all these A_v is k-uniform, has $(r-1)^k$ members and contains no weak Δ-system of size r.

For $r = 3$, Erdős, Milner and Rado observed that $g(2,3) = 5$, in particular, the family of the five edges of a 5-cycle does not contain any weak Δ-system of size 3. This together with (4) gives the bound. Abbott and Hanson [4] used this observation to derive the relation $g(k,3) \geq 5g(k-2,3)$ for $k \geq 2$ and, therefore, the bound

$$g(k,3) \geq 5^{\lfloor k/2 \rfloor} 2^{k-2\lfloor k/2 \rfloor}.$$

Construction 3 is better than Construction 1 in the sense that, for given k and r, it produces the family of the same cardinality but with the stronger property. Recall that due to (4), it is asymptotically (in r) optimal for every fixed k.

The only known exact value of $g(k,r)$ for $k \geq 3$ and $r \geq 3$ is $g(3,3) = 10$ (see [4]). The best known upper bound on $g(k,3)$ due to M. Axenovich, D. G. Fon-Der-Flaass and myself [6] is:

For every $\delta > 0$, there exists a constant $C = C(\delta)$ such that

$$g(k,3) < Cn!^{0.5+\delta}.$$

Abbott and Exoo [1] gave the lower bounds $g(k,4) \geq C \cdot 10^{n/2}$ and $g(k,5) \geq C \cdot 20^{n/2}$.

Δ-SYSTEMS IN SET SYSTEMS WITH A FIXED CARDINALITY OF THE GROUND SET

In [14], Erdős and Szemerédi showed

$$F(n,3) < 2^{n(1-\frac{1}{10\sqrt{n}})} \tag{5}$$

and stated that the probabilistic method implies that for each $r \geq 3$, there exists a constant $c_r > 0$, so that

$$F(n,r) > (1+c_r)^n$$

where $c_r \to 1$ as $r \to \infty$. Let

$$\beta_r = \lim_{n\to\infty} F(n,r)^{1/n}.$$

Abbott and Hanson [4] observed that β_r exists and that the probabilistic method mentioned above gives $\beta_r \geq 2(r+2)^{-1/r}$. They also presented a construction implying

$$\beta_r \geq \binom{2r-2}{r}^{1/(2r-2)} \sim 2^{(1-\frac{\log(2r)}{4r})}, \tag{6}$$

which is slightly better than the probabilistic bound.

The Erdős-Szemerédi proof [14] of (5) reveals relations between bounds for $f(k,r)$ and $F(n,r)$. It shows that good upper bounds for $f(k,r)$ yield satisfactory upper bounds for $F(n,r)$ and strong lower bounds (if found) for $F(n,r)$ might imply lower bounds for $f(k,r)$. W. A. Deuber, P. Erdős, D. S. Gunderson, A. G. Meyer and I [7] observed that the Erdős-Szemerédi argument together with (2) yields that for each r and sufficiently large n,

$$F(n,r) < 2^{n-\frac{\sqrt{n}\log\log n}{\log\log\log n}},$$

and that if there exists a constant C so that $f(k,3) < C^k$, then for n sufficiently large,

$$F(n,3) < 2^{n(1-0.65/C)}.$$

In particular, in this case, $\beta_3 \leq 2^{(1-1/2C)}$. It follows that if the Erdős–Rado conjecture is true, then there exists an $\epsilon > 0$ so that for large n, $F(n,3) < (2-\epsilon)^n$.

This motivates obtaining lower bounds on $F(n,r)$ and β_r. In [7], the following bound (improving (6)) is given: for every $r \geq 3$ and every n of the form $n = 2pr\lfloor \log r \rfloor$,

$$F(n,r) \geq 2^{n(1-\frac{\log\log r}{2r}-O(1/r))},$$

(and there are uniform families which witness this bound). In particular,

$$\beta_r \geq 2^{(1-\frac{\log\log r}{2r}-O(1/r))}.$$

It was also proved in [7] that for every n of the form $n = 48q + 2$, $F(n,3) \geq 1.551^{n-2}$; in particular, $\beta_3 \geq 1.551$.

WEAK Δ-SYSTEMS IN SET SYSTEMS WITH A FIXED CARDINALITY OF THE GROUND SET

Although Construction 3 gives an exponential (in k) lower bound on $g(k,3)$, it gives only linear (in n) lower bound on $G(n,3)$. In the middle of the seventies, Abbott asked if $G(n,3)$ is superlinear in n. Answering this question, Erdős and Szemerédi [14] proved that it is superpolynomial, namely,

$$G(n,3) \geq (1+o(1))n^{\log n/4\log\log n}. \tag{7}$$

To do this, they elaborated Construction 3 as follows.

Construction 4. Take $s = \lfloor \frac{\log_2 n}{2\log_2\log_2 n} \rfloor$ disjoint copies $T_t^1, \ldots, T_t^s$ of the complete binary tree T_t of height $t = \lfloor 0.5\log_2 n \rfloor$. For every $i = 2, \ldots, s$, replace every vertex of T_t^i by a set of cardinality $\lfloor(\log_2 n)^{i-1}\rfloor$ (all these sets are disjoint). Let $v_1, \ldots, v_s$ be some pendant vertices in $T_t^1, \ldots, T_t^s$, respectively. Define $B(v_1, \ldots, v_s)$ to be the union of the vertex sets of the paths connecting $v_1, \ldots, v_s$ with the corresponding roots, and let $\mathcal{F}$ be the family of the sets $B(v_1, \ldots, v_s)$ for all possible choices of $v_1, \ldots, v_s$. Clearly,

$$|\mathcal{F}| = (2^t)^s \geq (1+o(1))2^{\frac{\log_2^2 n}{4\log_2\log_2 n}},$$

and the cardinality of the ground set is at most

$$\sum_{i=1}^{s} 2^{t+1}(\log_2 n)^{i-1} < 2^{t+1}\cdot 2\cdot(\log_2 n)^{s-1} < 2\sqrt{n}\cdot 2\cdot\frac{\sqrt{n}}{\log_2 n} < n.$$

Thus, if we prove that no three members of $\mathcal{F}$ form a weak Δ-system, then (7) follows.

Assume that members B_1, B_2 and B_3 of $\mathcal{F}$ form a weak Δ-system and that i is the largest index such that B_1, B_2 and B_3 do not coincide on T_t^i. Then, due to the structure of the binary tree, we can reorder B_1, B_2 and B_3 so that

$$|B_1 \cap B_2 \cap T_t^i| > |B_1 \cap B_3 \cap T_t^i|. \tag{8}$$

If $i = 1$, then we are done. Let $i > 1$. Since T_t^i is obtained from T_t^1 by blowing every vertex into $\lfloor(\log_2 n)^{i-1}\rfloor$ vertices, (8) yields

$$|B_1 \cap B_2 \cap T_t^i| - |B_1 \cap B_3 \cap T_t^i| \geq \lfloor(\log_2 n)^{i-1}\rfloor. \tag{9}$$

But

$$|B_3 \cap \left(\bigcup_{j=1}^{i-1} T_t^j\right)| \leq (t+1)\sum_{j=1}^{i-1}(\log_2 n)^{j-1} = (1+o(1))0.5\log_2 n \cdot (\log_2 n)^{i-2} <$$

$$(\log_2 n)^{i-1}.$$

This together with (9) contradicts our assumption on B_1, B_2 and B_3.

Erdős and Szemerédi [14] also conjectured that for some $\epsilon > 0$,

$$G(n,3) \leq (2-\epsilon)^n.$$

This conjecture (as a consequence of a stronger result) was proved by Frankl and Rödl [15] for $\epsilon = 0.01$.

Recently, Rödl and Thoma [19] substantially improved (7) by showing that for sufficiently large n,

$$G(n,r) \geq 2^{\frac{1}{3}n^{1/5}\log_2^{4/5}(r-1)}. \tag{10}$$

To do this, they elaborated Construction 3 in a different manner than it was made in Construction 4. They replaced every vertex v in the $(r-1)$-nary tree $T_t(r)$ of height $t = \lceil 6n^{1/5}\log_2^{4/5}(r-1)\rceil$ by a set A_v of cardinality $m = \lfloor n^{3/5}\log_2^{2/5}(r-1)\rfloor$. In contrast with Construction 4, these sets A_v are not necessarily disjoint, but every two have a small intersection and the union of all A_v has the cardinality at most n. The members of the constructed family are the unions of the sets on the paths from pendant vertices of $T_t(r)$ to the root.

Later [17], this construction was elaborated to a random construction giving the bound

$$G(n,r) \geq r^{c(n\ln n)^{1/3}}.$$

Still, the gap between lower and upper bounds on $G(n,r)$ is challenging.

CONCLUDING REMARK

One of the aims of the present article was to show that there was some progress lately in studying every of the functions $f(k,r)$, $g(k,r)$, $F(n,r)$ and $G(n,r)$, but none of the main problems is solved.

References

[1] H. L. Abbott and G. Exoo, "On set systems not containing Delta systems", *Graphs and Combinatorics*, 8, 1992, 1–9.

[2] H. L. Abbott and B. Gardner, "On a combinatorial theorem of Erdős and Rado", in: *W. T. Tutte, ed., Recent progress in Combinatorics*, Academic Press, New York, 1969, 211-215.

[3] H. L. Abbott and D. Hanson, "On finite Δ-systems", *Discrete Math.*, 8, 1974, 1–12.

[4] H. L. Abbott and D. Hanson, "On finite Δ-systems, II", *Discrete Math.*, 17, 1977, 121–126.

[5] H. L. Abbott, D. Hanson, and N. Sauer, "Intersection theorems for systems of sets", *Journal of Combinatorial Theory, Series A*, 12, 1972, 381–389.

[6] M. Axenovich, D. G. Fon-Der-Flaass, and A. V. Kostochka, "On set systems without weak 3-Δ-subsystems", *Discrete Mathematics*, 138, 1995, 57-62.

[7] W. A. Deuber, P. Erdős, D. S. Gunderson, A. V. Kostochka, and A. G. Meyer, "Intersection statements for systems of sets", *Journal of Combinatorial Theory, Series A*, 79, 1997, 118-132.

[8] M. Deza, "Solution d'un probléme de Erdős–Lovász", *Journal of Combinatorial Theory, Series B*, 16, 1974, 166–167.

[9] P. Erdős, "Problems and results on finite and infinite combinatorial analysis", in: *Infinite and finite sets Colloq. Keszthely 1973, Vol. I, Colloq. Math. Soc. J. Bolyai*, 10, North Holland, Amsterdam, 1975, 403–424.

[10] P. Erdős, "Problems and results on set systems and hypergraphs", *Extended Abstract, Conf.on Extremal Problems for Finite Sets, 1991, Visegrad, Hungary*, 1991, 85-92.

[11] P. Erdős, E. C. Milner, and R. Rado, "Intersection theorems for systems of sets, III", *J. Austral. Math. Soc.*, 18, 1974, 22–40.

[12] P. Erdős and R. Rado, "Intersection theorems for systems of sets", *J.London Math. Soc.*, 35, 1960, 85-90.

[13] P. Erdős and R. Rado, "Intersection theorems for systems of sets, II", *J.London Math. Soc.* 44, 1969, 467-479.

[14] P. Erdős and E. Szemerédi, "Combinatorial properties of systems of sets", *Journal of Combinatorial Theory, Series A*, 24, 1978, 308–313.

[15] P. Frankl and V. Rödl, "Forbidden intersections", *Trans. Amer. Math. Soc.*, 300, 1987, 259–286.

[16] A. V. Kostochka, "An intersection theorem for systems of sets", *Random Structures and Algorithms*, 9, 1996, 213-221.

[17] A. V. Kostochka and V. Rödl, "On large systems of sets with no large weak Δ-subsystems", *Combinatorica*, 18, 1998, 235-240.

[18] A. V. Kostochka, V. Rödl and L. Talysheva, "On systems of small sets with no large Δ-subsystems", *Combinatorics, Probability and Computing*, 8, 1999, 265-268.

[19] V. Rödl and L. Thoma, "On the size of set systems on $[n]$ not containing weak r, Δ-systems", *Journal of Combinatorial Theory, Series A*, 80, 1997, 166-173.

[20] J. H. Spencer, "Intersection theorems for systems of sets", *Canad. Math. Bull.* 20, 1977, 249-254.

THE AVC WITH NOISELESS FEEDBACK AND MAXIMAL ERROR PROBABILITY: A CAPACITY FORMULA WITH A TRICHOTOMY

Rudolf Ahlswede and Ning Cai

Fakultät Mathematik, Universität Bielefeld
Postfach 100131, 33501 Bielefeld, Germany

Abstract: To use common randomness in coding is a key idea from the theory of identification. Methods and ideas of this theory are shown here to have also an impact on Shannon's theory of transmission. As indicated in the title, we determine the capacity for a classical channel with a novel structure of the capacity formula. This channel models a robust search problem in the presence of noise (see R. Ahlswede and I. Wegner, Search Problems, Wiley 1987).

INTRODUCTION

Let $\mathcal{X}, \mathcal{Y}$ be the finite input and output alphabets of an AVC defined by the class of $|\mathcal{X}| \times |\mathcal{Y}|$–stochastic matrices $\mathcal{W}$, which we assume to be finite . Eventhough our results hold for every $\mathcal{W}$, we assume here $\mathcal{W}$ to be finite, because already under this restriction the proofs are highly sophisticated and we don't want to burden the reader with additional technical, but known, approximation arguments (like i.e. in [2]).

It was assumed in [1] that $\mathcal{W}$ equals its row–convex hull $\overline{\overline{\mathcal{W}}}$ and it was shown that in the presence of noiseless feedback under the maximal error probability criterion its capacity $C_F(\overline{\overline{\mathcal{W}}})$ has the formula

$$C_F(\overline{\overline{\mathcal{W}}}) = \max_{P \in \mathcal{P}(\mathcal{X})} \min_{W \in \overline{\overline{\mathcal{W}}}} I(P, W), \quad \text{if the capacity is positive.} \tag{1}$$

Here $\mathcal{P}(\mathcal{X})$ is the set of probability distributions (PD) on $\mathcal{X}$ and I is the mutual information.

I. Althöfer et al. (eds.), Numbers, Information and Complexity, 151-176.

Actually, this result was shown with an explicit coding strategy. Clearly, the known (in [11]) exact condition for positivity in the absence of feedback, namely,

$$\overline{\mathcal{W}}(x) \cap \overline{\mathcal{W}}(x') = \varnothing \quad \text{for some} \quad x, x' \in \mathcal{X}, 1.2 \tag{2}$$

where $\overline{\mathcal{W}}(x) =$ convex hull $(\mathcal{W}(x))$ and $\mathcal{W}(x) = \{W(\cdot|x) : W \in \mathcal{W}\}$, is also sufficient for positivity in the presence of feedback.
However, it is not necessary for positivity of $C_F(\overline{\overline{\mathcal{W}}})$.
On the other hand (see Lemma 3 of [1]) condition (2) is necessary and sufficient for positivity of $C_F(\overline{\overline{\mathcal{W}}})$ (and also of $C_F(\mathcal{W})$), if $\mathcal{W}$ contains only 0–1–matrices. Furthermore, Example 2 of [1] shows that $C_F(\overline{\overline{\mathcal{W}}})$ and $C_F(\mathcal{W})$ can be different. This construction shows that in cases where (2) does not hold (for letters) its extension for feedback strategies can still hold.
In this paper we determine $C_F(\mathcal{W})$ completely. The formula distinguishes three cases and therefore we speak of a trichotomy. It is an absolute novelty for capacity formulas in Information Theory.
A *dichotomy occurred* — quite surprisingly at its time — for AVC without feedback under the average error criterion ([2]): $C_{av}(\mathcal{W})$ is zero or else equals the random code capacity $C_R(\mathcal{W}) = \max_P \min_{W \in \overline{\mathcal{W}}} I(P, W)$, where $\overline{\mathcal{W}}$ is the convex hull of $\mathcal{W}$.
We settle now the positivity problem for $C_F(\mathcal{W})$ and we prove the Trichotomy Theorem. The Positivity Theorem and the easy direction of its proof are presented in Section 2. The much harder direction is given in Section 6. It uses a Balanced Coloring Lemma, which we establish in Section 3.
The Trichotomy Theorem is stated in Section 4. It incorporates the Positivity Theorem and the Capacity Theorem for 0–1–matrices of [1], which also readily leads to the Converse of the Trichotomy Theorem. Its direct part, however, is far more complex. The main ingredients are the List Reduction Lemma of [1], the Elimination Technique of [2], and the Balanced Coloring Lemma (see [2], [7]) in the version of Section 3.
Finally we mention that the coding problem for the AVC with feedback has another appealing interpretation. One of the simplest search problems is to find an unknown element $x \in \mathcal{X}$ by sequentially "Yes–No" questions like "Is $x \in A$?" where A is any subset of $\mathcal{X}$. It is easy to see that the minimal number of such questions which specify x is in the worst case $\lceil \log |\mathcal{X}| \rceil$. Now, if the answers are false with probability ε, allowing an error probability λ, then this problem is equivalent to the coding problem for the BSC $W = \begin{pmatrix} 1-\varepsilon & \varepsilon \\ \varepsilon & 1-\varepsilon \end{pmatrix}$ with complete feedback. A proof can be found in the book mentioned in the abstract. More generally there is the same connection for a–ary questions with b–ary answers with noise, that is, the BSC can be replaced by a general DMC. In a robust noise model this DMC is to be replaced by an AVC.
Needless to say that channels with feedback links are of practical interest (see [13]) in error control coding (ARQ, FEC systems etc.). Here we settle the capacity problem for the robust channel model AVC.

POSITIVITY OF THE CAPACITY $C_F(\mathcal{W})$

We are given the set of transmission matrices

$$\mathcal{W} = \{W(\cdot|\cdot,s) : s \in \mathcal{S}\}, |\mathcal{S}| < \infty. \tag{3}$$

For a state sequence $s^n \in \mathcal{S}^n$ the n–length feedback transmission matrix $W_F^n(\cdot|\cdot,s^n)$ is an $|\mathcal{X}|^{\sum_{t=0}^{n-1}|\mathcal{Y}^t|} \times |\mathcal{Y}^n|$–stochastic matrix with entries $W(y_1|f_1,s_1)\times \prod_{t=2}^{n} W\big(y_t|f_t(y^{t-1}),s_t\big)$, where the feedback strategy $f^n = (f_1,\dots,f_n)$ is defined by $f_1 \in \mathcal{X}$ and $f_t : \mathcal{Y}^{t-1} \to \mathcal{X}$ for $t = 2,\dots,n$.
We denote the set of those strategies by $\mathcal{F}^n$ and then write $W_F^n(\cdot|\cdot,s^n) = (W^n(\cdot|f^n,s^n))_{f^n\in\mathcal{F}^n}$ and

$$\mathcal{W}_F^n = \{W_F^n(\cdot|\cdot,s^n) : s^n \in \mathcal{S}^n\} \tag{4}$$

and draw an immediate consequence of (2).
Lemma 1. *$C_F(\mathcal{W}) > 0$ iff for some n there are two n–length strategies $f^n, f'^n \in \mathcal{F}^n$ with disjoint corresponding convex hulls, that is, convex hull $\big(\{W^n(\cdot|f^n,s^n) : s^n \in \mathcal{S}^n\}\big) \cap$ convex hull $\big(\{W^n(\cdot|f'^n,s^n) : s^n \in \mathcal{S}^n\}\big) = \varnothing$.*
Next we need for our analysis two concepts, namely, for $x \in \mathcal{X}$

$$\mathcal{S}_x = \{s \in \mathcal{S} : \text{ for some } y \quad W(y|x,s) = 1\} \tag{5}$$

and

$$\mathcal{Y}_x = \{y \in \mathcal{Y} : \text{ for some } s \quad W(y|x,s) = 1\}. \tag{6}$$

Notice that both, $\mathcal{S}_x$ and $\mathcal{Y}_x$, can be empty and that $\mathcal{S}_x = \varnothing$ iff $\mathcal{Y}_x = \varnothing$.
Lemma 2. *If $C_F(\mathcal{W}) > 0$, then necessarily*

$$(i) \quad C_R(\mathcal{W}) > 0$$

$$\text{and} \quad (ii) \quad \mathcal{Y}_x \cap \mathcal{Y}_{x'} = \varnothing \text{ for some } x \neq x'.$$

Proof: If (i) does not hold, then there is a distribution P on $\mathcal{S}$ such that the matrix $\sum_s P(s)W(\cdot|\cdot,s)$ has identical rows. Therefore for all n and $P^n(s^n) = \prod_{t=1}^{n} P(s_t)$ also $\sum_{s^n} P^n(s^n)W_F^n(\cdot|\cdot,s^n)$ has identical rows and (as a special case of Lemma 1) $C_F(\mathcal{W}) = 0$.
If (ii) does not hold, then for all $x, x'(x \neq x')$ there are $y(x,x') \in \mathcal{Y}$ and $s(x,x')$, $s'(x,x') \in \mathcal{S}$ with the property $W\big(y(x,x')|x,s(x,x')\big) = W\big(y(x,x')|x',s'(x,x')\big) = 1$.
This implies that for all n and any two rows of W_F^n corresponding to the feedback strategies $f^n = (f_1, f_1, \dots, f_n)$ and $f'^n = (f'_1, f'_1, \dots, f'_n)$ we can choose

$y_1 = y(f_1, f_1')$, $s_1 = s(f_1, f_1')$, $s_1' = s'(f_1, f_1')$ and; for $t = 2, 3, \ldots, n$; $y_t = y\big(f_t(y^{t-1}), f_t'(y^{t-1})\big)$, $s_t = s\big(f_t(y^{t-1}), f_t'(y^{t-1})\big)$, and $s_t' = s\big(f_t(y^{t-1}), f_t'(y^{t-1})\big)$ such that
$W(y^n|f^n, s^n) = W(y^n|f'^n, s'^n) = 1$ and thus $C_F(\mathcal{W}) = 0$.
Quite remarkably also the converse of Lemma 2 holds. This is a much deeper result.

Positivity Theorem. *$C_F(\mathcal{W}) > 0$ iff (i) and (ii) in Lemma 2 hold.*

The rather sophisticated proof is based on the Coloring Lemma of Section 3, which is closely related to its predecessors in [3] and [7]. We give it in the last section so that readers, who are interested only in our coding scheme of Section 4 can skip it.

BALANCED COLORING

Lemma 3. *Let $\mathcal{Q} \subset \mathcal{P}(\mathcal{V})$ be a finite set of PD's on $\mathcal{V}$ and let there be associated with every $P \in \mathcal{Q}$ a family $\mathcal{E}(P)$ of subsets of $\mathcal{V}$ such that*

$$\alpha(P) \triangleq \max\left\{P(v) : v \in \bigcup_{E\in\mathcal{E}(P)} E\right\} < 1. \tag{7}$$

Now, if there are positive numbers $\eta(P)$ for all $P \in \mathcal{Q}$ such that for $k \geq 2$, $\delta \in (0,1)$ and all $E \in \mathcal{E}(P)$

$$\left(\frac{1}{\alpha(P)}\right)^{1-\delta}\left[\eta(P) - \frac{e}{2k}\alpha(P)^{\delta}P(E)\right] > ln\left\{2k\sum_{P\in\mathcal{Q}}|\mathcal{E}(P)|\right\}, \tag{8}$$

then there is a function $g : \mathcal{V} \to \{1, 2, \ldots, k\}$ which satisfies for all $P \in \mathcal{Q}, E \in \mathcal{E}(P)$, and $i \in \{1, 2, \ldots, k\}$

$$\left|P\big(g^{-1}(i) \cap E\big) - \frac{1}{k}P(E)\right| < \eta(P). \tag{9}$$

Furthermore, for $\delta = \frac{1}{4}$, $\eta(P) = 2\alpha(P)^{\frac{1}{4}}$, and $\alpha \triangleq \max_{P\in\mathcal{Q}}\alpha(P)$

$$\alpha^{-\frac{1}{2}} > ln\left[2k\sum_{P\in\mathcal{Q}}|\mathcal{E}(P)|\right] \tag{10}$$

implies (8) and thus (9) holds.

Proof: The idea behind the following probabilistic existence proof is to use a union bound argument to show that the probability of a randomly chosen coloring to be "bad" is less than 1. We color all $v \in \mathcal{V}$ at random independently and uniformly with k colors.

Next we introduce the RV's

$$\Psi_i(v) \quad = \begin{cases} 1, & \text{if } v \text{ gets color } i \\ 0 & \text{otherwise} \end{cases}$$

$$\text{and } Z_i^P(E) \quad = \textstyle\sum_{v \in E} P(v)\Psi_i(v) \quad \text{for} \quad P \in \mathcal{Q}.$$

With Bernstein's version of Chebyshev's inequality

$$\Pr\big(Z_i^P(E) > \frac{1}{k}P(E) + \eta(P)\big\}$$

$$\leq \exp_e\left\{-\alpha(P)^{-(1-\delta)}\left[\frac{1}{k}P(E)+\eta(P)\right]\right\}\cdot \mathbb{E}\exp_e\left\{\alpha(P)^{-(1-\delta)}\sum_{v\in E}P(v)\Psi_i(v)\right\}$$

$$= \exp_e\left\{-\alpha(P)^{-(1-\delta)}\left[\frac{1}{k}P(E)+\eta(P)\right]\right\}\cdot \prod_{v\in E}\mathbb{E}\exp_e\{\alpha(P)^{-(1-\delta)}P(v)\Psi_i(v)\}$$

$$= \exp_e\left\{-\alpha(P)^{-(1-\delta)}\left[\frac{1}{k}P(E)+\eta(P)\right]\right\}\times$$

$$\prod_{v\in E}\left(\frac{k-1}{k}+\frac{1}{k}\exp_e\{\alpha(P)^{-(1-\delta)}P(v)\}\right).$$

Using Lagrange's remainder formula for the Taylor series of the exponential function we continue with the upper bound

$$\exp_e\left\{-\alpha(P)^{-(1-\delta)}\left[\frac{1}{k}P(E)+\eta(P)\right]\right\}\times$$

$$\prod_{v\in E}\left\{1+\frac{1}{k}\left[\alpha(P)^{-(1-\delta)}P(v)+\frac{[\alpha(P)^{-(1-\delta)}P(v)]^2\cdot e}{2}\right]\right\}$$

and since $ln(1+x) < x$ for $x > 0$ with the upper bound

$$\exp_e\left\{-\alpha(P)^{-(1-\delta)}\left[\frac{1}{k}P(E)+\eta(P)-\frac{1}{k}\sum_{v\in E}P(v)-\frac{e}{2k}\alpha(p)^{-(1-\delta)}\sum_{v\in E}P^2(v)\right]\right\}$$

$$= \exp_e\left\{-\alpha(P)^{-(1-\delta)}\left[\eta(P)-\frac{e}{2k}\alpha(P)^{-(1-\delta)}\sum_{v\in E}P^2(v)\right]\right\}$$

$$\leq \exp_e\left\{-\alpha(P)^{-(1-\delta)}\left[\eta(P)-\frac{e}{2k}\alpha(P)^{-(1-\delta)}\cdot\sum_{v\in E}\alpha(P)P(v)\right]\right\},$$

because $P(v) \leq \alpha(P)$ for $v \in E$.
The last upper bound equals

$$\exp_e\left\{-\alpha(P)^{-(1-\delta)}\left[\eta(P)-\frac{e}{2k}\alpha(P)^{\delta}P(E)\right]\right\}.$$

Analogously, $\Pr\left\{Z_i^P(E) < \frac{1}{k}P(E) - \eta(P)\right\}$
$\leq \exp_e\left\{-\alpha(P)^{-(1-\delta)}\left[\eta(P) - \frac{e}{2k}\alpha(P)^{\delta}P(E)\right]\right\}$ for all $P \in \mathcal{Q}$, $E \in \mathcal{E}(P)$ and $i \in \{1, 2, \ldots, k\}$. This together with (8) implies (9).
Finally, since $\left(\frac{1}{\alpha(P)}\right)^{\frac{3}{4}}\left[2\alpha(P)^{\frac{1}{4}} - \frac{e}{2k}\alpha(P)^{\frac{1}{4}}P(E)\right] > \left(\frac{1}{\alpha(P)}\right)^{\frac{1}{2}} \geq \alpha^{-\frac{1}{2}}$
(10) implies (8).

THE TRICHOTOMY THEOREM

For the formulation of our main result we need a concept from [1].
With our set of matrices $\mathcal{W}$ we associate the set of stochastic $|\mathcal{X}| \times |\mathcal{Y}| - (0-1)$ matrices

$$\hat{\mathcal{W}} = \left\{\hat{W} : \hat{W}(\cdot|x) \in \mathcal{W}(x) \text{ for all } x \in \mathcal{X} \text{ and } \hat{W}(y|x) \in \{0, 1\} \text{ for all } y \in \mathcal{Y}\right\}, \tag{11}$$

where $\mathcal{W}(x) = \{W(\cdot|x, s), s \in \mathcal{S}$.
Let this set be indexed by the set $\hat{\mathcal{S}}$. Then we have that for all $\hat{s} \in \hat{\mathcal{S}}$ and $x \in \mathcal{X}$ there is an $s \in \mathcal{S}_x$ with

$$\hat{W}(\cdot|x, \hat{s}) = W(\cdot|x, s). \tag{12}$$

Of course, $\hat{\mathcal{W}}$ (and thus also $\hat{\mathcal{S}}$) can be empty. This happens exactly, if for some x $\mathcal{S}_x = \varnothing$ or (equivalently) $\mathcal{Y}_x = \varnothing$. These sets are defined in (5) and (6).
Shannon determined in [12] the zero–error feedback capacity $C_{0,F}(W)$ of a DMC W.
An alternate formula — called for by Shannon — was given in [1]. For

$$\hat{V}(\cdot|\cdot) = |\hat{\mathcal{S}}|^{-1} \sum_{\hat{s} \in \hat{\mathcal{S}}} \hat{W}(\cdot|\cdot, \hat{s})$$

this formula asserts

$$C_{0,F}(\hat{V}) = C_F(\hat{\mathcal{W}}) = \begin{cases} \max\limits_{P} \min\limits_{W \in \overline{\hat{\mathcal{W}}}} I(P, W), & \text{if } \mathcal{Y}_x \cap \mathcal{Y}_{x'} = \varnothing \text{ for some } x, x' \\ 0 & \text{otherwise.} \end{cases} \tag{13}$$

Moreover, we have an inequality for this quantity.
Lemma 4. *$C_F(\mathcal{W}) \leq C_F(\hat{\mathcal{W}})$, if $\hat{\mathcal{W}} \neq \varnothing$.*
Proof: It suffices to show that every feedback code with maximal error probability $\varepsilon < 1$ for $\mathcal{W}$ is a code for $\hat{\mathcal{W}}$. Indeed, otherwise there exists a feedback code for $\mathcal{W}$ with two encoding functions $f^n = (f_1, \ldots, f_n)$ and $f'^n = (f_1', \ldots, f_n')$ such that for some $y^n \in \mathcal{Y}^n$ and $\hat{s}^n, \hat{s}'^n \in \hat{\mathcal{S}}^n$

$$\hat{W}^n(y^n|f^n, \hat{s}^n) = \hat{W}^n(y^n|f'^n, \hat{s}'^n) = 1.$$

But then, if we choose s_t, s_t' corresponding to $\left(f_t(y^{t-1}), \hat{s}_t\right)$ and $\left(f_t'(y^{t-1}), \hat{s}_t'\right)$, respectively, according to (12), we get

$$W^n(y^n|f^n, s^n) = W^n(y^n|f'^n, s'^n) = 1,$$

a contradiction.
Clearly by averaging we see that an ε–code with feedback for the AVC $\mathcal{W}$ is an ε–code for the AVC with feedback and therefore $C_F(\mathcal{W}) = C_F(\overline{\mathcal{W}})$. Furthermore, since feedback does not increase the capacity of an individual DMC $W \in \overline{\mathcal{W}}$ we have that

Lemma 5. $C_F(\mathcal{W}) = C_F(\overline{\mathcal{W}}) \leq C_R(\mathcal{W})$.

We are now ready to state our main result.

Trichotomy Theorem.

$$C_F(\mathcal{W}) = \begin{cases} 0, & \textit{iff } C_R(\mathcal{W}) = 0 \textit{ or } \mathcal{Y}_x \cap \mathcal{Y}_{x'} \neq \varnothing \\ & \textit{for all } x, x' \in \mathcal{X} \qquad \text{(i)} \\ C_R(\mathcal{W}), & \textit{if } C_F(\mathcal{W}) > 0 \textit{ and } \mathcal{Y}_x = \varnothing \\ & \textit{for some } x \qquad \text{(ii)} \\ \min\{C_R(\mathcal{W}), C_F(\hat{\mathcal{W}})\}, & \textit{if } C_F(\mathcal{W}) > 0 \textit{ and } \mathcal{Y}_x \neq \varnothing \\ & \textit{for all } x. \qquad \text{(iii)} \end{cases}$$

Remark 1: There is almost no connection between the values of $C_R(\mathcal{W})$ and $C_F(\hat{\mathcal{W}})$.

Example 1:
Choose $\mathcal{X} = \mathcal{S} = \{1, 2, \ldots, a\}$, $\mathcal{Y} = \{1, 2, \ldots, a, b\}$, and $\mathcal{W}$ as set of matrices W with

$$W(y|x, s) = 1, \quad \text{if} \quad x \neq s \quad \text{and} \quad y = x \quad \textbf{or} \quad x = s, y = b.$$

Then $C_F(\hat{\mathcal{W}}) = 0$, but with P as uniform distribution on $\mathcal{X}$,

$$C_R(\mathcal{W}) \geq \min_{W \in \overline{\mathcal{W}}} I(P, W) = \left(1 - \frac{1}{a}\right) \log a$$

and this goes to infinity with a going to infinity.

Example 2:
Choose $\mathcal{X}' = \{0, 1, \ldots, a\}$, $\mathcal{S}' = \{1, 2, \ldots, a\}$, $\mathcal{Y}' = \{0, 1, \ldots, a, b\}$ and define $\mathcal{W}'$ as set of matrices with $W(y|x, s) = 1$, if $x = y = 0$ (for every s) **or** $x \neq 0$, $x \neq s$ and $y = x$ **or** $x = s$, $y = b$, $x \neq 0$.
Then $C_F(\hat{\mathcal{W}}') = \log 2 > 0$, however for $\mathcal{W}$ in Example 1 $C_R(\mathcal{W}') > C_R(\mathcal{W})$. So $C_R(\mathcal{W}')$ can be arbitrary large and much larger than a positive $C_F(\hat{\mathcal{W}})$.

Example 3:
Choose $\mathcal{X} = \mathcal{Y} = \mathcal{S} = \{0, 1\}$, $W(\cdot|\cdot, 0) = \begin{pmatrix} \frac{1}{2} & \frac{1}{2} \\ \frac{1}{2} & \frac{1}{2} \end{pmatrix}$, $W(\cdot|\cdot, 1) = \begin{pmatrix} 1 & 0 \\ 0 & 1 \end{pmatrix}$.
Then $C_R(\mathcal{W}) = 0$ and $C_F(\hat{\mathcal{W}}) = 1$.
Finally, we formulate the Trichotomy Theorem in a more elegant, but less informative way. For this we define

$$C_F^{\infty}(\hat{\mathcal{W}}) = \begin{cases} C_F(\hat{\mathcal{W}}), & \text{if } \hat{\mathcal{W}} \neq \varnothing \\ \infty, & \text{if } \hat{\mathcal{W}} = \varnothing. \end{cases} \tag{14}$$

Then Lemma 4 says that always

$$C_F(\mathcal{W}) \le C_F^\infty(\hat{\mathcal{W}})$$

and with Lemma 5 we conclude that

$$C_F(\mathcal{W}) \le \min\big(C_R(\mathcal{W}), C_F^\infty(\hat{\mathcal{W}})\big). \tag{15}$$

Furthermore, now (ii) and (iii) say that there is equality in (15), if $C_F(\mathcal{W}) > 0$. Finally, if $C_F(\mathcal{W}) = 0$, then by (i) and (13) either $C_R(\mathcal{W}) = 0$ **or** $C_F(\hat{\mathcal{W}}) = 0$. We summarize our findings.

Capacity Theorem. $C_F(\mathcal{W}) = \min\{C_R(\mathcal{W}), C_F^\infty(\hat{\mathcal{W}})\}$.

PROOF OF THE TRICHOTOMY THEOREM

It remains to be seen that for $C_F(\mathcal{W}) > 0$
(ii) $C_F(\mathcal{W}) \ge C_R(\mathcal{W})$, if $\mathcal{S}_x = \varnothing$ for some x,
and
(iii) $C_F(\mathcal{W}) \ge \min\{C_R(\mathcal{W}), C_F(\hat{\mathcal{W}})\}$ otherwise.
For the convenience of the reader we mention first that in the case, where $\mathcal{W}$ contains only 0–1–matrices, we are in the case (iii) and (13) gives the desired result.
In the other extreme case (ii) we have $\hat{\mathcal{W}} = \varnothing$ and can use Lemma 3 (to establish a common random experiment) in conjunction with the elimination technique of [2]. (This approach of [7] works here even for maximal errors, because the "edges E" are big enough, if 0–1–distributions are excluded. In contrast to the previous work now the sender cannot randomize!)
To be specific, for any $\gamma > 0$ choose $l \sim \frac{n}{2}\,\gamma\, C_R^{-1}(\mathcal{W})$, an $x_0 \in \mathcal{X}$ with $\mathcal{S}_{x_0} = \varnothing$, and the encoding

$$f_t(y^{t-1}) = x_0 \quad \text{for} \quad 1 \le t \le l. \tag{16}$$

Next, clearly for $x_0^l = (x_0, \dots, x_0)$ and all y^l, s^l

$$W^l(y^l|x_0^l, s^l) \le \omega^{*l} < 1, \tag{17}$$

where

$$\omega^* = \max\{W(y|x,s) : W(y|x,s) \ne 1, x \in \mathcal{X}, s \in \mathcal{S}, \text{ and } y \in \mathcal{Y}\}. \tag{18}$$

By applying Lemma 3 to $\mathcal{Q} = \{W^l(\cdot|x_0^l, s^l) : s^l \in \mathcal{S}^l\}$, $k = (n-l)^2$, $\mathcal{E}(P) = \{\mathcal{Y}^l\}$ for all P, $\alpha = w^{*^l}$ in (10) then when l is sufficiently large, so that $w^{*-\frac{1}{2}l} > ln(n-l)^2|\mathcal{S}|^l$, i.e. (10) holds, there is a coloring or equivalently a partition $\{A_i\}_{i=1}^{(n-l)^2}$ of $\mathcal{Y}^l$ such that for all $s^l \in \mathcal{S}^l$ and $i = 1, 2, \dots, (n-l)^2$

$$\left| W^l(A_i|x_0^l, s^l) - \frac{1}{(n-l)^2} \right| < 2^{-l\tau} \tag{19}$$

for a positive $\tau \left(= -\frac{1}{8}\log w^*\right)$, which is independent of l.
For this we have used l letters and for the remaining $n-l$ letters we use a random code with rate $C_R(\mathcal{W}) - \frac{\gamma}{2}$, maximum error probability $\frac{\lambda}{2}$, and with ensemble size $(n-l)^2$. Its existence is guaranteed by the elimination technique of [2].
Now, after having sent x_0^l and received $y^l \in A_i$, which is also known to the sender, because of the feedback, for any message m the m–th codeword in the i–th code of the ensemble is send next.
This n–length feedback code achieves a rate

$$\frac{1}{n}\left(n - \frac{n}{2}\,\gamma\, C_R^{-1}(\mathcal{W})\right)\left(C_R(\mathcal{W}) - \frac{\gamma}{2}\right) \geq C_R(\mathcal{W}) - \delta$$

and a maximum error probability less than $(n-l)^2 2^{-l\tau} + \frac{\lambda}{2} < \lambda$, when l is large enough.
The main issue is really to prove the direct part for the mixed case:

$$\hat{\mathcal{W}} \neq \varnothing \text{ and } \mathcal{W} \smallsetminus \hat{\mathcal{W}} \neq \varnothing,\ C_F(\mathcal{W}) > 0.$$

We design a strategy by compounding *four types* of codes. There germ is the iterative list reduction code of [1].
However, now we must achieve a higher rate by incorporating also codes based on common randomness. The detailed structure will become clear at the end of our description.
We begin with the codes announced.

1. List reducing or coloring code (LROCC)
As in [1] we start with $\mathcal{T}_P^l$, the set of P–typical sequences in $\mathcal{X}^l$, where $P \in \mathcal{P}_l(\mathcal{X}) = \{P \in \mathcal{P}(\mathcal{X}) : \mathcal{T}_P^l \neq \varnothing\}$.
However, right in the beginning we gain a certain freedom by deviating from [1] by choosing parameters such that $|\mathcal{T}_P^l|$ is much smaller than the size of the set of messages $\mathcal{M}$. An (l,ξ,c) LROCC (where the role of parameter ξ becomes clear in (21) and (22) is defined by a triple (g, L, K) of functions, which we now explain.

Function $g : \mathcal{L} \to \mathcal{T}_P^l$ (called *balanced partition function*) is chosen such that

$$\left||g^{-1}(x^l)| - |g^{-1}(x'^l)|\right| \leq 1 \text{ for all } x^l, x'^l \in \mathcal{T}_P^l. \tag{20}$$

Function $L : \mathcal{Y}^l \to 2^{\mathcal{L}}$
This function, which we call *list function*, assigns to every $y^l \in \mathcal{Y}^l$ a sublist of $\mathcal{L}$ as follows. Define first for $x^l \in \mathcal{X}^l$, $y^l \in \mathcal{Y}^l$, and $\mathcal{Y}_x$

$$\tilde{\delta}(x^l, y^l) = |\{t : y_t \notin \mathcal{Y}_{x_t}\}|, \tag{21}$$

the *discriminator*.

Then set

$$L(y^l) = \{v \in \mathcal{L} : \tilde{\delta}(g(v), y^l) < \xi\} \text{ for } y^l \in \mathcal{Y}^l. \tag{22}$$

We need later interpretations for the relation $v \in L(y^l)$. Since by our assumptions $\mathcal{Y}_x \neq \varnothing$ for all x, $\tilde{\delta}(x^l, y^l) < \xi$ implies that a $y'^l \in \mathcal{Y}^l$ can be found so that (in the Hamming distance)

$$d_H(y^l, y'^l) < \xi \tag{23}$$

and

$$y'_t \in \mathcal{Y}_{x_t} \quad \text{for all} \quad t = 1, 2, \ldots, l. \tag{24}$$

Equivalently, we can say that there is

$$\overline{\hat{W}} \in \overline{\hat{\mathcal{W}}} \quad \text{with} \quad y'^l \in \mathcal{T}^l_{\overline{\hat{W}}}(x^l).$$

Also, by (22) – (24) for all $y^l \in \mathcal{Y}^l$

$$\frac{1}{l} \log |L(y^l)| < \frac{1}{l} \log |\mathcal{L}| - \min_{\overline{\hat{W}} \in \overline{\hat{\mathcal{W}}}} I(P, \overline{\hat{W}}) + u(l, \xi), \tag{25}$$

where u is a function with

$$u(l, \xi) \to 0 \quad \text{as} \quad \frac{\xi}{l} \to 0 \quad \text{and} \quad l \to \infty. \tag{26}$$

(Notice: when $\xi = 1$, then L is a list reduction via $\hat{\mathcal{W}}$ as in [1].)
Function $K : \mathcal{Y}^l \to \{1, 2, \ldots, c\}$
In this *coloring function* we choose c of *polynomial growth in* l. Let $\mathcal{Q} = \{W^l(\cdot|x^l, s^l) : x^l \in \mathcal{X}^l, s^l \in \mathcal{S}^l\}$, $\mathcal{E}(W^l(\cdot|x^l, s^l)) = \{\{y^l : \tilde{\delta}(x^l, y^l) \geq \xi\}\}$ and $k = c$ in Lemma 3.
Then by Lemma 3 we can also assume that for all $x^l \in \mathcal{T}^l_P$, $s^l \in \mathcal{S}^l$, and $j \in \{1, 2, \ldots, c\}$

$$\left|W^l\left(K^{-1}(j) \cap \{y^l : \tilde{\delta}(x^l, y^l) \geq \xi\}|x^l, s^l\right) - c^{-1} W^l\left(\{y^l : \tilde{\delta}(x^l, y^l) \geq \xi\}|x^l, s^l\right)\right|$$
$$< 2\omega^{*\frac{1}{4}\xi}, \tag{27}$$

because $\tilde{\delta}(x^l, y^l) \geq \xi$ implies $W^l(y^l|x^l, s^l) \leq \omega^{*\xi}$ for all s^l (ω^* was defined in (18)) and consequently, $w^{*-\frac{1}{2}\xi} > \log[2c|\mathcal{X}|^l|\mathcal{S}|^l]$, i.e. (10) holds for sufficiently large ξ satisfying (26).

2. Index Code (IC)

This code has two codewords of length j and error probability μ. The codewords stand for messages L, K. They are used by the sender (based on the discriminator) to inform the receiver whether next he uses reducing the list, by sending L, or coloring on the output, by sending K.

3. Eliminated correlated code (ECC)

An m–length and (maximal) μ–error probability ECC is a *family*

$$\{\{(u_i^q, D_i^q) : 1 \leq i \leq M\} : 1 \leq q \leq m^2\}$$

of m^2 codes with the properties

$$m^{-2}\sum_{q=1}^{m^2} W^m(D_i^q|u_i^q, s^n) > 1-\mu \quad \text{for all} \quad s^n \in \mathcal{S}^n \quad \text{and all} \quad i=1,\dots,M \quad (28)$$

and

$$m^{-1}\log M > C_R(\mathcal{W}) - \delta'. \quad (29)$$

Their existence was proved in [2].

4. $(k, 2^{\gamma k}, \mu)$–Code

This is just an ordinary feedback code for $\mathcal{W}$ of length k, rate γ, and maximal error probability μ. Its existence is provided by $C_F(\mathcal{W}) > 0$.

Choice of parameters:

Before we present our coding algorithm we adjust the parameters. It is convenient to have the abbreviation

$$C = \min\big(C_R(\mathcal{W}), C_F(\hat{\mathcal{W}})\big). \quad (30)$$

a.) Let P attain the maximum in $\max_{P' \in \mathcal{P}_l(\mathcal{X})} \min_{\overline{\hat{W} \in \hat{\mathcal{W}}}} I(P', \overline{\hat{W}})$.

b.) Fix now any $\delta \in (0, C)$ and $\lambda \in (0,1)$.

c.) By our assumption $C_F(\mathcal{W}) > 0$ there is a positive number γ so that for large enough k and $\log M \le k \cdot \gamma$ (k, M, μ)–codes exist.

d.) Define

$$r_o = \left\lceil \frac{2}{\delta}\left(\frac{\log|\mathcal{X}|}{\gamma} + 2\right) C^2 \left(C - \frac{\delta}{2}\right)^{-1} \right\rceil \quad (31)$$

and let j be a fixed integer such that a j–length IC with error probability $\frac{\lambda}{4r_o}$ exists.

e.) Let ξ increase with l, but keep for sufficiently large l $\frac{\xi}{l}$ so small that for the u in (25)

$$u(l, \xi) < \frac{\delta}{4}. \quad (32)$$

f.) Insure

$$l > r_o j \quad (33)$$

and for the message set $\mathcal{M}$ set

$$n_0 = \log|\mathcal{M}| = \left\lceil \frac{2}{\delta}\left(\frac{\log|\mathcal{X}|}{\gamma} + 2\right) C^2 l \right\rceil. \quad (34)$$

g.) Require l and also ξ to be so large that the coloring function K for the LROCC can be obtained with Lemma 3 and still

$$n_0^2 \omega^{*\xi/4} < \frac{\lambda^2}{64 r_o}. \tag{35}$$

h.) Finally we make l so large that all codes in the following algorithm exist.

Encoding Algorithm
Begin:
Input: $v \in \mathcal{M}$

1. Set $i := 0$ and let $\mathcal{L}_i := \mathcal{M}$, go to 2.

2. If $|\mathcal{L}_i| \geq |\mathcal{T}_P^l|$, then let $m_i := \left\lfloor \frac{\log |\mathcal{L}_i|}{C_R(\mathcal{M}) - \frac{\delta}{2}} \right\rfloor$, encode $\mathcal{L}_i$ to an (l, ξ, m_i^2) LROCC (g, L, K) over $\mathcal{T}_P^l$, send $g(v) := x^l$ to the receiver, go to 3.

   Otherwise, go to 5.

3. Receive the output y^l and encode a j–length IC with $\frac{\lambda}{4r_o}$–error probability.

   If $\tilde{\delta}(x^l, y^l) < \xi$, send the word "$L$" of the IC to the receiver. Let $i := i+1$, $\mathcal{L}_i := L(y^l)$ and go to 2.

   Otherwise send the word "K" of the IC to the receiver, let $q = K(y^l)$, go to 4.

4. Encode $\mathcal{L}_i$ to an m_i–length ECC with $\frac{\lambda}{4}$–error probability and send the codeword u_v^q to the receiver, go to 6.

5. Encode $\mathcal{L}_i$ to a $\left(k, |\mathcal{L}_i|, \frac{\lambda}{4}\right)$–code with rate γ and send the codeword standing for v to the receiver, go to 6.

6. Stop.

End.

Decoding Algorithm
Begin:

1. Set $i := 0$ and let $\mathcal{L}_i = \mathcal{M}$, go to 2.

2. If $|\mathcal{L}_i| \geq |\mathcal{T}_P^l|$, go to 3.

   Otherwise go to 5.

3. Receive (y^l, y^j) and decode y^j for the j–length IC.

   If the decoding result is "L", let $i := i + 1$, $\mathcal{L}_i = L(y^l)$, go to 2.

Otherwise let $q = K(y^l)$ and go to 4.

4. Let $m_i := \left\lfloor \frac{\log|\mathcal{L}_i|}{C_R(\mathcal{M}) - \frac{\delta}{2}} \right\rfloor$, receive y^{m_i} and decode y^{m_i} for the q–th value–code of the m_i–length ECC, go to 6.

5. Receive y^k and decode it for the $\left(k, |\mathcal{L}_i|, \frac{\lambda}{4}\right)$ code with rate γ and length k, go to 6.

6. Stop

End.

Analysis

According to the choice of our P, by (25) and (32), for sufficiently large l we have

$$\frac{1}{l}\log|\mathcal{L}_{i+1}| < \frac{1}{l}\log|\mathcal{L}_i| - C_F(\hat{\mathcal{W}}) + \frac{\delta}{2}, \tag{36}$$

or in other words

$$\log|\mathcal{L}_i| < \log|\mathcal{M}| - ilC_F(\hat{\mathcal{W}}) + il\frac{\delta}{2} \leq \log|\mathcal{M}| - ilC + il\frac{\delta}{2}. \tag{37}$$

Thus, according to our encoding program, by (31), (34), and (37), at most r_o LROCC–IC–pairs may be encoded, and at most one "K". If it exists, it must be in the last IC. Therefore we can define the RV U as

$$U = \begin{cases} r, & \text{if } r \text{ LROCC–IC–pairs are sent and} \\ & \text{the last sent word of IC is "}K\text{"} \\ r_o + 1, & \text{if no "}K\text{" is sent,} \end{cases} \tag{38}$$

or in other words,

$$U = r \leq r_o$$

$\Leftrightarrow$ After the message set is reduced $r-1$ times, the "r–th output"is "colored" and then the message is sent by the value "with this color" in an ECC.

$$U = r_o + 1$$

$$\Leftrightarrow \quad \text{After the size of the message set is reduced to less than } |\mathcal{T}_P^l|, \\ \text{the message is sent by the ordinary (feedback) code with rate } \gamma. \tag{39}$$

The rate:

Although the encoding algorithm may produce sequences with different lengths, by obvious reasons, we only need their common bound, say b.

Moreover, we only have to show that

$$b \leq \left(C - \frac{\delta}{2}\right)^{-1} \log|\mathcal{M}| + \left(\frac{\log|\mathcal{X}|}{\gamma} + 2\right) l. \tag{40}$$

This is so, because by an elementary calculation, for any positive $a, aC^2 \leq \frac{\delta}{2}\log|\mathcal{M}|$ implies $\left(C - \frac{\delta}{2}\right)^{-1}\log|\mathcal{M}| + a \leq (C-\delta)^{-1}\log|\mathcal{M}|$ and then (34) and (40) imply that the lengths of the encoding sequences are bounded by $(C-\delta)^{-1}\log|\mathcal{M}|$.

Case $U = r \leq r_o$:

By (39), after having been reduced $r-1$ times, the "message list" with size at most $\log|\mathcal{M}| - (r-1)l\left(C - \frac{\delta}{2}\right)$ (by (37)), is encoded by an $\left\lfloor \left(C_R(\mathcal{M}) - \frac{\delta}{2}\right)^{-1}\left(\log|\mathcal{M}| - (r-1)l\left(C - \frac{\delta}{2}\right)\right)\right\rfloor$-length ECC.

Therefore the total length of the encoding sequences is not exceeding $r(l+j) + \left(C - \frac{\delta}{2}\right)^{-1}\left(\log|\mathcal{M}| - (r-1)l\left(C - \frac{\delta}{2}\right)\right) \leq \left(C - \frac{\delta}{2}\right)^{-1}\log|\mathcal{M}| + r_o j + l$ $\leq \left(C - \frac{\delta}{2}\right)^{-1}\log|\mathcal{M}| + 2l$ (by (33))

Case $U = r_o + 1$:

By (31), (33), (34), (39) and the wellknown fact that $|\mathcal{T}_P^l| \leq 2^{l\log|\mathcal{X}|}$, the total lengths of encoding sequences are bounded by

$$\begin{aligned} r_o(l+j) + \frac{\log|\mathcal{X}|}{\gamma}l &\leq \left[\left(l\left(C - \tfrac{\delta}{2}\right)\right)^{-1}\log|\mathcal{M}| + 1\right] l + r_o j + \tfrac{\log|\mathcal{X}|}{\gamma} l \\ &\leq \left(C - \tfrac{\delta}{2}\right)^{-1}\log|\mathcal{M}| + \left(2 + \tfrac{\log|\mathcal{X}|}{\gamma}\right) l, \end{aligned}$$

i.e. (40).

The error probability:

Denote by E, E_I, and E_γ, the events that errors occur *at any step*, at decoding an IC, and at the decoding of the ordinary code with rate γ, respectively, and by $\Pr(\cdot|v, s^n)$, $v \in \mathcal{M}$, $s^n \in \mathcal{S}^n$, the corresponding output probability, when v is sent and the channel is governed by s^n. Notice that E_I, $E_\gamma \subset E$. We have to upperbound $\Pr(E|v, s^n)$. For this we first notice that

$$\Pr(E_I|v, s^n) < \sum_{r=1}^{r_o} \Pr(U = r|v, s^n) \cdot r\frac{\lambda}{4r_o} \leq \frac{\lambda}{4} \tag{41}$$

and therefore

$$\Pr(E|v, s^n) < \frac{\lambda}{4} + \Pr(E|E_I^c, v, s^n). \tag{42}$$

We are left with upper bounding

$$\Pr(E|E_I^c, v, s^n) = \sum_{r=0}^{r_o+1} \Pr(U = r|E_I^c, v, s^n)\Pr(E|E_I^c, U = r, v, s^n). \tag{43}$$

Here the last summand is upper bounded by the error probability $\frac{\lambda}{4}$ in a $\left(k, |\mathcal{L}_r|, \frac{\lambda}{4}\right)$ –code, which is used for $r = r_o + 1$, because

$$\Pr(E|E_I^c, U = r_o + 1, v, s^n) = \Pr(E_\gamma|v, s^n) < \frac{\lambda}{4}, \tag{44}$$

Finally, for $r \leq r_o$ by our coding rules

$$W^l(\{y^l : \tilde{\delta}(x^l, y^l) \geq \xi\}|x^l, s^l(r)) \geq \Pr(U = r|E_I^c, v, s^n) \tag{45}$$

where $x^l \in \mathcal{T}_P^l$ is the value of the r–th $g(v)$, $s^l(r)$ is the segment of s^n corresponding to the r–th LROCC.
Therefore by (27), (28), and (35) in the case

$$\Pr(U = r|E_I^c, v, s^n) \geq \frac{\lambda}{4r_o}$$

and with the convention that $s^{m_r}(m_r)$ is the last part of s^n

$$\Pr(E|E_I^c, U = r, v, s^n) = \left(W^l(\{y^l : \tilde{\delta}(x^l, y^l) \geq \xi\}|x^l, s^l(r))\right)^{-1}$$

$$\times \sum_{q=1}^{m_r^2} W^l(K^{-1}(q) \cap \{y^l : \tilde{\delta}(x^l, y^l) \geq \xi\}|x^l, s^l(r)) W^{m_r}((D_v^q)^c|u_v^q, s^{m_r}(m_r))$$

$$\leq \sum_{q=1}^{m_r^2} m_r^{-2} W^{m_r}((D_v^q)^c|u_v^q, s^{m_r}(m_r)) + \left(\frac{\lambda}{4r_o}\right)^{-1} \cdot 2m_r^2 \omega^{*\frac{1}{4}\xi} < \frac{\lambda}{4},$$

This and (42) – (44) imply

$$\Pr(E|v, s^n) < \frac{\lambda}{4} + \frac{\lambda}{4} + \left(1 \cdot \frac{\lambda}{4} + \frac{\lambda}{4} \cdot 1\right) \leq \lambda.$$

PROOF OF THE POSITIVITY THEOREM

We shall, in this section, show that the conditions in Lemma 2 are also sufficient for the positivity. To this end we assume a contradiction, (i) and (ii) in Lemma 2 hold, that is,

$$C_R(\mathcal{W}) > 0 \tag{46}$$

and w.l.o.g. for $0, 1 \in \mathcal{X}$

$$\mathcal{Y}_0 \cap \mathcal{Y}_1 = \varnothing, \tag{47}$$

but that

$$C_F(\mathcal{W}) = 0. \tag{48}$$

We establish the desired result by deriving a contradiction. First we rewrite (46) in the form

$$\theta \triangleq \min_{\pi \in \mathcal{P}(\mathcal{S})} \max_{x, x', y} |\sum_s \pi(s) W(y|x', s) - \sum_s \pi(s) W(y|x, s)| > 0 \tag{49}$$

and with Lemma 1 (48) in the following form: for any two encoding functions f_0^n and f_1^n there exist PD's α^n and β^n on $\mathcal{S}^n$ such that for all $y^n \in \mathcal{Y}^n$

$$\sum_{s^n} \alpha^n(s^n) W^n(y^n | f_0^n, s^n) = \sum_{s^n} \beta^n(s^n) W^n(y^n | f_1^n, s^n). \tag{50}$$

The proof in this part is much harder than others in the paper and as well as in most papers in this direction, which contain only a few new ideas and techniques. So it may be hard to understand for some readers. Therefore, we first describe the main idea and give an outline of the proof.

For an input, a sequence of states (or a distribution on the sequences of states) governing the channel and a coloring of the output space, a subset in the output is said to be well colored if its members are colored with (nearly) uniform probability. We have seen that if one can find an input such that for all distributions on the sequences of states the output space is well colored (with a large probability), then the positivity follows. In fact, we shall see that by Lemma 1 any well colored subset is sufficient. However it cannot always be done, and actually it is not hard to see that one can never find such an input, if for all $x \in \mathcal{X}$ $S_x \neq \varnothing$ (unless (50) holds). To obtain the well colored subsets we have to construct 2 encoding functions f_0^n and f_1^n and to show that under the assumption (50) one is always able to find a well colored subset for both of them. Our functions consist of 3 blocks with lengths m_1, m_2 and 1, here m_1 and m_2 will be chosen carefully.

In the first two blocks and for both encoding functions, only letters "0" and "1" satisfying (47) are used. The first blocks of f_0^n and f_1^n are m_1 zeros and ones respectively. At the same time, the output space $\mathcal{Y}^{m_1}$ is colored by 2^{2m_2} colors, say $\{(b^{m_2}, b'^{m_2}) : b^{m_2}, b'^{m_2} \in \{0,1\}^{m_2}\}$. For the output y^{m_1} colored by (b^{m_2}, b'^{m_2}), the encoding functions f_0^n and f_1^n encode in the second block to b^{m_2} and b'^{m_2}, respectively. We use the Balanced Coloring Lemma 3, and color $\mathcal{Y}^m$ in the following way.

— Let $\delta^*(x^m, s^m) = |\{t : s_t \notin S_{x_t}\}|$. Then for 0^{m_1} and all s^{m_1} with $\delta^*(0^{m_1}, s^{m_1}) \geq l_1$ (i.e. the number of t's such that $s_t \in S_{x_t}$ is not "too large") for a properly chosen l_1, $\mathcal{Y}^{m_1}$ is well colored.

— For 1^{m_1} and all $s^{m_1} \in \mathcal{S}^{m_1}$ all subsets in $\mathcal{Y}^{m_1}$ of the form $A^{m_1} = \prod_{t=1}^{m_1} A_t$, $A_t \in \{\mathcal{Y}, \mathcal{Y}_0\}$, and $|\{t : A_t = \mathcal{Y}_0\}| = m_1 - l_1 + 1$, are well colored.

We shall show in Lemma 6 below that if for a probability measure μ on $\mathcal{S}^m$ and fixed $x^m \in \mathcal{X}^m$ $\mu\big(s^n : \delta^*(x^m, s^m) < l\big)$ is sufficiently small, then (for some coloring for x^m and μ), $\mathcal{Y}^{m_1}$ is well colored.

Thus,

Case 1: If $\alpha^n\big(s^n : \delta^*(0^{m_1}, s^{m_1}) < l_1\big)$ is sufficiently small, then for 0^{m_1} and α^m, $\mathcal{Y}^{m_1}$ (and $\mathcal{Y}^{m_1} \times L$ for all $L \subset \mathcal{Y}^{m_2+1}$) is well colored.

Moreover in Lemma 7 below we shall show

Case 2: If the condition in Case 1 does not hold, under condition (50) one can always find an A^{m_1} such that for 1^{m_1} and β^n, A^{m_1} (and $A^{m_1} \times L$ for all

$L \subset \mathcal{Y}^{m_2+1}$) is well colored. Thus in the first round of coloring at least for one input we can find a well colored subset.
Next we use the Balanced Coloring Lemma 3 again, but this time we color $\mathcal{Y}^{m_2}$ such that for 0^{m_1} and s^{m_1} with $\delta^*(0^{m_2}, s^{m_2}) \geq l_2$ (for suitable l_2) and for 1^{m_1} and s^{m_1} with $\delta^*(1^{m_2}, s^{m_2}) \geq l_2$, $\mathcal{Y}^{m_2}$ is well colored.
The hard kernel in the proof is Lemma 8, which we call the Crowd Lemma. It means that if the decoding functions (in the second block) take sufficiently many values and those values crowd the input space, one can always find "good pairs".
We shall show there that, because in the first block we can always for at least one encoding function find a well colored subset, we can always find a pair (b^{m_2}, b'^{m_2}) (as values for f_0^n and f_1^n, respectively, in the second block), such that for the probability distribution α^n or its conditional probability under certain conditions (probability distribution β^n or its conditional distribution under certain conditions), the probability of $\alpha^n\big(s^{m_2} : \delta^*(b^{m_2}, s^{m_2}) < l_2\big)$, $\big(\beta^n\big(s^{m_2} : \delta(b'^{m_2}, s^{m_2}) < l_2\big)\big)$ for suitable l_2 is sufficiently small.
Thus by Lemma 6 again, we show that for both, f_0^n and α^n and f_1^n and β^n, $\mathcal{Y}^{m_2}$ is well colored. This will complete our proof. Now let us start it.
First we define a pair (f_0^n, f_1^n) of encoding functions and then show that for them (49) and (50) cannot hold simultaneously. *The definition is given in four steps.*

1. Let $m_1 > l_1 > m_2 > l_2$ and $n = m_1 + m_2 + 1$ be (large) integers depending on a (small) real $\varepsilon > 0$, to be specified later, such that

$$\frac{l_2}{m_2}, \frac{m_2}{l_1}, \frac{l_1}{m_1} \sim \varepsilon. \tag{51}$$

2. Recall the definition of $\mathcal{S}_0, \mathcal{S}_1$ in (5). For $b^m \in \{0,1\}^m, s^m \in \mathcal{S}^m$ we introduce the "distance"

$$\delta^*(b^m, s^m) \triangleq |\{t : s_t \notin \mathcal{S}_{b_t}\}| \tag{52}$$

and for m_1 the sets of PD's

$$\mathcal{P}_1 \triangleq \{W^{m_1}(\cdot|0^{m_1}, s^{m_1}) : \delta^*(0^{m_1}, s^{m_1}) \geq l_1\}, \tag{53}$$

$$\mathcal{P}_2 \triangleq \{W^{m_1}(\cdot|1^{m_1}, s^{m_1}) : s^{m_1} \in \mathcal{S}^{m_1}\}, \tag{54}$$

and the set of output sets

$$\mathcal{A} \triangleq \{A^{m_1} = \prod_{t=1}^{m_1} A_t : A_t \in \{\mathcal{Y}, \mathcal{Y}_0\} \text{ and } |\{t : A_t = \mathcal{Y}_0\}| = m_1 - l_1 + 1\}. \tag{55}$$

We now apply the (balanced coloring) Lemma 3 for the choices $\mathcal{V} = \mathcal{Y}^{m_1}, \mathcal{Q} = \mathcal{P}_1 \cup \mathcal{P}_2$, and

$$\mathcal{E}(P) = \left\{ \begin{array}{ll} \{\mathcal{Y}^{m_1}\}, & \text{if } P \in \mathcal{P}_1 \\ \mathcal{A}, & \text{if } P \in \mathcal{P}_2. \end{array} \right\}, \tag{56}$$

and color $\mathcal{Y}^{m_1}$ with a coloring function $g = (\Phi_1, \Psi_1) : \mathcal{Y}^{m_1} \to \{0,1\}^{m_1} \times \{0,1\}^{m_1}$ with $k = 2^{2m_2}$ colors.

Let

$$w \triangleq \max\{W(y|x,s) : W(y|x,s) \neq 1, x = 0,1, s \in \mathcal{S} \text{ and } y \in \mathcal{Y}\}. \tag{57}$$

Denote the inverse image of the coloring function g for (b^{m_2}, b'^{m_2}) by

$$\Omega_1(b^{m_2}, b'^{m_2}) \triangleq g^{-1}(b^{m_2}, b'^{m_2}) = \Phi_1^{-1}(b^{m_2}) \cap \Psi_1^{-1}(b'^{m_2}) \tag{58}$$

and the subset of A^m colored by (b^{m_2}, b'^{m_2}) by

$$A^{m_1}(b^{m_2}, b'^{m_2}) \triangleq A^{m_1} \cap \Omega_1(b^{m_2}, b'^{m_2}), \tag{59}$$

(where $A^{m_1} \in \mathcal{A}$ is defined in (55)).

To apply Lemma 3, we check (10) i.e.

$\alpha^{-\frac{1}{2}} > ln\big[2k(1+|\mathcal{A}|)|\mathcal{S}^{m_2}|\big] \geq ln\left[2k \sum_{P\in Q} |\mathcal{E}(P)|\right]$, which is true when l_1, m_1 is sufficiently large (cf. 51) since by (52), (53) $\alpha(P) \leq w^{l_1}$ for $P \in \mathcal{P}_1$ and by (47), (54), (55) $\alpha(P) \leq w^{m_1 - l_1 + 1}$ for $P \in \mathcal{P}_2$.

Then by Lemma 3 we have that (c.f. the choices in (10))

$$|W^{m_1}(\Omega_1(b^{m_2}, b'^{m_2})|0^{m_2}, s^{m_1}) - \frac{1}{2^{2m_2}}| < 2w^{\frac{l_1}{4}} \tag{60}$$

for all $b^{m_2}, b'^{m_2} \in \{0,1\}^{m_2}$ and all s^{m_1} with

$$\delta^*(0^{m_1}, s^{m_1}) \geq l_1 \tag{61}$$

and

$$|W^{m_1}(A^{m_1}(b^{m_2}, b'^{m_2})|1^{m_1}, s^{m_1}) - \frac{W^{m_1}(A^{m_1}|1^{m_1}, s^{m_1})}{2^{m_2}}| < 2w^{\frac{1}{4}(m_1 - l_1 + 1)} \tag{62}$$

for all $b^{m_2}, b'^{m_2} \in \{0,1\}^{m_2}$, for all $A^{m_1} \in \mathcal{A}$, and for all $s^{m_1} \in \mathcal{S}^{m_1}$.

3. Apply Lemma 3 for the choices $\mathcal{V} = \mathcal{Y}^{m_2}, \mathcal{Q} = \mathcal{P}' = \{W^{m_2}(\cdot|b^{m_2}, s^{m_2}) : b^{m_2} \in \{0,1\}^{m_2}, s^{m_2} \in \mathcal{S}^{m_2}$, and $\delta^*(b^{m_2}, s^{m_2}) \geq l_2\}, \mathcal{E}(P) = \{\mathcal{Y}^{m_2}\}$ for all $P \in \mathcal{P}'$, $k = |\mathcal{X}|^2$ and $g' = (\Phi_2, \Psi_2) : Y^{m_2} \to \mathcal{X} \times \mathcal{X}$. Similarly as in 2. we have for

$$\Omega_2(x, x') \triangleq g'^{-1}(x, x') = \Phi_2^{-1}(x) \cap \Psi_2^{-1}(x') \tag{63}$$

$$|W^{m_2}(\Omega_2(x,x')|b^{m_2},s^{m_2}) - \frac{1}{|\mathcal{X}|^2}| < 2w^{\frac{l_2}{4}} \tag{64}$$

for all $x, x' \in \mathcal{X}, b^{m_2} \in \{0,1\}^{m_2}$, and $s^{m_2} \in \mathcal{S}^{m_2}$ with $\delta^*(b^{m_2}, s^{m_2}) \geq l_2$ since here $\alpha = w^{l_2}$ and the right hand side of (10) polynomially increases, i.e. (10) holds.

4. Finally define the announced encoding functions

$$f_0^n = (0^{m_1}, \Phi_1, \Phi_2) \text{ and } f_1^n = (1^{m_1}, \Psi_1, \Psi_2) \tag{65}$$

which lead to the desired contradiction. If they satisfy (50) for some α^n and β^n, then we can express this also by saying that for the pairs of *RV's* (S^n, Y^n) and (S'^n, Y'^n) with *PD*'s $\alpha^n(\cdot)W^n(\cdot|f_0^n,\cdot)$ and $\beta^n(\cdot)\cdot W^n(\cdot|f_1^n,\cdot)$, resp., Y^n and Y'^n have the same (marginal) distributions.

For the analysis of these *RV's* we need the following simple Lemmas 6 and 7 and finally the crucial Crowd Lemma 8.
In the sequel we write (with some abuse of notation) $S^{m_1}S^{m_2+1}$ or $S^{m_1}S^{m_2}S$ for S^n and $Y^{m_1}Y^{m_2+1}$ or $Y^{m_1}Y^{m_2}Y$ for Y^n.
We notice that Y^{m_1} or Y^{m_2} falling into $\Omega_1(b^{m_2}, b'^{m_2})$, i.e. it getting color (b^{m_2}, b'^{m_2}), implies that in the second block f_0^n and f_1^n will take values b^{m_2} and b'^{m_2}. A similar event will happen in the third block, when the output in the second block gets color (x, x'). These facts will repeatedly be used in our proof.

Lemma 6. *(i) Suppose that*

$$Pr\left(\delta^*(0^{m_1}, S^{m_1}) < l_1\right) < w^{l_1}, \tag{66}$$

then for all $b^{m_2}, b'^{m_2} \in \{0,1\}^{m_2}$ and $L \subset \mathcal{Y}^{m_2+1}$

$$|Pr(Y^{m_1} \in \Omega_1(b^{m_2}, b'^{m_2}), Y^{m_2+1} \in L) - \frac{1}{2^{2m_2}} \sum_{s^{m_2+1}} \Big[Pr(S^{m_2+1} = s^{m_2+1})$$

$$\cdot Pr(Y^{m_2+1} \in L|S^{m_2+1} = s^{m_2+1}, Y^{m_1} \in \Omega_1(b^{m_2}, b'^{m_2}))\Big]| < 2w^{\frac{l_1}{4}} + w^{l_1} \tag{67}$$

and one can choose l_1, m_1, and m_2 in (51) such that

$$|Pr(Y^{m_2+1} \in L|Y^{m_1} \in \Omega_1(b^{m_2}, b'^{m_2})) -$$

$$\sum_{s^{m_2+1}} Pr(S^{m_2+1} = s^{m_2+1}) Pr(Y^{m_2+1} \in L|S^{m_2+1} = s^{m_2+1}, Y^{m_1} \in \Omega_1(b^{m_2}, b'^{m_2})|$$

$$< w^{\frac{l_1}{8}} \tag{68}$$

(ii) Suppose that for some $b^{m_2} \in \{0,1\}^{m_2}$ and $E \subset \mathcal{Y}^{m_1}$

$$Pr(\delta^*(b^{m_2}, S^{m_2}) < l_2|Y^{m_1} \in E) < w^{l_2}, \tag{69}$$

then for all $x, x' \in \mathcal{X}, K \subset \mathcal{Y}$, and $b'^{m_2} \in \{0,1\}^{m_2}$

$$| \sum_{s^{m_2+1}} \Big[Pr(S^{m_2+1} = s^{m_2+1} | Y^{m_1} \in E)$$

$$\times Pr(Y^{m_2} \in \Omega_2(x,x'), Y \in K | S^{m_2+1} = s^{m_2+1}, Y^{m_1} \in \Omega_1(b^{m_2}, b'^{m_2})) \Big]$$

$$- \frac{1}{|\mathcal{X}|^2} \sum_{s \in S} Pr(S = s | Y^{m_1} \in E) W(K|x,s)| < 2w^{\frac{l_2}{4}} + w^{l_2}. \qquad (70)$$

Moreover, one can replace (S^{m_2}, Y^{m_1}) and $W(K|x,s)$ in (69) and (70) by (S'^n, Y'^n) and $W(K|x', s)$.

Proof: Let $L = \mathcal{Y}^{m_2+1}$ in (67). Then the resulting inequality

$$\left| \Pr\big(Y^{m_1} \in \Omega_1(b^{m_2}, b'^{m_2})\big) - \frac{1}{2^{2m_2}} \right| \le 2w^{\frac{l_2}{4}} + w^{l_1}$$

and (67) imply (68) (cf. (51)). We show now (67). By definition of (S^n, Y^n)

$$\Pr\big(Y^{m_1} \in \Omega_1(b^{m_2}, b'^{m_2}), Y^{m_2+1} \in L\big) =$$

$$\sum_{s^{m_1} s^{m_2}} \Big[\Pr(S^n = s^{m_1} s^{m_2+1}) W^{m_1}\big(\Omega_1(b^{m_2}, b'^{m_2}) | 0^{m_1} s^{m_1}\big)$$

$$\times \Pr\big(Y^{m_2+1} \in L | S^{m_2+1} = s^{m_2+1}, Y^{m_1+1} \in \Omega_1(b^{m_2}, b'^{m_2})\big)\Big]$$

and then the LHS of (67) does not exceed

$$\sum_{s^{m_1} s^{m_2+1}} \Big[Pr(S^n = s^{m_1} s^{m_2+1}) | W^{m_1}(\Omega_1(b^{m_1}, b'^{m_1}) | 0^{m_1}, s^{m_1}) - \frac{1}{2^{2m_2}} |$$

$$\times Pr(Y^{m_2+1} \in L | S^{m_2+1} = s^{m_2+1}, Y^{m_1+1} \in \Omega_1(b^{m_2}, b'^{m_2})) \Big],$$

which together with (60), (61) and (66) yields (67) (by splitting $\mathcal{S}^n$ to $\{s^{m_1+m_2+1} : \delta^*(0^{m_1}, s^{m_1}) \ge l_1\}$ and $\{s^{m_1+m_2+1} : \delta^*(0^{m_1}, s^{m_1}) < l_1\}$). Notice that by the definition of (Y^n, S^n) and (65) for $s^{m_2+1} = s^{m_2} s$ in (70)

$$Pr(Y^{m_2} \in \Omega_2(x,x'), Y \in K | S^{m_2+1} = s^{m_2+1}, Y^{m_1} \in \Omega_1(b^{m_2}, b'^{m_2}))$$

$$= W^{m_2}(\Omega_2(x,x') | b^{m_2}, s^{m_2}) W(K|x,s)$$

and hence (ii) can be established exactly like (i).
The importance of (67) and (68) (resp. (70)) is that S^{m_2+1} (resp. S) in the second terms (resp. term) is independent of $\Phi_1(Y^{m_1})$ (resp. $\Phi_2(Y^{m_2})$). Intuitively speaking, the jammer has very little knowledge about the output to come. The same phenomenon can be encountered in the next auxiliary result.

Lemma 7. *For all $A^{m_1} \in \mathcal{A}, b^{m_2}, b'^{m_2} \in \{0,1\}^{m_2}$ and $L \subset \mathcal{Y}^{m_2+1}$*

$$\begin{aligned}&|Pr(Y'^{m_1} \in A^{m_1}(b^{m_2}, b'^{m_2}), Y'^{m_2+1} \in L)\\ &-\tfrac{1}{2^{2m_2}} Pr(Y'^{m_1} \in A^{m_1}) \sum_{s^{m_2+1}} Pr(S'^{m_2+1} = s^{m_2+1} | Y'^{m_1} \in A^{m_1})\\ &\times Pr(Y'^{m_2+1} \in L | S'^{m_2+1} = s^{m_2+1}, \Psi_1(Y'^{m_1}) = b'^{m_2})|\\ &< 2w^{\frac{m_1-l_1+1}{4}}. \end{aligned} \tag{71}$$

Moreover, if (66) does not hold, one can always choose the parameters according to (51) and find an $A^{m_1} \in \mathcal{A}$ *in such a way that*

$$\begin{aligned}&|Pr(Y'^{m_2+1} \in L | Y'^{m_1} \in A^{m_1}(b^{m_2}, b'^{m_2}))\\ &- \sum_{s^{m_2+1}} \Big[Pr(S'^{m_2+1} = s^{m_2+1} | Y'^{m_1} \in A^{m_1})\\ &\times Pr(Y'^{m_2+1} \in L | S'^{m_2+1} = s^{m_2+1}, \Psi_1(Y'^{m_1}) = b'^{m_2}) \Big] | < w^{l_1}. \end{aligned} \tag{72}$$

Proof: (71) is proved analogously to (67). However, notice that here all $W^{m_1}(\cdot|1^{m_1}, s^{m_1})$ are contained in $\mathcal{P}_2 \subset \mathcal{Q}$ (see (54)) and therefore no condition analogous to (66) is necessary. To obtain (72) from (71) we let $L = \mathcal{Y}^{m_2+1}$ in (71) and get

$$|Pr(Y'^{m_1} \in A^{m_1}(b^{m_2}, b'^{m_2})) - \frac{1}{2^{2m_2}} Pr(Y'^{m_1} \in A^{m_1})| < 2w^{\frac{1}{4}(m_1-l_1+1)} \tag{73}$$

A difficulty now arises. In order to obtain a good bound w^{l_1} at the RHS of (2.27), we have to find an $A^{m_1} \in \mathcal{A}$ such that $Pr(Y'^{m_1} \in A^{m_1})$ is not too small. Assume then that (66) does not hold and we now look for our A^{m_1}. Since the set $\{s^m : \delta^*(0^{m_1}, s^{m_1}) < l_1\}$ is covered by the family of sets

$$\mathcal{B} \triangleq \left\{ \prod_{t=1}^{m_1} B_t : B_t \in \{\mathcal{S}_0, \mathcal{S}\} \text{ and } |\{t : B_t = \mathcal{S}_0\}| = m_1 - l_1 + 1 \right\},$$

$\sum_{B^{m_1} \in \mathcal{B}} \Pr(S^{m_1} \in B^{m_1}) \geq \Pr(\delta^*(0^{m_1}, S^{m_1}) < l_1) \geq w^{l_1}$ and therefore one member of $\mathcal{B}$, say $B^{m_1} = \mathcal{S}_0^{m_1-l_1+1} \times \mathcal{S}^{l_1-1}$, must have the probability

$$Pr(S^{m_1} \in B^{m_1}) \geq \binom{m_1}{l_1 - 1}^{-1} w^{l_1}, \tag{74}$$

if (66) does not hold since $|\mathcal{B}| = \binom{m_1}{l_1-1}$. We then choose $A^{m_1} = \mathcal{Y}_0^{m_1-l_1+1} \times \mathcal{Y}^{l_1-1}$. Notice that for all $s^{m_1} \in B^{m_1}$

$$W^{m_1}(A^{m_1}|0^{m_1}, s^{m_1}) = 1 \tag{75}$$

Recalling Y^n and Y'^n have the same distributions, we conclude from (65), (74), and (75) that

$$Pr(Y'^{m_1} \in A^{m_1}) = Pr(Y^{m_1} \in A^{m_1})$$

$$\geq \sum_{s^{m_1} \in B^{m_1}} Pr(S^{m_1} = s^{m_1}) W^{m_1}(A^{m_1}|0, s^{m_1}) \geq \binom{m_1}{l_1 - 1}^{-1} w^{l_1}.$$

With the above inequality and the relation $2^{2m_2+1}\binom{m_1}{l_1-1} w^{\frac{m_1-l_1+1}{4}-l_1} = o(1)$ (which follows from the assumption in (51)) and (73), (72) can be obtained by dividing (2.26) by $\Pr(Y'^{m_1} \in A^{m_1})$.
Now comes the kernel of the proof.

Crowd Lemma 8. *For suitable parameters in (51)*

(i) For all PD σ on $\mathcal{S}^{m_2}$ there exists a $b^{m_2} \in \{0,1\}^{m_2}$ such that

$$\sigma(s^{m_2} : \delta^*(b^{m_2}, s^{m_2}) < l_2) < w^{l_2}. \tag{76}$$

(ii) If (68) holds, then for all $b^{m_2} \in \{0,1\}^{m_2}$ there exists a $b'^{m_2} \in \{0,1\}^{m_2}$ such that

$$Pr(\delta^*(b'^{m_2}, S'^{m_2}) < l_2 | Y'^{m_1} \in \Omega_1(b^{m_2}, b'^{m_2})) < w^{l_2}. \tag{77}$$

(iii) If (72) holds, then for all $b'^{m_2} \in \{0,1\}^{m_2}$ there exists a $b^{m_2} \in \{0,1\}^{m_2}$ such that

$$Pr(\delta^*(b^{m_2}, S'^{m_2}) < l_2 | Y^{m_1} \in A^{m_1}(b^{m_2}, b'^{m_2})) < w^{l_2}. \tag{78}$$

Proof: Ad(i). Assume to the opposite that for some σ and all b^{m_2}

$$\sigma(s^{m_2} : \delta^*(b^{m_2}, s^{m_2}) < l_2) \geq w^{l_2}.$$

Then we add up these inequalities over all $b^{m_2} \in \{0,1\}^{m_2}$. Since for all $s^{m_2} \in \mathcal{S}^{m_2}$ there are at most $\sum_{j=0}^{l_{2-1}} \binom{m_2}{j} 2^j$ $b^{m_2'}s$ with $\delta^*(b^{m_2}, s^{m_2}) < l_2$ we obtain that

$$\sum_{j=0}^{l_2-1} \binom{m_2}{j} 2^j \geq \sum_{s^{m_2}} \sigma(s^{m_2}) |\{b^{m_2} : \delta^*(b^{m_2}, s^{m_2}) < l_2\}| =$$

$$\sum_{b^{m_2} \in \{0,1\}^{m_2}} \sigma(s^{m_2} : \delta^*(b^{m_2}, s^{m_2}) < l_2) \geq 2^{m_2} w^{l_2},$$

which cannot happen for sufficiently small ε and large l_2 in (51).
Ad (ii) and (iii). We only show that (77) holds under (68), because (iii) can be proved in the same way, whereas in (i) we dealt with one PD, we deal now with a family of PD's. This makes things harder. Define for all $b'^{m_2} \in \{0,1\}^{m_2}$ and $\tilde{\delta}$ in (21).

$$L^*(b'^{m_2}) \triangleq \{y^{m_2} \in \mathcal{Y}^{m_2} : \tilde{\delta}(b'^{m_2}, y^{m_2}) < l_2\}. \tag{79}$$

Then for all s^{m_2} with $\delta^*(b'^{m_2}, s^{m_2}) < l_2$ by the definitions of (S'^n, Y'^n) and $\mathcal{S}_x$,

$$Pr(Y'^{m_2} \in L^*(b'^{m_2}) | S'^{m_2} = s^{m_2}, Y'^{m_1} \in \Omega_1(b^{m_2}, b'^{m_2}))$$
$$= W^{m_2}(L^*(b'^{m_2}) | b'^{m_2}, s^{m_2}) = 1. \quad (80)$$

Consequently, if (77) is false, i.e. for *some* b^{m_2} and *all* b'^{m_2}.

$$Pr(\delta^*(b'^{m_2}, S'^{m_2}) < l_2 | Y'^{m_2} \in \Omega_1(b^{m_2}, b'^{m_2})) \geq w^{l_2},$$

then for such a b^{m_2} and all b'^{m_2}, by (80)

$$Pr(Y'^{m_2} \in L^*(b'^{m_2}) | Y'^{m_1} \in \Omega_1(b^{m_2}, b'^{m_2}))$$
$$= \sum_{s^{m_2}} \Big[Pr(S'^{m_2} = s^{m_2} | Y'^{m_2} \in \Omega_1(b^{m_2}, b'^{m_2}))$$
$$\times Pr(Y'^{m_2} \in L^*(b'^{m_2}) | S'^{m_2} = s^{m_2}, Y'^{m_1} \in \Omega_1(b^{m_2}, b'^{m_2}))\Big]$$
$$\geq \sum_{s^{m_2}:\delta^*(b^{m_2}, s^{m_2}) < l_2} \Big[\Pr(S'^{m_2} = s^{m_2} | Y^{m_2} \in \Omega_1(b^{m_2}, b'^{m_2})$$
$$\times \Pr(Y'^{m_2} \in L^*(b'^{m_2}) | S'^{m_2} = s^{m_2}, Y'^{m_2} \in \Omega_1(b^{m_2}, b'^{m_2}))\Big]$$
$$= \Pr\big(\delta^*(b'^{m_2}, S'^{m_2}) < l_2 | Y'^{m_2} \in \Omega_1(b^{m_2}, b'^{m_2})\big) > w^{l_2}.$$

Therefore, since Y^n and Y'^n have the same distributions,

$$Pr(Y^{m_2} \in L^*(b'^{m_2}) | Y^{m_1} \in \Omega_1(b^{m_2}, b'^{m_2})) \geq w^{l_2}. \quad (81)$$

Apply now (68) to $L = L^*(b'^{m_2})$ for all b'^{m_2}. Thus

$$\sum_{s^{m_2+1}} \Big[Pr(S^{m_2+1} = s^{m_2+1})$$
$$\times Pr(Y^{m_2} \in L^*(b'^{m_2}) | S^{m_2+1} = s^{m_2+1}, Y^{m_1} \in \Omega_1(b^{m_2}, b'^{m_2}))\Big] \geq w^{l_2} - w^{\frac{l_1}{8}}. \quad (82)$$

Finally, by adding both sides of (82) over $\{0,1\}^{m_2}$ and by using the fact that each $y^{m_2} \in Y^{m_2}$ is covered by at most $\sum_{j=0}^{l_2-1} \binom{m_2}{j} 2^j$ sets $L^*(b'^{m_2})$ in (79) we arrive at

$$\sum_{j=0}^{l_2-1} \binom{m_2}{j} 2^j \geq \sum_{s^{m_2+1}} Pr(S^{m_2+1} = s^{m_2+1})$$
$$\times \sum_{b'^{m_2} \in \{0,1\}^{m_2}} Pr(Y^{m_2} \in L^*(b'^{m_2}) | S^{m_2+1} = s^{m_2+1}, Y^{m_1} \in \Omega_1(b^{m_2}, b'^{m_2}))$$
$$\geq 2^{m_2}(w^{l_2} - w^{\frac{l_1}{8}}), \quad (83)$$

which contradicts (51).
The idea behind the Crowd Lemma is that an encoding function with enough different values has always "a good" value against the jamming.
Now it's time for the harvest.

Proof of Positivity Theorem: We use Lemmas 6-8 to obtain a contradiction to (49). This is done in two cases.

Case 1 (66) holds: Then by Lemma 6 also (68) holds. We apply Lemma 8 (i) to $\sigma = P_{S^{m_2}}$ and obtain a b^{m_2} such that (69) holds with $E = \mathcal{Y}^{m_1}$ (i.e. unconditional distribution). Fix this b^{m_2} and apply Lemma 6 (ii) for $E = \mathcal{Y}^{m_1}$. Thus we obtain (70) with $E = \mathcal{Y}^{m_1}$. Choose next $L = \Omega_2(x, x') \times K$ in (68) and combine it with (70) for $E = \mathcal{Y}^{m_1}$. Thus we get that for the fixed b^{m_2}, all $x, x' \in \mathcal{X}$, all $b'^{m_2} \in \{0,1\}^{m_2}$, and all $K \subset \mathcal{X}$

$$|Pr(Y^{m_2} \in \Omega_2(x,x'), Y \in K | Y^{m_1} \in \Omega_1(b^{m_2}, b'^{m_2}))$$

$$-\frac{1}{|\mathcal{X}|^2} \sum_s Pr(S=s) W(K|x,s)| < w^{\frac{l_1}{8}} + 2w^{\frac{l_2}{4}} + w^{l_2}. \tag{84}$$

On the other hand, since (68) holds, we can find a b'^{m_2} for the fixed b^{m_2} so that (77 holds by (ii) in Lemma 8. That is, after replacing (S^n, Y^n) by (S'^n, Y'^n), (69) holds for $E = \Omega_1(b^{m_2}, b'^{m_2})$ and therefore, by Lemma 6 (ii) again, but this time for (S'^n, Y'^n) (instead of (S^n, Y^n)) and $E = \Omega_1(b^{m_2}, b'^{m_2})$ we obtain for the fixed b^{m_2}, b'^{m_2}, all $x, x' \in \mathcal{X}$, and $K \subset \mathcal{Y}$

$$|Pr(Y'^{m_2} \in \Omega_2(x,x'), Y' \in K | Y'^{m_1} \in \Omega_1(b^{m_2}, b'^{m_2}))$$

$$-\frac{1}{|\mathcal{X}|^2} \sum_{s \in \mathcal{S}} Pr(S' = s | Y'^{m_1} \in \Omega_1(b^{m_2}, b'^{m_2})) W(K|x', s)| < 2w^{\frac{l_2}{4}} + w^{l_2}, \tag{85}$$

where we use the fact that

$$Pr(Y'^{m_2} \in \Omega_2(x,x'), Y' \in K | Y'^{m_1} \in \Omega_1(b^{m_2}, b'^{m_2}))$$

$$= \sum_{s^{m_2+1}} Pr(S'^{m_2+1} = s^{m_2+1} | Y'^{m_1} \in \Omega_1(b^{m_2}, b'^{m_2}))$$

$$\times Pr(Y'^{m_2} \in \Omega_2(x,x'), Y' \in K | S'^{m_2+1} = s^{m_2+1}, Y'^{m_1} \in \Omega_1(b^{m_2}, b'^{m_2})).$$

Finally, let l_1 and l_2 be sufficiently large, then from (84), (85), and the fact that Y^n and Y'^n have the same distributions we obtain that for θ in (49), all $x, x' \in \mathcal{X}$ and $K \subset \mathcal{Y}$,

$$|\sum_s Pr(S=s) W(K|x,s) - \sum_s Pr(S' = s | Y'^{m_1} \in \Omega_1(b^{m_2}, b'^{m_2})) W(K|x', s)| \leq$$

$$\leq \frac{\theta}{3},$$

or, for all $x, x'' \in \mathcal{X}$ and $K \subset \mathcal{Y}$.

$$\left| \sum_s Pr(S=s)W(K|x,s) - \sum_s Pr(S=s)W(K|x'',s) \right| < \frac{2\theta}{3} \tag{86}$$

which contradicts (49) (with $K = \{y\}$).

Case 2: (66) does not hold: Here by Lemma 7 we have (72) for an $A^{m_2} \in \mathcal{A}$. Fix this A^{m_2} by applying Lemma 8(i) for $\sigma = \Pr(\cdot | Y'^{m_2} \in A^{m_2})$, we obtain that for a (fixed) b'^{m_2} $Pr(\delta^*(b'^{m_2}, S'^{m_2}) < l_2 | Y'^{m_2} \in A^{m_2}) < w^{l_2}$, i.e. (69) in terms of the distribution (S'^n, Y'^n) and with $E = A^{m_2}$. Therefore we have (70) in terms of the distribution of (S'^n, Y'^n) with $E = A^{m_2}$ and then an inequality in terms of the distribution of (S'^n, Y'^n), analogous to (84), by combining (70) and (72). Next for the fixed b'^{m_2} (obtained by applying Lemma 8 (i) in this case), we find a b^{m_2} such that (78) holds. Now we set $E = \Lambda^{m_1}(b^{m_2}, b'^{m_2})$ in Lemma 6 (ii) and obtain an inequality, analogous to (85), but in terms of the distribution of (S^n, Y^n). Finally, we get an inequality analogous to (86), which contradicts (49).

References

[1] R. Ahlswede, "Channels with arbitrarily varying channel probability functions in the presence of noiseless feedback", *Z. Wahrsch. Verw. Gebiete*, vol. 25, 1973, 239–252.

[2] R. Ahlswede, "Elimination of correlation in random codes for arbitrarily varying channels", *Z. Wahrsch. Verw. Gebiete*, vol. 44, 1978, 159–175.

[3] R. Ahlswede, "Coloring hypergraphs: a new approach to multi–user source coding" , *J. Combin. Inform. System Sci.*, Part I, vol. 4, 1979, 76–115 and Part II, vol. 5, 1980, 220–268.

[4] R. Ahlswede and V.B. Balakirsky, "Identification under random processes", Preprint 95–098, SFB 343 "Diskrete Strukturen in der Mathematik", Universität Bielefeld, 1995, *Problemy peredachii informatsii*, (special issue devoted to M.S. Pinsker), vol. 32, no. 1, Jan-March 1996, 144–160.

[5] R. Ahlswede and N. Cai, "Two proofs of Pinskers conjecture concerning arbitrarily varying channels", *IEEE Trans. Inform. Theory*, vol. IT–37, 1991, 1647–1649.

[6] R. Ahlswede and N. Cai, "Correlated sources help the transmission over AVC", Preprint 95–106, SFB 343 "Diskrete Strukturen in der Mathematik", Universität Bielefeld, 1995, *IEEE Trans. Inf. Theory*, Vol. IT–43, No. 1, 1997, 37–67.

[7] R. Ahlswede and I. Csiszár, "Common randomness in information theory and cryptography, Part 1: Secret sharing", *IEEE Trans. Inform. Theory*, vol. IT–39, 1993, 1121–1132 and "Part 2: CR capacity", Preprint 95–101, SFB 343 "Diskrete Strukturen in der Mathematik", Universität Bielefeld, 1995, *IEEE Trans. Inf. Theory*, Vol. 44, No. 1, 1998, 55–62.

[8] R. Ahlswede and G. Dueck, "Identification via channels", *IEEE Trans. Inform. Theory*, vol. IT–35, 1989, 15–29.

[9] R. Ahlswede and G. Dueck, "Identification in the presence of feedback — a discovery of new capacity formulas", *IEEE Trans. Inform. Theory*, vol. IT–35, 1989, 30–39.

[10] R. Ahlswede and Z. Zhang, "New directions in the theory of identification via channels", *IEEE Trans. Inform. Theory*, vol. IT–41, 1995, 1040–1050.

[11] J. Kiefer and J. Wolfowitz, "Channels with arbitrarily varying channel probability functions", *Inform. and Control*, vol. 5, 1962, 44–54.

[12] C.E. Shannon, "The zero–error capacity of a noisy channel", *IRE Trans. Inform. Theory*, vol. IT–2, 1956, 8–19.

[13] S. Lin and D.J. Costello, Jr., *Error control coding: Fundamentals and Applications*, Prentice–Hall, Inc., Englewood Cliffs, N.J. 1983.

[14] J.M. Ooi, "A Framework for Low–Complexity Communication Channels with Feedback", Dissertation at MIT, *RLE Technical Report*, No. 617, Nov. 1997.

[15] J.M. Ooi and Gregory W. Wornell, "Fast iterative coding for feedback channels", *1997 Proceedings IEEE Int. Symp. on Inf. Theory*, Ulm, Germany, June 29 - July 4, 1997, 133.

CALCULATION OF THE ASYMPTOTICALLY OPTIMAL CAPACITY OF A T-USER M-FREQUENCY NOISELESS MULTIPLE-ACCESS CHANNEL

Leonid Bassalygo and Mark Pinsker*

Institute for Problems of Information Transmission, RAS,
19 Bolshoi Karetnii, 101447 Moscow, Russia

The statement of the problem is taken from [1]. Let $T(T \geq 2)$ be the number of users every of which transmits one symbol from the alphabet $\{1, 2, ..., M\}$, $M \geq 2$, at each time instant (the time is discrete); and the output is a binary sequence of length M where the symbol 0 is in the m-th position if and only if none user transmitted the symbol m. Such channel is referred to as an A-channel in [1]. Denote by $X = (X_1, ..., X_T)$ an M-ary sequence at the input of the channel and by $Y = (Y_1, ..., Y_M)$ a binary sequence at the output. Then (see [1]) the sum capacity of an A-channel is

$$C_{sum}(T, M) = maxH(Y),$$

where the maximum is taken over all product distributions on input random variables $X_1, \ldots, X_T$:

$$P(X) = P_1(X_1) \cdots P_T(X_T). \tag{1}$$

We shall also call the symbols of our alphabet frequences and call the users stations. Denote

$$C_{sum}(\lambda) = \lim_{M \to \infty} \frac{C_{sum}(\lambda M, M)}{M}, 0 < \lambda < \infty.$$

*This work was supported by the Russian Fundamental Research Foundation (grant 99-01-00828)

I. Althöfer et al. (eds.), Numbers, Information and Complexity, 177-180.

The existence of the limit and the convexity of the function $C_{sum}(\lambda)$:

$$C_{sum}(\alpha\lambda_1 + \beta\lambda_2) \geq \alpha C_{sum}(\lambda_1) + \beta C_{sum}(\lambda_2),\ \alpha + \beta = 1; \alpha, \beta \geq 0,$$

are easily proved by an apropriate partition of frequences. The cases $\lambda = 0, \infty$, are described in the end of the paper.

The formula for the output entropy $H_{unif}(Y)$ under the common uniform distribution of all $X_1, \ldots, X_T$:

$$P(X_t = m) = \frac{1}{M}, t = 1, \ldots, T, m = 1, \ldots, M, \tag{2}$$

was written in [1].

Asymptotic behavior of this entropy, i.e. the value $H_{unif}(\lambda) = \lim\limits_{M\to\infty} \frac{H_{unif}(Y)}{M}$ for $T = \lambda M,\ 0 < \lambda < \infty$, was calculated in [2]:

$$H_{unif}(\lambda) = h(1 - e^{-\lambda}), \quad h(u) = -ulogu - (1-u)log(1-u)$$

In the same paper it was observed that

$$C_{sum}(ln2) = H_{unif}(ln2) = 1.$$

An attempt to calculate $H_{unif}(\lambda)$ was made in [3], but formula (14) and, respectively, Theorem 2 from [3] are not right (the error is an effect of improper use of the approximation (12) for binomial coefficients).

In [1] it was also indicated that the uniform distribution is not good for $T > M$, and common distribution distorted for the benefit of one fixed frequence and equiprobable on the other frequences gives a better answer.

In [4] it was proposed to use the specific distorted distribution introduced in [5] for the analysis of some parameter of an A-channel, for fixed M (i.e. if $\lambda = \infty$) :

$$P_t(X_t = M) = 1 - \frac{(M-1)ln2}{T}, \quad P_t(X_t = m) = \frac{ln2}{T} \tag{3}$$

for all m from 1 till $M-1,\ t = 1, \ldots, T$.

Denote by $H_{distort}(Y)$ the entropy of Y for this distribution (we note that the uniform and the distorted distributions coincide for $T = Mln2$ and that the distorted distribution is defined only for $T \geq Mln2$). It is not difficult to calculate the asymptotical behavior of this entropy, i.e. to find the value $H_{distort}(\lambda) = \lim\limits_{M\to\infty} \frac{H_{distort}(Y)}{M}$ for $T = \lambda M$:

$$H_{distort}(\lambda) = 1, \quad ln2 \leq \lambda < \infty.$$

If we restrict ourselves to common imput distributions only (i.e. $P_1 = \cdots = P_T$ in (1)), then the asymptotical behavior of the right-hand side of (1) under this restriction (denote the corresponding value by $C_{com}(\lambda)$) is completely defined by the uniform (2) and the distorted (3) distribution.

Theorem 1. The equality

$$C_{com}(\lambda) = \begin{cases} H_{unif}(\lambda) & = & h(1-e^{-\lambda}) & \text{if} & 0 < \lambda \leq ln2, \\ H_{distort}(\lambda) & = & 1, & \text{if} & ln2 \leq \lambda < \infty. \end{cases}$$

holds.

Comment on Theorem 1. They assumed (see e.g. [1,3]) that the uniform distribution is optimal if $\lambda \leq 1$. Computer calculations (see, e.g. [4]) did not confirm it and Theorem 1 shows that this assumption could not be confirmed because the uniform distribution is certainly not asymptotically optimal if $\lambda > ln2$. But for $\lambda = ln2 = 0,693...$ it is such, and we presupposed (probably, as all other researchers) that it is such for all smaller $\lambda : 0 < \lambda \leq ln2$. Therefore we were very surprised when we discovered the uniform distribution to be asymptotically optimal for one λ only : $\lambda = ln2$; for a smaller λ, a better answer is given by the following input distribution (surely, not common; $t = 1, 2, ..., T; T < M$) :

$$P_t(X_t = m) = \begin{cases} 1/2 & \text{if } m = t, \\ \frac{1}{2(M-T)} & \text{if } m > T, \\ 0 & \text{otherwise.} \end{cases}$$

This distribution generates its own frequence at every station with probability $\frac{1}{2}$ and generates common $M - T$ frequences equiprobably. Denote the output entropy for this input distribution by $H_0(Y)$ and denote by $H_0(\lambda)$ the corresponding asymptotic value.

Theorem 2. The equality

$$H_0(\lambda) = \lambda + (1-\lambda)h(1 - e^{-\frac{\lambda}{2(1-\lambda)}}) \quad \text{if} \quad 0 < \lambda \leq \frac{2ln2}{1+2ln2} = 0,581...$$

holds.

Corollary 1. Since $H_0(\frac{2ln2}{1+2ln2}) = 1$ and $C_{sum}(\lambda)$ is a convex function, $C_{sum}(\lambda) = 1$ if $\lambda \geq \frac{2ln2}{1+2ln2}$.

Corollary 2. For other positive λ, the following lower and upper bounds of $C_{sum}(\lambda)$ hold :

$$C_{sum}(\lambda) \geq \lambda + (1-\lambda)h(1 - e^{-\frac{\lambda}{2(-\lambda)}}) \qquad \text{if} \quad 0 < \lambda \leq \frac{2ln2}{1+2ln2} = 0,581...$$

$$C_{sum}(\lambda) \leq \begin{cases} 1 & \text{if} \quad \frac{1}{2} \leq \lambda \leq \frac{2ln2}{1+2ln2}, \\ h(\lambda) & \text{if} \quad 0 < \lambda < 1/2. \end{cases}$$

It remains to consider two extreme points : $\lambda = 0$ and $\lambda = \infty$.

I. $\lambda = 0$, i.e. $\frac{T}{M} \to 0$ as $M \to \infty$. Then

$$C_{sum}(T, M) \sim Tlog\frac{M}{T}$$

(here and further, $f(n) \sim g(n)$ means that $\lim \frac{f(n)}{g(n)} = 1$ as $n \to \infty$).

II. $\lambda = \infty$, i.. $\frac{T}{M} \to \infty$ as $T \to \infty$. Then

$$C_{sum}(T, M) \sim \begin{cases} M & \text{if} \quad M \to \infty, \\ M - 1 & \text{if} \quad M \text{ is fixed.} \end{cases}$$

The case I follows in fact from [2], the case II was derived for fixed M in [4] and for $M \to \infty$ in [1] where it was proved that

$$C_{sum}(T, M) \geq M - 1 \text{ for } M \geq T - 1.$$

References

[1] S. C. Chang and J. K. Wolf, "On the T-user M-frequency noiseless multiple-access channels with and without intensity information", *IEEE Trans. Inform. Theory.*, 27, No. 1, 1981, 41-48.

[2] L. Wilhelmsson and K. Sh. Zigangirov, "On the asymptotical capacity of a multiple-access channel", *Probl. Inf. Trans.* 33, No. 1, 1997, 12-20.

[3] A. J. Grant and C. Schlegel, "Collision-type multiple-user communications", *IEEE Trans. Inform. Theory.* 43, No. 5, 1997, 1725-1736.

[4] P. Gober and A. J. Han Vinck " Note on "On the asymptotical capacity of a multiple-access channel" by L. Wilhelmsson and K. Sh. Zigangirov (Probl. Inf. Trans. 1997. Vol. 33, n.1, 9-16)" sunmitted Probl. Inf. Trans..

[5] A. J. Han Vinck and J. Keuning, "On the capacity of the asynchronous T-user M-frequency noiseless multiple-access channel without intensity information", *IEEE Trans. Inform. Theory.* 42, No. 6., 1996, 2235-2238.

A SURVEY OF CODING METHODS FOR THE ADDER CHANNEL

Gurgen H. Khachatrian

Institute for Problems of Informatics and Automation,
Armenian National Academy of Sciences, 375044 Yerevan, Armenia
gurgenkh@forof.sci.am

Abstract: In this survey the main results on coding for the noiseless multi-user adder channel are presented. The survey consists of two parts, where the coding methods for the 2-user adder channel and T-user adder channel are given respectively.

Dedicated to Rudolf Ahlswede on the occasion of his 60th birthday

PART I. Coding for 2-user adder channel.

I INTRODUCTION.

The problem of construction of uniquely decodable (UD) codes for the two-user binary adder channel (BAC) has been considered by many authors [1-13] . The problem can be formulated as follows: A pair of binary codes (C_1, C_2) of the same length is called to be UD, if and only if, for any two distinct pairs (u, v) and (u', v') $(u, u') \in C_1$ and $(v, v') \in C_2$ we have the property, that

$$u + v \neq u' + v'$$

where $u + v$ means the componentwise arithmetic sum of the binary components of the vectors u and v respectively, which is in fact a ternary vector. For an example if $u = (10100)$ and $v = (11101)$, then $u + v = (21201)$. The coding problem in most general form can be formulated as for given length n, rate R_1 of the code C_1, to construct UD pair of codes (C_1, C_2), such that the rate R_2 for the second code is maximum possible, where $R_i = \log_2(\sharp C_i)/n$. A less general problem would be for given n to construct UD pair of codes with maximum rate sum $R_1 + R_2$. Both problems are rather hard and the complete solution is not found yet.

I. Althöfer et al. (eds.), Numbers, Information and Complexity, 181-196.

II CAPACITY REGION

The average-error capacity region for the 2-user BAC has been established by R. Ahlswede in 1971 [1] as a special case of his multiple access channel coding theorem. It shows that the achievable rates are determined by $0 \leq R_1, R_2 \leq 1$, $R_1 + R_2 \leq 1.5$. A fortiori this is an upper bound for UD codes. Unfortunately all known constructions are still far away from the capacity bounds.

III CONSTRUCTION OF LINEAR UD CODES

Definition - A UD pair of codes (C_1, C_2) is called to be linear (LUD) if one of the codes, say C_1, is a linear (n, k) code. It was shown that unlike the case with ordinary block codes, the restriction that one of the codes is linear, essentially reduces the possiblity to construct good UD codes, due to the following theorem by Weldon in 1976 [3].

Theorem 1. . *Let C_1 have 2^k codewords and the property ,that some k-subset of n bits of the code takes all possible 2^k values. Then assuming, that (C_1, C_2) is UD, $|C_2|$ is upper bounded by*

$$|C_2| \leq 3^{n-k} \tag{1}$$

It can be shown, that the bound 1 can be easily achieved with $R_1 \geq 0.5$.

a) *Construction with* $R_1 = 0.5$. $C_1 = (00, 11)$ $C_2 = (00, 10, 01)$ - is UD and achieves the bound 1 . This construction can be repeated any m times to get codes for $n = 2m; |C_1| = 2^m$,$|C_2| = 3^m$

b) *Construction with* $R_1 > 0.5$. Now assume that we concatenate r positions to the previous code of length $2m$ to get the length $2m + r$. Obviously if in the extra r positions the code C_1 is arbitrary, and if C_2 is the all zero vector,then (C_1, C_2) for the length $(2m + r)$ will be again UD.

We get $|C_1| = 2^{m+r}$, $|C_2| = 3^m$ which means, that $|C_2|$ meets the upper bound 1. However , if $R_1 > 0.5$ and $R_2 = (1 - R_1)\log_2 3 < 0.5$, it can be shown,that if instead of the code with $R_2 < 0.5$ one takes the linear code with $R_1 < 0.5$, then he will get larger rate for the code C_2 . Therefore the construction of LUD codes is of interest with $R_1 < 0.5$. Kasami and Lin in 1978 [4] obtained an upper bound for

$$|C_2| \leq \sum_{j=0}^{k} \binom{n-k}{j} 2^j + 2^k \sum_{j=k+1}^{n-k} \binom{n-k}{j} \tag{2}$$

This bound is coming from the fact, that if the coset of an (n, k) code has maximum and mimimum weights $w_{\min}$ and $w_{\max}$, respectively, it can be shown, that at most $\min\{2^{n-w_{\max}}, 2^{w_{\min}})$ vectors can be chosen from each such coset for the code C_2.

The upper bound 2 is an improvement of 1 for the range $0 \leq R_1 < 0.4$. In an asymptotic form 2 for that range is:

$R_2 \leq 1$ if $0 \leq R_1 < 1/3$, $R_2 \leq R_1 + (1 - R_1)H(\rho) + O(1)$ if $1/3 \leq R_1 < 2/5$,where $H(\rho)$ is the entropy function, $\rho = R_1/(1 - R_1), O(1) \to 0$ when $n \to \infty$. This is the best known upper bound for LUD codes. The best known lower bound is obtained in the work by Kasami, Lin, Wei and Yamamura in 1983 [5] by using a graph-theoretical approach. The problem of LUD construction had been reduced to the computation of a maximum independent set of an undirected graph. The final result in an asymptotic form is as follows:

$$R_2 \geq 1 - O(1)\,\text{if}\,0 \leq R_1 < 1/4;$$

$$R_2 \geq 1/2(1 + H(2R_1)) - O(1)\,\text{if}\,1/4 \leq R_1 < 1/3; \tag{3}$$

$$R_2 \geq 1/2(\log_2 6) - R_1 - O(1),\ \text{if}\,1/3 \leq R_1 < 1/2$$

However the lower bound 3 is nonconstructive ,i.e it does not give a method of an explicit construction of codes.

c) *Constructions of LUD codes with* $R_1 < 0.5$

1) Construction (Shannon, 1961) (This idea is valid for any UD codes).The idea of construction is simply "time sharing" between two original UD codes. The users agree to use each of two UD pairs several times to get another UD pair with a longer length. Let (C_1, C_2) and (C_1', C_2') be UD pairs with rates (R_1, R_2), (R_1', R_2') and lengths n and n' recpectively. Then, if (C_1, C_2) is used a times, and then (C_1', C_2') is used b times, the resulting UD pair will have a length $(an + bn')$ and rates $(R_1'', R_2'') = (\frac{anR + bn' R_1'}{an+bn'}, \frac{anR_2 + bn' R_2'}{an+bn'})$. This construction will be further referred to as "time-sharing" technique(TS).

Definition 2. *Two pairs of UD codes* P_1 *and* P_2 *will be called equivalent if they can be constructed from each other by TS and this will be denoted by* $P_1 \sim P_2$. *It is easy to see, that if one applies TS to different pairs of UD codes with rates* (R_1, R_2) *and* (R_1', R_2'), $R_{\max} = \max\{(R_1, R_2, R_1', R_2')\}$, *it is not possible to get UD pair* (R_1'', R_2'') ,$R_{\max}' = \max\{R_1'', R_2''\}$ *with* $R_{\max}' > R_{\max}$. *From this observation it is natural to introduce the following partial order between different UD pairs:*

Definition 3. *It will be said that a UD pair* $P_1 = (R_1, R_2)$ *is superior to* $P_1' = (R_1', R_2')$ *denoted by* $P_1 \succeq P_1'$ *if* $R_1 + R_2 \geq R_1' + R_2'$ *and max* $\{R_1, R_2\} \geq \max\{R_1', R_2'\}$.

Definition 4. *It will be said that two different UD pairs* P_1, P_2 *are incomparable, if they are not equivalent or one of them is not superiour to the other. These three definitions give criteria how to compare different UD pairs.*

2) Construction 2 (Weldon, Yui, 1976). Let $C_1 = \{0^n, 1^n\}$ $C_2 = \{(0,1)^n \backslash 1^n\}$ Then (C_1, C_2) is UD. The proof is obvious, since if the sum vector has at least

one "2" then all one vector 1^n is transmitted by C_1, otherwise the all zero vector 0^n is transmitted.

Definition 5. *It is said that a vector $u = (u_1, u_2, \cdots u_n)$ does not cover a vector $v = (v_1, v_2, \cdots v_n)$ denoted by $u \not\to v$ if there is at least one i for which $v_i > u_i$. The following lemma plays an important role for the construction of LUD codes.*

Lemma 6. *(Kasami, Lin,1976 [4]). The code pair (C_1, C_2) is UD if and only if for any two distict pairs (u, v) and (u', v') in $C_1 \times C_2$ one of the following conditions holds:*
a) $u \oplus v \neq u' \oplus v'$ b) $u \oplus v = u' \oplus v'$ but $u \oplus v \not\to v \oplus v'$

Proof. Obviously, if two vectors are different modulo 2, they will be different modulo 3, i.e for the adder channel. Now let us have the second condition.,which means, that for some i , $v_i \oplus v_i' = 1$ and $u_i \oplus v_i = 0$ and hence $u_i' \oplus v_i' = 0$. Since $v_i \neq v_i'$, this implies, that $u_i + v_i \neq u_i' + v_i'$ and therefore $u + v \neq u' + v'$ Now let us apply lemma 6 for the construction of LUD codes. If C_1 is an (n, k) code, then evidently code vectors of C_2 must be chosen from the cosets of C_1 and the only common vector between C_1 and C_2 should be 0^n.

Lemma 7. *(Kasami, Lin, 1976 [4]). Let (C_1, C_2) be an LUD pair. Then two vectors v and v' from the same coset can be chosen as code vectors for the code C_2 if and only if $v \oplus v'$ can not be covered by any vector of that coset.*

Proof. Suppose that $v, v' \in C_2, u, u' \in C_1$ and $u \oplus v = u' \oplus v'$. According to the condition of the lemma, there is some i for which $v_i \oplus v_i' = 1$ and $u_i \oplus v_i = u_i' \oplus v_i' = 0$ and therefore as in Lemma 6 $u + v \neq u' + v'$. It is easy to see that the reverse statement of the lemma is also true. The Lemma 7 has been used by G.Khachatrian for the construction of LUD codes.

3) Construction (G.Khachatrian, 1981, 1982 [8], [9]). In [9] the following general construction of LUD codes is given. It is considered that the generator matrix of C_1 has the following form.

$$
\begin{array}{cccccccccccccccccccc}
 & 1 & \cdot & 1 & 0 & \cdot & \cdot & \cdot & \cdot & 0 & 1 & 1 & \cdot & \cdot & \cdot & \cdot & \cdot & \cdot & 1 & \\
 & 0 & \cdot & 0 & 1 & \cdot & 1 & 0 & 0 & 0 & \cdot & \cdot & \cdot & \cdot & \cdot & \cdot & \cdot & \cdot & \cdot & r^{(1)} \\
 & \cdot & \cdot & \cdot & \cdot & \cdot & \cdot & \cdot & \cdot & \cdot & 1 & 1 & \cdot & \cdot & \cdot & \cdot & \cdot & \cdot & 1 & \\
 & \cdot & \cdot & \cdot & \cdot & \cdot & \cdot & \cdot & \cdot & \cdot & 0 & 0 & 0 & 1 & 1\cdot & \cdot & \cdot & \cdot & 1 & \\
 & \cdot & \cdot & \cdot & \cdot & \cdot & \cdot & \cdot & \cdot & \cdot & \cdot 0 & 0 & 0 & \cdot & \cdot & \cdot & \cdot & \cdot & \cdot & r^{(2)} \\
I_k & \cdot & \cdot & \cdot & \cdot & \cdot & \cdot & \cdot & \cdot & \cdot & \cdot & \cdot & \cdot & 1 & 1 & \cdot & \cdot & \cdot & 1 & \\
 & \cdot & \cdot & \cdot & \cdot & \cdot & \cdot & \cdot & \cdot & \cdot & \cdot & \cdot & \cdot & \cdot & \cdot & \cdot & \cdot & \cdot & \cdot & \\
 & \cdot & \cdot & \cdot & \cdot & \cdot & \cdot & \cdot & \cdot & \cdot & 0 & 0 & 0 & 0 & 0 & 0 & 1 & 1 & 1 & \\
 & \cdot & \cdot & \cdot & \cdot & \cdot & \cdot & \cdot & \cdot & \cdot & \cdot & \cdot & \cdot & \cdot & \cdot & \cdot & \cdot & \cdot & \cdot & r^{(m)} \\
 & \cdot & \cdot & \cdot & \cdot & \cdot & \cdot & 1 & \cdot & 1 & 0 & 0 & 0 & 0 & 0 & 0 & 1 & 1 & 1 & \\
 & & l_1 & & & l_2 & & & l_k & & & r_1^{(1)} & & & & & & r_1^{(m_1)} & &
\end{array}
$$

where I_k is an identity matrix, $\sum_{j=1}^{m_1} r^{(j)} = k; \sum_{j=1}^{m_1} r_1^{(j)} = n - k - \sum_{i=1}^{k} l_i$; In [9] the following formula for the cardinality of C_2 is given with the restriction that $l_i = l(i = 1 \cdot \cdot k), r^{(j)} = r; (j = 1 \cdot m); r_1^{(j)} = r_1(i = 1 \cdot \cdot m_1)$ $|C_2| =$

R_1	R_2	n	R_1	R_2	n
0.125	0.99993	n = 120	0.2	0.99624	n = 210
0.13333	0.99981	n = 120	0.25	0.98458	n = 156
0.14285	0.99974	n = 252	0.2666	0.97957	n = 210
0.1666	0.99896	n = 144	0.3	0.9642	n = 100
0.1875	0.99729	n = 224	0.3333	0.9382	n = 30
0.4	0.8865	n = 60			

Table 1

$2^m \sum_{i=1}^{m_1} \sum_{j=0}^{i-1} F_1(i,j) * F(i) \quad (i = 0 \to j = 0)$ where
$F_1(i,j) = (-1)^j \times \binom{i}{j} \times \sum_{p=i}^{ir} \binom{(i-j)r}{p} \times (2^{l+1} - 2)^{k-p}$

$$F(i) = \sum_{j_1=0}^{m-i} \sum_{j_2=j_1+1}^{m-i+1} \cdot \sum_{j_i=j_{i-1}+1}^{m-1} 2^{j_1(r_1-1)} \times 2^{(j_2-j_1)(r_1-1)+1)} \times (2^{(m-j_i)(r_1-1)+1} - 1)$$

An analogous formula is obtained in [10] for arbitrary $r^{(i)}$,$r_1^{(i)}$, l_i which is more complicated and is not introduced here for the sake of space. The parameters of some codes obtained with the above consruction are presented in Table 1.

IV CONSTRUCTION OF NONLINEAR UNIQUELY DECODABLE CODES (NUD)

Construction 1.(H.Van Tilborg, P.C.Van den Braak, 1985 [11]). The idea of the construction is as follows: Let a code pair $(C, D \cup E)$ of the length n with partitions $C = C^0 \cup C^1$ and $D = D^0 \cup D^1$ be given, which is called a system of basic codes if (I) $C, D^i \cup E$ is UD for $i = 0, 1$, (II) $C^i, D \cup E$ is UD for $i = 0, 1$, (III) $\forall_{(c,d)\in C^0\times D^0} \forall_{(c',d')\in C^1\times D^1} [c + d \neq c' + d']$, (IV) there is a bijective mapping $\varphi : D^{(0)} \to D^{(1)}$ such that $\forall_{d\in D^0} \forall_{d'\in D^1} [d' = \varphi(d)$ if $\exists_{c,c'} \in C[c + d = c' + d']$, (V) $D \cap E = \ominus, C^{(0)} \neq \ominus, C^{(1)} \neq \ominus, D^{(1)} \neq \ominus$. Let Z be binary code of length s. Now consider a code $\Im$ of length ns which is obtained from the code Z by replacing each coordinate of $Z_i (i = 1 \cdots s)$ by the code vector from $C^{(i)} (i = 0, 1)$. $\Im$ will be considered to be the first code for the new UD pair of the length ns. Now the question is how many vectors from $(D \cup S)^s$ can be included in the second code. The following theorem gives an explicit answer about the cardinalities of both codes.

Theorem 8. . *Let $(C, D \cup E)$ be a system of basic codes of length n as defined above. Let Z be a code of length s, where $2 \leq w \leq s/2$,and $\Im$ be a code of length ns as defined above. Write $s = qw + r, 0 \leq r \leq w$ and define $N = sn$, $\delta = \max\{r, w - r\}, x =\mid D^{(0)} \mid \setminus \mid D^{(0)} \cup E \mid$ and $y =\mid C^{(0)} \mid \setminus \mid C \mid$. Then*

a) $\Im$ is a code of length N and size $|\Im|=|C|^s \times \sum_{k=0}^{q} \binom{s}{kw} y^{s-kw}(1-y)^{kw}$

b) there exists a code $\wp$ of length N s.t. $(\Im, \wp)$ is UD. The code $\wp$ has size $|\wp|=|D^{(0)} \cup E|^s \times \{w - \sum_{i=0}^{s} \binom{s}{i}(w-i-1)x^i(1-x)^{5-i} + \sum_{i=0}^{s} \binom{s}{i}(w-2-2i)x^{5-i}(1-x)^i + \sum_{i=w-\beta-1}^{s} \binom{s}{i}(\beta-1-i)x^{5-i}(1-x)^i\}$

For the numerical results a system of basic codes given by $C_i^{(0)} = D_i^{(0)} = \{0^{n_i}\}$ $C_i^{(1)} = D_i^{(1)} = \{1^{n_i}\}$ $E_i = \{0,1\}^{n_i} \setminus \{0^{n_i}, 1^{n_i}\}$ of length n_i is used which is in fact a system of UD codes given by construction 1. It is interesting to mention, that if Z is a parity check code correcting single erasures with $w = 2$ this construction coincides with the special case of construction 3, however it does not cover the construction 3 in more general form. The numerical results for the best UD code pairs obtained with this method will be presented in the final table. It is also interesting to mention, that in the paper [11] where the present construction is given it was also mentioned the construction of a UD pair of length 7 and sizes $|C|=12$ and $|D|=47$ found by C. Van den Braak in an entirely different way. Although no construction principle of that code has been explained it has the best known sum rate, namely $R_1 = 0.5121$, and $R_2 = 0.7935$, $R_1 + R_2 = 1.3056$.

Construction 2 (R.Ahlswede, V.Balakirski, 1997, [12]).

a) Construction of C_1: N -code length is $N = tn$, $|C_1| = \binom{t}{t/2}$. A code is constructed as follows: At first all $\binom{t}{t/2}$ vectors of the length t and weight $t/2$ are taken and each coordinate then is repeated exactly n times resulting in a code of the length tn and cardinality $\binom{t}{t/2}$.
b) Construction of C_2. The length tn is divided into t blocks of length n. It is obvious that if a block of length n is a vector $G = \{0,1\}^n \setminus \{0^n, 1^n\}$, then in that blocks C_1 and C_2 can be decoded uniquely (according to construction 1). In any r blocks where C_2 has elements from $B = \{0^n, 1^n\}, C_2$ may have one of the following $(r+1)$ possible vectors $\{\{0^n\}^i, \{1^n\}^{n-i}\}$ $(i = 0 \cdots r)$, therefore the cardinality of C_2 is defined by the formulae:
$|C_2| = \sum_{r=0}^{t} \binom{t}{r}(2^n-2)^{n-r}(1+r) = (2^n-1)^{n-1}(2^n-1+n)$. This construction gives relatively good codes with $n = 2$. The best sum rate is achieved with $t = 26, n = 2, R_1 = 0.4482$ $R_2 = .8554, R_1 + R_2 = 1.3036$. Although this construction does not give a significant improvement over previous NUD construction it gives by our opinion a very fruitful approach to the construction of better UD codes.

Construction 3 (G.Khachatrian,1997,[13]). The following construction is considered. Let N be the length of the codes C_1 and C_2, t is an arbitrary integer, $N = 2t$.

1) Construction of code C_1. We consider 2 cases,namely when t is odd and even. Vectors of C_1 have the form $(a_1 a_1 \cdots a_{ii} \cdots)$ where the number of nonzero elements a_i is equal to
i) $(t/2) \pm i (i = 0 \cdot \cdot r)$ if t is even, ii)$(t+1)/2 + i$ or $(t-1)/2 - i$ $(i = 0 \cdot \cdot r)$, if t

is odd. Therefore the cardinality of C_1 is equal to

$$| C_1 |= 2\sum_{j=0}^{r} \binom{t}{t/2+j}$$

if t is even

$$| C_1 |= 2\sum_{j=0}^{r} \binom{t}{\frac{t+1}{2}+j}$$

if t is odd.

2) Construction of the code C_2. The positions of C_2 are divided into t subblocks of the length 2. Let t_1 ($0 \leq t_1 \leq t$) be the number of subblocks of the length 2, where C_2 may have either (00) or (11), in the rest of $(t - t_1)$ subblocks C_2 has either (01) or (10). Now let's see what combinations of (00) and (11) specifically C_2 is allowed to have in these subblocks of the length t_1. C_2 will consist of vectors of the type$\{\{0^n\}^j, \{1^n\}^{n-j}\}$ where $j = (2r+1)k$, if t is even and $j = 2(r+1)k$, if t is odd . Therefore, the number of vectors corresponding to those t_1 subblocks is equal to $N(t_1) = \lceil (t_1+1)/(2r+1) \rceil$ if i is even $N(t_1) = \lceil (t_1+1)/(2(r+1)) \rceil$ if i is odd. We get the following formula for the cardinality of C_2; $| C_2 |= \sum_{r=0}^{t} \binom{t}{r} 2^{n-r} N(t_1)$ and we get that $| C_2 |\approx 3^{t-1}/2r * (t + 1.5(2r+1))$. The best code which is obtained according to this construction has the parameters: $t = 19, N = 38, r = 2, R_1 = 0.48305, R_2 = 0.82257, R_1 + R_2 = 1.30562$

PART II. Code Constructions for the T-User Noiseless Adder Channel

I INTRODUCTION

Let us consider a multiple-access communication system where T statistically independent sources which use binary block codes $C_1, \ldots, C_T$ of equal length n, simultaneously transmit information via one common channel, maintaining synchronization with respect to words and bits. The output of the adder channel is the $(T+1)$-ary vector which is the componentwise arithmetic sum of the transmitted binary vectors. The task of the decoder is to determine uniquely the messages of all users.

Let $u_1 = (u_1^1 \ldots u_n^1), \ldots, u_T = (u_1^T \ldots u_n^T)$ be the transmitted vectors, $u_i = C_i$, $i = 1, 2, \ldots, T$. Then the output of the channel will be the vector

$$u_1 + u_2 + \cdots + u_T = \left(\sum_{i=1}^{T} u_1^i, \cdots, \sum_{i=1}^{T} u_n^i \right).$$

The set of codes $C_1, \ldots, C_T$ is said to be **uniquely decodable(UD)** if for any two distinct sets $u_1, \ldots, u_T$ and $v_1, \ldots, v_T$; u_i, $v_i \in C_i$ $i = 1, 2, \ldots, T$ we have $u_1 + \cdots u_T \neq v_1 + \cdots v_T$. The rate of the ith user is given by

$$R_i = \frac{\log_2 |C_i|}{n}$$

where $|C_i|$ denotes the cardinality of the ith code. The problem in general is to construct UD set of codes $C_1, \ldots, C_T$ such that the point $(R_1, \ldots, R_T)$ will be as close as possible to the boundary of the capacity region for the channel. In this survey a more specific problem is considered, namely: the construction of a UD set of codes $C_1, \ldots, C_T$ so that the rate sum $R_{SUM}(T) = R_1 + \cdots + R_T$ is as large as possible. This problem has been considered in [15] -[20].

II CAPACITY REGION

Ulrey [14] generalized Ahlswede's MAC coding theorem to many senders. The capacity region for the noiseless T-user adder channel (AC) was calculated by Liao [15].

$$C = \{(R_1, R_2, \ldots, R_T): \quad 0 \le R_i \le 1,$$
$$0 \le R_1 + \cdots + R_t \le \sum_{i=0}^{t} \frac{\binom{t}{i}}{2^t} \log_2 \frac{2^t}{\binom{t}{i}}, \quad \text{for } t = 2, \ldots, T\}.$$

It can be shown that

$$C_{SUM}(T) = \sum_{i=0}^{T} \frac{\binom{T}{i}}{2^T} \log_2 \frac{2^T}{\binom{T}{i}}$$

and is asymptotically equal to $\frac{1}{2}\log_2(\pi e T/2)$ as $T \to \infty$.

III CONSTRUCTIONS

It should be mentioned here that the "time-sharing" technique proposed by Shannon in 1961 could also be applied to a channel with many users [7]. The first non-trivial construction of UD codes for the T-user AC was proposed by Chang and Weldon in 1979 [13], where a construction method for so called basic UD codes was given.

Definition 9. *A UD system $(C_1, \ldots, C_T)$ is called basic (BUD) if $|C_i| = 2$ $(i = 1, \ldots, T)$. For this case the sum rate of T-users is equal to T/n. Thus, the problem is to achieve the maximal rate for a fixed T or to have the maximal number of users for fixed n.*

Consider a BUD T-user code $C_1, \ldots, C_T$ where $C_i = \{x_i, y_i\}$.

Definition 10. *A matrix $D = (d_1, d_2 \ldots, d_T)$ where $d_i = x_i - y_i$ and $x_i - y_i$ means the componentwise arithmetic difference between the vectors x_i and y_i, is called to be the difference matrix (dm) of the BUD system $(C_1, \ldots, C_T)$.*

The difference matrix plays a central role in the construction of T -user BUD codes.

Theorem 11. *Let $(C_1, \ldots, C_T)$ be a T-user basic code. Let $m = (m_1, \ldots, m_T)$ $m_i \in \{0, 1, -1\}$. Then $(C_1, \ldots, C_T)$ is a BUD if and only if*

$$mD = 0^n,$$

where D is the dm of $(C_1, \ldots, C_T)$ implies that m is the all-zero T-tuple.

The proof of the theorem follows from the definition of *dm* and BUD.

According to Theorem 11 the construction of BUD is reduced to the construction of $T \times W$ matrices over $\{0, 1, -1\}$ such that the rows in the matrices are linearly independent over $\{0, 1, -1\}$. From a given matrix D we can construct more than one T-user BUD codes, since the lth components of x_i and y_i could be set either to "0" or to "1" when the lth component of d_i is a "0". The situation is the same when d_i is equal to "1" and " − 1". It is natural that all the T-user BUD codes constructed from the same *dm* of D will be said to be equivalent.

In paper [16] the following iterative construction of *dm* was proposed.

Theorem 12. *For any nonnegative integer j, the matrix*

$$D_j = \begin{bmatrix} D_{j-1} & D_{j-1} \\ D_{j-1} & -D_{j-1} \\ I_{j-1} & 0_{j-1} \end{bmatrix} \tag{4}$$

defines a $(j+2) \cdot 2^{j-1}$-user BUD of the length 2^j, where I_{j-1} is the 2^{j-1}-order identity matrix; 0_{j-1} is the $2^{j-1} \times 2^{j-1}$ zero matrix; $D_0 = [1]$.

Proof. The proof is by induction on j. For $j = 0$ $D_0 = [1]$ which specifies a trivial single-user code of length 1. Assume that D_{j-1} defines a $2^{j-2}(j+1)$-user BUD of length 2^{j-1}; $j \geq 1$. Let $m = (m_1, m_2, m_3)$ be a solution of $mD_j = 0^{N_j}$ over $(0, 1, 1)$, where $m_1 m_2 \in (0, 1, -1)^{T_{j-1}}$; $m_3 \in (0, 1, -1)^{n_{j-1}}$. From 4 we have

$$m_1 D_{j-1} + m_2 D_{j-1} + m_3 = 0^{N_{j-1}}$$

and

$$m_1 D_{j-1} - m_2 D_{j-1} = 0^{N_{j-1}}$$

which is reduced to

$$2m_1 D_{j-1} = -m_3. \tag{5}$$

It follows from (5) that $m_3 = 0^{n_{j-1}}$. Then we get that

$$m_1 D_{j-1} = 0^{N_{j-1}}$$
$$m_2 D_{j-1} = 0^{N_{j-1}}$$

and $m_i = 0^{N_{j-1}}$ $(i = 1, 2, \ldots)$, since D_{j-1} is assumed to be a dm for a T_{j-1}-user code. Thus $m = (m_1, m_2, m_3) = 0^{T_j}$ and by Theorem 12 D_j is a difference matrix of a T_j-user BUD of length N_j.

The rate $R_{SUM}(T_j)$ of a code described in Theorem 12 is

$$R_{SUM}(T_j) = \frac{T_j}{N_j} = 1 + \frac{j}{2} = 1 + \frac{1}{2}\log_2 N_j.$$

Since

$$\frac{1}{2}\log_2 T_j = \frac{1}{2}\log_2 N_j + \frac{1}{2}\log_2\left((\log_2 N_j) + 2\right) - \frac{1}{2},$$

it follows that

$$\lim_{N_j \to \infty} \frac{R_{SUM}(T_j)}{C_{SUM}(T_j)} = 1$$

This implies the following.

Corollary 13. *The Tj-user BUD defined by Theorem 12 has a sum rate $R_{SUM}(T_j)$ asymptotically equal to the maximal achievable sum rate $C_{SUM}(T_j)$ as T_j increases.*

Although this result looks very elegant, the coding problem of AC is rather interesting for the case when the number of users is fixed. The real goal would be the following: to get asymptotically optimum UD codes for fixed T as the length of the codes goes to infinity.

The construction given by Theorem 12 was generalized in the work by Ferguson [17] in 1982, where it was shown that instead of $(I_{j-1}\ 0_{j-1})$ in D_i could be used any $(A\ B)$ if $\overline{A+B}$ is an invertible binary matrix(in which the overbar refers to reduction *modulo* 2). The construction described in Theorem 12 gives codes with length $N = 2^i$.

In [18] a shortening technique pas proposed by Chang in 1984, which allows to construct BUD of arbitrary length. This result was improved in 1986 by Martirossian [19].

Theorem 14. *Let $\mathbf{m} = (m_1, m_2, \ldots, m_T)$ be an arbitrary vector with $m_1 \in \{0, 1, -1\}$. Then $C_1, C_2, \ldots, C_T$ is a uniquely decodable code with T users if the condition $\mathbf{m}\,\widetilde{D} = 0^n$ holds iff for $\mathbf{m} = 0^T$, where 0^n is the n-dimensional all-zero vector.*

For the code of length n we'll denote the difference matrix (dm) of a uniquely decodable code $C_1, C_2, \ldots, C_T$ by $\widetilde{D}_n = \{d_1^n, d_2^n, \ldots, d_n^n\}$ and the number of users by T_n, respectively.

Theorem 15. *If $\widetilde{D}_u$ and $\widetilde{D}_v$ are the dm of BUD codes of length u and v $(u \le v)$, respectively, then the matrix*

$$\widetilde{D}_{u+v} = \left|\begin{array}{c|cc} -\widetilde{D}_v^u & \widetilde{D}_v & \\ \widetilde{D}_u & \widetilde{D}_u & \widetilde{A} \\ \widetilde{I}_u & \widetilde{0}_u & \widetilde{B} \end{array}\right| \left|\begin{array}{ccc|cccc} -d_1^v - d_2^v & \cdots & d_u^v & d_1^v d_2^v & \cdots & & d_v^v \\ d_1^u d_2^u & \cdots & d_u^u & d_1^u d_2^u & \cdots & d_u^u & \widetilde{A} \\ e_1 e_2 & \cdots & e_u & 00 & \cdots & 0 & \widetilde{B} \end{array}\right|, \quad (6)$$

where $\widetilde{D}_v^u$ consists of the first columns of the matrix $\widetilde{D}_v$; $\widetilde{I}_u$ is the $u \times u$ identity matrix; $\widetilde{A}$, $\widetilde{B}$ are any two matrices with elements from $\{0, 1, -1\}$, is the difference matrix dm of a UD code of length $u + v$.

Theorem 15 allows us to construct D_u from the given $D_{u_1}, D_{u_2}, \ldots, D_{u_s}$, where $u = u_1 + u_2 + \cdots + u_s$ for any s.

Now we'll represent n as $n = \sum_{k=0}^{s} n_k 2^k$, $s = [\log_2^n]$, $n_k \in \{0,1\}$ and denote $n^{(j)} = \sum_{k=0}^{l} n_k 2^k$.

Thus, using Theorem 15 for the lengths $u = n_j n^{(j)}$, $v = \sum_{r=j+1}^{s} n_r 2^r$ and setting j equal to $s-1, s-2, \ldots, 1, 0$ successively, we'll reduce the construction of $\widetilde{D}_n$ to the one of constructing $\widetilde{D}_{2^0}, \widetilde{D}_{2^1}, \ldots, \widetilde{D}_{2^s}$. For this case the number of users is obtained successively from the relation $T_{u+v} = T_u + T_v + u$, i.e

$$
\begin{aligned}
T_n &= T_{2^s} + T_{n^{(s-1)}} + n^{(s-1)} \\
&= T_{2^s} + n_{s-1} T_{2^{s-1}} + T_{n^{(s-2)}} + n_{s-1} n^{(s-1)} + n^{(s-2)} \\
&= T_{2^s} + n_{s-1} T_{2^{s-1}} + \cdots + n_0 T_{2^0} + n_{s-1} n^{(s-1)} + n_{s-2} n^{(s-2)} + \cdots + n_0 = \\
&= \sum_{k=0}^{s} n_k T_{2^k} + \sum_{l=0}^{s-1} n_l \sum_{k=0}^{1} n_k 2^k
\end{aligned}
$$

or as $T_{2^k} = (k+2)\, 2^{k-1}$ (see [12], the same result is also obtained from Theorem 15) then

$$
T_n = \sum_{k=0}^{s} n_k (k+2)\, 2^{k-1} + \sum_{l=0}^{s-1} n_l \sum_{k=0}^{1} n_k 2^k. \tag{7}
$$

Let's denote the number of users of the code of length n constructed in [18] by T_n'. If we express n as $n = 2^l - j$, $0 <, j < 2^{l-1}$, then it will be given by the formula

$$
T_n' = (i+1)\, 2^{l-1} - j - \sum_{k=0}^{i-2} j_k (k+2)\, 2^{k-1}, \tag{8}
$$

where $j - 1 = \sum_{k=0}^{i-2} j_k 2^k, \quad j_k \in \{0,1\}$.

Lemma 16. $T_n \geq T_n'$.

Comparing 7 and 8, we have

$$
T_n - T_n' = \sum_{l=0}^{\alpha-1} n_l \sum_{k=0}^{1} n_k 2^k \geq 0. \tag{9}
$$

Now we will introduce the results obtained by Khachatrian and Martirossian (These results were reported during the First Armenian-Japanese Colloquium on Coding Theory, Diligan Armenia, September 1986 and are finally published in [20]). A construction of nonbasic UD codes constructed from BUD given in some special way is represented here. This construction is based on the following.

Lemma 17. *Let $C_1, \ldots, C_T$ be a UD set, and $\{\{u_1\}, \ldots, \{u_{T_1}\}\}$ be a split of this set into T_1 nonempty subsets. Then the system $\{C_1^1, \ldots, C_T^1\}$ will also be a UD, where C_i^1 is the set of all binary vectors that belong to the set of all possible sums*

$$\left(x_{r_1^{(i)}} + x_{r_2^{(i)}} + \cdots + x_{r^{(i)}_{|u_i|}}\right)$$

where

$$x_{r_j^i} \in C_{r_1^{(i)}} \quad j = \{1, \ldots, |u_i|\}, \{u_i\} = \left\{C_{r_1^{(i)}}, \ldots C_{r^{(i)}_{|u_i|}}\right\}$$

and $|u_i|$ is the cardinality of the set u_i

$$\sum_{i-1}^{T} |u_i| = T.$$

The proof of the Lemma follows directly from the definition of a UD system. The obtained T_1-user UD system will be called to be a T_1-*conjugate system* in respect to T-user $\{C_1, \ldots, C_T\}$ (in short $(T_1 - T)$ system).

The following 2 corollaries are deduced from Lemma 17.

Corollary 18. *Let $C_1, \ldots, C_T$ be a UD set and let*

$$\{C_{i_1} \cap C_{i_2} \cap \cdots \cap C_{i_r}\} \neq \emptyset.$$

Then the $(T - r + 1)$-user system $\left(C_0, C_{j_1} \ldots, C_{j_{T-r}}\right)$, $j_1, j_2, \ldots, j_{T-r} \in \{i_1, i_2, \ldots, i_r\}$ is also a UD, where $C_0 = C_{i_1} \cup C_{i_2} \cup \cdots \cup C_{i_r}$.

Corollary 19. *Let*

$$D = [d_{i_1}\ d_{i_2} \cdots d_{i_k}]^T$$

be the submatrix of dm for a BUD system. If each column in D has no more than one nonzero element then the corresponding $C_{i_1}, \ldots, C_{i_k}$ codes can be combined into one code with the cardinality equal to 2^k such that the obtained $(T - k + 1)$-user system is also UD.

The last corollary allows us to construct $(T - k + 1)$-user UD codes from T-user ones with the same sum rate, which is obviously more favorable since we have the same sum rate for a smaller number of users.

The UD codes will be constructed on the basis of some initial BUD codes and Lemma 17. Now we'll try to explain the problem if initial BUD codes of what kind are constructed. Two cases will be considered here.

First case $(n = 2^k)$.

The construction is implemented iteratively on k. On the kth step 2^{k-1} matrices $D_{2^k}^1, \ldots, D_{2^k}^{2^k-1}$ are constructed.

At the first step $(k = 1)$ $i = 1$.

$$D^1_{2^1} = \left|\begin{array}{cc} 1 & 1 \\ 1 & -1 \\ \hline 1 & 0 \end{array}\right| = \left|\begin{array}{c} A^1_{2^1} \\ \hline B^1_{2^1} \end{array}\right| = \left|\begin{array}{c} a_{21} \\ a_{22} \\ \hline b_{21} \end{array}\right|$$

At the second step ($k = 2$) there are two matrices.

$i = 1$

$$D^1_{2^2} = \left|\begin{array}{cccc} 1 & 1 & 1 & 1 \\ -1 & -1 & 1 & 1 \\ \hline 1 & -1 & 0 & 0 \\ 0 & 0 & 1 & -1 \\ 1 & 1 & 0 & 0 \\ 1 & 0 & 0 & 0 \\ 0 & 0 & 1 & 0 \end{array}\right| = \left|\begin{array}{c} A^1_{2^2} \\ \hline B^1_{2^2} \end{array}\right| = \left|\begin{array}{cc} a_{21} & a_{21} \\ -a_{21} & a_{21} \\ \hline a_{22} & 0 \\ 0 & a_{22} \\ a_{21} & 0 \\ B^1_2 & 0 \\ 0 & B^1_2 \end{array}\right|$$

$i = 2$

$$D^2_{2^2} = \left|\begin{array}{cccc} 1 & 1 & 1 & 1 \\ 1 & -1 & 1 & -1 \\ -1 & -1 & 1 & 1 \\ -1 & 1 & 1 & -1 \\ \hline 1 & 1 & 0 & 0 \\ 0 & 0 & 1 & -1 \\ 1 & 0 & 0 & 0 \\ 0 & 0 & 1 & 0 \end{array}\right| = \left|\begin{array}{c} A^2_{2^2} \\ \hline B^2_{2^2} \end{array}\right| = \left|\begin{array}{cc} a_{21} & a_{21} \\ a_{22} & a_{22} \\ -a_{21} & a_{21} \\ a_{22} & a_{22} \\ \hline a_{21} & 0 \\ 0 & a_{22} \\ B^1_2 & 0 \\ 0 & B^1_2 \end{array}\right|$$

At the kth step $i = 1, 2, \ldots, 2^{k-1}$

$$
D^i_{2^k} = \begin{array}{c|c|c|}
 & a_{2^{k-1}1} & a_{2^{k-1}1} \\
1 & \vdots & \vdots \\
 & a_{2^{k-1}i} & a_{2^{k-1}i} \\
\hline
 & -a_{2^{k-1}1} & a_{2^{k-1}1} \\
2 & \vdots & \vdots \\
 & -a_{2^{k-1}i} & a_{2^{k-1}i} \\
\hline
 & a_{2^{k-1}(i+1)} & 0 \\
3 & \vdots & \vdots \\
 & a_{2^{k-1}2^{k-1}} & 0 \\
\hline
 & 0 & a_{2^{k-1}(i+1)} \\
4 & \vdots & \vdots \\
 & 0 & a_{2^{k-1}2^{k-1}} \\
\hline
 & a_{2^{k-1}1} & 0 \\
5 & \vdots & \vdots \\
 & a_{2^{k-1}\lfloor \frac{i+1}{2} \rfloor^*} & 0 \\
\hline
 & 0 & a_{2^{k-1}\lfloor \frac{i+3}{2} \rfloor} \\
6 & \vdots & \vdots \\
 & 0 & 0 \\
\hline
7 & B^{2^{k-2}}_{2^{k-1}} & 0 \\
\hline
8 & 0 & B^{2^{k-2}}_{2^{k-1}}
\end{array}
= \begin{array}{|c|}
A^i_{2^k} \\
\hline
B^i_{2^k}
\end{array}
\qquad (10)
$$

For the sake of convenience the rows of the matrix $D^i_{2^k}$ are split into eight blocks and numbered. Let us denote the number of rows in $D^i_{2^k}$ (the number of users) by $T^i_{2^k}$. It is easy to see that for the matrices constructed by (10) the following recurrence relation holds:

$$
T^i_{2^k} = T^{2^{k-2}}_{2^{k-1}} + T^{2^{k-2}}_{2^{k-1}} + i. \qquad (11)
$$

It follows, particularly, from (11) that

$$
T^{2^{k-2}}_{2^k} = (k+2)\,2^{k-1} \text{ and } T^i_{2^k} = (k+1)\,2^{k-1} + i.
$$

The following theorems are proved in [20].

Theorem 20. *For all k and i $1 \le i \le 2^{k-1}$, the matrix $D^i_{2^k}$ is a dm for a BUD set of codes.*

T	$\sum R_i$	n	T	$\sum R_i$	n	T	$\sum R_i$	n	T	$\sum R_i$	n
2	1.2924	2	14	2.5680	16	25	3.0183	32	37	3.2683	32
3	1.5283	3	15	2.6250	16	26	3.0326	32	38	3.2826	32
4	1.6666	3	16	2.6666	12	27	3.0625	32	39	3.3125	32
5	1.8305	4	17	2.6930	16	28	3.0808	32	40	3.3308	32
6	2.0000	4	18	2.7500	16	29	3.0951	32	41	3.3451	32
7	2.0731	8	19	2.7586	16	30	3.1250	32	42	3.3750	32
8	2.1666	6	20	2.8180	16	31	3.1433	32	43	3.3933	32
9	2.2500	8	21	2.8750	16	32	3.1666	24	44	3.4076	32
10	2.3231	8	22	2.9116	16	33	3.1875	32	45	3.4375	32
11	2.3962	8	23	2.9430	16	34	3.2058	32	46	3.4358	32
12	2.5000	8	24	3.0000	16	35	3.2201	32	47	3.4701	32
13	2.5366	16				36	3.2500	32	48	3.5000	32

Now new BUD codes can be constructed by regrouping the rows of the matrix $D^i_{2^k}$ (see [20]). The results are summarized by

Theorem 21. *For UD codes $R_{SUM}(T)$ satisfies the following relation:*

$$1. R_{SUM}(T) \geq \frac{(k+2)2^k+2s}{2^{k+1}}, \qquad r=0$$
$$2. R_{SUM}(T) \geq \frac{(k+2)2^k+2s-1+\log_2 3}{2^{k+1}}, \qquad r=1$$
$$3. R_{SUM}(T) \geq \frac{(k+2)2^k+2s-1-k+\log_2\left(2^{k+1}+1\right)}{2^{k+1}}, \qquad r=2.$$

The table above gives the best known T-user UD codes based on the results in [20].

References

[1] R. Ahlswede, "Multi-way communication channels", *Proc. 2nd Int. Symp. Inform. Theory,* Thakhkadzor, Armenia, 1971, 23-25.

[2] T. Kasami and S. Lin, "Coding for a multiple-access channel", *IEEE Trans. Inform. Theory,* 1976, 129-137.

[3] E. J. Weldon, "Coding for a multiple-access channel", *Information and Control,* 1978, 256-274.

[4] T. Kasami and S. Lin, "Bounds on the achievable rates of block coding for a memoryless multiple-access channel", *IEEE Trans. Inform. Theory,* 1978, 186-187.

[5] T. Kasami, S. Lin, V.Wei and S. Yamamura, "Graph theoretic approach to the code construction of the two-user multiple-access binary adder channel", *IEEE Trans. Inform. Theory,* 1983, 114-130.

[6] H. van Tilborg, "An upper bound for codes in a two-access binaryerasure channel", *IEEE Trans. Inform. Theory*, 1978, 112-116.

[7] C. E. Shannon, "Two-way communication channels", *Proc. of 4th Berkley Symp. Math. Stat. Prob.*, Vol. N1, 611-644, 1961.

[8] G. Khachatrian, "Construction of uniquely decodable code pairs for two-user noiseless adder channel", *Problemi Peredachi Informasi*, 1981.

[9] G. Khachatrian, "On the construction of codes for noiseless synchronized 2-user channel", *Problems of Control and Inform. Theory*, 1982, 319-324.

[10] G. Khachatrian and H. Shamoyan, "The cardinality of uniquely decodable codes for two-user adder channel", *J. Inform. Process. Cybernet.*, EIK 27, 7, 1991, 351-355.

[11] P. Coebergh van den Braak and H. van Tilborg, "A family of good uniquely decodable code pairs for the two-access binary adder channel", *IEEE Trans. Inform. Theory* 31, 1985, 3-9.

[12] R. Ahlswede and V. B. Balakirsky, "Construction of uniquely decodable codes for the two-user binary adder channel", *Proc. 2nd INTAS Meeting on Inform. Theory and Combinatorics*, Essen, Germany, 1997, 1-2.

[13] G. Khachatrian, "New construction of uniquely decodable codes for two-user adder channel", *Colloquim dedicated to the 70-anniversary of prof. R. Varshamov*, Thakhkadzor, Armenia, October 1-7, 1997.

[14] M. L. Ulrey, "The capacity region of a channel with s senders and r receivers", *Information and Control* 29, 1975, 185-203.

[15] H. J. Liao, *Multiple-Access Channels*, PHD Dissertation, Dept. of Elect. Eng. University of Hawaii, 1972.

[16] S. C. Chang and E. J. Weldon, "Coding for t-user multiple-cccess channels", *IEEE Trans. on Inform. Theory*, 1979, 684-691.

[17] T. Ferguson, "Generalized T-user codes for multiple-access channels", *IEEE Trans. Inform. Theory* 28, 1982, 775-778.

[18] S. C. Chang, "Further results on coding for T-user multiple-access channels", *IEEE Trans. Inform. Theory* 30, 1984, 411-415.

[19] S. S. Martirossian, "Codes for noiseless adder channel", *X Prague Conference on Inform. Theory*, Abstracts of papers, Prague, 1986, 110-111.

[20] G. Khachatrian, S. S. Martirossian, "Code construction for the T-user noiseless adder channel", *IEEE Trans. Inform. Theory* 44, 1998, 1953-1957.

COMMUNICATION NETWORK WITH SELF-SIMILAR TRAFFIC

Boris Tsybakov

QUALCOMM Inc., 5775 Morehouse Drive, Room L-400G, San Diego, CA 92121, USA

borist@qualcomm.com

Abstract:

The paper is a review of some results on the discrete-time finite-buffer queueing system which models a communication network multiplexer fed by a self-similar cell traffic. The review includes also some new results. First, the definitions of second-order self-similar processes are given. Then, a queue model is introduced. It has a finite buffer, a number of servers with unit service time, and an input traffic which is an aggregation of independent source-active periods having Pareto-distributed lengths and arriving as Poisson batches. A source generates a Bernoulli sequence of cells. The asymptotic bounds to the buffer-overflow and cell-loss probabilities are given in some cases . The bounds show a true asymptotic behaviour of the probabilities. The bounds decay polynomially with buffer-size growth and exponentially with excess of channel capacity over traffic rate.

INTRODUCTION

A self-similar nature of traffic in high-speed communication networks was recently discovered by real-time measurements made in Bellcore and in other leading communication corparations (Leland, Taqqu, Willinger, and Wilson [10], [11], Crovella and Bestavros [4]). There is a widely-shared feeling, based on the experimental measurements, that an important performance measure of buffer overflow, the overflow probability, decays significantly slower with growing buffer size under self-similar traffic than under short-range dependent traffic such as the renewal or Markov-type traffics traditionally used in telecommunication models. The problem of developping an adequate mathematical approach to treat queues with long-range dependent traffic attracts now a lot of attention (Willinger, Taqqu, and Erromilli [26]). We believe that this problem is in the field of Prof. Dr. Rudolf Ahlswede's interest, with whom we enjoy having useful scientific contacts for several decades.

Primarily, the paper was intended as a review of some results on the finite-buffer queueing systems fed by self-similar traffic. However, in process of its writing, a few new results and generalizations were also included.

I. Althöfer et al. (eds.), Numbers, Information and Complexity, 197-219.

We begin our review with definitions of second-order self-similar processes and present some of their most important properties (Section 2). Pioneering mathematical work on these processes was done in [9] by A.N.Kolmogorov and later by B.B.Mandelbrot, M.S.Pinsker, A.M.Yaglom, Y.G.Sinai, D.R.Cox and many other famous mathematicians.

Then (in Section 3), it is introduced a queue model which has a finite buffer, a number of servers with unit service time, and an input traffic which is an aggregation of independent source-active periods having Pareto-distributed lengths and arriving as Poisson batches. A source generates a Bernoulli sequence of cells in its active period.

In Section 4, the definitions of buffer-overflow and cell-loss probabilities are given. After this, we present the relations between these probabilities.

The last Section 5 contains the asymptotic upper and lower bounds to overflow and loss probabilities. In the case of Bernoulli parameter being equal to 1, the bounds show a true asymptotic behaviour of the probabilities when the buffer size goes to infinity. The bounds decay algebraically with buffer-size growth and exponentially with excess of channel capacity over traffic rate. Such behaviour of the probabilities shows that one can better combat traffic losses in communication networks by increasing channel capacity rather than buffer size.

When we can give an appropriate reference related to a mentioned result, we do not give a proof of the result. In other cases, the proofs are given. Some of them are presented in the appendix of the paper.

SECOND-ORDER SELF-SIMILARITY

Here, the definitions of self-similarity of discrete-time stochastic processes are presented.

We begin with the introduction of $X = (X_1, X_2, \ldots)$, a semi-infinite segment of a second-order-stationary real-number stochastic process of discrete argument (time) $t \in \mathbf{N} \triangleq \{1, 2, \ldots\}$. Denote $\mu \triangleq EX_t < \infty$ and $\sigma^2 \triangleq \text{var}X_t < \infty$, the mean and the variance of X_t respectively. Denote $r(k) \triangleq \frac{E(X_{t+k}-\mu)(X_t-\mu)}{\sigma^2}$, $b(k) \triangleq \sigma^2 r(k)$, $k \in \mathbf{Z}_+ \triangleq \{0, 1, 2 \ldots\}$ the correlation coefficient and autocovariance of process X and denote by $f(l)$ its spectral density. The mean μ, the variance $\sigma^2 \equiv b(0)$, the correlation coefficient $r(k)$, and the autocovariance $b(k)$ do not depend on time t, and $r(k) = r(-k)$, $b(k) = b(-k)$.

Exact self-similarity

Definition A [3]: A process X is called *exactly second-order self-similar (es-s)* with the Hurst parameter $H = 1 - (\beta/2)$, $0 < \beta < 1$ if its correlation coefficient is

$$r(k) = \frac{1}{2}[(k+1)^{2-\beta} - 2k^{2-\beta} + (k-1)^{2-\beta}] \triangleq g(k), \; k \in \mathbf{N}. \tag{2.1}$$

The function $g(k)$ can be written as $g(k) = \frac{1}{2}\delta^2(k^{2-\beta})$ in terms of the central second difference operator $\delta^2(f(x))$ applied to a function $f(x)$. The function $g(k)$ is monotonically decreasing in k.

Before commenting the definition of es-s, we give an equivalent definition and present the most essential properties of es-s processes.

Definition B: A process X is called *es-s* with parameter $H = 1-(\beta/2)$, $0 < \beta < 1$ if

$$b_m(k) = b(k)m^{-\beta}, \quad k \in \mathbf{Z}_+, \quad m \in \{2,3,\ldots\} \tag{2.2}$$

where $b_m(k)$ is the autocovariance of the averaged (over blocks of length m) process $X^{(m)} = (X_1^{(m)}, X_2^{(m)}, \ldots)$ where $X_t^{(m)} = (X_{tm-m+1}+\ldots+X_{tm})/m;\ m, t \in \mathbf{N}$.

The properties of es-s processes are given by the following theorem.

Theorem 2.1 [17], [21]. *For a process X and $0 < \beta < 1$, the following are equivalent:*

a) *X is es-s in definition A, i.e., $r(k) = g(k)$,*

b) $b_m(0) = b(0)m^{-\beta}, \quad m \in \{2,3,\ldots\}$,

c) $f(l) = c \mid e^{2\pi il} - 1 \mid^2 \sum_{n=-\infty}^{\infty} \mid l+n \mid^{\beta-3}$, $-\frac{1}{2} \le l \le \frac{1}{2}$ *where $c > 0$ is a constant,*

d) *X is es-s in definition B, i.e. $b_m(k) = b(k)m^{-\beta}$, $k \in \mathbf{Z}_+$, $m \in \{2,3,\ldots\}$.*

Each of a)-d) implies that

$$r_m(k) = r(k), \quad k \in \mathbf{Z}_+,\ m \in \{2,3,\ldots\} \tag{2.3}$$

where $r_m(k)$ is the correlation coefficient of $X^{(m)}$.

We remark that each of $b_m(0) = b(0)m^{-\beta}$ in b) and $b_m(k) = b(k)m^{-\beta}$ in d) can be considered as a functional equation relative to autocovariance $b(k)$, since

$$b_m(k) = \frac{1}{m^2}[\sum_{i=0}^{m-1}(m-i)b(mk+i) + \sum_{i=1}^{m-1}(m-i)b(mk-i)],\ m \in \{2,3,\ldots\},\ k \in \mathbf{Z}_+. \tag{2.4}$$

The following theorem and its proof show that these equations have the same and unique solution.

Theorem 2.2. *For given $b(0)$ and parameter $\beta, 0 < \beta < 1$, the system of equations*

$$b_m(k) = b(k)m^{-\beta} \tag{2.5}$$

(where an individual equation corresponds to a particular choice of pair (k, m) with $k \in \mathbf{N}$ and $m \in \{2,3,\ldots\}$) with respect to $b(k)$ has the unique solution

$$b(k) = b(0)g(k), \quad k \in \mathbf{N}. \tag{2.6}$$

Proof. Since the function $b_m(k)$ is the autocovariance of process $X^{(m)}$, then, taking into account that $b_1(k) \equiv b(k)$ and using the notation $x_t \triangleq X_t -$

$\mu, \triangle_{t,m} \stackrel{\triangle}{=} x_{tm-m-1} + \ldots + +x_{t,m}, \triangle_{t,1} \stackrel{\triangle}{=} x_{t,1}$, the equation (2.5) can be written as

$$m^{-2} E[\triangle_{t,m} \triangle_{t+k,m}] = m^{-\beta} E[\triangle_{t,1} \triangle_{t+k,1}]. \tag{2.7}$$

Denote $\widehat{b}\ (k) \stackrel{\triangle}{=} b(k)$ and $\widehat{b}_m\ (k) \stackrel{\triangle}{=} E[\frac{\triangle_{t,m}}{m^H} \cdot \frac{\triangle_{t+k,m}}{m^H}]$. The equation

$$\widehat{b}_m\ (k) = \widehat{b}\ (k), \quad k \in \mathbf{Z}_+, \quad m \in \{2, 3, \ldots\} \tag{2.8}$$

goes from (2.7) by multiplication of (2.7) by m^β. This shows that the system of equations (2.8) with respect to $\widehat{b}\ (k)$ is a different record of the system (2.5).

To prove the theorem, it is now sufficient to show that the system (2.8) has the unique solution $\widehat{b}\ (k) \equiv b(k)$ given by (2.6). We shall do it showing that the system of equations

$$\widehat{b}_k\ (0) = \widehat{b}\ (0), \quad k \in \mathbf{N} \tag{2.9}$$

has the unique solution

$$\widehat{b}\ (k) (\equiv b(k)) = b(0) g(k), \quad k \in \mathbf{N} \tag{2.10}$$

and remarking that the solution of (2.9) satisfies the system (2.5) evidently.

The system (2.9) is written as

$$\widehat{b}\ (0) = \frac{1}{k^{\gamma_0}} E[\sum_{l=0}^{k-1} x_l]^2, \quad k \in \mathbf{N}, \quad \gamma_0 \stackrel{\triangle}{=} 2 - \beta. \tag{2.11}$$

(We note that the right-hand side of (2.11) can be expressed in terms of $\widehat{b}\ (k)$ to explain why (2.11) is an equation with respect to $\widehat{b}\ (k)$.)

For $k = 1$, (2.11) gives trivially

$$\widehat{b}\ (0) = E x_0^2 \stackrel{\triangle}{=} b(0). \tag{2.12}$$

For $k = 2$, (2.11) gives

$$\widehat{b}\ (0) = \frac{1}{2^{\gamma_0}} E[x_0 + x_1]^2 = \frac{1}{2^{\gamma_0 - 1}} (\widehat{b}\ (0) + \widehat{b}\ (1)). \tag{2.13}$$

The equation (2.13) is equivalent to the equation

$$\widehat{b}\ (1) = \frac{\widehat{b}\ (0)}{2} (2^{\gamma_0} - 2) \tag{2.14}$$

which is the same as (2.10) when it has $k = 1$. Thus, for system (2.9), any solution $\widehat{b}\ (k)$ is expressed by (2.14) when $k = 1$.

The remaining part of the proof uses an induction over k. We assume that, for system (2.9), any solution $\widehat{b}\ (k)$ is expressed by (2.10) when $k \in \{1, 2, \ldots, (K - 1)\}$ where $K \geq 2$. We have to show that the same statement is true when $k = K$.

Actually, for $k = K + 1$, the system (2.11) can be written as

$$(K+1)^{\gamma_0} \widehat{b}(0)[= E[\sum_{l=0}^{K-1} x_l + x_K]] = \widehat{b}(0)K^{\gamma_0} + \widehat{b}(0) + 2\widehat{b}(K) + 2E[x_K \sum_{l=1}^{K-1} x_l]$$

or as

$$2\widehat{b}(K) = (K+1)^{\gamma_0} \widehat{b}(0) - K^{\gamma_0} \widehat{b}(0) - \widehat{b}(0) - 2E[x_K \sum_{l=1}^{K-1} x_l]. \qquad (2.15)$$

For $k = K$, the comparison of (2.10) and (2.15) shows that we need to prove only that

$$2\sum_{k=1}^{K-1} \widehat{b}(k) = \widehat{b}(0)[K^{\gamma_0} - (K-1)^{\gamma_0} - 1] \qquad (2.16)$$

Let us sum the equations (2.10) over $k \in \{1, ..., (K-1)\}$. (These equations hold due to our induction assumption.) As a result, we get

$$2\sum_{k=1}^{K-1} \widehat{b}(k) = \widehat{b}(0) \sum_{k=1}^{K-1} [(k+1)^{\gamma_0} - 2k^{\gamma_0} + (k-1)^{\gamma_0}]. \qquad (2.17)$$

Since

$$\sum_{k=1}^{K-1} [(k+1)^{\gamma_0} - 2k^{\gamma_0} + (k-1)^{\gamma_0}] = K^{\gamma_0} - (K-1)^{\gamma_0} - 1,$$

(2.16) holds under the induction assumption.

Thus, (2.9) has the unique solution (2.10) that means (see that was said above when we introduced the system (2.9)) that we have proved the theorem. QED

The statement that the system of equations (2.11) has the unique solution

$$b(k) = \int_{-1/2}^{1/2} e^{2\pi i\lambda k} f(\lambda) d\lambda \qquad (2.18)$$

where $f(l)$ is given in c) of Theorem 2.1 was proved by Sinai [17, Theorem 2.1] under the condition that the ratio $b(k)/b(0)$ depends on k and γ_0 only. The equation (2.18) gives an expression different from (2.6) for the same unique solution of (2.11).

The equation (2.3) means that the es-s process does not change its correlation coefficient with averaging over blocks of any length m. This is a primary reason why X is called self-similar. The significance of the function $g(k)$ is in the fact that it gives a nondegenerate correlation coefficient of the limiting ($m \to \infty$) averaged process.

Since

$$g(k) \sim \frac{1}{2}(2-\beta)(1-\beta)k^{-\beta} = H(2H-1)k^{-\beta}, \quad k \to \infty$$

the es-s process has a heavy-tailed and even unsummable autocovariance function and correlation coefficient,

$$\sum_{k=0}^{\infty} r(k) = \infty.$$

(Here, $f(x) \sim h(x)$ means $f(x)/h(x) \to 1$ as $x \to \infty$.) This relates the es-s processes with the *long-range dependence (l-rd)* processes. The latter were defined (see [3]) as processes which have $r(k) \sim ck^{-\beta}$, $0 < \beta < 1$ where c is a constant. Thus we see that any es-s process is l-rd.

We note that if $1 < \beta < 2$ then X is the *short-range dependent (s-rd)* process which has $V_m \equiv b_m(0) \sim cm^{-1}$ as $m \to \infty$, summable $r(k)$, and uncorrelated $X_t^{(m)}$ as $m \to \infty$, whereas $0 < \beta < 1$ shows that X is l-rd process. When $0 < \beta < 1$, a value of β shows a level of long-range dependency in X, a lower β corresponds to higher dependency in X.

Asymptotic self-similarity

Definition D [3]: A process X is called *asymptotically second-order self-similar* (as-s) with parameter $H = 1 - (\beta/2), 0 < \beta < 1$ if

$$\lim_{m\to\infty} r_m(k) = g(k), \quad k \in \mathbf{N}. \tag{2.19}$$

Thus X is as-s if after averaging over blocks of length m and as $m \to \infty$, its correlational structure becomes identical to that of an es-s process. In other words, if $X^{(m)}$ es-s as $m \to \infty$, then X is as-s.

It is clear that an es-s process is as-s.

The following theorem gives a necessary and sufficient condition for X to be as-s in terms of the variance V_m of the averaged process $X^{(m)}$ and also gives a sufficient condition in terms of the correlation coefficient $r(k)$ of process X itself.

Theorem 2.3 [21]. *For a process X and $H = 1 - (\beta/2)$, $0 < \beta < 1$ the following are equivalent:*

e) *X is as-s, i.e.,* (2.4),

f) $(V_{km}/V_m) \sim k^{-\beta}$, *integer* $m \to \infty$, $k \in \mathbf{N}$.

The asymptotic equation

g) $r(k) \sim H(2H-1)L(k)k^{-\beta}$, *integer* $k \to \infty$

implies the asymptotic equation

h) $V_m \sim \sigma^2 L(m)m^{-\beta}$, *integer* $m \to \infty$,

(*where $L(k)$ is a slow varying function* [1]) *and each of* g) *and* h) *implies* e) *and* f).

The asymptotic equation f) is just a definition of the *index* $(-\beta)$ *regulary varying sequence* (*rvs*) V_m *with integer variable.* Thus Theorem 2.3 states that the asymptotic self-similarity of X is equivalent to the regular variation of variance of $X^{(m)}$.

According to g), each l-rd process is as-s.

Finally, we recall that there exist a concept of a *strictly self-similar (ss-s)* process and a concept of a *strictly asymptotically self-similar (sas-s)* process. The first of them is more widely known. Namely, a narrow-sense stationary X is ss-s with H if $m^{1-H}X^{(m)} \doteq X$, $m \in \mathbf{N}$ where $\doteq$ means equality in the sense of finite-dimensional distributions. Similarly, X is sas-s with H if $m^{1-H}X^{(m)}$ goes to ss-s as $m \to \infty$ in $\doteq$ sense. Here, $H = 1 - (\beta/2)$, $0 < \beta < 1$.

It is easy to see that if X is ss-s then it is es-s. But the opposite statement is not true. However, if X is Gaussian es-s with $EX_t = 0$ then it is ss-s.

Similarly, an sas-s process is as-s.

In this Section, it was used a segment $(X_1, X_2, \ldots)$ of stationary process $(\ldots, X_{-1}, X_0, X_1, \ldots)$. However, all definitions and statements hold true if we substitute the segment with the process and make some negligible changes.

MODELS OF INPUT TRAFFIC AND QUEUE

Communication system and its queueing model

Consider a discrete-time, $t \in \{\ldots, -1, 0, 1, \ldots\} \triangleq \mathbf{Z}$, communication system consisting of a finite buffer and a channel. An input traffic $G = (\ldots, G_{-1}, G_0,$- $G_1, \ldots)$, where $G_t \in \mathbf{Z}_+$ is the number of cells arriving at time $t \in \mathbf{Z}$, feeds the buffer. The buffer has a finite size h. This means that it can accommodate not more than h cells at a time. In any slot t, which is the interval $[t, t+1)$ containing only one time instant t from the discrete time axis $\mathbf{Z}$, the channel can transmit (serve) not more than C cells. C is a finite positive integer, $C \in \mathbf{N}$, and is called the channel capacity.

The communication system is considered as a queuing one. The following order of events is assumed at any time moment t : {end of service in slot $t-1$}, {end of slot $t-1$}, {new cell arrival if cells arrive at t}, {choice of next cells for service}, {loss of cells if it is required by discipline}, {putting non-lost cells into buffer}, {beginning of slot t}, {beginning of service in slot t}. The considered system is denoted as **G/D/C/h/d** where **G** denotes the input traffic G, **D** stands for the deterministic service time equal 1, C is the number of servers, h means that the buffer size is h, and **d** indicates that we take into account a discipline d in the system.

3.2. Discipline. In the considered queueing system at each time t and on the basis of available information, a discipline decides which one of the following alternatives should be applied to each cell (request) in the system: (1) To put the cell into service at time t, (2) To keep the cell in buffer till $t+1$, (3) To discard (to lose) the cell at time t.

The most important class of disciplines in this paper is denoted by $D_C(h)$. A discipline d is in $D_C(h)$ if it satisfies the following conditions [25], [19]: (i) If $G_t + Z_t > 0$ (where G_t is the number of new cells arrived at time t and

Z_t is the number of cells which already were in the buffer at time t), then $\min\{G_t + Z_t, C\}$ cells go into service at t. (ii) If $G_t + Z_t \leq h + C$, then no cells are discarded at t. If $G_t + Z_t > h + C$, instead, then $G_t + Z_t - h - C$ cells are discarded at t. Which cells are discarded and which cells go into service depend on a specific discipline $d \in D_C(h)$.

3.3 Input traffic. Here, a specific input traffic denoted as $Y = (\ldots, Y_{-1}, Y_0, Y_1, \ldots)$ is presented. This traffic Y is asymptotically self-similar and for it, the upper bounds to overflow and loss probabilities are found in Section 5.

The traffic Y is assumed to be a stream of cells. The cells have equal length 1. The cells are assigned to sources, so the traffic is an aggregation of cells generated by sources. The sources are enumerated by $s \in \mathbf{Z}$. A source s starts to generate its cells at time denoted by $\omega_s (\omega_s \leq \omega_{s+1})$. The moment ω_s is called the time of source s arrival.

At each time $\omega_s + i - 1$ in time interval $\omega_s, \ldots,\ \omega_s + \tau_s - 1,\ i \in \{1, \ldots, \tau_s\}$, the source s generates one cell with probability p and does not generate any cells with probability $1 - p$. The number of cells generated by source s at $t = \omega_s + j$ is denoted as $\Theta_s(t - \omega_s + 1),\ j \in \{0, \ldots, \tau_s - 1\}$. Given τ_s the variables $\Theta_s(i),\ i \in \{1, \ldots, \tau_s\}$ are i.i.d., $\Pr\{\Theta_s(i) = 1\} = 1 - \Pr\{\Theta_s(i) = 0\} = p$. The time interval $\omega_s, \ldots, \omega_s + \tau_s - 1$ is called the *active period* of source s; $\tau_s \in \mathbf{N}$ is called the length of the active period of source s. Thus, in its active period, a source generates a Bernoulli sequence of cells so that, given $\tau_s = m$, the number of cells generated by source s in its active period (this number is denoted as φ_s) is distributed as $\Pr\{\varphi_s = n \mid \tau_s = m\} = \binom{m}{n} p^n (1-p)^{m-n}$. Before time ω_s and after time $\omega_s + \tau_s - 1$, the source s does not generate any cells at all.

At any instant $t \in \mathbf{Z}$, more than one source arrival can occur. By ξ_t, we denote the number of sources arriving at t, that is, $\xi_t \in \mathbf{Z}_+$ is the number of sources started their active periods at t.

Thus,

$$Y_t = \sum_{s \in \mathbf{Z}} \Theta_s(t - \omega_s + 1), \quad t \in \mathbf{Z} \tag{3.1}$$

where $\Theta_s(i) = 0$ for $i \leq 0$ and $i \geq \tau_s + 1$. This means that Y_t is a total number of cells generated by all sources which are active at t.

It is assumed that $\tau_s,\ s \in \mathbf{Z}$ are i.i.d. for different s; the numbers of source arrivals, $\xi_t,\ t \in \mathbf{Z}$, are i.i.d. with $0 < \lambda \triangleq E\xi_t < \infty$ and $\Pr\{\xi_t = 0\} < 1$; the random variables τ_s are mutually independent of sequences ξ_t and ω_s. Let τ (let ξ) be a generic symbol for τ_s (for ξ_t). The sequences $(\Theta_s(1), \ldots, \Theta_s(\tau_s)),\ s \in \mathbf{Z}$ are i.i.d. and they are independent of sequences ξ_t and ω_s.

The most important case of traffic Y is the one that has Pareto-type distributed τ and Poissonian ξ,

$$\Pr\{\tau = l\} = c_0 l^{-\alpha-1}, \quad c_0 \triangleq (\sum_{l=1}^{\infty} l^{-\alpha-1})^{-1},\ 1 < \alpha < 2,\ l \in \mathbf{N}, \tag{3.2}$$

$$\Pr\{\xi = n\} = e^{-\lambda} \lambda^n / n!,\ 0 < \lambda < \infty,\ n \in \mathbf{Z}_+. \tag{3.3}$$

Such traffic Y is a stationary (in narrow sense) and ergodic process. Also, Y is asymptotically self-similar with $H = (3 - \alpha)/2$ [19, Statement A.1], [21, Theorem 6].

A special case of Y with (3.2), (3.3) and $p = 1$ was considered in [24].

RELATION BETWEEN OVERFLOW AND LOSS PROBABILITIES

In this Section, we get upper and lower bounds to the ratio $P_{\rm loss}/P_{\rm over}$ where $P_{\rm over}$ is the buffer-overflow probability and $P_{\rm loss}$ is the cell-loss probability in the queueing system. Each of these bounds is obtained for more general input traffic than Y defined in Subsection 3.3. However, among these bounds, the lower bound is for a more general traffic.

We start from the definitions of $P_{\rm over}$ and $P_{\rm loss}$. Let G_t be the number of new cells arrived at $t, 0 < EG_t < \infty$ and $G = (\ldots, G_{-1}, G_0, G_1, \ldots)$ be stationary and ergodic. Let L_t denote the number of cells lost at time t in a discrete-time queue $\mathbf{G/D/C/h}$ with $d \in D_C(h)$. According to [2] (see Theorem 2 in Section 4 of Chapter 4 in [2]) and since $L_t = \min\{0, \quad Z_t + G_t - h - C\}$, $Z_{t+1} = \min\{h, \max\{0, Z_t + G_t - C\}\}$ where Z_t is the number of cells which are in the buffer at time t just before a new cell arrival, the process $L = (\ldots, L_{-1}, L_0, L_1, \ldots)$ is, also, stationary and ergodic.

The overflow probability is defined as the stationary probability of event $\{G_t + Z_t - h - C > 0\}$ called the buffer overflow,

$$P_{\rm over} \stackrel{\triangle}{=} \Pr\{G_t + Z_t - h - C > 0\} \tag{4.1}$$

The loss probability is defined as

$$P_{\rm loss} \stackrel{\triangle}{=} \lim_{T\to\infty} (\sum_t L_t)/(\sum_t G_t) \tag{4.2}$$

where the limit is with probability 1 and $\sum_t$ means the sum over t in an interval of length T. According to Birkhoff's theorem,

$$P_{\rm loss} = EL_t/EG_t. \tag{4.3}$$

Theorem 4.1 [24]. *In* $\mathbf{G/D/C/h}$ *queue with* $d \in D_C(h)$ *and a stationary and ergodic* G *having* $0 < EG_t < \infty$, *the ratio* $P_{\rm loss}/P_{\rm over}$ *is lowerbounded as*

$$P_{\rm loss}/P_{\rm over} \geq 1/EG_t. \tag{4.4}$$

To get an upper bound to $P_{\rm loss}/P_{\rm over}$, we need to decrease the generality of input traffic to a queue. Namely, we assume here that the input traffic (denoted as $\tilde{Y}$) is defined as Y in all except an assumption that, now, $(\Theta(1), \ldots, \Theta(\tau))$ is a random sequence with only restriction that $\Theta(t)$ are identically distributed $(\Pr\{\Theta(t) = u\} = Q_u,\ u \in \mathbf{Z}_+)$ and $0 < E\Theta(t) < \infty$. We recall for lucidity that the active periods of different sources are still i.i.d. as in Y and, also as in Y, the conditional on $\tau = m$ distribution of $\Theta(t)$ does not depend on m.

The traffic $\tilde{Y}$ is stationary and ergodic.

In the below theorem which gives an upper bound to $P_{\text{loss}}/P_{\text{over}}, \eta_t$ denotes the number of cells generated at time t by sources arrived before time t, that is, $\eta_t = \tilde{Y}_t - \zeta_t$ where ζ_t is the number of cells generated at time t by sources arrived at t. The theorem uses the following restriction on the distribution of ζ_t :

$$\Lambda \triangleq \sup_{c \geq 0}(\sum_{l=1}^{\infty} l\Pr\{\zeta_t = l + c\})/(\sum_{l=1}^{\infty}\Pr\{\zeta_t = l + c\}) \leq \Lambda_0 \tag{4.5}$$

where Λ_0 is a finite constant which, generally, depends on λ and the distribution Q_u.

It is easy to check that (4.5) holds, for examples in the case of $\Theta(t) = R,\ 0 \leq t \leq \tau,\ R \in \mathbf{N}$ [24]; in the case of Y; and in the case of $b_1 e^{-al} \leq \Pr\{\zeta_t = l\} \leq b_2 e^{-al},\ l \in \mathbf{Z}_+$, where b_1, b_2, a are some positive constants.

It is easy to check also that (4.5) holds for $\Pr\{\zeta_t = l\} = c(l^{-(3+\varepsilon)} - (l + 1)^{-(3+\varepsilon)})$ where c is a normalization constant and $\varepsilon > 0$, and it does not hold for $\Pr\{\zeta_t = l\} = c(l^{-2} - (l+1)^{-2})$. This show that $E\zeta_t^2 < \infty$ is an important condition for satisfiability of (4.5).

Theorem 4.2. *In* $\tilde{\mathbf{Y}}/\mathbf{D}/\mathbf{C}/\mathbf{h}$ *queue with* $d \in D_C(h)$, *the ratio* $P_{\text{loss}}/P_{\text{over}}$ *is upperbounded as*

$$P_{\text{loss}}/P_{\text{over}} \leq (\max\{\Lambda_0, E\zeta_t\} + \sum_{u=1+C}^{\infty} \Pr\{\eta_t \geq u\})/E\tilde{Y}_t. \tag{4.6}$$

Proof. Since an overflow event is $\{G_t + Z_t \geq 1 + h + C\}$, (4.3) gives

$$P_{\text{loss}} = \frac{P_{\text{over}}}{EG_t} \sum_{m=1}^{\infty} \Pr\{G_t + Z_t \geq m + h + C \mid G_t + Z_t \geq 1 + h + C\} \tag{4.7}$$

where the summand is a stationary conditional probability.

Consider the sum

$$S \triangleq \sum_{m=1}^{\infty}\sum_{n=0}^{\infty} \Pr\{W_t = n\}\Pr\{\zeta_t \geq m+h+C-W_t \mid \zeta_t \geq 1+h+C-W_t,\ W_t = n\} \tag{4.8}$$

where

$$W_t = Z_t + \eta_t. \tag{4.9}$$

Since W_t does not depend on ζ_t and $\Pr\{X \geq x_1 \mid X \geq x_2\} = \Pr\{X \geq x_1\}/\Pr\{X \geq x_2\}$ for $x_1 \geq x_2$, then, denoting

$$S_{m,n} \triangleq \Pr\{\zeta_t \geq m + h + C - n\}/\Pr\{\zeta_t \geq 1 + h + C - n\}, \tag{4.10}$$

we get

$$S = S_1 + S_2,\ S_1 \triangleq \sum_{n=0}^{h+C} \Pr\{W_t = n\} \sum_{m=1}^{\infty} S_{m,n},$$

$$S_2 \triangleq \sum_{n=1+h+C}^{\infty} \Pr\{W_t = n\} \sum_{m=1}^{\infty} S_{m,n}. \tag{4.11}$$

In (4.11), we did not pay attention on the order of summation over m and n since the sums have nonnegative terms and, as it will be clear, S is finite.

Under the restriction (4.5), we have

$$S_1 \leq \Lambda_0 \sum_{n=0}^{h+C} \Pr\{W_t = n\}. \tag{4.12}$$

For S_2, we have

$$S_2 = \sum_{n=1+h+C}^{\infty} \Pr\{W_t = n\}(n - h - C + E\zeta_t). \tag{4.13}$$

Using (4.11)-(4.13) and the inequality $\Pr\{\zeta_t + \eta_t \geq n\} \leq \Pr\{\eta_t \geq n - h\}$, we get

$$\begin{aligned} S &\leq \max\{\Lambda_0, E\zeta_t\} + \sum_{n=1+h+C}^{\infty} \Pr\{\zeta_t + \eta_t \geq n\} \\ &\leq \max\{\Lambda_0, E\zeta_t\} + \sum_{u=1+C}^{\infty} \Pr\{\eta_t \geq u\}. \end{aligned} \tag{4.14}$$

Now, (4.7) and (4.14) give (4.6). QED

Corollary of Theorem 4.2. *In* **Y/D/C/h** *queue with $d \in D_C(h)$, the ratio $P_{\text{loss}}/P_{\text{over}}$ is upperbounded as*

$$P_{\text{loss}}/P_{\text{over}} \leq a(p\lambda, C)/p\lambda E\tau \tag{4.15}$$

where

$$a(p\lambda, C) = e^{2p\lambda} + e^{p\lambda(E\tau-1)} \frac{[p\lambda(E\tau - 1)]^{1+C}}{(1+C)!}. \tag{4.16}$$

In the case of $p = 1$, the bound (4.15) was obtained in [24].

BOUNDS TO OVERFLOW AND LOSS PROBABILITIES

In this Section, some known and new bounds to P_{over} and P_{loss} are presented. We discuss the bounds and compare them.

We start with the following theorem which states the upper bound to P_{over}.

Upper bounds.

Theorem 5.1. *The overflow probability, P_{over}, in* **Y/D/C/h/d** *queue with the Pareto-type τ, Poissonian ξ, $d \in D_C(h)$, and $C > p\lambda E\tau$ is upperbounded as*

$$P_{\text{over}} \leq \frac{(p\lambda c_0(\alpha-1)^{-\alpha}(C+2)^{\alpha-1})^k}{k!} h^{(-\alpha+1)k}, \; h \to \infty, \; k = 1 + \lfloor C - p\lambda E\tau \rfloor \tag{5.1}$$

where $g(x) \leq f(x),\ x \to \infty$ *means* $\lim_{x\to\infty} g(x)/f(x) \leq 1$.

Proof is given in Appendix. QED

This theorem is a generalization of a similar theorem proved in [24] for the case of $p = 1$. The paper [24] has also an extension of the theorem to the case of sources generating $\Theta(t) = R \in \mathbf{N}$ cells at each slot of their active periods, namely, this theorem states that

$$P_{\text{over}} \leq \frac{(\lambda c_0(\alpha-1)^{-\alpha}((C/R)+2)^{\alpha-1}R^{\alpha-1})^k}{k!} h^{(-\alpha+1)k},$$

$$h \to \infty,\ k = 1 + \lfloor \frac{C}{R} - \lambda E\tau \rfloor, C > \lambda R E\tau. \tag{5.2}$$

There are other papers which consider the problem of asymptotic evaluation of the buffer-occupancy distribution function, $F(x)$, in a queue with *infinite-size* buffer and input traffic Y having $p = 1$ (Parulekar and Makowski [14]- [16], Duffield [6], [7], Duffield and O'Connell [8], Liu, Nain. Towsley, and Zhang [13]). Those papers use a Large Deviation Principle (see Dembo and Zeitouni [5]) and the Gartner-Ellis theorem [5] which allows to apply the said Principle to the considered problem. An upper bound to $1 - F(x)$ can be interpreted as an upper bound to P_{over} with $h = x$ [20], [22]. Taking into account this interpretation, we give a review of related results of those papers.

By refining the Duffield-O'Connell theorem [8] and by using the Parulekar-Makowski results [14], Duffield [6] obtained the following large deviation upper bound:

$$\limsup_{x\to\infty} \frac{\log(1-F(x))}{\log x} \leq 1 - (\alpha-1)(C - \lambda E\tau),\quad C > \lambda E\tau \tag{5.3}$$

Since $(-\alpha+1)(1+\lfloor C - \lambda E\tau \rfloor) < 1 - (\alpha-1)(C-\lambda E\tau)$, the bound (5.1) is tighter than (5.3) for any λ, α, C in their set of values. In the case of $C = 1$, the bound (5.3) does not work in the sense that it is not better than the trivial bound $P_{\text{over}} \leq 1$. Concerning the bound (5.1) in the case of $C = 1$, it works and, even, gives the true asymptotic behaviour of P_{over},

$$\lim_{h\to\infty} \frac{\log P_{\text{over}}}{\log h} = -\alpha + 1,\quad \lambda E\tau < 1 \tag{5.4}$$

as it will be clear after presentation of the lower bounds below.

Liu, Nain, Towsley, and Zhang [13] proposed an alternative to the approach based on the Gartner-Ellis theorem, that yields the asymptotic lower and upper bounds to $1 - F(x)$. They derived the large deviation upper bound,

$$\limsup_{x\to\infty} \frac{\log(1-F(x))}{\log x} \leq -\alpha + 1,\quad C > \lambda E\tau. \tag{5.5}$$

This bound has the same exponent of h as the bound

$$P_{\text{over}} \leq \frac{\lambda c_0 R^\alpha h^{-\alpha+1}}{\alpha(\alpha-1)(C - \lambda R E\tau)},\quad h \to \infty,\quad C > \lambda R E\tau, \tag{5.6}$$

obtained in [20] (when $R = 1$) but does not reveal a factor which accompanies $h^{-\alpha+1}$.

Corollary of Theorem 5.1. *The loss probability, P_{loss}, in* **Y/D/C/h/d** *queue with the Pareto-type τ, Poissonian $\xi, d \in D_C(h)$, and $C > p\lambda E\tau$ is upperbounded as*

$$P_{\text{loss}} \leq \frac{(p\lambda c_0(\alpha-1)^{-\alpha}(C+2)^{\alpha-1})^k a(p\lambda, C)}{k!(p\lambda E\tau)} h^{(-\alpha+1)k},$$

$$h \to \infty,\ k = 1 + \lfloor C - p\lambda E\tau \rfloor \tag{5.7}$$

where $a(p\lambda, C)$ is given by (4.16).

Proof. The bound (5.7) follows from (4.15) and (5.1). QED

In [24], there is an extension of the bound (5.7) to the case of $\Theta(t) = R \geq 1$ when $p = 1$. In this case, the upper bound in [24] is the same as (5.7) with the only change that C is substituted with C/R.

5.2. Lower bounds. The lower bounds to P_{over} and P_{loss} are obtained only in the case of $p = 1$.

Theorem 5.2 [18]. *In* **Y/D/C/h/d** *queue with the Pareto-type τ, Poissonian ξ, $d \in D_C(h)$, $p = 1$, $\Theta(t) = R$, and $C \geq \lambda R E\tau$, the overflow and loss probabilities are asymptotically lower bounded as*

$$P_{\substack{\text{over}\\ \text{loss}}} \gtrsim b(c)_{\substack{\text{over}\\ \text{loss}}} h^{(-\alpha+1)k}, \quad h \to \infty, \quad k = 1 + \lfloor \frac{C}{R} - \lambda E\tau \rfloor \tag{5.8}$$

where $f(x) \gtrsim g(x), x \to \infty$ means $\liminf_{x\to\infty} f(x)/g(x) \geq 1$,

$$b(c)_{\substack{\text{over}\\ \text{loss}}} \triangleq \begin{cases} \frac{c_0^k R^{(\alpha-1)k}}{\alpha(\alpha-1)^k(E\tau+(1-e^{-\rho/E\tau})^{-1}-1)^{\alpha+k}} \triangleq \Gamma & \text{for overflow probability,} \\ \Gamma/\lambda R E\tau & \text{for loss probability,} \end{cases} \tag{5.9}$$

and $\rho = \lambda E\tau$ if $\lambda E\tau \leq 1$ and, if $\lambda E\tau > 1, \rho$ is any number such that

$$0 \leq \rho < \begin{cases} 1 + \delta - \Delta & \text{for } \Delta \geq \delta, \\ \delta - \Delta & \text{for } \Delta < \delta, \end{cases} \tag{5.10}$$

where

$$\delta \triangleq \lambda E\tau - \lfloor \lambda E\tau \rfloor, \quad \Delta \triangleq \frac{C}{R} - \lfloor \frac{C}{R} \rfloor. \tag{5.11}$$

In a relation in the theorem, one should ignore the subscript "loss" when considering the lower bound for the overflow probability and vice versa.

Thus, (5.2) and (5.8) reveal the function $h^{(-\alpha+1)k}$ which gives the asymptotic behaviour of $P_{\substack{\text{over}\\ \text{loss}}}$ with increasing buffer size. In particular, it can be shown that

$$\lim_{h\to\infty} \frac{\log P_{\substack{\text{over}\\ \text{loss}}}}{\log h} = (-\alpha+1)k, \quad C > \lambda R E\tau \tag{5.12}$$

where k is such as in (5.2) and (5.8).

An important feature of the probability decay here is that it is polynomially slow with buffer-size growth and exponentially fast with growth of excess of channel capacity over total traffic rate, $C - \lambda RE\tau$.

This result points to a tradeoff that can be important in the design of a communication system. For example, consider a system with $\lambda E\tau < 1$ and an integer $C/R > 1$. We have $k = C/R$ for this system. Now suppose we increase the channel capacity from C to bC. For simplicity, let us assume that $b > 1$ and bC/R is an integer. This increase in capacity reduces the main term $h^{(-\alpha+1)k}$ of overflow (loss) probability from $h^{(-\alpha+1)C/R}$ to $h^{(-\alpha+1)bC/R}$. To achieve the same reduction of $h^{(-\alpha+1)k}$ but now at the expense of buffer size, we need to increase the buffer size from h to h^{b-1}. To take an illustrative example, suppose that $b = 2.5$ and we start with $h = 10^4$. The reduction of $h^{(-\alpha+1)k}$ by increasing the capacity from C to $2.5C$ will be the same as what will be achieved by increasing the buffer size from $h = 10^4$ to $h = 10^{10}$.

Note also, that an increase in capacity is accompanied by *decrease* in transmitted-cell delay whereas an increase in buffer size is accompanied by *increase* in transmitted-cell delay. Thus, to combat traffic losses, one can better increase channel capacity rather than buffer size. This conclusion, however, does not take into account any other practically important factors such as availability, cost etc.

The problem of finding the lower bounds to $P_{\substack{\text{over}\\ \text{loss}}}$ was consider also in [19], [22], and [23]. In [19] and [22], it was considered the case of $R = C = 1$. In [19], it was proved that $P_{\substack{\text{over}\\ \text{loss}}} \geq \underline{c}_{\substack{\text{over}\\ \text{loss}}} h^{(-\alpha+1)}$ for each $h \in \mathbf{Z}_+$ (but not only asymptotically as $h \to \infty$) and without the restriction $C \geq \lambda RE\tau$. Thus when $h \to \infty$, the result of [19] is a special case of (5.8). In [22], $\underline{c}_{\substack{\text{over}\\ \text{loss}}}$ was increased making more precise the bounds from [19]. Also, [22] gives a numerical and analytical comparison of lower bounds, upper bounds and exact values (in a singular case of $h = 0$) of $P_{\substack{\text{over}\\ \text{loss}}}$. A brief proof of results of [23] is given in the appendix of [18].

APPENDIX. PROOF OF THEOREM 5.1

The theorem 5.1, first, is proved under the additional restriction that $C > 1 + p\lambda E\tau$. Then it is proved when $p\lambda E\tau < C \leq 1 + p\lambda E\tau$.

Thus, let $C > 1 + p\lambda E\tau$. The following proof is based on the three lemmas which are presented below.

Let us consider the $\mathbf{Y/D/C/h/d}$, $d \in D_C(h)$ queue (introduced in Section 3) with $C \in \mathbf{N}$, the Poisson ξ and the Pareto-type τ. For a given $0 \leq \gamma \leq 1$, we split the $\mathbf{Y/D/C/h/d}$ queue into two queues $\mathbf{Y^{(i)}/D/C^{(i)}/h^{(i)}/d^{(i)}}$, $d(Q_i) \in D_{C^{(i)}}(h^{(i)})$, $i = 1, 2$ denoted as Q_1 and Q_2 respectively, where $d(Q_i)$ is a discipline in $Q_i, Y^{(i)} = (\ldots, Y^{(i)}_{-1}, Y^{(i)}_0, Y^{(i)}_1, \ldots)$,

$$Y_t^{(1)} \triangleq \sum_{s:\tau_s > \gamma h, s \in \mathbf{Z}} \Theta_s(t - \omega_s + 1), \; Y_t^{(2)} \triangleq \sum_{s:\tau_s \leq \gamma h, s \in \mathbf{Z}} \Theta_s(t - \omega_s + 1) \quad (A.1)$$

$$Y_t^{(1)} + Y_t^{(2)} = Y_t, \; C^{(1)} + C^{(2)} = C, \; h^{(1)} = 0, \; h^{(2)} = h;$$
$$C^{(i)}, C \in \mathbf{N}; \; Y_t^{(i)}, Y_t \in \mathbf{Z}_+ \;\; i = 1, 2. \quad (A.2)$$

Thus, the input traffic $Y^{(1)} = (\ldots, Y_{-1}^{(1)}, Y_0^{(1)}, Y_1^{(1)}, \ldots)$ in Q_1 is composed of the traffic Y sources which have long active periods (with lengths which are greater than γh), and the input traffic $Y^{(2)} = (\ldots, Y_{-1}^{(2)}, Y_0^{(2)}, Y_1^{(2)}, \ldots)$ in Q_2 is composed of the traffic Y sources which have bounded active periods (with lengths which are not greater than γh). The Q_i-queue has $C^{(i)}$ servers. The Q_1-queue has a zero-size buffer, $h^{(1)} = 0$ (that is, Q_1 has no buffer); this means that, if $Y_t^{(i)} \leq C^{(i)}$, then all $Y_t^{(1)}$ new cells go into service at t, and if $Y_t^{(1)} > C^{(1)}$, then $C^{(1)}$ new cells go into service at time t and the rest $Y_t^{(1)} - C^{(1)}$ cells are discarded. The Q_2-queue has a buffer of size $h^{(2)} = h$; this is the size of buffer in the initial $\mathbf{Y/D/C/h/d}$ queue also.

We note that the numbers of new sources which come at time t in $Y^{(1)}$ and $Y^{(2)}$ are the Poisson random variables with parameters $\lambda_1 \triangleq \lambda \Pr\{\tau > \gamma h\}$ for $Y^{(1)}$ and $\lambda_2 \triangleq \lambda \Pr\{\tau \leq \gamma h\}$ for $Y^{(2)}$. The probability distributions $\Pr\{Y_t = n\}$, $\Pr\{Y_t^{(1)} = n\}$, and $\Pr\{Y_t^{(2)} = n\}$, $n \in \mathbf{Z}_+$ are also Poissonian with parameters denoted as $\mu_0 \triangleq EY_t$, $\mu_1 \triangleq EY_t^{(1)}$, and $\mu_2 \triangleq EY_t^{(2)}$ respectively. All traffics, $Y, Y^{(1)}$, and $Y^{(2)}$ are stationary and ergodic.

Denote the overflow probability in the $\mathbf{Y/D/C/h/d}$ queue by P_{over} and, in Q_i, by $P_{\text{over}}(Q_i)$. The probabilities $P_{\text{over}}, P_{\text{over}}(Q_1)$, and $P_{\text{over}}(Q_2)$ do not depend on the disciplines in their queues since $d \in D_C(h)$ and $d(Q_i) \in D_{C^{(i)}}(h^{(i)})$ [19].

The following Lemma gives a relation between $P_{\text{over}}, P_{\text{over}}(Q_1)$, and $P_{\text{over}}(Q_2)$. In spite of the difference in input traffics, here and in [24] (where $p = 1$), the proof of the Lemma is the same as in [24].

Lemma A.1.

$$P_{\text{over}} \leq a(p\lambda_1, C^{(1)}) P_{\text{over}}(Q_1) + a(p\lambda_2, C^{(2)}) P_{\text{over}}(Q_2) \tag{A.3}$$

where $a(\lambda, C)$ *is defined in* (4.16).

Now, to upperbound P_{over}, we want to obtain the upper bounds to $P_{\text{over}}(Q_1)$ and $P_{\text{over}}(Q_2)$. We shall get the bounds under the following specific choice of $C^{(1)}$ and $C^{(2)}$:

$$C^{(1)} = C - C^{(2)} = \lfloor C - \varepsilon - p\lambda E\tau \rfloor \geq 1, \; C^{(2)} = \lceil \varepsilon + p\lambda E\tau \rceil \tag{A.4}$$

where $\lceil x \rceil$ denotes the minimum integer which is greater than or equal to x and $\varepsilon \geq 0$. The condition $C^{(1)} \geq 1$ holds if $C > 1 + p\lambda E\tau$ and ε is sufficiently small.

First, we get an upper bound to $P_{\text{over}}(Q_1)$.

Lemma A.2.

$$P_{\text{over}}(Q_1) \leq \frac{(p\lambda c_0 (\alpha - 1)^{-1} \gamma^{-\alpha+1})^{1+C^{(1)}}}{(1 + C^{(1)})!} h^{(-\alpha+1)(1+C^{(1)})} \tag{A.5}$$

where c_0 *is defined in* (3.2).

Proof of Lemma A.2. In Q_1, $\{t \text{ is an overflow moment}\} = \{Y_t^{(1)} \geq 1 + C^{(1)}\}$ since $h^{(1)} = 0$.

The number of active periods existing at time t is the Poisson random variable with parameter

$$\mu_1 = \lambda \Pr\{\tau > \gamma h\} E[\tau \mid \tau > \gamma h] = \lambda c_0 \sum_{i > \gamma h} i^{-\alpha} \leq \frac{\lambda c_0 (\gamma h)^{-\alpha+1}}{\alpha - 1}. \qquad (A.6)$$

The distribution $\Pr\{Y_t = l\}$ is Poissonian with parameter $p\mu_1$ since

$$\Pr\{Y_t = l\} = \sum_{m=l}^{\infty} e^{-\mu_1} \frac{\mu_1^m}{m!} \binom{m}{l} p^l (1-p)^{m-l} =$$

$$= \frac{e^{-\mu_1} p^l}{l!(1-p)^l} \sum_{m=l}^{\infty} \frac{(\mu_1(1-p))^m}{(m-l)!} = e^{-p\mu_1} \frac{(p\mu_1)^l}{l!}.$$

Thus, we have

$$P_{\text{over}}(Q_1) \leq \sum_{l=1+C^{(1)}}^{\infty} e^{-p\mu_1} \frac{(p\mu_1)^l}{l!} \leq \frac{(p\mu_1)^{1+C^{(1)}}}{(1+C^{(1)})!}. \qquad (A.7)$$

The statement (A.5) follows from (A.6) and (A.7). QED

In [24], Lemma A.2 was proven for $p = 1$.

The following lemma is proved for traffic $\tilde{Y}$ (introduced in Section 4 after (4.4)) with the additional restriction that $\Theta(t)$ takes its values on $\{0, \ldots, J\}$, $1 \leq J < \infty$. However, the lemma will be used later only for traffic Y. The lemma uses $C^{(2)} = \lceil \varepsilon + \lambda(E\Theta)(E\tau) \rceil$ instead of $C^{(2)}$ given by (A.4). For traffic Y, we have $E\Theta = p$ that gives (A.4).

Lemma A.3. *If γ and ν are such that*

$$0 < \gamma \leq \frac{\alpha - 1}{(C+2)J} - \nu, \quad 0 < \nu < \frac{\alpha - 1}{(C+2)J} \qquad (A.8)$$

then

$$P_{\text{over}}(Q_2) \leq \frac{e^{\phi}}{(C+1)\varepsilon} h^{-(1+C-(C+2)\gamma C^{(2)})} \qquad (A.9)$$

for any $\phi > 0$, $C^{(2)} = \lceil \varepsilon + \lambda J(E\Theta)(E\tau) \rceil$, $\varepsilon > 0$, and a sufficiently large h.

Proof of Lemma A.3. We have [20],

$$P_{\text{over}}(Q_2) \leq \Pr\{\sup_{n \geq 1}(T_n - nC^{(2)}) > h\} \leq \sum_{n=1}^{\infty} u_n \qquad (A.10)$$

where

$$u_n \triangleq \Pr\{T_n > h + nC^{(2)}\}, \; \overline{T}_n = \sum_{u \in \{t-n, \ldots, t-1\}} \tilde{Y}_u^{(2)} \leq \overline{T}_n \triangleq \sum_{m=1}^{\lfloor \gamma h \rfloor} \sum_{v=0}^{mJ} v\vartheta(m, v) \qquad (A.11)$$

where $\vartheta(m,v)$ is the number of active periods with length $\tau_s = m$ and also such that they have v cells each and $\omega_s \in S \triangleq \{s : \tau_s \leq \gamma h,\ \omega_s \in \{t-n-\lfloor\gamma h\rfloor, \dots, t-1\},\ s \in \mathbf{Z}\}$. The random variables $\vartheta(m,v)$ with different (m,v), $m \in \mathbf{N}$, $0 \leq v \leq mJ$ are independent and Poissonian with parameters $\lambda_{m,v} = \lambda N P_{m,v}$ where $N = n + \lfloor\gamma h\rfloor$ is the length of the interval S, $P_{m,v} = \Pr\{(\tau,\psi) = (m,v)\}$, and (τ,ψ) =(length of the source's active period, number of cells in this active period).

To upperbound u_n, we use the Chernoff bound,

$$\log u_n \leq -r(h + nC^{(2)}) + g_n(r) \triangleq U_n, \quad r > 0 \tag{A.12}$$

where $g_n(r)$ is the semi-invariant moment generating function of the random variable $\overline{T}_n$,

$$g_n(r) \triangleq \log Ee^{r\overline{T}_n} = \lambda(n + \lfloor\gamma h\rfloor) \sum_{m=1}^{\lfloor\gamma h\rfloor} \sum_{v=0}^{mJ} P_{m,v}(e^{rv} - 1). \tag{A.13}$$

Now we want to obtain an upper bound to U_n. We have from (A.12) and (A.13) that

$$U_n \leq -r(h - \lfloor\gamma h\rfloor C^{(2)}) - \varepsilon r(n - \lfloor\gamma h\rfloor) + w_n \tag{A.14}$$

where

$$w_n \triangleq \lambda N \sum_{m=1}^{\lfloor\gamma h\rfloor} \sum_{v=0}^{mJ} P_{m,v}(e^{rv} - 1 - rv). \tag{A.15}$$

For w_n, we obtain with the help of the inequality $e^x - 1 - x \leq x^2 + x^3 e^x$, $x > 0$ that

$$\begin{aligned} w_n &\leq \lambda N \sum_{m=1}^{\lfloor\gamma h\rfloor} \sum_{v=0}^{mJ} P_{m,v}(r^2 v^2 + r^3 v^3 e^{rv}) \\ &\leq \lambda N \sum_{m=1}^{\lfloor\gamma h\rfloor} (r^2 m^2 J^2 + r^3 m^3 J^3 e^{rmJ}) \Pr\{\tau = m\}. \end{aligned} \tag{A.16}$$

Also in (A.16), it was noticed that

$$\sum_{v=0}^{mJ} vP_{m,v} = \Pr\{\tau = m\} \sum_{v=0}^{mJ} v\Pr\{\psi = v \mid \tau = m\} = m(E\Theta)\Pr\{\tau = m\}, \tag{A.17}$$

$$N^{-1} \sum_{\mu=1}^{\lfloor\gamma h\rfloor} \sum_{v=0}^{mJ} \lambda_{m,v} v \leq \lambda \sum_{m=1}^{\infty} (\sum_{v=0}^{mJ} vP_{m,v}) = \lambda(E\Theta)(E\tau). \tag{A.18}$$

In our next step in upperbounding w_n, we use the Pareto-type distribution $\Pr\{\tau = m\} = c_0 m^{-\alpha-1}$, $1 < \alpha < 2$ and we use a specific $r > 0$, namely, $r = (C+2)h^{-1}\log h$, $h > 1$. So, we have

$$\sum_{m=1}^{\lfloor\gamma h\rfloor} r^2 m^2 \Pr\{\tau = m\} \leq c_0 r^2 (2 - \alpha)^{-1} \lfloor\gamma h\rfloor^{-\alpha+2}$$

$$\leq c_0(C+2)^2(2-\alpha)^{-1}\gamma^{-\alpha+2}h^{-\alpha}\log^2 h, \qquad (A.19)$$

$$\sum_{m=1}^{\lfloor\gamma h\rfloor} r^3 m^3 J e^{rmJ} \leq c_0 r^3 J \lfloor\gamma h\rfloor^{-\alpha+2} \sum_{m=1}^{\lfloor\gamma h\rfloor} e^{rmJ} \leq$$

$$\leq c_0(C+2)^2\gamma^{-\alpha+2}h^{-\alpha+(C+2)Jh^{-1}+(C+2)\gamma J}\log^2 h. \qquad (A.20)$$

Above, in (A.19) and (A.20) were used the inequalities

$$\sum_{m=1}^{y} m^{-x} \leq 1 - \frac{1}{1-x} + \frac{y^{-x+1}}{1-x}, \quad x > 0$$

and

$$\sum_{m=1}^{y} m^b e^{xm} \leq m^y \int_1^{y+1} e^{xz} dz, \quad b > 0, \quad x > 0, \ y \geq 1.$$

The bounds (A.19) and (A.20) give

$$w_n \leq c_1(n+\gamma h)h^{-\alpha+(C+2)Jh^{-1}+(C+2)\gamma J}\log^2 h, \qquad (A.21)$$

where

$$c_1 \triangleq c_0\lambda(C+2)^2 J^2\gamma^{-\alpha+2}(3-\alpha)(2-\alpha)^{-1}.$$

Now (A.14) and (A.21) give

$$U_n \leq -(C+2)(1-\gamma C^{(2)})\log h - (C+2)\varepsilon n h^{-1}\log h + (C+2)\varepsilon h^{-1}\log h +$$

$$+ c_1(n+\gamma h)h^{-\alpha+(C+2)Jh^{-1}+(C+2)\gamma J}\log^2 h \qquad (A.22)$$

where it was used the following inequalities:

$$\varepsilon N(C+2)h^{-1}\log h \geq (\varepsilon(C+2)nh^{-1}\log h) - \varepsilon(C+2)h^{-1}\log h,$$

$$(C+2)(h-\lfloor\gamma h\rfloor)C^{(2)}h^{-1}\log h \geq (C+2)(1-\gamma C^{(2)})\log h.$$

In order to obtain a simpler expression, we weaken the bound toU_n, namely

$$U_n \leq -(C+2)(1-\gamma C^{(2)})\log h - n\varepsilon(C+1)h^{-1} + \phi \qquad (A.23)$$

for any $\phi > 0$ and a large h.

In the derivation of (A.23), we noticed that $h^{1-\alpha+(C+2)Jh^{-1}+(C+2)\gamma J}\log^2 h$ (when $\nu > 0$) and $(C+2)\varepsilon h^{-1}\log h$ can be made less than any given positive number by large enough h. Also, we used the inequality $((C+2)\varepsilon\log h) - c \geq \varepsilon(C+1)$ for a sufficiently large h and $0 < c < \infty$.

It follows from (A.23) that

$$\sum_{n=1}^{\infty} u_n \leq e^{\phi} h^{-(C+2)(1-\gamma C^{(2)})} \sum_{n=1}^{\infty} (e^{-\varepsilon(C+1)h^{-1}})^n \leq$$

$$\leq \frac{e^{\phi}h^{-(1+C-(C+2)\gamma C^{(2)})}}{\varepsilon(C+1)} \qquad (A.24)$$

for a sufficiently large h. QED

Lemma A.3 is a generalization of a lemma from [24] proved for traffic Y with $p = 1$.

The statement (5.1) under the restriction $C > 1 + p\lambda E\tau$ now follows from Lemmas A.1, A.2, and A.3 if we take $\gamma = \frac{\alpha-1}{C+2} - \nu$ and $\varepsilon = \varepsilon(h)$ such that $\varepsilon(h) \to 0$, $h \to \infty$ and $\varepsilon^{-1}h^{-(1+C-(C+2)\gamma C^{(2)})-(-\alpha+1)(1+C^{(1)})} \to 0$, $h \to \infty$ where $C^{(1)}$ and $C^{(2)}$ are given by (A.4); and notice that, as $h \to \infty$, we have $a(p\lambda_1, C^{(1)}) \to 1$ (since $\lambda_1 \to 0$) and $a(p\lambda_2, C^{(2)})$ goes to a finite value $a(p\lambda, C^{(2)})$ which is independent of h (since $\lambda_2 \to \lambda$). In this argument, we took into account that $1+C-(C+2)\gamma C^{(2)} > (\alpha-1)(1+C^{(1)})$ since $1+C-(C+2)\gamma C^{(2)} = 1+C-(C+2)(\frac{\alpha-1}{C+2} - \nu)C^{(2)} = 1 + C - C^{(2)}[(\alpha-1) - \nu(C+2)] > (\alpha-1)(1+C-C^{(2)}) = (\alpha-1)(1+C^{(1)})$ where it was used that $0 < (\alpha-1) - \nu(C+2) < 1$ for $\nu < (\alpha-1)/(C+2)$ and $0 < \alpha - 1 < 1$.

Let $\lambda E\tau < C \leq 1 + \lambda E\tau$. For this we need an extension of Lemma A.1 to **Y/D/C/h** queue which is split into two queues Q_1 and Q_2 having non-integer $C^{(1)}$ and $C^{(2)}$ respectively. First we explain what we mean under "$\mathbf{Y^{(i)}/D/C^{(i)}/h^{(i)}}$ queue with non-integer $C^{(j)}$" [24].

Let us consider a **G/D/C/h** queue denoted by Q_3. The Q_3-queue has the discrete time $t \in \mathbf{Z}$, a general stationary and ergodic input traffic of requests $G = (\ldots, g_{-1}, g_0, g_1, \ldots)$ where g_t is the number of requests arrived at time t, the service time 1, C servers ($C \in \mathbf{N}$), a buffer of size h (h-buffer), which can keep up to h requests, and a discipline $d(Q_3) \in D_C(h)$.

The discipline $d(Q_3)$ is specified by the following way.

($1Q_3$) If $0 < N_t + g_t \leq h + C$, where N_t (like Z_t in Subsection 3.2 where the traffic was Y) is the number of requests in the h-buffer at time t (just before the new request arrival), then $K_t = \min(C, N_t + g_t)$ requests go into service at time t ($t \in \mathbf{Z}$).

($2Q_3$) If $N_t + g_t > h + C$, then $N_t + g_t - h - C$ requests are discarded (lost) without any service, h non-lost requests occupy the h-buffer (that is the buffer of size h), and the remaining non-lost C requests go into the service at time t ($t \in \mathbf{Z}$).

($3Q_3$) A request which goes into service at time t ($t \in \mathbf{Z}$) gets full service by $t+1$ and leaves the system at time $t+1$.

For certainty, here and in the sequel, it is assumed that a request loss can be only at time $t \in \mathbf{Z}$ and only the requests arrived at t can be lost at t. This means, in particular, that a request, which is in the h-buffer, can not be lost and leaves the h-buffer only to get a service.

Notice also, that

$$N_{t+1} = \min\{h, \max\{0, N_t + g_t - C\}\}, t \in \mathbf{Z}. \qquad (A.25)$$

Now, we need to give a different interpretation to Q_3. The interpretation is denoted by Q_4 and it shows how to understand the Q_3-queue with a non-integer C.

The Q_4-queue has the same input traffic G as Q_3. It has one server (but not C servers as Q_3). The server in Q_4 has the rate C; this means that a request gets a full service for the time $1/C$. In other words, $1/C$ is the service time in Q_4. The Q_4-queue has two buffers, a h-buffer of size h and a C-buffer of size C. (We note that the last buffer is not included into a calculation of the system buffer size.) The Q_4-queue is a discrete arrival but continuous departure time queue.

The discipline $d(Q_4)$ in Q_4 is the following.

($1Q_4$) If $0 < N_t + g_t \le h + C$, then $K_t = \min(C, N_t + g_t)$ requests go into the C-buffer and the remaining $N_t + g_t - K_t$ requests go into the h-buffer at time t ($t \in \mathbf{Z}$).

($2Q_4$) If $N_t + g_t > h + C$, then $N_t + g_t - h - C$ requests are discarded, h requests occupy the h-buffer, and C requests occupy the C-buffer at time t ($t \in \mathbf{Z}$).

($3Q_4$) At each time moment (from the set of real numbers $\mathbf{R}$) when the server finishes a service, it begins a service of another request from the C-buffer if the C-buffer is not empty.

For certainty, here and in the sequel, it is assumed that a request from the h-buffer can go only into the C-buffer and a request from the C-buffer can leave this C-buffer only to get a service.

We note that

$$P_{\text{over}}(Q_3) = P_{\text{over}}(Q_4) \qquad (A.26)$$

The Q_4-queue is presented above as a different interpretation of Q_3. This means, in particular, that, in the Q_4-queue, C is a positive integer. However, we can extend Q_4-queue to any non-integer $C > 0$. For $C \in \mathbf{R}_+$ ($\mathbf{R}_+$ is the set of non-negative real numbers), the presentation of Q_4-queue is the same as above in the case of $C \in \mathbf{N}$ with the only change that the C-buffer should be replaced by a variable C_t-buffer where, at time t, C_t is the maximum number of requests which can go into the service in the interval $[t, t+1)$ after time t^* when the server finishes the service of a request which has gone into the service before t and is still under the service after t ($t \in \mathbf{Z}$). (If the server has no such requests under the service at time t then $t^* = t$.) We note that C_t can take one of two values $\lfloor C \rfloor$ or $\lceil C \rceil$., depending on the amount of unfinished work at the server at time t ($t \in \mathbf{Z}$).

Thus, after this explanation, we are able to consider the queues with C servers, where C is not an integer.

In the following Lemma A.1a, we consider $\mathbf{G/D/C/h}$ queues with $C \in \mathbf{R}_+$ using the above explanation.

Let $\mathbf{Y/D/C/h/d}$ with $C \in \mathbf{N}$ be the queue considered at the beginning of the theorem proof. Let, like there, the queue be split into $\mathbf{Y^{(i)}/D/C^{(i)}/h^{(i)}/d^{(i)}}$,

$i = 1, 2$ (denoted, as earlier, by Q_1 and Q_2 respectively) with (A.1) and (A.2) where, instead of $C^{(i)} \in \mathbf{N}$, we assume that $C^{(i)} \in \mathbf{R}_+$.

Lemma A.1a. *The inequality (A.3) with $a(\lambda_i, C^{(i)})$ changed to $a(\lambda_i, \lfloor C^{(i)} \rfloor)$ holds in the case of $C^{(i)} \in \mathbf{R}_+$.*

The proof of Lemma A.1a is omitted since, basically, it repeats the proof of Lemma 1, which also was omitted above.

Next we choose

$$C^{(1)} = C - C^{(2)}, \quad C^{(2)} = \epsilon + \lambda E\tau \tag{A.27}$$

where $\epsilon \geq 0$ is sufficiently small. Unlike (A.4), (A.27) allows the non-integer $C^{(i)}, i = 1, 2$.

To upper bound $P_{\text{over}}(Q_1)$, we use

Lemma A.2a. *If $\lambda E\tau < C \leq 1 + \lambda E\tau$, then for a sufficiently large h,*

$$P_{\text{over}}(Q_1) \leq \frac{p\lambda c_0 \gamma^{-\alpha+1} h^{-\alpha+1}}{\alpha - 1} \tag{A.28}$$

where c_0 is defined in (3.2).

Proof of Lemma A.2a. In Q_1, the event $\{$ t is an overflow moment $\}$ implies the event $\{Y_t^{(1)} > 0$ $\}$. Taking it into account and using (A.6) and (A.7), we get (A.28). QED

Now, we notice that Lemma 3 holds, for a non-integer $C^{(2)}$ given by (A.27), without any changes.

Finally, the statement (5.1) in the case of $\lambda E\tau < C \leq 1 + \lambda E\tau$ follows from Lemmas 1a, 2a, and 3 with the same ϵ, γ, and ν as above under the restriction $C > 1 + \lambda E\tau$. QED

References

[1] N.H.Bingham, C.M.Goldie, and J.L.Teugels, *Regular Variation,* Cambridge, New York, Melburn: Cambridge Univ. Press, 1987.

[2] A.A.Borovkov, *"Asymptotic Methods in Queueing Theory,"* Wiley, 1984.

[3] D.R.Cox, "Long-Range Dependence: A Review," in *Statistics: An Appraisal,* H.A.David and H.T.David, eds. Ames, IA: The Iowa State University Press, 1984, 55-74.

[4] M.E.Crovella and A.Bestavros, "Self-Similarity in Word Wide Web Traffic: Evidence and possible causes," *Proceedings of the 1996 ACM SIGMETRICS. International Conference on Measurement and Modeling of Computer Systems,* May, 1996 and *IEEE/ACM Trans. on Networking* 5, No.6, 1997, 835-846.

[5] A.Dembo and O.Zeitouni, *"Large Deviation Techniques and Applications,"* Jones and Bartlett, Boston (MA), 1993.

[6] N.G.Duffield, "On the Relevance of Long-Tailed Durations for the Statistical Multiplexing of Large Aggregations," *Proc. 34-th Annual Allerton*

Conf. on Communication, Control, and Computing, Oct. 2-4, 1996, 741-750.

[7] N.G.Duffield, "Queueing at Large Resources Driven by Long-Tailed $M/D/\infty$-modulated Processes", a manuscript, 1996, December 30.

[8] N.G.Duffield and N.O'Connell, "Large Deviations and Overflow Probabilities for the General Single-server Queue with Applications," *Math. Proc. Cam. Phil. Soc.* 118, 1995, 363-374.

[9] A.N.Kolmogorov, "Wiener's Spiral and Some Other Interesting Curves in Hilbert's Space," *Dokl. Akad. Nauk USSR* 26, No.2, 1940, 115-118 (in Russian).

[10] W.E.Leland, M.S.Taqqu, W.Willinger and D.V.Wilson, "On the Self-Similar Nature of Ethernet Traffic," *Proc. ACM SIGCOMM'93,* San Fransisco, CA, 1993, 183-193.

[11] W.E.Leland, M.S.Taqqu, W.Willinger, and D.V.Wilson, "On the Self-Similar Nature of Ethernet Traffic (Extended version)," *IEEE/ACM Trans. on Networking* 2, No. 1, 1994, 1-15.

[12] N.Likhanov, B.Tsybakov, and N.D.Georganas, "Analysis of an ATM Buffer with Self-Similar ("Fractal") Input Traffic", *Proc. IEEE INFOCOM'95,* Boston, MA, 1995, 985-992.

[13] Z.Liu, P.Nain, D.Towsley, and Z.-L.Zhang, "Asymptotic Behavior of a Multiplexer Fed by a Long-Range Dependent Process," *CMPSCI Technical Report 97-16,* University of Massachusetts at Amherst, 1997.

[14] M.Parulekar and A.M.Makowski, "Tail Probabilities for a Multiplexer with Self-Similar Traffic," *Proc. IEEE INFOCOM'96 Conf.,* Mar. 26-28, 1996, 1452-1459.

[15] M.Parulekar and A.M.Makowski, "Tail Probabilities for $M/G/\infty$ Input Processes (I): Preliminary Asymptotics," Preprint, University of Maryland, 1996.

[16] M.Parulekar and A.M.Makowski, "$M/G/\infty$ Input Processes: A Versatile Class of Models for Network Traffic", Preprint, University of Maryland, 1996.

[17] Y.G.Sinai, "Automodel Probability Distributions," *Probab. Theory and its Applic.* 21, No.1, 1976,63-80 (in Russian).

[18] B.Tsybakov, "Decay of Loss Probabilities in a Network with Self-Similar Input," submitted *IEEE Trans. Inform. Theory.*

[19] B.Tsybakov and N.D.Georganas, "On Self-Similar Traffic in ATM Queues: Definitions, Overflow Probability Bound and Cell Delay Distribution", *IEEE/ACM Trans. on Networing* 5, No.3, 1997, 397-409.

[20] B.Tsybakov and N.D.Georganas, "Self-Similar Traffic and Upper Bounds to Buffer-Overflow Probability in ATM Queue," *Performance Evaluation* 32, 1998, 57-80.

[21] B.Tsybakov and N.D.Georganas, "Self-Similar Processes in Communications Networks," *IEEE Trans. Inform. Theory* 44, 1998, 1713-1725.

[22] B.Tsybakov and N.D.Georganas, "Overflow and Loss Probabilities in a Finite ATM Buffer Fed by Self-Similar Traffic", *Queueing Systems*, 32, 1999, 233-256.

[23] B.Tsybakov and N.D.Georganas, "Buffer Overflow under Self-Similar Traffic," *Proceedings of SPIE (SPIE - The International Society for Optical Engineering), Performance and Control of Network Systems III*, Eds. R.D. van der Mei, D.P. Heyman, Vol. 3841, 1999, 172-183.

[24] B.Tsybakov and N.D.Georganas, "Overflow and Losses in a Network Queue with Self-Similar Input", submitted *IEEE Trans. Inform. Theory.*

[25] B.Tsybakov and P.Papantoni-Kazakos, "The Best and Worst Packet Transmission Policies", *Problems of Information Transmission,* Vol. 32, No. 4, 1996, 365-382.

[26] W.Willinger, M.S.Taqqu, and A.Erramilli, "A Bibliographical Guide to Self-Similar Traffic and Performance Modeling for Modern High-Speed Networks" in *Stochastic networks: Theory and applications",* Ed. F.P.Kelly, S.Zachary, and I.Ziedins, Clarendon Press (Oxford University Press), Oxford, 1996, 339-366.

ERROR PROBABILITIES FOR IDENTIFICATION CODING AND LEAST LENGTH SINGLE SEQUENCE HOPPING *

Edward C. van der Meulen

Dept. of Math., Catholic University of Leuven,
Celestijnenlaan 200B, 3001 Heverlee, Belgium

ecvdm@gauss.wis.kuleuven.ac.be

Sándor Csibi †

Dept. of Telecom., Techn. Univ. of Budapest,
Stoczek utca 2, 1111 Budapest, Hungary.
csibi@hit.bme.hu

Abstract: Upper and lower bounds on the probabilities of the missed and the false identification are proved for Poisson population, for multiple access with least length single sequence hopping, and identification plus transmission coding at each potential source. False identification due to possible worst pairs of identifiers is considered. It is shown, how can one drastically suppress the probability of this event provided not just a single code word but at least a

*Presented in part at IEEE Intern. Workshop on Inform. Theory, Haifa, Israel, June 9-14, 1996. Research of both authors was partially supported by 1995-98, Project Math. Inform. Theory of the Royal Belgian Ac. Sc., Letts. and Fine Arts, and the Hung. Ac. of Sc.
†Research was partially supported also by the Hung. Nat. Sc. Res. Found. Grant No. OTKA 11601-206.

I. Althöfer et al. (eds.), Numbers, Information and Complexity, 221-238.

couple of code words might be sent from each source, following each demand, consecutively. An approriate kind of randomization is assumed for this purpose, frequently needed anyhow. The combination of identification plus transmission coding and single sequence hopping might be appealing for certain tasks of identification through a multiple access channel. This might be the case, e.g., for certain public emergency services, meant to convey within some area many kinds of occasional demands from a vast population of potential sources, each sending a very short message following a demand, very infrequently.

Index terms – Identification, hopping, Poisson population, single sequence, least length, probability bounds.

INTRODUCTION

Consider Poisson population and multiple access with least length single sequence hopping [9, 14]. Sources can not any more be identified under such circumstances at the output of the multiple access channel by the single common hop sequence itself (even if the separation was successful). The well known way of identification might still be possible, namely assigning members of a finite set of identifiers to the potential users at the outset, adding these as headings to the messages to be sent next to a demand, and decoding the identifier in the same way as the message itself.

Nevertheless, it is also known from fundamental results due to Ahlswede, Dueck, Han, Verdú, and Wei ([1, 2, 3]), that even a vast amount of possible identifiers can be made available for the sources, without lengthening unduely the total message. This can be done by using codes especially designed for identification plus transmission. Remarkable capabilities of identification codes for this purpose were first discovered and proved by Ahlswede and Dueck ([1]). For an overview of the place of identification coding ideas within a general theory of information transfer, see Ahlswede ([4]).

The first well-implementable explicit constructions of asymptotically optimal identification plus message transmission codes (IT codes) are due to Verdú and Wei ([3]).

Using a single common control sequence for all potential users was already kept in mind by Abramson ([6]) for the Spread Aloha principle, prior to [9, 14], and other studies on least length single sequence hopping by the second author (cited in [14]). For more on the Spread Aloha principle see Abramson ([7]), and for regarding single sequence hopping see Pap ([8]).

The worst possible identification error probabilities are of our present interest, inherently due to (i) multiple access by Poisson population with least length single sequence hopping, and (ii) identification plus transmission (IT) coding.

The choice of single sequence hopping, as a simple kind of multiple access, might be particularly appealing for public emergency services, with a huge but precisely unknown number of potential sources, each with a demand to send some short message occurring extremely infrequently. This is definitely the case if each source would come up with any one out of very many possible demands. Identification codes , in the aforementioned sense, might offer broad perspectives for serving particularly many potential users economically, the precise number of whom might unexpectedly increase in the future. One might want to offer, under such circumstances, tolerable reliability in advance at some highest admissible average number of demands per unit time (called demand rate). One might want to do so even under possible worst circumstances of false identification.

THE MODEL

A temporary homogeneous Poisson source population is assumed, with total demand rate λ, and demands occurring at each Poisson arrival at one of the sources. ν Q-ary message blocks, each of length k, are sent successively, and also an identifier, next to each demand ($Q = 2^{\mu}$, μ some positive integer not less than 2).

Assume $N_a < \infty$ potential sources to be served in the underlying real-life task, each of these sources associated with one of N_a distinct identifiers. Suppose there is a slotting in time with equispaced nodes. Consider, in the model, the slot duration as the unit of time.

A ν-tuple of message blocks is sent from one of these identified sources following a demand only if no demand occurred at the very source throughout previous T slots. Assume otherwise a ν tuple of message blocks is sent from one of the sources of an unidentified source population of infinite size. (Unidentified sources are considered in the mathematical model only, and are not attributed to the real-life task motivating the study in any respect).

The unidentified sources are introduced to avoid any contradiction in the model between assuming a finite set of admissible identifiers and a Poisson demand process at the same time. (For more see Appendix III.)

A multiple access erasure channel is supposed, which is memoryless and noiseless (and, for simplicity only, of no delay). Slotted access is considered (as in [9, 14]) and exclusively time hopping (see, e.g., [13]). By slotted it is meant that the very same temporary slotting is available at each source and at the common output. More distinctly, that the transmission of any Q-ary symbol can start at one of the nodes of this slotting only, and the same constraint holds also for the arrival of these symbols at the output of the multiple access channel.

The ν message blocks at the source just activated are fed successively into the message input of an identification plus transmission (IT) encoder in the sense of Verdú and Wei ([3]), controlled by the identifier associated with the source, provided this is an identified one. For any unidentified source, the IT encoder is controlled by an additional 'identifier' common to all such sources. (Unidentified, fictitious sources contribute to the total demand rate in the model, but are of no interest to any other party.)

Let us consider first the transmission of any message block from any source u.

A single symbol "one" of a binary codeword of weight n is sent from the output of the IT code, following an (n, k) code for error contol, to a control unit controlled by a copy of a binary sequence s_0, called hop sequence. It is essential that the same s_0 is assigned to each potential source (no matter whether it is identified or not).

This control unit is called enhopper; s_0 is a binary sequence of weight n and length $N \gg n$ (and of additional properties to be assumed in the sequel).

A multiple access erasure channel is supposed that is memoryless and noiseless (and, for simplicity only, of no delay.) Because of the asumptions of the model, the same slotting with corresponding nodes at the same instances is available not only at each source but also at the common output of the multiple access channel. Hence the transmission of any Q-ary symbol can start at each source at one of the nodes of this slotting only, and the same is the case at the single common output of the channel.

Let us consider next the identification plus transmission (IT) code in more detail.

Of the well-known versions of Verdú-Wei IT codes that is assumed per message block ([3]), which is generated by the concatenation $\mathcal{C}_{IT}$ of a binary code $\mathcal{C}_1$ of constant weight M_{IT}, and of maximal correlation K_{IT}, and two Reed-Solomon codes $\mathcal{C}_2$ and $\mathcal{C}_3$ (for a concise account on these notions see Appendix I). The code layers of the concatenation are enumerated from inside on, starting at the output of the encoder of the binary code $\mathcal{C}_{IT}$. The same copy of $\mathcal{C}_{IT}$ is assumed at each potential source, and also at the output of the multiple access channel.

Each of the code words of $\mathcal{C}_{IT}$ might be associated with an identifier. Thus the possible number $N_a + 1$ of identifiers equals at most the number N_{IT} of the binary codewords in $\mathcal{C}_{IT}$ ($N_{IT} :=| \mathcal{C}_{IT} |$.)

The same binary codeword of $\mathcal{C}_{IT}$ is sent ν-times successively following each demand.

$\mathcal{C}_{IT}$ is a constant weight binary code, with M_{IT} as its weight.

The content of the lth block of the message, just to be sent, is conveyed by a single "one" out of the M_{IT} "ones" of the binary codeword of $\mathcal{C}_{IT}$. (The latter has already been assigned as an identifier of the actual demand.) Distinct "ones" of the codeword of $\mathcal{C}_{IT}$ stand for the distinct possible messages to be transmitted.

Assume the instant of the demands and the ν-tuples of the message block contents from distinct sources to be independent random variables.

Assume, more distinctly, as an oversimplified model of scrambling, each (scrambled) source block content uniformly distributed over M_{IT} integers, with probability $\frac{1}{M_{IT}}$, and the consecutive ν (scrambled) message input block contents to be independent. (Notice, however, that no model including the well-known techniques of scrambling and descrambling themselves are treated within the scope of the present paper.)

It was already pointed out in [3] that, if no message is sent at all, the symbol "one" of the binary codeword c_{IT} to be transmitted should be drawn for transmission from the "ones" of c_{IT} uniformly. Let us assume also independence for the ν positions of the "ones" of the copies of the same codeword c_{IT} sent ν-times successively next to a demand.

Recall that the identification part of the IT code found by Verdú and Wei is optimal as $\tau \to \infty$, $q \to \infty$, $\kappa \to \infty$, and $\frac{\kappa}{q} \to \infty$ (in the sense of [3], obtained from properties, oriented toward code construction, relying upon the theoretical fundamentals in [1]).

One out of the M_{IT} ones of c_{IT} selected by the just considered (scrambled) message block is encoded into one of the codewords of an (n, k) Reed-Solomon code $\mathcal{C}_0$. This code is shortened to $n = p - 1 < Q - 1$ (p stands for the greatest prime less than Q. (For the meaning of shortening $\mathcal{C}_0$ to $n = p-1$, see Appendix IV.) $\mathcal{C}_0$ is defined over $GF(Q)$.

Thus one of the M_{IT} distinct message blocks can be sent.

The codeword c_{IT}, obtained in this way, is sent into the enhopper, (and from this over the multiple access channel). $\mathcal{C}_0$ is the Reed-Solomon code considered usually for forward-erasure correction outside of enhopping and dehopping ([9, 14]).

ν codewords of $\mathcal{C}_0$ (each of n Q-ary symbols) are sent successively, following each demand, for conveying the ν consecutive message blocks via ν consecutive frames.

The consecutive ν code words, each of n Q-ary symbols, are placed into ν consecutive frames, each of $N \gg n$ slots. More distinctly, the jth ($j = 1, 2, \ldots, n$) Q-ary symbol of the lth codeword from C_0 is placed into that very slot of the lth frame at which s_0 takes the value 1 the jth time (counted from the beginning of the sequence). By this, each source might give a fair chance to other simultaneously active sources. (These are the sources of which at least one of the message carrying slots covers the just considered lth frame in question.) This is so, as merely $n \ll N$ out of the N slots of s_0 take the value 1 only.

Recall that, by the definition of the model, the frame ν-tuples initiated by distinct demands, start from distinct instants on (apart from the exceptional case when more than one demand occurs within the same slot). Thus, while the multiple access is slotted, it is frame asynchonous.

Obviously, the single common sequence s_0 should be such as to give really chance under these circumstances, at least within the objectives of the designer. Obviously, s_0 should be subject to certain constraints.

C.1 s_0 should be of full cyclic order (in the sense that the s_0 and $S^l s_0$ (mod N) should not match, for any cyclic shift S^l of s_0 of l slots).

C.2 The cyclic correlation c should equal 1. (c stands for the maximum possible mutual covers of active slots of s_0 and $S^l s_0$, for any $l = 1, 2, \ldots N - 1$. Obviuosly the no shift, $l = 0$, is excluded.)

Let us denote by N' the least sequence length: $N' \leq N$. (It is easy to show that a least sequence length exists.) Denote by s_0' any hop sequence of length N'.

$c = 1$ is chosen also in the present paper (as was did in [14]). One should know that not just the number of simultaneously admissible sources, but also N', take in this case their possible greatest values at the same time. (For a precise notion of simultaneous activity, see Definition 2 in the sequel.)

All frame ν-tuples arriving from distinct sources at the common output of the multiple access channel are fed to the same dehopper. In order to focus on the essential principles only, let us consider the following simplified model sufficient for the present study:

The n Q-ary symbols of each frame ν-tuple form each active source are marching through the same $Q + 2$-ary shift register of N cells, called message register $\mathcal{MR}$. Assume an erasure any time a slot with symbols from at least two sources enter the very same cell at the same time. (The additional two of the $Q + 2$ states are to take, in a simple way, the silence symbol Λ and the erasure symbol Δ into account.)

Let us call the content of $\mathcal{MR}$, at any instant t, the N-tuple of the $Q+2$-ary symbols within $\mathcal{MR}$, at t. Number the cells in $\mathcal{MR}$ from its output end on. Let us call a cell in $\mathcal{MR}$ active at t if it is just carrying a $Q + 2$-ary symbol distinct from Λ.

Assume that a copy of s_0 is stored also at the single common ouput of the channel in a binary register of length N, with s_0 as a fixed content, called reference register $\mathcal{RR}$. Number the cells in $\mathcal{RR}$ also from its output end on. Call a cell in $\mathcal{RR}$ active if s_0 takes the binary value 1 at this cell.

Assume that a frame from some source u, the quality of service of which we are just interested in, is touching at instant t the output end of $\mathcal{MR}$ with its front. (Such an u is usually called the tagged source.)

It is a question of interest under what condition one might separate (and decode without erasure) a frame from a tagged source u the front of which is at the output end of $\mathcal{MR}$ at t. How to do so even if other sources are also present at t in $\mathcal{MR}$? Recall that $\mathcal{C}_0$ is an (n, k) $R-S$ code, thus at most $n - k$ erasures in the separated code word can be corrected by $\mathcal{C}_0$.

In order to answere these questions concisely, let us introduce some notions, and assert some basic facts ([14]).

Definition 1. *Assume there is a frame front from source u at the output end of $\mathcal{MR}$ at t. Let us say there is frame front coincidence at t if a frame front from at least another source, distinct from u, is also at the output end of $\mathcal{MR}$ at t.*

Definition 2. *Call any source u window-active at t if at least one slot of the frame ν-tuple from u is just within the N-tuple of cells of $\mathcal{MR}$ at t.*

Definition 3. *Given some positive integer A, assume there are M_t sources window- active at t. We say there is overflow with respect to the activity threshold A, if $M_t > A$.*

Definition 4. *We say that a frame from u and s_0 match at t, provided (i) the front of the frame within $\mathcal{MR}$ is at t, and (ii) all active slots of $S^t s_0 \pmod N$ cover all active slots of s_0 in $\mathcal{RR}$.*

As next other kind of matches, between arriving and stored identifiers, will also be of our interest, let us call the match between a frame and s_0, according Definition 4, dehopper match any time a distinction seems necessary.

Lemma 5. *(see, e.g., [14]) Assume an enhopper as already defined in the present section, and an s_0 according to C.1 and C.2. The frame front from any source u can match s_0 at t if and only if its front is at the output end of $\mathcal{MR}$ at t.*

Definition 6. *[14] Assume a frame from a tagged source u arrives at the output end of $\mathcal{MR}$ at t, and $\nu \geq 2$. Call the positive number A_0 the highest admissible activity threshold provided (i) all erasures of this frame, due to covers from other sources, can be corrected by C_0 if $M_t \leq A_0$; but (ii) at least one erasure can not be corrected by C_0, if $M_t = A_0 + 1$, and the configuration of the fronts of the frame ν-tuples from the M_t window-active sources is possible worst.*

Lemma 7. *[14] Consider s_0 with a cyclic correlation $c = 1$, and $\nu \geq 2$. Assume (i) a frame front from u, just considered as a tagged source, is at the output end of $\mathcal{MR}$ at t, (ii) s_0 is according to C.1 and C.2, and (iii) that neither frame front coincidence no overflow with respect to A_0 occurs at t. Then the considered frame from source u can be separated at t, and the frame decoded without error.*

Remark 1 [14]: *It can be easily seen that, for $\nu \geq 2$, cyclic instead of conventional shifts can be considered for any worst front configuration of the frame ν-tuples that are just window active.*

Recall again that at most $n - k$ erasures can be corrected by C_0. By this it follows (Lemma 3 in [14]) that, for $c = 1$, and $k \geq 2$, $A = A_0 = n - k + 1$.

Let us choose, for simplicity,

$$k = k' := \lfloor \frac{n+1}{2} \rfloor = \lfloor \frac{p}{2} \rfloor.$$

(For the meaning of this choice see Appendix IV.) For the choice of $c = 1$ and $k = k'$:

$$A_0 = k' + 1$$

(see Appendix I in [14]).

As a next step, we want to decide at the common output of the multiple access channel, whether identifier a is just sent or not, following a demand; and if so, how to recover the scrambled message sent at the place of the output of the channel. (Assume for doing so that the way of scrambling is known also at the output of the multiple access channel by means of some helper.)

Assume that a copy of the codeword $c_{IT} \in C_{IT}$, assigned to identifier a, is stored for this purpose, at the output of the multiple access channel. Let this be done at the output of the decoder of C_0 (placed at the common output of the multiple access channel). Declare as identifier a, the position of the binary symbol "one" of c'_{IT} obtained after decoding of the incoming codeword c'_0 (assigned to the position of $c'_{IT} \in C_{IT}$ by C_0), provided the actually transmitted single binary symbol "one" of the codeword c'_{IT} covers any of the "ones" of the codeword c_{IT}, stored. (Recall that c_{IT} is standing for a at the output of the multiple access channel).

Call this event an identifier match. (The superscript prime of c'_0 is just to warn that the identifier b actually sent can be $b = a$ as well as $b \neq a$.)

Notice that the content of the (scrambled) input block sent is recovered only if $b = a$, and c'_0 is decoded successfully. The message block content, just transmitted by the position of a single symbol one of c'_{IT}, is conveyed via the codeword $c'_0 \in C_0$ corresponding to this. By that, one of the possible message blocks, actually sent is recovered together with the identifier b in this case.

Observe that while no common clock has been assumed for receiving the consecutive ν frames from distinct sources, one can still immediately read out the decoded codeword c'_0 at this register step t. Thus one can compare, symbol by symbol, the inverse image c'_{IT} with the copy of c_{IT} stored at this place. This is because one can use, without any modification, the code C_{IT} (due to [3]), designed to compare c'_{IT} and c_{IT} under the circumstances of frame synchronism, for identification even if frame asynchronous multiple access is inserted between the output of the encoder and the input of the decoder of C_0. Thus the original restriction of IT-codes to frame synchronism (by virtue of the the well-known appealingly simple form, introduced by Verdú and Wei) no longer holds if IT-codes are combined with time hopping (as is the case considered in this paper). This greater freedom is of particular interest for the kind of actual networking tasks kept in mind in this paper.

We have confined ourselves, at the beginning of this section, to slotted access (as in [14]). The unslotted version of the same model of single sequence hopping is left, for simplicity, outside the present study. (One should notice, however, that by an appropriate modification of the present model to unslotted access, separation and decoding without error is possible also up to the same highest admissible activity threshold $A_0 = k' + 1$, provided there is no frame front coincidence, at t [10, 15]. (For more see Remark 4 in the section on error probabilities after Theorem 11.) Notice, however, that the notion of frame front coincidence should be somewhat modified, with respect to the slotted case, under the circumstances of single sequence hopping with unslotted access (see [15]).

Observe that the distinct paths of the encoding and decoding for the identifier and that of the message makes C_{IT} a code especially suited for efficient identification. This fact justifies to call, in our present context, C_{IT} itself identification plus transmission (IT) code. Notice, however, that the term identification plus transmission code has been introduced, originally in [3], not for the code C_{IT} itself but for the code meant between (i) the input of the identifier and message block pair of C_{IT} and (ii) the output of the channel code C_0 (both notations C_{IT} and C_0 understood in our present sense).

ON THE ERROR PROBABILITIES OF INTEREST

Consider, according to the model of the previous section, an IT code $\mathcal{C}_{IT}$ ([3]), with the following parameters: input block length $q^\tau - 1$ of the outer Reed-Solomon code $\mathcal{C}_3$, defined over $GF(q^\kappa)$, input block length κ of the inner Reed-Solomon Code $\mathcal{C}_2$, defined over $GF(q)$. ($\tau < \kappa$). All these parameters of $\mathcal{C}_{IT}$ are chosen to be consistent with well-known concatenation constraints, and also with the value of p, assuming for $\mathcal{C}_3$ and $\mathcal{C}_2$ primitive R-S codes with $q = p$. For more on the slightly revised definitions of some of the code parameters, and also on some changes of the notations necessary in the present context to be consistent with [9] and [14], see Appendix I.

Lemma 8. *Assume $q \geq 3$, $\tau \geq 1$, and $\kappa - \tau > 1$. Then:*

$$\frac{\kappa}{q}(1-\frac{1}{\kappa})(1-\frac{1}{q^{\kappa-\tau}}) < \frac{K_{IT}}{M_{IT}} < \frac{\kappa}{q}\,\frac{1+\frac{1}{\kappa q^{\kappa+\tau-1}}}{1-\frac{1}{q}-\frac{1}{q^\kappa}}$$

Proof See Appendix II.

Corollary 9.

$$\frac{K_{IT}}{M_{IT}} \asymp \frac{\kappa}{q} \to 0,$$

as $\tau \to \infty$, $q \to \infty$, $\kappa \to \infty$, $\frac{\kappa}{q} \to 0$. (Recall that $\kappa - \tau > 1$, thus $\frac{1}{q^{\kappa-\tau-1}} \to 0$.)

Remark 2: *Notice that the conditions for Corollary 9 are the same as in Proposition 3 in [3] (taking the already mentioned changes in the notation into account).*

Proof This follows obviously from Lemma 8.

Recall (from the section on the model) that all codewords of $\mathcal{C}_{IT}$ are used for identification. Let (a, b) stand for any possible identifier pair, and (a', b') for any worst possible identifier pair (the latter with the corresponding codeword pair in $\mathcal{C}_{IT}$ at the minimum possible distance apart).

Assume identifier a, stored at the single common output of the multiple access channel (next to the decoder), is to decide at any step t with dehopper match, whether identifier $b = a$ did arrive or not.

Define next, particularly for identifier $b = a$, incoming at any such t, the probability of missed classification by

$$\mathbf{P}(missed) := \mathbf{P}(\{a\ missed\}_t \mid \{a\ arrived\}_t\ ).$$

Define, at any such t and for any identifier pair (a, b), with $b \neq a$ the probability of false identification by

$$\mathbf{P}(false, (a, b)) := \mathbf{P}(\{b\ detected\}_t \mid \{a\ arrived\}_t\ ).$$

(It obviously follows from the model defined in the previous section that $\mathbf{P}(missed)$ takes the very same value, at any t with dehopper match and any identifier a, and $\mathbf{P}(false, (a, b))$ takes the very same value for any such t, given any pair (a, b) of distinct identifiers, i.e., $b \neq a$.)

Obviously, for any t, (a', b') and (a, b) :

$$\mathbf{P}(false)' := \mathbf{P}(false, (a', b')) \geq \mathbf{P}(false, (a, b)),$$

$\mathbf{P}(false)'$, for any worst identifier pair (a', b'), takes the very same value at any considered step t. Next, concerning false identification, particularly the worst probability of misclassification $\mathbf{P}(false)'$ will be of our interest.

Recall, from Section IV of [14], the definition of decoding error $\mathbf{P}(dec\ err)$, at any step t with dehopper match, for least length single sequence hopping (the latter meant in precisely the same same way as in [14]). It obviously follows, from the model of Section IV of [14], that $P(dec\ err)$ also takes the very same value at any step t with dehopper match.

Lemma 10. *Consider any step t with dehopper match. Then (i) for any admissible identifier pair* (a, b) :

$$\mathbf{P}(missed) = \mathbf{P}(dec\ err),$$

and (ii) for any admissible worst identifier pair (a', b') *of distinct identifiers* $b \neq a$:

$$\mathbf{P}(false)' = (1 - \mathbf{P}(dec\ err))\ \frac{K_{IT}}{M_{IT}}.$$

Proof Assertion (i) readily follows from the fact that the detection of the identifier, incoming at t, is missed only if the incoming codeword in code $\mathcal{C}_0$ (corresponding to the codeword in $\mathcal{C}_{IT}$, assigned to identifier a) is not decoded without error. Assertion (ii) follows from the model and from notions concerning $\mathcal{C}_{IT}$. Namely, it follows partly from the fact that false identification can occur only if the codeword in $\mathcal{C}_0$, just incoming from source u, is received without error; and partly from the definition of the weight M_{IT} and that of the possible worst correlation K_{IT} (the latter obviously occurring for some worst identifier pair (a', b').) □

Denote by

$$1 + \delta := \frac{A_0}{\mathbf{E}M_0}.$$

($1+\delta$ stands for a design parameter, called in the present study, peak-to-average ratio. $\mathbf{E}M_t$ denotes, at any decoding instant t, the expectation of the number M_t of simultaneously active sources. $\mathbf{E}M_t = \mathbf{E}M_0$.)

Recall that the symbol "one" of the selected codeword of $\mathcal{C}_{IT}$ is drawn, according to the previous section, randomly.

Theorem 11. *Given* $\nu \geq 3$, $Q = 2^{\mu}$, *for some* $\mu \geq 2$. *Let* $p < Q$ *stand for the largest prime less than* Q, *and assume a shortening of the word length of an* (n, k) $R - S$ *code to* $n = p - 1 < Q - 1$. *Consider a threshold* $C \geq 1$, *for constraining the peak-to-average ratio* $1 + \delta$ *by* $\delta \leq C$ *(see Appendix III,*

[14]). Choose single sequence hopping according to [14] with highest admissible activity threshold $A = A_0 = k' = k'(n) = \lfloor \frac{n+1}{2} \rfloor$. *Let* $q = p \geq 2^2$, $\tau \geq 1$, *and* $\kappa - \tau > 1$. *Then,*

$$\mathbf{P}(dec\ err)_{LB} \leq \mathbf{P}(missed) \leq \mathbf{P}(dec\ err)_{UB}, \tag{1}$$

and

$$(1 - \mathbf{P}(dec\ err)_{UB})\frac{\kappa}{q}(1 - \frac{1}{\kappa})(1 - \frac{1}{q^{\kappa-\tau}}) < \mathbf{P}(false)' < \frac{\kappa}{q}\frac{1 + \frac{1}{\kappa q^{\kappa-\tau-1}}}{1 - \frac{1}{q} - \frac{1}{q^\kappa}}. \tag{2}$$

Here

$$\mathbf{P}(dec\ err)_{LB} := (1 - g_1)\frac{1}{4(1+\delta)(1+\frac{1}{\nu})}\frac{1}{k'}, \tag{3}$$

$$\mathbf{P}(dec\ err)_{UB} := (1 + g_2)(1 + g_3)(1 + g_4)\frac{1}{e(1+\delta)(1+\frac{1}{\nu})}\frac{1}{k'}, \tag{4}$$

$$g_1 := (1 + g_2)\frac{1}{(1+\delta)(\nu+1)e}\frac{1}{k'},$$

$$g_2 := \frac{1}{(1 - f_1)(1 - \frac{1}{(\nu+1)ek'(1-f_1)})} - 1,$$

$$g_3 := \frac{1}{1 - \frac{1}{(1+\delta)(\nu+1)4}\frac{1}{k'}} - 1,$$

$$g_4 := \frac{e(1+\delta)(1+\frac{1}{\nu})k'}{(1+g_2)(1+g_3)} \exp -k'\frac{\delta^2}{(1+\delta)}\frac{\ln(1+C)}{2C}, \quad f_1 := \frac{1.6}{k'}.$$

Remark 3: *The bounds from both sides on* $\mathbf{P}(dec\ err)$ *(namely* $\mathbf{P}(dec\ err)_{LB}$ *and* $\mathbf{P}(dec\ err)_{UP}$*) are the same as given in Theorem 1 in [14].*

Remark 4: *The present Theorem 11 could readily be carried through, along the lines of [10, 15], even to unslotted multiple access. For the highest admissible activity threshold, for* $c = 1$ *and* $k = k'$, *still* $A = A_0 = k' + 1$ *holds for the unslotted version. (For the tightening of the constraint to* $\nu \geq 3$, *also for frame asynchronous access, see Appendix IV in the present paper, and [15].)*

Proof See the subsequent section.

Up to this point both error probabilities of interest have been considered for selecting one of the possible identifiers by framewise identifier match. Consider next the identifier declared at the common output of the multiple access channel not framewise but messagewise, using the very same identifier and a simple kind of joint evaluation of framewise decisions over all ν frames (see Theorem 11). More distinctly, detect an identifier b, if b was selected over all ν consecutive frames unanimously.

Denote the probabilities of missed and false identification, obtained in this way over all ν frames, by $\mathbf{P}(missed)_\nu$ and $\mathbf{P}(false)'_\nu$, respectively. (Recall that the corresponding error probabilities, obtained by framewise identifier matching, are denoted, simply without subscripts, by $\mathbf{P}(missed)$ and $\mathbf{P}(false)'$, respectively.)

Theorem 12. *Assume $\nu\mathbf{P}(dec\ err)_{UB} \leq 1$. Let $Q = 2^{\mu}$, $\mu \geq 2$, $\nu \geq 3$, for any threshold C such that $1 \leq C$ and $\delta \leq C$. Let $q = p \geq 2^2$, $\tau > 1$, and $\kappa - \tau \geq 1$ (as in Theorem 11). Assume scrambling done over the ν consecutive frames independently. Then*

$$\mathbf{P}(dec\ err)_{LB} \leq \mathbf{P}(missed)_{\nu} \leq \nu\mathbf{P}(dec\ err)_{UB}, \tag{5}$$

and

$$(1 - \nu\mathbf{P}(dec\ err)_{UB})(\frac{\kappa}{q}(1 - \frac{1}{\kappa})(1 - \frac{1}{q^{\kappa-\tau}}))^{\nu}$$
$$\leq \mathbf{P}(false)'_{\nu} \leq (\frac{\kappa}{q}\frac{1 + \frac{1}{\kappa q^{\kappa-\tau-1}}}{1 - \frac{1}{q} - \frac{1}{q^{\kappa}}})^{\nu}. \tag{6}$$

(For $\mathbf{P}(dec\ err)_{LB}$ and and $\mathbf{P}(dec\ err)_{UB}$ see Theorem 11)

Proof of Theorem 12 See the subsequent section.

THE PROOFS OF BOTH THEOREMS

Proof of Theorem 11 – The first pair of assertions of Theorem 11 (bounds (1)) on $\mathbf{P}(missed)$ follow from Assertion (i) of Lemma 10. (The lower and higher bounds on $\mathbf{P}(dec\ err)$, given by bounds (3) and (4), are the same as given by Theorem 1 of [14].)

The second pair of assertions of Theorem 11 (bounds (2)) on $\mathbf{P}(false)'$ follow from Assertion (ii) of Lemma 10. By this the proof of Theorem 11 is complete. □

Proof of Theorem 12 – Let us start with proving the first assertion of Theorem 12 (see inequalities (5) on $\mathbf{P}(missed)_{\nu}$).

Recall that an identifier is accepted, in this case, only if the very same identifier was declared unequivocally over all ν frames. Accordingly, at some step t with dehopper match, incoming $b' = a'$ is missed (i.e., event $\{missed\}_{\nu}$ occurs) if event $\{dec\ err\}_l$ occurs in any of the $l = 1, 2, \ldots \nu$ consecutive frames. Accordingly:

$$\mathbf{P}(missed)_{\nu} = \mathbf{P}(\cup_{l=1}^{\nu}\{dec\ err\}_l) \tag{7}$$

Consider just a single term, say $\mathbf{P}(dec\ err)_l$, on the right side of equation (7) as a lower bound, and the union bound as an upper bound (on the probability of the union event in Equation (7)). Notice that, for all frames l, $\mathbf{P}(dec\ err)_l = \mathbf{P}(dec\ err)$. Thus:

$$\mathbf{P}(dec\ err)_{\nu} < \mathbf{P}(missed)_{\nu} < \nu\mathbf{P}(dec\ err). \tag{8}$$

Let us refer, this time again, to $\mathbf{P}(dec\ err)_{LB}$ and $\mathbf{P}(dec\ err)_{UB}$ as given in Theorem 1 (by Bounds (3) and (4)). The second pair of the assertions of Theorem 12 on $\mathbf{P}(missed)$ (i.e., Bounds (6)) follow from $\mathbf{P}(dec\ err)_{UB}$, Assertion (ii) of Lemma 10, and Lemma 8. By this the proof of Theorem 12 is complete. □

Appendix I

$\mathcal{C}_{IT}$ is defined as in [3]. Recall that both Reed-Solomon codes within $\mathcal{C}_{IT}$ have been confined , for the sake of definiteness, in the Model to primitive codes (the codeword length of which equals the underlying alphabet size minus 1). Assume, for any positive integers $q = p \geq 2^2$, $\tau \geq 1$, $\kappa - \tau > 1$, input length $q^\tau - 1$ and κ for $\mathcal{C}_3$ and $\mathcal{C}_2$, respectively. Accordingly, $\mathcal{C}_3$ stands for a $(q^\kappa - 1, q^\tau - 1)$ and $\mathcal{C}_2$ for a $(q - 1, \kappa)$ Reed-Solomon code (differing from those in [3] in just the slightly different parameter definitions).

Following our present definitions, each of the binary code words of $\mathcal{C}_{IT}$ is of length

$$S_{IT} = (q^\kappa - 1)(q - 1)q,$$

and is of the following weight (the latter meant to be the number of "ones" per codeword):

$$M_{IT} = (q^\kappa - 1)(q - 1).$$

It is well known that correlation K_{IT} is the maximum possible number of mutual covers of "ones" of any binary codeword pair in $\mathcal{C}_{IT}$.

Consider also the widely used notion of correlation K of any non-binary code. This is the maximum possible number of positions at which the symbols of any pair of distinct codewords are equal (i.e., cover each other at most at K positions). It is well known that for any $(\tilde{n}, \tilde{k})$ Reed-Solomon code: $K = \tilde{k} - 1$, (as Reed-Solomon codes are minimum distance codes [5]).

In Appendix II we will confine ourselves to codeword pairs of both $\mathcal{C}_3$ and $\mathcal{C}_2$ being just the code distance apart. (Namely, such pairs will be assigned to worst pairs of identifiers (a', b'). Obviously, for any identifier a' such an identifier exists as Reed-Solomon codes are linear codes.)

Lemma 13. *For any pair of codewords at minimum distance of any minimum distant $(\tilde{n}, \tilde{k})$ code the number of mutually covered binary symbols equals precisely $K = \tilde{k} - 1$.*

Proof Obvious, from the definition of the notion of minimum distance codes, and that of code distance.

Appendix II

Proof of Lemma 10 – As readily seen from Appendix I, for any worst identifier pair (a', b') the following properties hold:

(**P.1**) The correlation of Reed-Solomon code $\mathcal{C}_3$ (that is a q^κ-ary code of input block length $q^\tau - 1$) equals:

$$(q^\tau - 1) - 1 = q^\tau - 2$$

(see Lemma 13 of Appendix I). Thus altogether $q^\tau - 2$ symbols of the codewords $c_{3,a}$ and $c_{3,b}$ (standing, within $\mathcal{C}_3$, for some worst choice (a', b') of the identifier pair) are equal; the rest of the symbols is distinct. (Recall, when considering this, that both $c_{3,a}$ and $c_{3,b}$ are of length $q^\kappa - 1$. Take for both codewords c_{3a} and c_{3b} the subset of the positions with distinct symbols. Notice also that the symbols within this subset are mapped into distinct codewords of $\mathcal{C}_2$.)

(**P.2**) For any worst pair (a', b'), however, it follows by Lemma 13 that just $\kappa - 1$ of the symbols of each of the corresponding codeword pairs in $\mathcal{C}_2$ are equal. (Notice, in this respect, that the input block length of $\mathcal{C}_2$ equals κ.)

Observe that each of the $q - 1$ consecutive $(q^\kappa - 1)(q - 1)$-ary codewords, mapping the consecutive symbols of $c_{3,a}$ and $c_{3,b}$, respectively, are mapped into consecutive binary codewords of $\mathcal{C}_1$ (q of which generate consecutively a codeword of $\mathcal{C}_{IT}$). By this, and Properties (P.1) and (P.2):

$$K_{IT} = (q^\tau - 2)(q - 1) + (q^\kappa - 1 - (q^\tau - 2))(\kappa - 1). \tag{9}$$

Next let us restrict the remainder of the proof to $q = p \geq 2^2$, and $\kappa - \tau > 1$. By equation (9):

$$K_{IT} < \; q^{\tau+1} + q^\kappa \kappa, \tag{10}$$

and

$$\begin{aligned} K_{IT} > \; & (\kappa - 1)q^\kappa + q^\tau((q - 1) - (\kappa - 1)) - 2(q - 1) \\ & = (\kappa - 1)(q^\kappa - q^\tau) + q^\tau(q - 1) - 2(q - 1) \\ & = \kappa q^\kappa (1 - \frac{1}{\kappa})(1 - \frac{1}{q^{\kappa-\tau}}) + \; q^\tau(q - 1)(1 - \frac{2}{q^\tau}). \end{aligned} \tag{11}$$

Consider next inequalities (10) and (11), and the bounds $K_{IT,LB}$ and $K_{IT,UB}$, defined in the following way:

$$K_{IT} < \; K_{IT,UB} := \; q^{\tau+1} + q^\kappa \kappa, \tag{12}$$

and

$$K_{IT} > \; K_{IT,LB} := \; \kappa q^\kappa (1 - \frac{1}{\kappa})(1 - \frac{1}{q^{\kappa-\tau}}). \tag{13}$$

By the expression of M_{IT} (see Appendix I):

$$M_{IT} < \; M_{IT,UB} := q^\kappa q, \;\; M_{IT} > \; M_{IT,LB} := q^\kappa(q - 1) - q. \tag{14}$$

By equations (12) through (14):

$$\frac{K_{IT,UB}}{M_{IT,LB}} = \frac{\kappa q^{\kappa} + q^{\tau+1}}{q^{\kappa}(q-1) - q}$$
$$< \frac{\kappa(1 + \frac{1}{\kappa q^{\kappa-\tau-1}})}{q - 1 - \frac{1}{q^{\kappa-1}}}$$
$$= \frac{\kappa}{q} \frac{(1 + \frac{1}{\kappa q^{\kappa-\tau-1}})}{1 - \frac{1}{q} - \frac{1}{q^{\kappa}}} =: (\frac{K_{IT}}{M_{IT}})_{UB}, \tag{15}$$

and

$$\frac{K_{IT,LB}}{M_{IT,UB}} = \frac{\kappa q^{\kappa}(1 - \frac{1}{\kappa})(1 - \frac{1}{q^{\kappa-\tau}})}{q^{\kappa+1}}$$
$$> \frac{\kappa}{q}(1 - \frac{1}{\kappa})(1 - \frac{1}{q^{\kappa-\tau}}) =: (\frac{K_{IT}}{M_{IT}})_{LB}. \tag{16}$$

Notice that both $(\frac{K_{IT}}{M_{IT}})_{LB}$ and $(\frac{K_{IT}}{M_{IT}})_{UB}$ are of our interest for $q = p \geq 2^2$ $\tau \geq 1$, and $\kappa - \tau > 1$. Lemma 8 immediately follows from equations (15) and (16). □

Appendix III

One might even consider for a model of Poisson population, which admits in any given time interval of positive duration an unlimited number of demands, a finite set of identifiers.

Let us recollect from the Model the definition of the frame, that of the frame ν-tuple, and the fact that any demand is followed by the front of the frame ν-tuple of the message block, to be sent at the beginning of the following slot. Recollect from the same section the definition of the highest admissible activity threshold A_0.

Recall that in the Model a constraint was introduced in terms of T slots, just with a warning that T should be appropriately great.

We want to choose an appropriately large T to exclude, in the mathematical model, a demand occurring at an identified source u at θ, any time within its immediate past of T slots during which a demand already occured at the same identified source u.

Obviously this objective is met if $T > \nu n + 1$. It can be readily seen that all frame ν-tuples arriving from sources window-active at t arrive from distinct sources and are, therefore, independent.

Obviously, by inserting (fictitious) unidentified sources occasionally, the total demand rate, due to the N_a identified sources only, is increased. However, given λ and T, the increment is negligible if N_a is appropriately great.

We have not defined the rules in all detail for drawing one of the identified sources over the N_a sources following a demand, any time such choice is admissible. The omitted details are, however, of no interest for our present study.

Appendix IV

1 Recall, from Remark 1 on the Model that k is confined to a single fixed value in our present study, namely

$$k = k(n) = k' := \lceil \frac{n+1}{2} \rceil = \lfloor \frac{p}{2} \rfloor$$

(as done for example in [14]).

Is this choice associated with any property which is meaningful for the multiple access task considered? In a sense, yes.

To show this, consider a hop sequence of unit correlation $c = 1$, given Q, and $n = p - 1$. Consider the total number $r(k)$ of the p-ary blocks of length k, conveyed jointly by the greatest admissible number $A_0 = n - k + 1$ of window active sources (for which all erasures can be corrected even under any worst frame configuration of the just window-active sources). As $A_0 = n - k + 1$ for $c = 1$,

$$r(k) := (n - k + 1)k.$$

As $n+1 = p$ is odd, $r(k)$ takes, for any value of n, its maximum at $k = k' = \lfloor \frac{p}{2} \rfloor$. (For more see Section III and Appendix II in [14].)

2 Recall that $\mathcal{C}_0$ stands for the $R - S$ code meant for forward error control at the output of the multiple access channel. Notice that this code is shortened to $n = p - 1$. (p stands for the greatest prime not exceeding $Q = 2^{\mu}$.)

The shortening to p is just to obtain an appropriate upper bound on N' by the well known design approach of cyclically permutable sequences due to Nguyen, Győrfi and Massey, based on censoring an $R - S$ code appropriately [13].

3 A lower bound on N' is obtained along the lines of a well-known basic counting approach due to Bassalygo and Pinsker for estimating, for a finite number of distinct sequences and for frame synchronism, the possible shortest sequence length N'. One should, however, still settle in our present context a difficulty concerning this approach, that occurs specifically for single sequence hopping, by an additional idea ([12, 14]). Namely, that the aforementioned counting approach itself does not lead, in this latter case, to an explicit lower bound on N', but to an inequality including implicitly N'. (For the solution of this problem see [14].)

The proof of (3) and (4) in the section on the error probabilities relies, among other considerations, on bounds on N' obtained in this way.

For a detailed study of the extremal additive set problem, underlying the lower estimates on N', and the correponding bounds on $\mathbf{P}(dec\ err)$, see [14].

4 Tightening the constraint from $\nu \geq 2$ to $\nu \geq 3$, in Theorems 1 and 2, is needed to admit the use of cyclic instead of conventional shifts for posing the extremal additive set problem for N', and also in the proof of the lower bound on N'. (See Section VI, and Theorem I in Section V in [14].)

Acknowledgement

The authors wish to thank the reviewers for their comments, helping them to improve the paper, and also make it more self-supportive.

References

[1] R. Ahlswede and G. Dueck, "Identification via channels," *IEEE Trans. on Inform Theory,* IT-35, no.1, 1989, pp. 15-29.

[2] T.S. Han and S. Verdú, "New results in the theory of identification via channels," *IEEE Trans. Inform. Theory,* IT-38, no. 1, 1992, pp. 14-25.

[3] S. Verdú and V.K. Wei, "Explicit constructions of constant-weight codes for identification via channels," *IEEE Trans. on Inform. Theory,* IT-39, no. 1, 1993, pp. 30-36, 1993.

[4] R. Ahlswede, "General Theory of Information Transfer," *Preprint 97-118, Sonderforschungsbereich 343, Diskrete Strukturen in der Mathematik* Universität Bielefeld, D, 1997.

[5] R.E. Blahut, *Theory and Practice of Error Control Codes.* Reading, MA: Addison-Wesley Publ. Co., 1983.

[6] N. Abramson "Development of ALOHANET," *IEEE Trans. on Inform. Theory,* vol. 31, 1985, pp. 119-123.

[7] N. Abramson "Multiple access in wireless digital networks," *Proc. IEEE,* vol. 82, 1994, pp. 1360-1370.

[8] L. Pap, "Performance analysis of DS unslotted packet radio networks with given auto- and crosscorrelation sidelobes," *Proc. IEEE Third Internat. Symp. Spread Spectrum Techniques and Applications,* Oulu, Finland, 1994, pp. 343-345.

[9] S. Csibi, "Two-sided bounds on the decoding error probability for structured hopping, single common sequence and Poisson population," *Proc. 1994 IEEE Internat. Symp. on Inform. Theory,* Trondheim, 1994, p. 290.

[10] S. Csibi, "On the least decoding error probability for truly asynchronous single sequence hopping," *Proc. 1995 IEEE Internat. Symp. on Inform. Theory, Whistler,* 1995, p. 385.

[11] E. C. van der Meulen and S. Csibi, "Identification coding for least length single sequence hopping," *Abstracts, 1996 IEEE Information Theory Workshop,* Dan-Carmel, Haifa, 1996, p. 67.

[12] L.A. Bassalygo and M.S. Pinsker, "Limited multiple-access to an asynchronous channel," *Problems of Information Transmission,* (in Russian) Vol. 19, 1983, pp. 92 - 96.

[13] Q.A. Nguyen, L. Győrfi, and J.L.Massey, "Constructions of binary constant weight cyclic codes and cyclically permutable codes," *IEEE Trans. on Inform Theory,* IT-38, 1992, pp. 940-949.

[14] S. Csibi, "On the decoding error probability of slotted asynchronous access and least length single sequence hopping," *Preprint,* 1997.

[15] S. Csibi, "On the decoding error probability of truly asynchronous least length single sequence hopping," *Preprint,* 1997.

A NEW UPPER BOUND ON CODES DECODABLE INTO SIZE-2 LISTS

Alexei Ashikhmin

Los Alamos National Laboratory
Mail Stop P990, Los Alamos, NM 87545, USA

alexei@c3serve.c3.lanl.gov

Alexander Barg

Bell Laboratories, Lucent Technologies
600 Mountain Avenue 2C-375, Murray Hill, NJ 07974, USA

abarg@research.bell-labs.com

Simon Litsyn*

Department of Electrical Engineering–Systems, Tel Aviv University,
Ramat Aviv 69978, Israel

litsyn@eng.tau.ac.il

DEDICATED TO R. AHLSWEDE ON THE OCCASION OF HIS 60-TH BIRTHDAY

Abstract: A new asymptotic upper bound on the size of binary codes with the property described in the title is derived. The proof relies on the properties of the distance distribution of binary codes established in earlier related works of the authors.

INTRODUCTION

Let $C \in \mathbb{Z}_2^n$ be a binary block code. One says that C corrects r errors if every sphere of radius r in $\mathbb{Z}_2^n$ contains at most one codevector and r is the maximal number with such property. Relaxing this definition, one may require that every such sphere contain at most m vectors from the code. Then if r or fewer errors occur in the channel, the transmitted vector can be isolated by compiling

*Research done while visiting DIMACS Center, Rutgers University, Piscataway, NJ 08854

I. Althöfer et al. (eds.), Numbers, Information and Complexity, 239-244.

a list of m codevectors closest to the received vector. If these conditions hold true, one says that C corrects r errors under list decoding. For brevity, we call such a code C an (m, r) code. The number r will be called the size-m list radius of C.

Let C be an (m, r) code of rate $R(C) = \log_2 |C|/n$. We assume that $r = \rho n$, i.e., the number of errors depends linearly on n, and m is a constant. The main asymptotic problem for (m, r) list codes is to determine the value of

$$R(m, \rho) = \limsup_{n\to\infty} R(C),$$

where the limit is computed over all sequences of codes whose size-m list radius converges to ρ.

The concept of list decoding was introduced by Elias [6] and Wozencraft [12]. Ahlswede [1] showed that it enables one to determine capacity of a wide class of communication channels. Some 30 years after (m, r) codes had been introduced, Blinovsky [4] (see also [5]) derived lower and upper asymptotic bounds on their size for any given value of m. Since in this paper we deal only with the case of $m = 2$, in the theorem below we quote only the relevant bounds from [4].

Let $H(x) = -x \log_2 x - (1 - x) \log_2(1 - x)$ be the entropy function and $H^{-1}(x)$ its inverse.

Theorem 1. [4] *We have* $\overline{R}_2(\rho) \geq R(2, \rho) \geq \underline{R}_2(\rho)$, *where for* $0 \leq h < \infty$, $\underline{R}_2(\rho)$ *is defined parametrically as follows:*

$$\underline{R}_2 = 1 - \tfrac{1}{2}\Big[h\rho + \log_2\Big(1 + 3 \cdot 2^{-h/3}\Big)\Big], \tag{1}$$

$$\rho = \frac{2^{-h/3}}{1+3\cdot 2^{-h/3}}. \tag{2}$$

Further,

$$\overline{R}_2 \leq 1 - H\big(\frac{1}{2} - \frac{1}{2}\sqrt{1 - 4\rho}\big) \tag{3}$$

Lower bounds on (m, r) codes for finite n were derived in [7].

Note that formally the upper bound (3) coincides with the well-known Bassalygo-Elias bound on the size of error-correcting codes. Technically it will be more convenient to us to study the function

$$\rho(m, R) = \limsup_{n\to\infty} \frac{1}{n} r(m, C),$$

where $r(m, C)$ is the size-m list radius of the code C. In this paper we are concerned with upper bounds on $\rho(2, R)$ (typically, any such bound also gives an upper bound on $R(2, \rho)$). Eq. (3) implies the bound

$$\rho(2, R) \leq H^{-1}(1 - R)(1 - H^{-1}(1 - R)). \tag{4}$$

In this paper we derive an improvement of this bound (and so also of (3)).

The principal technical tool of obtaining the new bound is an application of Delsarte's linear programming method to deriving lower bounds on code invariants, found recently in [10], [2]. In particular, in [10] it is proved that in every code of rate $R > 0$ and sufficiently large length n, there necessarily exists an exponentially large component of the weight distribution. This theorem was used in [3] together with bounds on constant-weight codes to prove sharp estimates of the distance distribution of codes meeting the MRRW upper bound [9] (provided that such exist). These results are also used below. More details and notation are given in Section 20. Section 20 is devoted to the new bound.

NOTATION AND PRELIMINARIES

Let

$$\delta(R) = \limsup_{n\to\infty} \operatorname{dist}(C),$$

where dist C is the distance of the code C and the limsup is computed over all sequences of codes of rate R. In other words, $\delta(R) = 2\rho(1, R)$. We shall use the upper (linear programming) bound on $\delta(R)$ [9], which has the form

$$\delta(R) \le \delta^{lp}(R) = \min_{\substack{0\le\beta\le\alpha\le 1/2\\ H(\alpha)-H(\beta)=1-R}} 2\frac{\alpha(1-\alpha)-\beta(1-\beta)}{1+2\sqrt{\beta(1-\beta)}} \tag{5}$$

Likewise, let C be a binary code of distance $d = \delta n$ and constant weight $w = \alpha n$. Define

$$\begin{aligned}\delta(R,\alpha) &= \limsup_{n\to\infty} \frac{d(Rn, \alpha n)}{n},\\ R(\delta,\alpha) &= \limsup_{n\to\infty} R(C).\end{aligned}$$

By [9], we have

$$\delta(R,\alpha) \le \delta^{lp}(R,\alpha) = 2\frac{\alpha(1-\alpha)-H^{-1}(R)(1-H^{-1}(R))}{1+2\sqrt{H^{-1}(R)(1-H^{-1}(R))}}, \quad H^{-1}(R) \le \alpha \le \frac{1}{2} \tag{6}$$

In a certain range of parameters this bound can be improved. The improvement is based on a result in [8] and appears in an explicit form in [11]. In the form convenient to us it is given in [3]. Let $\alpha_m(R)$ be the value of α that furnishes the minimum to the right-hand side of (5)[1]. Then

$$\delta(R,\alpha) \le \delta^{lp}(1+R-H(\alpha)), \quad \alpha_m(R) \le \alpha \le \frac{1}{2}.$$

Let us summarize these results in the following theorem.

[1]Note that β in (5) is a dummy variable whose value is determined uniquely given α and R.

Theorem 2.

$$\delta(R,\alpha) \le \delta^{up}(R,\alpha) = \begin{cases} \delta^{lp}(R,\alpha), & H^{-1}(R) \le \alpha \le \alpha_m(R), \\ \delta^{lp}(1+R-H(\alpha)), & \alpha_m(R) \le \alpha \le \frac{1}{2} \end{cases} \qquad (7)$$

The second ingredient that we need is the following theorem, which gives a lower bound on the components of the distance distribution of the code. Let

$$A_i = \frac{1}{|C|}\{(c',c'') \in C^2 : \mathrm{dist}\,(c',c'') = i\}.$$

Theorem 3. [10] *For every code of rate R and sufficiently large length n there exists a number ξ,*

$$\xi \in \Big(0, 2\frac{\alpha(1-\alpha)-\beta(1-\beta)}{1+2\sqrt{\beta(1-\beta)}}\Big], \qquad (8)$$

such that

$$\frac{1}{n}\log_2 A_{\xi n} \ge R-1+H(\beta)+2H(\alpha)-2q(\alpha,\beta,\xi/2)-\xi-(1-\xi)H\left(\frac{\alpha-\xi/2}{1-\xi}\right),$$

where α and β are arbitrary numbers satisfying

$$0 \le \beta \le \alpha \le 1/2, \quad H(\alpha)-H(\beta) \ge 1-R, \qquad (9)$$

and

$$q(\alpha,\beta,\gamma) = H(\beta) + \int_0^\gamma \log_2\Bigg(\frac{\alpha(1-\alpha)-y(1-2y)-\beta(1-\beta)}{2(\alpha-y)(1-\alpha-y)} + \frac{\sqrt{(\alpha(1-\alpha)-y(1-2y)-\beta(1-\beta))^2-4(\alpha-y)(1-\alpha-y)y^2}}{2(\alpha-y)(1-\alpha-y)}\Bigg)dy. \qquad (10)$$

THE NEW BOUND

Theorem 4. *Let C be a $(2,\rho n)$ code of rate R, $0 \le R \le 1$. Then*

$$\rho \le \rho(2,R) \le \frac{1}{2}\min_{\alpha,\beta}\max_{\xi}\delta^{up}(R'(\alpha,\beta,\xi),\delta^{lp}(R)),$$

where

$$R'(\alpha,\beta,\xi) = R-1+H(\beta)+2H(\alpha)-2q(\alpha,\beta,\xi/2)-\xi-(1-\xi)H\left(\frac{\alpha-\xi/2}{1-\xi}\right),$$

ξ,α,β satisfy (8)-(9), *and $q(\alpha,\beta,\gamma)$ is defined in* (9).

Proof. By Theorem 3, there exists ξ in the interval (8) such that the number of codevectors on the sphere of radius ξn centered at a certain codevector a satisfies (3). We can translate the space $\mathbb{Z}_2^n$ by a; then this claim is equivalent to the existence of a constant-weight code of rate R'. This code has relative minimum distance at most $\delta^{up}(R,\xi)$ (cf. (7)). Take two codevectors c', c'' at a distance $n\delta^{up}(R,\xi)$ and consider them together with the center of the sphere (0, for that matter). It is easy to see that the center of the sphere of minimal radius that contains c', c'' and 0 has weight $\frac{1}{2}n\delta^{up}(R'(\alpha,\beta,\xi),\xi)$. □

Optimization carried out in [3] leads to the following corollary.

Corollary 5.

$$\rho(2,R) \le \begin{cases} \frac{1}{2}\delta^{lp}(R-1+H(\delta^{lp}(R)),\delta^{lp}(R)), & 0 \le R \le R_0, \\ \frac{1}{2}\delta^{lp}(R), \quad R_0 \le R \le 1, \end{cases} \tag{11}$$

where $R_0 = 0.421\ldots$ is the root of the equation $\alpha_m(R) = \delta^{lp}(R)$.

Bound (11) is plotted in Fig. 1 together with bounds (1)-(2) and (3). Computations show that it is better than (3) for all $R \in (0,1)$. Note that the second segment in (11) coincides with the best known upper bound (5) on $\rho(1,R) = \delta(R)$. We wish to stress a difference between this result and the upper bound in Theorem 1. The upper bound in Theorem 1 is the same for the cases of $m = 1$ and $m = 2$ simply because the way of counting the contribution to the weight of the center of the sphere in [4] cannot tell between $m = 2i-1$ and $m = 2i$. Corollary 5, in contrast, indicates a geometric property of (hypothetical) codes meeting the MRRW upper bound (5), namely, that for some pairs of vectors at a distance $n\delta^{lp}(R)$ apart, there is a third vector at the same distance from each of them.

For reference purposes we also give a short table of values of the bounds.

Table 1 Bounds on $\rho(2,R)$.

R	0.1	0.2	0.3	0.4	0.5	0.6	0.7	0.8	0.9
Lower bound (1)-(2)	0.168	0.133	0.105	0.082	0.063	0.046	0.031	0.018	0.0079
Elias bound (4)	0.216	0.184	0.153	0.125	0.098	0.073	0.050	0.030	0.0128
New bound (11)	0.196	0.165	0.138	0.114	0.091	0.069	0.048	0.029	0.0127

References

[1] R. Ahlswede, "Channel capacities for list codes", *J. Appl. Probability*, 10, 1973, 824–836.

[2] A. Ashikhmin and A. Barg, "Binomial moments of the distance distribution: Bounds and applications", *IEEE Trans. Inform. Theory*, 45, 1999, 438–452.

[3] A. Ashikhmin, A. Barg, and S. Litsyn, "New upper bounds on generalized distances", *IEEE Trans. Inform. Theory*, 45, 1999, 1258–1263.

[4] V. Blinovsky, "Bounds for codes decodable in a list of finite size", *Problems of Information Transmission*, 22(1), 1986, 11–25.

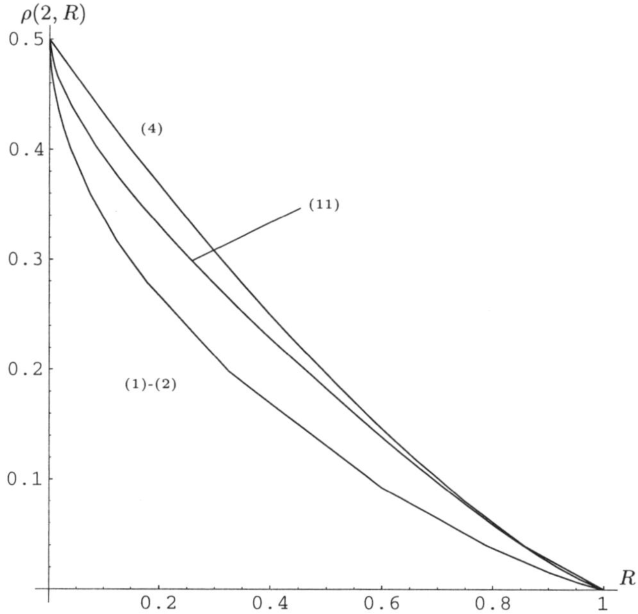

Figure 1 Bounds on the size-2 list radius of a code of rate R

[5] V. Blinovsky, "Asymptotic Combinatorial Coding Theory", Kluwer Academic Publishers, Boston, 1997.

[6] P. Elias, "List decoding for noisy channels", Rep. No. 335 Research Laboratory of Electronics, Massachusetts Institute of Technology, Cambridge, Mass. MR 20 #5702, 1957.

[7] P. Elias, "Error correcting codes for list decoding", *IEEE Trans. Inform. Theory*, 37, 1991, 5–12.

[8] V. I. Levenshtein, "Upper-bound estimates for fixed-weight codes", *Problemy Peredachi Informatsii*, 7(4), 1971, 3–12, in Russian. English translation in *Probl. Inform. Trans.* 7, 281–287.

[9] R. J. McEliece, E. R. Rodemich, H. Rumsey, and L. R. Welch, "New upper bound on the rate of a code via the Delsarte-MacWilliams inequalities", *IEEE Trans. Inform. Theory*, 23, 1977, 157–166.

[10] S. Litsyn, "New bounds on error exponents", *IEEE Trans. Inform. Theory*, 45, 1999, 385–398.

[11] A. Samorodnitsky, "On the optimum of Delsarte's linear program", *J. Combinatorial Theory, Ser. A*, to appear, 1999.

[12] J. M. Wozencraft, "List decoding", Quarterly Progr. Rep., Res. Lab. Electronics, MIT, 48, 1958, 90–95.

CONSTRUCTIONS OF OPTIMAL LINEAR CODES

Stefan Dodunekov

Institute of Mathematics and Informatics, Bulgarian Academy of Sciences
8. G. Bonchev Str, 1113 Sofia, Bulgaria
stedo@moi.math.bas.bg

Juriaan Simonis

Delft University of Technology, Faculty of Information Technology and Systems
P.O.Box 5031, 2600 GA Delft, the Netherlands
J.Simonis@twi.tudelft.nl

Abstract: The goal of this paper is to present an overview on known constructions of length-optimal linear codes. First we discuss the interrelation between various definitions of optimality in terms of the basic parameters of a linear code: length, dimension and minimum distance.. Then we give some general constructions of Griesmer codes based on the anticode technique. Constructions using the correspondences between codes and projective multisets are also considered. A survey on quasi-cyclic and quasi-twisted optimal codes is included.

INTRODUCTION

An $[n, k, d]_q$ = [length, dimension, minimum distance]$_q$-code is defined to be a k-dimensional subspace of the n-dimensional standard vector space $\mathbb{F}_q^n$ over the field $\mathbb{F}_q$ of prime power size q and minimum nonzero Hamming distance at least d. Since the basic parameters of a code are its length, dimension and minimum

I. Althöfer et al. (eds.), Numbers, Information and Complexity, 245-263.

distance, it is natural to study those codes that optimize one parameter for fixed values of the two others. This leads to the following definition:

Definition 1.1. An $[n, k, d]_q$-code is said to be

- *length-optimal* (*N-optimal*) if no $[n-1, k, d]_q$-code exists,
- *dimension-optimal* (*K-optimal*) if no $[n, k+1, d]_q$-code exists, and
- *distance-optimal* (*D-optimal*) if no $[n, k, d+1]_q$-code exists.

One can take a different point of view. There are three basic ways of creating new codes. Let $S := \{1, 2, \ldots, n\}$ be the coordinate index set of $\mathbb{F}_q^n$. We can identify $\mathbb{F}_q^n$ with $\mathbb{F}_q^S$, the $\mathbb{F}_q$-vector space of the mappings $S \to \mathbb{F}_q$. If T is an m-subset of S, then $\mathbb{F}_q^T$ can be identified with the subspace

$$(\mathbb{F}_q^n)^T := \{\mathbf{x} \in \mathbb{F}_q^n \mid \operatorname{supp}(\mathbf{x}) \subseteq T\}$$

of $\mathbb{F}_q^n$, where $\operatorname{supp}(\mathbf{x})$, the support of a vector $\mathbf{x} = (x_1, x_2, \ldots, x_n)$, is the subset

$$\operatorname{supp}(\mathbf{x}) := \{i \mid x_i \neq 0\}.$$

Any bijection $T \to \{1, 2, \ldots, m\}$ induces an isomorphism between $\mathbb{F}_q^T$ and $\mathbb{F}_q^m$.

Let $\mathbf{x}_T$ denote the *restriction* of $\mathbf{x} \in \mathbb{F}_q^n$ to T. More generally, if U is any subset of $\mathbb{F}_q^n$, then $U_T = \{\mathbf{x}_T \mid \mathbf{x} \in U\}$.

Let $\overline{T}$ denote the complement of T in S.

Definition 1.2. Let $\mathcal{C}$ be a an $[n, k, d]_q$-code.

- The restriction $\mathcal{C}_T$ of $\mathcal{C}$ to T is said to be obtained by *puncturing* $\mathcal{C}$ with respect to $\overline{T}$.
- The code
$$\mathcal{C}^T := \{\mathbf{c} \in \mathcal{C} \mid \operatorname{supp}(\mathbf{c}) \subseteq T\}_T$$
is said to be obtained by *shortening* $\mathcal{C}$ with respect to $\overline{T}$.

So both $\mathcal{C}_T$ and $\mathcal{C}^T$ have length m. The minimum distance of $\mathcal{C}^T$ is at least $d(\mathcal{C})$ and the minimum distance of $\mathcal{C}_T$ is at least $d(\mathcal{C}) - n + m$, where $d(\mathcal{C}) = d$ is the minimum distance of $\mathcal{C}$.

The inverse process of puncturing is *lengthening*. Let $\varphi : \mathcal{C} \to \mathbb{F}_q^m$ be any linear mapping. Then the linear code

$$\mathcal{C}' := \{(\mathbf{c}, \varphi(\mathbf{c})) \mid \mathbf{c} \in \mathcal{C}\}$$

has length n and dimension k. If $\mathcal{D} := \varphi(\mathcal{C})$ is an $[m, k, e]_q$-code, then $\mathcal{C}'$ is an $[n+m, k, d+e]_q$-code which is called a *juxtaposition* of $\mathcal{C}$ and $\mathcal{D}$.

Using lengthening, puncturing or shortening with respect to one coordinate position we obtain the following useful result.

Proposition 1.3. If an $[n, k, d]_q$-code exists with $0 < k < n$, then codes with parameters $[n+1, k, d]_q$, $[n-1, k, d-1]_q$, and $[n-1, k-1, d]_q$ exist.

We consider an $[n,k,d]_q$-code to be *optimal* with respect to lengthening, puncturing or shortening if no code with these parameters can be obtained by such a construction. In the case of lengthening this kind of optimality is length-optimality, but in the cases of puncturing and shortening new definitions of optimality emerge.

Definition 1.4. An $[n,k,d]_q$-code is said to be:

- P-optimal if no $[n+1,k,d+1]_q$-code exists,
- S-optimal if no $[n+1,k+1,d]_q$-code exists, and
- strongly-optimal if it is N-optimal, P-optimal and S-optimal.

Let us compare the five basic types of optimality. First of all, it is clear that N-optimality implies both K-optimality and D-optimality, that P-optimality implies D-optimality and that S-optimality implies K-optimality. In the other cases, there is independence, as the following table shows. The examples are drawn from the extremely useful table of bounds for binary D-optimal codes maintained by A.E. Brouwer (on-line version:
http://www.win.tue.nl/math/dw/voorlincod.html).

	N	P	S	K	D
$[10,5,4]$	+	+	−		
$[30,6,14]$	+	−	+		
$[69,4,36]$	−	+	+		
$[33,5,16]$		+		−	
$[12,4,6]$		−		+	
$[32,6,15]$			+		−
$[15,4,8]$			−		+

These facts suggest that length-optimality is the most important basic type of optimality. That is why in what follows the word optimal will be reserved for length-optimal codes.

Let us introduce a fundamental function.

Definition 1.5.

$$N_q(k,d) := \min\{n \mid \text{an } [n,k,d]_q\text{-code exists}\}.$$

So the optimal codes are precisely the $[N_q(k,d),k,d]_q$-codes. A nice feature of the function $N_q(k,d)$ is that it is strongly increasing in both arguments.

Theorem 1.6. *(The Griesmer bound)*

$$N_q(k,d) \geq g_q(k,d) := \sum_{j=0}^{k-1} \lceil \frac{d}{q^j} \rceil. \tag{1}$$

Inequality (1) was proved for $q=2$ by Griesmer [21] and Varshamov in his thesis [72] (unpublished) and was later generalized to any q by Solomon and

Stiffler [67]. For historical reasons we will refer to it as to the Griesmer bound. Codes with parameters meeting (1) with equality will be called Griesmer codes.

The Griesmer bound is achievable. Two important examples of Griesmer codes are the simplex code $\mathcal{S}_k(q)$ with parameters

$$\left[\frac{q^k-1}{q-1}, k, q^{k-1}\right]_q \tag{2}$$

and the MacDonald code $\mathcal{M}_k^u(d)$ with parameters

$$\left[\frac{q^k-q^u}{q-1}, k, q^{k-1}-q^{u-1}\right]_q, \quad 1 \leq u \leq k-1. \tag{3}$$

The following two natural problems are still open in general.

Problem 1.7. Determine $N_q(k,d)$ for all values of q, k and d. Given q, k and d, characterize up to equivalence all $[N_q(k,d), k, d]_q$-codes.

Problem 1.8. Find all values of q, k and d for which $N_q(k,d) = g_q(k,d)$ (i.e. for which there exists a Griesmer code). In case $N_q(k,d) = g_q(k,d)$ characterize up to equivalence all Griesmer codes.

Even Problem 1.8., which is weaker, is far away from a complete solution. An optimistic observation is that for any given k and q there exists a constant $D(k,q)$ such that $d > D(k,q)$ implies $N_q(k,d) = g_q(k,d)$ (Baumert and McEliece [1] for $q = 2$, Hamada and Tamari [37], Dodunekov [14] and Hill [47] for any q). In other words, for any fixed k and q, $N_q(k,d)$ is known for all but a finite number of cases.

However, for any given q, $d > 2$ and integer l, there exists a constant $K(d,q,l)$ such that $k > K(d,q,l)$ implies $N_q(k,d) \geq l + g_q(k,d)$ (Dodunekov [15]). The history of $N_2(8,d)$ nicely illustrates the difficulties. Helleseth [44] proved that $N_2(8,d) = g_2(8,d)$ for any $d \geq 128$ but for $d < 128$ still there is at least one open case.

The paper is organized as follows. In Section 2 we present general constructions of Griesmer codes. In Section 3 we describe a general approach for optimal code construction based on the interrelation between codes and projective multisets. Finally, in Section 4 we summarize results about quasi-cyclic optimal codes.

The authors are fully aware of the existence of many more construction methods of optimal linear codes. Many of these techniques, however have been adequately surveyed elsewhere. First of all we should mention Brouwer's chapter in the Handbook of Coding Theory [64]. The powerful max- and min-hyper approach of Hamada et al. has been surveyed in [38], [68] and [41]. See also Hill and Kolev's forthcoming paper [49]. For algebraic geometry codes, we refer to the chapter by Høholdt, van Lint and Pellikaan in [64] and its list of references. The special issue [55] of the IEEE Transactions on Information Theory is also an excellent source.

As a general reference about notions and facts from coding theory which are not defined here, we refer to [62].or [64].

CONSTRUCTIONS OF GRIESMER CODES

Some simple constructions

In this section we shall consider several general constructions of Griesmer codes. First we mention that certain juxtapositions of the simplex codes $\mathcal{S}_k(q)$ (2) and the MacDonald codes $\mathcal{M}_k^u(q)$ (3) are Griesmer codes. We give two examples.

Example 2.1. For any integer $t > 0$ and any $[g_q(k,d),k,d]_q$-code $\mathcal{D}$, a juxtaposition

$$\mathcal{C} := (\mathcal{D} \mid t\mathcal{S}_k(q))$$

is a Griesmer code.

Example 2.2. If the integers a_i, $i = 1,2,\ldots,k-1$, satisfy the condition $0 \le a_i \le q-1$, the code

$$\mathcal{C} := (a_1\mathcal{M}_k^{k-1}(q) \mid a_2\mathcal{M}_k^{k-2}(q) \mid \cdots \mid a_{k-1}\mathcal{M}_k^1(q))$$

is a Griesmer code.

Next we observe that using puncturing we get some Griesmer codes for free.

Proposition 2.3. Suppose that $q \mid d$ and that $N_q(k,d-b) = g_q(k,d-b)$ for some b with $0 \le b \le q-1$. Then

$$N_q(k,d-a) = g_q(k,d-a)$$

for all a with $b \le a \le q-1$.

Codes and projective multisets

The coordinate index set of any full length code (i.e. a code without an all-zero coordinate) can be interpreted as a projective multiset.

Definition 2.4. Let $\mathcal{C}$ be an $[n,k,d]_q$-code of full length and let

$$G := [\ \mathbf{g}_1 \quad \mathbf{g}_2 \quad \cdots \quad \mathbf{g}_n\ ]$$

be a generator matrix of $\mathcal{C}$. Then the multiset

$$\gamma_\mathcal{C} := (\langle \mathbf{g}_i \rangle, i = 1,2,\ldots,n)$$

in the projective space $\mathbb{P}(\mathbb{F}_q^k)$ is called the projective multiset associated with $\mathcal{C}$.

A nonzero codeword $\mathbf{c} := \xi G$ of $\mathcal{C}$ corresponds to the linear form $\sum_{i=1}^k \xi_i x_i$ of the vector space $\mathbf{F}_q^k$ and hence to a hyperplane $H_\mathbf{c}$ of $\mathbb{P}(\mathbb{F}_q^k)$. Then the weight of $\mathbf{c}$ is the size of the complement of $H_\mathbf{c}$ in the multiset $S_\mathcal{C}$:

$$\mathrm{wt}(\mathbf{c}) = |\gamma_\mathcal{C}| - |\gamma_\mathcal{C} \cap H_\mathbf{c}|.$$

This leads to the following interpretation of the minimum distance.

Proposition 2.5. Let α be the maximum multiplicity of $\mathcal{C}$ (or $\gamma_\mathcal{C}$). Put

$$T := \alpha\mathbb{P}(\mathbb{F}_q^k) \setminus \gamma_\mathcal{C}.$$

Then

$$d(\mathcal{C}) = \alpha q^{k-1} - |T| + \min_H |T \cap H|,$$

where H runs through all hyperplanes of $\mathbb{P}(\mathbb{F}_q^k)$.

A promising strategy to construct optimal codes is by starting with a good code $\mathcal{C}$ and puncturing it with respect to a suitable submultiset $T \subset \gamma_{\mathcal{C}}$. What kind of T is suitable? Since

$$d(\mathcal{C}_{\overline{T}}) \geq d(\mathcal{C}) - \max\{\mathbf{x} \mid \mathbf{x} \in \mathcal{C}_T\},$$

we would like the maximum distance of $\mathcal{C}_T$ to be as small as possible. For this reason the code $\mathcal{C}_T$ is sometimes called an anticode [19]. An excellent choice for T is a projective space, because then the code $\mathcal{C}_T$ is a simplex code, and in a simplex code the minimum distance and the maximum distance coincide.

As an example, we apply these observations to Griesmer codes.

Proposition 2.6. Let $\mathcal{C}$ be a $[g_q(k,d),k,d]_q$-code, with $d = sq^{k-1} - \sum_{i=1}^{k-1} a_i q^{i-1}$, $0 \leq a_i \leq q-1$ for all i. Suppose that an integer t and a $(t-1)$-dimensional projective subspace $L \subset \mathbb{P}(\mathbb{F}_q^k)$ exist such that $a_t < q-1$ and $L \subseteq S_{\mathcal{C}}$. Then $\mathcal{C}_{\overline{L}}$ is a $[g_q(k, d-q^{t-1}, k, d-q^{t-1}]_q$-code.

Belov's theorems

Solomon and Stiffler [67] were the first to apply the idea of Proposition 2.5. recursively, using as starting code an s times replicated simplex code. The best general result was obtained by Belov, Logachev and Sandimirov [2] in the binary case and generalized to arbitrary field size in [17] and [47].

Theorem 2.7. Let $(u_i | i = 1, 2, \ldots, t)$ be a nonincreasing sequence of integers between $k-1$ and 1, and such that no value is taken more than $q-1$ times. Then successive puncturing of $s\mathcal{S}_k(q)$ with respect to projective subspaces of dimension $u_i - 1$ can yield a k-dimensional Griesmer code with minimum distance

$$d := sq^{k-1} - \sum_{i=1}^{t} q^{u_i}$$

if and only if

$$\sum_{i=1}^{\min(s+1,t)} u_i \leq sk.$$

Another idea, which can be already found in Belov [2], is to add small Griesmer codes to larger ones. As a straightforward consequence we formulate the following result.

Proposition 2.8. Let $\mathcal{C}$ be a $[g_q(k,d),k,d]_q$-code with $d = sq^{k-1}$ - $\sum_{i=1}^{k-1} a_i q^{i-1}$, $0 \leq a_i \leq q-1$ for all i, and such that its multiset $S_{\mathcal{C}}$ contains an $(l-1)$-dimensional subspace L of $\mathbb{P}(\mathbb{F}_q^k)$ with multiplicity $s' \leq q-1-a_i$. Also, let $\mathcal{D}$ be a $[g_q(l,e),l,e]_q$-code with $e = s'q^{l-1} - \sum_{i=1}^{l-1} b_i q^{i-1}$, $0 \leq b_i \leq q-1$ for all i, and such that $b_i \leq q-1-a_i$, $i = 1, 2, \ldots, l-1$. Then there exists a $[g_q(k,d'),k,d']_q$-code with $d' := d - s'q^{l-1} + e$.

Example 2.9. Take for $\mathcal{C}$ the simplex code $\mathcal{S}_k(q)$ and for $\mathcal{D}$ any $[[g_q(l,e),k,e]_q$-code with $l < k$ and

$$e = q^{l-1} - \sum_{i=1}^{l-1} a_i q^{i-1}, \quad 0 \leq a_i \leq q-1 \text{ for all } i.$$

Then we get a $[g_q(k,d'),k,d'_q]_q$-code $\mathcal{C}'$ with minimum distance

$$d' := q^{k-1} - \sum_{i=1}^{l-1} a_i q^{i-1}.$$

This example can be used to create families of Griesmer codes.

Theorem 2.10. Suppose that a $[g_q(l,e),l,e]_q$-code exists with $l < k$ and $e = q^{l-1} - \sum_{i=1}^{l-1} a_i q^{i-1}$, $0 \leq a_i \leq q-1$ for all i. Then for any sequence of integers a_i, $i = l, l+1, \ldots, k-1$ with $0 \leq a_i \leq q-1$ for all i, there exists a $[g_q(k,d),k,d]_q$-code with

$$d := (1 + \sum_{i=1}^{k-1} a_i) q^{k-1} - \sum_{i=1}^{k-1} a_i q^{i-1}.$$

Example 2.11. For $e = 1, 2$ we have

$$N_q(l,e) = g_q(l,e)$$

for all l. So we can use all optimal codes of minimum distance ≤ 2 for the construction given by Theorem 2.10. Consider the binary case. Then

$$e = 2^{l-1} - \sum_{i=e}^{l-1} 2^{i-1}.$$

Hence binary k-dimensional Griesmer codes with minimum distance

$$d := (1 + \sum_{i=l+1}^{k-1} a_i) 2^{k-1} - \sum_{i=l+1}^{k-1} a_i 2^{i-1} - \sum_{i=e}^{l-1} 2^{i-1}$$

exist for $e = 1, 2$ and for all $l \geq e$ and all $a_i \in \{0, 1\}$, $i = l, l+1, \ldots, k-1$.

Let us introduce the following notation.

Notation 2.12.

1. $\mathfrak{C}(k,d;q)$ is the set $\{C \mid C \text{ is a } [g_q(k,d)k,d]_q\text{-code}\}$,
2. $\mathfrak{C}^{(1)}(k,d;q)$ is the subset of $\mathfrak{C}(k,d;q)$ obtainable by the construction of Theorem 2.7., and
3. $\mathfrak{C}^{(2)}(k,d;q)$ is the subset of $\mathfrak{C}(k,d;q)$ obtainable by the construction of Theorem 2.10.

Remark 2.13. Problem 1.8. can be rephrased as follows: to identify the parameters for which $\mathfrak{C}(k,d;q)$ is nonempty and to describe up to equivalence its elements. This task is easy if $k \leq 2$, for then all optimal codes are Griesmer codes. The sets $\mathfrak{C}(1,d;q)$ and $\mathfrak{C}(2,d;q)$ are nonempty for all d and all codes within a set are equivalent.

Let us now specify the constructions of Theorem 2.7. and Example 2.11. for the binary case.

Theorem 2.14. [2] Let $s = \lceil \frac{d}{2^{k-1}} \rceil$ and define $k > u_1 > \cdots > u_m \geq 1$ such that

$$s2^{k-1} - d = \sum_{i=1}^{m} 2^{u_i - 1}.$$

Then there exists a $[g_2(k,d), k, d]_2$-code if

$$\sum_{i=1}^{\min(s+1,m)} u_i \leq sk$$

or $u_{i+1} = u_i - 1$ for $i = s, s+1, \ldots, m-1$ and $u_m = 1$ or 2.

It is easy to check that for $d \leq 2^{k-1}$ the conditions of Theorem 2.14. are satisfied for all values of d outside the intervals $J(k,i) = [2^{k-1} - 2^{k-i} + 3, 2^{k-1} - 2^{k-i-1} - 2^i]$, $i = 1, 2, \ldots, \lfloor \frac{k-2}{2} \rfloor$.

Belov [2] conjectured that if $d \in J(k,i)$ then $N_2(k,d) \geq g_2(k,d) + 1$, i.e. that for $s = 1$ the conditions of Theorem 2.13. are necessary. We shall call the $J(k,i)$ the *Belov intervals*.

The Belov conjecture was proved by Logachev [57] for $i = 1$, by van Tilborg [71] for $i = 2$ and by Helleseth [42] in general. In fact, Helleseth proved a stronger result.

Theorem 2.15. [42]. If $d \leq 2^{k-1}$ then

$$\mathfrak{C}(k,d;2) = \mathfrak{C}^{(1)}(k,d;2) \cup \mathfrak{C}^{(2)}(k,d;2).$$

For some cases it is possible to find the exact value of $N_2(k,d)$ even if d is in the Belov intervals.

Theorem 2.16. [14] Let

$$d_0 := 2^{k-1} - 2^{k-i-1} - 2^i.$$

Then

$$N_2(k,d) = g_2(k,d) + 1$$

for $d := d_0$ if $1 \leq i \leq \lfloor (k-2)/2 \rfloor$ and for $d := d_0 - 2$ if $2 \leq i \leq \lfloor (k-2)/2 \rfloor$.

Remark 2.17. There exist general constructions of Griesmer codes which are not of Solomon-Stiffler or Belov type. For $q = 2$, $d > 2^{k-1}$ such constructions were suggested by Helleseth and Van Tilborg [42], Helleseth [44] and Logachev [58, 59, 60, 61]. More recently, Hamada, Helleseth and Ytrehus constructed new codes meeting the Griesmer bound over $\mathbb{F}_{q^t}$ from Solomon-Stiffler codes over $\mathbb{F}_q$. The resulting codes are generally not equivalent to Solomon-Stiffler codes. (See also Hamada and Helleseth [39] for the quaternary version of this construction.)

There are also many sporadic Griesmer codes which do not belong to any known general class of Griesmer codes, cf. Helleseth's survey paper [45].

DUAL TRANSFORMS OF MULTISETS

In this section we consider a general approach to constructive coding theory which is based on the interrelation between codes and projective multisets mentioned in Subsection 2.2. The first one to use this relationship was Slepian [66], see also [63], who used the term modular representation. A lot of work has been done to study the relation between projective two-weight codes and projective (n, k, h_1, h_2) sets (Delsarte [13], Hill [46] and others). These are subsets of size n of $\mathbb{P}(\mathbb{F}_q^k)$ such that every hyperplane is met in h_1 or h_2 points. A nice survey on two-weight codes is the paper by Calderbank and Kantor [8].

The spanning subsets $K \subseteq \mathbb{P}(\mathbb{F}_q^{N+1})$ of size n and such that all s-dimensional projective subspaces of $\mathbb{P}(\mathbb{F}_q^{N+1})$ intersect K in at most s points, called $(n; r, s; N, q)$-sets, are surveyed by Hirschfeld and Storme [50]. The $(n; k-2, n-d; k-1, q)$-sets correspond to linear $[n, k, d]_q$-codes for which the columns of any generator matrix are pair-wise independent. Another good reference is the survey paper by Landjev [56]. Recently, Brouwer and van Eupen [6] used a correspondence between projective codes and two-weight codes to construct optimal codes and to prove the uniqueness of certain codes. Their idea - a generalization of a result by Hill [46] - is to transform subsets of a finite projective space Π into multisets of the dual space Π^*. The dual transform of its full generality is described in [18]. Variations on this theme can be found in [52].

Projective multisets revisited

Formally, a multiset γ in $\mathbb{P}(\mathbb{F}_q^k)$ is nothing but a mapping $\gamma: \mathbb{P}(\mathbb{F}_q^k) \to N$, and the size of γ is the integer $\sum_{p \in \mathbb{P}(\mathbb{F}_q^k)} \gamma(p)$. Then a generator matrix

$$G := [\ \mathbf{g}_1 \quad \mathbf{g}_2 \quad \cdots \quad \mathbf{g}_n\ ]$$

for a full-length $[n, k, d]_q$-code $\mathcal{C}$ determines a projective multiset $\gamma_\mathcal{C}$

$$\gamma_\mathcal{C}(\langle \mathbf{x} \rangle) := |\{i \mid \langle \mathbf{g}_i \rangle = \langle \mathbf{x} \rangle\}|.$$

in the projective space $\mathbb{P}(\mathbb{F}_q^k)$. This definition depends on the choice of the generator matrix, but other choices yield projectively equivalent multisets. Conversely, any multiset γ in $\mathbb{P}(\mathbb{F}_q^k)$ that spans $\mathbb{P}(\mathbb{F}_q^k)$ determines a full-length $[n, k, d]_q$-code up to code equivalence. Let us denote any code from this equivalence class by $\mathcal{C}_\gamma$.

Definition 3.1. Let γ be a projective multiset on $\mathbb{P}(\mathbb{F}_q^k)$.

- The *multiplicity set* of γ (and of the corresponding code $\mathcal{C}_\gamma$) is the set

$$M_\gamma := \mathrm{Im}\gamma.$$

- The *weight function* of γ is the function

$$\mu_\gamma : \mathbb{P}(\mathbb{F}_q^k) \to N, \quad \mu_\gamma(\langle \mathbf{x} \rangle) := \sum_{\substack{\langle \mathbf{y} \rangle \in \mathbb{P}(\mathbb{F}_q^k), \\ \mathbf{x}\cdot\mathbf{y}=0}} \gamma(\langle \mathbf{y} \rangle),$$

where $\mathbf{x} \cdot \mathbf{y} := \sum x_i y_i$ is the standard scalar product on $\mathbb{F}_q^k$.

Let us describe the connection between the weights of codewords in $\mathcal{C}$ and the weight function of $\gamma_{\mathcal{C}}$.

Definition 3.2. The *weight distribution* of a code $\mathcal{C} \subseteq \mathbb{F}_q^k$ is the sequence $A_0(\mathcal{C})$, $A_1(\mathcal{C})$, ..., $A_n(\mathcal{C})$ defined by

$$A_i(\mathcal{C}) := |\{\mathbf{c} \mid \mathbf{c} \in \mathcal{C} \wedge |\mathbf{c}| = i\}|, \; i = 0, 1, \ldots, n.$$

The *weight set* of $\mathcal{C}$ is the set

$$W_{\mathcal{C}} := \{i \mid i \in \{1, 2, \ldots, n\} \wedge A_i(\mathcal{C}) \neq 0\}$$

Proposition 3.3. If the projective multiset γ is constructed by means of the generator matrix G of the full-length $[n, k, d]_q$-code $\mathcal{C}$, then

$$\mathrm{wt}(\mathbf{x}G) = \mu_\gamma(\langle \mathbf{x} \rangle), \; \mathbf{x} \in \mathbb{F}_q^k \setminus \{\mathbf{o}\}.$$

Hence

$$A_i(\mathcal{C}) = (q-1)\left|\mu_\gamma^{-1}(i)\right|, \; i = 1, 2, \ldots, n,$$

and

$$W_{\mathcal{C}} = \mathrm{Im}\mu_\gamma.$$

Dual transforms

Let $\mathcal{C} \subseteq \mathbb{F}_q^S$ be a k-dimensional full-length code, and let σ be a any function that takes integer values on the weight set W of $\mathcal{C}$. We extend this function to a polynomial function

$$\sigma(i) := \sum_{y \in W} \sigma(y) \frac{\prod\limits_{w \in W \setminus y} (i - w)}{\prod\limits_{w \in W \setminus y} (y - w)}$$

on $\mathbb{Q}$ by Lagrange interpolation. Note that the degree $g := g_\sigma$ of the polynomial σ does not exceed $|W| - 1$.

For each σ, we shall construct from γ a new multiset on $\mathbb{P}(\mathbb{F}_q^k)$.

Definition 3.4. The *dual transform* of the projective multiset $\gamma := \gamma_{\mathcal{C}}$ with respect to σ is the multiset

$$\gamma^\sigma : \mathbb{P}(\mathbb{F}_q^k) \to \mathbb{N}, \; p \longmapsto \sigma(\mu(p)).$$

The *dual transform* of the code $\mathcal{C}$ with respect to σ is the code $\mathcal{C}^\sigma := \mathcal{C}_{\gamma^\sigma}$.

Let us describe a matrix that generates the code $\mathcal{C}^\sigma$.The nonzero codewords fall into sets of $q-1$ pairwise dependent codewords. Now take from each set $\sigma(w)$ copies, where w is the weight of the codewords in the sets, and put all these vectors as *columns* in a matrix. The *row space* of this matrix is $\mathcal{C}^\sigma$.

It might happen that the multiset γ^σ does not span $\mathbb{P}(\mathbb{F}_q^k)$. In the sequel we assume that this is not the case, i.e. that the dimension of the dual transform $\mathcal{C}^\sigma$ is equal to k. We now look at the other parameters. Let us express the polynomial σ in the Krawtchouk polynomials

$$K_i(x) := \sum_{m=0}^{i}(-1)^m(q-1)^{i-m}\binom{x}{m}\binom{n-x}{i-m},$$

cf. [54]. There are – uniquely determined – rational numbers $a_0, a_1, \ldots, a_g$ such that

$$\sigma(j) = \sum_{l=0}^{g} a_l K_l(j).$$

Proposition 3.5. The length of $\mathcal{C}^\sigma$ is equal to

$$\sum_{i=0}^{g} a_i\{\frac{q^k}{q-1}A_i(\mathcal{C}^\perp) - (q-1)^{i-1}\binom{n}{i}\}. \tag{4}$$

So the length of $\mathcal{C}^\sigma$ depends on the weight distribution of the dual code $\mathcal{C}^\perp$. For the weights in $\mathcal{C}^\sigma$ we need more information on $\mathcal{C}^\perp$. This is the kernel of the mapping

$$\varphi : \mathbb{F}_q^n \to \mathbb{F}_q^k, \quad \mathbf{y} \longmapsto G\mathbf{y}^T,$$

where G is a generator matrix of $\mathcal{C}$.

Definition 3.6. The *reduced distribution matrix* of $\mathcal{C}^\perp$ is the $\frac{q^{n-k}-1}{q-1} \times (n+1)$ matrix $\bar{\mathbf{D}}$ parametrized by $\mathbb{P}(\mathbb{F}_q^k) \times \{0, 1, \ldots, n\}$ and having

$$\bar{\mathbf{D}}_{\langle \mathbf{x}\rangle, i} := |\{\mathbf{c} \in \mathcal{C}^\perp \mid d(\mathbf{c}, \varphi^{-1}(\mathbf{x})) = i\}| - A_i(\mathcal{C}^\perp)$$

as its $(\langle \mathbf{x}\rangle, i)$ entry.

Proposition 3.7. The weight function of the projective multiset γ^σ is given by

$$\mu_{\gamma^\sigma}(p) = -q^{k-1}\sum_{i=0}^{g} a_i \bar{\mathbf{D}}_{p,i}. \tag{5}$$

Hence to determine the weight distribution, and more specifically the minimum distance, of the dual transform $\mathcal{C}^\sigma$, we need to know the first $g+1$ columns of the reduced distribution matrix of $\mathcal{C}^\perp$.

Example 3.8. Let $\mathcal{C}$ be the unique binary $[48, 8, 22]$-code. (Cf. [16] for a construction and [51] for a computerized uniqueness proof.) The weight set of $\mathcal{C}$ is {22,24,30,32}. If we choose for σ the function with $\sigma(22) = \sigma(30) = 1$ and

$\sigma(24) = \sigma(32) = 0$, then the dual transform $\mathcal{C}^\sigma$ turns out to be a $[192, 8, 96]$-code which in fact is optimal. Another, record breaking, example is the $[245, 9, 120]$ code described in [52]. D. Jaffe found this example (and several others that happen to improve the table [5]) by means of an extensive computer search. The basic problem here is to develop a theory that predicts which input codes $\mathcal{C}$ and which transform functions σ produce record-breaking output codes $\mathcal{C}^\sigma$.

Dual transforms of degree one

Let $\mathcal{C} \subseteq \mathbb{F}_q^n$ be k-dimensional full-length code, and let $\gamma := \gamma_\mathcal{C}$ be the corresponding projective multiset. In this section, we study dual transforms $\mathcal{C}^\sigma$ under the assumption that the transform function σ has *degree one:* $\sigma(j) := aj + b$. Let W be the weight set of $\mathcal{C}$. Two choices for σ are particularly useful: If $\Delta := \gcd W$, $d := \min W$ and $D := \max W$, then the functions σ_+ and σ_- defined by

$$\sigma_+(j) := \frac{j-d}{\Delta}, \ \sigma_-(i) := \frac{-j+D}{\Delta}$$

indeed take nonnegative integer values on $W_\mathcal{C}$.

Expressing the polynomial σ in the Krawtchouk polynomials $K_0(j) := 1$ and $K_1(j) := (q-1)n - qj$, we get

$$\sigma(j) = (b + \frac{(q-1)an}{q})K_0(j) + (-\frac{a}{q})K_1(j).$$

Let $\mathcal{D} := \mathcal{C}^\sigma$ be the dual transform of $\mathcal{C}$ with respect to σ. Since the code $\mathcal{C}$ is of full length, i.e. $A_1(\mathcal{C}^\perp) = 0$, the formula for the length of $\mathcal{D}$ reduces to

$$n_\mathcal{D} = nq^{k-1}a + \begin{bmatrix} k \\ 1 \end{bmatrix} b. \tag{6}$$

Now we consider the weight function of γ^σ. Formula (4) gives us

$$\mu_{\gamma^\sigma}(p) = \alpha\gamma(p) + \beta, \tag{7}$$

with

$$\alpha := q^{k-2}a, \quad \beta := \frac{(q-1)n_\mathcal{D} + b}{q}.$$

Remark 3.9. Note that the weight set $W_\mathcal{D}$ of $\mathcal{D}$ is equal to

$$\{\alpha m + \beta \mid m \in M_\gamma\} \setminus \{0\}.$$

If, in particular, $\mathcal{C}$ is *projective*, then $\mathcal{D}$ is a (≤ 2)-weight code. This case is the main subject matter of Brouwer and Van Eupen's paper [6].

Formula (5) immediately gives the *minimum distance* of $\mathcal{D}$. If $k_\mathcal{D} = k$, then

$$d_\mathcal{D} = \begin{cases} (an+b)q^{k-1} + a(\min M_\gamma - n)q^{k-2} & \text{if} \quad a > 0, \\ (an+b)q^{k-1} + a(\max M_\gamma - n)q^{k-2} & \text{if} \quad a < 0. \end{cases}$$

QUASI-CYCLIC OPTIMAL CODES

In this section we will consider the class of quasi-cyclic (QC) codes which turns out to contain many optimal codes. It is a natural generalization of the class of cyclic codes. QC-codes were introduced by Townsend and Weldon [69]. They achieve a Gilbert-Varshamov type bound [53].

Definition 4.1 A linear code of length $n = pm$ is called *p-quasi-cyclic* (*p*-QC) if it is invariant under a coordinate permutation which is a product of p $m - cycles$.

Let us order the coordinate places in such a way that the permutation in the definition takes the form

$$(1, 2, \ldots, m)(m+1, m+2, \ldots 2m) \cdots (n-m+1, n-m+2, \ldots n).$$

The best studied QC-codes are those that possess a generator matrix which consists of circulant matrices.

Definition 4.2. An $m \times m$ matrix $\mathbf{C}$ over $\mathbb{F}_q$ is said to be *circulant* if $\mathbf{P}^{-1}\mathbf{CP} = \mathbf{C}$, where $\mathbf{P}$ is the permutation matrix corresponding to $(1, , \ldots, m)$.

$$\mathbf{P} := \begin{bmatrix} 0 & 1 & . & . & & \\ 0 & & 1 & & & \\ . & & & . & & \\ . & & & & . & \\ 0 & & & & & 1 \\ 1 & 0 & . & . & . & 0 \end{bmatrix}.$$

Example 4.3. Let $\mathbf{C}_1$, $\mathbf{C}_2$,... , $\mathbf{C}_p$ be $m \times m$ circulant matrices over $\mathbb{F}_q$. Then the row space of the matrix

$$\mathbf{G} := [\mathbf{C}_1 \ \mathbf{C}_2 \ \cdots \ \mathbf{C}_p] \tag{8}$$

is a p-QC-code $\mathcal{C}$ with length mp and dimension $k \leq m$. This type of QC code is called by Séguin and Dralet [65] a 1-*generator QC-code*. The case $k = m$ is well researched, and in fact the older literature reserved the term quasi-cyclic for this type of codes. Note that for these codes the rate is $1/p$ and the rate of the dual is $(p-1)/p$.

From Definition 4.2. it is clear that the $m \times m$ circulant matrices constitute an algebra. In fact, this algebra is isomorphic to the algebra of polynomials $\mathbb{F}_q[x]/(x^m - 1)$. Let us identify a vector $\mathbf{c} = (c_0, c_1, \ldots, c_m) \in \mathbb{F}_q^m$ with the polynomial $c(x) := c_0 + c_1 x + \cdots + c_m x^{m-1}$. Then an isomorphism between the circulant algebra and $\mathbb{F}_q[x]/(x^m - 1)$ is given by

$$\mathbf{C} \mapsto C^1(x),$$

where C^1 is the first row vector of $\mathbf{C}$.

Let us go back to Example 4.3. and denote the polynomials corresponding to the circulant matrices $\mathbf{C}_i$ by $c_i(x)$. These polynomials are called the *defining polynomials* of the QC-code $\mathcal{C}$. The dimension of the code $\mathcal{C}$ can be determined in a very simple way. Following [65], we define the order of the 1-generator QC-code $\mathcal{C}$ to be the polynomial

$$h(x) := \frac{x^m - 1}{\gcd(x^m - 1, c_1(x), c_2(x), \dots c_p(x))}.$$

Then $\dim \mathcal{C} = \deg h(x)$.

A good description of QC-codes can be found in Greenough and Hill [20]. Special cases of QC-codes, i.e. QC-codes of rates $1/p$, $(m-1)/pm$, $(p-1)/p$ and $2/p$ were considered by many authors: van Tilborg [70], Gulliver [22, 28, 30], Gulliver and Bhargava [23, 24, 25, 26, 27, 29, 31, 32, 33, 34], Gulliver and Östergård [35, 36], Boukliev [4], Daskalov [10, 11], Daskalov and Gulliver [12].

In [24, 31, 34, 11], the authors considered the special case when $\gcd(x^m - 1, c_1(x), c_2(x), \dots, c_p(x)) = x - 1$ and found many good binary [24, 34], ternary and quaternary [31, 11] QC-codes. As a rule, good QC-codes typically are obtained if there are no cyclic conjugates among the defining polynomials $c_1(x), c_2(x), \dots, c_p(x)$.

In [48], Hill and Greenough introduced the concept of *quasi-twisted codes.* A constacyclic code (or α-twisted code) of length m (see [3]) is a linear code over $\mathbb{F}_q$ which is invariant under the transformation

$$\mathbf{c} \mapsto \mathbf{c}\mathbf{Q}_m,$$

where $\mathbf{Q}_m$ is the $m \times m$ matrix

$$\mathbf{Q}_m := \begin{bmatrix} 0 & 1 & . & . & & \\ 0 & & 1 & & & \\ . & & & . & & \\ . & & & & . & \\ 0 & & & & & 1 \\ \alpha & 0 & . & . & . & 0 \end{bmatrix} \tag{9}$$

for some given nonzero $\alpha \in \mathbb{F}_q$.

Definition 4.4. [48] A linear code of length $n = pm$ is called *p-quasi-twisted* (p-QT) if it equivalent to a code which is invariant under a transformation of the form

$$\mathbf{c} \mapsto \mathbf{c}(\mathbf{I}_p \otimes \mathbf{Q}_m).$$

In order to define a class of quasi-twisted codes corresponding to those of Example 4.3, we need the notion of twistulant matrices.

Definition 4.5. An $m \times m$ matrix $\mathbf{T}$ over $\mathbb{F}_q$ is said to be *α-twistulant* if $\mathbf{Q}^{-1}\mathbf{T}\mathbf{Q} = \mathbf{T}$, where $\mathbf{Q}$ is the matrix (9).

Example 4.6. Let $\mathbf{T}_1$, $\mathbf{T}_2$, ... , $\mathbf{T}_p$ be $m \times m$ α-twistulant matrices over $\mathbb{F}_q$. Then the row space of the matrix

$$\mathbf{G} := [\mathbf{T}_1 \ \mathbf{T}_2 \ \cdots \ \mathbf{T}_p]$$

is a p-QT-code $\mathcal{C}$ with length mp and dimension $k \leq m$.

The theory of quasi-twisted codes is similar to that of quasi-cyclic ones, because the algebra of the twistulant $m \times m$ matrices over $GF(q)$ is isomorphic to the algebra of polynomials $\mathbb{F}_q[x]/(x^m - \alpha)$.

To conclude this section, note that many of the papers on QC- and QT-codes contain results of computer searches. Two algorithms for searching good binary QC-codes were developed in [73]. This thesis presents also an overview on computer searches for QC-codes as well as tables of the best found binary QC-codes. It is worth to mention here some of the results. In [70], van Tilborg considered binary $[pk, k]$-QC-codes for small values of p and with dimension seven and eight up to code length 120. By a computer search he computed the best possible minimum distances of such codes. For $k = 7$, he found $[42, 7, 19]$, $[56, 7, 26]$, $[63, 7, 31]$, $[70, 7, 33]$, $[105, 7, 52]$, $[112, 7, 56]$ and $[119, 7, 59]$ QC-codes, which are optimal.

Chen, Peterson and Weldon [9] carried out an exhaustive search for the best possible rate 1/2 binary QC-codes up to code length 42. For k = 3, 4, 5, 8, 9, 10, 11, 12, 13, 14, 15 there are optimal QC-codes. For k = 18, 19, 20, 21 the QC-codes found in [9] are the best known.

References

[1] L.D. Baumert, R.J. McEliece, "A note on the Griesmer bound" *IEEE Trans. Inform. Theory* 19, 2, 1973, 134–135.

[2] B.I. Belov, "A conjecture on the Griesmer boundary", *Optimization methods and their applications All-Union Summer Sem., Khakusy, Lake Baikal, 1972 Russian,* 182. *Sibirsk. Energet. Inst. Sibirsk. Otdel. Akad. Nauk SSSR, Irkutsk,* 1974, 100–106.

[3] E.R. Berlekamp, *Algebraic coding theory,* McGraw-Hill Book Co., New York - Toronto, Ont. - London, 1968, xiv+466.

[4] I.G. Boukliev, "New bounds for the minimum length of quaternary linear codes of dimension five", *Discrete Math.* 169, no. 1-3, 1997, 185–192.

[5] A.E. Brouwer, T. Verhoeff, "An updated table of minimum-distance bounds for binary linear codes", *IEEE Trans. Inform. Theory* 39, no. 2, 1993, 662–676.

[6] A.E. Brouwer, M. van Eupen, "The correspondence between projective codes and 2-weight codes", *Des. Codes Cryptogr.* 11, no. 3, 1997, 262–266.

[7] A.E. Brouwer, "Bounds on the size of linear codes", *Handbook of Coding Theory,* eds. V. S. Pless and W. C. Huffman, Elsevier, Amsterdam etc., 1998, ISBN: 0-444-50088-X.

[8] A.R. Calderbank, W.M. Kantor, "The geometry of two-weight codes", *Bull. London Math. Soc.* 18, 1986, 97–122.

[9] C.L. Chen, W.W. Peterson, E.J. Jr. Weldon, "Some results on quasi-cyclic codes", *Information and Control* 15, 1969, 407–423.

[10] R.N. Daskalov, "Ten good quasi-cyclic 10-dimensional quaternary linear codes", *Proc. Int. Workshop on Optimal Codes and Related Topics, Sozopol, Bulgaria, May 26 - June 1,* 1995, 45–49.

[11] R.N. Daskalov, "Some good rate $m - 1/pm$ quaternary quasi-cyclic codes of dimension ten", *Mathematics and Education in Mathematics* , Sofia, 1996, 104–108.

[12] R.N. Daskalov, T.A. Gulliver, "New good quasi-cyclic ternary and quaternary linear codes", *IEEE Trans. Inform. Theory* 43, no. 5, 1997, 1647–1650.

[13] P. Delsarte, "Weights of linear codes and strongly regular normed spaces", *Discrete Math.* 3, 1972, 47–64.

[14] S.M. Dodunekov, "The minimum block length of a linear q-ary code with given dimension and code distance", *Problemy Peredachi Informatsii* 20, no. 4, 1984, 11–22.

[15] S.M. Dodunekov, " A note on the Griesmer bound", *C. R. Acad. Bulgare Sci.* 37, no. 9, 1984, 1177–1178.

[16] S.M. Dodunekov, ", N.L. Manev, An improvement of the Griesmer bound for some small minimum distances", *Discrete Appl. Math.* 12, no. 2, 1985, 103–114.

[17] S.M. Dodunekov, *Optimal linear codes*, Doctor Thesis, Sofia, 1985.

[18] S.M. Dodunekov, J. Simonis, "Codes and projective multisets", *Electron. J. Combin.* 5, 1998, no. 1, Research Paper 37, 23 pp. electronic.

[19] P.G. Farrell, "Linear binary anticodes", *Electron. Lett.* 6, 1970, 419–421.

[20] P.P. Greenough, R. Hill, "Optimal ternary quasi-cyclic codes", *Des. Codes Cryptogr.* 2, no. 1, 1992, 81–91.

[21] J. Griesmer, "A bound for error-correcting codes", *IBM J. Res. Develop.* 4, 1960, 532–542.

[22] T.A. Gulliver, *Construction of quasi-cyclic codes*, PhD Thesis, Univ. of Victoria, Canada, 1989.

[23] T.A. Gulliver, V.K. Bhargava, "Some best rate $1/p$ and rate $p - 1/p$ systematic quasi-cyclic codes", *IEEE Trans. Inform. Theory* 37, no. 3, 1991, part 1, 552–555.

[24] T.A. Gulliver, V.K. Bhargava, "Nine good rate $m - 1/pm$ quasi-cyclic codes", *IEEE Trans. Inform. Theory* 38, 1992, no. 4, 1366–1369.

[25] T.A. Gulliver, V.K. Bhargava, "Some best rate $1/p$ and rate $p-1/p$ systematic quasi-cyclic godes over $GF3$ and $GF4$", *IEEE Trans. Inform. Theory* 38, 1992, no. 4, 1369–1374.

[26] T.A. Gulliver, V.K. Bhargava, *New good rate $m - 1/mp$ ternary and quaternary quasi-cyclic codes*, Technical report SCE-93-18, Carlton University, 1993.

[27] T.A. Gulliver, V.K. Bhargava, "Two new rate $2/p$ binary quasi-cyclic codes", *IEEE Trans. Inform. Theory* 40, no. 5, 1994, 1667–1668.

[28] T.A. Gulliver, "New optimal ternary linear codes of dimension 6", *Ars Combin.* 40, 1995, 97–108.

[29] T.A. Gulliver, V.K. Bhargava, "An updated table of rate $1/p$ binary quasi-cyclic codes", *Appl. Math. Lett.* 8, no. 5, 1995, 81–86.

[30] T.A. Gulliver, "Two new optimal ternary two-weight codes and strongly regular graphs", *Discrete Math.* 149, no. 1–3, 1996, 83–92.

[31] T.A. Gulliver, V.K. Bhargava, "New good rate $m - 1/pm$ ternary and quaternary quasi-cyclic codes", *Des. Codes Cryptogr.* 7, 1996, no. 3, 223–233.

[32] T.A. Gulliver, V.K. Bhargava, "Some best rate $1/p$ quasi-cyclic codes over $GF5$", *Information theory and applications II. Proceedings of the fourth Canadian Workshop Lac Delaye, Québec, Canada, 1995,* 28–40, Lecture Notes in Comput. Sci., 1133, Springer, Berlin - New York, 1996.

[33] T.A. Gulliver, V.K. Bhargava, "Improvements to the bounds on optimal binary linear codes of dimensions 11 and 12", *Ars Combin.* 44, 1996, 173–181.

[34] T.A. Gulliver, V.K. Bhargava, "New optimal binary linear codes of dimensions 9 and 10", *IEEE Trans. Inform. Theory* 43, no. 1, 1997, 314–316.

[35] T.A. Gulliver, P.R.J. Östergård, "Improved bounds for ternary linear codes of dimension 7", *IEEE Trans. Inform. Theory* 43, 1997, 1377–1381.

[36] T.A. Gulliver, P.R.J. Östergård, "Improved bounds for quaternary linear codes of dimension 6", *Appl. Algebra Engrg. Comm. Comput.* 9, no. 2, 1998, 153–159.

[37] N. Hamada, F. Tamari, "Construction of optimal codes and optimal fractional factorial designs using linear programming", Combinatorial mathematics, optimal designs and their applications Proc. Sympos. Combin. Math. and Optimal Design, Colorado State Univ., Fort Collins, Colo., 1978. *Ann. Discrete Math.* 6, 1980, 175–188.

[38] N. Hamada, M. Deza, "A survey of recent works with respect to a characterization of an n, k, d, q-code meeting the Griesmer bound using a min-hyper in a finite projective geometry", *Discrete Math.* 77, no. 1-1, 1989, 75–87.

[39] N. Hamada, T. Helleseth, "A characterization of some linear codes over $GF4$ meeting the Griesmer bound", *Math. Japon.* 37, no. 2, 1992, 231–242.

[40] N. Hamada, T. Helleseth, O. Ytrehus, "A new class of nonbinary codes meeting the Griesmer bound", *Discrete Appl. Math.* 47, no. 3, 1993, 219–226.

[41] N. Hamada, "A survey of recent work on characterization of minihypers in PGt, q and nonbinary linear codes meeting the Griesmer bound", *J. Combin. Inform. System Sci.* 18, no. 3-4, 1993, 161–191.

[42] T. Helleseth, "A characterization of codes meeting the Griesmer bound", *Inform. and Control* 50, no. 2, 1981, 128–159.

[43] T. Helleseth, H.C.A. van Tilborg, "A new class of codes meeting the Griesmer bound", *IEEE Trans. Inform. Theory* 27, no. 5, 1981, 548–555.

[44] T. Helleseth, "New constructions of codes meeting the Griesmer bound", *IEEE Trans. Inform. Theory* 29, no. 3, 1983, 434–439.

[45] T. Helleseth, "Projective codes meeting the Griesmer bound", *Discrete Math.*, 106/107, 1992, 265–271.

[46] R. Hill, "Caps and codes", *Discrete Math.* 22, no. 2, 1978, 111–137.

[47] R. Hill, "Optimal linear codes", *Cryptography and coding, II Cirencester, 1989,* 75-104, Inst. Math. Appl. Conf. Ser. New Ser., 33, Oxford Univ. Press, New York, 1992.

[48] R. Hill, P.P. Greenough, "Optimal quasi-twisted codes", *Proc. Third Int. Workshop on Algebraic and Combinatorial Coding Theory, Voneshta Voda, Bulgaria, June 22-28,* 1992, 92–97.

[49] R. Hill, E. Kolev, "A survey of recent results on optimal linear codes", to appear in *Comb. Designs and their Appl.*.

[50] J. W.P. Hirschfeld, L. Storme, "The packing problem in statistics, coding theory and finite projective spaces", *J. Stat. Planning and Inference* , 72, 1998, 355–380.

[51] D. B. Jaffe, "Binary linear codes: new results on nonexistence", *Draft version accessible through the author's web page http://www.math.unl.edu/~djaffe,* 11/10/1997 Version 0.5. Dept. of Math. and Statistics, University of Nebraska, Lincoln.

[52] D. B. Jaffe, J. Simonis, "New binary linear codes which are dual transforms of good codes",to be published to *IEEE Trans. Inform. Theory.*

[53] T.A. Kasami, "Gilbert-Varshamov bound for quasi-cyclic codes of rate 1/2", *IEEE Trans. Inform. Theory* 20, 1974, 679.

[54] M. Krawtchouk, "Sur une généralisation des polynomes d'Hermite", *Comptes Rendus* 189, 1929, 620–622.

[55] G. Lachaud, M.A. Tsfasman, J. Justesen, V.K. Wei, "Special Issue on Algebraic Geometry Codes", *IEEE Trans. Inform. Theory* 41, 1975.

[56] I.N. Landgev, "Linear codes over finite fields and finite projective geometries" to appear in *Discrete Math.*

[57] V.N. Logachev, "An improvement of the Griesmer bound in the case of small code distances", *Optimization methods and their applications All-Union Summer Sem., Khakusy, Lake Baikal, 1972 Russian,* 182, *Sibirsk. Energet. Inst. Sibirsk. Otdel. Akad. Nauk SSSR, Irkutsk,* 1974, 107–111.

[58] V.N. Logachev, "A construction of a class of codes meeting the Varshamov-Griesmer bound", *Russian Modelling and optimization in large energy systems,* 116–120, *Sib. Otd. AN SSSR, Irkutsk,* 1975, 116–120.

[59] V.N. Logachev, "A construction of a class of optimal anticodes", *Proc. Eighth All-Union Conference on Coding Theory and Inf. Transmission. Abstracts, part 2,* 1981, *Moscow - Kuibyshev,* 95–97.

[60] V.N. Logachev, "New sufficient conditions for the existence of codes attaining the Varshamov-Griesmer bound", *Problemy Peredachi Informatsii* 22, no. 2, 1986, 3–26.

[61] V.N. Logachev, "Characterization and existence conditions for codes that meet the Varshamov-Griesmer bound", *Problems Inform. Transmission* 24, no. 3, 1988, 24–41, *translated from Problemy Peredachi Informatsii* 24, no. 3, 1988, 189-204 Russian.

[62] F.J. MacWilliams, N.J.A. Sloane, *The theory of error-correcting codes*, 2nd reprint, North-Holland Mathematical Library, Vol. 16, North-Holland Publishing Co., Amsterdam - New York - Oxford, 1983, xx+762 pp. ISBN: 0-444-85009-0 and 0-444-85010-4.

[63] W.W. Peterson, E.J. Jr. Weldon, *Error-correcting codes*, Second edition. The M.I.T. Press, Cambridge, Mass. - London, 1972. xi+560 pp.

[64] V.S. Pless, W.C. Huffman, *Handbook of Coding Theory*, Elsevier, Amsterdam, 1998, ISBN: 0-444-50088-X.

[65] G.E. Séguin, G. Drolet, *The theory of* 1*-generator quasicyclic codes*, Preprint, Royal Military College of Canada, Kingston, ON, June 1990

[66] F. Slepian, "A class of binary signaling alphabets", *Bell System Tech. J.* 35, 1956, 203–234.

[67] G. Solomon, J.J. Stiffler, "Algebraically punctured cyclic codes", *Inform. and Control* 8, 1965, 170–179.

[68] F. Tamari, "A construction of some $[n, k, d, q]$-codes meeting the Griesmer bound", *Discrete Math.* 116, 1993, 269–287.

[69] R.L. Townsend, E.J. Jr. Weldon, "Self-orthogonal quasi-cyclic codes", *IEEE Trans. Inform. Theory* 13, no. 2, 1967, 183–195.

[70] H.C.A. van Tilborg, "On quasi-cyclic codes with rate $1/m$", *IEEE Trans. Inform. Theory* 24, no. 5, 1978, 628–630.

[71] H.C.A. van Tilborg, "On the uniqueness resp. nonexistence of certain codes meeting the Griesmer bound", *Inform. and Control* 44, no. 1, 1980, 16–35.

[72] R.R. Varshamov, *Problems of the general theory of linear coding*, Extended abstract of a PhD Thesis, Moscow State University, 1959.

[73] S. Weijs, *A computer search for quasi-cyclic codes*, Master's thesis, Dept. Math. and Comp. Sci., Eindhoven Univ. Technol., Eindhoven, The Netherlands, 1997.

NEW APPLICATIONS AND RESULTS OF SUPERIMPOSED CODE THEORY ARISING FROM THE POTENTIALITIES OF MOLECULAR BIOLOGY

Arkadii G. D'yachkov, Anthony J. Macula and Vyacheslav V. Rykov

State University of New York, College at Geneseo,
Department of Mathematics, Geneseo, NY, 14454, USA.
dyachkov@nw.math.msu.su, macula@uno.cc.geneseo.edu, rykov@rvv.dnttm.ru

Abstract: Superimposed codes (SC) were introduced by Kautz-Singleton (1964) [1], who worked out the important constructive methods. Dyachkov-Rykov [2, 3, 4, 5, 6] and Erdos-Frankl-Furedi [7] obtained upper and lower bounds on the rate of SC. Dyachkov-Macula-Rykov [8, 9, 10, 11] investigated the development of constructions for SC (nonadaptive pooling designs) intended for the clone-library screening problem. (See Balding-Torney [12] and Knill-Bruno-Torney [13]). In this paper, we give an introduction to the problem and a detailed survey of our recent results on constructive methods of SC. We discuss superimposed distance codes and list-decoding superimposed codes.

APPLICATION TO DNA LIBRARY SCREENING

To understand what a DNA library is, think of *several copies* of an identical but incredibly long word (of length $\sim 10^8$, e.g., a chromosome) from letters of the quaternary alphabet $\{A, C, G, T\}$. Each copy of the word has been cut in thousands of contiguous pieces (of length $\sim 10^4$, e.g., chromosome fragments). Take those pieces and copy those letter strings onto their own separate small piece of paper. The thousands of little pieces of paper (i.e., clones) that result essentially constitute a DNA library. In other words, each *clone* represents some contiguous subpiece of a contiguous superpiece of DNA. The *DNA library*, or the *clone-library* consists of thousands separate clones.

I. Althöfer et al. (eds.), Numbers, Information and Complexity, 265-282.

A unique and contiguous sub-subpiece of DNA (of length $\sim 10^2$) is called a *sequenced tagged site* (STS). For a fixed STS, a clone is called positive (negative) for that STS if it contains (does not contain) that given STS.

Example. Let the following $s = 4$ copies of the DNA superpiece be given and $\{C_1, C_2, C_3, C_4, C_5\}$ be the library of 5 clones.

$$\overbrace{\boxed{\mathrm{AAA}}\,GCGTCT\boxed{\mathrm{TAA}}}^{C_1}\,\overbrace{CCGATAGGCAAC}^{C_3}\,TTG,$$

$$\boxed{\mathrm{AAA}}\,GC\,\overbrace{GTCT\boxed{\mathrm{TAA}}}^{C_2}\,CCGATAGGCAACTTG,$$

$$\boxed{\mathrm{AAA}}\,GCGT\,\overbrace{CT\boxed{\mathrm{TAA}}\,CCGATAGGC}^{C_4}\,AACTTG,$$

$$\boxed{\mathrm{AAA}}\,GCGTCT\boxed{\mathrm{TAA}}\,CCGAT\,\overbrace{AGGCAACTTG}^{C_5}.$$

Clones $\{C_1, C_3\}$ could be taken from the same copy of the DNA superpiece. Clones $\{C_2, C_4\}$ are taken from different copies. Let $\mathrm{STS}_1 = \boxed{\mathrm{AAA}}$ and $\mathrm{STS}_2 = \boxed{\mathrm{TAA}}$. Then C_1 is positive for $\boxed{\mathrm{AAA}}$ and C_1, C_2 and C_4 are positive for $\boxed{\mathrm{TAA}}$. Note that C_1 is positive for both $\boxed{\mathrm{AAA}}$ and $\boxed{\mathrm{TAA}}$. Clones C_3, C_5 are negative for both $\boxed{\mathrm{AAA}}$ and $\boxed{\mathrm{TAA}}$.

A *pool* is a subset of clones. Each pool is tested as a group by exposing that entire group to a chemical probe (e.g. polymerase chain reaction [12]) which can detect a given STS. A pool is called *positive for the STS* if the probe indicates that some member of that group contains the given STS. In other words, if the tests are error-free, then a pool is positive for an STS if that pool contains at least one clone that contains the given STS.

Let $1 \leq s < t$, $N > 1$ be integers. Mathematically, *clone-library screening* for positive clones is modeled by searching a t-set of objects (clone-library) for a particular p-subset, $p \leq s$, called a *subset of positive clones.* A nonadaptive pooling design is a series of N apriori group tests that often be carried out simultaneously. Every parallel pooling design is nonadaptive.

A *pool outcome* (result of the group testing) is said to be positive if one of the pool's clones is positive, negative otherwise. Using this binary N-sequence of outcomes, an investigator *has to identify* the p-subset, $p \leq s$, of positive clones.

When screening clone-libraries for positive clones, there are the following features [13, 14] which determine the cost of finding the positive clones.

1. The same library is screened with many different probes. Each probe is associated with the subset of clones which are positive for the unique STS that that probe detects.

2. It is expensive to prepare a pool for testing the first time, although once the pool is prepared, it can be screened many times with different probes.

3. Screening one pool at a time is expensive. Screening many pools in parallel with the same probe is cheaper.

4. It is common practice to individually test potential positive clones for confirmation. It means that clone-library screening consists of:

   (a) the first screening stage,

   (b) a confirmatory screening stage.

   These confirmatory tests can be relatively costly.

5. The screening results are not always reliable. Tests may be *false positive* ($\leq 7\%$), that is they identify a positive clone in a pool when there does not exist any. Similarly, tests may be *false negative* ($\leq 10\%$), that is, they fail to identify a positive clone in a pool that contains positive clones. Therefore, errors must be tolerated.

6. There are constraints on pool sizes. Pools can't be too large because if a pool containing a positive clone(s) contains too many other clones, then that pool can become too dilute and the probe may not be sensitive enough to detect the presence of the positive clone(s). This could lead to a positive pool being mislabeled as negative. This loss of information could result in some positive clones remaining unidentified.

The goal of our paper is to construct a class of efficient nonadaptive pooling designs and their two-stage modifications which give the possibility to identify any p-subset, $p \leq s$, of positive clones in a clone-library of size t. These designs (called superimposed codes) are based on the combinatorial and algebraic methods of Coding Theory [15]. We also investigate the error-correcting abilities of these pooling designs.

In Sect. 2, we introduce the definitions of superimposed codes which yield mathematical models for pooling designs.

In Sect. 3, we consider the constructive superimposed codes called *incidence matrix codes.* These codes give the optimal pooling designs for the Renyi (1965) [16] group testing model in which the *size of a testing group* (or the *size of a pool*) *is restricted.*

In Sect. 4, we discuss superimposed codes which are based on the q-ary Reed-Solomon codes (RS-codes) [15]. They were suggested by Kautz-Singleton [1]. We introduce some generalizations of the Kautz-Singleton codes and identify the parameters of the best known superimposed codes.

SUPERIMPOSED CODES & POOLING DESIGNS

Notations and definitions

We will use the terminology of combinatorial *coding theory* and the following collection of notations. Let

- $1 \leq s < t,\ 1 \leq k < t,\ N > 1$ be integers;

- t – code (clone-library) size, N – code length (number of pools);
- code (pooling design) $X = \|x_i(u)\|$, $i = 1, 2, \ldots, N$, $u = 1, 2, \ldots, t$, be a binary $(N \times t)$–matrix, $x_i(u) = 1$, if the u-th clone is in the i-th pool and $x_i(u) = 0$, otherwise;
- $\mathbf{x}(u) = (x_1(u), x_2(u), \ldots, x_N(u))$, $u = 1, 2, \ldots t$, be columns (codewords), and $\mathbf{x}_i = (x_i(1), x_i(2), \ldots, x_i(t))$, $i = 1, 2, \ldots N$, be rows (pools);
- $w = \min_u \sum_{i=1}^N x_i(u)$ be the minimal weight of codewords;
- $\lambda = \max_{u,v} \sum_{i=1}^N x_i(u) x_i(v)$ be the maximal dot product of codewords;
- $k = \max_i \sum_{u=1}^t x_i(u)$ be the maximal weight of rows;

We say that the binary column $\mathbf{x}$ *covers* binary column $\mathbf{y}$ if the boolean sum $\mathbf{x} \vee \mathbf{y} = \mathbf{x}$.

Definition 1 [1, 3, 17]. The code X is called a *superimposed* (s, N, t)-code, or *s-disjunct* code if the boolean sum of any s-subset of columns of X covers those and only those columns of X which are the terms of the given boolean sum.

Let $p \leq s$ be the number of positive clones in a clone-library of size t. To identify an unknown p-subset of positive clones, we apply the pooling design X which satisfies Definition 1, i.e. X is the superimposed (s, N, t)-code. Obviously, the binary N-sequence $\mathbf{y}$ of pool outcomes is the boolean sum of the unknown p-subset of columns of X. Definition 1 means that the unknown p-subset is represented by all columns which are below $\mathbf{y}$. Thus, we need to carry out $\leq t$ successive comparisons of the boolean sum $\mathbf{y}$ with codewords of X. Hence, the *identification complexity* of (s, N, t)-code does not exceed t.

Let $|\mathbf{x}| = \sum_{i=1}^N x_i$ denote the Hamming weight of a binary column $\mathbf{x} = (x_1, x_2, \ldots, x_N)$ and $\vee$ denote the boolean sum symbol. Define the value

$$\mathcal{D}(\mathbf{x}\|\mathbf{y}) \stackrel{\text{def}}{=} |\mathbf{x} \vee \mathbf{y}| - |\mathbf{y}|$$

which will be called a *superimposed distance* from a binary column $\mathbf{x}$ to a binary column $\mathbf{y}$. Note that $\mathcal{D}(\mathbf{x}\|\mathbf{y}) \neq \mathcal{D}(\mathbf{y}\|\mathbf{x})$.

Let $D = 1, 2, \ldots, N$ and $s = 1, 2, \ldots, t-1$ be arbitrary fixed integers and

$$\mathbf{u} = \{u_1, u_2, \ldots, u_s, u_{s+1}\}, \quad u_i \neq u_j, \quad u_i \in [t],$$

be any $(s+1)$-collection of integers.

Definition 2. The number

$$\mathcal{D}_s(X) \stackrel{\text{def}}{=} \min_{\mathbf{u}} \mathcal{D}\left(\mathbf{x}(u_{s+1})\| \bigvee_{i=1}^{s} \mathbf{x}(u_i)\right) =$$

$$\min_{\mathbf{u}} \left\{ \left| \mathbf{x}(u_{s+1}) \bigvee \left(\bigvee_{i=1}^{s} \mathbf{x}(u_i) \right) \right| - \left| \bigvee_{i=1}^{s} \mathbf{x}(u_i) \right| \right\},$$

is called a *superimposed s-distance* of code X.

Definition 3. Code X is called a superimposed $(s,\ N,\ t)$-code (or s-disjunct code) *with distance* D if $\mathcal{D}_s(X) = D$.

Remark 1. For the case $D = 1$, Definition 3 coincides with Definition 1.

Remark 2. Let the number of positives $p = s$. It is easy to understand [4] that superimposed $(s,\ N,\ t)$-code with distance D corrects any combination of $\leq D - 1$ errors distorting N-sequence of pool outcomes.

Lower bound

In this section, we discuss the lower bound on the length N of codes X intended for the group testing model of Renyi (1965) [16] in which the size of a testing group (or the size of a pool) is restricted. It means that the maximal row weight of X should be given.

Definition 4. Let $t > k$ and $N > D \geq 1$ be an arbitrary fixed integers. Code X is called a *superimposed* $(s,\ N,\ t)_k$*-code with distance* D*, or superimposed* (s, t, D, k)*-code*, if $\mathcal{D}_s(X) = D$ and the maximal row weight of X is equal to k.

The following proposition is a generalization of the corresponding bound for the case $D = 1$ from [9, 10, 11].

Proposition 1. *Let* $t > k$, $Dk \geq s + D$ *and* $N > 1$ *be integers.*

1. *For any superimposed* (s, t, D, k)*-code* X*, the length*

$$N \geq \left\lceil \frac{(s+D)t}{k} \right\rceil .$$

2. *If* $Dk \geq s+D+1$, $(s+D)t = kN$ *and there exists the optimal superimposed* (s, t, D, k)*-code* X *of length* $N = (s+D)t/k$*, then*

   (a) *code* X *is a constant-weight code of weight* $w = s + D$*, for any* $i = 1, 2, \ldots, N$*, the weight of row* $|\mathbf{x}_i| = k$ *and the maximal dot product* $\lambda = 1$;

   (b) *the following inequality is true*

$$k^2 - \frac{k(k-1)}{s+D} \leq t.$$

Proof. 1). Consider an arbitrary superimposed (s, t, D, k)-code X of length N. Let t', $0 \leq t' \leq t$ be the number of codewords of X having a weight $\leq s + D - 1$. From definition of (s, t, D, k)-code it follows that $t'D \leq N$ and the following inequality is true

$$(t - t')(s + D) \leq (N - t'D)k \iff t(s + D) \leq Nk - t'[Dk - (s + D)].$$

Since $Dk \geq s + D$, the statement 1 is proved.

2). The proof of statement 2 is based on the Johnson inequality [15]. The arguments are similar to those given in [9, 10, 11] for the case $D = 1$.

Denote by $N(s,t,D,k)$ the minimal possible length of a superimposed (s,t,k,D)-code. From Proposition 1 it follows

$$N(s,t,D,k) = \begin{cases} = Dt, & \text{if } Dk \le s+D, \\ \ge \frac{t(s+D)}{k}, & \text{if } Dk \ge s+D+1. \end{cases}$$

SUPERIMPOSED CODES BASED ON INCIDENCE MATRICES

Notations and definitions

Let $m \ge 2,\ l \ge 1,\ n \ge 2m+l+1$ be arbitrary integer, $[n] = \{1,2,\ldots,n\}$ be the set of integers from 1 to n and $\mathcal{E}(m,n)$ be the collection of all $\binom{n}{m}$ m-subsets of $[n]$.

Let $X = \|x_B(A)\|$, $B \in \mathcal{E}(m,n)$, $A \in \mathcal{E}(m+l,n)$, be the binary code, where

$$x_B(A) \stackrel{\text{def}}{=} 1 \quad \text{if and only if} \quad B \subset A.$$

This code will be called incidence matrix code (IM-code) with codewords (columns) $\mathbf{x}(A)$, $A \in \mathcal{E}(m+l,n)$. One can easily understand that IM-code $X = \|x_B(A)\|$ is the constant-weight code with parameters:

$$t = \binom{n}{m+l}, \quad N = \binom{n}{m}, \quad k = \binom{n-m}{l},$$

$$w = \binom{m+l}{m}, \quad \lambda = \binom{m+l-1}{m},$$

with t–code size, N–code length, $N < t$, w–weight of columns (codewords), k–weight of rows and λ–the maximal dot product of codewords.

List-Decoding Superimposed Codes & Two-Stage Screening Pooling Design

Let $A_1, A_2, \ldots, A_{m+1}$, where $A_i \in \mathcal{E}(m+l,n)$, be an arbitrary $(m+1)$-collection of $(m+l)$-subsets of the set $[n]$. Denote by $\mathbf{y} \stackrel{\text{def}}{=} \mathbf{x}(A_1) \vee \mathbf{x}(A_2) \vee \cdots \vee \mathbf{x}(A_{m+1})$ the boolean sum of the corresponding $(m+1)$ codewords of IM-code $\|x_B(A)\|$. Let $L(m,l) \ge m+1$ be the maximal possible number of codewords of $\|x_B(A)\|$ covered by $\mathbf{y}$. The detailed description of the function $L(m,l)$ was obtained by Vilenkin (1998) [18]. As a particular case of his result, we give the following important property of an IM-code $\|x_B(A)\|$.

Proposition 1. *If* $1 \le l \le m$, *then* $L(m,l) - (m+1) = 2l$.

Let $1 \le l \le m$. ¿From Proposition 1 it follows that the Boolean sum of any $(m+1)$-subset of codewords of X can cover not more than $2l$ codewords that are not components of the $(m+1)$-subset. This yields the possibility to apply IM-code X as the pooling design at the first screening stage. If $s = m+1$,

$l \leq m$ and the number of positive clones $p \leq s = m+1$, then $\leq s+2l$ candidates are confirmed individually in a confirmatory screening stage.

Using the terminology of superimposed codes [5], IM-code $X = \|x_B(A)\|$ is called the *list-decoding superimposed code* of size $t = \binom{n}{m+l}$, length $N = \binom{n}{m}$, strength $s = m+1$, constraint $k = \binom{n-m}{l}$ and list-size $L = 2l$, $l \leq m$.

Example. Let $m = 2$, $l = 1$, $n = 16$. We have

$$t = 560,\ N = 120,\ s = 3,\ k = 14,\ L = 2.$$

Hence, if the number of positive clones $p \leq 3$, then the two-stage list decoding algorithm needs to carry out $\leq 120 + 5 = 125$ pools. On the other hand, Proposition 1 from Sect. 2 says that for a one-stage algorithm with $D = 1$, we need at least $(s+1)t/k = 4 \cdot 560/14 = 160$ pools.

Superimposed s*-distance of IM-codes*

Let $m \geq 2$, $l \geq 1$, $n \geq 2m + l + 1$ be fixed parameters of an IM-code X.

Proposition 2. *For any* s, $1 \leq s \leq m$, *the superimposed* s*-distance of* an IM-*code* is

$$\mathcal{D}_s(X) = \binom{l+m-s}{m-s}.$$

Proof. 1.) Let $A_0, A_1, \ldots, A_s$, where $A_i \in \mathcal{E}(m+l, n)$, be an arbitrary $(s+1)$-collection of different $(m+l)$-subsets. Since $A_0 \neq A_i$, for any $i = 1, 2 \ldots, s$, there exists an element $a_i \in A_0$ and $a_i \notin A_i$. Hence, there exists an s-subset $B = \{a_1, a_2, \ldots, a_s\}$, $B \subset A_0$ and for any $i = 1, 2, \ldots, s$, $B \not\subset A_i$.

Consider the $(l+m-s)$-set $A_0 \setminus B$. There exist $\binom{l+m-s}{m-s}$ distinct $(m-s)$-subsets of $A_0 \setminus B$. Hence, for any $i = 1, 2 \ldots, s$, there exist at least $\binom{l+m-s}{m-s}$ distinct m-subsets of A_0 which do not belong to A_i. It means that

$$\mathcal{D}_s(X) \geq \binom{l+m-s}{m-s}. \tag{1}$$

2.) Now we show that the lower bound (1) is true with the sign of equality. Let $A = \{a_0, a_1, \ldots, a_s\}$ be an arbitrary $(s+1)$-subset and B be an $(m+l-s)$-subset, $B \bigcap A = \emptyset$. Consider the collection of $(m+l)$-subsets $A_0, A_1, \ldots, A_s$, where $A_i = B \bigcup A \setminus \{a_i\}$, $i = 0, 1, \ldots, s$. They have the following form:

$$\begin{array}{l} A_0 = \{a_1, a_2, a_3, \ldots, a_s\} \bigcup B, \\ A_1 = \{a_0, a_2, a_3, \ldots, a_s\} \bigcup B, \\ A_2 = \{a_0, a_1, a_3, \ldots, a_s\} \bigcup B, \\ \ldots\ldots\ldots\ldots\ldots\ldots\ldots\ldots \\ A_s = \{a_0, a_1, a_2, \ldots, a_{s-1}\} \bigcup B. \end{array}$$

Note that $A_0 \setminus A_i = \{a_i\}$.

Lemma. *Let* $C \in \mathcal{E}(m, n)$ *be an arbitrary* m*-subset of the set* $[n]$. *Then* $C \in A_0$ *and for any* $i = 1, \ldots, s$, $C \notin A_i$ *if and only if* C *has the following* *form*

$$C = \{a_1, \dots, a_s\} \bigcup B', \tag{2}$$

where B' is an $(m-s)$-subset of B.

Proof of lemma. One can see that if C has the form (2), then $C \subset A_0$ and $C \not\subset A_i$ for any i. Conversely, if $C \subset A_0$ and $C \not\subset A_i$, then C intersects with $A_0 \backslash A_i = \{a_i\}$. Since it is true for any $i = 1, \dots, s$, then C has the form (2) and the lemma is proved.

The superimposed distance $\mathcal{D}\left(A_0 \| A_1 \bigvee \cdots \bigvee A_s\right)$ is the number of subsets C having the form (2). This value is equal to the number of different $(m-s)$-subsets B' of the set B, i.e. $\left|\mathcal{E}(m-s, m+l-s)\right| = \binom{m+l-s}{m-s}$. Hence, the definition of superimposed s-distance and inequality (1) yield the statement of Proposition 2.

For the case $l = 1$, we have $w = m+1$, $\lambda = 1$, $k = n-m$ and the superimposed s-distance $D = \mathcal{D}_s(X) = m-s+1$. Thus, the parameters (m, n) of the IM-code could be written in the form $m = D + s - 1$, $n = k + m = k + D + s - 1$. Therefore, from Proposition 1 of Sect. 2 it follows

Proposition 3. *Let $s \geq 2$, $D \geq 1$, $k \geq s + D + 1$ be fixed integers. Then the optimal length*

$$N\left(s, \binom{k+D+s-1}{s+D}, D, k\right) = \binom{k+D+s-1}{s+D-1}.$$

Superimposed s-distance of generalized IM-codes

We need the following notations. Let

$$2 \leq m < w < n,\ 0 \leq \lambda < w,\ d = w - \lambda, \quad \binom{n}{m} < t \leq \binom{n}{w}$$

be integers and let there exist a t-family $\mathcal{K} = \{K_1, K_2, \dots, K_t\}$ of subsets of $[n]$, where

$$|K_u| = w,\ K_u \subset [n],\ u = 1, 2, \dots, t,$$

and

$$\max_{u \neq v} |K_u \cap K_v| = \lambda, \quad \min_{u \neq v} |K_u \setminus K_v| = d = w - \lambda.$$

Let $X = \|x_B(u)\|$, $B \in \mathcal{E}(m, n)$, $u = 1, 2, \dots, t$, be the binary code of size t and length $N = \binom{n}{m}$, where an element $x_B(u) \overset{\text{def}}{=} 1$ if and only if $B \subset K_u$.

The binary code X will be called a *generalized* IM-code.

Proposition 4. *The following statements are true.*

1. *For any s, $2 \leq s \leq \min\{m, d\}$, the superimposed s-distance of a generalized IM-code*

$$\mathcal{D}_s(X) \geq \left\lceil \frac{\binom{d}{s} \cdot \binom{w-s}{m-s}}{\binom{m}{s}} \right\rceil. \tag{3}$$

2. *For $s = m = 2$, lower bound* (3) *could be improved and the following inequality is true*

$$\mathcal{D}_2(X) \geq \begin{cases} d^2 - \frac{(d-\lambda)(d-\lambda+1)}{2}, & \text{if } \lambda \leq d; \\ d^2, & \text{if } \lambda \geq d. \end{cases} \tag{4}$$

Proof. 1). Denote by $D_s(d, m, w)$ the right-hand side of (3). Let (without loss of generality) $K_1, K_2, \ldots, K_s, K_{s+1}$ be an arbitrary fixed $(s+1)$-collection of elements of t-family $\mathcal{K}$. We need to check that there exist at least $D \stackrel{\text{def}}{=} D_s(d, m, w)$ different subsets $B_1, B_2, \ldots, B_D$ of the set $[n]$ where for any $i = 1, 2, \ldots, D$, the following conditions take place

$$|B_i| = m, \; B_i \subset K_{s+1}, \; B_i \not\subset K_u, \; u = 1, 2, \ldots, s. \tag{5}$$

It is easy to see that $D_s(d, m, w)$ could be written in the form

$$D_s(d, m, w) =$$
$$\left\lceil \frac{1}{m!} \cdot d(d-1)\cdots(d-s+1) \cdot [w-s][w-s-1]\cdots[w-s-(m-s)+1] \right\rceil .$$

This implies the existence of at least $D = D_s(d, m, w)$ different ways to choose m-sets $B = B_i, \quad i = 1, 2, \ldots, D$, satisfying (5), in the following form

$$B = (a_1, a_2, \ldots, a_s, b_1, b_2, \ldots, b_{m-s}),$$

where

$$a_1 \in (K_{s+1} \setminus K_1), \quad a_u \in (K_{s+1} \setminus K_u) \setminus \{a_1, a_2, \ldots, a_{u-1}\}, \; u = 2, 3, \ldots, s,$$

$$b_1 \in K_{s+1} \setminus \{a_1, a_2, \ldots, a_s\}, \quad b_j \in K_{s+1} \setminus \{a_1, a_2, \ldots, a_s, b_1, b_2, \ldots, b_{j-1}\},$$
$$j = 2, 3, \ldots, m-s.$$

2). Let (without loss of generality) K_1, K_2, K_3 be an arbitrary fixed triple of elements of t-family $\mathcal{K}$. Using the standard notation of the complement of a set, define nonintersecting subsets

$$P_3 = K_3 \bar{K}_1 \bar{K}_2, \; P_1 = K_3 K_1 \bar{K}_2, \; P_2 = K_3 \bar{K}_1 K_2.$$

We have

$$|K_3 \setminus K_2| = |P_3| + |P_1| \geq d, \; |K_3 \setminus K_1| = |P_3| + |P_2| \geq d.$$

It is not difficult to see that for each $i = 1, 2$, there exist

$$\binom{|P_3|}{2} + |P_3| \cdot (w - |P_3|) + |P_1| \cdot |P_2|$$

distinct 2-subsets of K_3 which do not belong to K_i. Hence, the superimposed 2-distance

$$\mathcal{D}_2(X) \geq \min_{0 \leq v \leq d} \left\{ \binom{v}{2} + v(w-v) + (d-v)^2 \right\}.$$

Let $\lambda \overset{\text{def}}{=} w-d$. If $\lambda \le d$, then the minimum is achieved at $v = d-\lambda$. If $\lambda \ge d$, then the minimum is achieved at $v = 0$. The corresponding minimal values are given by the right-hand side of (4).

Corollary. *Let*

$$f_d = \frac{d}{n}, \quad f_w = \frac{w}{n}, \quad f_D = \frac{\mathcal{D}_s(X)}{N}, \, N = \binom{n}{m}$$

be the corresponding parameter fractions for a t-family (constant weight code) $\mathcal{K} = \{K_1, K_2, \ldots, K_t\}$ and code X. The inequality (3) *yields*

$$f_D \ge \left(\frac{f_d}{f_w}\right)^s \cdot (f_w)^m = (f_d)^s (f_w)^{m-s}, \quad 2 \le s \le m.$$

Open problems

1. For IM-code $X = \|x_B(A)\|$, find the maximal possible number of codewords covered by any fixed $(m+2)$-collection of codewords of X.
2. Let $m = 3$, $2 \le s \le 3$. Is it possible to improve lower bound (3)?
3. Let $s = m = 2$. Do there exist any "nontrivial" t-families (constant weight codes) $\mathcal{K} = \{K_1, K_2, \ldots, K_t\}$ for which the lower bound (4) is achieved?

SUPERIMPOSED CODES BASED ON REED-SOLOMON CODES

Generalized Kautz-Singleton codes

Let $\mathcal{P}$ be the set of all primes or prime powers ≥ 2, i.e.,

$$\mathcal{P} \overset{\text{def}}{=} \{2, 3, 4, 5, 7, 8, 9, 11, 13, 16, 17, 19, 23, 25, 27, 29, 31, 32, 37, \ldots\}.$$

Let $q_0 \in \mathcal{P}$ and $2 \le k_0 \le q_0 + 1$ be fixed integers for which there exists the q_0-ary Reed-Solomon code (RS-code) B of size $q_0^{k_0}$, length (q_0+1) and the Hamming distance $d_0 = q_0 - k_0 + 2 = (q_0+1) - (k_0-1)$ [15]. We will identify the code B with a $\left((q_0+1) \times q_0^{k_0}\right)$–matrix whose columns, (i.e., (q_0+1)-sequences from the alphabet $\{0, 1, 2, \ldots, q_0-1\}$) are the codewords of B. Therefore, the maximal possible number of positions (rows) where its two codewords (columns) can coincide, called a *coincidence* of code B, is equal to $k_0 - 1$.

Fix an arbitrary integer $r = 0, 1, 2, \ldots, k_0 - 1$ and introduce the *shortened* RS-code $\tilde{B}$ of size $t = q_0^{k_0-r}$, length $n_0 = q_0 + 1 - r$ that has the same Hamming distance $d_0 = q_0 - k_0 + 2$. Code $\tilde{B}$ is obtained by the *shortening* of the *subcode* of B which contains $0's$ in the first r positions (rows) of B. Obviously, the coincidence of $\tilde{B}$ is equal to

$$\lambda_0 \overset{\text{def}}{=} n_0 - d_0 = (q_0 + 1 - r) - d_0 = q_0 + 1 - r - (q_0 - k_0 + 2) = k_0 - r - 1. \quad (1)$$

Consider the following standard transformation of the q_0-ary code $\tilde{B}$, when each symbol of the q_0-ary alphabet $\{0, 1, 2, \ldots, q_0 - 1\}$ is substituted for the corresponding binary column of length q_0 and weight 1, namely:

$$0 \Leftrightarrow \underbrace{(1,0,0,\ldots,0)}_{q_0}, \quad 1 \Leftrightarrow \underbrace{(0,1,0,\ldots,0)}_{q_0}, \quad \ldots \quad q_0 - 1 \Leftrightarrow \underbrace{(0,0,0,\ldots,1)}_{q_0}.$$

As a result we have the binary constant-weight code X of size t, length N and weight w, where

$$t = q_0^{k_0 - r}, \quad N = n_0 q_0 = (q_0 + 1 - r) q_0, \quad w = n_0 = q_0 + 1 - r.$$

¿From (1) it follows that for the obtained binary code X, the maximal dot product of codewords is $\lambda = \lambda_0 = k_0 - r - 1$.

Let X be a binary code with parameters w and λ. Kautz-Singleton [1] suggested the following evident sufficient condition of the s-disjunct property: $s\lambda \leq w - 1$. Hence, by virtue of (1), code X is the s-disjunct code if

$$s(k_0 - r - 1) = s\lambda_0 \leq w - 1 = n_0 - 1 = q_0 - r.$$

For the particular case $r = 0$, this construction of s-disjunct codes was given in [1].

Let $m \geq 1$ and $2 \leq s < 2^m$ be arbitrary fixed integers. We look for the parameters q_0, k_0 and r yielding the s-disjunct code X of size t, $2^m \leq t < 2^{m+1}$, having the minimal possible length N. In paper [19], we proved

Proposition 1. *If there exists the solution of the given extreme problem, then the optimal parameters q_0, k_0, r and N are connected by the following formulas*:

$$q_0 \geq s\lambda_0, \quad \textit{where} \quad \lambda_0 \stackrel{\text{def}}{=} \left\lceil \frac{m}{\log_2 q_0} \right\rceil - 1, \tag{2}$$

$$r = q_0 - s\lambda_0 \geq 0, \quad k_0 = r + \lambda_0 + 1, \tag{3}$$

$$n_0 = q_0 + 1 - r = s\lambda_0 + 1, \tag{4}$$

$$w = n_0 = s\lambda_0 + 1, \quad N = q_0(s\lambda_0 + 1), \quad t = q_0^{\lambda_0 + 1}, \quad 2^m \leq t < 2^{m+1}. \tag{5}$$

Table 1, which was computed in [19], gives the numerical values of the optimal parameters q_0, λ_0 and N, when $s = 2, 3, \ldots, 7$, $m = 5, 6, \ldots, 20$.

Example. For the case $s = 3$, $m = 10$, Table 1 gives $q_0 = 11$, $\lambda_0 = 2$, $N = 77$. It means that there exists 3-disjunct constant-weight code with

$$\lambda = \lambda_0 = 2, \quad w = s\lambda_0 + 1 = 7, \quad t = q_0^{\lambda_0 + 1} = 11^3 = 1331, \quad N = 11 \cdot 7 = 77.$$

This code is obtained from shortened RS-code with $q_0 = 11$, $k_0 = 7$ and $r = 4$.

Remark 1. If $\lambda = \lambda_0 = 1$, then the length N of the corresponding code from Table 1 coincides (for the case $D = 1$) with the lower bound from Sect. 2.2.

Remark 2. In Table 1, we marked by boldface type the example of the superimposed code parameters which were known from [1, 3].

Table 1 Parameters of constant-weight (s, N, t)-codes of strength s, $2 \le s \le 7$, length N, size t, $2^m \le t < 2^{m+1}$, $5 \le m \le 20$, based on the q_0-ary shortened Reed-Solomon codes.

s	2	3	4	5	6	7
m	q_0, λ_0, N	q_0, λ_0, N	q_0, λ_0, N	q_0, λ_0, N	q_0, λ_0, N	q_0, λ_0, N
5	–	7, 1, 28	7, 1, 35	7, 1, 42	7, 1, 49	
6	4, 2, 20	8, 1, 32	8, 1, 40	8, 1, 48	8, 1, 56	9, 1, 72
7	–	–	13, 1, 65	13, 1, 78	13, 1, 91	13, 1, 104
8	7, 2, 35	7, 2, 49	–	16, 1, 96	16, 1, 112	16, 1, 128
9	8, 2, 40	8, 2, 56	8, 2, 72	–	23, 1, 161	23, 1, 184
10	–	11, 2, 77	11, 2, 99	11, 2, 121	–	–
11	7, 3, 49	–	13, 2, 117	13, 2, 143	13, 2, 169	–
12	8, 3, 56	9, 3, 90	16, 2, 144	16, 2, 176	16, 2, 208	16, 2, 240
13	–	11, 3, 110	–	23, 2, 253	23, 2, 299	23, 2, 345
14	–	13, 3, 130	13, 3, 169	–	27, 2, 351	27, 2, 405
15	8, 4, 72	–	–	–	–	32, 2, 480
16	–	16, 3, 160	16, 3, 208	16, 3, 256	19, 3, 361	–
17	11, 4, 99	–	–	–	–	–
18	13, 4, 117	13, 4, 169	–	23, 3, 368	23, 3, 437	23, 3, 506
19	–	–	–	27, 3, 432	27, 3, 513	27, 3, 594
20	11, 5, 121	16, 4, 208	**16, 4, 272**	–	32, 3, 608	32, 3, 704

Superimposed concatenated codes

A further extention [19] of the Kautz-Singleton superimposed (s, N, t)-codes is based on the following *concatenated codes.*

Consider the q_0-ary shortened Reed-Solomon code with parameters (2)-(4), where q_0 is a prime power. Let q_0-ary *symbols* of this code *be substituted, i.e., be coded, for the binary codewords of a known constant-weight s-disjunct code of size* $q' \ge q_0$, *length* $q \le q_0$ *and weight* $w' < q$. Denote this binary superimposed code as an (s, q, q')-code. *We will apply* (s, q, q')*-codes which are the standard binary constant-weight* (n, d, w')*-codes of size* $A(n, d, w') = q'$, *length* $n = q$, *weight* $w' = s\lambda' + 1$, *the Hamming distance* $d = 2(w' - \lambda')$ *and the maximal dot product* λ'. Proposition 1 can be generalized as follows.

Proposition 2 [19]. *The given substitution yields the concatenated code which is the binary constant-weight superimposed* $(s,\ qn_0,\ q_0^{\lambda_0+1})$*-code of weight* $w = w'n_0$.

In Proposition 1, we used only the *trivial* substitution, where $q_0 = q = q'$ and $w' = 1$.

Remark 3. If we apply the trivial substitution $q_0 = q = q'$, i.e., $w' = 1$, then we obtain the *concatenated code* which *is a standard constant-weight* $(N,\ d,\ w)$*-code of size* t, where

$$N = q_0 n_0, \quad w = n_0, \quad t = q_0^{k_0 - r}, \quad d = 2(n - \lambda_0) = 2(q_0 - k_0 + 2) = 2d_0.$$

Remark 4. If $q < q_0$ and $w' > 1$, then one knows only the weight $w = w'n_0$ of the constant-weight concatenated code. We cannot identify its distance d and the maximal dot product $\lambda \geq w'\lambda_0$.

Let $d = 2, 4, 6, \ldots$, $d \leq n$ and $w \leq n$ be arbitrary integers. Denote by $A(n, d, w)$ the *maximal size* (known up to now) *of constant-weight binary code* of length n, distance d and weight w. The tables of $A(n, d, w)$ called Standard Tables (ST) are available [20] and:

http://www.research.att.com/~njas/codes/Andw/index.html

On the base of Standard Tables, we calculated [19] the numerical values of optimal parameters for superimposed concatenated (s, N, t)-codes, when $s = 2$ and $s = 3$.

Superimposed s*-distance for concatenated codes*

Let $s \geq 2$, $m \geq 1$ and $D \geq 1$ be arbitrary fixed integers and we look for a binary code X whose superimposed s-distance $\mathcal{D}_s(X) \geq D$ and size t, $2^m \leq t < 2^{m+1}$.

Parameters of s-distance superimposed codes. It is easy to understand that such binary code X can be constructed on the base of the q_0-ary shortened RS-codes if the following generalizations of (2)-(5) are true

$$q_0 \geq s\lambda_0 + (D-1), \quad \text{where} \quad \lambda_0 \stackrel{\text{def}}{=} \left\lceil \frac{m}{\log_2 q_0} \right\rceil - 1, \tag{6}$$

$$\begin{aligned} k_0 &\stackrel{\text{def}}{=} q_0 + s - (s-1)\left\lceil \frac{m}{\log_2 q_0} \right\rceil - (D-1), \\ r &\stackrel{\text{def}}{=} k_0 - \left\lceil \frac{m}{\log_2 q_0} \right\rceil = q_0 - s\lambda_0 - (D-1) \geq 0, \end{aligned} \tag{7}$$

$$n_0 \stackrel{\text{def}}{=} q_0 + 1 - r = s\lambda_0 + D, \quad t \stackrel{\text{def}}{=} q_0^{k_0 - r} = q_0^{\lambda_0+1}, \quad 2^m \leq t < 2^{m+1} \tag{8}$$

In addition, if there exists an (s, q, q')-code, where $q \leq q_0 \leq q'$, then the code X has the length

$$N = q[s\lambda_0 + D] = q[(s\lambda_0 + 1) + (D-1)]. \tag{9}$$

It is known [4] that X corrects any combination of $\leq D-1$ errors distorting the boolean sum of s codewords. Let f, $0 < f < 1/q$, be the *error-correction fraction* of X. We have

$$f \leq \frac{D-1}{N} = \frac{D-1}{q[(s\lambda_0+1)+(D-1)]} \iff D-1 \geq \frac{fq}{1-fq}(s\lambda_0+1).$$

Hence, (6) gives the following *upper bound* on the error-correction fraction of X:

$$q_0 - s\lambda_0 \geq \frac{fq}{1-fq}(s\lambda_0+1) \iff f \leq f_0 \stackrel{\text{def}}{=} \frac{q_0 - s\lambda_0}{q(q_0+1)}. \tag{10}$$

We can summarize as follows.

Proposition 3. *Consider the class $C_f(s, m)$ of codes which have the given fixed error-correction fraction f, $0 < f \leq f_0$, where f_0 is defined by (10). For an arbitrary code X from $C_f(s, m)$, the minimal possible length N_f and the maximal possible rate $R_f \overset{def}{=} m/N_f$ are defined by formulas*

$$N_f = q[s\lambda_0 + D_f], \quad R_f = \frac{m}{q[s\lambda_0 + D_f]}, \quad \text{where } D_f \overset{def}{=} 1 + \left\lceil \frac{fq}{1 - fq}(s\lambda_0 + 1) \right\rceil$$

The tight upper bound on the rate R_f takes place

$$R_f \leq \bar{R}_f = \frac{m(1 - fq)}{q(s\lambda + 1)}, \quad \text{where} \quad \frac{m}{q(1 + q_0)} \leq \bar{R}_f \leq \frac{m}{q(1 + s\lambda_0)}, \; f_0 \geq f \geq 0.$$

Superimposed 2-distance for concatenated codes. Let there exist a constant-weight $(2, q, q')$-code of weight w', $m \geq 3$ be an arbitrary fixed integer and the RS-code base q_0 satisfy the following conditions

$$q \leq q_0 \leq q' \quad q_0 \in \mathcal{P}, \quad 2 \leq k_0 \leq q_0 + 1,$$

$$q_0 > 2\lambda_0, \quad \text{where} \quad \lambda_0 \overset{\text{def}}{=} \left\lceil \frac{m}{\log_2 q_0} \right\rceil - 1.$$

In formulas (6)-(9), we assign $r = 0$, $n_0 = q_0 + 1$, $k_0 = \lambda_0 + 1$ and obtain a constant-weight concatenated $(2, N, t)$-code X whose length N, superimposed 2-distance D, weight w, size t, error-correction fraction f and code rate R_f are defined as follows

$$N = q(q_0 + 1), \quad D = q_0 - 2\lambda_0 + 1, \quad w = w'(q_0 + 1), \tag{11}$$

$$t = q_0^{\lambda_0 + 1}, \quad 2^m \leq t < 2^{m+1}, \tag{12}$$

$$f = \frac{D - 1}{N} = \frac{q_0 - 2\lambda_0}{q(q_0 + 1)}, \quad R_f = \frac{m}{q(q_0 + 1)} = f \cdot \frac{m}{q_0 - 2\lambda_0}. \tag{13}$$

For several codes X, numerical values (11)-(13) are are given in Table 2.

Two last rows of Table 2 contain the values of the maximal possible random coding rate $R_2^{ran}(f)$ and the corresponding optimal random weight fraction $Q_2^{ran}(f) = w^{ran}/N$ [4].

The comparison shows that the *rate R_f of the given concatenated code exceeds the random coding rate $R_2^{ran}(f)$, if $0 < f < .065$.*

List-decoding characteristics of generalized Kautz-Singleton codes

Let the *random* p-collection $1 \leq p \leq t - 1$ of positives has the uniform distribution on the $\binom{t}{p}$-set of all p-subsets of the set $[t]$.

Table 2 Parameters of constant-weight concatenated $(2, N, t)$-codes of weight w, length N and size t, $2^m \leq t < 2^{m+1}$, $10 \leq m \leq 18$, with superimposed 2-distance $\mathcal{D}_2(X) = D$

	Auxiliary parameters								
q_0	7	8	11	9	13	11	13	17	11
λ_0	3	3	4	3	4	3	3	3	2
q	7	8	9	9	10	9	10	11	9
w'	1	1	3	1	3	3	3	3	3
q'	7	8	12	9	13	12	13	17	12
	Parameters of superimposed 2-distance codes								
N	56	72	108	90	140	108	140	198	108
w	8	9	36	10	42	36	42	54	36
D	2	3	4	4	6	6	8	12	8
m	11	12	17	13	18	13	14	16	10
t	7^4	8^4	11^5	9^4	13^5	11^4	13^4	17^4	11^3
f	.0179	.0278		.0333	.0357	.0463	.05	.0555	.0648
R_f	.1964	.1667	.1574	.1333	.1286	.1203	.10	.0808	.0926
$R_2^{ran}(f)$	.1251	.1031		.094	.0880	.0703	.0648	.0571	.0452
$Q_2^{ran}(f)$	.272	.277		.279	.282	.287	.289	.292	.297

To identify the p-collection, we use the constant-weight (s, N, t)-code X of strength s, weight w, length N, size t, $2^m \leq t < 2^{m+1}$, $m = 5, 6, \ldots$ and the maximal dot product λ, based on q_0-ary shortened RS-codes with parametes (2)-(5). For the given code X having parameters (s, q_0, k_0, r), denote by $\mathcal{L}(p)$ the *average number of extra codewords*, i.e., the average value of the list-decoding size, covered by the boolean sum of the corresponding random p-collection of codewords of X. Obviously $\mathcal{L}(p) = 0$, if $p \leq s$ and one can prove [22] that $1 \leq \mathcal{L}(p) \leq t - p$, if $p \geq s + 1$.

Let us apply code X of length N as the pooling design at the first screening stage. Then $p + \mathcal{L}(p)$ is the average number of potential positives which are confirmed individually in the second confirmatory screening stage. Therefore, the number $N + p + \mathcal{L}(p)$ is the average length of the two-stage screening pooling design, based on the shortened RS-codes.

To simplify the subsequent notations, we define the new parameter

$$K \stackrel{\text{def}}{=} k_0 - r = \lambda_0 + 1, \quad K \geq 1,$$

and consider the shortened RS-code $\tilde{B}$ as an q_0-ary *maximum-distance separable* code (MDS-code) [15, 21], which is identified by its length n_0, $K < n_0 \leq q_0 + 1$, size $t = q_0^K$ and coincidence $\lambda_0 = K - 1$. Formulas (2)-(5) take the form

$$q_0 \geq s\lambda_0 = s(K-1), n_0 = s\lambda_0 + 1, \quad 2^m \leq t = q_0^K < 2^{m+1}, \quad K \stackrel{\text{def}}{=} \left\lceil \frac{m}{\log_2 q_0} \right\rceil.$$

$$\lambda = \lambda_0 = K - 1, \qquad w = n_0 = s\lambda + 1, \qquad N = q_0 n_0 = q_0(s\lambda_0 + 1).$$

Hence, for an arbitrary fixed integer p, $s+1 \leq p \leq t = q_0^K$, the average value $\mathcal{L}(p)$ depends also on the MDS-code parameters (n_0, q_0, K) and

$$\mathcal{L}(p) = q_0^K \frac{\binom{q_0^K - 1}{p} - C(n_0, p, q_0, K)}{\binom{q_0^K}{p}},$$

$$C(n_0, p, q_0, K) = \sum_{v=1}^{n_0} (-1)^{v+1} \binom{n_0}{v} D_v(p, q_0, K)$$

$$D_v(p, q_0, K) = \begin{cases} \binom{q_0^{K-v}(q_0-1)^v}{p}, & \text{if } v \leq K, \\ \binom{A_v(q_0, K)}{p}, & \text{if } v \geq K+1, \end{cases}$$

$$A_v(q_0, K) = (q_0 - 1) \sum_{j=0}^{K-1} (-1)^j \binom{v-1}{j} q_0^{K-j-1}.$$

These formulas are obtained in [22].

For a given threshold $L \geq 0$, define the *averaged list-of-L decoding strength* $S(L)$:

$$S(L) \Leftrightarrow \{\mathcal{L}(S(L)) \leq L \quad \text{and} \quad \mathcal{L}(S(L)+1) > L\}.$$

Note that $S(0) = s$. Table 3 is similar to Table 1. It gives the optimal parameters of (s, N, t)-codes of the minimax strength s, $15 \leq s \leq 20$, weight w, size t, $2^m \leq t < 2^{m+1}$, $9 \leq m \leq 19$, based on the q_0-ary shortened RS-codes. In addition, Table 3 contains the numerical values of the averaged list-of-L decoding strength $S(L)$, when $L = 0.1$ and $L = 1$.

Example. For the case $s = 16$, $m = 11$, Table 3 gives $q_0 = 47$, $\lambda_0 = 1$ and $N = 799$. It means that there exists a 16-disjunct constant-weight binary code with

$$\lambda = 1, \quad w = s\lambda + 1 = 17, \quad t = 47^2 = 2209, \quad N = q_0 w = 47 \cdot 17 = 799.$$

The averaged list-decoding strengths $S(.1) = 43$ and $S(1) = 52$ essentially exceed the minimax strength $s = 16$.

Open problem

Find the parameters of superimposed codes based on the q_0-ary shortened RS-codes which yield efficient possibilities for the minimax combinatorial constructions of list-decoding superimposed codes. This problem is similar to that that we considered in Sect. 3 for IM-codes.

ACKNOWLEDGMENT

The authors wish to acknowledge Prof. Ahlswede for his permanent interest and support of their investigations in the superimposed code theory. In a recent paper [23], superimposed codes play a big role in so-called k-identification.

Table 3 Averaged list-of-L decoding strength $S(L)$, $L = .1, 1$, of constant-weight (s, N, t)-codes of the minimax strength s, $15 \leq s \leq 20$, size t, $2^m \leq t < 2^{m+1}$, $9 \leq m \leq 19$, length N, based on the q_0-ary shortened RS-codes.

s	15	16	17	18	19	20
m	q_0, λ_0, N $S(.1), S(1)$	q_0, λ_0, N $S(.1), S(1)$	q_0, λ_0, N $S(.1), S(1)$	q_0, λ_0, N $S(.1), S(1)$	q_0, λ_0, N $S(.1), S(1)$	q_0, λ_0, N $S(.1), S(1)$
9	23, 1, 368 25, 30	23, 1, 391 26, 31	23, 1, 414 27, 32	23, 1, 437 29, 33	23, 1, 460 30, 35	23, 1, 483 30, 36
10	32, 1, 512 31, 38	32, 1, 544 33, 39	32, 1, 576 34, 41	32, 1, 608 36, 42	32, 1, 640 37, 44	32, 1, 672 39, 45
11	47, 1, 752 41, 49	47, 1, 799 43, 52	47, 1, 846 45, 54	47, 1, 893 47, 56	47, 1, 940 50, 58	47, 1, 987 51, 60
12	– –	– –	67, 1, 1206 60, 71	67, 1, 1273 62, 74	67, 1, 1340 65, 77	67, 1, 1407 67, 80
13	– –	– –	– –	– –	– –	– –
14	31, 2, 961 34, 39	– –	– –	– –	– –	– –
15	32, 2, 992 36, 40	**32,2,1056** 37, 42	37, 2, 1295 44, 49	37, 2, 1369 45, 51	39, 2, 1521 49, 55	– –
16	41, 2, 1271 34, 39	41, 2, 1353 36, 41	41, 2, 1435 38, 43	41, 2, 1517 40, 44	41, 2, 1599 42, 46	41, 2, 1681 43, 48
17	53, 2, 1643 53, 60	53, 2, 1749 56, 63	53, 2, 1855 58, 66	53, 2, 1961 60, 68	53, 2, 2067 63, 70	53, 2, 2173 65, 73
18	64, 2, 1984 61, 70	64, 2, 2112 65, 74	64, 2, 2240 68, 77	64, 2, 2368 70, 80	64, 2, 2496 74, 82	64, 2, 2624 76, 85
19	– –	– –	– –	81, 2, 2997 87, 97	81, 2, 3159 93, 104	81, 2, 3321 97, 107

References

[1] W.H. Kautz, R.C. Singleton, "Nonrandom Binary Superimposed Codes," *IEEE Trans. Inform. Theory* 10 (4), 1964, 363-377.

[2] A.G. D'yachkov, V.V. Rykov, "Bounds on the Length of Disjunctive Codes," *Problemy Peredachi Inform.* 18 (3) 1982, 7-13 (in Russian).

[3] A.G. D'yachkov, V.V. Rykov, "A Survey of Superimposed Code Theory," *Problems of Control and Inform. Theory* 12 (4), 1983, 229-242.

[4] A.G. D'yachkov, V.V. Rykov, A.M. Rashad, "Superimposed Distance Codes", *Problems of Control and Inform. Theory* 18 (4), 1989, 237-250.

[5] A.G. D'yachkov, V.V. Rykov, "On Superimposed Codes," *Fourth International Workshop "Algebraic and Combinatorial Coding Theory"*, Novgorod, Russia, September 1994, 83-85.

[6] A.G. D'yachkov, "Designing Screening Experiments", *Lectures in the Bielefeld University"*, Bielefeld, Germany, Jan.-Feb., 1997.

[7] P. Erdos, P. Frankl, Z. Furedi, "Families of Finite Sets in which No Set Is Covered by the Union of r Others", *Israel Journal of Math.* 51, no. 1-2, 1985, 75-89.

[8] A.J. Macula, "A Simple Construction of d-Disjunct Matrices with Certain Constant Weight," *Discrete Mathematics* 162, 1996, 311-312.

[9] A.G. D'yachkov, V.V. Rykov, "Some Constructions of Optimal Superimposed Codes," *Conference "Computer Science & Information Technologies"*, Yerevan, Armenia, September 1997, 242-245.

[10] A.G. D'yachkov, V.V. Rykov, "Optimal Superimposed Codes and Designs for Renyi's Search Model", *Preprint 97-062, SFB 343*, University of Bielefeld, Germany, 1997.

[11] A.G. D'yachkov, A.J. Macula, V.V. Rykov, "On Optimal Parameters of a Class of Superimposed Codes and Designs", *1998 IEEE International Symposium on Information Theory*, MIT, Cambridge, MA USA, 16-21 August 1998, p. 363.

[12] D.J. Balding, D.C. Torney, "Optimal Pooling with Detection", *Journal of Combinatorial Theory, Ser. A* 74, 1996, 131-140.

[13] E. Knill, W.J. Bruno, D.C. Torney, "Non-adaptive Group Testing in the Presence of Error", *Discrete Applied Mathematics* 88, 1998, 261-290.

[14] E. Knill, S. Muthukrishnan, "Group Testing Problems in Experimental Molecular Biology", *Los Alamos National Laboratory, Preliminary Report*, Los Alamos, 1995.

[15] F.J.MacWilliams, N.J.A.Sloane, "The Theory of Error-Correcting Codes", North Holland, 1983.

[16] A. Renyi, "On the Theory of Random Search", *Bull. Amer. Math. Soc.* 71 (6), 1965, 809-828.

[17] D.-Z. Du, F.K. Hwang, *Combinatorial Group Testing and its Applications*, World Scientific, Singapore-New Jersey-London-Hong Kong, 1993.

[18] P.A. Vilenkin, "On Constructions of List-Decoding Superimposed Codes", *Sixth International Workshop "Algebraic & Combinatorial Coding Theory"*, Pskov, Russia, September 1998, 228-231.

[19] A.G. D'yachkov, A.J. Macula, V.V. Rykov, "New Constructions of Superimposed Codes", *IEEE Trans. Inform. Theory*, to appear.

[20] A.E. Brouwer, J.B. Shearer, N.J.A. Sloane, W.D. Smitt, "A New Table of Constant-Weight Codes", *IEEE Trans. Inform. Theory* 36 (6), 1990, 1334-1380.

[21] R.S. Singleton, "Maximum Distance Q-Nary Codes", *IEEE Trans. Inform. Theory* 10 (2), 1964 116-118.

[22] V.V. Rykov, S.M. Yekhanin, "On the Averaged List-Decoding Size for Superimposed Codes Based on RS-codes", submitted.

[23] R. Alswede, "General Theory of Information Transfer", *Preprint 97-118, SFB 343*, University of Bielefeld, 1997.

RUDIFIED CONVOLUTIONAL ENCODERS*

Rolf Johannesson

Department of Information Technology, Information Theory Group, Lund University
P.O. Box 118, S-221 00 LUND, Sweden

rolf@it.lth.se

Abstract: In this semi-tutorial paper convolutional codes and their various encoders are presented. The terminology *rudified* convolutional encoders is introduced for convolutional encoders that are both systematic and polynomial. It is argued that these rudified convolutional encoders—contrary to common belief—are sometimes the best choice.

I. Introduction

It is well-known that convolutional codes encoded by nonsystematic encoders or by systematic, rational (feedback) encoders have a larger free distance than convolutional codes encoded by systematic, polynomial encoders. This latter class of encoders are therefore considered inferior to the former. However, in this semi-tutorial paper we will argue that the systematic, polynomial convolutional encoders—contrary to common belief—are the best choice in some situations. Due to their excellent performance we call these encoders *rudified convolutional encoders.*

After having defined convolutional codes and their various encoders in Section II we define the free distance and discuss briefly some free distance bounds in Section III. In the following two sections we compare the performances of Viterbi and list decoding of convolutional codes encoded by general and rudified encoders. We conclude with a challenge for Rudi and an envoi. No proofs are given, instead we refer to [1].

*This research was supported in part by the Swedish Research Council for Engineering Sciences under Grants 97–235 and 97–723.

I. Althöfer et al. (eds.), Numbers, Information and Complexity, 283-293.

II. Convolutional Codes and Their Encoders

Convolutional codes are often thought of as nonblock linear codes over a finite field, but it can be an advantage to treat them as block codes over certain infinite fields. For simplicity we consider only binary convolutional codes.

First we define a convolutional transducer.

Definition: A rate $R = b/c$ (binary) *convolutional transducer* over the field of rational functions $\mathbb{F}_2(D)$ is a linear mapping

$$\begin{aligned}\tau: \mathbb{F}_2^b((D)) &\rightarrow \mathbb{F}_2^c((D)) \\ \boldsymbol{u}(D) &\mapsto \boldsymbol{v}(D),\end{aligned}$$

which can be represented as

$$\boldsymbol{v}(D) = \boldsymbol{u}(D)G(D), \tag{1}$$

where $G(D)$ is a $b \times c$ *transfer function matrix* of rank b with entries in $\mathbb{F}_2(D)$ and the Laurent series $\boldsymbol{v}(D)$ is called a *code sequence* arising from the *information sequence* $\boldsymbol{u}(D)$. □

Obviously we must be able to reconstruct the information sequence $\boldsymbol{u}(D)$ from the code sequence $\boldsymbol{v}(D)$. Therefore we require that the transducer map is injective, i.e., the transfer function matrix $G(D)$ has rank b over the field $\mathbb{F}_2(D)$.

Next we have the following

Definition: A rate $R = b/c$ *convolutional code* $\mathcal{C}$ over $\mathbb{F}_2$ is the image set of a rate $R = b/c$ convolutional transducer with $G(D)$ of rank b over $\mathbb{F}_2(D)$ as its transfer function matrix. □

It follows immediately from the definition that a rate $R = b/c$ convolutional code $\mathcal{C}$ over $\mathbb{F}_2$ with the $b \times c$ matrix $G(D)$ of rank b over $\mathbb{F}_2(D)$ as a transfer function matrix can be regarded as the $\mathbb{F}_2((D))$ row space of $G(D)$. Hence, it can also be regarded as the rate $R = b/c$ block code over the infinite field of Laurent series encoded by $G(D)$.

A transfer function matrix (of a convolutional code) is called a *generator matrix* if it (has full rank and) is realizable, that is every entry consists of a rational function with a constant term 1 in the denominator polynomial.

Definition: A rate $R = b/c$ *convolutional encoder* of a convolutional code with generator matrix $G(D)$ over $\mathbb{F}_2(D)$ is a realization by a linear sequential circuit of a rate $R = b/c$ convolutional transducer whose transfer function matrix $G(D)$ (has full rank and) is realizable. □

A given convolutional code can be encoded by many essentially different encoders.

Example 2.1:

Consider the rate $R = 1/2$, binary convolutional code with the basis vector $\boldsymbol{v}_0(D) = (1 + D + D^2 \quad 1 + D^2)$. The simplest encoder for this code has the generator matrix

$$G_0(D) = (1 + D + D^2 \quad 1 + D^2). \tag{2}$$

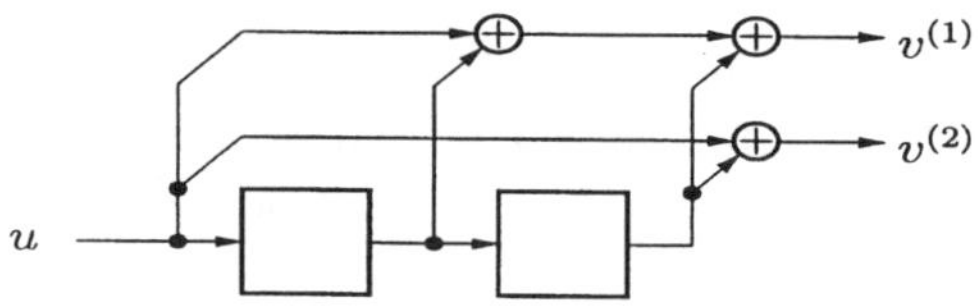

Figure 1 A rate $R = 1/2$ convolutional encoder with generator matrix $G_0(D)$.

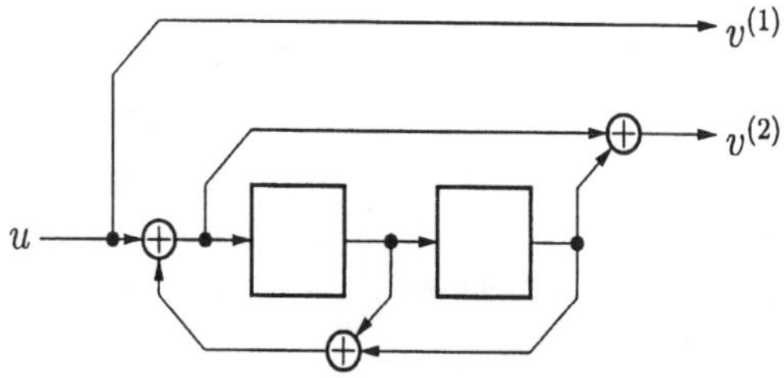

Figure 2 A rate $R = 1/2$ systematic convolutional encoder with feedback and generator matrix $G_1(D)$.

A realization in *controller canonical form* is shown in Fig. 1. □

An encoder which realizes a polynomial generator matrix is called a *polynomial encoder.*

Example 2.1 (cont.):

If we choose the basis to be $\boldsymbol{v}_1(D) = a_1(D)\boldsymbol{v}_0(D)$, where the scalar $a_1(D)$ is the rational function $a_1(D) = 1/(1 + D + D^2)$, we obtain the generator matrix

$$G_1(D) = \begin{pmatrix} 1 & \frac{1+D^2}{1+D+D^2} \end{pmatrix} \tag{3}$$

for the same code. The output sequence $\boldsymbol{v}(D) = (v^{(1)}(D)\ v^{(2)}(D))$ of the encoder with generator matrix $G_1(D)$ shown in Fig. 2 can be written as

$$\begin{aligned} v^{(1)}(D) &= u(D) \\ v^{(2)}(D) &= u(D)\tfrac{1+D^2}{1+D+D^2}. \end{aligned} \tag{4}$$

The input sequence appears unchanged among the two output sequences. □

Definition: A rate $R = b/c$ convolutional encoder whose b information sequences appear unchanged among the c code sequences is called a *systematic encoder* and its generator matrix is called a *systematic generator matrix.* □

If a convolutional code $\mathcal{C}$ is encoded by a systematic generator matrix we can always permute its columns and obtain a generator matrix for an *equivalent* convolutional code $\mathcal{C}'$ such that the b information sequences appear unchanged *first* among the code sequences. Thus, without loss of generality a systematic generator matrix can be written as

$$G(D) = (I_b \ R(D)), \tag{5}$$

where I_b is a $b \times b$ identity matrix and $R(D)$ a $b \times (c-b)$ matrix whose entries are rational functions of D.

Being 'systematic' is a generator matrix property, not a code property. Every convolutional code has both systematic and *nonsystematic* generator matrices.

III. The Free Distance and Heller's Upper Bounds

Let $\mathcal{C}$ be a convolutional code. The *free distance* is the principal determiner for the error correcting capability of a convolutional code when we are communicating over a channel with small error probability and use maximum-likelihood (or nearly so) decoding. It is defined as the minimum Hamming distance between any two differing codewords,

$$d_{\text{free}} \stackrel{\text{def}}{=} \min_{\boldsymbol{v} \neq \boldsymbol{v}'} \{d_{\text{H}}(\boldsymbol{v}, \boldsymbol{v}')\}. \tag{6}$$

Let $\mathcal{E}_t$ be the set of all error patterns with t or fewer errors. Then a convolutional code $\mathcal{C}$ can correct all error patterns in $\mathcal{E}_t$ if and only if $d_{\text{free}} > 2t$.

Let $G(D) = (g_{ij}(D))$ be a generator matrix. Then the *memory* of $G(D)$ is

$$m \stackrel{\text{def}}{=} \max_{1 \le i \le b} \max_{1 \le j \le c} \{g_{ij}(D)\}. \tag{7}$$

Heller used Plotkin's bound on the minimum distance for block codes to derive a surprisingly tight bound on the free distance for convolutional codes [2]:

Theorem 1. *The free distance for any binary, rate $R = b/c$ convolutional code encoded by a generator matrix of memory m satisfies*

$$d_{free} \le \min_{i \ge 1} \left\{ \left\lfloor \frac{(m+i)c}{2(1-2^{-bi})} \right\rfloor \right\}. \tag{8}$$

□

For convolutional codes encoded by rudified encoders, that is encoders that are both systematic and polynomial, we have the corresponding bound:

Theorem 2. *The free distance for any binary, rate $R = b/c$ convolutional code encoded by a rudified generator matrix of memory m satisfies*

$$d_{free} \le \min_{i \ge 1} \left\{ \left\lfloor \frac{(m(1-R)+i)c}{2(1-2^{-bi})} \right\rfloor \right\}. \tag{9}$$

□

For the ensemble of periodically, time-varying convolutional codes Costello [3] proved the following lower bound on the free distance.

Theorem 3. *There exists a binary, periodically time-varying, rate $R = b/c$ convolutional code with a polynomial generator matrix of memory m that has a free distance satisfying the inequality*

$$\frac{d_{free}}{mc} \geq \frac{R}{-\log(2^{1-R}-1)} + O\left(\frac{\log m}{m}\right). \tag{10}$$

□

For convolutional codes encoded by rudified encoders we have the following counterpart:

Theorem 4. *There exists a binary, periodically time-varying, rate $R = b/c$ convolutional code with a rudified generator matrix of memory m that has a free distance satisfying the inequality*

$$\frac{d_{free}}{mc} \geq \frac{R(1-R)}{-\log(2^{1-R}-1)} + O\left(\frac{\log m}{m}\right). \tag{11}$$

□

By comparing these bounds we notice that in order to obtain the same value of the bound for rudified encoders as for general encoders we have to increase the memory for the rudified encoders by the factor $(1-R)^{-1}$. Rudified encoders are *inferior* to general encoders from the free distance point of view.

IV. Maximum-likelihood (Viterbi) Decoding

For convolutional encoders, it is sometimes useful to draw the *state-transition diagram.* If we ignore the labeling, the state-transition diagram is a *de Bruijn graph* [4]. In Fig. 3, we show a simple convolutional encoder and its state-transition diagram.

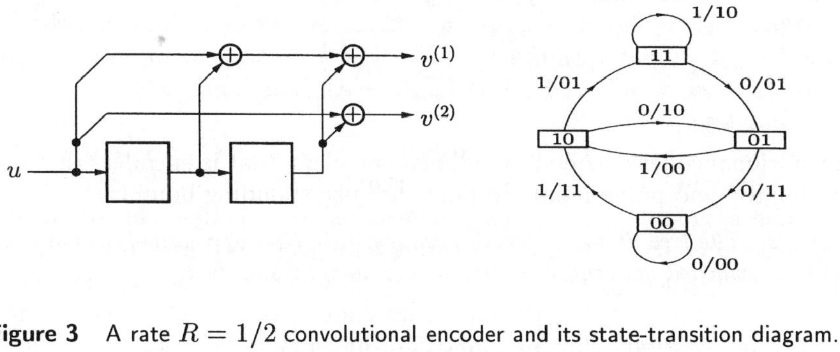

Figure 3 A rate $R = 1/2$ convolutional encoder and its state-transition diagram.

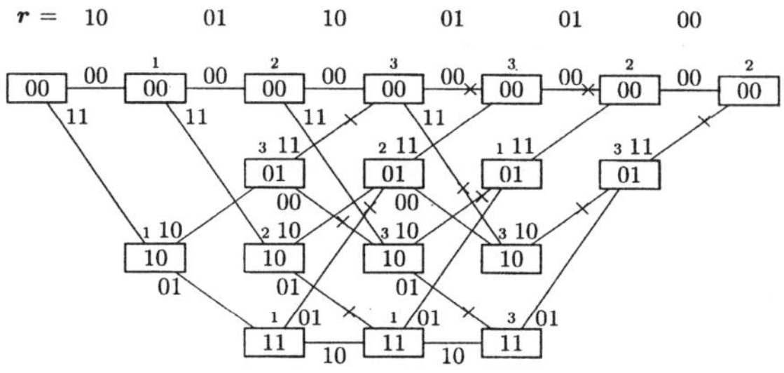

Figure 4 An example of Viterbi decoding—hard decisions.

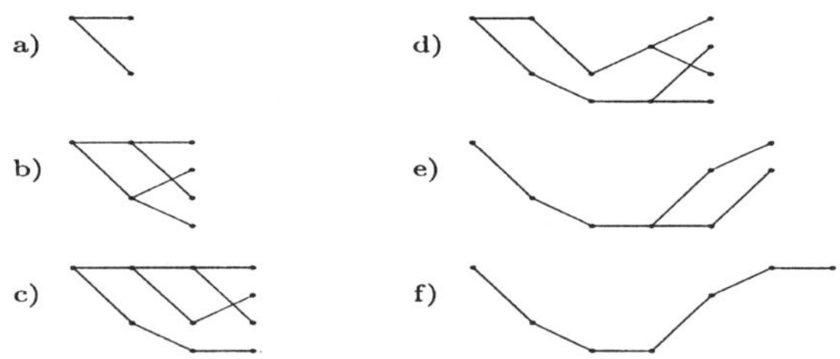

Figure 5 Development of subpaths through the trellis.

Assume that we start in the 00 state and draw the states successively to the right as time progresses. Then we obtain the *trellis* representation of the convolutional code shown in Fig. 4 [5].

The Viterbi algorithm is an efficient procedure to obtain a maximum-likelihood estimate of the codeword. When comparing the subpaths leading to each state, the Viterbi algorithm discard all subpaths except the one closest (in Hamming distance) to the received sequence, since those discarded subpaths cannot possibly be the initial part of the path $\widehat{\boldsymbol{v}}$ that minimizes $d_{\mathrm{H}}(\boldsymbol{r},\boldsymbol{v})$, i.e.,

$$\widehat{\boldsymbol{v}} = \arg\min_{\boldsymbol{v}}\{d_{\mathrm{H}}(\boldsymbol{r},\boldsymbol{v})\}. \tag{12}$$

This is the *principle of nonoptimality.* In case of a tie, we can arbitrarily choose one of the closest subpaths as the survivor. If we are true to the principle of nonoptimality when we discard subpaths the path remaining at the end must be the optimal one.

The Hamming distances and discarded subpaths at each state determined by the Viterbi algorithm are shown in Fig. 4 (the discarded subpaths are marked with $\times$. The estimated information sequence is $\widehat{\boldsymbol{u}} = 1110$. The successive development of the surviving subpaths through the trellis is illustrated in Fig. 5.

It can be shown that (see, e.g., [1]) that there exists a binary rate $R = b/c$, periodically time-varying convolutional code encoded by a polynomial, periodically time-varying generator matrix of memory m and period T, where

$T = O(m^2)$, such that the error probability from a Viterbi decoder is upper-bounded by

$$P_B^T \leq 2^{-(E_C(R)+o(1))mc}, \quad 0 \leq R \leq C, \tag{13}$$

where $E_C(R)$ is the convolutional coding exponent shown in Fig. 6 and C is the channel capacity.

Furthermore, there exists a binary rate $R = b/c$, periodically time-varying convolutional code encoded by a rudified, periodically time-varying encoding matrix of memory m and period T, where $T = O(m^2)$, such that the error probability from a Viterbi decoder is upper-bounded by

$$P_B^T \leq 2^{-(E_C^{\text{sys}}(R)+o(1))mc}, \quad 0 \leq R < C, \tag{14}$$

where

$$E_C^{\text{sys}}(R) = E_C(R)(1-R) \tag{15}$$

is the convolutional coding exponent for convolutional codes encoded by rudified generator matrices. (See Fig. 6.)

We also have a corresponding lower bound. For any rate $R = b/c$ convolutional code $\mathcal{C}$ encoded by a generator matrix of memory m that is used to communicate over a binary symmetric channel (BSC) with crossover probability ϵ the error probability is lower-bounded by

$$P_B > 2^{-\left(E_C^{\text{low}}(R)+o(1)\right)mc}, \tag{16}$$

where $E_C^{\text{low}}(R)$ is the convolutional lower bound exponent

$$E_C^{\text{low}}(R) = \min\left\{E_C^{\text{sph}}(R), -\frac{1}{2}\log\left(2\sqrt{\epsilon(1-\epsilon)}\right)\right\} \tag{17}$$

and $0 \leq R < C$. The sphere-packing exponent $E_C^{\text{sph}}(R)$ is the upper curve in Fig. 7.

In the region $R_0 \leq R < C$ the exponent E_C is optimal. For rates $0 \leq R < R_0$ the true value of the exponent is somewhere between the values of $E_C(R)$ and $E_C^{\text{low}}(R)$.

V. List Decoding

List decoding is a nonbacktracking breadth-first search of a code tree. At each depth only the L most promising subpaths are extended, not all, as is the case of Viterbi decoding. These subpaths form a *list* of size L. Since not all paths are extended it can happen that the correct path is lost. This is a serious kind of error event that is typical for list decoding.

For a given list size L the *list weight* w_{list} of the convolutional code $\mathcal{C}$ is

$$w_{\text{list}} = \min_t \min_{\boldsymbol{v}_{[0,t]}} \{\delta_L(\boldsymbol{v}_{[0,t]})\}, \tag{18}$$

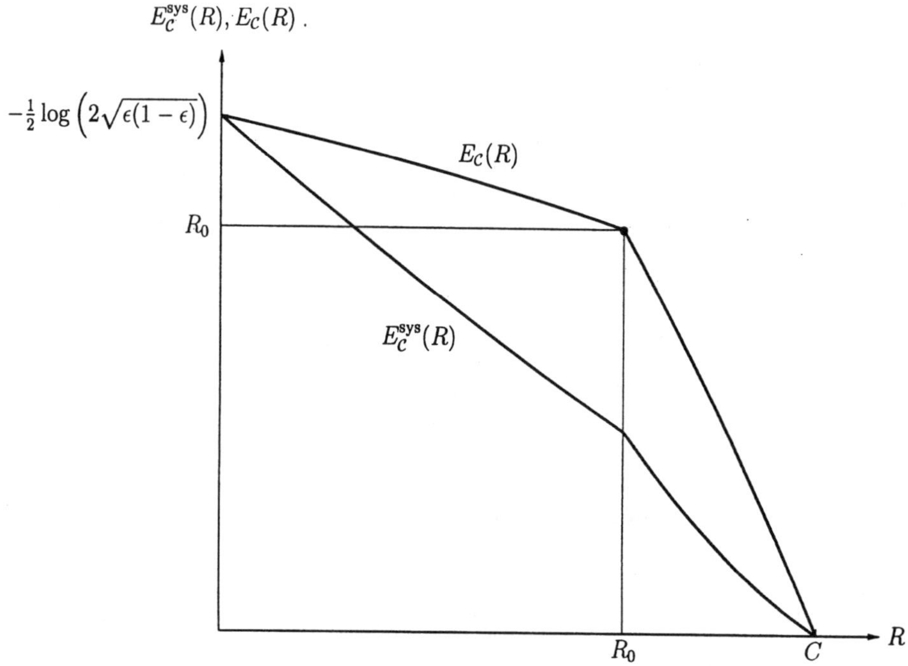

Figure 6 The convolutional coding exponents $E_C^{\rm sys}(R)$ and $E_C(R)$ for the binary symmetric channel (BSC) with crossover probability $\epsilon = 0.045$. R_0 is the so-called computational cutoff rate.

where $\boldsymbol{v}_{[0,t]}$ is the initial part of the codeword $\boldsymbol{v} \in \mathcal{C}$ and $\delta_L(\boldsymbol{v}_{[0,t]})$ is the largest radius of a sphere with center $\boldsymbol{v}_{[0,t]}$ such that the number of codewords in the sphere is less than or equal to L. The importance of the list weight is seen from the following

Theorem 5. *Given a list decoder of list size L and a received sequence with at most $\lfloor \frac{1}{2} w_{list} \rfloor$ errors. Then the correct path will not be forced outside the list of L survivors.* □

If the number of errors exceeds $\lfloor \frac{1}{2} w_{\rm list} \rfloor$, then it depends on the code $\mathcal{C}$ and on the received sequence $\boldsymbol{r}$ whether the correct path is not forced outside the list.

For the list weight we have [1]:

Theorem 6. *There exist binary, rate $R = b/c$, infinite memory, time-invariant convolutional codes with nonsystematic and rudified generator matrices used with list decoding of list size L, having a list weight w_{list}, and satisfying the*

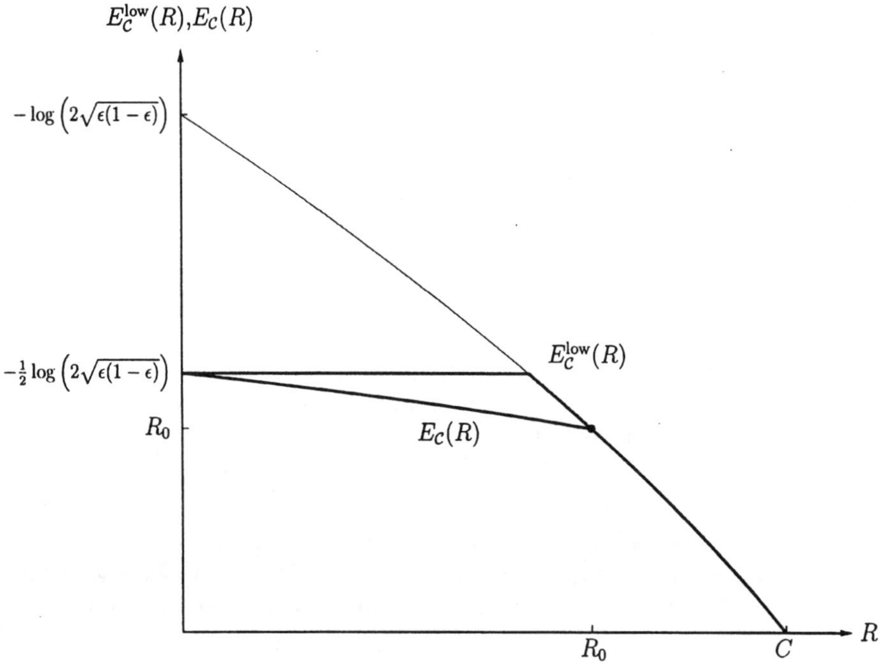

Figure 7 The convolutional coding exponent $E_C(R)$ and the lower bound exponent $E_C^{\rm low}$ for the BSC with crossover probability $\epsilon = 0.045$.

inequality

$$w_{list} > \frac{\log L}{-\log(2^{1-R}-1)} + O(1), \tag{19}$$

where

$$O(1) = \frac{\log((2^R-1)(2^{1-R}-1))}{-\log(2^{1-R}-1)}. \tag{20}$$

□

If we choose the list size L equal to the number of encoder states, i.e., $L = 2^{bm}$, then our lower bound on $w_{\rm list}$ (19) coincides with Costello's lower bound on $d_{\rm free}$ (10).

It follows from Theorem 6 that convolutional codes encoded by both nonsystematic and rudified generator matrices have principle determiners of the correct path loss probability that are lower-bounded by the same bound! For the free distance, which is the principal determiner of the error probability with Viterbi decoding, Costello's lower bounds on the free distance differ by the factor $(1 - R)$ depending on whether the convolutional code is encoded by a nonsystematic or by a rudified generator matrix. This different behavior reflects a fundamental and important difference between list and Viterbi decoding.

Theorem 7. *For a list decoder of list size L there exist infinite memory, time-invariant, binary convolutional codes of rate $R = b/c$ with nonsystematic and rudified generator matrices such that the probability of correct path loss is upper-bounded by the inequality*

$$P(\mathcal{E}_t^{cpl}) \leq 2^{-(E_C(R)+o(1))(\log L)/R}, \tag{21}$$

where $E_C(R)$ is the convolutional coding exponent shown in Fig. 6. □

If we choose the list size L equal to the number of encoder states, i.e., $L = 2^{bm}$, then our upper bound on the probability of correct path loss (21) coincides with the upper bound on the error probability (13).

We can prove a corresponding lower bound [1],

$$P(\mathcal{E}_t^{\mathrm{cpl}}) > 2^{-\left(E_C^{\mathrm{sph}}(R)+o(1)\right)(\log L)/R}, \tag{22}$$

where $E_C^{\mathrm{sph}}(R)$ is shown as the upper curve in Fig. 7.

For the ensemble of general, nonlinear trellis codes it can be shown that for list decoding the exponent of (22) is somewhat surprisingly correct for all rates, $0 \leq R < C$ [7]!

We conjecture that this also holds for the exponent for the ensemble of convolutional codes encoded by rudified encoders and, hence, that also list decoding of convolutional codes encoded by rudified encoders is superior to Viterbi decoding of convolutional codes encoded by nonsystematic encoders.

Our conjecture is given strong support by the experiments reported in [1].

VI. Comments

Rudified encoders are inferior to nonsystematic ones if we consider the free distance. Hence, since the free distance is the principle determiner of the error probability when Viterbi decoding is used at high signal-to-noise ratios, rudified encoders should not be used together with Viterbi decoding. If we use list decoding then the principle determiner of the correct path loss, viz., the list weight, is the same for both nonsystematic and rudified encoders. It has recently been shown [6], [1] that rudified encoders support a spontaneous recovery of a lost correct path which for list decoding leads to a superior performance for these encoders. Somewhat surprisingly, for a given *decoder* complexity a rudified encoder together with list decoding outperforms a nonsystematic encoder together with Viterbi decoding. The explanation is that with the list decoder we can use a more powerful code, that is a code whose encoder has a state space that is larger than the decoder state space. Finally, we remark—as conjectured by J. L. Massey twentyfive years ago—that together with sequential decoding rudified encoders perform as well as nonsystematic ones.

VII. A Challenge for Rudi

Prove the conjecture in Section V and collect SEK 500!

VIII. Envoi

Happy Birthday, Rudi!

References

[1] R. Johannesson and K. Sh. Zigangirov, *Fundamentals of convolutional coding*, Piscataway, N. J., IEEE Press, 1999.

[2] J. A. Heller, "Sequential decoding: Short constraint length convolutional codes", Jet Propulsion Lab., California Inst. Technol., Pasadena, Space Program Summary 37–54, vol. 3, Dec. 1968, 171–174.

[3] D. J. Costello, Jr., "Free distance bounds for convolutional codes", *IEEE Trans. Inform. Theory*, vol. 20, 1974, 356–365.

[4] S. W. Golomb, *Shift Register Sequences*, Holden-Day, San Fransisco, 1967. Revised ed., Aegean Park Press, Laguna Hills, Cal., 1982.

[5] G. D. Forney, Jr., (1967), *Review of random tree codes* (NASA Ames. Res. Cen., Contract NAS2-3637, NASA CR 73176, Final Rep.;Appx A). See also Forney, G. D., Jr. (1974), Convolutional codes II: Maximum-likelihood decoding and convolutional codes III: Sequential decoding. *Inform Contr.*, 25:222–297.

[6] H. Osthoff, J. B. Anderson, R. Johannesson, and C.-f. Lin, "Systematic feed-forward convolutional encoders are better than other encoders with an M-algorithm decoder", *IEEE Trans. Inform. Theory*, vol. 44, 1998, 831–838.

[7] K. Sh. Zigangirov and V. D. Kolesnik, "List decoding of trellis codes", *Problems of Control and Information Theory* 9, 1980, 347–364.

ON CHECK DIGIT SYSTEMS USING ANTI–SYMMETRIC MAPPINGS

Ralph–Hardo Schulz

FB Mathematik und Informatik, Freie Universität Berlin
Arnimallee 3, 14195 Berlin, Germany

schulz@math.fu-berlin.de

Abstract: We consider check digit systems over a group G with check equation $T(a_1)T^2(a_2)\cdots T^n(a_n) = e$ (for codewords $a_1a_2\ldots a_n \in G^n$) with $e \in G$ and permutation T of G. Such a system detects all single errors (i.e. errors in only one component); and it detects adjacent transpositions (i.e. errors of the form $\ldots ab\ldots \longrightarrow \ldots ba\ldots$) iff T is anti–symmetric that means that T fulfills the condition $x\,\mathrm{T}(y) \neq y\,\mathrm{T}(x)$ for all $x, y \in G$ with $x \neq y$. In this survey we shall report on the existence of groups with anti–symmetric mappings, define equivalence relations between check digit systems and describe, in the special case of the dihedral group D_5, the equivalence classes.

INTRODUCTION

A check digit system with one check character is a systematic error detecting code over an alphabet A which arises by appending a *check digit* a_n to every word $a_1a_2\ldots a_{n-1} \in A^{n-1}$:

$$c : \begin{cases} A^{n-1} & \longrightarrow & A^n \\ a_1a_2\ldots a_{n-1} & \longmapsto & a_1a_2\ldots a_{n-1}a_n. \end{cases}$$

The aim of using such a system is to discover transmission errors. Empirical investigations by VERHOEFF [27], and BECKLEY [1] (see Table 1) show that *single errors*, i.e. errors in only one component, and *adjacent transpositions* (also called *neighbour transposition errors*), i.e. errors of the form $\ldots ab\ldots \longrightarrow \ldots ba\ldots$, are the most important errors made by human operators; (for insertion and deletion errors a detection rate of 100% is achieved by adding leading zeros to make all codewords of equal length).

Note that the numbers of Table 1 can vary from sample to sample and may depend on the location of the affected digits; e.g. the rightmost two digits may be affected by single errors more than the other digits together ([27] p. 14).

Choosing $G = (A, \cdot)$ as a set endowed with a group structure one can determine the check digit a_n by a "check"-equation

$$(\diamond) \qquad \delta_1(a_1)\delta_2(a_2)\ldots\delta_n(a_n) = e$$

I. Althöfer et al. (eds.), Numbers, Information and Complexity, 295-310.

Table 1 Error types and their frequencies

Error type			*Relative frequency* Verhoeff	Beckley
single error	$\dots a \dots$	$\longrightarrow \dots a' \dots$	79.0% (60-95)	86%
adjacent transposition	$\dots a\, b \dots$	$\longrightarrow \dots b\, a \dots$	10.2 %	8%
jump transposition	$\dots acb \dots$	$\longrightarrow \dots bca \dots$	0.8%	
twin error	$\dots aa \dots$	$\longrightarrow \dots bb \dots$	0.6%	6%
phonetic error (a≥2)	$\dots a0 \dots$	$\longrightarrow \dots 1a \dots$	0.5%	
jump twin error	$\dots aca \dots$	$\longrightarrow \dots bcb \dots$	0.3%	
other error			8.6%	

Source: Verhoeff [27](12,112 pairs, 6 digits), Beckley [1].

for fixed permutations δ_i of G, $i = 1, \dots, n$ and $e \in G$; a usual choice is $e = e_G$ where e_G denotes the neutral element of G. Often, δ_i is chosen such that $\delta_i = T^i$ for a fixed permutation T of G. A check digit system with this check equation detects all *single errors;* and it detects all adjacent transpositions iff $\delta_{i+1}\delta_i^{-1}$ is *anti–symmetric* for $i = 1, \dots, n-1$; here we use the following definition.

1.1 Definition : Anti-symmetric mapping. A bijection T of a group G onto itself is called **anti-symmetric** iff it fulfills the condition

$$(**) \qquad x\,\mathrm{T}(y) \neq y\,\mathrm{T}(x) \quad \text{for all} \quad x, y \in G \ \text{ with } x \neq y.$$

(This is a generalization of a condition of SCHAUFFLER [17]).

In this survey we shall report on the existence of groups with anti–symmetric mappings, define equivalence relations between check digit systems over the same group and describe, in the special case of the dihedral group D_5 of order 10, the equivalence classes with their error detection capacities. Note that we do not discuss check digit systems using general quasigroups and schemes using two or more check digits.

1.2 First examples. Examples of check digit systems are (see for instance [3], [19], [22]): •• the **European Article Number code (EAN)** and (after adding 0 as first digit) the **Universal Product Code (UPC)** with $G = (\mathbb{Z}_{10}, +)$, $n = 13, \delta_{2i-1} = \mathrm{id} = \delta_{13}$ and $\delta_{2i}(a) = 3a$ for $e = 0, i = 1, \dots, 6$; this system does not detect adjacent transpositions $\dots ab \dots \longrightarrow \dots ba \dots$ for $|a-b| = 5$ (see 2.4(ii)); •• the **International Standard Book Number code (ISBN)** with $G = (\mathbb{Z}_{11}, +), n = 10$ and $\delta_i(a) = ia$ for $e = 0$, $i = 1, \dots, 10$; this system detects all adjacent transpositions but needs an element $X \notin \{0, \dots, 9\}$; •• the system of the **serial numbers of German banknotes** which uses an anti–symmetric mapping T_0 (found by VERHOEFF [27], see 4.2) of the dihedral group $G = D_5$ with $e = e_G$, $\delta_i = {T_0}^i$ for $i = 1, \dots 10$ and $\delta_{11} = \mathrm{id}$.

1.3 General assumption: *From now on we consider check digit systems over a group G of order q with codewords of length $n \geq 3$ and check equation*

$$(*) \qquad T(a_1)T^2(a_2)\cdots T^n(a_n) = e = e_G.$$

1.4 Detection of other errors The following Table 2, concerning the detection of other important errors, is a variation of VERHOEFF's table [27] by GIESE [9]. The numbers are under the assumption that all error locations and digits are of equal probability. In section 6 we shall use the detection rates of these errors to compare different anti–symmetric mappings of the same group.

Table 2 Detection of other errors

Errortype	*Detection set*	*Percentage of detection*
twin errors	$M_{\text{TE}}=\{(x,y)\in G^2 \mid xT(x)\neq yT(y)\}$	$\lvert M_{\text{TE}}\rvert/q(q-1)$
jump transpositions	$M_{\text{JT}}=\{(x,y,z)\in G^3 \mid xyT^2(z)\neq zyT^2(x)\}$	$\lvert M_{\text{JT}}\rvert/q^2(q-1)$
jump twin errors	$M_{\text{JZ}}=\{(x,y,z)\in G^3 \mid xyT^2(x)\neq zyT^2(z)\}$	$\lvert M_{\text{JZ}}\rvert/q^2(q-1)$

ORTHOMORPHISMS OF ABELIAN GROUPS

If G is an abelian group the condition $(**)$ is equivalent to $x\ \mathrm{T}(x)^{-1} \neq y\ \mathrm{T}(y)^{-1}$ for all $x, y \in G$ with $x \neq y$. We remind of the following.

2.1 Definition For a group $(G,\cdot)$, a mapping $f : G \longrightarrow G$ of is denoted as **orthomorphism** (or *perfect difference mapping*) if both f and g with $g(x) := x \cdot f(x)^{-1}$ are permutations. In this case, $\mathrm{inv} \circ f : x \mapsto f(x)^{-1}$ is called a **complete mapping** [15]. (Here "inv" denotes the mapping $x \mapsto 1/x$.) Hence h is complete iff h is a permutation with $xh(x) \neq yh(y)$ for all x, y in G with $x \neq y$. Thus we have

2.2 Proposition *If G is an abelian group and T a permutation of G then T is anti–symmetric. $\Longleftrightarrow$ T is an orthomorphism. $\Longleftrightarrow$ $\mathrm{inv} \circ T$ is a complete mapping.*

The theory of complete mappings is well developed. So we know e.g.:

2.3 Theorem a) *A finite abelian group G admits a complete mapping iff G has odd order m or contains more than one involution;* (PAIGE [16]). *Note that, for m odd, $x \mapsto 2x$ is anti–symmetric.* **b)** *A necessary condition for a finite group of even order to admit complete mappings is that its Sylow 2-subgroups be non-cyclic. For soluble groups this condition is also sufficient; (HALL and PAIGE [13]).*

Surveys on orthomorphism are given in [14] and [2]. The existence of complete mappings has been proved for many classical groups. In 1989, DENES and KEEDWELL [7] conjectured that all non-soluble groups admit a complete mapping. Consequences of HALL and PAIGE's theorem are the following. (For a proof see SIEMON [25] and, shortened by using groups with signum, DAMM [6]; here, a homomorphism $\mathrm{sgn}_G : G \longrightarrow \{1, -1\}$ is called signum, see DAMM [6] and ŠIRÁŇ and ŠKOVIERA; such a signum exists iff G contains a subgroup of index 2.)

2.4 Corollary (i) *A group of order $m = 2u$ with u odd does not admit a complete mapping and thus, in the abelian case, no anti–symmetric mapping.* **(ii)** *There does not exist a check digit system over $\mathbb{Z}_{10}$ which detects all adjacent transpositions.* More general: *The cyclic group G admits an anti–symmetric mapping iff $|G|$ is odd;* (see as well 2.5). **(iii)** *Groups of order $m = 2u$ with u odd, especially D_5 and $\mathbb{Z}_{10}$, don't admit a check digit system which detects all twin errors or all jump twin errors.*
PROOF of (iii). Otherwise, according to Table 2, the mappings T or $L_y \circ T^2$ would be complete; (here $L_y(z) := yz$ denotes the left multiplication with y).□
2.5 Examples. a) inv $= 1/\text{id}$ *is an anti–symmetric mapping of a finite abelian group G iff $|G|$ is odd;* (variation of SIEMON [25] 3.16; see as well 3.2.)
b) *For a finite cyclic group of order m the mapping $x \mapsto x^k$ is anti–symmetric iff gcd $(k, m) = 1 =$ gcd $(k-1, m)$;* (cf. SIEMON l.c. 3.18, [8] L. 4.3).
c) *If G is abelian and $T \in AutG$ then T is anti–symmetric iff T is fixed point free on G;* (cf. [23] 1.4, see as well 5.3).
PROOF. (a) If $|G| = 2k+1$, then $x = x^{2k+2} = (x^{k+1})^2$; thus $x \mapsto x^2$ is surjective, hence bijective. If $|G|$ is even this mapping is not injective.
(b) $x \mapsto x^k$ and $x \mapsto x^{k-1}$ have to be bijective; hence g^k and g^{k-1} must have order m if g is a generating element. □
2.6 Remarks. 1.) A system using the mapping "inv" does not detect any twin error. 2.) If m is even then there is no k satisfying the conditions of b). 3.) If $\gcd(k, m) = 1$ then $x \mapsto x^k$ is an automorphism of G.

THE GENERAL CASE

We come back to anti–symmetric mappings of not necessary commutative groups. Note that in [11], [28], [10] and in [6] the condition (**) is replaced by (**') $\phi(x)y \neq \phi(y)x$ for all $x, y \in G$ with $x \neq y$. The reason is that these authors use, for codewords $x_{n-1}x_{n-2}\dots x_1x_0$, the check equation

$$(*') \quad \phi^{n-1}(x_{n-1})\phi^{n-2}(x_{n-2})\cdots\phi(x_1)x_0 = e.$$

By putting $T = \phi^{-1}$ and $\bar{x} = \phi(x), \bar{y} = \phi(y)$ one comes back to (**). Having this in mind we have to reformulate results of these authors.
3.1 Examples. a) Examples for dihedral groups are given in sections 4 and 6. **b)** *Let G be the group of all $m \times m$* **triangular matrices** *over $K = \text{GF}(q)$* **with diagonal 1...1**. *Define T by*

$$T : \begin{pmatrix} 1 & & & a_{ij} \\ & 1 & & \\ & & \cdot & \\ & & & \cdot \\ 0 & & & 1 \end{pmatrix} \longmapsto \begin{pmatrix} 1 & & & f_{ij}(a_{ij}) \\ & 1 & & \\ & & \cdot & \\ & & & \cdot \\ 0 & & & 1 \end{pmatrix}$$

where the f_{ij}'s are orthomorphisms of (GF$(q), +$). *Then T is anti-symmetric;*(cf. [23] 1.2b). For instance, f_{ij} can be chosen to be the mapping $\hat{d_{ij}} : K \longrightarrow K$ with $x \mapsto d_{ij}x$ for $d_{ij} \in K \setminus \{0, 1\}$ and $j > i$. **c)** *Let $q = 2^m > 2$*

and $K = \mathrm{GF}(q)$*; put* $u_{ac} = 1$ *if* $a^2 \neq c$ *and* $u_{ac} = u$ *for a fixed* $u \in K \setminus \{0,1\}$ *otherwise. Then the mapping* $T : \begin{pmatrix} a & 0 \\ b & c \end{pmatrix} \longmapsto \begin{pmatrix} a^2 & 0 \\ u_{ac} \cdot b & c^2 \end{pmatrix}$ *is an anti–symmetric mapping of the group* $G_0 = \{\begin{pmatrix} a & 0 \\ b & c \end{pmatrix} \mid a, b \in K \wedge a \cdot c \neq 0\}$ *of all regular* 2×2- **triangular matrices** *over* GF (q) (see [21] 3.1). **d)** *In the same way there can be defined an anti–symmetric mapping of the* **affine group** $A(1, q) = \{\begin{pmatrix} 1 & 0 \\ b & c \end{pmatrix} \mid b, c \in \mathrm{GF}(q) \wedge c \neq 0\}$ *for* $q = 2^m > 2$; (see [21] 3.2). **e)** *Let* $q = p^m > 2$ *and* t *a prime with* $t|(q-1)$ *and* $t > 2$*; choose* l *from* $\{2, 3, \ldots, t-1\}$ *and* $u_0 \in \mathrm{GF}(q) \setminus \{0, 1\}$*; furthermore, put* $u_j = ({d_1}^j)^{(l-2)} {d_2}^j$ *for* $j \in \{1, \ldots, t-1\}$ *where* d_1, d_2 *are fixed elements of* $K = \mathrm{GF}(q)$ *with* $d_1 \neq d_2$ *and* ${d_1}^t = 1 = {d_2}^t$*. Then the mapping* $T : \begin{pmatrix} {d_1}^j & 0 \\ k & {d_2}^j \end{pmatrix} \longmapsto \begin{pmatrix} {d_1}^{jl} & 0 \\ u_j k & {d_2}^{jl} \end{pmatrix}$ *is an anti–symmetric mapping of the* **Îto group** $G = \{\begin{pmatrix} {d_1}^j & 0 \\ k & {d_2}^j \end{pmatrix} \mid j = 0, \ldots, t-1; k \in K\}$*. Choosing e.g.* $q = 23$, $t = 11$, $l = 2$, $u_0 = 2 \in GF(23)$ *or* $q = 29$, $t = 7$, $l = 2$, $u_0 = 4 \in GF(29)$ *one gets a check digit system detecting all single, twin, jump twin errors and adjacent transpositions; the alphabet then contains* 253 *and* 203 *elements respectively;* (see [21] 3.4 and 3.6). **f)** *Taking* G *as the group* H *of all* $4 \times 4-$*matrices over* $K = GF(q)$ *of the form* $[x, y, z] :=$

$$\begin{pmatrix} 1 & x & y & z \\ 0 & 1 & 0 & y \\ 0 & 0 & 1 & -x \\ 0 & 0 & 0 & 1 \end{pmatrix}$$

with $x, y, z \in K$*, we get an anti–symmetric mapping by* $T : [x, y, z] \mapsto [f(x), g_x(y), h_{xy}(z)]$ *if* f, g_x, h_{xy} *are orthomorphisms of* $(K, +)$ *for all* $x, y \in K$; (see [23] 1.2c). **g)** For $m \geq 2$, the group $Q_m := < a, b \mid a^{2m} = b^4 = e,\ b^2 = a^m,\ ab = ba^{-1} >$ is called a **dicyclic group** or (for m a power of 2) a **generalized quaternion group**; it is a group of order $4m$. *One gets an anti–symmetric mapping* φ *by putting* (cf. [10] Th.2.1 ii) $\varphi(a^i) = a^{-i}$ (*for* $0 \leq i \leq m-1$), $\varphi(a^i) = b \cdot a^{i-1}$(*for* $m \leq i \leq 2m-1$), $\varphi(ba^i) = ba^{i-1}$ (*for* $0 \leq i \leq m-1$) *and* $\varphi(ba^i) = a^{-i}$ (*for* $m \leq i \leq 2m-1$). For Q_2 and Q_3, there exist results of a computer search by S. UGAN ([26], [24]). **h)** For anti–symmetric mappings of the **semi-dihedral groups** of order $8m$ with m even, see [10] 2.1 (iii), (iv).

3.2 Theorem (GALLIAN and MULLIN) *Let* G *be a group and* $g \in G$*. The mapping* φ *with* $\varphi(x) = gx^{-1}$ *is anti–symmetric iff* g *commutes with no element of order 2;* (cf. [10] Th.3.1).

The proof is technical and shows that g commutes with the involution $y^{-1}x$ if $x\varphi(y) = y\varphi(x)$ for $x \neq y$. □

3.3 Corollary (i) *All groups of odd order admit anti–symmetric mappings;* ([10] 3.2). (ii)*For* $m > 2$*, the symmetric group* $\boldsymbol{S_m}$ *and the alternating group* $\boldsymbol{A_m}$ *have anti–symmetric mappings;* [10] 3.3).

PROOF. (i) 3.2 . (ii) As the element g of 3.2, choose an $m-$cycle when m is odd and an $(m-1)-$cycle when m is even. □

Using the classification of finite simple groups and applying Theorem 3.2, GALLIAN and MULLIN state the first part of the following.

3.4 Theorem *(a) Every finite simple group except $\mathbb{Z}_2$ has an anti–symmetric mapping;* ([10]). *(b) Every non–trivial finite p–group which is not a cyclic 2–group has an anti–symmetric mapping;* ([10] Th. 7.1).

PROOF of (b)(Idea). If p is odd one can apply 3.3. If $p = 2$ then there exist two elements of order 2 generating a group of order 4; now one considers a maximal subgroup containing this group and constructs a normal subgroup with non-cyclic factor group for which one gets anti–symmetric mappings by induction. □

An important tool to construct anti–symmetric mappings is the following.

3.5 Extension–Theorem (GALLIAN and MULLIN) *If H is a normal subgroup of G and there exist anti–symmetric mappings φ and ψ of H and G/H respectively then there exists an anti–symmetric mapping of G;* (cf. [10]).

PROOF(Sketch). Put $\gamma(u_i h) = \varphi(h)\psi^*(u_i)$ where ψ^* is the mapping induced by ψ on a set of representatives $\{u_i\}$ of the cosets of H . □

Especially, the direct product of groups with anti–symmetric mappings has an anti–symmetric mapping; this was known already to GUMM [11] and, implicitely, to VERHOEFF. So one can extend the results on the existence of anti–symmetric mappings from $p-$ groups: *Nilpotent groups with trivial or non–cyclic Sylow 2-subgroup admit anti–symmetric mappings.* This leeds to the following conjecture.

3.6 Conjecture of Gallian and Mullin *All non-abelian groups have anti–symmetric mappings;* ([10])

This conjecture has been confirmed by HEISS [12] for soluble groups.

3.7 Theorem (HEISS) *Every finite non-abelian solvable group admits an anti–symmetric mapping.*

PROOF (Idea): Recursive construction of anti–symmetric mappings starting from a normal subgroup of odd order and a cyclic 2-subgroup of a minimal counter–example. □

In a lecture given at the DMV-ÖMG meeting 1997, HEISS announced to have proved the full conjecture of Gallian and Mullin.

There exist as well an upper bound for the size of Ant(G), the set of anti–symmetric mappings of a finite group G (cf. DAMM [6] p.38 Th.9):

3.8 Theorem *For a group G of order m the following inequality holds (with e the Eulerian number).* $|\text{Ant}(G)| \leq m! - m\lceil (m-1)!(e-1)/e \rceil \leq m!/e + m/2$. This bound is sharp for $m = 2, 3, 4$ but not e.g. for $m = 10$ (bound $1,334,960$ for $|\text{Ant}(D_5)| = 34,040$, see section 6).

ANTI–SYMMETRIC MAPPINGS OF DIHEDRAL GROUPS

4.1 Representations of dihedral groups a) The *dihedral group* of order $2m$ is the symmetry group of the regular $m-$gon. Denoting the rotation through angle $2\pi/m$ by d and a reflection by s one has $D_m =< d, s \mid e = d^m = s^2 \wedge ds = sd^{-1} >$. The $2m$ elements are of the form $d^i s^j$ for $i = 0, \ldots, m-1$ and $j = 0, 1$.

b) If m is odd then, by defining $d = \begin{pmatrix} 1 & 0 \\ -1 & 1 \end{pmatrix}$ and $s = \begin{pmatrix} -1 & 0 \\ 0 & 1 \end{pmatrix}$, the dihedral group D_m can be represented as a matrix group (see e.g. [11]), namely $D_m \cong \{\begin{pmatrix} a & 0 \\ b & 1 \end{pmatrix} \mid a, b \in \mathbb{Z}_m \wedge a \in \{1, -1\}\}$. **c)** More general, for any $m > 2$, we have $D_m \cong \{(f, x) \mid f \in \{1, -1\} \wedge x \in \mathbb{Z}_m\}$ with operation $(f_1, x) \cdot (f_2, y) = (f_1 f_2, x f_2 + y)$ (cf. [11]). **d)** For any natural number m one can identify the element $d^i s^j \in D_m$ with the integer $j \cdot m + i$ $(j = 0, 1, i = 0, \ldots, m-1)$ (or $(1, -i) \mapsto i$ and $(-1, i) \mapsto m + i$ for the description according to c). Thus one gets a representation of D_m on $\{0, \ldots, 2m-1\}$ with induced operation $*$. In case $m = 5$ this operation has the following composition table (see e.g. [27], [8], [11], [19], [22], [28]); here k MOD m denotes the remainder of k under division by m.

$i * j$	$0 \leq j \leq 4$	$5 \leq j \leq 9$
$0 \leq i \leq 4$	$(i+j)$MOD 5	$5 + (i+j)$MOD 5
$5 \leq i \leq 9$	$5 + (i-j)$MOD 5	$(i-j)$MOD 5

4.2 Verhoeff's anti-symmetric mappings of dihedral groups (i) For the system of serial numbers of German banknotes, the anti–symmetric mapping used is the one found by VERHOEFF [27] p.95:

$$T_0 = \begin{pmatrix} 0 & 1 & 2 & 3 & 4 & 5 & 6 & 7 & 8 & 9 \\ 1 & 5 & 7 & 6 & 2 & 8 & 3 & 0 & 9 & 4 \end{pmatrix} = (01589427)(36).$$

In this scheme, there is used the check equation ($\diamond$) (see the introduction) with $\delta_i = {T_0}^i$ for $i = 1, \ldots, 10$ and $\delta_{11} = \mathrm{id}$; (cf. e.g. [19]). Furthermore, the alpha–numeric alphabet is encoded as in Table 3. **(ii)** Further anti-symmetric permutations found by computer search are, among others, (07319854)(26) and (03986215)(47) ([27] p.95). **(iii)** For D_m with m odd and $r \neq 0$ MOD m , the following mapping is anti–symmetric (cf. [27] p.91); $T(d^k) = d^{-k}$ and $T(d^j s) = d^{j+r} s$; for $m = 5$ this yields the permutations $\rho = (14)(23)(56789)$, which is mentioned again in [28], and (14)(23)(58697), see as well [11].

Table 3 Encoding the letters of the serial numbers of German banknotes

A	D	G	K	L	N	S	U	Y	Z
0	1	2	3	4	5	6	7	8	9

4.3 Other anti–symmetric mappings of $\mathbf{D}_m$ In the following we mention several other anti–symmetric mappings of D_m. That they have this property is proved by direct and exhaustive calculation. **a)** For m odd the mapping $T\begin{pmatrix} a & 0 \\ b & 1 \end{pmatrix} := \begin{pmatrix} a & 0 \\ h_a(b) & 1 \end{pmatrix}$ is anti–symmetric if h_a is injective and fulfills $bk - la \neq h_a(b) - h_k(l)$ for $(a, b) \neq (k, l)$; (see [20] 3.7). It is sufficient to put $h_a(b) = u_a - ab$ with $u_1 \neq u_{-1}$. Choosing $u_a = -at - c$ with $c, t \in \mathbb{Z}_m$ and $t \neq 0$ one gets the system of GUMM ([11] p.103), namely $T(d^k) = d^{c+t-k}$ and $T(d^j s) = d^{t-c+j} s$, especially for $t = r/2 = -c$ the system 4.2

(iii); and putting $u_{-1} = 0$ and $u_1 = 1 - m$ (or $c = t = (m-1)/2$ in GUMM l.c.) one has the systems of BLACK ([4]) for $m = 5$ and ECKER and POCH ([8] Th.4.4) $T(d^k) = d^{m-k-1}$ and $T(d^j s) = d^j s$. For $m = 5$, this mapping can be expressed as (04)(13). Choosing $u_1 = 0$ and $u_{-1} = -1$ (or $c = 1/2 = -t$ in Gumm l.c.) yields the scheme of WINTERS (again for m odd): $T = (0)(1\ m-1)(2\ m-2)\ldots(\frac{m-1}{2}\ \frac{m+1}{2})(2m-1\ \ 2m-2\ldots m+1\ \ m)$ or $T(d^k) = d^{-k}$ and $T(d^j s) = d^{j-1}s$; this is the system of VERHOEFF for $r = -1$. For $m = 5$ one gets the mapping ρ of 4.2(iii). Putting $c = t = 1$ in GUMM's system, one gets the scheme of GALLIAN and MULLIN ([10]Th.2.1 (i)) for m odd: $T(d^k) = d^{2-k}$ and $T(d^j s) = d^j s$. **b)** For m odd, the mapping $D_m \to D_m$ with $x \mapsto ax^{-1}b$ is anti–symmetric if $a \in \{d, \ldots, d^{m-1}\}$; (see 3.2 and 6.2 b). Choosing $a = d^t$ and $b = d^c$ yields the system of GUMM, see part (a). **c)** GALLIAN and MULLIN observed that for $m = 2k$ and $G = D_m$ the following mapping is anti–symmetric; ([10] l.c.;see as well [6]p.22). $T(s) = e; T(d^{-1}s) = ds;\ T(d^j) = d^{1-j}s\,(1 \le j \le k); T(d^j) = d^{1-j}(k+1 \le j \le m); T(d^j s) = d^{j+1}s(1 \le j \le k-1); T(d^j s) = d^{j+1}(k \le j \le m-2)$.
What can be said about the detection of other errors with D_m ? An important answer gives the following theorem of DAMM (cf.[6] p.55).

4.4 Theorem *For $m \ge 3$ odd there does not exist a check digit system over D_m which detects* (i) *all jump transpositions or* (ii) *all twin errors or all jump twin errors* (HALL/PAIGE).

PROOF (Sketch). In order to prove that D_m does not admit a jump transposition detecting mapping T one shows that the mappings $L_y \circ T^2$ can not be anti–symmetric (see Table 2) for all $y \in D_m$: Using the terminology of 4.1c we define $T^2(f, x) = (g_1(f, x), g_2(f, x))$. There exists an element $(-1, x) \in D_m$ such that the component function g_1 of T^2 fulfills $-g_1(1, 0) = g_1(-1, x)$; otherwise there would be $m + 1$ elements with the same signum, in contradiction to the fact that the positive elements of D_m form a subgroup of index 2. Then $L_c \circ T^2$ with $c = (1, \frac{1}{2}g_1(1,0)(g_2(-1,x) - xg_1(1,0) - g_2(1,0)))$ is not anti–symmetric (in the sense of (**)) as a straight-forward calculation shows. For twin errors and jump twin errors the statement is part of 2.4(iii). □

Therefore we are going to search for other groups with better detection rates. In connection with dihedral groups, there are still to mention the following results involving group–(anti–)automorphisms (– for definitions see 5.1).

4.5 Theorem (DAMM) (i) *D_m allows no anti–symmetric automorphism for* $m > 2$. (ii) *D_m admits an anti–symmetric anti–automorphism iff m is odd.*

PROOF (cf. [6]Th.28). One can show that an automorphism of D_m has a fixed conjugacy class and hence can't be anti–symmetric, see 5.3 and [23] 1.5. If m is even then (1, m/2) is a fixed point of any (signum and order preserving) anti–automorphism. If m is odd then $\psi : x \mapsto (1, -1)x^{-1}(1, 1)$ is a fixed point free anti–automorphism. Now, the assertion follows from 5.2(b). □

ANTI-SYMMETRIC (ANTI–)AUTOMORPHISMS

As seen in 2.5 and 3.2, the mapping inv: $x \longmapsto x^{-1}$ is, under certain conditions, an anti-symmetric mapping. On the other hand " inv" is, for every group, an anti-automorphism.

5.1 Definition A bijection $\psi : G \longrightarrow G$ of a group G is called **anti-automorphism** if $\psi(xy) = \psi(y) \cdot \psi(x)$ for all $x, y \in G$. The set of all anti-automorphisms of G is denoted by Antaut G.

Note that Antaut G = Aut G $\circ$ inv. In [6], DAMM uses anti-automorphisms to construct anti-symmetric mappings. He states:

5.2 Theorem (DAMM) **(a)** *If φ is anti-symmetric and ψ an anti-automorphism then $\psi \circ \varphi^{-1} \circ \psi^{-1}$ is anti-symmetric.* **(b)** *For an anti–automorphism ψ holds: ψ is anti-symmetric $\Longleftrightarrow$ ψ is fixed point free $\Longleftrightarrow$ $\varphi^{-1} \circ \psi \circ \varphi$ is fixed point free for any (anti-) automorphism φ.*

An overview on conditions for error detection using anti–automorphisms is given in Table 4a). The proofs are straight forward calculations. We continue with group–automorphisms.

Table 4 Error detection for anti–automorphisms ψ and automorphisms T

	Error type	*a) Conditions on ψ* (for all $x, y \in G$, $x \neq e$)	*b) Conditions on T* (for all $x, y \in G$, $x \neq e$)
1.	single error	none	none
2.a)	adjacent transpos.	$\psi(x) \neq x$	$T(x) \neq y^{-1}xy$
2.b)	jump transposition	$\psi^2(x) \neq y^{-1}xy$	$T^2(x) \neq y^{-1}xy$
3.a)	twin error	$\psi(x) \neq x^{-1}$	$T(x) \neq y^{-1}x^{-1}y$
3.b)	jump twin error	$\psi^2(x) \neq y^{-1}x^{-1}y$	$T^2(x) \neq y^{-1}x^{-1}y$
4.	phonetic error ($0 \hat{=} e, 1 \hat{=} g$)	$g^{-1}a \neq \psi(a) \neq ag^{-1}$ (for $a \hat{=} 2, \ldots, 9$)	$T(a) \neq g^{-1}a$ (for $a \hat{=} 2, \ldots 9$)

Source: [6][5]

5.3 Proposition. (i) *Let G be a finite group and $T \in$ Aut G. Then T is anti-symmetric iff T does not fix any conjugacy class of $G \setminus \{e\}$ (where e denotes the neutral element of G). When G is abelian, then this is the case iff T operates fixed point freely on G;* (see [23] 3.1 and 2.5 c).**(ii)** Sufficient (and for $n > 4$ also necessary) conditions on the automorphism T for the detection of errors are stated in Table 4 b) (cf. [5]).

PROOF (i)T is anti-symmetric iff $T(x)T(y)^{-1} = T(xy^{-1}) \neq x^{-1}(xy^{-1})x$ for all $x, y \in G$ with $x \neq y$. (ii) The condition for adjacent transpositons follows from (i). A twin error is detected iff $T^i(a)T^{i+1}(a) \neq T^i(b)T^{i+1}(b)$ which is equivalent to $T(ba^{-1}) \neq b^{-1}(ba^{-1})^{-1}b$. The other conditions follow similarly. □

5.4 Definition Let G be a finite group. An automorphism T of G is called **good** provided $T(x)$ is not conjugate to x or x^{-1} and $T^2(x)$ is not conjugate to x or x^{-1} for all $x \in G, x \neq e$; (cf.[5]).

5.5 Remarks. a) *A good automorphism is anti-symmetric and detects single errors, adjacent transpositions, jump transpositions, twin errors and jump twin errors;* (see 5.3). **b)** *If G is abelian then the automorphism T admits to detect single errors, adjacent transpositions, jump transpositions and twin errors if T^2 is fixed point free; and T is good if T^4 is fixed point free.*
c) *For any group G and automorphism T of odd order t already condition* 2a) *of* Table4b) *implies that T is good.*
PROOF.c) Since $\gcd(4,t) = 1$ there are integers r, s with $4r + st = 1$; any conjugacy class fixed by T^4 must be fixed by $T = T^{4r+st}$ too. □
5.6 An example *Choose $q = 2^m > 2$ and G as the* **Sylow 2–subgroup of the unitary group $\mathbf{SU(3,q^2)}$** *of order q^3, formed by the matrices $Q(x,y) = \begin{pmatrix} 1 & x & y \\ 0 & 1 & x^q \\ 0 & 0 & 1 \end{pmatrix}$ with $x, y \in GF(q^2)$ and $y + y^q + x^{q+1} = 0$. The automorphism $T : Q(x,y) \longmapsto Q(x\lambda^{2q-1}, y\lambda^{q+1})$, induced by conjugation with $H_\lambda = \begin{pmatrix} \lambda^{-q} & 0 & 0 \\ 0 & \lambda^{q-1} & 0 \\ 0 & 0 & \lambda \end{pmatrix}$ for $\lambda \in GF(q^2) \setminus \{0\}$, is good iff the multiplicative order of λ is not a divisor of $q+1$;* (BROECKER following a hint of G. STROTH). *The check character system using the automorphism T of order $q-1$ detects all single errors, adjacent-transpositions, twin errors, jump transpositions and jump-twin errors.* Generalization:
5.7 Good automorphisms on p-groups *Let P be a p-group and T be an element of* Aut*P. Suppose* gcd $(o(T), p(p-1)) = 1$. *Then T is good iff T is fixed point free on P;* (cf.[5]).
PROOF (Sketch). Take $P_1 := \Omega_1(Z(P))$ and define P_i inductively such that $P_i/P_{i-1} = \Omega_1(Z(P/P_{i-1}))$. (Here $\Omega_1(G)$ denotes the subgroup of G generated by the elements of order p.) One gets a T-invariant chain $P_0 = \{e\} < P_1 < \ldots < P_n = P$. If T is fixed point free on P then it acts fixed point freely on each P_i/P_{i-1}. Choose $x \in P$ such that $T(x)$ is conjugate to x and let i be minimal with $x \in P_i$. Suppose $i > 0$ then one can show $T(< xP_{i-1} >) =< xP_{i-1} >$. As Aut $(< xP_{i-1} >)$ is cyclic of order $p-1$ this shows $T(xP_{i-1}) = xP_{i-1}$, a contradiction. So $i = 0$ and $x = e$. Hence T is good by 5.5 (iii). □
5.8 Corollary *Let S be the* **Sylow 2–subgroup of PSL $(2,q)$** , $q = 2^m, m > 1$, *defined by $S = \{\begin{pmatrix} 1 & 0 \\ v & 1 \end{pmatrix} \mid v \in GF(q)\}$; then $T = \begin{pmatrix} t & 0 \\ 0 & t^{-1} \end{pmatrix}$ with $t \in GF(q) \setminus \{0,1\}$ acts fixed point freely on S. Therefore S admits a good automorphism hence a check digit system which detects all single errors, adjacent transpositions, twin errors, jump transpositions and jump-twin errors;* (cf.[5]). *Similarly, the* **Sylow 2–subgroups of the Suzuki group $\mathbf{Sz(q)}$** *(for $q = 2^{2t+1}, q > 2$) admit a good automorphism.* More general
5.9 Theorem *The* **Sylow 2–subgroup of a Chevalley group** *over* GF $(q), q = 2^m$, *admits a good automorphism T with $o(T) \mid (q-1)$ provided q is large enough;* (cf.[5] Result 2).

EQUIVALENCE OF CHECK DIGIT SYSTEMS

Although the systems over Chevalley groups admit to detect all single errors, adjacent transpositions, twin errors, jump transpositions and jump-twin errors we concentrate now on the dihedral group of order 10 since their elements can be interpreted as $0, 1, \ldots, 9$ and used in the decimal system. Because there are (exactly) 34,040 anti–symmetric mappings over D_5 (VERHOEFF [27] p.92, DAMM [6] p.44 with sieve methods , GIESE [9]) we want to define equivalences between these schemes. But there are several possibilities to do so. In the whole section, let G be a group and T_1, T_2 permutations of G.

6.1 Definition T_1 and T_2 are called **weak equivalent** if there exist elements a, b and an automorphism α of G such that $T_2 = R_a \circ \alpha^{-1} \circ T_1 \circ \alpha \circ L_b$. Here $R_a(x) := x \cdot a$ and, as before, $L_b(y) := by$; (cf. [27], [6], [18]).

6.2 Proposition. a) *Weak equivalence is an equivalence relation (i. e. reflexive, symmetric and transitive).* **b)** *If T_1 and T_2 are weak equivalent and if T_1 is anti–symmetric, then T_2 is anti-symmetric;* ([6] p.30, [27]). **c)** *If T_1 and T_2 are weak equivalent permutations of G then they detect the same percentage of twin errors;* ([18]). **d)** *If T_1 is an* **automorphism** *of G and T_2 is weak equivalent to T_1 then T_1 and T_2 detect the same percentage of jump transpositions and the same percentage of jump twin errors;* ([18]).

PROOF. a) Straight forward calculation (cf.[6] p.31). b) $xT_2(y) = yT_2(x)$ implies $xR_a \circ \alpha^{-1} \circ T_1 \circ \alpha \circ L_b(y) = yR_a \circ \alpha^{-1} \circ T_1 \circ \alpha \circ L_b(x)$, therefore $\alpha(b)\alpha(x)\alpha R_a \alpha^{-1} T_1 \alpha L_b(y) = \alpha(b)\alpha(y)\alpha R_a \alpha^{-1} T_1 \alpha L_b(x)$, hence $\alpha(bx)T_1(\alpha(by))$ $= \alpha(by)T_1\big(\alpha(bx)\big)$ showing $\alpha(bx) = \alpha(by)$, so $x = y$. c)We have $xT_2(x) \neq yT_2(y) \iff x\alpha^{-1}T_1\alpha(bx)a \neq y\alpha^{-1}T_1\alpha(by)a \iff \overline{x}T_1(\overline{x}) \neq \overline{y}T_1(\overline{y})$ for $\overline{x} = \alpha(bx)$ and $\overline{y} = \alpha(by)$; therefore the detection sets $M_{\mathrm{TE}}(T_1)$ and $M_{\mathrm{TE}}(T_2)$ (see Table 2) have the same cardinality. d) We get $xyT_2^2(z) \neq zyT_2^2(x) \iff \overline{x}\,\overline{y}\,T_1^2(\overline{z}) \neq \overline{z}\,\overline{y}\,T_1^2(\overline{x})$ for $\overline{x} = \alpha(bx), \overline{z} = \alpha(bz)$ and $\overline{y} = \alpha(y)T_1(\alpha(b))$; hence $|M_{\mathrm{JT}}(T_1)| = |M_{\mathrm{JT}}(T_2)|$ (see Table 2). A similar argument holds for jump twin errors. □

6.3 Weak equivalence and detection rates. The assertion of 6.2 d) might be wrong if T_1 and T_2 are not automorphisms; see the following counterexample (cf. [9], [18]). Let T_0 be VERHOEFF's anti–symmetric mapping (see 4.2) $T_0 = (01589427)(36)$. It detects 94.22 % of jump transpositions and 94.22 % of jump twin errors. Consider the weak equivalent permutation $T_1 := R_4 \circ \mathrm{id} \circ T_0 \circ \mathrm{id} \circ L_3$, namely $T_1 = (079482)(36)$. This mapping detects only 87.56 % of all jump transpositions and jump twin errors respectively.

6.4 Weak equivalence in the case of $\mathbf{D_5}$. According to GIESE [9] and DAMM [6] p.32, there exist exactly 20 equivalence classes with respect to weak equivalence; one of it contains 40 elements (with (01)(24) as representative); and 4 further classes have 1,000 elements each; the other 15 classes all are of cardinality 2,000. Since weak equivalence might not respect all error detecting capabilities, see 6.3, we restrict ourselves to stronger relations.

6.5 Definition T_1 and T_2 are called **automorphism equivalent** if there exists an $\alpha \in \mathrm{Aut}\ G$ such that $T_2 = \alpha^{-1} \circ T_1 \circ \alpha$; ([18]).

6.6 Proposition (i) *Automorphism-equivalence is an equivalence relation; and if T_1 and T_2 are automorphism equivalent then T_1 and T_2 are weak equivalent.* (ii) *If T_1 and T_2 are automorphism equivalent, then T_1 and T_2 detect the same percentage of adjacent transpositions, jump transpositions, twin errors and jump twin errors;*([18], [9]).

PROOF of (ii). The detection sets $M_{AT} = \{(x,y) \in G^2 \mid xT(y) \neq yT(x)\}, M_{JT}$ and M_{JZ} of $T = T_1$ can be mapped bijectively onto the corresponding sets of $T_2 = \alpha^{-1} \circ T_1 \circ \alpha$; for instance $(x,y) \in M_{AT}(T_2) \iff xT_2(y) \neq yT_2(x) \iff \alpha(x)\alpha\alpha^{-1}T_1(\alpha(y)) \neq \alpha(y)\alpha\alpha^{-1}T_1(\alpha(x)) \iff (\alpha(x), \alpha(y)) \in M_{AT}(T_1)$. For twin errors, (ii) follows from (i) and 6.2 c). □

Table 5 Types of anti–symmetric mappings of D_5 and their detection rates in %

	Type I	IIa	IIb	III	IV	V	VIa	VIb
single errors	100%	100%	100%	100%	100%	100%	100%	
adjacent transpos.	100%	100%	100%	100%	100%	100%	100%	
twin errors	95.56	95.56	91.11	91.11	91.11		55.56	
jump transpos.	94.22	92.00	94.22	92.00	90.22	2)	66,67	
jump twin errors	94.22	92.00	94.22	92.00	90.22		66.67	
Detection rate of all 5 error types[1)]	99.90	99.87	99.87	99.84	99.82	99.85–99.42	99.30	
Number of equivalence classes	2	44	8	160	16	1470	1	5
Size of classes	20	20	20	20	20	20	20	4

Source: [9].

1) weighted with the relative frequencies (without phonetic errors)
2) at least one rate below 90%

6.7 Types of equivalence classes over D_5 According to computations by GIESE with the program package MAGMA there are 1,706 equivalence classses of anti-symmetric mappings with respect to automorphism-equivalenće; [9]. Giese distinguishes 8 types of classes according to the rate of detection of errors and the size of equivalence-classes, see Table 5. In type V and VI, there are contained all classes which have at least one detection rate below 90%. Class VI is distinguisted since it contains many of the systems already known before. There exist classes of type V with a detection rate of 95.56 % for twin errors and of 89.78% for jump transpositions and for jump twin errors, thus giving an over all detection rate of 99.85 %. (The detection rates of systems of type I are in accordance with [27]p.95, those of type VI with [28] p.304). To give another point of view, GIESE has calculated as well the unweighted error detection rates for some codeword lengths, see [9].

6.8 Remarks on Table 5. (i) The phonetic error detecting capability may alter between automorphism equivalent systems (see 6.11). Therefore, this error type is not considered in Table 5. **(ii)** The relative frequency of errors used for the computation of detection rates of all 5 errors together, see Table 5, is based on the occurence among the non-coincidental errors without phonetic errors according to VERHOEFF's list (Table 1). So single errors are weighted with 86.909 %, adjacent transpositions with 11.221%, twin errors with 0.66%, jump transpositions with 0.88% and jump twin errors with 0.33% (of errors of these five types).

6.9 Description of equivalence classes over D_5 An overview on the number of classes and their sizes is given in Table 5. Type I contains 2 equivalence classes with 20 elements each; a representative of one class is the anti-symmetric mapping (0 7 3 1 9 8 5 4) (2 6) found by VERHOEFF; the second class contains the mapping T_0 of 4.2 (i) used for the German banknotes and (0 3 9 8 6 2 1 5) (4 7). The equivalence class of type VIa is represented by (0849) (1735) (26) ; the 5 classes of type VIb with 4 elements each contain all systems of the equivalent schemes of GUMM [11] and SCHULZ [20]. One of these classes has (04) (13) as a representative, the mapping found by BLACK [4], see as well ECKER & POCH [8], one other consists of 4 mappings given by VERHOEFF, namely (14)(23)(56789), (14)(23)(58697) and their inverses, see 4.2 (iii).

6.10 Phonetic errors The calculation of phonetic error detection rates is problematic in several ways. **(i)** The distibution of these errors depends on their position in the codeword. In VERHOEFF's statistics [27], this distribution is $15, 0, 9, 1$ and 34 for positions $(j, j+1)$ with $j = 1, \ldots, 5$ respectively. Verhoeff explains this by the habit of quoting the words in pairs of decimals. When partitioning a word in blocks of size 3 the position is likewise important (e.g. 15,000 is taken for **50**,000 more easily than for **11,5**00 , cf. DAMM). But at other places, the error probability may be different from that which one gets by partitioning in blocks of size 2 completely. Therefore we consider unweighted phonetic detection rates mainly. **(ii)** In VERHOEFF's random sample, the distribution of the errors $1x \longrightarrow x0$ and $x0 \longrightarrow 1x$ over x is that of Table 6. This shows how strongly these errors depend on the language and the phonetic resemblance of pairs in it; for Dutch, the low frequency of 8 is typical. So there should be made an extra statistics for each language.

Table 6 Distribution of phonetic errors $1x \longrightarrow x0$ and $x0 \longrightarrow 1x$

x	2	3	4	5	6	7	8	9
%	5.1	20.3	16.9	15.3	6.8	11.9	1.7	22.0

Sources: [27], [18].

6.11 Detection of phonetic errors As mentioned before, the detection rate of phonetic errors may vary in an automorphism equivalent class. Taking the class of T_0 and word length $n = 10$ as example, the number of recognizable phonetic errors out of 72 possible errors is 69 (for T_0 and 4 other mappings) 61, 60 and 57 (as well for 5 mappings each). Furthermore, for a permutation T which is anti-symmetric in the sense of (**), the detection rate of phonetic errors using check equation (*) may be different from that of T^{-1} when using check equation (*'). While the inverse mappings $\phi = T^{-1}$ of the examples T of Type I, III and VI of Table 7 have the same detection rates (for $n \leq 10$) as the corresponding mapping T, the permutation (146389725) given by DAMM ([6]) with check equation $(*')$ has phonetic detection rates of 87.5%, 85%, 85.42%, 87.5%, 87.5% and 87.5% for $n = 5, 6, 7, 8, 9$ and 10 respectively. (For Type VI, the percentages differ from WINTERS assertion [28]).

Table 7 Detection rates of phonetic errors (in %) for some representatives using D_5, check equation $(*)$ and word length n

	Type I	II a	III	VI b
T	$T_0=$ (01589427)(36)	(152798364)	(175)(238694)	(14)(23)(59876)
a) Unweighted phonetic error detection rate for T	n=5 96.9 n=6 95.0 n=7 95.8 n=8 96.4 n=10 95.8	n=5 87.5 n=6 90.0 n=7 87.5 n=8 85.7 n=10 87.5	100 %	n=5 56.3 n=6 62.5 n=7 56.3 n=8 60.7 n=10 59.7
b) Detection rate for all non – random errors ($n = 6$)	99.87 %	99.82 %	99.84 %	99.10 %
c) Unweighted phonetic error detection rate for T (without $12 \longleftrightarrow 20$)	n=5 96.4 n=6 97.1 n=7 97.6 n=8 98.0 n=10 96.8	n=5 89.3 n=6 91.4 n=7 90.5 n=8 89.8 n=10 90.5	100 %	n=5 50.0 n=6 57.1 n=7 50.0 n=8 55.1 n=10 54.0

Sources: [27], [9], [18].

If one wants to compare the error detection rates of some representatives one can take all non–coincidental errors as a base, so the weight–percentages are slightly different from those in 6.8(ii): 86.433, 11.160, 0.656, 0.875, 0.328 and (for phonetic errors) 0.547 %; (here the error $\ldots 12 \ldots \longleftrightarrow \ldots 20 \ldots$ is included as in Verhoeffs statistics). One gets detection rates according to Table 7b). Since the phonetic resemblance of 12 and 20 is, in German or English, not very large, DAMM does not count the error $\ldots 12 \ldots \longleftrightarrow \ldots 20 \ldots$. Using $(*')$, he gets a (weighted) phonetic error detection probability of 90.48% for $\phi = (146389725)$ and of 96.83% for $\phi = (07249851)(36)$. As well GIESE has calculated the (unweighted) detection rates of phonetic errors with $\ldots 12 \ldots \longleftrightarrow \ldots 20 \ldots$ excluded, see Table 7 c).

6.12 Remarks (i) *A similar investigation on eqivalence of anti–symmetric mappings of dicyclic groups and generalized quaternion groups has been made by Sehpanuhr UGAN [26] for her diploma thesis.* **(ii)** *Note that check digit systems using so called total anti–symmetric mappings of the quasi-groups* $(\mathbb{Z}_{10}, *)$ *with* $x * y = (x + y)\ MOD 10$ *if* x *is even and* $x * y = (x - y - 2)\ MOD 10$ *if* x *is odd may have an error detecting rate of 99,89 % for all* 6 *non–random error types; ([6]) . In view of Theorem 5.2 a) we define*

6.13 Definition T_1 and T_2 are called **strongly equivalent** if there exists an $\alpha \in \text{Aut } G$ such that $T_2 = \alpha^{-1} \circ T_1 \circ \alpha$ or a $\psi \in \text{Antaut } G$ with $T_2 = \psi^{-1} \circ {T_1}^{-1} \circ \psi$.

6.14 Proposition. a) *Strong equivalence is an equivalence relation; and if T_1, T_2 are strongly equivalent then T_1 is anti-symmetric iff T_2 is anti-symmetric.* **b)** *If T_1 and T_2 are strongly equivalent then T_1 and T_2 detect the same percentage of adjacent transpositions, jump transpositions, twin errors and jump-twin errors;* ([18]).

PROOF. a) 6.2 b) and 5.2 a). b) In view of 6.6 it suffices to consider $T_2 = \psi^{-1} \circ {T_1}^{-1} \circ \psi$ for $\psi \in \text{Antaut } G$; one gets e.g. $(x, y) \in M_{AT}(T_2) \iff x\psi^{-1}(T_1^{-1} \circ \psi(y)) \neq y\psi^{-1}(T_1^{-1} \circ \psi(x)) \iff T_1^{-1} \circ \psi(y)\, T_1\, T_1^{-1}\, \psi(x) \neq T_1^{-1} \circ \psi(x) T_1 T_1^{-1} \psi(y) \iff (T_1^{-1} \circ \psi(x), T_1^{-1} \circ \psi(y)) \in M_{AT}(T_1)$. The other cases can be handled similarly. □

6.15 Strong equivalence of schemes over D_5 . According to computer calculations by GIESE [9] (again using MAGMA) there are 911 equivalence classes of anti-symmetric mappings of D_5 with respect to strong equivalence; 115 classes, containing 4,600 systems, belong to type I to IV (see 6.7). Type I consists now of 1 equivalence class; for types II to IV as well, two equivalence classes with respect to automorphism equivalence fuse to one class with respect to strong equivalence (with 40 elements each). But the classes of Type VI remain unchanged.

References

[1] D.F. Beckley, "An optimum system with modulus 11", *The Computer Bulletin*, 11, 1967, 213–215.

[2] D. Bedford, "Orthomorphisms and near orthomorphisms of groups and orthogonal Latin squares", *Bulletin of the ICA*, 15, 1995, 13–33. Addendum to orthomorphisms .... Bulletin of the ICA, 18, 1996, p.86.

[3] A. Beutelspacher, "Vertrauen ist gut, Kontrolle ist besser! Vom Nutzen elementarer Mathematik zum Erkennen von Fehlern", in *Jahrbuch Überblicke Mathematik 1995*, Vieweg, 1995, 27–37.

[4] W.L. Black, "Error detection in decimal numbers", *Proc IEEE (Lett.)*, 60, 1972, 331–332.

[5] C. Broecker, R.-H. Schulz, and G. Stroth, "Check character systems using Chevalley groups", *Designs, Codes and Cryptography*, 10, 1997, 137–143.

[6] H.M. Damm, "Prüfziffersysteme über Quasigruppen", Diplomarbeit Universität Marburg, März 1998.

[7] J. Dénes and A.D. Keedwell, "A new conjecture concerning admissibility of groups", *Europ. J. of Combin.*, 10, 1989, 171–174.

[8] A. Ecker and G. Poch, "Check character systems", *Computing*, 37 (4), 1986, 277–301.

[9] S. Giese," Äquivalenz von Prüfzeichensystemen am Beispiel der Diedergruppe D_5", Staatsexamensarbeit FU Berlin, 1999.

[10] J.A. Gallian and M.D. Mullin, "Groups with antisymmetric mappings", *Arch.Math.*, 65, 1995, 273–280.

[11] H.P. Gumm, "A new class of check-digit methods for arbitrary number systems", *IEEE Trans. Inf. Th. IT*, 31, 1985, 102–105.

[12] S. Heiss, "Antisymmetric mappings for finite solvable groups", *Arch. Math.*, 69(6), 1997, 445–454.

[13] M. Hall and L.J. Paige, "Complete mappings of finite groups", *Pacific J. Math.*, 5, 1955, 541–549.

[14] D.M. Johnson, A.L. Dulmage, and N.S. Mendelsohn, "Orthomorphisms of groups and orthogonal Latin squares I", *Canad. J. Math.*, 13, 1961, 356–372.

[15] H.B. Mann, "The construction of orthogonal Latin squares", *Ann. Math. Statistics*, 13, 1942, 418–423.

[16] L.J. Paige, "A note on finite abelian groups", *Bull. AMS*, 53, 1947, 590–593.

[17] R. Schauffler, " Über die Bildung von Codewörtern", *Arch. Elektr. Übertragung*, 10(7), 1956, 303–314.

[18] R.-H. Schulz, "Private communication with S.Giese", 1997/98.

[19] R.-H. Schulz, *Codierungstheorie. Eine Einführung*, Vieweg Verlag, Braunschweig/Wiesbaden, 1991.

[20] R.-H. Schulz, "A note on check character systems using Latin squares", *Discr. Math.*, 97, 1991, 371–375.

[21] R.-H. Schulz, "Some check digit systems over non-abelian groups", *Mitt. der Math. Ges. Hamburg*, 12(3), 1991, 819–827.

[22] R.-H. Schulz, "Informations- und Codierungstheorie – eine Einführung", in R.-H. Schulz (editor), *Mathematische Aspekte der angewandten Informatik*, BI, Mannheim etc. 1994, 89–127.

[23] R.-H. Schulz, "Check character systems over groups and orthogonal Latin squares", *Applic. Algebra in Eng., Comm. and Computing, AAECC*, 7, 1996, 125–132.

[24] R.-H. Schulz, "Equivalence of check digit systems over the dicyclic groups of order 8 and 12", *Geburtstagsband für Harald Scheid*, To appear.

[25] H. Siemon, *Anwendungen der elementaren Gruppentheorie in Zahlentheorie und Kombinatorik*, Klett–Verlag, Stuttgart, 1981.

[26] S. Ugan, "Prüfzeichensysteme über dizyklischen Gruppen der Ordnung 8 und 12", Diplomarbeit FU Berlin, 1999.

[27] J. Verhoeff, *Error detecting decimal codes*, volume 29 of *Math. Centre Tracts*, Math. Centrum Amsterdam, 1969.

[28] S.J. Winters, "Error detecting schemes using dihedral groups", *The UMAP Journal*, 11(4), 1990, 299–308.

SWITCHINGS AND PERFECT CODES *

Faina I. Solov'eva

Sobolev Institute of Mathematics, pr. Koptyuga 4
Novosibirsk 630090, Russia

sol@math.nsc.ru

Dedicated to Rudolf Ahlswede on the occasion of his 60th birthday

Abstract: Let C be a code (or a design or a graph) with some parameters. Let A be a subset of C. If the set $C' = (C \setminus A) \cup B$ is a code (a design or a graph) with the same parameters as C we say that C' is obtained from C by *a switching*. Special switchings for perfect binary codes are considered. A survey of all nontrivial properties of perfect codes given by the switching approach is presented. Some open questions are discussed.

INTRODUCTION

Investigating perfect codes is one of the most fascinating subjects in coding theory. It is well known [43-45], [39] that nontrivial perfect q-ary single-error-correcting codes (briefly perfect codes) exist only for length $n = (q^k - 1)/(q-1)$, $k \geq 2$, for length 23 (the binary Golay code) and for length 11 (the ternary Golay code). Both Golay codes are unique up to equivalence. Many problems regarding perfect codes are still open, for example, the main problem of the construction and enumeration of perfect codes remains unsolved. Especially in recent years, a lot of papers have been devoted to the construction and investigation of properties of perfect codes. Several approaches were developed for studying these questions. The switching approach appeared to be the most fruitful. It allows a series of problems to be solved. The aim of the paper is to survey all known nontrivial properties of perfect binary codes given by the switching approach. We present a short summary of other nontrivial properties

*This research was supported by the Russian Foundation for Basic Research under grant 97-01-01104

I. Althöfer et al. (eds.), Numbers, Information and Complexity, 311-324.

of perfect codes and give a list of references concerning the properties and constructions of perfect codes. Some open problems will be considered.

NECESSARY DEFINITIONS

A *q-ary code* C of length n is a subset of the vector space E_q^n of dimension n over the Galois field $GF(q)$. The elements of C are called *codewords* or *vectors.* The best progress in studying perfect codes was made for $q = 2$.

Recall the necessary definitions and notions for binary codes. We denote the vector space of dimension n over $GF(2)$ by E^n. Two codes $C, C' \subset E^n$ are said to be *isomorphic* if there exists a permutation π such that $C' = \pi(C) = \{\pi(x) : x \in C\}$. Codes $C, C' \subset E^n$ are *equivalent* if there exists a vector $b \in E^n$ and a permutation π such that $C' = b \oplus \pi(C) = \{b \oplus \pi(x) : x \in C\}$. *The Hamming distance* $d(x, y)$ between vectors $x, y \in C$ is the number of coordinates in which x and y differ. *The Hamming weight* of $x \in C$ is given by

$$wt(x) = d(x, \mathbf{0}),$$

where $\mathbf{0}$ is the all-zero vector. A code distance is given by $d = min\, d(x, y)$ for any different codewords $x, y \in C$. *A neighborhood* $K(M)$ of a set M in E^n is the union of spheres of radius 1 with centers at the vectors of M. A set $C \subseteq E^n$ is called *a perfect code* of length n if $K(C) = E^n$ and for any $x, y \in C$ one has $K(x) \cap K(y) = \emptyset$. Let $M \subset C$. Exchanging the bit in the i'th coordinate of all vectors of a set M with the opposite bit we obtain a new set, denoted by $M \oplus i$. A set M is an *i-component* of the perfect code C if $K(M) = K(M \oplus i)$. It is not difficult to see that the set $C' = (C \setminus M) \cup (M \oplus i)$ is a perfect code. We say that C' is obtained from the code C by *a switching* (or a *translation,* see [9]) of an i-component M.

SHORT SUMMARY OF PROPERTIES

It is known that there are many interesting properties concerning perfect codes especially perfect binary codes. The linear perfect codes called *Hamming codes* are unique up to equivalence.

A code is *distance-invariant* if the number $A_i(n)$ of all codewords on distance i from the fixed codeword does not depend on the choice of the codeword. In 1957 Lloyd [20] and in 1959 Shapiro and Slotnik [31] proved a perfect binary code to be distance-invariant. Abdurahmanov [1] showed the same result for any q-ary perfect code. A binary code of length n is *distance-regular* if for any codewords α, β and any integers $i, j \in \{1, \dots, n\}$ the number of codewords γ such that $d(\alpha, \gamma) = i, \;\; d(\beta, \gamma) = j$, does not depend upon the choice of α, β but only depends on $d(\alpha, \beta)$. In [10] it is proved that among the perfect binary codes with distance 3 only Hamming codes of length 3 and 7 are distance-regular. A subset F of all vectors in E^n with fixed $n - k$ coordinates is called a k-dimensional *face.* Every perfect binary code of length n has uniform distribution in k-dimensional faces of E^n, $k \geq (n + 1)/2$. The result is proved by Delsarte [14] in 1972 and independently by Pulatov [29] in 1973. In [30] Pulatov generalized the result for any q-ary perfect codes. Spectral properties of perfect binary codes generalizing results of Shapiro and Slotnik, Delsarte, Pulatov were developed by Vasil'eva [43, 46]. In [45] the concept of a centered characteristic function of a perfect code is introduced and it is established that the centered characteristic function of a perfect code is presented as a linear combination of the centered characteristic functions of an arbitrary class of equivalent perfect codes.

Many papers are concerned with the construction of perfect codes. A survey of perfect binary codes is given in [36] and one of q-ary perfect codes in [21]. All constructions can be divided into two parts, the former being concatenation constructions, the latter being switching constructions. We discuss switching constructions in Sections 4, 5 and 7 below. In 1962 Vasil'ev [40] discovered the first class of nonequivalent perfect binary codes. Vasil'ev's construction is a switching construction. It can be found in Section 4. In 1986 Mollard [24] generalized Vasil'ev's construction, see Section 5 below. The general switching construction can be found in [9], see also Section 7.

Every finite group is isomorphic to the full permutation automorphism group of some perfect binary code. Hence there exist perfect binary codes with the trivial permutation automorphism group. This was proved in 1986 by Phelps [25].

In 1995 Avgustinovich [4] showed that every perfect binary code was uniquely determined by its codewords of weight $(n-1)/2$.

Let $C \subseteq E^n$ be a code. The set K of all vectors $x \in E^n$, for which $C \oplus x = C$ is called the kernel of C. Bayer, Ganter and Hergert [13] developed algebraic techniques for nonlinear perfect binary codes and investigated their kernels. Heden [17] found three perfect binary codes of length 15 which have kernels of dimension 1, 2 and 3. For all $k \geq 4$ there exists a nonlinear perfect binary code of length $n = 2^k - 1$ which had a kernel of dimension j if and only if $j \in \{1, 2, \ldots, 2^k - k - 3\}$. This result was established by Phelps and LeVan [26].

Etzion and Vardy [15] presented a perfect binary code of full rank for every $n = 2^k - 1, k \geq 4$, see Section 8 below.

In [8] it is proved that there exist nonsystematic perfect binary codes of length n for every $n = 2^k - 1,\ \ k \geq 8$. For $5 \leq k \leq 7$ such codes were found by Phelps and LeVan [27]. A class of nonsystematic perfect binary codes of length $n > 127$ with a trivial automorphism group is presented in [11]. An analogous result is found in [22] by Malyugin for a systematic case for all admissible lengths greater than 15.

The intersection number was investigated by Etzion and Vardy in [15, 16] and Vasil'eva in [44]. In [15] it is proved that the smallest nonempty intersection of two perfect binary codes of length n consists of two codewords for all admissible n, see Section 12 below.

A mapping $\phi : C \to E_q^n$ is called an isometry from the code C to the code $\phi(C)$ if $d(x, y) = d(\phi(x), \phi(y))$ for all codewords $x, y \in C$. A code C in E_q^n is called *metrically rigid* if every isometry $\phi : C \to E_q^n$ with respect to the Hamming metric is extendable to an isometry of the whole space E_q^n. The metrical rigidity of perfect codes with the exception of the binary Hamming code of length 7 and the ternary Hamming code of length 4 was proved in [3, 35]. Two codes C_1 and C_2 are *weakly isometric* if there is a map $J : C_1 \to C_2$ such that the equality $d(\alpha, \beta) = 3$ holds iff $d(J(C_1), J(C_2)) = 3$. It is clear that isometric codes are weakly isometric. In [28] Phelps and LeVan ask whether perfect codes with isomorphic minimum distance graphs are always equivalent. It means: are two weakly isometric perfect codes equivalent? In [12] Avgustinovich proves that any two weakly isometric perfect binary codes are equivalent.

Exact upper and lower bounds on the number of i-components of an arbitrary perfect binary code were found in [32, 33]. According to [32] there exist nonextremal cardinality i-components of perfect binary codes of length n for all admissible $n > 7$. A perfect binary code of length $n, n > 7$, with i-components of different structures and cardinalities was presented in [5]. A class of perfect binary codes of length n with nonextremal cardinality i-components is constructed for all admissible $n > 7$ and the existence of maximal cardinality nonisomorphic i-components of different perfect binary codes of length n for all $n = 2^k - 1, k > 3$, was proved, see [37, 38].

VASIL'EV CODES

From now on we consider only perfect binary codes (briefly perfect codes). Let V^p be a perfect code of length $p = 2^k - 1, k \geq 2$. Let λ be an arbitrary function from V^p to the set $\{0, 1\}$. For $\gamma \in E^p$ let $|\gamma| = \gamma_1 + \cdots + \gamma_p \pmod 2$, where $\gamma = (\gamma_1, \ldots, \gamma_p)$. Set $n = 2p + 1$.

Theorem 1. (Vasil'ev, [40].) *The set $V^n = \{(\gamma, \gamma \oplus \beta, |\gamma| \oplus \lambda(\beta)) : \gamma \in E^p, \beta \in V^p\}$ is a perfect code of length n.*

Since λ is an arbitrary function, we obtain (taking the previous iterative steps into account) the following lower bound on the number of different perfect codes:

$$N(V^n) \geq 2^{2^{\frac{n+1}{2} - \log(n+1)}} \cdot 2^{2^{\frac{n+1}{4} - \log(n+1)}},$$

where $N(V^n)$ denotes the number of Vasil'ev codes of length n. This bound has been the best lower bound for a long time.

The concept of i-components (in terminology of disjunctive normal forms) was introduced by Vasil'ev [40, 41].

It is easy to see that the set $M_n = \{(\gamma, \gamma, |\gamma|) : \gamma \in E^p\}$ is the n-component of V^n of cardinality $2^{\frac{n-1}{2}}$, $n = 2p + 1$, and Vasil'ev's construction is the switching construction. Let $K(M_n)$ and $K(M_n \oplus n)$ be neighborhoods of M_n and $M_n \oplus n$ respectively. It is true that $K(M_n) = K(M_n \oplus n)$. Therefore M_n is an n-component by the definition and $(V^n \setminus M_n) \cup (M_n \oplus n)$ is a perfect code. Analogously

$$V^n \setminus (\bigcup_{\beta \in V_1^p} M_n^\beta)) \cup (\bigcup_{\beta \in V_1^p} (M_n^\beta \oplus n)$$

is a perfect binary code of length n, where V_1^p is a subcode of the code V^p and $M_n^\beta = M_n \oplus (\mathbf{0^p}, \beta, \mathbf{0})$, $\beta \in V^p$.

An i-component is *minimal* if it cannot be subdivided into smaller i-components. In [33] it was proved that an i-component of cardinality $2^{(n-1)/2}$ is minimal i-component with minimal cardinality. It is not difficult to see that minimal i-component is unique up to equivalence. In [32, 42] the concept of i-components was developed and other switching constructions of perfect binary codes were found. The lower bound given there is of the form $2^{2^{\frac{n+1}{2}(1-\varepsilon_n)}}$, where $\varepsilon_n \to 0$ if $n \to \infty$.

MOLLARD CODES

Some unessential improvement of $N(V^n)$ can be obtained by Mollard's construction [24], which we shall present now.

Let C^r and C^m be two perfect codes of length r and m respectively. Let

$$\alpha = (\alpha_{11}, \alpha_{12}, \ldots, \alpha_{1m}, \alpha_{21}, \ldots, \alpha_{2m}, \ldots, \alpha_{r1}, \ldots, \alpha_{rm}) \in E^{rm}.$$

The generalized parity functions $p_1(\alpha)$ and $p_2(\alpha)$ are defined by $p_1(\alpha) = (\sigma_1, \sigma_2, \ldots, \sigma_r) \in E^r$, $p_2(\alpha) = (\sigma'_1, \sigma'_2, \ldots, \sigma'_m) \in E^m$, where $\sigma_i = \sum_{j=1}^m \alpha_{ij}$ and $\sigma'_j = \sum_{i=1}^r \alpha_{ij}$. Let f be an arbitrary function from C^r to E^m.

Theorem 2. (Mollard, [24].) *The set*

$$M^n = \{(\alpha, \beta \oplus p_1(\alpha), \gamma \oplus p_2(\alpha) \oplus f(\beta)) : \alpha \in E^{rm}, \beta \in C^r, \gamma \in C^m\}$$

is a perfect code of length $n = rm + r + m$.

In the case $m = 1$ Mollard's and Vasil'ev's constructions coincide. In [34] the existence of Mollard codes which are not Vasil'ev codes was demonstrated.

STRUCTURE OF I-COMPONENTS

The next problem concerning perfect codes is the analysis of the cardinality and the investigation of the structure of i-components. In this section we consider the progress in the study of these questions. In [9] were proved the following Propositions.

Proposition 1. *Let M be an i-component of any perfect code C. Then the set $C \setminus M$ is an i-component of the perfect code C too.*

Proposition 2. *Let M_1 and M_2 be i-components of a perfect code C. Then the sets $M_1 \cup M_2$, $M_1 \cap M_2$, $M_1 \setminus (M_1 \cap M_2) = M_1 \setminus M_2$ are i-components of the perfect code C.*

Proposition 3. *Let M be an i-component of a perfect code C and for some perfect code D it is true that $M \subset D$. Then M is an i-component of the code D.*

Theorem 3. (See [32, 33]) *The exact upper and lower bounds on the number of minimal i – components of a perfect code of length n,* $n = 2^q - 1$, *are*

$$2 \leq L_n \leq 2^{\frac{n+1}{2}}/(n+1),$$

where L_n *is the number of minimal i-components.*

Consequence. *The cardinality of the minimal i-components can vary from* $2^{(n-1)/2}$ *to* $2^{n-1}/(n+1)$.

Theorem 4. (See [5]) *For any* $n = 2^q - 1$, $q \geq 4$, *there exists a perfect code of length n such that the set of minimal i-components of the code contains i-components with different structures and cardinalities for some i.*

Theorem 5. (See [37, 38]) *There exist maximal cardinality nonisomophic minimal i-components of different perfect codes of length n for all* $n = 2^k - 1, k > 3$.

Theorem 6. (See [37, 38]) *There exists a perfect code of length n with minimal i-components cardinality* $(t+1)2^{n-t}/(n+1)$ *for every* $n = 2^k - 1, k > 3$ *and* $t = 2^s - 1$, *where* $s = 2, \ldots, log(n+1)/2$.

However, the problem of enumerating all possible sizes of minimal i-components of perfect binary codes remains open.

$\tilde{\alpha}$-COMPONENTS, LOWER BOUND

We further identify a vector $x = (x_1, \ldots, x_n) \in E^n$ with its support $\{i : x_i = 1\}$. Let $\tilde{\alpha} \subseteq N = \{1, \ldots, n\}$. The set M is called an *$\tilde{\alpha}$-component* of the perfect code C if it is an i-component for every $i \in \tilde{\alpha}$.

Proposition 4. *Let* M *be the* $\tilde{\alpha}$*-component of a perfect code* C, $i \in \tilde{\alpha}$, *and let the set* $M' \subseteq M$ *be the i-component of the code* C. *Then* $M^* = (M \setminus M') \cup (M' \oplus i)$ *is the* $\tilde{\alpha}$*-component of the code* $C^* = (C \setminus M') \cup (M' \oplus i)$.

Given a perfect code C of the length n. Let $\tilde{\alpha} = \{\alpha_1, \ldots, \alpha_t\}$ be the vector of weight t with only the α_1'th, ..., α_t'th coordinates equal to 1. Let $M^1_{\alpha^1}, \ldots, M^k_{\alpha^k}$ be mutually disjoint subsets of the code C such that $M^s_{\alpha^s}$ is the α^s-component of C, where $\alpha^1, \ldots, \alpha^k \subset \{1, \ldots, n\}$ are not all necessarily different and let $\beta^s \subseteq \alpha^s$.

Theorem 7. (See [9].) *The set*

$$C' = (C \setminus (\bigcup_{s=1}^{k} M^s_{\alpha^s})) \cup (\bigcup_{s=1}^{k} (M^s_{\alpha^s} \oplus \beta^s))$$

is a perfect binary code of length n.

Define (the *switching class*) the *single switching class* of a perfect code C as the set of all perfect codes obtained from C by (a sequence of) $\tilde{\alpha}$-component switches. Phelps and LeVan [28] presented a perfect code of length 15 and showed that it does not belong to the switching class of the Hamming code. Hence for any n there exist switching classes of perfect codes and it is interesting to clarify the number of classes for every $n = 2^k - 1, k > 3$. A classification of all perfect codes of length 15 formed from the Hamming code of length 15 by single switchings is presented in [23].

Hamming codes are unique up to equivalence therefore for any two different Hamming codes H_1^n and H_2^n of length n there exists a vector b and a permutation π such that $H_1^n = b \oplus \pi(H_2^n)$. By the definition of a switching $b \oplus H^n$ belongs to the switching class of the Hamming code H^n of length n. It is not difficult to prove that a transposition $(j,k)(H^n)$ of coordinates j and k of H^n switches exactly a half of i-components of H^n, where $(i,j,k) \in H^n$. Therefore $\pi(H^n)$ and H^n are switching equivalent and we have than

Proposition 5. *Any two Hamming codes H_1^n and H_2^n of length n are switching equivalent.*

Now we give a short description of the construction of Avgustinovich and Solov'eva [6, 9]. Consider the Hamming code H^n of length n. Let $\{i,j,k\}$ be the vector of H^n of weight 3. It means that only the i'th, j'th and k'th coordinates are equal to 1. Let $N_1 = 2^{\frac{n+1}{4} - \log(n+1)}$, $N_2 = 2^{\frac{n-3}{4}}$.

Proposition 6. *The Hamming code H^n can be partitioned in $\{i,j,k\}$-components R_{ijk}^t :*

$$H^n = \bigcup_{t=1}^{N_1} R_{ijk}^t.$$

Proposition 7. *Every $\{i,j,k\}$-component R_{ijk}^t, $t = 1,\ldots,N_1$, can be partitioned in i-components R_i^l :*

$$R_{ijk}^t = \bigcup_{l=1}^{N_2} R_i^l.$$

We now choose one of the coordinates i,j or k for every $\{i,j,k\}$-component R_{ijk}^t and divide the $\{i,j,k\}$-component into the components in the chosen coordinate. Thus the code H^n is split into the i-, j- and k-components with minimal cardinalities. This partition of the Hamming code allows us to construct a large class of different perfect binary codes.

Theorem 8. (See [6, 9].) *There are at least*

$$2^{2^{\frac{n+1}{2} - \log(n+1)}} \cdot 6^{2^{\frac{n+1}{4} - \log(n+1)}}$$

different perfect binary codes of length n.

This bound is better than the other known lower bounds. A full proof can be found in [9]. It is easy to see that this construction method is possible for the Hamming code divided into some $\tilde{\alpha}$-components, where every $\tilde{\alpha}$-component is divided into $\tilde{\alpha'}$-components, $\tilde{\alpha'} \subseteq \tilde{\alpha}$. Such partitions yield complicated classes of perfect codes. We restrict ourselves to the case which gave us the maximal factor in the lower bound of Theorem 8.

From Section 5 it is not difficult to see that Mollard's construction can be described by the method of $\tilde{\alpha}$-components, see also [6].

RANKS OF PERFECT CODES

The rank $r(C)$ of a code $C \subset E^n$ is the maximum number of linearly independent vectors in the code C. Ranks of perfect binary codes were investigated by Hergert [19], Heden [17], Etzion and Vardy [15, 16]. A code C^n of length n is of *full rank* if $r(C^n) = n$. Using switchings of i-components Etzion and Vardy [15] constructed full rank perfect code of length n from the Hamming code for all admissible n.

Consider the Hamming code H^n as a set of all vectors $\alpha = (\alpha_1, \dots, \alpha_n)$ such that $\bigoplus_{i=1}^n \alpha_i h_i = \mathbf{0}^k$, where $h_i \in E^k \setminus \mathbf{0}^k$ and h_i is the binary presentation of i, $k = \log(n+1)$. A set $\{i_1, \dots, i_k\} \subset \{1, \dots, n\}$ of numbers such that $\{h_{i_1}, \dots, h_{i_k}\}$ are independent vectors is called the set of *independent points.*

Lemma 1. (See Lemma 6.1 in [15] and Lemma 5 in [26].) *Let H^n be the Hamming code of length $n = 2^k - 1$, $k \geq 4$ with the set $\{1, \dots, k\}$ as the set of its independent points. Then there are k minimal i-components $M_1, \dots, M_k$ with minimal cardinality in H^n such that $M_i \cap M_j = \emptyset$ for any distinct $i, j \in \{1, \dots, k\}$.*

Theorem 9. (See [15].) *The set*

$$D^n = (H^n \setminus (\bigcup_{i=1}^k M_i)) \cup (\bigcup_{i=1}^k (M_i \oplus i))$$

is a full rank perfect binary code of length n for every $n = 2^k - 1, \;\; k \geq 4$.

In [15] Etzion and Vardy proved the following result

Theorem 10. *For all $k \geq 4$ there exists a nonlinear perfect binary code of length $n = 2^k - 1$ with a rank of dimension t if and only if $t \in \{2^k - k, 2^k - k + 1, \dots, 2^n\}$.*

KERNELS OF PERFECT CODES

Let $C \subseteq E^n$ be a code. The set $Ker(C)$ of all vectors $x \in E^n$, for which $C \oplus x = C$ is called the kernel of C. In 1994 Heden [17] constructed three perfect codes of length 15 which had kernels of dimension 1, 2 and 3. In 1995 Phelps and LeVan [26] established the following result

Theorem 11. *The dimension of a kernel $Ker(D^n)$ of the code D^n given in Theorem 9 is equal to 1.*

By multiple special switchings Phelps and LeVan obtained perfect codes with kernels of all possible sizes.

Theorem 12. *For all $k \geq 4$ there exists a nonlinear perfect binary code of length $n = 2^k - 1$ which has a kernel of dimension j if and only if $j \in \{1, 2, \ldots, 2^k - k - 3\}$.*

It is interesting to clarify the connection between ranks and kernels. Which pairs (r, k) are attainable as the rank r and kernel dimension k of a perfect code of length $2^k - 1$? The question was posed by Etzion and Vardy in [16]. The first connection between the rank $r(C^n)$ and the kernel $Ker(C^n)$ of a perfect code (C^n) is established by Hergert [19].

Theorem 13. *For any perfect binary code C^n of length n it is true*

$$1 + n - r(C^n) \leq Ker(C^n).$$

Hence, if $Ker(C^n) = 1$ then the rank $r(C^n)$ coincides with the dimension n of E^n regardless of the size of the permutation automorphism group of the code C^n. Some pairs (r, k) are admissible, see [16] and Section 11 below. A full rank perfect code of length $n = 2^k - 1$ can also be constructed by induction on $k, k \geq 4$. According to Lemma 2.2 in [15], if we use a code V^p of rank $r(V^p)$ in Vasil'ev's construction we will obtain a perfect code V^n of length $n = 2p + 1$ of rank $r(V^n) = r(V^p) + p + 1$ as a resulting code. If $r(V^p) = p$ then $r(V^n) = n$ and V^n is a full rank perfect code. As the first full rank perfect code one can use, for example, Heden's full rank perfect code of length 15 from [17].

NONSYSTEMATICY

Avgustinovich and Solov'eva [7, 8] constructed a class of nonsystematic perfect binary codes of length n for every $n = 2^k - 1,\ \ k \geq 8$. The question about the existence of nonsystematic perfect codes was posed by Hergert [19]. A perfect code C of length n is systematic if there are $n - \log(n+1)$ coordinates such that the code C deleted in the remaining $\log(n+1)$ coordinates coincides with $E^{n-\log(n+1)}$.

Proposition 8. *Let $n = 2^k - 1,\ \ k \geq 8$. There are n minimal components $M_1, \ldots, M_n$ with minimal cardinalities in the Hamming code H^n such that the i'th component M_i is an i-component and the distance between two components M_i and M_j is greater than 4 if $i \neq j$.*

This property allows us to switch every i-component M_i in the i'th coordinate. Thus we obtain

Theorem 14. (See [7, 8].) *The set*

$$C = (H^n \setminus (\bigcup_{i=1}^{n} M_i)) \cup (\bigcup_{i=1}^{n} (M_i \oplus i))$$

is a nonsystematic perfect binary code of length n *for every* $n = 2^k - 1,\ \ k \geq 8$.

The existence of nonsystematic perfect codes of length $n = 2^k - 1,\ \ k \leq 7$, was proved by Phelps and LeVan [27].

TRIVIAL AUTOMORPHISM GROUPS

Define the automorphism of a perfect code C of length n as an (not necessarily linear) isometry of the n-dimensional vector space E^n over $GF(2)$ with respect to the Hamming metric which leaves C invariant. Every isometry of E^n can be represented as a mapping $A_\pi^v : x \to \pi(x)$, where π is a permutation of the n coordinate positions and v is a vector of E^n (cf. [18], p.50). We denote the identity permutation by e, the all-one vector by $\mathbf{1}$. We denote the kernel respectively the symmetry subgroup of the automorphism group $Aut(C)$ by $Ker(C) = \{A_e^v : A_e^v(C) = C\}$ and $Sym(C) = \{A_\pi^{\mathbf{0}} : A_\pi^{\mathbf{0}}(C) = C\}$, here $\mathbf{0}$ is the all-zero vector as above. The automorphism group of a perfect code C is called trivial, if $Aut(C) = Ker(C) = \{A_e^0, A_e^1\}$, i.e. if the identity permutation and the replacement of the codeword by its complement are the only automorphisms of C.

It should be noted that $Sym(C) \times Ker(C) = Aut(C)$ is not true for every code C. Hence it is not sufficient to investigate $Sym(C)$ and $Ker(C)$ separately.

Let H^n be the Hamming code of length n. An integer vector $\alpha = (\alpha_1, \ldots, \alpha_n)$ is called *heterogeneous* if α_i is odd, greater than 0 for $i = 1, \ldots, n$ and $\alpha_i \neq \alpha_j$ for $i \neq j$. Assume that there exist minimal components $M_{i_1}^{\gamma_1}, \ldots, M_{i_m}^{\gamma_m}$, $m = \sum_1^n \alpha_i$, of minimal cardinality in the code H^n such that the distance between two components $M_{i_j}^{\gamma_j}$ and $M_{i_t}^{\gamma_t}$ is greater than 6 for $j \neq t$ and such that there are exactly α_i i-components, $i = 1, \ldots, n$. We call a code C *α-heterogeneous* if it is obtained from H^n by a translation of the components $M_{i_1}^{\gamma_1}, \ldots, M_{i_m}^{\gamma_m}$ (every i-component is exchanged in the i'th coordinate).

Theorem 15. (See [11].) *There exists a perfect α-heterogeneous code of length* n *for every* $n = 2^k - 1,\ \ k \geq 8$.

In particular we can choose the vector $(1, 3, \ldots, 2n-1)$ of length n as the vector α. A code C is called a code of *full t-rank* if every vector from E^n is a linear combination of not more than t vectors from C. It is evident that a code of full rank is a code of full t-rank for some t. We have $t \geq 3$ for the codes of full rank with distance greater than 1.

Theorem 16. (See [11].) *A perfect α-heterogeneous code is a perfect nonsystematic code of full 3-rank and has a trivial automorphism group.*

An analogous result holds for systematic perfect codes.

Theorem 17. (See [22].) *There exists a perfect full rank systematic code of length n with a trivial automorphism group for all $n = 2^k - 1,\ \ k \geq 5$.*

The construction of such codes was done again using special switchings of minimal i-components with minimal cardinality. The question if there is a perfect binary code of length 15 with a trivial automorphism group remains open.

INTERSECTION NUMBERS

The *intersection number* of two binary codes C_1 and C_2 is defined as $\eta(C_1, C_2) = |C_1 \cap C_2|$. Etzion and Vardy [15, 16] established the following result

Theorem 18. *If C_1, C_2 are two distinct perfect codes of length $n = 2^k - 1,\ \ k \geq 3$, then*

$$2 \leq \eta(C_1, C_2) \leq 2^{n-log(n+1)} - 2^{\frac{n-1}{2}}.$$

Both bounds are tight. For all $k \geq 3$ there exist perfect codes C_1, C_2 of length $n = 2^k - 1$ such that $\eta(C_1, C_2) = 2^{n-log(n+1)} - 2^{\frac{n-1}{2}}$. The bound was established using a switch of one i-component in Vasil'ev's construction. Moreover using multiple switchings they obtained intersection numbers of the form

$$t 2^{\frac{n-1}{2}}$$

for all $t = 1, 2, \ldots, 2^{\frac{n-1}{2} - log(n+1)} - 1$, see [16].

The lower bound for $\eta(C_1, C_2)$ was constructed in [16] exploring a switch for the concatenation construction of the Hamming code. Using induction Etzion and Vardy gave a complete solution of the intersection number problem for Hamming codes.

Theorem 19. *For each $k \geq 3$ there exist two Hamming codes H_1^n, H_2^n of length $n = 2^k - 1$, such that*

$$\eta(H_1, H_2) = 2^{n-t}$$

for $t = log(n+1) + 1, \ldots, 2log(n+1)$.

There is a close connection between an intersection number of two perfect codes C_1 and C_2 and a distance $d(C_1, C_2) = |(C_1 \backslash C_2) \cup (C_2 \backslash C_1)|$ between them $d(C_1, C_2) = |C_1| + |C_2| - 2\eta(C_1, C_2)$. A difference of numbers of codewords of C_1 and C_2 in any k-dimensional face of E^n is investigated and the lower bound for the distance $d(C_1, C_2)$ using the difference is established in [44].

The problem of enumerating all possible intersection numbers of distinct perfect binary codes is still open.

CONCLUDING REMARK

We have verified that the switching approach gave unexpected progress in investigating perfect binary codes. It may also be fruitful for studying and constructing (not necessarily perfect) q-ary codes. Recently Ahlswede, Aydinian and Khachatrian [2] introduced and analyzed the new concept of *diameter* perfect codes.

References

[1] J.K. Abdurahmanov, *On geometrical structure of codes correcting errors*, PhD Thesis, Tashkent, Usbekiston (1991), 66 p.

[2] R. Ahlswede, H. Aydinian and L. Khachatrian, "On perfect codes and related concepts", *Designs, Codes, and Cryptography*, to appear.

[3] S.V. Avgustinovich, "On nonisometry of perfect binary codes", *Proc. of Institute of Math. SB RAN* 27, 1994, 3-5.

[4] S.V. Avgustinovich, "On a property of perfect binary codes", *Discrete Analysis and Operation Research* 2 (1), 1995, 4-6.

[5] S.V. Avgustinovich and F.I. Solov'eva, "On projections of perfect binary codes", *Proc. Seventh Joint Swedish-Russian Workshop on Information Theory*, St.-Petersburg, Russia, June 1995, 25-26.

[6] S.V. Avgustinovich and F.I. Solov'eva, "Construction of perfect binary codes by sequential translations of the i-components", *Proc. of Fifth Int. Workshop on Algebraic and Comb. Coding Theory. Sozopol*, Bulgaria, June 1996, 9-14.

[7] S.V. Avgustinovich and F.I. Solov'eva, "Existence of nonsystematic perfect binary codes", *Proc. of Fifth Int. Workshop on Algebraic and Comb. Coding Theory*, Sozopol, Bulgaria, June 1996, 15-19.

[8] S.V. Avgustinovich and F.I. Solov'eva, "On the nonsystematic perfect binary codes", *Probl. Inform. Transmission* 32 (3), 1996, 258-261.

[9] S.V. Avgustinovich and F.I. Solov'eva, "Construction of perfect binary codes by sequential translations of an $\tilde{\alpha}$-components", *Probl. Inform. Transmission* 33 (3), 1997, 202-207.

[10] S.V. Avgustinovich and F.I. Solov'eva, "On distance regularity of perfect binary codes", *Probl. Inform. Transmission* 34 (3), 1998, 247-249.

[11] S.V. Avgustinovich and F.I. Solov'eva, "Perfect binary codes with trivial automorphism group", *Proc. of Int. Workshop on Information Theory*, Killarney, Ireland. June 1998, 114-115.

[12] S.V. Avgustinovich, "To minimal distance graph structure of perfect binary $(n, 3)$-codes", *Discrete Analysis and Operation Research* 1 (5) 4, 1998, 3-5 (in Russian).

[13] H. Bauer, B. Ganter, and F. Hergert, "Algebraic techniques for nonlinear codes", *Combinatorica* 3, 1983, 21-33.

[14] P. Delsarte, "Bounds for unrestricted codes by linear programming", *Philips Res. Report* 27, 1972, 272-289.

[15] T. Etzion and A. Vardy, "Perfect binary codes: Constructions, properties and enumeration", *IEEE Trans. Inform. Theory* 40 (3), 1994, 754-763.

[16] T. Etzion and A. Vardy, "On perfect codes and tilings: problems and solutions", *SIAM J. Discrete Math.* 11 (2), 1998, 205-223.

[17] O. Heden, "A binary perfect code of length 15 and codimension 0", *Designs, Codes and Cryptography* 4, 1994, 213-220.

[18] W. Heise and P. Quattrocchi, *Informations- und Codierungtheorie*, 3. Aufl., Springer-Verlag, 1995.

[19] F. Hergert, "Algebraische Methoden fur Nichtlineare Codes", Thesis Darmstadt, 1985.

[20] S.P. Lloyd, "Binary block coding", *Bell Syst. Techn. J.* 36, 1957, 517-535.

[21] G. Cohen, I. Honkala, A. Lobstein and S. Litsyn, *Covering codes*, Chapter 11, Elsevier, 1998.

[22] S.A. Malyugin, "Perfect codes with trivial automorphism group", *Proc. II Int. Workshop on Optimal Codes*, Sozopol, Bulgaria, June 1998, 163-167.

[23] S.A. Malyugin, "On counting of perfect binary codes of length 15", *Discrete Analysis and Operation Research*, submitted (in Russian).

[24] M. Mollard, "A generalized parity function and its use in the construction of perfect codes", *SIAM J. Alg. Disc. Meth.* 7 (1), 1986, 113-115.

[25] K.T. Phelps, "Every finite group is the automorphism group of some perfect code", *J. of Combin. Theory* Ser. A 43 (1), 1986, 45-51.

[26] K.T. Phelps and M.J. LeVan, "Kernels of nonlinear Hamming codes", *Designs, Codes and Cryptography* 6, 1995, 247-257.

[27] K.T. Phelps and M.J. LeVan, "Non-systematic perfect codes", *SIAM Journal of Discrete Mathematics* 12 (1), 1999, 27-34.

[28] K.T. Phelps and M.J. LeVan, "Switching equivalence classes of perfect codes", *Designs, Codes and Cryptography* 16 (2), 1999, 179 - 184.

[29] A.K. Pulatov, "On geometric properties and circuit realization of subgroup in E^n", *Discrete Analysis* 23, 1973, 32-37 (in Russian).

[30] A.K. Pulatov, "On structure of close-packed (n,3)-codes", *Discrete Analysis* 29, 1976, 53-60 (in Russian).

[31] G.S. Shapiro and D.L. Slotnik, "On the mathematical theory of error correcting codes", *IBM J. Res. and Devel.* 3 (1), 1959, 25-34.

[32] F.I. Solov'eva, "Factorization of code-generating disjunctive normal forms", *Methody Discretnogo Analiza* 47, 1988, 66-88 (in Russian).

[33] F.I. Solov'eva, "Exact bounds on the connectivity of code-generating disjunctive normal forms", *Inst. Math. of the Siberian Branch of Acad. of Sciences USSR*, Preprint 10, 1990, 15 (in Russian).

[34] F.I. Solov'eva, "A combinatorial construction of perfect binary codes", *Proc. of Fourth Int. Workshop on Algebraic and Comb. Coding Theory*, Novgorod, Russia, September 1994, 171-174.

[35] F.I. Solov'eva, S.V. Avgustinovich, T. Honold T. and W. Heise, "On the extendability of code isometries", *J. of Geometry*, 61, 1998, 3-16.

[36] F.I. Solov'eva, "Perfect binary codes: bounds and properties", *Discrete Mathematics*, to appear.

[37] F.I. Solov'eva, "Perfect binary codes components", *Proc. of Int. Workshop on Coding and Cryptography*, Paris, France. January, 1999, 29-32.

[38] F.I. Solov'eva, "Structure of i-components of perfect binary codes", *Discrete Appl. of Math.*, submitted.

[39] A. Tietäväinen, "On the nonexistence of perfect codes over finite fields", *SIAM J. Appl. Math.* 24, 1973, 88-96.

[40] Y.L. Vasil'ev, "On nongroup close-packed codes", *Problems of Cybernetics* 8, 1962, 375-378 (in Russian).

[41] Y.L. Vasil'ev, "On comparing of complexity of deadlock and minimal disjunctive normal forms", *Problems of Cybernetics* 10, 1963, 5-61 (in Russian).

[42] Y.L. Vasil'ev and F.I. Solov'eva, "Codegenerating factorization on n-dimensional unite cube and perfect codes", *Probl. Inform. Transmission* 33 (1), 1997, 64-74.

[43] A.Y. Vasil'eva, "Spectral properties of perfect binary (n,3)-codes", *Discrete Analysis and Operation Research* (2) 2, 1995, 16-25 (in Russian).

[44] A. Y. Vasil'eva, "On distance between perfect binary codes", *Discrete Analysis and Operation Research* 1 (5) 4, 1998, 25-29 (in Russian).

[45] A.Y. Vasil'eva, "On centered characteristic functions of perfect binary codes", *Proc. of Sixth Int. Workshop on Algebraic and Combin. Coding Theory*, Pskov, Russia, September 1998, 224-227.

[46] A.Y. Vasil'eva, "Local spectrum of perfect binary codes", *Discrete Analysis and Operation Research* 1 (6) 1, 1999, 3-11 (in Russian).

[47] V.A. Zinov'ev and V.K. Leontiev, "A theorem on nonexistence of perfect codes over Galois fields", *Inst. of Problems Information Transmission*, Preprint, 1972 (in Russian).

[48] V.A. Zinov'ev and V.K. Leontiev, "On perfect codes", *Probl. Control and Inform. Theory* 1, 1972, 26-35.

[49] V.A. Zinov'ev and V.K. Leontiev, "Nonexistence of perfect codes over Galois fields", *Probl. Control and Inform. Theory* 2 (2), 1973, 123-132.

ON SUPERIMPOSED CODES

A.J. Han Vinck and Samuel Martirossian

Institute for Experimental Mathematics
University of Essen, Ellernstrasse 29, D-45326 Essen, Germany
vinck@exp-math.uni-essen.de

Abstract: We introduce the concept of q-ary superimposed codes. These codes are to be used in a multi-user concept where the set of active users of size m is small compared to the total amount of users T. The active transmitters use signatures of q-ary symbols to be transmitted over a common channel and the channel output is equal to the active set of input values. We give a class of codes that can be used to uniquely determine the set of active users from the composite signature at the channel output.

INTRODUCTION

We discuss the transmission of information over the so called T-user M-frequency noiseless multiple access channel without intensity information. The users have the same channel input alphabet of M integers from a q-ary alphabet. As defined by Chang and Wolf [2], the channel output at each time instant is a symbol which identifies which subset of integers occurred as inputs to the channel, but not how many of each integer occurred. As a practical example, in Pulse Positioning Modulation (PPM) format each integer is transmitted as a single pulse positioned in one of q disjoint sub slots. The detector output after each slot is equal to the positions where a pulse is detected. Hence, for a q-ary input we have 2^{q-1} possible outputs. This channel model is equivalent to the T-User M-Frequency Multi Access channel. It is the purpose of this paper to describe a signaling method that allows m users to use the q-ary input channel simultaneously.

We extend and modify the class of binary Superimposed Codes (SIC) introduced by Kautz-Singleton [1]. A Superimposed code $\mathrm{SIC}(n, N, 2, m)$ consists of N binary code words of length n, with the property that from the Boolean sum of any m-subset we are able to uniquely determine the individual code words from the m-subset. Proposition 1 gives a relation between N, m and n.

I. Althöfer et al. (eds.), Numbers, Information and Complexity, 325-331.

It follows directly from the property of SICs.

Proposition 1:

$$2^n \geq \sum_{i=1}^{m} \binom{N}{i}.$$

We extend the definition of SICs to the situation where code words have q-ary symbols and the channel output is a symbol which identifies which subset of integers occurred as input to the channel (no intensity information). We first have to give some additional definitions.

Definition 1: The q-ary "$\cup$", $\cup(a, b, \cdots, c)$ is defined as the set of different symbols of the argument $(a, b, \cdots, c)$.

Example: $\cup(1, 2, 3, 3, 2) = \{1, 2, 3\}$.

Example: $\cup(0, 1, 1, 0, 0) = \{0, 1\}$.

Let $V \subset \{0, 1, \cdots, (q-1)\}^n, | V |= N$, represent an $N \times n$ matrix V.

Definition 2: The q-ary "$\underline{\cup}$" of m code words in $V, \underline{\cup}(\underline{r}, \underline{s}, \cdots, \underline{t})$ is defined as the component wise $\cup$ of the symbols.

Example: $\underline{\cup}(1223, 1321, 1111) = (\{1\}, \{1, 2, 3\}, \{1, 2\}, \{1, 3\})$.

Definition 3: The $\underline{\cup}$ of m code words $(\underline{r}, \underline{s}, \cdots, \underline{t})$ cover a code word $\underline{v}$ if

$$\underline{\cup}((\underline{r}, \underline{s}, \cdots, \underline{t}) = \underline{\cup}((\underline{r}, \underline{s}, \cdots, \underline{t}), \underline{v}).$$

Example: The vector $(\{1\}, \{1, 2, 3\}, \{1, 3\})$ covers the code word $\underline{v} = (1, 3, 3)$.

Definition 4: A q-ary-Superimposed Code (q-SIC) V with parameters n, N, q, m contains N q-ary code words of q-ary code words of length n with the property that the $\underline{\cup}$ of any set S containing m or less code words does not cover any code word not in S. Proposition 2 again follows from the definition 4.

Proposition 2:

$$(2^q)^n \geq \sum_{i=1}^{m} \binom{N}{i}. \tag{1}$$

For large values of N and constant m,

$$n \approx \frac{m}{q} log_2 N.$$

In the next theorem we give a more explicit bounding technique for the length of a q-SIC.

Theorem 1: For a q-SIC (n, N, q, m) the following inequalities hold

i) for $m < n = ms + r, 0 \leq r < m,$
$N \leq (m-r)(q^s - 1) + r(q^{s+1} - 1);$

ii) for $n \leq m \leq n(q-1)$, the maximum number of code words
$N_{max} = n(q-1).$

Proof:
i) $m < n$. Consider a particular partition of the code words of a q-SIC(n, N, q, m) in m non-empty parts of size $n_1, n_2, \cdots, n_m$, where $\sum_{j=1}^{m} n_j = n$. Every code word from the q-SIC must have at least one part different from the corresponding part of all other code words. This part contains at least one symbol, called special element, that can be used to distinguish a code word from the $\uplus$ of any set S of m or less code words. If the number of special elements in a particular column is exactly q^{n_i} we have $N = q^{n_i}$. We must therefore assume that every column contains at most $q^{n_i} - 1$ special elements. The maximum number of different parts we can choose is an upper bound for the number of different code words in the q-SIC, and thus

$$N \leq \begin{matrix} \textit{minimum over} \\ \textit{all partitions} \end{matrix} \sum_{i=1}^{m} (q^{n_i} - 1). \tag{2}$$

The minimum is obtained for $n_i = s$ or $n_i = s + 1$ for $r > 0$. We thus obtain as an upper bound

$$N \leq (m-r)(q^s - 1) + r(q^{s+1} - 1) \tag{3}$$

ii) Let $m \geq n$. In this case, every code word must have a special element in at least one of its columns. If one of the columns contains exactly q special elements, then $N = q$. Therefore, every column must contain no more than $q - 1$ special elements. Hence, we obtain as an upper bound

$$N \leq n(q-1). \tag{4}$$

Example: The following example gives a q-SIC$(n = 5, N = 5 * 3, 4, m)$, where $n \leq m < n(q-1)$, that equals the upperbound in (4). The example can easily be generalized to other values of n and q.
The q-SIC$(5, 15, 4, m)$ contains the following code words

10000 01000 00100 00010 00001
20000 02000 00200 00020 00002
30000 03000 00300 00030 00003

In section II we give some of the properties of q-SICs and we develop some code constructions. In section III we give an asymptotic construction.

PROPERTIES AND CONSTRUCTIONS

In this section we consider the construction of q-ary SICs using error correcting codes, such as Reed Solomon codes. We first give a general relation between the minimum distance of a code and the existence of a q-ary SIC.

Theorem 2: Let $V \subset \{0, 1, \cdots, q-1\}^n$ be an error correcting code with minimum distance d and cardinality N. If

$$d > \frac{m-1}{m} n, \tag{5}$$

then V is also q-SIC(n, N, q, m).

Proof: The number of agreements between two code words is less than or equal to $n - d$. For the $\uplus$ of any set S of m code words the number of agreements with a specific code word not in S is thus less than or equal to $m(n-d)$. For $m(n-d) < n$, there must be at least one special element in any other code word not in S. Hence, the members of the set S can be determined uniquely.
Remark: For linear codes we can use the Plotkin upper bound to limit the value for m as

$$\frac{m}{m-1} > \frac{n}{d} \geq \frac{q^k - 1}{q^{k-1}(q-1)}.$$

It is easy to check that for $m \leq q$ the conditions are fulfilled.

Corollary 1: Let V be a q-ary MDS code with parameters $(n, k, d = n - k + 1)$. Then, for $k = \lceil \frac{n}{m} \rceil$ the code V is q-SIC(n, q^k, q, m).

Proof: $d = n - k + 1 = n - \lceil \frac{n}{m} \rceil + 1 > n - \frac{n}{m} = \frac{m-1}{m} n$.
This construction is the first step in the well known Kautz-Singleton construction [1].

Corollary 2: The extended Reed-Solomon $(n = q^s, k = q^{s-1}, d = q^s - q^{s-1} + 1)$ code over $GF(q^s)$, where q is any prime power and $m \leq q$, defines a q-SIC(q^s, q^{sk}, q^s, m).

Proof: It is easy to check that for $m \leq q$, the condition of theorem 2 is fulfilled.
Example: For $m = 3$ and $q = 9$, the shortened RS-code with parameters $(n, k, d) = (7, 3, 5)$ gives a q-SIC$(7, 9^3, 9, 3)$ and the shortened RS-code with

parameters $(n, k, d) = (4, 2, 3)$ gives a q-SIC$(4, 9^2, 9, 3)$.

Remark: The condition (5) in Theorem 2 is a sufficient but not a necessary condition for the existance of a q-SIC. This follows from the next example.

Example: Let $q = 3, m = 2$ and $n = 4$. The following code has distance 2. The corresponding $q-$SIC does not satisfy condition (5).

	0000	0110	0221	1122
q-SIC(4, 12, 3, 2)	1201	1010	2211	2021
	2101	2220	0012	2202

Example: The code $B = (100, 010, 001, 111)$ with minimum distance 2 and length 3 is a q-SIC for $m = 2$, since $d = 2 > n(m-1)/m = 3/2$. The code $A = (100, 010, 001, 111, 110)$ is not a q-SIC according to the definition. However, it can be verified that the $\underline{\cup}$ of any set of 2 code words can be identified uniquely. As an example, the $\underline{\cup}(010, 111) = (\{0,1\}, \{1\}, \{0,1\})$ covers $(1,1,0)$. However, for $m = 2$, the code word $(1,1,0)$ in combination with $(0,1,1)$ gives $(\{0,1\}, \{1\}, \{0,1\})$, which is not a member of the code.

Theorem 3: If there exists a q-SIC(n_0, N_0, q_0, m) and a q-SIC(n_1, N_1, q_1, m), where $q_1 \leq N_0$ then there also exists a q-SIC$(n_0 n_1, N_1, q_0, m)$.

Proof: Assign to each symbol $\{0, 1, \cdots, q_1 - 1\}$ a different code word from q-SIC(n_0, N_0, q_0, m). Replace the symbols in q-SIC(n_1, N_1, q_1, m) by these code words. Since we replaced all q_1-ary elements by different code words from q-SIC(n_0, N_0, q_0, m) we thus obtain a q-SIC$(n_0 n_1, N_1, q_0, m)$.

Corollary 3: If there exists a SIC$(n_0, N_0, 2, m)$ and a q-SIC(n_1, N_1, q_1, m), where $q_1 \leq N_0$, then there also exists a SIC$(n_0 n_1, N_1, 2, m)$.

Proof: Assign to each symbol $\{0, 1, \cdots, q_1 - 1\}$ a different code word from SIC$(n_0, N_0, 2, m)$. Replace the symbols in q-SIC(n_1, N_1, q_1, m) by these code words. Since we replaced all q_1-ary elements by different code words from SIC$(n_0, N_0, 2, m)$ we thus obtain a SIC$(n_0 n_1, N_1, 2, m)$.
The codes constructed in Corollary 3 can be seen as a generalization of the Kautz-Singleton codes.

Example: Suppose that we have the following starting code q-SIC$(3, 4, 2, m = 2)$ with the 4 code words

$$\{100, 010, 001, 111\} \equiv \{0, 1, a, b\}.$$

The second code to be used is a RS code over $GF(2^2)$ with parameters $(n, k, d) = (3, 2, 2)$. This code has a distance $d = 2 > n(m-1)/m = 3/2$. Hence, we can construct a q-SIC$(3, 16, 2^2, 2)$. We can replace every element with a code word from the first code and obtain a SIC$(9, 16, 2, 2)$ with 16 code words

$$q\text{-SIC}(9,16,2,2) = \begin{matrix} 000, 01a, 0ab, 0b1, 1a0, ab0, b10, a01 \\ \\ b0a, 10b, 1ba, a1b, ba1, 111, aaa, bbb \end{matrix}$$

As a third code we construct a RS code over $GF(2^4)$ with parameters $(n = 15, k = 8, d = 8)$, where $d > n(m-1)/m = 15/2$. From this code we obtain a q-SIC$(15, 2^{32}, 2^4, m = 2)$. Combining with the second code we obtain a q-SIC$(9 * 15 = 135, 2^{32}, 2, 2)$. This example shows that we can construct a series of codes. We will use this later fact to predict the asymptotic behavior of a particular construction.

Example: Let $q = 4$ and $m = 3$. The first code we use is a RS code over $GF(2^2)$ with parameters $(n = 4, k = 2, d = 3)$. Since $d > 2n/3$, we obtain a q-SIC$(4, 2^4, 2^2, 3)$. The second code we choose is a shortened RS code over $GF(2^4)$ with parameters $(n = 13, k = 5, d = 9)$. Since $9 > 26/3$, we obtain a q-SIC$(13, 2^{20}, 16, 3)$. Combining both codes, we obtain a q-SIC$(4*13 = 42, N = 2^{20}, q = 4, m = 3)$.

AN ASYMPTOTIC CONSTRUCTION

We give an algorithm for constructing arbitrary long codes based on Theorem 3 and Corollary 2.

Step 0. Suppose that we have a q-SIC$(n_0, N_0 = q^i, q_0, m)$ for arbitrary $i > 1$ and q is a prime power, $q \geq m$.

Step 1. Using corollary 2, we obtain a q-SIC(q^i, q^{ik}, q^i, m), where $k = q^{i-1}$. From Theorem 3 we then construct a q-SIC$(n_0 q^i, q^{ik}, q_0, m)$.

Suppose that from step $l-1$ we have a q-SIC$(n_{l-1}, N_{l-1}, q_0, m)$. Using corollary 2 we construct a q-SIC$(N_{l-1}, N_l, N_{l-1}, m)$, where

$$N_l = N_{l-1}^{N_{l-1}/q}. \tag{6}$$

From Theorem 3 we then obtain a q-SIC$(n_l = n_{l-1}N_{l-1}, N_l, q_0, m)$. For this construction we easily see that

$$n_l = \frac{n_0}{\log_q N_0} q^l \log_q N_l. \tag{7}$$

The asymptotic behavior of (7) can be estimated as follows. Taking the base-q logarithm of N_l, l times, we obtain for $N_0 = q^i, l < i$

Index

$$\log_q \log_q \ldots \log_q N_l > (N_0/q) \log_q N_0 = iq^{i-1} > q^l, \tag{8}$$

where we used the fact that

$$\log_q \log_q N_l > log_q N_{l-1}.$$

For $i - 1 < l \leq iq^{i-1}$ we take the base-q logarithm of N_l, $l - 1$ times to obtain

$$\log_q \log_q \ldots \log_q N_l > (N_1/q) \log_q N_1 > q^l,$$

where

$$N_1 = N_0^{N_0/q}.$$

Hence, using (7), we can say that asymptotically,

$$n_l \approx \log_q N_l. \tag{9}$$

This is exactly what we expect from the bound as given in Section I.

CONCLUSIONS

We extended the binary superimposed codes to the case where q-ary symbols are used. We derive bounds on the cardinality of the codes and give an asymptotic code construction that behaves according to the upper bound. These codes can be used for random access systems using multiple frequency shift keying, or in systems where pulse position plays the key role.

References

[1] W.H. Kautz and R.C. Singleton, “Nonrandom Binary Superimposed Codes,” *IEEE Trans. Inform. Theory* 10, 1964, 363-377.

[2] Shin-Chun Chang and J.K. Wolf, “On the T-User M-Frequency Noiseless Multiple Access Channel with and without Intensity Information,” *IEEE Trans. Inform. Theory* 27 (1), 1981, 41-48.

[3] A.G. Dyachkov and V.V. Rykov, "A Survey of Superimposed Code Theory," *Problems of Control and Inform. Theory* 12 (4), 1-13, English Translation.

THE MACWILLIAMS IDENTITY FOR LINEAR CODES OVER GALOIS RINGS

Zhe-Xian Wan

Department of Information Technology, Lund University
Box 118, S-221 00 Lund, Sweden

Abstract: The MacWilliams identity relating the weight enumerators of a binary linear code and its dual code is generalized to linear codes over Galois rings.

Index terms – Galois ring, linear code, MacWilliams identity.

INTRODUCTION

Let R be a Galois ring of characteristic p^e and cardinality p^{em}, where p is a prime and e and m are positive integers. Without of generality we can assume that $R = \mathbb{Z}_{p^e}[\xi]$, where ξ is a root of a monic basic irreducible polynomial $h(x)$ of degree m over $\mathbb{Z}_{p^e}$. For the rudiments of Galois rings, see [3] Nechaev (1989).

Let n be a positive integer and R^n be the set of all n-tuples over R. R^n is an R-module of rank n under the componentwise addition and scalar multiplication of n-tuples and $|R^n| = p^{emn}$. Any R-submodule of R^n is called a *linear code* of *length* n over R. Elements of R^n are called *words* and those of a linear code are called its *codewords*.

For any $\boldsymbol{x} = (x_1, x_2, \ldots, x_n)$, $\boldsymbol{y} = (y_1, y_2, \ldots y_n) \in R^n$ define

$$\boldsymbol{x} \cdot \boldsymbol{y} = x_1 y_1 + x_2 y_2 + \ldots + x_n y_n,$$

which is called the *dot product* of $\boldsymbol{x}$ and $\boldsymbol{y}$. If $\boldsymbol{x} \cdot \boldsymbol{y} = 0$, $\boldsymbol{x}$ and $\boldsymbol{y}$ are said to be *orthogonal.*

For any linear code $\mathcal{C}$ of length n over R define

$$\mathcal{C}^{\perp} = \{\boldsymbol{x} \in R^n : \boldsymbol{x} \cdot \boldsymbol{y} = 0 \ \forall \boldsymbol{y} \in \mathcal{C}\} .$$

I. Althöfer et al. (eds.), Numbers, Information and Complexity, 333-338.

It is easy to see that $C^{\perp}$ is also a linear code of length n over R. $C^{\perp}$ is called the *dual code* of C.

In the present paper, the MacWilliams identity relating the weight enumerators of a binary linear code and its dual is generalized to linear codes over Galois rings.

MACWILLIAMS IDENTITY

Let C be a linear code of length n over R and α be an element of R. For any $\boldsymbol{x} = (x_1, x_2, \ldots, x_n) \in R^n$ define the *weight* of $\boldsymbol{x}$ *at* α to be

$$w_\alpha(\boldsymbol{x}) = |\{i : x_i = \alpha\}| \ .$$

For simplicity, let $r = p^{em} - 1$, the $r+1$ elements of R be written as $\alpha_0 = 0$, $\alpha_1, \ldots, \alpha_r$, and

$$w_i(\boldsymbol{x}) = w_{\alpha_i}(\boldsymbol{x}) \ .$$

Then the *complete weight enumerator* of C is defined to be the homogeneous polynomial of degree n in p^{em} indeterminates $X_0, X_1, \ldots X_r$

$$W_C(X_0, X_1, \ldots X_r) = \sum_{\boldsymbol{c} \in C} X_0^{w_0(\boldsymbol{c})} X_1^{w_1(\boldsymbol{c})} \ldots X_r^{w_r(\boldsymbol{c})} \ .$$

Let α be an element of R, then α can be expressed uniquely as

$$\alpha = \alpha_0 + \alpha_1 \xi + \ldots + \alpha_{m-1} \xi^{m-1}, \text{ where } \alpha_0, \alpha_1, \ldots \alpha_{m-1} \in \mathbb{Z}_{p^e} \ .$$

Let ζ be a primitive p^e-th root of unity in the complex field.

Lemma 1: Let H be an R-submodule of R. Then

$$\sum_{\alpha \in H} \zeta^{\alpha_0} = \begin{cases} 1 & \text{if } H = \{0\} \ , \\ 0 & \text{otherwise.} \end{cases}$$

Proof: If $H = \{0\}$, $\sum_{\alpha \in H} \zeta^{\alpha_0} = \zeta^0 = 1$. Now let $H \neq \{0\}$. For any $a \in \mathbb{Z}_{p^e}$ let

$$S_a = \{\alpha \in H : \ \alpha_0 = a\} \ .$$

For $0 \in \mathbb{Z}_{p^e}$, we have $0 \in H$ such that $0_0 = 0$. Therefore S_0 is non-empty. Since $(\alpha + \beta)_0 = \alpha_0 + \beta_0$ for all $\alpha, \beta \in R$, S_0 is an additive subgroup of H. Moreover, if $S_a \neq \phi$ for some $a \in \mathbb{Z}_{p^e}$, there is an $\alpha \in H$ such that $\alpha_0 = a$ and consequently $S_a = S_0 + \alpha$. Consider the group homomorphism

$$\begin{array}{rccc} g: & H & \longrightarrow & \mathbb{Z}_{p^e} \\ & \alpha & \longmapsto & \alpha_0 \ . \end{array}$$

Then $\operatorname{Im} g$ is a subgroup of the additive group $\mathbb{Z}_{p^e}$ and

$$H = \bigcup_{a \in \operatorname{Im} g} S_a$$

is the coset decomposition of H relative to the subgroup S_0. Thus $|S_a| = |S_0|$ for all $a \in \operatorname{Im} g$. Then

$$\sum_{\alpha \in H} \zeta^{\alpha_0} = \sum_{a \in \operatorname{Im} g} |S_a| \zeta^a = |S_0| \sum_{a \in \operatorname{Im} g} \zeta^a .$$

We assert that $\operatorname{Im} g \neq \{0\}$. Since $H \neq \{0\}$, there is a non-zero element $\alpha \in H$. Let $\alpha = \alpha_0 + \alpha_1 \xi + \ldots + \alpha_{m-1} \xi^{m-1}$, where $\alpha_0, \alpha_1, \ldots, \alpha_{m-1} \in \mathbb{Z}_{p^e}$. We can assume that $\alpha_0 = \alpha_1 = \ldots = \alpha_{k-1} = 0$ and $\alpha_k \neq 0$, where $0 \leq k \leq m-1$. Since H is an R-submodule of R, $\xi^{-k}\alpha \in H$. Clearly $(\xi^{-k}\alpha)_0 = \alpha_k \neq 0$. Our assertion is proved. Being a subgroup of the cyclic group $\mathbb{Z}_{p^e}$, $\operatorname{Im} g$ is also cyclic and is generated by an element of $\mathbb{Z}_{p^e}$, say c. Clearly, $0 < c < p^e$ and $c \,|\, p^e$. Then $\operatorname{Im} g = \{0, c, 2c, \ldots, ((p^e/c) - 1)c\}$. Thus

$$\sum_{a \in \operatorname{Im} g} \zeta^a = \sum_{i=0}^{(p^e/c)-1} \zeta^{ci} = (1 - \zeta^{p^e})/(1 - \zeta^c) = 0 .$$

Consequently

$$\sum_{\alpha \in H} \zeta^{\alpha_0} = |S_0| \sum_{a \in \operatorname{Im} g} \zeta^a = 0 .$$

□

Lemma 2: Let $\mathcal{C}$ be a linear code of length n over R. Then

$$\sum_{\boldsymbol{x} \in \mathcal{C}} \zeta^{(\boldsymbol{x} \cdot \boldsymbol{y})_0} = \begin{cases} |\mathcal{C}| & , \quad \text{if } \boldsymbol{y} \in \mathcal{C}^\perp , \\ 0 & , \quad \text{if } \boldsymbol{y} \notin \mathcal{C}^\perp . \end{cases}$$

Proof: Consider first the case $\boldsymbol{y} \in \mathcal{C}^\perp$. We have $\boldsymbol{x} \cdot \boldsymbol{y} = 0 \ \forall \boldsymbol{x} \in \mathcal{C}$ and $\zeta^{(\boldsymbol{x} \cdot \boldsymbol{y})_0} = \zeta^0 = 1$. Hence

$$\sum_{\boldsymbol{x} \in \mathcal{C}} \zeta^{(\boldsymbol{x} \cdot \boldsymbol{y})_0} = \sum_{x \in \mathcal{C}} 1 = |\mathcal{C}| .$$

Now suppose that $y \notin \mathcal{C}^\perp$. For any $\alpha \in R$ let

$$\mathcal{C}_\alpha = \{\boldsymbol{x} \in \mathcal{C} : \boldsymbol{x} \cdot \boldsymbol{y} = \alpha\} .$$

It is easy to verify that $\mathcal{C}_0$ is an R-submodule of $\mathcal{C}$ and that if $\mathcal{C}_\alpha$ is non-empty, $\mathcal{C}_\alpha = \mathcal{C}_0 + \boldsymbol{x}$ where $\boldsymbol{x} \in \mathcal{C}$ with $\boldsymbol{x} \cdot \boldsymbol{y} = \alpha$. Consider the R-homomorphism

$$\begin{aligned} f_{\boldsymbol{y}} : \ & \mathcal{C} \longrightarrow R \\ & \boldsymbol{x} \longmapsto \boldsymbol{x} \cdot \boldsymbol{y} . \end{aligned}$$

Then $\operatorname{Im} f_{\boldsymbol{y}}$ is an R-submodule of R. Since $\boldsymbol{y} \notin \mathcal{C}^\perp$, $\operatorname{Im} f_{\boldsymbol{y}} \neq \{0\}$. Clearly,

$$\mathcal{C} = \bigcup_{\alpha \in \operatorname{Im} f_{\boldsymbol{y}}} \mathcal{C}_\alpha$$

is the coset decomposition of $\mathcal{C}$ relative to $\mathcal{C}_0$. Thus $|\mathcal{C}_\alpha| = |\mathcal{C}_0|$ for all $\alpha \in \operatorname{Im} f_{\boldsymbol{y}}$. Then

$$\sum_{\boldsymbol{x}\in\mathcal{C}} \zeta^{(\boldsymbol{x}\cdot\boldsymbol{y})_0} = |\mathcal{C}_0| \sum_{\alpha\in\operatorname{Im} f_{\boldsymbol{y}}} \zeta^{\alpha_0} = 0 \ ,$$

where the last equality follows from Lemma 1. □

Let f be a function defined on R^n with values in $\mathbb{C}[X_0, X_1, \dots X_r]$. The *Hadamard transform* of f, denoted by $\hat{f}$, is defined by

$$\hat{f}(\boldsymbol{x}) = \sum_{\boldsymbol{y}\in R^n} \zeta^{(\boldsymbol{x}\cdot\boldsymbol{y})_0} f(\boldsymbol{y}) \ \ \forall \ \ \boldsymbol{x} \in R^n \ .$$

Lemma 3: Let $\mathcal{C}$ be a linear code of length n over R. Then

$$\sum_{\boldsymbol{x}\in\mathcal{C}^\perp} f(\boldsymbol{x}) = \frac{1}{|\mathcal{C}|} \sum_{\boldsymbol{x}\in\mathcal{C}} \hat{f}(\boldsymbol{x}) \ .$$

Proof: By definition of Hadamard transform and Lemma 2

$$\begin{aligned} \sum_{\boldsymbol{x}\in\mathcal{C}} \hat{f}(\boldsymbol{x}) &= \sum_{\boldsymbol{x}\in\mathcal{C}} \sum_{\boldsymbol{y}\in R^n} \xi^{(\boldsymbol{x}\cdot\boldsymbol{y})_0} f(\boldsymbol{y}) \\ &= \sum_{\boldsymbol{y}\in R^n} f(\boldsymbol{y}) \sum_{x\in\mathcal{C}} \xi^{(\boldsymbol{x}\cdot\boldsymbol{y})_0} \\ &= \sum_{y\in\mathcal{C}^\perp} f(\boldsymbol{y})|\mathcal{C}| \ . \end{aligned}$$

□

Now we can prove the MacWilliams identity for linear codes over a Galois ring. When the Galois ring is a Galois field, see [2] MacWilliams and Sloane (1977) and when it is $\mathbb{Z}_4$, see [1] Klemm (1987).

Theorem 4: Let $\mathcal{C}$ be a linear code of length n over R and ζ be a primitive p^e-th root of unity in the complex field. Then
$W_{\mathcal{C}^\perp}(X_0, X_1, \dots X_r)$

$$= \frac{1}{|\mathcal{C}|} W_{\mathcal{C}} \left(\sum_{s=0}^{r} \zeta^{(\alpha_0\alpha_s)_0} X_s, \ \sum_{s=0}^{r} \zeta^{(\alpha_1\alpha_s)_0} X_s, \dots, \sum_{s=0}^{r} \zeta^{(\alpha_r\alpha_s)_0} X_s \right) \ .$$

□

Proof: Let

$$f(\boldsymbol{x}) = X_0^{w_0(\boldsymbol{x})} X_1^{w_1(\boldsymbol{x})} \dots \ X_r^{w_r(\boldsymbol{x})} \ \ \forall \boldsymbol{x} \in R^n \ .$$

Then

$$\hat{f}(\boldsymbol{x}) = \sum_{y\in R^n} \zeta^{(\boldsymbol{x}\cdot\boldsymbol{y})_0} f(\boldsymbol{y}) = \sum_{y\in R^n} \zeta^{(x_1y_1+x_2y_2+\cdots+x_ny_n)_0} X_0^{w_0(\boldsymbol{y})} X_1^{w_1(\boldsymbol{y})} \dots \ X_r^{w_r(\boldsymbol{y})} \ .$$

But for $i = 0, 1, \dots, r$,

$$w_i(\boldsymbol{y}) = \delta_{\alpha_i, y_1} + \delta_{\alpha_i, y_2} + \cdots + \delta_{\alpha_i, y_n} \ ,$$

where δ is the Kronecker delta. Then $\hat{f}(\boldsymbol{x})$ can be written as

$$\begin{aligned}
\hat{f}(\boldsymbol{x}) &= \sum_{\boldsymbol{y} \in R^n} \left(\zeta^{(x_1 y_1)_0} \prod_{i=0}^{r} X_i^{\delta_{\alpha_i, y_1}} \right) \dots \left(\zeta^{(x_n y_n)_0} \prod_{i=0}^{r} X_i^{\delta_{\alpha_i, y_n}} \right) \\
&= \left(\sum_{y_1 \in R} \zeta^{(x_1 y_1)_0} \prod_{i=0}^{r} X_i^{\delta_{\alpha_i, y_1}} \right) \dots \left(\sum_{y_n \in R} \zeta^{(x_n y_n)_0} \prod_{i=0}^{r} X_i^{\delta_{\alpha_i, y_n}} \right) \\
&= \left(\sum_{s=0}^{r} \zeta^{(x_1 \alpha_s)_0} X_s \right) \dots \left(\sum_{s=0}^{r} \zeta^{(x_n \alpha_s)_0} X_s \right) \\
&= \prod_{t=0}^{r} \left(\sum_{s=0}^{r} \zeta^{(\alpha_t \alpha_s)_0} X_s \right)^{w_t(\boldsymbol{x})} . \qquad (1)
\end{aligned}$$

The last equality follows from the observation that $\sum_{s=0}^{r} \zeta^{(x_\ell \alpha_s)_0} X_s = \sum_{s=0}^{r} \zeta^{(\alpha_t \alpha_s)_0} X_s$ when $x_\ell = \alpha_t$ and that there are $w_t(\boldsymbol{x})$'s x_ℓ equal to α_t, which contributes together $\left(\sum_{s=0}^{r} \zeta^{(\alpha_t \alpha_s)_0} X_s\right)^{w_t(\boldsymbol{x})}$.

By Lemma 3 and (1) we have

$$W_{\mathcal{C}^\perp}(X_0, X_1, \dots, X_r) = \sum_{\boldsymbol{c} \in \mathcal{C}^\perp} X_0^{w_0(\boldsymbol{c})} X_1^{w_1(\boldsymbol{c})} \dots X_r^{w_r(\boldsymbol{c})}$$

$$= \sum_{\boldsymbol{c} \in \mathcal{C}^\perp} f(\boldsymbol{c})$$

$$= \frac{1}{|\mathcal{C}|} \sum_{\boldsymbol{c} \in \mathcal{C}} \hat{f}(\boldsymbol{c})$$

$$= \frac{1}{|\mathcal{C}|} \sum_{\boldsymbol{c} \in \mathcal{C}} \left(\sum_{s=0}^{r} \zeta^{(\alpha_0 \alpha_s)_0} X_s \right)^{w_0(\boldsymbol{c})} \left(\sum_{s=0}^{r} \zeta^{(\alpha_1 \alpha_s)_0} X_s \right)^{w_1(\boldsymbol{c})} \dots$$

$$\dots \left(\sum_{s=0}^{r} \zeta^{(\alpha_r \alpha_s)_0} X_s \right)^{w_r(\boldsymbol{c})}$$

$$= \frac{1}{|\mathcal{C}|} W_{\mathcal{C}} \left(\sum_{s=0}^{r} \zeta^{(\alpha_0 \alpha_s)_0} X_s, \sum_{s=0}^{r} \zeta^{(\alpha_1 \alpha_s)_0} X_s, \dots, \sum_{s=0}^{r} \zeta^{(\alpha_r \alpha_s)_0} X_s \right) .$$

□

Now for any $\boldsymbol{x} = (x_1, x_2, \ldots, x_n) \in R^n$ define the *Hamming weight* of $\boldsymbol{x}$ to be

$$w_H(\boldsymbol{x}) = |\{i | x_i \neq 0\}| \ .$$

Then the *Hamming weight enumerator* of a linear code $\mathcal{C}$ of length n over R is defined as

$$\mathrm{Ham}_{\mathcal{C}}(X,Y) = \sum_{\mathbf{c}\in\mathcal{C}} X^{n-w_H(\mathbf{c})} Y^{w_H(\mathbf{c})} \ .$$

Clearly,

$$\mathrm{Ham}_{\mathcal{C}}(X,Y) = W_{\mathcal{C}}(X,Y,Y,\ldots,Y) \ .$$

Then from Theorem 4 we deduce immediately the following MacWilliams identity for $\mathrm{Ham}_{\mathcal{C}}$.

Theorem 5: Let $\mathcal{C}$ be a linear code of length n over R. Then

$$\mathrm{Ham}_{\mathcal{C}^\perp}(X,Y) = |\mathcal{C}|^{-1}\mathrm{Ham}_{\mathcal{C}}(X + rY, X - Y) \ .$$

□

Specializing the Galois ring R to be $\mathbb{Z}_{p^e}$, we obtain the following Corollaries 6 and 7 of Theorems 4 and 5, respectively.

Corollary 6: Let $\mathcal{C}$ be a linear code of length n over $\mathbb{Z}_{p^e}$ and ζ be a primitive p^e-th root of unity in the complex field. Then

$$W_{\mathcal{C}^\perp}(X_0, X_1, \cdots, X_{p^e-1}) =$$

$$= |\mathcal{C}|^{-1} W_{\mathcal{C}} \left(\sum_{s=0}^{p^e-1} \zeta^{0\cdot s} X_s \, , \sum_{s=0}^{p^e-1} \zeta^{1\cdot s} X_s \, , \ldots , \sum_{s=0}^{p^e-1} \zeta^{(p^e-1)s} X_s \right) \ .$$

□

Corollary 7: Let $\mathcal{C}$ be a linear code of length n over $\mathbb{Z}_{p^e}$. Then

$$\mathrm{Ham}_{\mathcal{C}^\perp}(X,Y) = |\mathcal{C}|^{-1} \ \mathrm{Ham}_{\mathcal{C}}(X + (p^e - 1)Y, X - Y) \ .$$

□

References

[1] M. Klemm, "Über die Identität von MacWilliams für die Gewichtsfunktion von Codes", *Arch. Math.*, 49, 1987, 400–406.

[2] F. J. MacWilliams and N. J. A. Sloane, *The Theory of Error-Correcting Codes*, North Holland, 1977.

[3] A. A. Nechaev, "Kerdock-code in a cyclic form", *Diskretnaya Mat. (USSR)*, 1, 1989, 123–139 (in Russian), English translation: *Discrete Math. Appl.*, 1, 1991, 365–384.

ON THE STRUCTURE OF A COMMON KNOWLEDGE CREATED BY CORRELATED OBSERVATIONS AND TRANSMISSION OVER HELPING CHANNELS

Vladimir B. Balakirsky*

Electrical Engineering Department,
Eindhoven University of Technology,
P.O.Box 513, 5600 MB Eindhoven, the Netherlands
on leave from the Data Security Association "Confident",
193060 St.-Petersburg, Russia

vbal@ei1.ei.ele.tue.nl

INTRODUCTION AND STATEMENT OF THE PROBLEM

Suppose that two individuals, person X and person Y, communicate with each other in such a way that X sends one of M_X messages to Y and, simultaneously, Y sends one of M_Y messages to X. The messages are numbered by the integers $1, \ldots, M_X$ and $1, \ldots, M_Y$. Assuming the numbers to be the identifiers for the corresponding messages, we consider the pairs of the exchanged messages $(i,j) \in \{1, \ldots, M_X\} \times \{1, \ldots, M_Y\}$ as possible common values of X and Y which describe their common knowledge. Suppose also that there is another person, called the source, who gives the same binary vector $\mathbf{x}$ of length n to the individuals. Then X and Y update their knowledge by including this vector,

*The work was supported by the University of Bielefeld (Germany) and the Eindhoven University of Technology (the Netherlands). The author is grateful to Professor Rudolf Ahlswede and Professor Imre Csiszár for helpful and stimulating discussions, which essentially affected this research and presentation of the results. The help of Dr. Roger Bultitude in the preparing of the manuscript is also highly appreciated.

I. Althöfer et al. (eds.), Numbers, Information and Complexity, 339-352.

which means that now they have a triple $(i, j, \mathbf{x})$ in common and if 2^n is much greater than $M_X M_Y$, then the total number of possible common values is also much greater. However, if the source changes the rules in such a way that $\mathbf{x}$ is given to X and $\mathbf{y}$ is given to Y, where the vectors $\mathbf{x}$ and $\mathbf{y}$ do not coincide, but correlated, then this updating of the transmitted pair of messages is not possible any more, and the individuals can revert the situation in which they may agree on $M_X M_Y$ common values. An alternative algorithm can be fixed as follows : X and Y compute their messages using deterministic functions of the observations and each individual, based on the vector given by the source and the message received from the other person, constructs a value belonging to some "virtual" space, which is assumed to be common to both of them and can be formally presented as a finite set Ω. The algorithm should be assigned in such a way that the values are also common. We will investigate this possibility and demonstrate the example in which one of 20 pairs of messages is exchanged, one of 60 pairs of vectors is given by the source, while X and Y construct one of 50 common values.

The three participants may have many reasons for communication under the rules described above; in fact, these reasons come as corollaries from saying "another person called the source". We will mention those reasons, which are relevant to the foregoing formal discussion. The source considers the communication system as a system of control: he knows how many messages can be exchanged and controls the common knowledge of X and Y by sending them sequences having a certain correlation. This knowledge is bounded from above, since the individuals cannot agree on more than the fixed number of common values, which is defined by the correlation of the source sequences and can be achieved if X and Y form their messages in an optimal way. ¿From a technical perspective, X and Y are interested in the possibility of communication using the source sequences to establish the result of a common random experiment for other purposes, like cryptography and identification. For example, they have a table of binary sequences and want to use one of these sequences as a secret key; an agreement on the particular sequence is achieved by constructing the common pointer to some of the rows of this table. Another possibility can be viewed as matching through the source : person X communicates with many individuals and he wants to discover which of them is the one who receives the correlated sequence from the source and sends his messages using the algorithm expected by X; a similar problem has to be solved by Y. In other words, in analyzing the source sequences and communicating over the channels the individuals investigate each other, and the source presents data for this study. Note also that, in a general context, any message sent by a person is the value of some function of his observations, and any discussion about the reasons would be rather artificial in a sense that the person does not have any choice.

We consider the class of problems described above as belonging to the multi–user direction of information theory started by Shannon [1]. The development of this direction in the 1970s was essentially initiated by Ahlswede [2], [5] who determined the achievable rate region for memoryless multiple access channels under the condition of arbitrary small average decoding error probability. The statements of the problems studied for multiple access channels include the situation when the decoder wants to recover the sequence at the output of a random generator based on the message of the encoder and another message, which was formed as a function of a correlated sequence and transmitted by the "helper" [4], [6], [7], where the role of the helper in this case is to present some side information to the decoder. The asymptotic characterization of the achievable rates of encoding these sequences was found by Wyner [6] and Ahlswede–Körner [7], but a generalization of their approach to the case of several helpers is a difficult problem related to the analysis of the dependent partitions of the spaces containing sequences of the helpers, and this problem is still open [8]. The point that there are interesting applications when the source sequences play some auxiliary role in the communication process was discovered by Ahlswede–Dueck [9], who showed that the noise in the channel gives the randomization that can be effectively used in identification schemes. The role of the common randomness in the communication systems where the participants try to take advantage from the observations of public random processes was also studied in [10], [11], [12], [13], and other papers. There also exists a notion of so–called "common information" introduced by Gács–Körner [3], which measures randomness contained in the variables that can be independently constructed using correlated random sequences; the authors showed that the desired random variables exist only if the source has a "special structure". Note that the possibility to communicate allows X and Y to represent the source as a collection of sub–sources having such a structure (if they have the capabilities to do so).

FORMAL STATEMENT OF THE PROBLEM

Given $M_X, M_Y \geq 1$ and the set $\overline{\mathcal{X}\mathcal{Y}} \subseteq \mathcal{X} \times \mathcal{Y}$, construct four functions f, g, K, L defined by the values,

$$\big(f(\mathbf{x}) \in [M_X] \big)_{\mathbf{x}\in\mathcal{X}}, \quad \big(g(\mathbf{y}) \in [M_Y] \big)_{\mathbf{y}\in\mathcal{Y}}$$

$$\big(K(\mathbf{x}|g(\mathbf{y})) \in \Omega \big)_{(\mathbf{x},\mathbf{y})\in\overline{\mathcal{X}\mathcal{Y}}}, \quad \big(L(\mathbf{y}|f(\mathbf{x})) \in \Omega \big)_{(\mathbf{x},\mathbf{y})\in\overline{\mathcal{X}\mathcal{Y}}}$$

where $[M_X] = \{1, \ldots, M_X\}$, $[M_Y] = \{1, \ldots, M_Y\}$, and Ω is a finite set, in such a way that

$$K(\mathbf{x}|g(\mathbf{y})) = L(\mathbf{y}|f(\mathbf{x})), \quad \text{for all } (\mathbf{x},\mathbf{y}) \in \overline{\mathcal{X}\mathcal{Y}} \tag{1}$$

and

$$|\,\Omega(f,g,K,L)\,| \longrightarrow \max$$

where

$$\Omega(f,g,K,L) = \big\{ K(\mathbf{x}|g(\mathbf{y})) : (\mathbf{x},\mathbf{y}) \in \overline{\mathcal{X}\mathcal{Y}} \big\} = \big\{ L(\mathbf{y}|f(\mathbf{x})) : (\mathbf{x},\mathbf{y}) \in \overline{\mathcal{X}\mathcal{Y}} \big\}. \tag{2}$$

The notations above are illustrated in Figure 1.

EXAMPLES

Let $M_X = 4$, $M_Y = 5$ and let X and Y have access to a source generating pairs of binary vectors $(\mathbf{x}, \mathbf{y})$ of length 6 in such a way that the vector $\mathbf{x}$ has 3 ones, the vector $\mathbf{y}$ has the 2 ones, and the Hamming distance between $\mathbf{x}$ and $\mathbf{y}$ is equal to 1,

$$\begin{aligned} \mathcal{X} \times \mathcal{Y} &= \{0,1\}_3^6 \times \{0,1\}_2^6 \\ \overline{\mathcal{X}\mathcal{Y}} &= \{\, (\mathbf{x}, \mathbf{y}) \in \mathcal{X} \times \mathcal{Y} : \; d_H(\mathbf{x}, \mathbf{y}) = 1 \,\}. \end{aligned} \tag{3}$$

The set $\overline{\mathcal{X}\mathcal{Y}}$ can be also specified by the matrix shown in Table 1, where the $*$ symbols mark all pairs belonging to the set $\overline{\mathcal{X}\mathcal{Y}}$ and we use the octal representation for the binary vectors (07, 13,... denote the vectors 000111, 001011,...). Thus, the source generates one of

$$\binom{6}{3}\binom{3}{1} = \binom{6}{2}\binom{4}{1} = 60$$

pairs of vectors. Person X partitions the set $\mathcal{X}$ into M_X subsets, determines the number of the subset $f(\mathbf{x})$ containing $\mathbf{x}$, and sends this number to Y. An example of the partitioning is presented in Table 2, where

$$\begin{aligned} \mathbf{x} \in \{07, 13, 15, 16, 23\} &\implies f(\mathbf{x}) = 1 \\ \mathbf{x} \in \{25, 26, 31, 32, 34\} &\implies f(\mathbf{x}) = 2 \\ \mathbf{x} \in \{43, 45, 46, 51, 52\} &\implies f(\mathbf{x}) = 3 \\ \mathbf{x} \in \{54, 61, 62, 64, 70\} &\implies f(\mathbf{x}) = 4. \end{aligned}$$

At the same time, person Y partitions the set $\mathcal{Y}$ into M_Y subsets, determines the number of the subset $g(\mathbf{y})$ containing $\mathbf{y}$, and also sends this number to X. Thus, the participants represent the source as a collection of $M_X M_Y$ sub–sources having the alphabets $\mathcal{X}_i \times \mathcal{Y}_j$, where

$$\mathcal{X}_i = \{\, \mathbf{x} \in \mathcal{X} : \; f(\mathbf{x}) = i \,\}, \;\; \mathcal{Y}_j = \{\, \mathbf{y} \in \mathcal{Y} : \; g(\mathbf{y}) = j \,\}$$

for all $i \in [M_X]$ and $j \in [M_Y]$.

In the considerations below we assume the partitioning of the sets $\mathcal{X}$ and $\mathcal{Y}$ specified in Table 2. Suppose that $\mathbf{x} = 45$. If X receives $g(\mathbf{y}) = 1$, then he knows that $\mathbf{y} = 05$. Person Y receives $f(\mathbf{x}) = 3$ in this case and he knows that $\mathbf{x} = 45$. Therefore the pair of vectors $(45, 05)$ describes the common knowledge of X and Y. Let person X receive $g(\mathbf{y}) = 5$ and know that $\mathbf{y} = 44$. In this case, Y having received the message 3 only knows that the vector observed by X belongs to the set $\{45, 46\}$. Person X imagines that he is Y and also knows this. Since X wants to establish a common knowledge with Y, he replaces the vector 45 by the set $\{45, 46\}$, and the pair $(\{45, 46\}, 44)$ becomes the common value. At last, if $g(\mathbf{y}) = 4$, then several iterations of the estimating procedure lead to a description of the common knowledge by the pair of sets $(\{43, 45, 46, 51, 52\}, \{41, 42\})$. Inspecting this procedure for all pairs $(\mathbf{x}, \mathbf{y}) \in \overline{\mathcal{X}\mathcal{Y}}$, we come to the conclusion that the participants can agree on one of 26 common values, and if these values are associated with the capital letters of the Latin alphabet, we write $\Omega(f,g,K,L) = \{\mathrm{A},\mathrm{B},\ldots,\mathrm{Z}\}$; the common values corresponding to the observed vectors are shown in Table 2. Another partitioning of the sets $\mathcal{X}$ and $\mathcal{Y}$ given in Table 3 leads to 50 possible common values, which can be associated with the letters $\mathrm{A},\mathrm{B},\ldots,\mathrm{Z},\mathrm{a},\mathrm{b},\ldots,\mathrm{x}$. In this case,

$$\begin{aligned}
\mathcal{Y}_1 &= \{03,14,60\} = \{000011,001100,110000\}\\
\mathcal{Y}_2 &= \{06,30,41\} = \{000110,011000,100001\}\\
\mathcal{Y}_3 &= \{05,12,24\} = \{000101,001010,010100\}\\
\mathcal{Y}_4 &= \{21,42,50\} = \{010001,100010,101000\}\\
\mathcal{Y}_5 &= \{11,22,44\} = \{001001,010010,100100\}.
\end{aligned}$$

Thus, $\mathcal{Y}_1,\mathcal{Y}_2$, and $\mathcal{Y}_5$ are binary block codes having the minimum distance 4, and the vector $\mathbf{y}$ is uniquely determined by X based on the vector $\mathbf{x}$ if he receives the messages 1, 2, or 5. Furthermore, there is only one vector $\mathbf{x} = 010101$ when X has the ambiguity about $\mathbf{y}$ if the message is 3 and only one such a vector $\mathbf{x} = 101010$ if the message is 4.

The general procedure for constructing the common values can be described in at least two different ways. Suppose that there is a matrix of dimension $|\mathcal{X}| \times |\mathcal{Y}|$ containing the $*$ symbols at the positions corresponding to the pairs of vectors $(\mathbf{x},\mathbf{y}) \in \overline{\mathcal{X}\mathcal{Y}}$ and gaps at all other positions. We permute the rows and the columns of this matrix in accordance with the functions f,g and split the resulting matrix into $M_X M_Y$ rectangles. Then any two pairs of vectors generate the same common value if and only if there exists a "path" connecting the corresponding $*$ symbols, which completely belongs to this rectangle and may turn by 90 degrees passing through any of the $*$ symbols. Another way of representing this procedure relates to bipartite graphs (see Figure 3) : given $(i,j) \in [M_X] \times [M_Y]$, we introduce a bipartite graph having left and right sides; the vertices at the left side of the graph are the vectors $\mathbf{x} \in \mathcal{X}_i$ and the vertices at the right side are the vectors $\mathbf{y} \in \mathcal{Y}_j$; any two vertices $\mathbf{x}$ and $\mathbf{y}$ are connected by an edge if and only if $(\mathbf{x},\mathbf{y}) \in \overline{\mathcal{X}\mathcal{Y}}$. In the next section we prove the statement that the different values of the functions K,L can be assigned only to distinct connected components of the bipartite graphs constructed for all $i \in [M_X]$ and $j \in [M_Y]$, where the term "connected component" is taken in the classical graph theory sense : any two vertices belong to the same connected component if and only if there is a path connecting these vertices. In further extensions of the problem these considerations should be represented in a more general form; in particular, empty sets can be the set of vertices belonging to the connected components.

SEPARABILITY OF A COMMON KNOWLEDGE

Any two pairs of vectors $(\mathbf{x}_1,\mathbf{y}_1),(\mathbf{x}_2,\mathbf{y}_2) \in \overline{\mathcal{X}\mathcal{Y}}$ can be separated by the participants either based on the description of the source when $(\mathbf{x}_1,\mathbf{y}_2),(\mathbf{x}_2,\mathbf{y}_1) \notin \overline{\mathcal{X}\mathcal{Y}}$ or based on different pairs of messages corresponding to these vectors. If both criteria cannot be used, then X and Y have to come to the same common value in the cases when they are given $(\mathbf{x}_1, \mathbf{y}_1)$ and $(\mathbf{x}_2, \mathbf{y}_2)$. However we speak about the "separability of the knowledge" in another context and present the discussion in the end of the section.

For all $i \in [M_X]$, $j \in [M_Y]$, and sets $\varphi \subseteq \mathcal{X}$, $\psi \subseteq \mathcal{Y}$, let

$$\begin{aligned}\mathcal{F}_j(\varphi) &= \{\, \mathbf{y} \in \mathcal{Y}_j : (\mathbf{x}, \mathbf{y}) \in \overline{\mathcal{X}\mathcal{Y}}, \text{ for some } \mathbf{x} \in \varphi \,\} \\ \mathcal{G}_i(\psi) &= \{\, \mathbf{x} \in \mathcal{X}_i : (\mathbf{x}, \mathbf{y}) \in \overline{\mathcal{X}\mathcal{Y}}, \text{ for some } \mathbf{y} \in \psi \,\}.\end{aligned}$$

Definition :

(X) *For all $i \in [M_X]$, $j \in [M_Y]$, and $\mathbf{x} \in \mathcal{X}_i$, the set $\varphi(\mathbf{x}|j)$ is the j–th ghost of the vector $\mathbf{x}$ if and only if $\varphi(\mathbf{x}|j) = \varphi$, where $\varphi = \emptyset$ when $\mathcal{F}_j(\{\mathbf{x}\}) = \emptyset$ and φ is the non–empty set of minimal cardinality which satisfies the conditions*

$$\begin{cases} \mathbf{x} \in \varphi \\ \varphi = \mathcal{G}_i\,(\,\mathcal{F}_j(\varphi)\,) \end{cases} \tag{4}$$

when $\mathcal{F}_j(\{\mathbf{x}\}) \neq \emptyset$.

(Y) *For all $i \in [M_X]$, $j \in [M_Y]$, and $\mathbf{y} \in \mathcal{Y}_j$, the set $\psi(\mathbf{y}|i)$ is the i–th ghost of the vector $\mathbf{y}$ if and only if $\psi = \emptyset$ when $\mathcal{G}_i(\{\mathbf{y}\}) = \emptyset$ and ψ is the non–empty set of minimal cardinality which satisfies the conditions*

$$\begin{cases} \mathbf{y} \in \psi \\ \psi = \mathcal{F}_j\,(\,\mathcal{G}_i(\psi)\,) \end{cases} \tag{5}$$

when $\mathcal{G}_i(\{\mathbf{y}\}) \neq \emptyset$.

Lemma :

(a) *For all $\mathbf{x} \in \mathcal{X}$ and $\mathbf{y} \in \mathcal{Y}$, the sets $\varphi(\mathbf{x}|1), \ldots, \varphi(\mathbf{x}|M_Y)$ and $\psi(\mathbf{y}|1), \ldots,$ $\psi(\mathbf{y}|M_X)$ are uniquely defined by* **(X)**, **(Y)**.

(b) *If $(\mathbf{x}, \mathbf{y}) \in \overline{\mathcal{X}\mathcal{Y}}$, then*

$$\begin{aligned}(\,\varphi(\mathbf{x}|g(\mathbf{y})), \psi(\mathbf{y}|f(\mathbf{x}))\,) &= (\,\varphi(\mathbf{x}|g(\mathbf{y})), \mathcal{F}_{g(\mathbf{y})}(\varphi(\mathbf{x}|g(\mathbf{y}))\,) &(6)\\ &= (\,\mathcal{G}_{f(\mathbf{x})}(\psi(\mathbf{y}|f(\mathbf{x})), \psi(\mathbf{y}|f(\mathbf{x}))\,) &(7)\end{aligned}$$

i.e., the pair of sets $(\,\varphi(\mathbf{x}|g(\mathbf{y})), \psi(\mathbf{y}|f(\mathbf{x}))\,)$ is uniquely determined both by $(\mathbf{x}, g(\mathbf{y}))$ and $(\mathbf{y}, f(\mathbf{x}))$.

(c) *For all $\mathbf{x}' \in \mathcal{X}$ and $\mathbf{y}' \in \mathcal{Y}$,*

$$\begin{aligned}\mathbf{x}' \in \varphi(\mathbf{x}|g(\mathbf{y})) &\implies K(\mathbf{x}'|g(\mathbf{y})) = K(\mathbf{x}|g(\mathbf{y})) &(8)\\ \mathbf{y}' \in \psi(\mathbf{y}|f(\mathbf{x})) &\implies L(\mathbf{y}'|f(\mathbf{x})) = L(\mathbf{y}|f(\mathbf{x})). &(9)\end{aligned}$$

Theorem : *Let*

$$\overline{\Phi\Psi}(f,g) = \{\,(\,\varphi(\mathbf{x}|g(\mathbf{y})), \psi(\mathbf{y}|f(\mathbf{x}))\,) \in 2^{\mathcal{X}} \times 2^{\mathcal{Y}} : (\mathbf{x}, \mathbf{y}) \in \overline{\mathcal{X}\mathcal{Y}}\,\}.$$

The functions f, g, K, L satisfy the restriction (1) if and only if there exists a function $\Theta : \overline{\Phi\Psi}(f,g) \to \Omega$ defined by the values

$$(\, \Theta\,(\varphi, \psi) \in \Omega\,)_{(\varphi,\psi)\in\overline{\Phi\Psi}(f,g)}$$

such that

$$K(\mathbf{x}|g(\mathbf{y})) = \Theta\,(\,\varphi(\mathbf{x}|g(\mathbf{y})), \psi(\mathbf{y}|f(\mathbf{x}))\,)\,, \quad \textit{for all } (\mathbf{x},\mathbf{y}) \in \overline{\mathcal{X}\mathcal{Y}} \tag{10}$$

and

$$L(\mathbf{y}|f(\mathbf{x})) = \Theta\,(\,\varphi(\mathbf{x}|g(\mathbf{y})), \psi(\mathbf{y}|f(\mathbf{x}))\,)\,, \quad \textit{for all } (\mathbf{x},\mathbf{y}) \in \overline{\mathcal{X}\mathcal{Y}}. \tag{11}$$

Corollary : *Given functions f and g,*

$$\max_{K,L} |\, \Omega(f,g,K,L)\,| = |\, \overline{\Phi\Psi}(f,g)\,| \tag{12}$$

where the maximum is taken over the functions K, L satisfying (1).

The set $\varphi(\mathbf{x}|j)$ satisfying $(\mathbf{X})$ can be constructed by the following recurrent procedure. We introduce the sequence of sets $\varphi^{(0)} = \{\mathbf{x}\}, \varphi^{(1)}, \ldots \subseteq \mathcal{X}_i$, where

$$\varphi^{(t)} = \mathcal{G}_i\left(\, \mathcal{F}_j(\varphi^{(t-1)})\,\right), \quad t = 1, 2, \ldots \tag{13}$$

Then all but a finite number of sets in this sequence coincide with some set φ, and $\varphi(\mathbf{x}|j) = \varphi$. In other words, we consider $\mathbf{x}$ as the 1–element set $\{\mathbf{x}\}$ and use the possibility of extending this set. Returning to the applications, one can say that the individuals have to replace the vectors received from the source with their ghosts and extend these ghosts until they become common. The theorem claims that this is the only way of creating a common knowledge in the sense of (1).

The result of the theorem can be interpreted using the scheme given in Figure 2 where X and Y are given the vectors generated by the source, but they do not communicate to each other. Different values of the function g correspond to different partitions of the space $\mathcal{X}$. If X knows g, but does not know the value of $g(\mathbf{y})$, then he considers all possible values and outputs the vector $\boldsymbol{\varphi}$ containing M_Y ghosts of the vector $\mathbf{x}$. For example, if the functions f and g are defined by Table 2 and $\mathbf{x} = 45$ (see also Figure 3), then $\boldsymbol{\varphi} = (\{45\}, \emptyset, \emptyset, \{43, 45, 46, 51, 52\}, \{45, 46\})$. By $(\mathbf{X})$, at least one component of the vector $\boldsymbol{\varphi}$ is a non–empty set and the value of the function f is a constant $f(\mathbf{x})$ for every non–empty set. Similarly, Y outputs the vector $\boldsymbol{\psi}$ containing M_X ghosts of the vector $\mathbf{y}$. The pair of vectors $(\boldsymbol{\varphi}, \boldsymbol{\psi})$ goes to the person called the decoder (in the situation considered before, the decoder is simultaneously X and Y), and the decoding algorithm is fixed as follows : the decoder first extracts $f(\mathbf{x})$ from $\boldsymbol{\varphi}$ and uses this value as a pointer to the vector $\boldsymbol{\psi}$, then he extracts $g(\mathbf{y})$ from $\boldsymbol{\psi}$, uses this value as a pointer to the vector $\boldsymbol{\varphi}$, and finally outputs $(\,\varphi(\mathbf{x}|g(\mathbf{y})), \psi(\mathbf{y}|f(\mathbf{x}))\,)$. If $(\mathbf{x},\mathbf{y}) \in \overline{\mathcal{X}\mathcal{Y}}$, then both components of this vector are non–empty sets, and this fact can be viewed as the matching of φ and ψ. In this case, the result of the decoding is the pair of elements of dependent partitions of the sets $\mathcal{X}$ and $\mathcal{Y}$. Note that the same result can be also obtained if X knows $g(\mathbf{y})$ and Y knows $f(\mathbf{x})$: then each person outputs only one subset, and the decoder combines them into a pair. Such a conclusion is possible only because the subsets presented by the persons are not arbitrarily chosen, but uniquely defined by the functions f and g in accordance with **(X)**, **(Y)**. In other words, the individuals establish a common knowledge only if they construct and present certain vectors φ and ψ that can be viewed as their individual contributions to the common knowledge. That is, the "common knowledge" can be achieved with the use of the source only if there exist separated individual knowledges matching each other.

PROOFS

Proof of Lemma

Statement (a) If there exist two sets, φ_1 and φ_2, satisfying (4), then their intersection, $\varphi_1 \cap \varphi_2$, also satisfies (4). Since $\varphi(\mathbf{x}|j)$ is the set of minimal cardinality satisfying (4) and the total number of subsets of $\mathcal{X}$ is finite, this set is uniquely defined by **(X)**. Similar considerations with the use of (5) prove that the set $\psi(\mathbf{y}|i)$ is uniquely defined by **(Y)**.

Statement (b) Let $(\mathbf{x},\mathbf{y}) \in \overline{\mathcal{X}\mathcal{Y}}$ and $\mathbf{x} \in \mathcal{X}_i$, $\mathbf{y} \in \mathcal{Y}_j$. We refer to the algorithm for constructing the sequence of sets $\varphi^{(0)} = \{\mathbf{x}\}, \varphi^{(1)}, \ldots \subseteq \mathcal{X}_i$, which is recurrently defined by (13), and construct another sequence $\varphi^{(0)\prime} = \mathcal{G}_i(\{\mathbf{y}\}), \varphi^{(1)\prime}, \ldots \subseteq \mathcal{Y}_j$ by

$$\varphi^{(t)\prime} = \mathcal{G}_i\left(\mathcal{F}_j(\varphi^{(t-1)\prime})\right), \quad t = 1, 2, \ldots$$

If φ' is the set belonging to this sequence such that all but a finite number of sets differ from φ', then

$$\{\mathbf{y}\} \subseteq \mathcal{F}_j(\{\mathbf{x}\}) \implies \mathcal{G}_i(\{\mathbf{y}\}) = \varphi^{(0)\prime} \subseteq \varphi^{(1)} \implies \varphi' \subseteq \varphi.$$

However φ is the set of minimal cardinality satisfying (4). Thus $\varphi' = \varphi$, and (6) follows. The proof of (7) is similar.

Statement (c) By the requirement for $\varphi(\mathbf{x}|g(\mathbf{y}))$ to be the set of minimal cardinality satisfying (4), $\mathbf{x}' \in \varphi(\mathbf{x}|g(\mathbf{y}))$ implies the existence of a sequence of $\lambda \geq 2$ vectors $\mathbf{x}_1, \ldots, \mathbf{x}_\lambda \in \varphi(\mathbf{x}|g(\mathbf{y}))$ such that $\mathbf{x}_1 = \mathbf{x}$, $\mathbf{x}_\lambda = \mathbf{x}'$, and

$$\left\{\, \mathbf{y} \in \mathcal{Y}_j : \, (\mathbf{x}_\mu, \mathbf{y}), (\mathbf{x}_{\mu-1}, \mathbf{y}) \in \overline{\mathcal{X}\mathcal{Y}} \,\right\} \neq \emptyset, \quad \text{for all } \mu \in \{2, \ldots, \lambda\}.$$

By (1),

$$K(\mathbf{x}_{\mu-1}|g(\mathbf{y})) = L(\mathbf{y}|f(\mathbf{x}_{\mu-1})), \;\; K(\mathbf{x}_\mu|g(\mathbf{y})) = L(\mathbf{y}|f(\mathbf{x}_\mu))$$

for any vector $\mathbf{y}$ belonging to this set and $f(\mathbf{x}_{\mu-1}) = f(\mathbf{x}_\mu)$ by **(X)**. Thus $K(\mathbf{x}_{\mu-1}|g(\mathbf{y})) = K(\mathbf{x}_\mu|g(\mathbf{y}))$. Using this argument for $\mu = 2, \ldots, \lambda$ we obtain $K(\mathbf{x}_1|g(\mathbf{y})) = K(\mathbf{x}_\lambda|g(\mathbf{y}))$ and prove (8). The proof of (9) is similar.

Proof of Theorem

Direct statement : *if there is a function* Θ *such that (10) and (11) hold, then the functions* K, L *satisfy (1).*

By (10) and (11), we write

$$K(\mathbf{x}|g(\mathbf{y})) = \Theta\left(\varphi(\mathbf{x}|g(\mathbf{y})), \psi(\mathbf{y}|f(\mathbf{x}))\right) = L(\mathbf{y}|f(\mathbf{x}))$$

for all $(\mathbf{x},\mathbf{y}) \in \overline{\mathcal{X}\mathcal{Y}}$ and obtain (1).

Converse statement : *if the functions* K, L *satisfy (1), then there is a function* Θ *such that (10) and (11) hold.*

By (8), $K(\mathbf{x}|g(\mathbf{y}))$ is a function of $\varphi(\mathbf{x}|g(\mathbf{y}))$ and $g(\mathbf{y})$. The integer $g(\mathbf{y})$ is available from any of the sets $\psi(\mathbf{y}|1),\dots,\psi(\mathbf{y}|M_X)$ since

$$\mathbf{y}' \in \psi(\mathbf{y}|i) \implies g(\mathbf{y}') = g(\mathbf{y}), \quad \text{for any } i \in [M_X].$$

In particular, $g(\mathbf{y})$ can be extracted from the set $\psi(\mathbf{y}|f(\mathbf{x}))$, and $K(\mathbf{x}|g(\mathbf{y}))$ can be represented as a function Θ_K depending on $\varphi(\mathbf{x}|g(\mathbf{y}))$ and $\psi(\mathbf{y}|f(\mathbf{x}))$, i.e.,

$$K(\mathbf{x}, g(\mathbf{y})) = \Theta_K\left(\varphi(\mathbf{x}|g(\mathbf{y})), \psi(\mathbf{y}|f(\mathbf{x}))\right). \tag{14}$$

Similar considerations with the use of (9) prove that $L(\mathbf{y}|f(\mathbf{x}))$ can be represented as a function Θ_L depending on the same arguments, i.e.,

$$L(\mathbf{y}|f(\mathbf{x})) = \Theta_L\left(\varphi(\mathbf{x}|g(\mathbf{y})), \psi(\mathbf{y}|f(\mathbf{x}))\right). \tag{15}$$

The coincidence of $K(\mathbf{x}|g(\mathbf{y}))$ and $L(\mathbf{y}|f(\mathbf{x}))$ for all $(\mathbf{x},\mathbf{y}) \in \overline{\mathcal{X}\mathcal{Y}}$ and (14), (15) imply the coincidence of the functions Θ_K and Θ_L. We denote $\Theta = \Theta_K = \Theta_L$ and using (2) obtain (10).

Corollary : Let $\Omega^*(f,g)$ denote the set of values of the function Θ. Then

$$|\Omega^*(f,g)| \le |\overline{\Phi\Psi}(f,g)| \tag{16}$$

since Θ is a mapping $\overline{\Phi\Psi}(f,g) \to \Omega$. On the other hand, by (10) and (11),

$$|\Omega^*(f,g)| = |\Omega(K,L,f,g)| \tag{17}$$

where we use the notations (2). Combining (16) and (17) we obtain

$$|\Omega(K,L,f,g)| \le |\overline{\Phi\Psi}(f,g)|. \tag{18}$$

If Θ is a one-to-one mapping, then (16) holds with the equality, which also implies the equality in (18), and (12) follows.

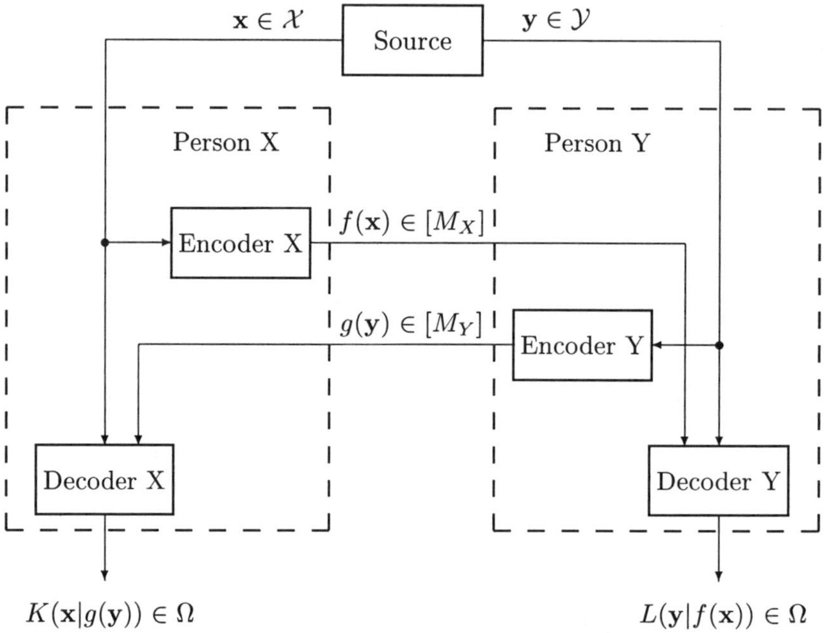

Figure 1 Model of creating a common knowledge by correlated observations and transmission over helping channels.

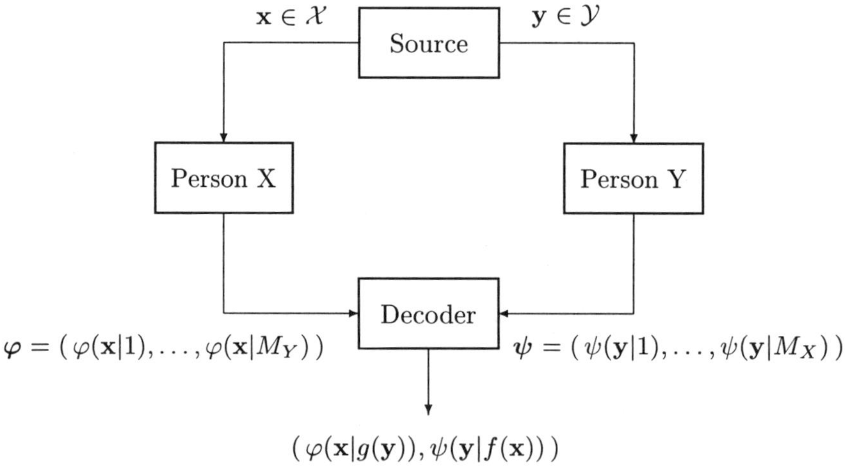

Figure 2 Logical scheme of creating a common knowledge.

Table 1 Structure of the set $\overline{\mathcal{X}\mathcal{Y}}$ defined in (3), where the binary vectors of length 6 are given in the octal representation.

	03	05	06	11	12	14	21	22	24	30	41	42	44	50	60
07	*	*	*												
13	*			*	*										
15		*		*		*									
16			*		*	*									
23	*						*	*							
25		*					*		*						
26			*					*	*						
31				*			*			*					
32					*			*		*					
34						*			*	*					
43	*										*	*			
45		*									*		*		
46			*									*	*		
51				*							*			*	
52					*							*		*	
54						*							*	*	
61							*				*				*
62								*				*			*
64									*				*		*
70										*				*	*

Table 2 The 26 common values A, . . . ,Z that can be constructed by the participants when X transmits one of 4 messages, Y transmits one of 5 messages, and the set $\overline{\mathcal{X}\mathcal{Y}}$ is defined in (3).

	03	05	06	11	12	14	21	22	24	30	41	42	44	50	60
07	A	A	A												
13	A			B	B										
15		A		B		B									
16			A		B	B									
23	A						C	C							
25		D					I		I						
26			E					I	I						
31				F			I			J					
32					G			I		J					
34						H			I	J					
43	K										P	P			
45		L									P		Q		
46			M									P	Q		
51				N							P			R	
52					O							P		R	
54						S							Z	Z	
61							T				X				Z
62								U				Y			Z
64									V				Z		Z
70										W				Z	Z

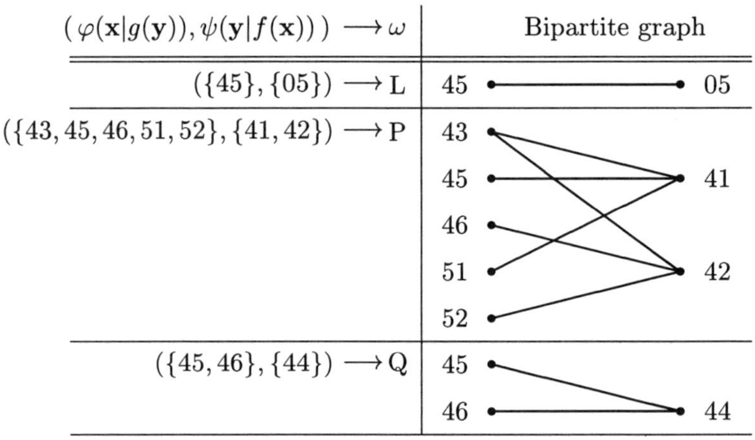

Figure 3 The common values ω and the corresponding bipartite graphs when $\mathbf{x} = 45$; the set $\overline{\mathcal{X}\mathcal{Y}}$ and the functions f, g are specified in Table 2.

Table 3 The 50 common values A, . . . ,Z that can be constructed by the participants when X transmits one of 4 messages, Y transmits one of 5 messages, and the set $\overline{\mathcal{X}\mathcal{Y}}$ is defined in (3).

	03	14	60	06	30	41	05	12	24	21	42	50	11	22	44
07	A			D			G								
25							G		G	I					
32					E			H						K	
54		B										J			L
61			C			F				I					
16		M		O				S							
34		M			P				T						
45						Q	R								Y
51						Q						V	W		
62			N								U			X	
26				b					f					j	
31					c					g			i		
43	Z					d					h				
52								e			h	h			
64			a						f						k
13	l							r					v		
15		m					q						v		
23	l									s				w	
46				o							t				x
70			n		p							u			

References

[1] C. E. Shannon, "Two–way communication channels," in *Claude Elwood Shannon : Collected Papers.* N. J. A. Sloane and A. D. Wyner (eds.). New York: IEEE Press, 1993, 351–384.
The paper was published in the *Proc. 4-th Berkley Symp. Math. Stat. and Prob.,* 1961, 611–644.

[2] R. Ahlswede, "Multi-way communication channels," in *2nd Int. Symp. Inform. Theory; Tsahkadzor, Armenian SSR, 1971.* Publishing House of the Hungarian Academy of Sciences, 1973, 23–52.

[3] P. Gács, J. Körner, "Common information is far less than mutual information," *Probl. Inform. Control,* 2(2), 1973, 149–162.

[4] D. Slepian, J. K. Wolf, "Noiseless coding of correlated information sources," *IEEE Trans. Inform. Theory,* 19(4), 1973, 772–777.

[5] R. Ahlswede, "The capacity region of a channel with two senders and two receivers," *Ann. Prob.,* 2(5), 1974, 805–814.

[6] A. D. Wyner, "On source coding with side information at the decoder," *IEEE Trans. Inform. Theory,* 21(3), 1975, 294–300.

[7] R. Ahlswede, J. Körner, "Source coding with side information and a converse for degraded broadcast channels," *IEEE Trans. Inform. Theory,* 21(6), 1975, 629–637.

[8] J. Körner, K. Marton, "How to encode modulo-two sum of binary sources," *IEEE Trans. Inform. Theory,* 25(2), 1979, 219–221.

[9] R. Ahlswede, G. Dueck, "Identification in the presence of feedback – A discovery of new capacity formulas," *IEEE Trans. Inform. Theory,* 35(1), 1989, 30–36.

[10] U. Maurer, "Secret key agreement by public discussion from common information," *IEEE Trans. Inform. Theory,* 39(3), 1993, 733–742.

[11] R. Ahlswede, I. Csiszár, "Common randomness in information theory and cryptography – Part I : Secret sharing," *IEEE Trans. Inform. Theory,* 39(5), 1993, 1121–1131.

[12] R. Ahlswede, V. B. Balakirsky, "Identification under random processes," *Problemy Peredachi Informatsii* (special issue honoring Mark S. Pinsker), 32(1), 1996, 144–160 (in Russian). English translation : *Probl. Inform. Transmission,* 32, 1996, 123–138.

[13] R. Ahlswede, I. Csiszár, "Common randomness in information theory and cryptography – Part II : CR capacity," *IEEE Trans. Inform. Theory,* 44(1), 1998, 225–240.

HOW TO BROADCAST PRIVACY: SECRET CODING FOR DETERMINISTIC BROADCAST CHANNELS

Ning Cai and Kwok Yan Lam

School of Computing, National University of Singapore,
Lower Kent Ridge Road, Singapore 119260

{ncai,lamky}@comp.nus.edu.sg

Abstract: We consider a broadcast channel, a channel with one sender and two receivers, and introduce a new model in which we require that each receiver not only can correctly (with a probability close to one) decode his/her own message but also obtains no (significant amount of) information about the message for the other receiver. We determine the capacity region for the deterministic broadcast channel in the presence of randomization at the sender's side. In the case that randomization is not allowed, we reduce the coding problem to an open problem in Combinatorics.

INTRODUCTION

People today have more and more privacy, e. g. the amount of their annual salaries or the balances in their bank account. So, it is natural for us to study how to protect it in communication. For example, when a company wants to adjust the salaries of its employees, how does it broadcast its decision so that everyone only knows the amount of his/her own new salary?

In this paper, we study a communication model for a broadcast channel for which both receivers should not obtain any knowledge about the message dedicated for the other receiver. The broadcast channel was introduced by T. M. Cover in 1972 [6]. It consists of one sender (or encoder) E and two receivers (or decoders) D_l, $l = 1, 2$. The sender E is required to send the messages m_1 and m_2 from the message sets $\mathcal{M}_1$ and $\mathcal{M}_2$ to D_1 and D_2, respectively, correctly with probability close to one. In general, the capacity regions for this kind of channels are still unknown. To determine them probably is one of the hardest open problems in Shannon Theory. So in this paper we focus on

I. Althöfer et al. (eds.), Numbers, Information and Complexity, 353-368.

deterministic broadcast channels whose capacity regions were determined by M. S. Pinsker (1978, [7]). In our model, we require not only that m_l, $l = 1, 2$ are correctly decoded by the corresponding receivers, respectively, but also that D_1 (D_2) is not allowed to obtain any (significant) knowledge about m_2 (m_1), the message for the other receiver. We call the code with the desired properties a secret code for the broadcast channel.

Another related model is the wire-tap channel (A. D. Wyner 1975 [8] and I. Csiszár-J. Körner 1978 [5]). The wire-tap channel has the same statistical properties as the broadcast channel. The difference is that one of the receivers, say D_2, now is assumed to be an eavesdropper and there is only one message from $\mathcal{M}_1$ to be transmitted (i. e. $\mathcal{M}_2$ does not exist at all). The requirement for the wire-tap channel is that the legal receiver D_1 should be able to recover the message from $\mathcal{M}_1$ correctly with high probability whereas the eavesdropper D_2 should obtain no significant knowledge about the message. The capacity regions for wire-tap channels were determined ([8] and [5]). Moreover a sharper result for wire-tap channels was obtained by I. Csiszár, (1996 [4]) by applying a lemma from [3]. In some sense our model can be understood as a "double wire-tap channel". That is, each receiver is both a legal receiver (in respect to the message for himself/herself) and an eavesdropper (in respect to the message for the other). However, it is not hard to see that the behavior of our secret code is more like that of a code for a broadcast channel because for both receivers there are two messages to be decoded.

Usually, for a wire-tap channel, randomization (at the sender's side) is allowed. In this paper, we consider both cases: with and without randomization. We show that the secret coding problem for the deterministic broadcast channels is equivalent to an open problem in Combinatorics when randomization is not allowed. We determine the capacity regions for secret codes with randomization for deterministic broadcast channels, which is our main result in this paper. An example shows that randomization can improve the performance.

Our model is formulated in Section 2. The case that randomization is not allowed is disscussed in Section 3. Our main result is stated and proved in Section 4. We conclude our paper by an example in Section 5.

DEFINITIONS

Let us first recall the definitions of the broadcast channel and the wire-tap channel.

Let $\mathcal{X}$, $\mathcal{Y}$, and $\mathcal{Z}$ be finite sets which will serve as the input alphabet, the output alphabet for the first receiver D_1, and the output alphabet for the second receiver D_2, respectively. Let us consider a (memoryless) broadcast channel described by a pair of stochastic matrices : $W_1 : \mathcal{X} \longrightarrow \mathcal{Y}$ and $W_2 : \mathcal{X} \longrightarrow \mathcal{Z}$. When an $x^n := (x_1, \ldots, x_n) \in \mathcal{X}^n$ is fed into the channel, the receivers D_1 and D_2 receive $y^n := (y_1, \ldots, y_n) \in \mathcal{Y}^n$ and $z^n := (z_1, \ldots, z_n) \in \mathcal{Z}^n$ with the probabilities

$$W_1^n(y^n|x^n) = \prod_{t=1}^{n} W_1(y_t|x_t) \tag{1}$$

and

$$W_2^n(z^n|x^n) = \prod_{t=1}^{n} W_2(z_t|x_t) \tag{2}$$

respectively. A rate pair (R_1, R_2) of non-negative reals is achievable iff for all positive reals λ and ϵ and for sufficiently large n (depending on λ and ϵ), there exists a code of lenght n, a system $\{(u_{i,j}, \mathcal{D}_i, \mathcal{D}'_j) : 1 \leq i \leq M_1, \quad \text{and} \quad 1 \leq j \leq M_2\}$ with $u_{i,j} \in \mathcal{X}^n, \mathcal{D}_i \subset \mathcal{Y}^n$ and $\mathcal{D}'_j \subset \mathcal{Z}^n$, for $1 \leq i \leq M_1$ and $1 \leq j \leq M_2$,

$$\mathcal{D}_i \cap \mathcal{D}_{i'} = \emptyset \quad \text{for} \quad i \neq i', \quad \mathcal{D}'_j \cap \mathcal{D}'_{j'} = \emptyset \quad \text{for} \quad j \neq j', \tag{3}$$

$$\frac{1}{n}\log M_1 \geq R_1 - \epsilon, \quad \text{and} \quad \frac{1}{n}\log M_2 \geq R_2 - \epsilon, \tag{4}$$

such that for all $i \in \{1, \ldots, M_1\}$ and $j \in \{1, \ldots, M_2\}$,

$$W_1^n(\mathcal{D}_i|u_{i,j}) \geq 1 - \lambda, \tag{5}$$

and

$$W_2^n(\mathcal{D}'_j|u_{i,j}) \geq 1 - \lambda. \tag{6}$$

A wire-tap channel can also be described by stochastic matrices W_1 and W_2 through (1) and (2). The secrecy capacity $C_s(1)$, in the case that D_1 is the legal receiver and D_2 is the eavesdropper, is the supremum of the reals R such that for all positive reals λ, μ and ϵ, and for sufficiently large n, there exits a code, a system $\{(Q, \mathcal{D}_i) : 1 \leq i \leq M\}$, where Q is a stochastic matrix, $Q : \mathcal{M} = \{1, \ldots, M\} \longrightarrow \mathcal{X}^n$, and $\mathcal{D}_i$, $i = 1, \ldots, M$ are pairwise disjoint susets of $\mathcal{Y}^n$, with $\frac{1}{n}\log M \geq R - \epsilon$ such that for all $u \in \mathcal{M}$

$$\sum_{x^n \in \mathcal{X}} Q(x^n|u) W_1^n(\mathcal{D}_u|x^n) \geq 1 - \lambda, \tag{7}$$

and

$$I(U \wedge Z^n) \leq n\mu. \tag{8}$$

Here U is the random variable with uniform distribution over $\mathcal{M}$ and Z^n is the output random variable of the channel W_2^n, when the input x^n of the channel is chosen with the probability $\sum_{u\in\mathcal{M}} P_U(u)Q(x^n|u)$. We notice that the factor n in front of μ in (8) is quite standard but it was shown in [4] that it can be removed without changing the secrecy capacity. Analogously, we denote by $C_s(2)$ the secrecy capacity in the case that D_2 is the legal receiver and D_1 is the eavesdropper.

Next, we define our secret code for the broadcast channel described by W_1 and W_2, through (1) and (2). An $(n, M_1, M_2, \lambda, \mu)$ secret code with randomization (at the sender's side) for the broadcast channel is a system $\{(Q, \mathcal{D}_i, \mathcal{D}'_j) : 1 \leq i \leq M_1 \text{ and } 1 \leq j \leq M_2\}$, where Q is a stochastic matrix $Q : \mathcal{M}_1 \times \mathcal{M}_2 \longrightarrow \mathcal{X}^n$, $\mathcal{M}_l = \{1, \dots, M_l\}$ for $l = 1,\ 2$ and $\mathcal{D}_i,\ i = 1, \dots M_1$ and $\mathcal{D}'_j,\quad j = 1, \dots, M_2$ are pairwise disjoint susets of $\mathcal{Y}^n$ and $\mathcal{Z}^n$ respectively, such that for all $u \in \mathcal{M}_1$ and $v \in \mathcal{M}_2$,

$$\sum_{x^n\in\mathcal{X}^n} Q(x^n|u,v)W_1^n(\mathcal{D}_u|x^n) \geq 1 - \lambda, \tag{9}$$

$$\sum_{x^n\in\mathcal{X}^n} Q(x^n|u,v)W_2^n(\mathcal{D}'_v|x^n) \geq 1 - \lambda, \tag{10}$$

$$I(U \wedge Z^n) \leq n\mu, \tag{11}$$

and

$$I(V \wedge Y^n) \leq n\mu, \tag{12}$$

where the random variables U and V are uniformly distributed and independently take values in $\mathcal{M}_1$ and $\mathcal{M}_2$ respectively, and Y^n and Z^n are output random variables of the channels W_1^n and W_2^n (observed by D_1 and D_2), respectively, when the random variable X^n with distribution $P_{X^n}(x^n) = \sum_{u\in\mathcal{M}_1, v\in\mathcal{M}_2} P_U(u)P_V(v)Q(x^n|u,v)$ is the input.

A pair (R_1, R_2) of non-negative reals is said to be achievable by secret codes with randomization if for all positive λ, μ and ϵ and a sufficiently large n, there exists an $(n, M_1, M_2, \lambda, \mu)$ secret code with randomization and rates $\frac{1}{n}M_l \geq R_l - \epsilon$ for $l = 1,\ 2$. The set of pairs achievable by secret codes with randomization is called capacity region for secret codes with randomization, or for probability-type secret codes, denoted by $C_s(1, 2)$. Here we also refer to our codes as probability-type secret codes to emphasize the contrast with combinatorics-type secret codes, the codes without randomization defined below. We shall show that the factors n in front of μ in (11) and (12) can be dropped without changing the capacity region.

A secret code without randomization, or a combinatorics-type secret code, is just a code for the broadcast channel (satisfying (5) and (6)), with the additional properties

$$I(U \wedge Z^n) = I(V \wedge X^n) = 0 \tag{13}$$

(where U, V, Y^n and Z^n are defined as before). Its capacity region (defined in the standard way), denoted by $C_s^*(1,2)$, is called capacity region for secret codes without randomization, or for combinatorics-type secret codes. Notice that instead of the condition "$\leq n\mu$" or "$\leq \mu$" we here require the condition "$= 0$". This is because we would like to model our problem "purely combinatorially".

In the sequel, we always consider deterministic broadcast channels, or noiseless broadcast channels, whose capacity regions (for classical codes) were determined by M. S. Pinsker [7]. For such a channel, there exists a pair of functions $\phi : \mathcal{X} \longrightarrow \mathcal{Y}$ and $\psi : \mathcal{X} \longrightarrow \mathcal{Z}$ such that $W_1(y|x) = 1$ iff $y = \phi(x)$ and $W_2(z|x) = 1$ iff $z = \psi(x)$. Furthermore, x and x' play the same role in the communication and neither D_1 nor D_2 can distinguish them if there are $x, x' \in \mathcal{X}$ with $x \neq x'$ such that $\phi(x) = \phi(x')$ and $\psi(x) = \psi(x')$. In this case, we can delete one of the two letters without making any difference. Thus, w. l. o. g. we assume that there are no such pairs of input letters. For the convenience of the notation, we assume that "0" is not a letter in $\mathcal{X}$. Thus, under our assumption we can define a function $\tau : \mathcal{Y} \times \mathcal{Z} \longrightarrow \mathcal{X} \cup \{0\}$ such that for $x \in \mathcal{X}$,

$$\tau(y,z) = x \quad \text{iff} \quad \phi(x) = y \quad \text{and} \quad \psi(x) = z, \tag{14}$$

and

$$\tau(y,z) = 0 \quad \text{iff there is no} \quad x \in \mathcal{X} \quad \text{with } \phi(x) = y \quad \text{and} \quad \psi(x) = z \tag{15}$$

Obviously,

$$\tau(y,z) = \tau(y',z') \quad \text{and} \quad (y,z) \neq (y',z') \Longrightarrow \tau(y,z) = \tau(y',z') = 0, \tag{16}$$

$$\begin{aligned} &\tau(y,z) = x \in \mathcal{X} \Longleftrightarrow \phi(x) = y \quad \text{and} \quad \psi(x) = z \\ &\Longleftrightarrow W_1(y|x)W_2(z|x) > 0 \Longleftrightarrow W_1(y|x)W_2(z|x) = 1, \end{aligned} \tag{17}$$

and

$$\tau(y,z) = 0 \Longleftrightarrow \text{for all} \quad x \in \mathcal{X}, W_1(y|x)W_2(z|x) = 0. \tag{18}$$

W. l. o. g., we also assume that for all $y \in \mathcal{Y}(z \in \mathcal{Z})$ there is an $x \in \mathcal{X}$ with $\phi(x) = y(\psi(x) = z)$, otherwise the output letter is useless and therefore can be deleted. For the deterministic broadcast channel notice, that if (5) and (6) hold for any $\lambda > 0$ then they hold for all $\lambda' \geq 0$.

THE COMBINATORIAL MODEL

We shall first state a problem from Combinatorics and then show that the combinatorially secret coding problem is equivalent to it. For any matrix A, we denote by $A^{\otimes n}$ its n-th Kronecker power (in the field where it is defined). Then

Problem : What is the largest $m = m(n, B)$ (or $\lim_{n\to\infty} \frac{1}{n}\log m$) for a given $(0,1)$-matrix B and any fixed n such that $B^{\otimes n}$ has an $l_1 \times l_2 = m$ all-one submatrix?

This problem has been studied by different groups of people but is still open. So far very little is known when the size of B is large, e. g. larger than 6×6, say. One motivation to study the problem is the search for Yao - type lower bounds ([9]) in the communication complexity of vector-valued functions (for example, c.f. [2]).

For a fixed deterministic broadcast channel, we let A_n be a $|\mathcal{Y}^n| \times |\mathcal{Z}^n|$ matrix whose rows and columns are labelled by $y^n \in \mathcal{Y}^n$ and $z^n \in \mathcal{Z}^n$ respectively and whose (y^n, z^n)-th entry is $\tau_n(y^n, z^n) := (\tau(y_1, z_1), \dots, \tau(y_n, z_n))$ if $\tau(y_t, z_t) \in \mathcal{X}$ for $t = 1, \dots, n$ and $\tau_n(y^n, z^n) := 0$, if there is a $t \in \{1, \dots, n\}$ with $\tau(y_t, z_t) = 0$. Let J be the operator acting on matrices by changing all non-zero entries to "ones" (and keeping the zero entries unchanged). We formally define the "(n-th) product" of the elements in $\mathcal{X} \cup \{0\}$ such that

$$x_1 \times \dots \times x_n = (x_1, \dots, x_n) \quad \text{for} \quad x_t \in \mathcal{X}, t = 1, \dots, n, \tag{19}$$

and

$$w_1 \times \dots \times w_n = 0 \quad \text{if there exists a} \quad t \in \{1, \dots, n\} \quad \text{with} \quad w_t = 0, \tag{20}$$

and then formally the "Kronecker power" $A_1^{\otimes n}$ of A_1 with the definition of the (formal) product. Then, we have that

$$A_n = A_1^{\otimes n}, \tag{21}$$

and

$$J(A_n) = J(A_1)^{\otimes n}. \tag{22}$$

Moreover,

$$\text{The } (y^n, z^n)\text{−th entry of } J(A_n) = J(A_1)^{\otimes n} \text{ is } 1 \iff \text{the } (y^n, z^n)\text{−th entry of } A_n = A_1^{\otimes n}, \tau_n(y^n, z^n) \in \mathcal{X}^n \iff \text{There is an } x^n \in \mathcal{X}^n \text{ s. t. } W_1^n(y^n|x^n)W_2^n(z^n|x^n) = 1 \text{ (and therefore } x^n = \tau_n(y^n, z^n)), \tag{23}$$

$$\text{The } (y^n, z^n)\text{−th entry of } J(A_n) = J(A_1)^{\otimes n} \text{ is } 0 \iff \text{there is no } x^n \in \mathcal{X}^n \text{ with } W_1^n(y^n|x^n)W_2^n(z^n|x^n) > 0. \tag{24}$$

Proposition 1. *The deterministic broadcast channel has a combinatorics-type secret code of length n and rates* $(\frac{1}{n}\log M_1,\ \frac{1}{n}\log M_2)$ *iff* $J(A_1)^{\otimes n}$ *has an* $M_1 \times M_2$ *all-one submatrix.*

Proof :

"If part": Suppose $J(A_1)^{\otimes n} = J(A_n)$ has an all-one submatrix whose rows and columns are labeled by $y^n(1), \ldots, y^n(M_1)$ and $z^n(1), \ldots, z^n(M_2)$ respectively. Let $u_{i,j}$ be the $(y^n(i), z^n(j))$-th entry of the submatrix, $\mathcal{D}_i = \{y^n(i)\}$ for $i = 1, \ldots, M_1$, and $\mathcal{D}'_j = \{z^n(j)\}$ for $j = 1, \ldots, M_2$. Then by (23),

$$W_1^n(\mathcal{D}_i|u_{i,j}) = 1 \quad \text{for all} \quad i,\ j, \tag{25}$$

and

$$W_2^n(\mathcal{D}'_j|u_{i,j}) = 1 \quad \text{for all} \quad i,\ j, \tag{26}$$

that is (5), (6) and (13) hold, or in other words, $\{(u_{i,j}, \mathcal{D}_i, \mathcal{D}'_j) : 1 \leq i \leq M_1 \text{ and } 1 \leq j \leq M_2\}$ is a combinatorics-type secret code.

"Only if" part: Let $\{(u_{i,j}, \mathcal{D}_i, \mathcal{D}'_j) : 1 \leq i \leq M_1 \text{ and } 1 \leq j \leq M_2\}$ be a combinatorics-type secret code of length n. Notice that all elements in $\mathcal{X}^n$, especially $u_{i,j}$ $i = 1, \ldots, M_1$, $j = 1, \ldots, M_2$ are located at A_n and the corresponding entries in $J(A_n) = J(A_1)^{\otimes n}$ are "1"'s. It is easy to see that for all (fixed) $i \in \{1, \ldots, M_1\} := \mathcal{M}_1$, $u_{i,j}$, $j \in \{1, \ldots, M_2\} := \mathcal{M}_2$ must be in the same row. Otherwise one could find a row of A_n, say the x^n-th row, and a proper non-empty subset of $\mathcal{M}_2$, say $\mathcal{M}'_2$ such that $u_{i,j}$ is in the x^n-th row iff $j \in \mathcal{M}'_2$. Thus, when a $u_{i,j}$, $j \in \mathcal{M}'_2$ is sent, the receiver D_1 receives x^n with probability one and therefore knows a message in $\mathcal{M}'_2$ is sending to the receiver D_2. This is a contradiction to (13). Thus, all codewords of the code are located in M_1 rows of A_n and by the same reason, they are located in M_2 columns. In other words, all codewords are located in an $M_1 \times M_2$ submatrix of A_n. However, the number of entries in the submatrix is only $M_1 M_2$, which is equal to the total number of codewords. So it cannot contain a zero entry. Thus the corresponding submatrix in $J(A_1)^{\otimes n}$ is an $M_1 \times M_2$ all-one submatrix.

THE MAIN RESULT

In this section, we state and prove our main result. First we need an auxiliary result. Intuitively, the following lemma says that the rows, the columns, and the non-zero entries in each row and each colmun of a given matrix satisfying certain conditions can be almost uniformly colored by a pair of coloring functions for rows and columns, respectively. We have the pleasure to point out that coloring-type lemmas were introduced to Information Theory by R. Ahlswede in [1] and they have played and will play important roles in Shannon Theory and related topics. We believe that it is one of Rudi Ahlswede's many remarkable and important contributions in Information Theory.

Lemma 2. *Let* $B = (b_{ij})_{ij}$ *be an* $N_1 \times N_2$ *matrix such that each of its row contains at least* L_2 *non-zero entries and each of its column contains at least*

L_1 non-zero entries respectively. Let K_1 and K_2 be two positive integers and δ be a positive real such that

$$4K_2(N_1+1) < e^{\frac{L_2\delta^2}{2K_2}}, \tag{27}$$

and

$$4K_1(N_2+1) < e^{\frac{L_1\delta^2}{2K_1}}. \tag{28}$$

Then there exists a pair (α, β) of coloring functions coloring the rows and columns of B, $\alpha : \{1, \ldots, N_1\} \longrightarrow \mathcal{K}_1 := \{1, \ldots, K_1\}$ and $\beta : \{1, \ldots, N_2\} \longrightarrow \mathcal{K}_2 := \{1, \ldots, K_2\}$ such that

$$\frac{N_1}{K_1}(1-2\delta) < |\alpha^{-1}(k)| < \frac{N_1}{K_1}(1+2\delta) \quad \text{for all} \quad k \in \mathcal{K}_1, \tag{29}$$

$$\frac{N_2}{K_2}(1-2\delta) < |\beta^{-1}(k')| < \frac{N_2}{K_2}(1+2\delta) \quad \text{for all} \quad k' \in \mathcal{K}_2, \tag{30}$$

$$\frac{B'_j}{K_1}(1-2\delta) < |\alpha_j^{-1}(k)| < \frac{B'_j}{K_1}(1+2\delta) \quad \text{for all} \quad k \in \mathcal{K}_1 \quad \text{and} \quad j = 1, \ldots, M_2, \tag{31}$$

and

$$\frac{B_i}{K_2}(1-2\delta) < |\beta_i^{-1}(k')| < \frac{B_i}{K_2}(1+2\delta) \quad \text{for all} \quad k' \in \mathcal{K}_2 \quad \text{and} \quad i = 1, \ldots, M_1, \tag{32}$$

where α^{-1} and β^{-1} are inverse images of α and β respectively, B_i and B'_j are the numbers of non-zero entries in the i-th row and the j-th column respectively, $\alpha_j^{-1}(k) = \{i : \ b_{i,j} \neq 0 \ \text{ and } \ \alpha(i) = k\}$, and $\beta_i^{-1}(k') = \{j : b_{i,j} \neq 0 \ \text{ and } \ \beta(j) = k'\}$.

Proof: We color the rows and columns of B with K_1 and K_2 colors randomly and independently with uniform distributions over $\mathcal{K}_1$ and $\mathcal{K}_2$ respectively. For any fixed color $k \in \mathcal{K}_1$, let for all $i \in \{1, \ldots, M_1\}$ the random variable

$$S_i = \begin{cases} 1 & \text{if the } i\text{-th row of } B \text{ is colored by the color } k \\ 0 & \text{else} \end{cases}$$

Then,

$$S_i = S_i^2 (a.\ s.) \quad \text{and} \quad \mathbf{E}S_i = \frac{1}{K_1} \quad \text{for all} \quad i = 1, \ldots, N_1 \tag{33}$$

where $\mathbf{E}(.)$ is the operator of the expectation.

Thus for any fixed $j \in \{1, \ldots, M_2\}$,

$$\begin{aligned}
&\Pr \quad \{|\alpha_j^{-1}(k)| \geq \frac{B'_j}{K_1}(1+2\delta)\} \\
&= \Pr\{ \sum_{i \in \{i' : b_{i'j} \neq 0\}} S_i \geq \frac{B'_j}{K_1}(1+2\delta)\} \\
&\leq e^{-\delta[\frac{B'_j}{K_1}(1+2\delta)]} \mathbf{E}\{e^{\delta \sum_{i \in \{i' : b_{i'j} \neq 0\}} S_i}\} \\
&= e^{-\delta[\frac{B'_j}{K_1}(1+2\delta)]} \prod_{i \in \{i' : b_{i'j} \neq 0\}} \mathbf{E} e^{\delta S_i} \\
&\leq e^{-\delta[\frac{B'_j}{K_1}(1+2\delta)]} \prod_{i \in \{i' : b_{i'j} \neq 0\}} \mathbf{E}(1 + \delta S_i + \frac{e\delta^2}{2} S_i^2) \\
&= e^{-\delta[\frac{B'_j}{K_1}(1+2\delta)]} \prod_{i \in \{i' : b_{i'j} \neq 0\}} (1 + \delta \mathbf{E} S_i + \frac{e\delta^2}{2} \mathbf{E} S_i^2) \\
&= e^{-\delta[\frac{B'_j}{K_1}(1+2\delta)]} \prod_{i \in \{i' : b_{i'j} \neq 0\}} [1 + \delta(1 + \frac{e\delta}{2}) \mathbf{E} S_i] \\
&= e^{-\delta[\frac{B'_j}{K_1}(1+2\delta)]} [1 + \delta(1 + \frac{e\delta}{2}) \frac{1}{K_1}]^{B'_j} \\
&\leq e^{-\delta[\frac{B'_j}{K_1}(1+2\delta)] + B'_j \delta (1 + \frac{e\delta}{2}) \frac{1}{K_1}} \\
&= e^{-\frac{B'_j \delta^2}{K_1} \frac{4-e}{2}} < e^{-\frac{L_1 \delta^2}{2K_1}}.
\end{aligned} \tag{34}$$

Here the first inequality follows from $\Pr\{T \geq a\} = \Pr\{e^{-\delta a} e^{\delta T} \geq 1\} \leq \mathbf{E}[e^{-\delta a} e^{\delta T}]$, for a random variable T, real a and positive real δ. The second equality holds by the independence. The second inequality follows from the inequality $e^t \leq 1 + t + \frac{1}{2} e t^2$ for $0 \leq t \leq 1$. The fourth and fifth equalities hold by (33). The third inequality follows from the inequality $1 + t \leq e^t$ for non-negative t.

In the analogous way, instead of the inequalities $e^t \leq 1 + t + \frac{1}{2} e t^2$ and $1 + t \leq e^t$ we apply the inequalities $e^{-t} \leq 1 - t + \frac{1}{2} e t^2$ and $1 - t \leq e^{-t}$, and obtain that

$$\Pr\{|\alpha_j^{-1}(k)| \leq \frac{B'_j}{K_1}(1-2\delta)\} = \Pr\{ \sum_{i \in \{i' : b_{i'j} \neq 0\}} S_i \leq \frac{B'_j}{K_1}(1-2\delta)\} < e^{-\frac{L_1 \delta^2}{2K_1}}. \tag{35}$$

Therefore, for all $k \in \mathcal{K}_1$ and all $j = 1, \dots, N_2$,

$$\Pr\{\frac{B'_j}{K_1}(1-2\delta) < |\alpha_j^{-1}(k)| < \frac{B'_j}{K_1}(1+2\delta)\} > 1 - 2e^{-\frac{L_1\delta^2}{2K_1}}. \tag{36}$$

Next we take summation of the random variables S_i over $\{1, \dots, N_1\}$, instead over $\{i' : b_{i'j} \neq 0\}$ in (34) and (35), and have that for all $k \in \mathcal{K}_1$

$$\Pr\{\frac{N_1}{K_1}(1-2\delta) < |\alpha^{-1}(k)| < \frac{N_1}{K_1}(1+2\delta)\} > 1 - 2e^{-\frac{N_1\delta^2}{2K_1}} \geq 1 - 2e^{-\frac{L_1\delta^2}{2K_1}}, \tag{37}$$

(since $L_1 \leq N_1$). Finally, we exchange the roles of rows and columns and by symmetry have that for all $k' \in \mathcal{K}_2$ and $i = 1, \dots, N_1$

$$\Pr\{\frac{B_i}{K_2}(1-2\delta) < |\beta_i^{-1}(k')| < \frac{B_i}{K_2}(1+2\delta)\} > 1 - 2e^{-\frac{L_2\delta^2}{2K_2}} \tag{38}$$

and

$$\Pr\{\frac{N_2}{K_2}(1-2\delta) < |\beta^{-1}(k')| < \frac{N_2}{K_2}(1+2\delta)\} > 1 - 2e^{-\frac{L_2\delta^2}{K_2}} \tag{39}$$

Thus by (27), (28), and (36) - (39), the probability of existence of the pair of the coloring functions satisfying (29) - (32) is positive, and this completes our proof.

Let $\mathcal{Q}$ be the set of triples (X, Y, Z) of random variables with joint distributions $P_{XYZ}(x, y, z) = P_X(x)W_1(y|x)W_2(z|x)$ for all $x \in \mathcal{X}, y \in \mathcal{Y}$, and $z \in \mathcal{Z}$, where P_X is an arbitrary probability distribution over $\mathcal{X}$, and $\mathcal{R}(X, Y, Z) = \{(R_1, R_2) : 0 \leq R_1 \leq H(X|Z) \text{ and } 0 \leq R_2 \leq H(X|Y)\}$. Denote by $Conv(A)$, the closed convex hull of the set A. Let T^n_{XYZ} be the set of triples (x^n, y^n, z^n) of sequences of length n with joint type P_{XYZ} and T^n_X, T^n_Y, T^n_Z, $T^n_{X|Y}(.)$ and $T^n_{X|Z}(.)$ are defined analogously. Then

Theorem 3. *For a deterministic broadcast channel,*

$$C_s(1,2) = Conv\{\cup_{(X,Y,Z)\in\mathcal{Q}}\mathcal{R}(X, Y, Z)\}. \tag{40}$$

Proof:

The Direct Part: For any fixed $(X, Y, Z) \in \mathcal{Q}$ and a sufficiently large n specified later (such that $T^n_{XYZ} \neq \emptyset$), let $B_n(XYZ) = (b_{y^n z^n})_{y^n z^n}$ be a $|T^n_Y| \times |T^n_Z|$ matrix whose rows and columns are labeled by $y^n \in T^n_Y$ and $z^n \in T^n_Z$, respectively, such that

$$b_{y^n z^n} = \begin{cases} x^n & \text{if there exists an } x^n \text{ such that } (x^n, y^n, z^n) \in T^n_{XYZ} \\ 0 & \text{else} \end{cases} \tag{41}$$

Notice that the x^n in (41) is unique by our assumption in Section 2 if it exists. Moreover all $x^n \in T_X^n$ are located in the matrix. In other words, $T_X^n = \{b_{y^n z^n} : y^n \in T_Y^n, z^n \in T_Z^n \text{ and } b_{y^n z^n} \neq 0\}$. Since $(x^n, y^n, z^n) \in T_{XYZ}^n$ iff $x^n \in T_X^n$ and for $t = 1, \ldots, n$, $y_t = \phi(x_t)$ and $z_t = \psi(x_t)$, by the definitions, $0 \neq b_{y^n z^n} = x^n(\text{say})$, implies that $W_1^n(y^n|x^n)W_2^n(z^n|x^n) = 1$.

APPLYING THE LEMMA: Since for all $z^n \in T_Z^n$, $|T_{X|Z}^n(z^n)|$ have the same value, we can denote this quantity by $t_{X|Z}^n$. Analogously, the common value of $|T_{X|Y}^n(y^n)|$, $y^n \in T_Y^n$ is denoted by $t_{X|Y}^n$. For an arbitrary small but positive ϵ, we choose $M_1 = \lfloor 2^{-\frac{n\epsilon}{2}} t_{X|Z}^n \rfloor$ and $M_2 = \lfloor 2^{-\frac{n\epsilon}{2}} t_{X|Y}^n \rfloor$. Then for sufficiently large n

$$\frac{1}{n}\log M_1 > H(X|Z) - \epsilon \quad \text{and} \quad \frac{1}{n}\log M_2 > H(X|Y) - \epsilon. \tag{42}$$

By the definition of $B_n(XYZ)$ its y^n-th row has exactly $|T_{X|Y}^n(y^n)| = t_{X|Y}^n$ non-zero entries and its z^n-th column has exactly $|T_{X|Z}^n(z^n)| = t_{X|Z}^n$ non-zero entries. Thus we substitute $K_l = M_l$ for $l = 1, 2$, $B = B_n(XYZ)$, and correspondingly the other parameters in Lemma 4.1 and find that the right hand sides of (27) and (28) are $e^{\frac{\delta^2}{2} 2^{\frac{n\epsilon}{2}}}$ whereas their left hand sides are growing exponentially with n. So, the conditions of the lemma are satisfied and a pair (α, β) of coloring functions with the desired properties exist.

TO DEFINE THE CODE: For $u \in \mathcal{M}_1 := \{1, \ldots, M_1\}$ and $v \in \mathcal{M}_2 := \{1, \ldots, M_2\}$let $Q(.|u,v)$ be the uniform distribution over $\{b_{y^n z^n} : b_{y^n z^n} \neq 0,\ \alpha(y^n) = u, \text{ and } \beta(z^n) = v\}$, $\mathcal{D}_u = \alpha^{-1}(u)$, and $\mathcal{D}'_v = \beta^{-1}(v)$. Then a code $\{(Q, \mathcal{D}_i, \mathcal{D}'_j) : 1 \leq i \leq M_1 \text{ and } 1 \leq j \leq M_2\}$ is defined. We have to show that it is a probability-type secret code (or a secret code with randomization), i. e. (9)-(12) must be satisfied.

THE ANALYSIS: By definition of the code, for all $u \in \mathcal{M}_1, v \in \mathcal{M}_2$ and x^n with $Q(x^n|u,v) > 0$, $W_1^n(\mathcal{D}_u|x^n)W_2^n(\mathcal{D}'_v|x^n) = 1$. So (9) and (10) hold for all non-negative λ. Next we show that (11) and (12) hold even when the factors n in front of μ are dropped. For this purpose, we let (U, V, X'^n, Y'^n, Z'^n) be the quintuple of random variables with the joint distribution

$$P_{UVX'^nY'^nZ'^n}(u, v, x^n, y^n, z^n) = \frac{1}{M_1}\frac{1}{M_2}Q(x^n|u,v)W_1^n(y^n|x^n)W_2^n(z^n|x^n)$$

for all $u \in \mathcal{M}_1$, $v \in \mathcal{M}_2$, $x^n \in \mathcal{X}^n$, $y^n \in \mathcal{Y}^n$ and $z^n \in \mathcal{Z}^n$. It is obvious that Z'^n takes values in T_Z^n with probability one. Further for all fixed $u \in \mathcal{M}_1$, and $z^n \in T_Z^n$, $\beta(z^n) = v$ (say), we have that

$$P_{UZ'^n}^n(u, z^n) = \frac{1}{M_1}\sum_{v'=1}^{M_2}\frac{1}{M_2}\sum\nolimits_{x^n \in T_X^n} Q(x^n|u,v')W_2^n(z^n|x^n)$$

$$= \frac{1}{M_1 M_2}Q(\{b_{y^n z^n} : b_{y^n z^n} \neq 0 \text{ and } \alpha(y^n) = u\}|u,v)$$

$$= \frac{1}{M_1M_2}\frac{|\{b_{y^nz^n}:\ b_{y^nz^n}\neq 0 \quad \text{and} \quad \alpha(y^n)=u\}|}{|\{b_{y^nz'^n}:b_{y^nz'^n}\neq 0,\ \alpha(y^n)=u \quad \text{and} \quad \beta(z'^n)=v\}|}$$
$$= \frac{1}{M_1M_2}\frac{|\alpha_{z^n}^{-1}(u)|}{|\{b_{y^nz'^n}:b_{y^nz'^n}\neq 0,\ \alpha(y^n)=u, \quad \text{and} \quad \beta(z'^n)=v\}|}. \tag{43}$$

The second equality holds because

$$Q(x^n|u',v')>0 \quad \text{iff} \quad x^n\in\{b_{y^nz^n}:\ b_{y^nz^n}\neq 0,\ \alpha(y^n)=u' \quad \text{and} \quad \beta(z^n)=v'\} \tag{44}$$

and

$$W_2^n(z^n|x^n)=\begin{cases}1 & \text{if } x^n \text{ is in } z^n\text{-th column of } B_n(XYZ)\\ 0 & \text{else}\end{cases} \tag{45}$$

The third equality follows from the definition of Q and the last equality follows from the definition of $\alpha_{z^n}^{-1}$ (in Lemma 4.1). Notice that for all $z^n\in T_Z^n$, B'_{z^n} in Lemma 4.1 now is $t_{X|Z}^n$. By (30), (31), we have that

$$\frac{|T_Z^n|t_{X|Z}}{M_1M_2}(1-2\delta)^2\leq|\{b_{y^nz'^n}:b_{y^nz'^n}\neq 0,\ \alpha(y^n)=u, \quad \text{and} \quad \beta(z^n)=v\}|$$
$$\leq\frac{|T_Z^n|t_{X|Z}}{M_1M_2}(1+2\delta)^2. \tag{46}$$

We now apply (31) and (46) to (43), and obtain that for all $u\in\mathcal{M}_1$, and $z^n\in T_Z^n$,

$$\frac{1}{M_1}\frac{1}{|T_Z^n|}\frac{1-2\delta}{(1+2\delta)^2}\leq P_{UZ'^n}(u,z^n)\leq\frac{1}{M_1}\frac{1}{|T_Z^n|}\frac{1+2\delta}{(1-2\delta)^2}. \tag{47}$$

By summing up the above inequality over $u\in\mathcal{M}_1$, we have that

$$\frac{1}{|T_Z^n|}\frac{1-2\delta}{(1+2\delta)^2}\leq P_{Z'^n}(z^n)\leq\frac{1}{|T_Z^n|}\frac{1+2\delta}{(1-2\delta)^2} \quad \text{for all } z^n\in T_Z^n, \tag{48}$$

or for all $u\in\mathcal{M}_1$ and $z^n\in T_Z^n$

$$\frac{1}{M_1}\frac{1}{|T_Z^n|}\frac{1-2\delta}{(1+2\delta)^2}\leq P_U(u)P_{Z'^n}(z^n)\leq\frac{1}{M_1}\frac{1}{|T_Z^n|}\frac{1+2\delta}{(1-2\delta)^2}, \tag{49}$$

which with (47) yields that for all $u\in\mathcal{M}_1$ and all $z^n\in T_Z^n$

$$\frac{(1-2\delta)^3}{(1+2\delta)^3}\leq\frac{P_{UZ'^n}(u,z^n)}{P_u(u)P_{Z'^n}(z^n)}\leq\frac{(1+2\delta)^3}{(1-2\delta)^3}. \tag{50}$$

Thus for any positive μ, one can choose sufficiently small δ (and consequently sufficiently large n) such that

$$I(U\wedge Z'^n)=\mathbf{E}\log\frac{P_{UZ'^n}(U,Z'^n)}{P_U(U)P_{Z'^n}(Z'^n)}<\mu. \tag{51}$$

In the same way, one can show that for any positive μ, sufficiently small δ, and sufficiently large n,

$$I(V \wedge Y'^n) < \mu. \tag{52}$$

Finally our proof of the direct part is completed by time sharing.

The Converse Part: Let $\{(Q, \mathcal{D}_i, \mathcal{D}'_j) : 1 \leq i \leq M_1 \text{ and } 1 \leq j \leq M_2\}$ be a code satisfying (9) - (12), random variables U, V, X^n, Y^n, and Z^n be defined as in (11) and (12). Then for the rate R_1 of D_1,

$$\begin{aligned}
&nR_1 = H(U) \leq H(U) - I(U \wedge Z^n) + n\mu = H(U|Z^n) + n\mu \\
&\leq H(UX^n|Z^n) + n\mu = H(X^n|Z^n) + H(U|X^nZ^n) + n\mu \\
&= H(X^n|Z^n) + H(U|X^n) + n\mu = H(X^n|Z^n) + H(U|X^nY^n) + n\mu \\
&\leq H(X^n|Z^n) + H(U|Y^n) + n\mu \leq H(X^n|Z^n) + n\theta(\lambda) + n\mu \\
&= \sum_{t=1}^{n} H(X_t|Z_t) + n[\theta(\lambda) + \mu],
\end{aligned} \tag{53}$$

where $\theta(\lambda) \to 0$ as $\lambda \to 0$. By (11) the first inequality holds. The fourth and the fifth equalities follow from the Markovity of $U \leftrightarrow X^n \leftrightarrow Y^nZ^n$. The fourth inequality is Fano's inequality under the condition (9). The last equality holds because the channel is memoryless. By the same reason, for the rate R_2 of D_2,

$$nR_2 \leq \sum_{t=1}^{n} H(X_t|Y_t) + n[\theta(\lambda) + \mu]. \tag{54}$$

(52) and (54) complete our proof of the converse part.

AN EXAMPLE

Let $\mathcal{X} = \{x_1, x_2, x_3, x_4, x_5, x_6\}$, $\mathcal{Y} = \mathcal{Z} = \{1, 2, 3\}$. Let us use the notation (for the deterministic broadcast channels) in Section 2 to define a deterministic broadcast channel as follows. Let

$$\phi(x_1) = \phi(x_2) = 1,\ \phi(x_3) = \phi(x_4) = 2, \quad \text{and} \quad \phi(x_5) = \phi(x_6) = 3, \tag{55}$$

and

$$\psi(x_1) = \psi(x_5) = 1,\ \psi(x_2) = \psi(x_3) = 2, \quad \text{and} \quad \psi(x_4) = \psi(x_6) = 3. \tag{56}$$

Thus the matrices A_1 and $J(A_1)$ defined in Section 3 are

$$A_1 = \begin{pmatrix} x_1 & x_2 & 0 \\ 0 & x_3 & x_4 \\ x_5 & 0 & x_6 \end{pmatrix}, \tag{57}$$

and

$$J(A_1) = \begin{pmatrix} 1 & 1 & 0 \\ 0 & 1 & 1 \\ 1 & 0 & 1 \end{pmatrix}. \tag{58}$$

It is very easy to see by direct observation or by the capacity formula in [5] that for all deterministic wire-tap channels (under our assumption for deterministic channels in Section 3),

$$C_s(1) = \log \max_{z \in \mathcal{Z}} |\{x : \psi(x) = z\}| \quad \text{and} \quad C_s(2) = \log \max_{y \in \mathcal{Y}} |\{x : \phi(x) = y\}|. \tag{59}$$

We leave it to the reader as an easy exercise. Thus for our example,

$$C_s(1) = C_s(2) = \log 2 = 1. \tag{60}$$

Moreover, for any input random variable X and output random variables Y and Z via the channel, we have that for all $y \in \mathcal{Y}$ and $z \in \mathcal{Z}$

$$|\{x : P_{X|Y}(x|y) > 0\}| \leq 2 \quad \text{and} \quad |\{x : P_{X|Z}(x|z) > 0\}| \leq 2, \tag{61}$$

and therefore

$$H(X|Y) \leq 1 \quad \text{and} \quad H(X|Z) \leq 1. \tag{62}$$

On the other hand, by taking uniform distribution over $\mathcal{X}$ we get a triple (X, Y, X) of random variables, the input and the output random variables for the channel, with $H(X|Y) = H(X|Z) = 1$. Thus by Theorem 4.2, the capacity region for probability-type secret codes of the example is

$$C_s(1,2) = \{(R_1, R_2) : \ 0 \leq R_1 \leq 1 \quad \text{and} \quad 0 \leq R_2 \leq 1\}, \tag{63}$$

the unit square. This is already interesting. By (60) and (63), $C_s(1,2) = [0, C_s(1)] \times [0, C_s(2)]$. We can send information to the legal receiver D_1 with a rate at most $C_s(1) = 1$ if we use the channel as a wire-tap channel for which D_2 is the eavesdropper. But if we want to use the same channel to send the messages to both receivers privately, the rate 1 can be achieved for both receivers too. That is sending an additional secret message to D_2 does not reduce the optimal rate for D_1. Our "double wire-tap" channel has the same optimal rate as the wire-tap channel.

For this simple example, the answer to the problem at the beginning of Section 3 and therefore the derivation of $C_s^*(1,2)$ via Proposition 3.1 are not hard. Our answer is based on the fact that for any submatrix S of a matrix $A = (a_{ij})_{ij}$, $i \neq i'$, and $j \neq j'$,

$$a_{ij} \quad \text{and} \quad a_{i'j'} \quad \text{are in} \quad S \Longrightarrow a_{ij'} \quad \text{and} \quad a_{i'j} \quad \text{are in} \quad S. \tag{64}$$

and the observation for $J(A_1) := (a'_{yz})_{yz}$ in (58), that

$$y \neq y',\ z \neq z', \quad \text{and} \quad a'_{yz} = a'_{y'z'} = 1 \Longrightarrow a'_{y'z} = 0 \quad \text{or} \quad a'_{yz'} = 0. \tag{65}$$

Denoting by $a^{(n)}_{y^n z^n}$ the (y^n, z^n)-th entry of $J(A_1)^{\otimes n}$ we claim

Claim: For any $a^{(n)}_{y^n z^n}$ and $a^{(n)}_{y'^n z'^n}$ in an all-one submatrix S of $J(A_1)^{\otimes n}$ the set $\{1, \ldots, n\}$ can be decomposed into the disjoint union of its subsets $T'_Y := \{t : 1 \leq t \leq n,\ y_1 = y_{t'} \quad \text{and} \quad z_t \neq z_{t'}\}$, $T'_Z := \{t : 1 \leq t \leq n\ y_t \neq y_{t'} \quad \text{and} \quad z_t = z_{t'}\}$ and $T' = \{t : 1 \leq t \leq n\ y_t = y_{t'} \quad \text{and} \quad z_t = z_{t'}\}$ Therefore there exists a $T^* \subset \{1, \ldots, n\}$ such that all entries $a^{(n)}_{y^n z^n}$ in S have the same y_t for $t \in T^*$ and the same z_t for $t \notin T^*$.

To prove the claim we assume a contradiction. Then there are an all-one submatrix S, its two entries $a^{(n)}_{y^n z^n}$ and $a^{(n)}_{y'^n z'^n}$ and a $t \in \{1, \ldots, n\}$ such that $y_t \neq y'_t$ and $z_t \neq z'_t$. By (65), w. l. o. g. assume that $a'_{y_t z'_t} = 0$. Then by the definition of the Kronecker product, $a^{(n)}_{y^n z'^n} = \prod_{t'=1}^{n} a'_{y_{t'} z'_{t'}} = 0$. However, by (64) $a^{(n)}_{y^n z'^n}$ is in S, a contradiction. To see the existence of T^*, we fix an $a^{(n)}_{y^n z^n}$ and consider the decompositions (T'_Y, T'_Z, T') for $(a^{(n)}_{y^n z^n}, a^{(n)}_{y'^n z'^n})$ and $(T"_Y, T"_Z, T")$ for $(a^{(n)}_{y^n z^n}, a^{(n)}_{y"^n z"^n})$. We find that $T'_Y \cap T"_Z = T'_Z \cap T"_Y = \emptyset$ because otherwise there is no decompsition for $(a^{(n)}_{y'^n z'^n}, a^{(n)}_{y"^n z"^n})$. Thus we can choose the union of the T_Y-type components of the decompositions as our T^*.

Let S be an $M_1 \times M_2$ all-one submatrix of $J(A_1)^{\otimes n}$ and T^* be the subset in the claim. Then $M_1 \leq 2^{|T*|}$ and $M_2 \leq 2^{n-|T^*|}$. Thus by Proposition 3.1 we have that the capacity region for the combinatorics-type secret codes is the triangle $C^*_s(1,2) = \{(R_1, R_2) : R_1 \geq 0,\ R_2 \geq 0, \quad \text{and} \quad R_1 + R_2 \leq 1\}$. So, for this example the capacity region for secret codes without randomization is a proper subset of the capacity region for the secret codes with randomization. Hence, randomization here improves the performance.

References

[1] R. Ahlswede, "Coloring hypergraphs: A new approach to multi-user source coding", *J. Comb. Inform. Syst. Sci.*, Part I, vol.4, 1979, 76-115: Part II, vol. 5, 1980, pp. 220-268.

[2] R. Ahlswede and N. Cai, "On communication complexity of vector-valued functions", *IEEE Trans. on Inform. Theory*, vol. IT-40, 1994, 2062-2067.

[3] R. Ahlswede and I. Csiszár, "Common randomness in information theory and cryptography", Part I: Secret sharing, *IEEE Trans on Inform. Theory*, vol. IT-39, 1993,1121-1132.

[4] I. Csiszár, "Almost independence and secrecy capacity", *Probl. Inform. Trans.*, vol. 32, 1996, 40-47.

[5] I. Csiszár and J. Körner, "Broadcast channels with confidential messages", *IEEE Trans. on Inform. Theory*, vol. IT-24, 1978, 339-348.

[6] T. M. Cover, "Broadcast channels", *IEEE Trans. on Inform. Theory*, vol. IT-18, 1972, 2-14.

[7] M. S. Pinsker, "Capacity region of noiseless broadcast channels", *Prob. Inform. Trans.*, vol. 14, 1978, 28-32.

[8] A. D. Wyner, "The wire-tap Channels", *Bell System Tech. J.*, vol. 54, 1975, 1355-1387.

[9] A. Yao, "Some complexity questions related to distributive computing", *Proc. 11th ACM Symp. Theory Comput.*, 1979, 209-213.

ASYMPTOTICALLY TIGHT BOUNDS ON THE KEY EQUIVOCATION RATE FOR ADDITIVE–LIKE INSTANTANEOUS BLOCK ENCIPHERERS

Zhaozhi Zhang

Institute of Systems Science, Academia Sinica, Beijing 100080

INTRODUCTION

In [1] R. Ahlswede and G. Dueck investigate secrecy systems with additive–like instantaneous block (ALIB) encipherers subject to the error probability criterion. They give asymptotically tight bounds on the probability of correct decryption for ALIB encipherers. But there are many criteria for secrecy systems. The important one is the key equivocation criterion. In this paper, we give asymptotically tight bounds on the key equivocation rate for ALIB encipherers.

DEFINITIONS AND NOTATION

Let $\mathcal{X}, \mathcal{K}, \mathcal{Y}$ be finite sets with

$$|\mathcal{X}| = |\mathcal{K}| = |\mathcal{Y}|$$

where the number of elements in a set $\mathcal{X}$ is denoted by $|\mathcal{X}|$. Let $(X_i)_{i=1}^{\infty}$ be a message source, where all the X_i, $i = 1, 2, \dots$ are independent replicae of a random variable X with values in $\mathcal{X}$. The probability distribution of $X^n = (X_1, \dots, X_n)$ is given by

$$\Pr(X^n = x^n) = \prod_{i=1}^{n} \Pr(X = x_i)$$

I. Althöfer et al. (eds.), Numbers, Information and Complexity, 369-374.

for all $x^n = (x_1, \ldots, x_n) \in \mathcal{X}^n$. Let $f : \mathcal{X} \times \mathcal{K} \to \mathcal{Y}$ be a function, where $f(x, \cdot)$ is bijective for each $x \in \mathcal{X}$ and $f(\cdot, k)$ is bijective for each $k \in \mathcal{K}$. $f^n : \mathcal{X}^n \times \mathcal{K}^n \to \mathcal{Y}^n$ denotes the n–fold product of f.

An (n, R) ALIB encipherer is a subset $\mathcal{C} \subset \mathcal{K}^n$ with $|\mathcal{C}| \leq 2^{nR}$. Given a pair $(f, \mathcal{C})$, we define a secrecy system which works as follows. A key word k^n is generated by a random key generator K^n according to the uniform distribution on $\mathcal{C}$. Using f^n and k^n, the sender encrypts the output x^n of the message source to the cryptogram $y^n = f^n(x^n, k^n)$ and sends it to the receiver over a noiseless channel. The receiver uses the same key word k^n and f^{-1} to decrypt the message $x^n = (f^{-1})^n(y^n, k^n)$, where the key word k^n is given to the receiver separately over a secure channel. The cryptanalyst intercepts the cryptogram y^n and attempts to decrypt x^n. Since the cryptanalyst does not know the actual key word k^n being used he has to search for a correct key word by using his knowledge of the system. Suppose that the random key K^n and the source output X^n are mutual independent. Let $Y^n = f^n(X^n, K^n)$. Then the average uncertainty about the key when the cryptanalyst intercepts a cryptogram is the conditional entropy $H(K^n|Y^n)$. The quantity $H(K^n|Y^n)/n$ which is called key equivocation rate is used as a security criterion for the secrecy system $(f, \mathcal{C})$. Define a function

$$\alpha(n, R) = \max_{\mathcal{C}} H(K^n|Y^n)/n$$

where the maximum is taken over all (n, R) ALIB encipherers $\mathcal{C} \subset \mathcal{K}^n$. Our aim is to derive a computable expression for $\lim_{n\to\infty} \alpha(n, R)$.

UPPER BOUNDS FOR $\alpha(N, R)$

Lemma 1. *For* $0 \leq R \leq \log |\mathcal{K}|$

$$\alpha(n, R) \leq R.$$

Proof. For any (n, R) ALIB encipherer $\mathcal{C} \subset \mathcal{K}^n$, $H(K^n|Y^n)/n \leq \frac{H(K^n)}{n} = \frac{1}{n} \log |\mathcal{C}| \leq R$. Then the lemma follows from the definition of $\alpha(n, R)$.

Lemma 2. *For* $0 \leq R \leq \log |\mathcal{K}|$

$$\alpha(n, R) \leq H(X).$$

Proof. It is well known that the key equivocation is related to the message equivocation by

$$H(K^n|Y^n) = H(X^n|Y^n) + H(K^n|X^n, Y^n).$$

The definition of the function f implies that K^n is a function of (X^n, Y^n). Then $H(K^n|X^n, Y^n) = 0$. Therefore, $H(K^n|Y^n) = H(X^n|Y^n) \leq H(X^n)$. Since the inequality is valid for all (n, R) encipherer $\mathcal{C} \subset \mathcal{K}^n$, the lemma follows from the definition of $\alpha(n, R)$.

Theorem 1. *For* $0 \le R < H(X)$

$$\limsup_{n\to\infty} \alpha(n,R) \le R.$$

For $H(X) \le R \le \log|\mathcal{K}|$

$$\limsup_{n\to\infty} \alpha(n,R) \le H(X).$$

Proof. The theorem is an immediate consequence of Lemma 1 and 2.

ASYMPTOTICAL LOWER BOUNDS FOR $\alpha(N,R)$

By the definition of the secrecy system, the joint probability distribution of X^n, K^n, Y^n is

$$\Pr(X^n = x^n, K^n = k^n, Y^n = y^n) = \\ \Pr(X^n = x^n)\Pr(K^n = k^n)\delta\big(y^n, f^n(x^n, k^n)\big)$$

where

$$\delta\big(y^n, f^n(x^n,k^n)\big) = \begin{cases} 1 & y^n = f^n(x^n,k^n) \\ 0 & \text{otherwise} \end{cases}$$

Then the conditional probability

$$\Pr(K^n = k^n | Y^n = y^n) = \sum_{x^n} \Pr(X^n = x^n)\delta\big(y^n, f^n(x^n,k^n)\big) \text{ for } k^n \in \mathcal{C}.$$

Define a discrete memoryless channel with transmission probability matrix $W = (W_{y|k}, k \in \mathcal{K}, y \in \mathcal{Y})$, where

$$W_{y|k} = \sum_x \Pr(X = x)\delta\big(y, f(x,k)\big).$$

Then the transmission probabilities for n–words k^n, y^n are

$$W^n_{y^n|k^n} = \prod_{i=1}^n W_{y_i|k_i} = \prod_{i=1}^n \sum_{x_i} \Pr(X_i = x_i)\delta\big(y_i, f(x_i,k_i)\big)$$

$$\sum_{x_1} \cdots \sum_{x_n} \prod_{i=1}^n \Pr(X_i = x_i)\delta\big(y_i, f(x_i,k_i)\big) = \sum_{x^n} \Pr(X^n = x^n)\delta\big(y^n, f^n(x^n,k^n)\big).$$

Therefore, an (n,R) ALIB encipherer $\mathcal{C} \subset \mathcal{K}^n$ can be regarded as an (n,R) code for the memoryless channel W. Furthermore, the random cryptogram Y^n is the output of the channel W^n when the input is the random key K^n. By this observation, we can use a result on the secrecy capacity of a wire–tap channel (broadcast channel) which was proved by Csiszár and Körner [2].

A broadcast channel is a memoryless channel with one input S and two putputs U and V. Its transmission probability is the conditional probability

$P_{UV|S}$. Two memoryless channels $W_1 = P_{U|S}$ and $W_2 = P_{V|S}$ are determined by $P_{UV|S}$. In the model of a wire–tap channel, W_1 is the receiver's channel and W_2 is the cryptanalyst's channel. Here S, U and V assume values in $\mathcal{S}$, $\mathcal{U}$ and $\mathcal{V}$ respectively. An (n, R) code is a subset $\mathcal{C} \subset \mathcal{S}^n$ with $|\mathcal{C}| \leq 2^{nR}$. Let C^n be uniformly distributed over C. The secrecy capacity of a broadcast channel $P_{UV|S}$ is defined as the maximum R for which for every $\varepsilon > 0$ and all sufficiently large n, there exists an $(n, R-\varepsilon)$ (possibly random) code $\mathcal{C} \subset \mathcal{S}^n$ such that for C^n uniformly distributed over $\mathcal{C}$ the following two conditions are satisfied:

1) there exists a decoding function $d; \mathcal{U}^n \to \mathcal{C}$ such that $\Pr\big(d(U^n) \neq C^n\big) < \varepsilon$

2) $H(C^n|V^n)/n > R - \varepsilon$

where U^n is the output of channel W_1 and V^n is the output of channel W_2 when the input is C^n.

A known result [2] [3]. If a broadcast channel $P_{UV|S}$ satisfies the condition that $I(S;U) \geq I(S;V)$ for all choices of probability distributions P_S, then the secrecy capacity of the broadcast channel $P_{UV|S}$ is

$$C_S(P_{UV|S}) = \max_{P_S}\big(I(S;U) - I(S;V)\big)$$

where $I(S;U)$ is the mutual information of S and U.

We use this result for a special broadcast channel $P_{KY|K}$, where K is a random variable in $\mathcal{K}$ with the probability distribution P_K, the receiver's channel $W_1 = P_{K|K}$ is a noiseless channel and the cryptanalyst's channel $W_2 = P_{Y|K} = W = (W_{y|k}, k \in \mathcal{K}, y \in \mathcal{Y})$ which is induced by the secrecy system.

Lemma 3. *The secrecy capacity of the broadcast channel $P_{KY|K}$ is*

$$C_S(P_{KY|K}) = H(X)$$

where X is the random output of the message source.

Proof. Evidently, the broadcast channel $P_{KY|K}$ satisfies the condition of the known result. Using the known result, we obtain

$$C_S(P_{KY|K}) = \max_{P_K}\big[H(K) - I(K;Y)\big].$$

The definition of the function f implies that any one of the random variables X, K, Y is a function of the remaining two others. Then $H(X|K,Y) = H(K|X,Y) = 0$. Therefore

$$H(K) - I(K;Y) = H(K,Y) - H(Y) = H(X,K,Y) -$$

$$H(X|K,Y) - H(Y) = H(X,Y) + H(K|X,Y) - H(Y)$$
$$= H(X,Y) - H(Y) = H(X|Y) \leq H(X).$$

It remains to prove that the equality

$$H(K) - I(K;Y) = H(X)$$

is achieved by some choice of the distribution P_K. By the definition of the channel W and the function f, we see that

$$W_{y|k} = \Pr(X = x) \text{ for } f^{-1}(y,k).$$

Furthermore, the channel W is a symmetric channel. Hence, the channel capacity

$$C = \max_{P_K} I(K;Y) = \log|\mathcal{K}| - H(X)$$

is achieved by the uniform distribution P_K. This proves that the equality $H(K) - I(K;Y) = H(X)$ is valid for the uniform distribution P_K. The lemma is proved.

Theorem 2. *For* $0 \le R < H(X)$

$$\liminf_{n\to\infty} \alpha(n,R) \ge R.$$

For $H(X) \le R \le \log|\mathcal{K}|$

$$\liminf_{n\to\infty} \alpha(n,R) \ge H(X).$$

Proof. If $R < H(X)$, then for every sufficiently small $\varepsilon > 0$, $R+\varepsilon < H(X)$. From Lemma 3, we have $C_S(P_{KY|K}) = H(X)$. According to the definition of the secrecy capacity of a broadcast channel, for every $\varepsilon > 0$, for all sufficiently large n, there exists an $(n, R+\varepsilon-\varepsilon) = (n,R)$ code $\mathcal{C} \subset \mathcal{K}^n$ such that for K^n uniformly distributed over $\mathcal{C}$, $H(K^n|Y^n)/n > R+\varepsilon-\varepsilon = R$. Where Y^n is the output of channel W^n when the input is K^n. We have noted before, by the definition of the channel W, Y^n is just the random cryptogram when the random key is K^n. This proves that

$$\liminf_{n\to\infty} \alpha(n,R) \ge R.$$

Next, if $H(X) \le R \le \log|\mathcal{K}|$, then for every $R' < H(X)$, $\alpha(n,R) \ge \alpha(n,R')$. Hence, by the first part of the theorem, we have for every $R' < H(X)$

$$\liminf_{n\to\infty} \alpha(n,R) \ge \liminf_{n\to\infty} \alpha(n,R') \ge R'.$$

This implies that

$$\liminf_{n\to\infty} \alpha(n,R) \ge H(X).$$

Combining Theorem 1 and Theorem 2, we obtain

Theorem 3. *For* $0 \le R < H(X)$

$$\lim_{n\to\infty} \alpha(n,R) = R.$$

For $H(X) \le R \le \log|\mathcal{K}|$

$$\lim_{n\to\infty} \alpha(n,R) = H(X).$$

Corollary. $\alpha(n,R)$ *is an increasing, continuous function of* $R \in [0, \log|\mathcal{K}|]$.

References

[1] R. Ahlswede and G. Dueck, "Bad codes are good ciphers", *Problems of Control and Information Theory* 11, 1982, 337–351.

[2] I. Csiszár and J. Körner, "Broadcast channels with confidential messages", *IEEE Trans. Inform. Theory* 24, 1978, 339–348.

[3] U. M. Maurer, "Secret key agreement by public discussion from common information", *IEEE Trans. Inform. Theory* 39, 1993, 733–742.

SPACE EFFICIENT LINEAR TIME COMPUTATION OF THE BURROWS AND WHEELER-TRANSFORMATION

Stefan Kurtz

Technische Fakultät, Univ. Bielefeld, Postfach 100 131, 33501 Bielefeld, Germany*

kurtz@techfak.uni-bielefeld.de

Bernhard Balkenhol

Fakultät für Mathematik, Univ. Bielefeld, Postfach 100 131, 33501 Bielefeld, Germany

bernhard@mathematik.uni-bielefeld.de

INTRODUCTION

In [4] a universal data compression algorithm (BW-algorithm, for short) is described which achieves compression rates that are close to the best known rates achieved in practice. Due to its simplicity, the algorithm can be implemented with relatively low complexity. Recently [2] modified the BW-algorithm to improve the compression rate even further. For a thorough discussion on the information theoretic background of the BW-algorithm and more references, see [1]. The most time and space consuming part of the BW-algorithm is the Burrows and Wheeler-Transformation (BWT, for short), which permutes the input string in such a way that characters with a similar context are grouped

*partially supported by DFG-grant Ku 1257/1-1

I. Althöfer et al. (eds.), Numbers, Information and Complexity, 375-383.

together. In [4], it was observed that for an input string of length n, this transformation can be computed in $O(n)$ time and space using suffix trees. However, suffix trees have a reputation of being very greedy for space, and therefore most researchers resorted to alternative non-linear methods for computing the BWT: The algorithm of [9] runs in $O(n \log n)$ worst case time and it requires $8n$ bytes of space. The algorithm of [3] is based on *Quicksort*. It is fast on average, but the worst case running time is $O(n^2)$. The Benson-Sedgewick algorithm requires $4n$ bytes. Its running time can be improved in practice, for the cost of $4n$ extra bytes. Recently, [11] showed how to combine the Manber-Myers Algorithm with the Bentley-Sedgewick Algorithm, to achieve a method running in $O(n \log n)$ worst case time and using $9n$ bytes.

With the recently developed implementation technique of [7], suffix trees can be represented more space efficiently, so that the space advantage of the non-linear methods is considerably reduced. In this paper, we further improve on [7], and show that a suffix tree based method requires on average about the same amount of space as the non-linear methods mentioned above. The improvement is achieved by exploiting the fact, that in practice, the BW-algorithm processes long input strings in blocks of a limited size (for this reason some researchers use the notion of "Block-Sorting"-algorithm). Assuming a maximal block size of $2^{21} - 1 = 2{,}097{,}151$, we show that the suffix tree can be implemented in $8.83n$ bytes on average for the files of the Calgary Corpus. This is $0.6n$ and $9.77n$ bytes less than the implementation technique of [7] and of [10], respectively. The worst case space requirement of our implementation technique is $16n$ bytes, compared to $20n$ bytes for [7] and $28n$ bytes for [10]. The reduction of the space requirement due to an upper bound on n seems trivial. However, we will see that it involves a considerable amount of engineering work to achieve the improvement, while retaining the linear worst case running time for constructing the BWT.

PRELIMINARIES

Let Σ be a finite ordered set, the *alphabet*. k denotes the size of Σ. We assume that x is a string over Σ of length $n \geq 1$ and that $\$ \in \Sigma$ is a character such that for any $i \in [1, n]$ we have $x_i < \$$. For any $i \in [1, n+1]$, let $S_i = x_i \dots x_n\$$ denote the ith non-empty suffix of $x\$$. Let $S_{j_1}, S_{j_2}, \dots, S_{j_{n+1}}$ be the sequence of all non-empty suffixes of $x\$$ in lexicographic order. This gives a bijective mapping $\varphi : [1, n+1] \to [1, n+1]$ defined by $\varphi(i) = j_i$. φ is the *suffix order on* $x\$$. Note that $\varphi(n+1) = n+1$, since $S_{n+1} = \$$. The *Burrows and Wheeler Transformation of* x is the string $\widetilde{x}$ of length $n+1$ such that for any $i \in [1, n+1]$ we have $\widetilde{x}_i = \$$ if $\varphi(i) = 1$, and $\widetilde{x}_i = x_{\varphi(i)-1}$ otherwise.

A Σ^+-*tree* T is a finite rooted tree with edge labels from Σ^+. For each $a \in \Sigma$, a node u in T has at most one a-edge $u \xrightarrow{av} w$ for some string v and some node w. Let u be a node in T. We denote u by $\overline{w}$ if and only if w is the concatenation of the edge labels on the path from the *root* to u. The node $\overline{\varepsilon}$ is the *root*. $depth(\overline{w}) := |w|$ is the *depth* of $\overline{w}$. A string s *occurs* in T if T contains a node $\overline{sv}$, for some string v.

Figure 1 The suffix tree for $x = abab$. Leaves are annotated with leaf numbers and branching nodes with head positions.

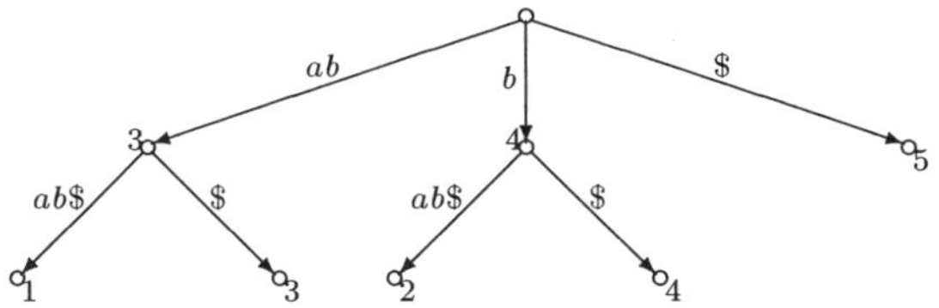

SUFFIX TREES AND THEIR IMPLEMENTATION

The *suffix tree* for x, denoted by ST, is the Σ^+-tree T with the following properties: (i) each node is either a leaf, a branching node, or the *root*, and (ii) a string w occurs in T if and only if w is a substring of $x\$$.

ST can be constructed and represented in linear time and space using one of the algorithms described in [13, 10, 12, 5]. See also [6] which reviews [13, 10, 12] and reveals relationships between these algorithms much closer than one would think. The *suffix link* for a node $\overline{aw}$ in ST is an unlabeled directed edge from $\overline{aw}$ to the node $\overline{w}$. Note that the latter exists in ST, whenever $\overline{aw}$ exists. We consider suffix links to be a part of the suffix tree, since they are required for most of the linear time suffix tree constructions (see [13, 10, 12]). For any branching node $\overline{aw}$ in ST, $\mathit{suffixlink}(\overline{aw})$ refers to node $\overline{w}$.

The *raison d'etre* of a branching node $\overline{w}$ in ST is the first branching occurrence of w in t, i.e., the first occurrence of wa, for some $a \in \Sigma$, such that w occurs to the left, but not wa. We therefore introduce the notions *head* and *head position*: Let $\mathit{head}_1 = \varepsilon$ and for $i \in [2, n+1]$ let head_i be the longest prefix of S_i which is also a prefix of S_j for some $j \in [1, i-1]$. For each branching node $\overline{w}$ in ST, let $\mathit{headposition}(\overline{w})$ denote the smallest integer $i \in [1, n+1]$ such that $w = \mathit{head}_i$. If $\mathit{headposition}(\overline{w}) = i$, then we say that the *head position* of $\overline{w}$ is i. Since there is a one-to-one correspondence between the *heads* and the branching nodes in ST (see [7]), the notion of head positions is well defined. Figure 1 shows the suffix tree for $x = abab$.

The head position j of some branching node $\overline{wu}$ tells us that the leaf $\overline{S_j}$ occurs in the subtree below node $\overline{wu}$. Hence wu is the prefix of S_j of length $\mathit{depth}(\overline{wu})$, i.e., the equality $wu = x_j \dots x_{j+\mathit{depth}(\overline{wu})-1}$ holds. As a consequence, the label of the incoming edge to node $\overline{wu}$ can be obtained by dropping the first $\mathit{depth}(\overline{w})$ characters of wu, where $\overline{w}$ is the predecessor of $\overline{wu}$: If $\overline{w} \xrightarrow{u} \overline{wu}$ is an edge in ST and $\overline{wu}$ is a branching node, then we have $u = x_i \dots x_{i+l-1}$ where $i = \mathit{headposition}(\overline{wu}) + \mathit{depth}(\overline{w})$ and $l = \mathit{depth}(\overline{wu}) - \mathit{depth}(\overline{w})$. Similarly, the label of the incoming edge to a leaf is determined from the leaf number and the depth of the predecessor: If $\overline{w} \xrightarrow{u} \overline{wu}$ is an edge in ST and $\overline{wu} = \overline{S_j}$ for some $j \in [1, n+1]$, then $u = x_i \dots x_n\$$ where $i = j + \mathit{depth}(\overline{w})$.

It is straightforward to show that for any branching node $\overline{aw}$ in ST either $\mathit{headposition}(\overline{aw}) + 1 = \mathit{headposition}(\overline{w})$ or $\mathit{headposition}(\overline{aw}) > \mathit{headposition}(\overline{w})$

holds, see [7]. As a consequence, we can discriminate all non-root nodes accordingly: $\overline{aw}$ is a *small* node if and only if $headposition(\overline{aw})+1 = headposition(\overline{w})$. $\overline{aw}$ is a *large* node if and only if $headposition(\overline{aw}) > headposition(\overline{w})$. The *root* is neither small nor large.

Let $b_1, b_2, \ldots, b_q$ be the sequence of branching nodes ordered by their head position, i.e., $headposition(b_i) < headposition(b_{i+1})$ for any $i \in [1, q-1]$. Obviously, b_1 is the *root*. One can show that a small node in this sequence is always immediately followed by another branching node, and that b_q is a large node, see [7]. We can thus partition the sequence $b_2, \ldots, b_q$ of branching nodes into *chains* of zero or more consecutive small nodes followed by a single large node. More precisely, a *chain* is a contiguous subsequence $b_l, \ldots, b_r$, $r \geq l$, of $b_2, \ldots, b_q$ such that *(i)* b_{l-1} is not a small node, *(ii)* $b_l, \ldots, b_{r-1}$ are small nodes, and *(iii)* b_r is a large node.

One easily observes that any non-root branching node in ST is a member of exactly one chain. The following lemma, which is proved in [7], shows an interesting relationship between the small nodes and the large node of a chain:

Lemma 1. *Let $b_l, \ldots, b_r$ be a chain. The following properties hold for any $i \in [l, r-1]$:*

(1) $suffixlink(b_i) = b_{i+1}$

(2) $depth(b_i) = depth(b_r) + (r - i)$

(3) $headposition(b_i) = headposition(b_r) - (r - i)$

According to this observation, it is not necessary to store $suffixlink(b_i)$, $depth(b_i)$, and $headposition(b_i)$ for any small node b_i. $suffixlink(b_i)$ refers to the next node in the chain, and if the distance $r - i$ of b_i to the large node b_r (denoted by $distance(b_i)$) is known, then $depth(b_i)$ and $headposition(b_i)$ can be obtained in constant time. This observation allows the following implementation technique: ST is represented by two tables T_{leaf} and T_{branch} which store the following values: For each *leaf number* $j \in [1, n+1]$, $T_{leaf}[j]$ stores a reference to the right brother of leaf $\overline{S_j}$. If there is no such brother, then $T_{leaf}[j]$ is a nil reference. Leaf $\overline{S_j}$ is referenced by leaf number j. Table T_{branch} stores the information for the small and the large nodes: For each small node $\overline{w}$, there is a *small record* which stores $distance(\overline{w})$, $firstchild(\overline{w})$, and $rightbrother(\overline{w})$. The latter two are references to the first child of $\overline{w}$ and to the right brother of $\overline{w}$, respectively. If there is no such brother of $\overline{w}$, then $rightbrother(\overline{w})$ is a nil reference. For any large node $\overline{w}$ there is a *large record* which stores $firstchild(\overline{w})$, $rightbrother(\overline{w})$, $depth(\overline{w})$, and $headposition(\overline{w})$. It also stores $suffixlink(\overline{w})$, whenever $depth(\overline{w}) \leq 2^{11} - 1$. The successors of a branching node are therefore found in a list whose elements are linked via the *firstchild*, *rightbrother*, and T_{leaf} references. To speed up the access to the successors, each such list is ordered according to the first character of the edge labels.

To guarantee constant time access from a small node b_i to the large node b_r, all records consist of integers (the general assumption is that an integer occupies 4 bytes or equivalently 32 bits). The integers are stored in table T_{branch}, ordered by the head positions of the corresponding branching nodes. All branching nodes are referenced by their *base address* in T_{branch}. The base address is the index of the first integer of the corresponding record. Since there are at most n large nodes in ST, the maximal base address is $3n-3$. A reference is either a base address or a leaf number. To distinguish these, we store a base address as an integer with offset $n+1$, i.e., base address i is stored as $n+1+i$. So a reference is smaller than $4n$, and if $n \leq 2^{21}-1$, then it occupies 23 bits. Each depth and each head position occupies at most 21 bits.

Consider the range of the distance values. In the worst case, take e.g. $x = a^n$, there is only one chain of length $n-1$, i.e., the maximal distance value is $n-2$. However, this case is very unlikely to occur. To save space, we delimit the maximal length of a chain to 65536. As a consequence, after at most 65535 consecutive small nodes an "artificial" large node is introduced, for which we store a large record. In this way, we delimit the distance value to be at most 65535, and thus the distance occupies 16 bits, which are stored with the two integers occupied by a small record. Thus we trade a delimited distance value for the saving of one integer for each small record.

Now let us consider how to store the values of a large record. The first two integers of a large record store the *firstchild* reference and the *rightbrother* reference, as in a small record. We need just one extra integer to store the remaining values of a large record: Consider some large node, say $\overline{w}$, and let $\overline{v}$ be the rightmost child of $\overline{w}$. There is a sequence consisting of one *firstchild* reference and at most $k-1$ *rightbrother*/T_{leaf} references which link $\overline{w}$ to $\overline{v}$. If $\overline{v} = \overline{S_j}$ for some $j \in [1, n+1]$, then $T_{leaf}[j]$ is a nil reference. Otherwise, if $\overline{v}$ is a branching node, then *rightbrother*$(\overline{v})$ is a nil reference. Of course, it only requires one bit to mark a reference as a nil reference. Hence the integer used for the nil reference contains unused bits, in which we store *suffixlink*$(\overline{w})$. As a consequence, retrieving the suffix link of $\overline{w}$ requires traversing the list of successors of $\overline{w}$ until the nil reference is reached, which encodes the suffix link of $\overline{w}$. This *linear retrieval* of suffix links takes $O(k)$ time in the worst case. However, despite linear retrieval, the suffix tree can still be constructed in $O(kn)$ time, since suffix links are retrieved at most n times during suffix tree construction (see [10, 7]).

Experiments show that linear retrieval may slow down suffix tree construction in practice. For this reason, we use the following method which makes linear retrieval of suffix links an exception: Whenever the depth of a large node does not exceed $2^{11}-1 = 2047$, we mark this fact and use the remaining bits of the corresponding large record to also store the suffix link. This can later be retrieved in constant time. For those large nodes whose depth exceeds 2047, linear traversal of suffix links is required. But those nodes are usually very rare, and if they occur, then the number of their successors is expected to be small. Hence the linear retrieval of suffix links is expected to be fast.

A small record stores two references ($2 \cdot 23$ bits), a distance value (16 bits), one *small/large bit* to mark whether the first integer is part of a small or a large record, and one *nil bit* to mark a reference as a nil reference. Altogether, a small record occupies 64 bits which fit into two integers. A large record, say for a large node $\overline{w}$, stores two references, one nil bit, one small/large bit, and one *small depth bit* which tells whether the depth is at most $2^{11}-1$. Moreover, there are 21 bits required for the head position, and 11 or 21 bits for the depth, depending on whether the small depth bit is set or not. Thus a large record requires 81 or 91 bits, which fit into three integers. If the depth of $\overline{w}$ is at most $2^{11}-1$, there are 15 unused bits in the large record. These are used to store the suffix link. The remaining 8 bits of the suffix link for $\overline{w}$ are stored in the integer $T_{leaf}[headposition(\overline{w})]$. Recall that this stores a reference (23 bits) and one nil bit.

Let σ be the number of small records and λ be the number of large records. Thus table T_{branch} requires $2\sigma + 3\lambda$ integers. Table T_{leaf} occupies n integers, and hence the space requirement of our implementation technique is $n+2\sigma+3\lambda$ integers. The implementation technique of [7] requires $n+2\sigma+4\lambda$ integers (for $n \leq 2^{27}-1$), while a previous implementation technique (see [10]) requires $2n+5(\sigma+\lambda)$ integers. In the worst case $\lambda = n$ and $\sigma = 0$.

The proposed suffix tree representation can be constructed in linear time, using the algorithm of [10]. The basic observation is that this algorithm constructs the branching nodes of ST in order of their head positions, which is compatible with our implementation technique. For details, see [7].

An alternative representation of the suffix tree uses a hash table to store the edges, as recommended in [10]. Unfortunately, this representation does not directly allow the depth first traversal to run in linear time. As already remarked in [8], an additional step is required to sort the edges lexicographically. This can be done by a bucket sorting algorithm, and thus requires linear time. In [7] it is shown that in practice this approach requires about 60% more space than the proposed linked list implementation, and it leads to a faster sorting procedure only if the alphabet is very large.

DEPTH FIRST TRAVERSAL

Due to the one-to-one correspondence between the leaves of ST and the non-empty suffixes of $x\$$, the BWT can be read from ST by a simple depth first traversal. This processes the edges outgoing from some branching node $\overline{w}$ in order $<_{\overline{w}}$ which is defined by $\overline{w} \xrightarrow{au} \overline{wau} \quad <_{\overline{w}} \quad \overline{w} \xrightarrow{cv} \overline{wcv} \iff a < c$. It is obvious that such a depth first traversal visits leaf $\overline{S_i}$ before leaf $\overline{S_j}$ if and only if $S_i < S_j$. Thus the suffix order $\varphi(1), \varphi(2), \ldots, \varphi(n+1)$ on $x\$$ is just the list of suffix numbers encountered at the leaves during the traversal. The linked list implementation of Section 31 allows the depth first traversal to run in $O(n)$ time. The only extra space required is for a stack storing references to the predecessors of a branching node. The stack occupies at most $r_{\max}$ integers where $r_{\max}$ is the length of the longest repeated substring of x.

The depth first traversal constructs $\widetilde{x}$ from left to right. Whenever it visits a leaf $\overline{S_j}$, $j > 1$, it has found the next character x_{j-1} of $\widetilde{x}$. It stores this character and proceeds with the right brother of $\overline{S_j}$ (if it exists). Thus x_{j-1} is accessed immediately before $T_{leaf}[j]$. Now recall that the integer $T_{leaf}[j]$ stores a reference and a nil bit, occupying 24 bits together. The 8 bits storing a part of the suffix link of the father (if this is a large node and $\overline{S_j}$ is the rightmost child) are not needed during the depth first traversal. For this reason, we store character x_{j-1} (which occupies 8 bits) in the unused bits of $T_{leaf}[j]$. This can be done very efficiently in one sweep over x and T_{leaf} before the depth first traversal. As a consequence, x is no longer accessed in a "random" fashion, which improves the cache coherence of the program and therefore its running time in practice. Moreover, during the traversal the space for the input string x can be reclaimed to store $\widetilde{x}$.

EXPERIMENTAL RESULTS

We used the programming language C to implement the techniques proposed here. The resulting program computes the BWT, and is referred to by *stbwt*. In order to compare *stbwt* with the Manber-Myers and the Benson-Sedgewick algorithm, we modified the original code of [9] and [3], since these only compute the suffix order. The program derived from [9], referred to by *mamy*, requires $8n$ bytes. We developed two programs based on [3]: *bese1* applies the Benson-Sedgewick algorithm to all suffixes of the input string. It requires $4n$ bytes. *bese2* first uses bucket sort to presort all suffixes according to their first $l = \lfloor \log_k n \rfloor$ characters. Then it applies the Benson-Sedgewick algorithm independently to all groups of suffixes whose prefix of length l is identical. This presorting step runs in linear time, but it requires $4n$ extra bytes. Thus the space requirement of *bese2* is $8n$ bytes. Unfortunately, the program of Sadakane is not available, and so we cannot compare it to *stbwt*. However, experiments in [11] show that Sadakane's algorithm is on average slightly slower than a suffix tree based method implemented by Larsson.

We applied all four programs to the 14 files of the Calgary Corpus. Table 1 shows the lengths and the alphabet sizes of the files and the running times in seconds on a computer with a Pentium MMX Processor (166 MHz, 32 MB RAM). The last column shows the total space requirement for *stbwt* in bytes per input character. In each row, the shortest running time is shown in a grey box. The last row gives the total file length, the total running times, and the average space requirement for *stbwt*. The table shows that *mamy* is the slowest program. Except for the file *pic* it is always considerably slower than the other programs. *bese1* is always slower than *bese2*. Both are faster than *stbwt* for the same 9 files, but the advantage is small (mostly within a factor of two). However, *bese1* and *bese2* are very slow for the file *pic* which contains long repeated substrings. This clearly reveals the poor worst case behavior of the Benson and Sedgewick algorithm. For most files, *stbwt* requires about n bytes more space than *mamy* and *bese2*. For *pic* and *obj1* it requires even less space.

Acknowledgements. We thank Gene Myers for providing a copy of his program code.

			mamy	*bese1*	*bese2*	*stbwt*	
file	*length*	*k*	*time*	*time*	*time*	*time*	*space*
bib	111261	81	4.13	0.60	0.49	0.71	8.87
book1	768771	82	35.72	6.08	4.39	8.62	8.92
book2	610856	96	28.93	4.45	3.30	5.67	8.96
geo	102400	256	2.38	0.36	0.30	1.87	6.83
news	377109	98	27.39	2.80	2.24	4.54	8.84
obj1	21504	256	0.39	0.21	0.20	0.11	7.14
obj2	246814	256	10.99	1.56	1.33	2.46	8.80
paper1	53161	95	1.15	0.20	0.17	0.28	9.09
paper2	82199	91	2.45	0.34	0.27	0.51	9.01
pic	513216	159	29.61	190.86	192.18	2.44	8.67
progc	39611	92	0.73	0.15	0.12	0.20	8.93
progl	71646	87	2.32	0.48	0.43	0.34	9.69
progp	49379	89	1.52	0.53	0.50	0.21	9.81
trans	93695	99	6.35	1.03	0.96	0.44	10.06
	3141622		154.04	209.66	206.87	28.40	8.83

Table 1 Running times (in seconds) and Space Requirement (bytes/input character)

References

[1] B. Balkenhol, S. Kurtz, "Universal Data Compression Based on the Burrows and Wheeler Transformation: Theory and Practice", *Technical Report, Sonderforschungsbereich: Diskrete Strukturen in der Mathematik, Universität Bielefeld, 98-069*, 1998, **http://www.mathematik.uni-bielefeld.de/sfb343/preprints/**.

[2] B. Balkenhol, S. Kurtz and Y. Shtarkov, "Modification of the Burrows and Wheeler Data Compression Algorithm", In *Proceedings of the IEEE Data Compression Conference, Snowbird, Utah*, IEEE Computer Society Press, 1999, 188–197.

[3] J. Bentley, R. Sedgewick, "Fast Algorithms for Sorting and Searching Strings", In *Proceedings of the ACM-SIAM Symposium on Discrete Algorithms*, 1997, 360–369. **http://www.cs.princeton.edu/~rs/strings/**.

[4] M. Burrows, D. Wheeler, "A Block-Sorting Lossless Data Compression Algorithm", *Research Report 124*, Digital Systems Research Center, 1994 **http://gatekeeper.dec.com/pub/DEC/SRC/research-reports/abstracts/src- rr-124.html**.

[5] M. Farach, "Optimal Suffix Tree Construction with Large Alphabets". In *Proceedings of the 38th Annual Symposium on the Foundations of Computer Science, FOCS 97*, New York. IEEE Comput. Soc. Press, 1997. **ftp://cs.rutgers.edu/pub/farach/Suffix.ps.Z**.

[6] R. Giegerich, S. Kurtz, "From Ukkonen to McCreight and Weiner: A Unifying View of Linear-Time Suffix Tree Construction". *Algorithmica*, **19**, 1997, 331–353.

[7] S. Kurtz, "Reducing the Space Requirement of Suffix Trees". *Report 98–03*, Technische Fakultät, Universität Bielefeld, 1998. **http://www.TechFak.Uni-Bielefeld.DE/techfak/~kurtz/publications. html**.

[8] N. Larsson, "The Context Trees of Block Sorting Compression". In *Proceedings of the IEEE Data Compression Conference, Snowbird, Utah, March 30 - April 1*, IEEE Computer Society Press, 1998, 189–198.

[9] U. Manbar, E. Myers, "Suffix Arrays: A New Method for On-Line String Searches", *SIAM Journal on Computing*, **22**(5), 1993, 935–948.

[10] E. McCreight, "A Space-Economical Suffix Tree Construction Algorithm", *Journal of the ACM*, **23**(2), 1976, 262–272.

[11] K. Sadakane, "A Fast Algorithm for Making Suffix Arrays and for Burrows-Wheeler Transformation". In *Proceedings of the IEEE Data Compression Conference, Snowbird, Utah, March 30 - April 1*, IEEE Computer Society Press, 1998, 129–138.

[12] E. Ukkonen, "On-line Construction of Suffix-Trees", *Algorithmica*, **14**(3), 1995.

[13] P. Weiner, "Linear Pattern Matching Algorithms". In *Proceedings of the 14th IEEE Annual Symposium on Switching and Automata Theory*, The Univsersity of Iowa, 1973, 1–11.

SEQUENCES INCOMPRESSIBLE BY SLZ (LZW), YET FULLY COMPRESSIBLE BY ULZ

Larry A. Pierce II and Paul C. Shields *

Mathematics Department, The University of Toledo, Toledo OH 43606

lpierce@math.utoledo.edu, pshields@math.utoledo.edu

Abstract: Binary sequences are constructed that are fully compressible by one infinite memory form of Lempel-Ziv, yet cannot be compressed by other infinite memory forms. The constructions make use of de Bruijn sequences.

Three versions of the Lempel-Ziv data compression algorithm are considered in this paper, simple Lempel-Ziv (SLZ), Lempel-Ziv-Welch (LZW), and unrestricted Lempel-Ziv (ULZ). All three algorithms parse sequences sequentially into words that have occurred in some way in the past; the words are then encoded by describing where they occurred in the past. They differ in the way the next word is defined.

1. SLZ, also known as LZ'78, [7], defines the next word to be the shortest block that has not appeared as a prior word.

2. ULZ, a version of LZ'77, [6], defines the next word to be the shortest block that does not start anywhere in the past.

3. LZW, [5], defines the next word as the longest block that is a prior word plus the symbol that follows it.

Nice descriptions of each of these algorithms and how next words are encoded can be found in [2, 3].

All sequences in this paper are assumed to be binary, unless stated otherwise. The finite sequence $x_m, x_{m+1}, \ldots, x_n$ is denoted by x_m^n, and product notation is used for concatenation of finite sequences, e. g., uv is the concatenation of u

*Supported in part by joint NSF-Hungarian Academy grant INT-9515485.

I. Althöfer et al. (eds.), Numbers, Information and Complexity, 385-390.

and v, and u^n is the concatenation of n copies of u. Infinite binary sequences are denoted by single letters, such as x or y. As in [7], the limiting compression ratio for SLZ is defined by

$$\mathrm{SLZ}(x) = \limsup_{n\to\infty} \frac{\mathrm{SLZ}(x_1^n)}{n}$$

where $\mathrm{SLZ}(x_1^n)$ denotes the length of the binary code word assigned to x_1^n by SLZ; the corresponding limiting compression ratios $\mathrm{LZW}(x)$ and $\mathrm{ULZ}(x)$ have similar definitions.

The principal goal of this paper is to establish the following, which answers some questions raised in [4].

Theorem.
There are binary sequences x and y such that

$$\mathrm{SLZ}(x) = \mathrm{LZW}(y) = 1 \text{ and } \mathrm{ULZ}(x) = \mathrm{ULZ}(y) = 0$$

It is easy to construct sequences that are not compressible by SLZ, namely, just concatenate all 1-blocks in some order, followed by all 2-blocks in some order, then all 3-blocks in some order, ..., [7]. Sequences constructed by this method will be called Champerknowne sequences as they first appeared in [1]. The new feature in this paper is that by carefully choosing the ordering of the k-blocks at each stage, one can force full compression by ULZ. A modification of the idea then provides a sequence incompressible by LZW and fully compressible by ULZ.

Both constructions utilize de Bruijn cycles. For each k, let $d(k)$ denote a de Bruijn k-cycle, that is, a binary sequence of length 2^k with the property that every member of $\{0,1\}^k$ starts at exactly one place in the first 2^k places of the concatenation $d(k)d(k)$. Let S denote the (circular) shift operator on binary sequences of length 2^k, that is, the mapping defined by

$$S(b_1, b_2, \ldots, b_{2^k}) = (b_2, b_3, \ldots, b_{2^k}, b_1).$$

The key to our first construction is the following lemma.

Lemma 1.
There are integers $\{\phi(j) \in [0,k) \colon 1 \le j < k\}$ such that

$$x(k) = d(k)S^{\phi(1)}d(k)S^{\phi(2)}d(k)\cdots S^{\phi(k-1)}d(k) \tag{1}$$

is a concatenation $b(1)b(2)\cdots b(2^k)$ of distinct k-blocks.

To see how the lemma gives the desired SLZ result, let x be the concatenation

$$x = x(1)x(2)\cdots x(k)\cdots$$

where $x(k)$ is given by (1) for each k. The lemma guarantees that x is a Champerknowne sequence, hence $\mathrm{SLZ}(x) = 1$. To show that $\mathrm{ULZ}(x) = 0$ first note that if $j > 0$ and $w(j)$ denotes the first $2^k - \phi(j)$ terms of $S^{\phi(j)}d(k)$, then $w(j)$ starts at the $(1 + \phi(j))$-th position of the first block $S^{\phi(0)}d(k) = d(k)$. In particular, the sequence $w(j)$ started earlier so at most one ULZ phrase can start in $w(j)$. This means, however, that $\mathrm{ULZ}(x) = 0$, since the fraction of $x(k)$ covered by the $w(j)$, $0 < j < k$, goes to 1 as $k \to \infty$.

Proof of Lemma 1. The idea is to create shifts so the set of successive *nonoverlapping* k-blocks in $x(k)$ is the same as the set of distinct *overlapping* k-blocks that start in the first 2^k places of $d(k)d(k)$. Towards this end, let Z_k denote the (additive) group of integers (mod k), choose $0 \le r < k$ such that $2^k = nk + r$, and let $G(r)$ be the subgroup of Z_k generated by r, represented as $G(r) = \{0, \beta, \dots, (\alpha - 1)\beta\}$, where α is the order of $G(r)$ and $\beta = k/\alpha$.

The desired $x(k)$ is defined as the concatenation

$$x(k) = [d(k)]^\alpha [Sd(k)]^\alpha [S^2 d(k)]^\alpha \cdots [S^{\beta-1}d(k)]^\alpha, \tag{2}$$

that is, a concatenation of β blocks, the j-th one being the concatenation of α copies of $S^j d(k)$. The length of $x(k)$ is $k2^k$, so it is a concatenation $b(1)b(2)\cdots b(2^k)$ of k-blocks. The proof that these k-blocks are distinct is given in the following two paragraphs.

Let Z_{2^k} denote the (additive) group of integers (mod 2^k), and let $\mathrm{H}(k)$ denote the subgroup of Z_{2^k} generated by k. Also let $h = |\mathrm{H}(k)|$ and $t = 2^k/|\mathrm{H}(k)|$, so that $\mathrm{H}(k)$ can be represented as

$$\mathrm{H}(k) = \{0, t, 2t, \dots, (h-1)t\}$$

Let $w = (d(k))^\alpha$. The k-block w_{ik+1}^{ik+k} is equal to the k-block $x_{\phi(ik)+1}^{\phi(ik)+k}$, where $\phi(ik)$ is the member of $\{0, t, 2t, \dots, (h-1)t\}$ that is congruent to ik (mod 2^k). In other words, the successive nonoverlapping k-blocks in w are exactly the k-blocks that start in $d(k)d(k)$ in the positions $\ell + 1$ for which ℓ belongs to the subgroup $\mathrm{H}(k)$. Likewise, the successive nonoverlapping k-blocks in $(S^j(d(k))^\alpha$ are exactly the k-blocks that start in $d(k)d(k)$ in the positions $\ell + 1$ for which ℓ belongs to the coset $j + \mathrm{H}(k)$. Since the cosets of $\mathrm{H}(k)$ are disjoint, it follows from the de Bruijn property that the sequence $x(k)$ defined by (2) indeed factors into distinct k-blocks. This completes the proof of Lemma 1. □

The SLZ parsing of a Champerknowne sequence has the property that all the k-blocks appear before any $(k+1)$-block appears. In SLZ parsing each word appears at most once, while in LZW parsing each word can appear twice, once followed by 0 and once followed by 1. The key to our LZW result is to force each k-block to appear two times in the LZW parsing before any $(k+1)$-block appears. A bit more care is needed to make this happen.

In the next lemma $\hat{S}$ denotes the circular shift on sequences of length 2^{k+1} and $d_0(k+1)$ denotes a de Bruijn $(k+1)$-cycle of length 2^{k+1} whose first $k+1$

coordinates are 0's and whose last $k+1$ coordinates are 1's (the existence of such cycles is easy to establish).

Lemma 2.
There are integers $\{\phi(j) \in [0,k) \colon 1 \le j < k\}$ *such that*

$$y(k) = d_0(k+1)[\hat{S}^{\phi(1)}d_0(k+1)][\hat{S}^{\phi(2)}d_0(k+1)]\cdots[\hat{S}^{\phi(k-1)}d_0(k+1)] \quad (3)$$

is a concatenation of k*-blocks* $b(1)b(2)\cdots b(2^{k+1})$ *such that*

1. *Each member of* $\{0,1\}^k$ *appears twice among the* $b(m)$.
2. *If* b_m *denotes the symbol that follows* $b(m)$ *in* $y(k)$, *then*
   (a) $b(m)b_m \neq b(m')b_{m'}$, *for* $m \neq m'$.
   (b) *If* $b(m) = b(2^{k+1})$ *with* $m < 2^{k+1}$, *then* $b_m = 1$.

To see how the lemma yields the desired LZW example, let y be the concatenation

$$y = y(1)y(2)\cdots y(k)\cdots,$$

where, $y(k)$ is given by the lemma, for each k. The conditions of the lemma and the definition of $d_0(k+1)$ imply that every word appears twice in the LZW parsing of y, which immediately implies that $\mathrm{LZW}(y) = 1$. The argument used for the SLZ case also shows that $\mathrm{ULZ}(y) = 0$.

Proof of Lemma 2. The principal difference between this and Lemma 1 is that here the focus is on the k-block parsing of sequences of length $k2^{k+1}$, rather than $k2^k$. Again Z_k denotes the additive group of integers (mod k) and $G(r)$ denotes the subgroup of Z_k generated by r, but now the remainder r is defined by $2^{k+1} = nk + r$, $0 \le r < k$. Again we can write $G(r) = \{0, \beta, \ldots, (\alpha-1)\beta\}$, where α is the order of $G(r)$ and $\beta = k/\alpha$.

The desired $y(k)$ is defined as the concatenation

$$y(k) = [d_0(k+1)]^\alpha[\hat{S}d_0(k+1)]^\alpha[\hat{S}^2d_0(k+1)]^\alpha\cdots[\hat{S}^{\beta-1}d_0(k+1)]^\alpha. \quad (4)$$

The length of $y(k)$ is $k2^{k+1}$, so it is a concatenation $b(1)b(2)\cdots b(2^{k+1})$ of blocks of length k. The proof that properties 1, 2(a), and 2(b) hold is given in the following two paragraphs.

In this new setting $\mathrm{H}(k)$ denotes the subgroup of $Z_{2^{k+1}}$ generated by k, and $\alpha = |\mathrm{H}(k)|$, $\beta = 2^{k+1}/|\mathrm{H}(k)|$. The earlier argument extends to show that the successive nonoverlapping k-blocks in $(\hat{S}^j(d_0(k+1))^\alpha$ are exactly the k-blocks that start in $d_0(k+1)d_0(k+1)$ in the positions $\ell+1$, for ℓ belonging to the coset $j + \mathrm{H}(k)$. Since each k-block starts in exactly two places in the first 2^{k+1} positions in $d_0(k+1)d_0(k+1)$ it follows that the sequence $y(k)$ defined by (4) has the first property of the lemma.

To establish property 2(a) it is enough to prove the following.

(i) The term that follows a k-block in the nonoverlapping k-block parsing of $y(k)$ is the same as the term that follows the corresponding k-block in $d_0(k+1)d_0(k+1)$.

This is obvious for those nonoverlapping k-blocks in $y(k)$ that are not the final block in one of the $[\hat{S}^j d_0(k+1)]^\alpha$, for $0 \le j < \beta - 1$. For final blocks we use the assumption that $d_0(k+1)$ begins with $k+1$ 0's, for it guarantees that first term of $[\hat{S}^{j+1} d_0(k+1)]^\alpha$ is a 0, which is exactly the term that follows the k-block in $d_0(k+1)d_0(k+1)$ that corresponds to the final k-block of $[\hat{S}^j d_0(k+1)]^\alpha$.

To establish property 2(b) first note that $b(2^{k+1}) = 1^{k-\beta+1}0^{\beta-1}$. The $(k+1)$-block $1^{k-\beta+1}0^\beta$ starts at position $2^{k+1} - k + \beta$ in $d_0(k+1)d_0(k+1)$. Suppose $m < 2^{k+1}$ and

$$b(m)b_m = 1^{k-\beta+1}0^\beta. \tag{5}$$

The definition, (4), of $y(k)$ then implies that $b(m)$ cannot be interior to any of the blocks $\hat{S}^j d_0(k+1)$, and hence there must be a $j \le \beta - 1$ such that $b(m)b_m$ is equal to the k-block that starts at position $2^{k+1} - r + 1 + j$ in $d_0(k+1)d_0(k+1)$, where $\alpha r \equiv 0 \pmod{k}$. The de Bruijn property implies that $2^{k+1} - r + 1 + j$ must be equal to $2^{k+1} - k + \beta$, that is,

$$k - \beta = r - 1 + j.$$

Multiplying this by α then shows $\alpha(1+j)$ is divisible by k, that is, $j + 1 = \beta$, which, in turn, cannot be true unless $m = 2^{k+1}$. This shows that property 2(b) is also true and completes the proof of Lemma 2. □

Remark 3. Most Champerknowne sequences have limiting ULZ compression close to 1, for there are $2^k!$ ways to order the k-blocks at each stage, and hence the number of such sequences grows at the same rate as the number of all sequences. To our surprise, the explicit k-block orderings we have tried produce small ULZ compression; in fact, we have not been able to find any simple way, analogous to the Champerknowne construction, to create sequences incompressible by ULZ.

Remark 4. A number of questions about the performance of LZ-algorithms on individual infinite sequences remain unsolved. It is easy to see that $\text{ULZ}(x) \le \text{SLZ}(x)$ and $\text{SW}(x) \le \text{LZW}(x)$ always hold, where SW is sliding-window Lempel-Ziv with unbounded look-back, see [4], where slightly different terminology is used. It is not known, however, whether there is any relationship between $\text{SLZ}(x)$ and $\text{LZW}(x)$, or between $\text{ULZ}(x)$ and $\text{SW}(x)$. Such relationships appear to be quite difficult to determine, for in each case one algorithm looks for longest "old" words, while the other looks for shortest "new" words.

Another question of interest is stationarity, that is, the relation between the compression ratios of x and its shift Tx. It is easy to see $\text{ULZ}(x) = \text{ULZ}(Tx)$ and that $\text{SW}(x) = \text{SW}(Tx)$, since neither algorithm restricts where it looks in the past. Nothing is known about stationarity for SLZ and LZW, both of which restrict where they look in the past.

Remark 5. We close by making a disclaimer. The algorithms discussed in this paper all compress almost every sequence drawn from an ergodic process to the entropy of the process. This paper is concerned only with individual sequences and no probability model is assumed; in fact, the set of Champerknowne sequences has measure 0 with respect to any ergodic process.

References

[1] D. G. Champerknowne, "The construction of decimals normal in the scale of ten", *Journal of the London Math. Soc.*, vol. 8, 1933, 254-260.

[2] S. A. Savari, "Redundancy of the Lempel-Ziv incremental parsing rule", *IEEE Trans. Inform. Theory*, vol. IT-43, 1997, 9-21.

[3] S. A. Savari, "Redundancy of the Lempel-Ziv string matching code", *IEEE Trans. Inform. Theory*, vol. IT-44, 1998, 787-791.

[4] P. Shields, "Finite-state coding of individual sequences", *IEEE Trans. Inform. Theory*, to appear.

[5] T. A. Welch, "A technique for high-performance data compression", *IEEE Computer*, vol. 17, no. 6, 1984, 8-19.

[6] J. Ziv and A. Lempel, "A universal algorithm for sequential data compression", *IEEE Trans. Inform. Theory*, vol. IT-23, 1978, 337-343.

[7] J. Ziv and A. Lempel, "Compression of individual sequences via variable rate coding", *IEEE Trans. Inform. Theory*, vol. IT-24, 1978, 530-536.

UNIVERSAL CODING OF NON-PREFIX CONTEXT TREE SOURCES*

Yuri M. Shtarkov

Institute for Problems of Information Transmission, RAS,
19 Bolshoi Karetnii, 101447 Moscow, Russia

shtarkov@iitp.ru

INTRODUCTION

The efficiency of data compression with the help of universal coding depends on the used *model* or *set of models* of the source. By expanding the set of models and/or increasing their complexity we can improve the approximation of the statistical properties of messages. However, this entails a higher redundancy and (usually) a higher complexity of coding. For this reason, the development of comparatively simple models capable of improving the statistical description of messages is of great importance. Not surprisingly, this problem has attracted much attention.

The present paper considers non-prefix context tree source models, which were discussed in [1] and [2] (the latter reference is taken from [1]). A general description of the models is given, followed by a discussion of a number of particular cases and universal coding problems.

THE MAIN DEFINITIONS AND CONCEPTS

Let A be a discrete alphabet of α letters, $\alpha \geq 2$; $x^k = x_1, \dots, x_k$, $x_i \in A$, be the first k letters of the message; $p(x^k|\omega)$ be the probability of appearance of x^k at the output of source ω, and $\varphi^{(n)}$ be a uniquely decodable binary code for blocks x^n of length n with codewords $\varphi^{(n)}(x^n)$ of length $|\varphi^{(n)}(x^n)| \leq -\log q(x^n|\varphi^{(n)}) + c$, where $|\chi|$ is the length of the sequence χ or the cardinality of the set χ, and $\{q(x^n|\varphi^{(n)}), x^n \in A^n\}$ is any "coding" probability distribu-

*This work was partly supported by the Russian Foundation of Basic Research (project number 96-01-0084) and by INTAS (project number 94469)

I. Althöfer et al. (eds.), Numbers, Information and Complexity, 391-402.

tion (the value of c can be added to any estimate of the redundancy and in what follows is not taken into account). The cumulative (per block) individual redundancy of the coding of message x^n at the output of source ω with code $\varphi^{(n)}$ is equal to

$$\rho(x^n|\varphi^{(n)},\omega) \triangleq |\varphi^{(n)}(x^n)| + \log p(x^n|\omega) \le \rho_n(\varphi^{(n)},\omega) \triangleq \max_{x^n\in A^n} \rho(x^n|\varphi^{(n)},\omega),$$

where $\log(.) = \log_2(.)$. The average redundancy $r_n(\varphi^{(n)},\omega)$ is equal to $E_\omega\{\rho(x^n|\varphi^{(n)},\omega)\}$, where $E_\omega\{\zeta(x^n)\}$ is the average value of the real function $\zeta(x^n)$ over $\{p(x^n|\omega), x^n \in A^n\}$.

The efficiency of universal coding $\varphi^{(n)}$ for any set Ω of the known sources ω is assessed by the maximal individual redundancy

$$\rho(\varphi^{(n)},\Omega) \triangleq \max_{x^n\in A^n} \sup_{\omega\in\Omega} \rho(x^n|\varphi^{(n)},\omega) = \max_{x^n\in A^n}\left[\log\frac{p(x^n|\Omega)}{q(x^n|\varphi^{(n)})}\right]$$

$$\ge \frac{\sigma(\Omega)}{2}\log n + c(\Omega) \tag{1}$$

or by the maximal average redundancy $r(\varphi^{(n)},\Omega) = \sup\{r(\varphi^{(n)},\omega), \omega \in \Omega\}$, where $p(x^n|\Omega) = \sup\{p(x^n|\omega), \omega \in \Omega\}$, $\sigma(\Omega)$ is the number of unknown parameters in the expressions for conditional probabilities and $c(\Omega)$ is independent of n. The maximal probability (MP) code [3,4] is optimal according to the first criterion (usually it achieves the lower bound in (1)) and, as a rule, is asymptotically optimal according to the second one.

Sequential arithmetic codes for the sequences of any length n (in particular, one unknown in advance) are considered below. The codes are denoted as φ rather than $\varphi^{(n)}$,

The above expressions primarily hold for the sets $\Omega = \Omega_m$ described by one particular model m, i.e. by a known method of calculation of probabilities $p(x^n|\omega)$ for a given parameter vector $\Theta = \Theta(\omega)$. Let now M be a set of models m, φ_m be *any* universal arithmetic code for Ω_m and $\Omega = \Omega(M)$ be the union of all Ω_m (usually the Ω set can be described by different sets of models). The codeword lengths $|\varphi_m(x^n)| = -\log q(x^n|\varphi_m)$ depend on m , which is why it is natural to use "the most convenient" model for the description of x^n (see [3, 5-8]). Therefore, *the multimodel properties* of any code $\varphi = \varphi_M$ for the set $\Omega(M)$ are estimated by the set of values $\delta\rho_n(m|M)$ which satisfy the inequalities

$$\delta\rho(x^n|M) \triangleq |\varphi_M(x^n)| - \min_{m'\in M} |\varphi_{m'}(x^n)| \le \delta\rho_n(m|M), \tag{2}$$

where $m = m(x^n)$ is a model for which a minimum of $|\varphi_{m'}(x^n)|$, is achieved, so that it is desirable to maximally reduce the values of $\delta\rho_n(m|M)$ (for the maximal average redundancy criterion, the problem is formulated similarly).

An optimal solution of this problem for a given n (see [3,8]) does not allow to use the arithmetic coding. Therefore, the weighting algorithm proposed in [5,6], which makes use of the coding probabilities

$$q^{(w)}(x^n|\varphi_M) = \sum_{m\in M} w(m)q(x^n|\varphi_m) \ge \max_{m\in M}\left[w(m)q(x^n|\varphi_m)\right], \tag{3}$$

where $\{w(m), m \in M\}$ is any probability distribution, is more preferable. The advantages of weighting include simple estimations $\delta\rho_n(m|M) \leq -\log w(m)$ which follow from the inequality in (3), and the possibility of arithmetic coding.

Sequential estimation of an (unknown) source model, proposed in [9] for a particular set M, agrees with arithmetic coding as well (see also [8]). Such an estimation consists in using a unique mapping

$$x^k \to e(x^k) = m_k \in M, \qquad \forall x^k \in A^k, \quad k = 1, 2, \ldots, \tag{4}$$

and conditional probabilities

$$\vartheta(a|x^k) = \vartheta(a|x^k, \varphi_{m_k}) = q(x^k a|\varphi_{m_k})/q(x^k|\varphi_{m_k}), \qquad \forall a \in A, \tag{5}$$

corresponding to code φ_{m_k}, for the encoding of the next letter x_{k+1} of the message. To obtain the upper bounds of $\delta\rho_n(m|M)$ for this natural approach is very difficult.

SOME SETS OF MODELS

Let U be a set of "segments" $u \in A^d$, $0 \leq d \leq D$, i.e. a set of nodes of a uniform α-ary tree T^* of depth D, including the root λ.

1) The Markov chain of connectedness (depth, order) d is described by the conditional probabilities $\theta(a|x^k) = \theta(a|u)$, where $u = x_k, \ldots, x_{k-d+1}$ and Markov models $m = d$ with $\sigma(d) = (\alpha - 1)\alpha^d$ (see (1)) differ only in the values of d. The set $\{d, 0 \leq d \leq D\}$ contains only $D + 1$ models having values of $\sigma(.)$, which differ from one another by at least a factor of α. Therefore, the minimum (over d) of the sum of two redundancy components that are due to an inaccurate approximation of the real source and to the unknown values of model parameters, respectively, is usually rather big.

2) The latter fact requires that the set of Markov chain models should be expanded. An important step in solving this problem was the introduction in [9] of Markov context tree (FSMX) models. Later, in [10-13], context tree (CT) models (lacking the Markov property), were proposed and investigated.

Definition 1. *A CT-source with memory depth $d \leq D$ is a source described by the complete and proper set S of contexts (segments s from U), the set of conditional probability distributions $\{\Theta_s, s \in S\} = \{\{\theta_s(a), a \in A\}, s \in S\}$ and the probability distribution of the first D letters of the message.*

The completeness and properness of the set S mean that, for any $x^k \in A^k$, $k \geq D$, the equality $x_k, \ldots, x_{k-d+1} = s_k \in S$ is valid for one and only one value of $d \leq D$. The conditional probability $\theta(a|x^k, \omega)$ of the appearance of the next letter $a = x_{k+1}$, $k \geq D$, is equal to $\theta_s(a)$, where $s = s_k$. The Markov property is defined by the condition $|s_{k+1}| \leq |s_k| + 1$ for all $x^{k+1} \in A^{k+1}$ and $k = D + 1, D + 2, \ldots$

The set S or the corresponding complete and proper α-ary tree T_S is the model of an CT-source with $\sigma(S) = (\alpha - 1)|S|$. A number of parameters decreases (relative to $(\alpha - 1)\alpha^D$) since all the segments of length D with the same "beginning" $s \in S$ have *the same* conditional probability distributions Θ_s. This is a "grouping" of segments.

Thus, CT-models are in better agreement with the properties of messages which have *contexts of various lengths* (for example, texts) than are Markov chains. Furthermore, the set $M(D)$ of CT-models is much wider than the set of Markov chains ($|M(D)|$ is a double exponent of D). Finally, the complexity of a universal coding for $M(D)$ is comparable with the complexity of coding for the set of Markov chains with $d \leq D$ [10-13].

One of the disadvantages of CT-models is a fixed rule of segment grouping. Therefore we will consider more general models.

3) Let $\bar{g}$ is a partition of the set A^D of segments of length D into a set of groups. This set is a model of a source with *grouped contexts* (GC-model) such that the conditional probability distributions are equal for all the segments of the same group and $\sigma(\bar{g}) = (\alpha - 1)|\bar{g}|$. The set $G(D)$ of such models corresponds to *all possible partitions* $\bar{g}$. This set was first mentioned in [14] and later discussed by F. M. J. Willems. It is obvious that CT-model S is a particular case of a partition $\bar{g}$. The following proposition is valid for this general case.

Theorem 1. *The maximal individual redundancy of the universal MP-coding for the set of GC-sources with the known model $\bar{g}$ is equal to the right-hand side of (1) with $\sigma(\bar{g}) = (\alpha - 1)|\bar{g}|$, and the multimodel redundancy (2) of the weighted coding for the set $G(D)$ is upperbounded by a constant.*

The first statement can be proved in the same way as for the set $M(D)$ [10-12], whereas the second statement follows from (3) since $|G(D)| = \text{const} < \infty$.

A significant expansion of the set $M(D)$ to $G(D)$ results in an increased redundancy (2) and an increased coding complexity. However, only a small fraction of models $\bar{g}$ is useful; usually the segments with equal conditional probability distributions are not grouping in an arbitrary way. Therefore it is important to introduce and study models which are intermediate between CT and GC-models.

NON-PREFIX CONTEXT TREE MODELS (NCT)

We will start by explaining the drawbacks of the fixed grouping rule for segments of CT-models (the drawbacks of arbitrary grouping were mentioned above).

Usually the coding probability for the universal coding of the set Ω_m of all CT-sources with a known model $m = S$ is equal to the product of the coding probability for the first D letters and of $q(x^k(s)|\varphi_0) = q_0(x^k(s))$ over all $s \in S$, where φ_0 is a universal code for memoryless sources and $x^k(u)$ is a subsequence of letters x_i of x^k, such that $x_{i-1}, \ldots, x_{i-|u|} = u$, $u \in U$ [4,10-12]. For any $u \in U$ the asymptotically optimal code φ_0 is described by the conditional probabilities

$$\vartheta_0(a|x^k(u)) = \frac{t_k(a|u) + 1/2}{k_u + \alpha/2}, \tag{6}$$

where $t_k(a|u) = t(a|x^k(u))$ is a number of appearances of a in $x^k(u)$ and $k_u = |x^k(u)|$. The corresponding coding probability is equal to

$$q_0(x^k(u)) = \frac{\Gamma(\alpha/2)}{\pi^{(\alpha-1)/2}\,\Gamma(k_u+\alpha/2)} \prod_{a\in A} \Gamma(t_k(a|u)+1/2) \approx \frac{\Gamma(\alpha/2)}{\sqrt{\pi}\, k_u^{(\alpha-1)/2}} e^{-\,k_u H_u}, \tag{7}$$

where $\Gamma(.)$ is a Gamma-function and H_u is the entropy of the "empirical probability distribution" $\{t_k(a|u)/k_u\}$ in nats under the condition that $0\ln 0 = 0$.

Let us assume that the only difference between models S_1 and S_2 is that the context $s \in S_1$ is replaced in S_2 by a proper and complete (with respect to $u = s$) subset $S(u)$ of contexts v for which the first d letters are equal to $u = s$. Then the model S_1 describes x^k better than S_2 if $q_0(x^k(u))$ is greater than the product of $q_0(x^k(v))$ over all $v \in S(u)$. Considering (7) and the fact that $\{x^k(v), v \in S(u)\}$ is a "splitting" of $x^k(u)$ the above inequality may be re-written after taking the logarithm as

$$k_u H_u + \frac{\alpha-1}{2}\log k_u - \sum_{v\in S(u)} \left[k_v H_v + \frac{\alpha-1}{2}\log k_v - c_\alpha\right] =$$

$$= \sum_{v\in S(u)} k_v \left\{ \left[\sum_{a\in A} \frac{t_k(a|v)}{k_v} \left(\ln\frac{t_k(a|v)}{k_v} - \ln\frac{t_k(a|u)}{k_u} \right) \right] - \right.$$

$$\left. - \frac{\alpha-1}{2}\left(\frac{\ln k_v}{k_v} - \frac{\ln k_u}{k_u}\right) + \frac{c_\alpha}{k_v} \right\} < c_\alpha, \tag{8}$$

where $c_\alpha = \ln\Gamma(\alpha/2) - (\ln\pi)/2$ (see, e.g. [8]).

If for some v the expression in braces is positive then $x^k(v)$ should be encoded in the node v; otherwise in the node u. Such an approach, which allows to increase the coding probability for x^k (i.e. to reduce the description length), is only possible under the condition that we withdraw the requirement of properness for the set S. We will consider one type of non-proper (non-prefix) context tree (NCT) models based on the CT-model S.

Let $\bar{\nu} = \{\nu(s), s \in S\}$, $0 \le \nu(s) \le \min(|s|, \nu_0)$ be the index set over S and $S_u(\bar{\nu})$ be the set of contexts $s \in S$ for which the first $|u| = |s| - \nu(s)$ letters coincide with segment u.

Definition 2. *The model of an NCT source is described by the complete and proper set S, by the index set $\bar{\nu}$ and by the set $\{\bar{q}_u\}$ of groupings of contexts $s \in S_u(\bar{\nu})$ for all internal nodes u of tree T_S with $|S_u(\bar{\nu})| > 1$.*

Any group of the NCT model consists of segment subsets (rather than "individual" segments as in the GC model); any such subset contains all the segments with the first $|s|$ letters coinciding with $s \in S_u(\bar{\nu})$. Such groupings are more "intelligent" than arbitrary ones, and their number is less than $|G(D)|$. With $\nu_0 = 0$ such a model coincides with the CT model S and with $|s| = \nu(s) = \nu_0 = D$ it conicides with the GC model.

In the NCT model, the prefix tree T_S is replaced by a nonprefix (but still complete) tree T_S^* since for the segments $uya\ldots$ and $uyb\ldots$, $|y| \le \nu_0$, $|u|+|y| <$

D and $a \neq b$ the node u can be considered as the leave and the internal node, respectively (which is what we need in (8)). A similar consideration was used in [1] for introducing NCT models with $\nu_0 = 1$ and $|g_u| = 1$ (in our notations). Thus, Definition 2 only contains a generalization of the main idea of [1]. If ν_0 equals 1, the prefix requirement is eliminated, while an increase in ν_0 make the NCT models more promising and flexible.

It is convenient to assign the values of $\nu(s)$ to the leaves s of the tree T_S stored in the memory of the encoder and the decoder. At the $(k+1)$-th step of the universal coding of sources with a known NCT model we successively define the current context s_k, the value $\nu(s_k)$, the node $u = u_k$ which satisfies the condition $s_k \in S_u(\bar{\nu})$, and the group of $S_u(\bar{\nu})$ containing s_k. It is obvious that Theorem 1 is valid in this case also and that the complexity is slightly larger than for the known CT model.

As usual, the most essential problems arise when the NCT model is unknown.

Let us stress that Definition 2 describes only one class of NCT models. Different NCT models correspond to different statistical properties of data.

If, for example, the message is the text file then for large $|u|$, $u \in U$, the subsequences $x^k(ua)$, $a \in A$, usually contain a small number of different letters. Some of these subsequences are repetitions of the same letter and it is natural to propose that the conditional probabilities of this letter are *equal* for all such $x^k(ua)$. Therefore it is reasonable to encode all such subsequences *together*; this corresponds to *grouping* of all such ua together (another subsequences can be encoded together or independently).

This simplified NCT model explains the rather high coding efficiency of the Burrows-Wheeler Transform (see, e. g., [15]) and corresponds to a generalization of the PPM* algorithm. It needs more attentive consideration. Therefore only the Definition 2 is discussed below.

WEIGHTING FOR THE SUBSET OF NCT MODELS

Theorem 1 is valid for the set $M^*(D)$ of NCT models, and the main problem of coding is the complexity which is significantly larger than for $M(D)$. The known algorithms for $M(D)$ use the mutual "partial embedding" of CT models [10-14]. However, for any set S various sets $\bar{\nu}$ and $S_u(\bar{\nu})$ exist, and for any $S_u(\bar{\nu})$ there exist various groupings $\bar{g}_u$. Therefore it is hardly possible to order the set $M^*(D)$ in a way convenient for coding.

Hence, it is necessary to introduce constraints which could help reduce the complexity. We will consider the constraints that do not obstruct the minimization of the left-hand side of (8).

1) The decision not to use the grouping of $v \in S_u(\bar{\nu})$ means that $|\bar{g}_u| = 1$ for all $u \in U$ (this is the starting case in [1]). Now the grouping is only provided by the choice of the model S and the set $\bar{\nu}$ so that its arbitrariness decreases (as compared to the general NCT model).

2) Even with this constraint it is necessary to take into account all possible sets $S_u(\bar{\nu})$ and index sets $\bar{\nu}$. To avoid the weighting of all such cases, it is

sufficient to use in u the coding conditional probabilities which are *independent* of the sets $S_u(\bar{\nu})$.

Let $q(x^k(v)|u)$ be the coding probability for the subsequence $x^k(v)$ encoded in the node u which satisfies this condition. Then, following [10-12] and taking into account the first constraint we can represent the weighted probability for $x^k(v)$ as

$$q^{(w)}(x^k(v)) = \sum_{\nu=0}^{\nu_0} w(\nu) q(x^k(v)|u(v,\nu)) + w(\nu_0+1) \prod_{a \in A} q^{(w)}(x^k(va)), \tag{9}$$

where $u = u(v,\nu)$ is the beginning of the segment v of length $|v| - \nu$ and $\{w(\nu), 0 \le \nu \le \nu_0+1\}$ is a probability distribution. Following [10-12] it is easy to prove that the weighted coding probability for x^k is equal to $q^{(w)}(x^k(\lambda)) = q^{(w)}(x^k)$. The first sum permits us to take into account that the value of $\nu(v)$ (if v is a leaf of an unknown model S) is unknown and that it can take the values varying from 0 to ν_0. At $\nu_0 = 0$ and $\nu_0 = 1$, this expression coincides with the original one for $M(D)$ and with the main expression in [1] respectively.

The probabilities $q(x^k(v)|u)$ are independent of $S_u(\nu)$ if the coding conditional probabilities depend only on $\{t_k(a|u)\}$ (e.g. as in (6)) and, possibly, on $\{t_k(a|v)\}$. This condition is equivalent to an assumption that $x^k(u)$ is the sequence of *independent identically distributed* (i.i.d.) letters (i.e. u is a leaf (?) of an unknown model) or that the real conditional probabilities for all $v \in S_u(\bar{\nu})$ are equal to *average* (?) values of the conditional probabilities in u (see [1,2,13]). In both cases, the scope of values of parameters is reduced but the subset of NCT models is not.

The following assertion helps to choose and analyze the efficiency of coding conditional probabilities for the calculation of the value of $q(x^k(v)|u)$.

Theorem 2. *If $-\log \psi(x^k)$ is an aim function (a desirable length of codeword for x^k), where $\psi(x^k) > 0$ is an arbitrary function defined over all $x^k \in A^k$ and $k = 1, 2, \ldots$ and $\psi(x^0) = 1$ then for any coding method $q(x^k)$ the redundancy introduced at the $(k+1)$-step is equal to*

$$\log N(x^k) = \log \left[\frac{\psi(x^{k+1})}{\psi(x^k)} \frac{q(x^k)}{q(x^{k+1})} \right] = \log \left[\frac{\psi(x^{k+1})}{\psi(x^k)} \frac{1}{\vartheta(x_{k+1}|x^k)} \right]. \tag{10}$$

In fact, after $k+1$ and k steps the cumulative redundancies are equal to $\log[\psi(x^{k+1})/q(x^{k+1})]$ and $\log[\psi(x^k)/q(x^k)]$, respectively, and the difference of these values is equal to the change in the cumulative redundancy at the $(k+1)$-th step. Equality (10) is valid for arbitrary $\psi(.)$ and $\vartheta(.)$. Earlier (see, e.g. [8]) only the local optimization was considered, for which $\vartheta(x_{k+1}|x^k) = \psi(x^{k+1})[\sum_{a \in A} \psi(x^k a)]^{-1}$ and $N(x^{k+1}) = N(x^k)$ is independent of x_{k+1}.

Considering the constraints introduced, it is natural to choose for the problem at hand

$$\psi(x^k(v)|u) = \frac{1}{Q(k_u, k_v)} \prod_{a \in A} \left[\frac{t_k(a|u)}{k_u} \right]^{t_k(a|v)}, \tag{11}$$

where $Q(k_u, k_v)$ can be introduced as a "normalizing" factor that brings ψ closer to the probability measure. The conditional probability (6) was used in [1] and [2] (reference from [1]). If local optimization is used for function (11) with any $Q(.)$ we obtain

$$\vartheta(a|x^k(v), u) \approx \frac{t_k(a|u) + t_k(a|v) + 1}{k_u + k_v + \alpha}, \tag{12}$$

where approximation $(1+t)(1+t^{-1})^\tau \approx t + \tau + 1$, $0 \leq \tau \leq t$, $t > 0$, is used (introducing the exponent $e^{\tau/t}$ increases the accuracy but complicates the calculation and estimation of the denominator). In contrast to (6) $t_k(a|v)$ is twice present in (12): "inside" $t_k(a|u)$ and outside of it, but for $t_k(a|u) = t_k(a|v)$ and $k_u = k_v$ (12) coincides with (6). Note that (12) is an example of *frequency weighting*, helpful for certain problems; it is sometimes useful to multiply $t_k(a|v)$ and k_v by the weight factor $w \neq 1$.

For any set V of segments with a common initial part u the conditional probabilities (6) and (12) produce the equality

$$\prod_{v \in V} q(x^k(v)|u) = q(x^k(V)|u), \tag{13}$$

where $x^k(V)$ is the union of $x^k(v)$ over all $v \in V$ (in the order of appearance of their letters in $x^k(u)$). Equality (13) determines the independence of the encoding from $S_u(\bar{\nu})$. The coding redundancy for $x^k(V)$ equals to the sum of redundancies for $x^k(v)$ over all $v \in V$. If $x^k(V) = x^k(u)$ then (6) provides the minimal redundancy of the coding of $x^k(u)$.

The substitution of (11) (with $Q(.) = 1$) in (10) gives for (6) and (12)

$$\ln N_1(x^{k+1}) = \ln\left[\frac{(t_u+1)(k_u+\alpha/2}{(t_u+1/2)(k_u+1)}\right] + t_v \ln\left(1+\frac{1}{t_u}\right) - k_v \ln\left(1+\frac{1}{k_u}\right) \tag{14}$$

and

$$\ln N_2(x^{k+1}) = \ln\left[\frac{(t_u+1)(k_u+k_v+\alpha)}{(t_u+t_v+1)(k_u+1)}\right] + t_v \ln\left(1+\frac{1}{t_u}\right) - k_v \ln\left(1+\frac{1}{k_u}\right) \tag{15}$$

respectively, where $t_u = t_k(x_{k+1}|u)$ and $t_v = t_k(x_{k+1}|v)$. If $t_v/k_v = t_u/k_u$ then the difference between the second and the third terms, which are the same in (14) and (15), is close to zero (the codings in nodes u and v are almost the same). The first term in (14) is independent of t_v and k_v and approximately equals $(\alpha-2)/(2k_u) + [1/(2t_u) - 1/(2k_u)]$; it may be only slightly larger than $(\alpha-1)/(2k_u)$.

The redundancies of codes (6) and (12) depend on the *arrangement* of letters of $x^n(v)$ in $x^n(u)$. Therefore it is useful to introduce a coding efficiency criterion, which generalizes the maximal individual redundancy criterion and can be applied to the problem at hand.

Let $T_v = \{t_n(a|v), a \in A\}$, $T_u = \{t_n(a|u), a \in A\}$ and $X_n(T_v, T_u)$ be a set of sequences $(x^n(v), x^n(u))$ with given T_v, T_u and equal probabilities of occurrence *for any values of parameters* of the NCT model.

Definition 3. *For any T_v and T_u the combinatorial redundancy is equal to*

$$\rho_n(T_v, T_u) = \frac{1}{|X_n(T_v, T_u)|} \sum_{X_n(T_v, T_u)} \rho(x^n(v)|x^n(u)), \tag{16}$$

where summing is performed over all $(x^k(v), x^k(u)) \in X_n(T_v, T_u)$ and $\rho(x^n(v)| x^n(u))$ is the redundancy of coding of $x^n(v)$ as a part of subsequence $x^n(u)$.

The introduced value is no more dependent on the location of letters of $x^n(v)$ in $x^n(u)$ and the values of conditional probabilities parameters. If $\rho(x^n(v)| x^n(u))$ are equal for all $(x^n(v), x^n(u)) \in X_n(T_v, T_u)$, it is equal to the maximal individual redundancy, otherwise it assumes an intermediate value between the maximal individual and the maximal average redundancy. The introduced criterion can be used for (comparative) analysis of different algorithms and for the choice of factors $Q(.)$ in (11).

SEQUENTIAL ESTIMATION OF NCT MODEL

Sequential estimation of an unknown source model was proposed in [9] and was studied for FSMX and CT models in [13] and [14] respectively (see also [8]). It can be also applied to the general NCT model (see Definition 2). Estimation of some of the components of the NCT-model (in particular, of groupings $\{\bar{g}_u\}$) is simpler than their weighting but it remains to be rather complicated. Therefore, firstly the same constrains as in (9), are considered.

Let $z(x^k) = x_k, \ldots, x_{k-D+1}$ be the "current" context branch and Z_k be the set of all nodes at this branch. If the criterion of minimal description length (MDL) is used for a current estimation of the NCT-model (with the above constraints), then the encoding of the next letter of the subsequence $x^k(v)$, $v \in Z_k$, has to be made in the node $u_k(v)$, $u \in Z_k$ such that

$$q^*(x^k(v)) = q(x^k(v)|u_k(v)) \triangleq \max_u q(x^k(v)|u), \qquad 0 \le |v| - |u| \le \nu_0. \tag{17}$$

It is now necessary to choose the best (for coding) node $v \in Z_k$. As the lengths of subsequences $x^k(v)$ are *different*, it makes no sense to compare the values of (17). The estimation rule in [13] allows to avoid this difficulty. However, the meaning of the rule is not entirely clear. So we need a new criterion of estimation.

Definition 4. *For any set $\{q(x^k(v)), v \in V\}$ the criterion of minimal description rate (MDR) corresponds to the choice of $v_0 \in V$, which minimizes the coding rate*

$$R(v) = -\frac{1}{|x^k(v)|} \log q(x^k(v)) \tag{18}$$

over all $v \in V$.

This criterion, which is a natural generalization of MDL for the sequences of varying lengths, allows to fully define the estimation procedure for the encoding of NCT sources with an unknown model (from the above subset). At the first step the node $u(v)$ and probability $q^*(x^k(v))$ are defined with the help of (17) for any $v \in Z_k$. At the second step MDR criterion (18) is used to determine the best node $v = v_0$. Then the conditional probabilities that correspond to the coding of $x^k(v_0)$ in the node $u(v_0)$ are used for the coding of the $(k+1)$-th letter of the message.

As was mentioned in the end of Section 5, the probabilities $q(x^k(v)|u)$ are strongly dependent on the arrangement of letters of $x^k(v)$ in $x^k(u)$. We can avoid this dependence by substituting functions $\psi(.)$ for the probabilities $q(.)$ in (17) and (18). For example, the left-hand side of (8) is minimized by the choice of function (11) with

$$Q(k_u, k_v) = e^{c(u,v)} \left[k_u^{(\alpha-1)/2} \right]^{k_v/k_u}, \tag{19}$$

where $c(v,v) = c_\alpha$ and $c(u,v) = 0$ otherwise. The resulting estimation rule (17) is similar to the rule used in [13]. It should however be noted that in most cases the values of the function (19) are much smaller than the ones which could render this function a "normalizing factor".

Despite the fact that model estimation procedures for the set $M(D)$ and the subset of $M^*(D)$ introduced above are rather close, the generalization of the upper bound of the maximal individual redundancy for $M(D)$ (see [8]) to the subset of $M^*(D)$ has to be considered in details.

The current estimation of the only group (all $S_u(\bar{\nu})$) of nodes v, encoded in the node u, allows us to withdraw the second (rather contradictory) constraint of Section 5. The estimation rules can be different. In particular, to minimizing the left-hand side of (8) we can use the following estimation (sorting) rule: v is an element of $S_u(\bar{\nu})$, if and only if

$$\sum_{a \in A} \frac{1}{2\xi^2(a|v)} \left(\frac{t_k(a|u)}{k_u} - \frac{t_k(a|v)}{k_v} \right)^2 \leq \frac{\alpha - 1}{2} \left(\frac{\ln k_v}{k_v} - \frac{\ln k_u}{k_u} \right) - \frac{c_\alpha}{k_v} \tag{20}$$

where $\xi(a|v)$ assumes a value between $t_k(a|u)/k_u$ and $t_k(a|v)/k_v$ (see [8]). It is important that the result is independent of another v and of the unknown set $S_u(\bar{\nu})$.

If $|u| \leq D - \nu_0$ then *at any step* the condition (20) has to be checked for all α^{ν_0} nodes v with $|u|$ letters coinciding with u. To reduce the complexity of such sorting we can reduce ν_0 (up to $\nu_0 = 1$), or use a weak dependence of the left-hand side of (20) on the next ($(k+1)$-th) letter, or introduce a few rather weak constraints, etc.

The complicated structure of NCT models permits us to combine the weighting and estimation in the same coding algorithm. For example, for any $v \in Z_k$ we can update the probability $q(x^k(v))$ according to the rule

$$\hat{q}(x^{k+1}(v)) = \hat{q}(x^k(v))\vartheta(x_{k+1}|x^k(v), x^k(u_k(v))), \tag{21}$$

where $u_k(v)$ is defined in (17), and any conditional probability can be chosen as $\vartheta(.|.)$ in (21), for example, (6) or (12) (it should be reminded that $u_k(v)$ is a function of $x^k(v)$ and $x^k(u)$). Now only one probability is associated with any node v, and we can replace the estimation of the best $v = v_0 \in Z_k$ (see (18)) by CTW for probabilities $\hat{q}(x^k(v))$.

To conclude, we would like to note that the universal coding with fuzzy MDR-estimation (see [8]) of NCT-model is close to a well-known and efficient algorithm PPM (for data compression), but PPM uses simplified rules.

References

[1] P.A.J. Volf and F.M.J. Willems, "A Context-Tree Branch-Weighting Algorithm,", *Proc. of 18th Symp. on Inform. Theory in the Benelux*, 1997, 115-122.

[2] M. J. Weinberger, J. J. Rissanen and R. B. Arps, "Applications of Universal Context Modeling to Losseless Compression of Gray-Scale Images", *IEEE Trans. Image Processing*, vol. 5, no. 4, 1996, 575-586.

[3] Yu.M. Shtarkov, "Coding of discrete sources with unknown statistics", *Topics in Inform. Theory (Second Colloquium, Keszely, 1975)*, Colloquia Mathematica Sosietatis Janos Bolyai, Amsterdam, North Holland, vol. 16, 1977, 559-574.

[4] Yu.M. Shtarkov, "Universal Sequential Coding of Single Messages", *Probl. Inform. Trans.*, vol. 23, no. 3, 1987, 3-17.

[5] B.Ya. Ryabko, "Twice-Universal Coding", *Probl. Inform. Trans.*, vol. 20, no. 4, 1984, 396-402.

[6] B.Ya. Ryabko, "Prediction of Random Sequences and Universal Coding", *Probl. Inform. Trans.*, vol. 24, no. 2, 1988, 3-14.

[7] J.J. Rissanen, *Stochastic Complexity in Statistical Inquiry*, New Jersey: World Scientific Publ. Co., 1989.

[8] Yu.M. Shtarkov, "Aim Functions and Sequential Estimation of Source Model for Universal Coding", *Probl. Inform. Trans.*, vol. 35, no. 3, 1999.

[9] J.J. Rissanen, "Complexity of Strings in the Class of Markov Sources", *IEEE Trans. Inform. Theory*, vol. 32, no. 4, 1986, 526-532.

[10] F.M.J. Willems, Yu.M. Shtarkov and Tj.J. Tjalkens, "Context Tree Weighting: A Sequential Universal Coding Procedure for FSMX Sources", *Proc. 1993 IEEE Intern. Symp. Inform. Theory*, USA, 1993, 59.

[11] F.M.J. Willems, Yu. M. Shtarkov and Tj. J. Tjalkens, "The Context Tree Weighting Method: Basic Properties", *IEEE Trans. Inform. Theory*, vol. 41, no. 3, 1995, 653-664.

[12] Yu.M. Shtarkov, Tj.J. Tjalkens and F.M.J. Willems, "Multialphabet Weighted Universal Coding of Context Tree Sources", *Probl. Inform. Trans.*, vol. 33, no. 1, 1997, 3-11.

[13] M.J. Weinberger, J.J. Rissanen and M. Feder, "A Universal Finite Memory Source" *IEEE Trans. Inform. Theory*, vol. 41, no. 3, 1995, 643-652.

[14] M.J. Weinberger, A. Lempel and J. Ziv, “A Sequential Algorithm for the Universal Coding of Finite Memory Sources” *IEEE Trans. Inform. Theory*, vol. 38, no. 3., 1992, 1002-1014.

[15] B. Balkenhol, S. Kurtz, and Yu.M. Shtarkov, “Modifications of the Burrows and Wheeler Data Compression Algorithm”, *Proc. of Data Compression Conference*, 1999, 188-197.

HOW MUCH CAN YOU WIN WHEN YOUR ADVERSARY IS HANDICAPPED?

Ludwig Staiger

Martin-Luther-Universität Halle-Wittenberg, Institut für Informatik
Kurt-Mothes-Str. 1, D–06120 Halle, Germany

staiger@cantor.informatik.uni-halle.de

Abstract: We consider infinite games where a gambler plays a coin-tossing game against an adversary. The gambler puts stakes on heads or tails, and the adversary tosses a fair coin, but has to choose his outcome according to a previously given law known to the gambler. In other words, the adversary is not allowed to play all infinite heads-tails-sequences, but only a certain subset F of them.

We present an algorithm for the player which, depending on the structure of the set F, guarantees an optimal exponent of increase of the player's capital, independently on which one of the allowed heads-tails-sequences the adversary chooses.

Using the known upper bound on the exponent provided by the maximum Kolmogorov complexity of sequences in F we show the optimality of our result.

It is well-known that random sequences do not admit successful gambling strategies. Here we consider a game where a player bets at fixed odds, but with unlimited amount, on the tosses of a coin. We further agree on the fact that the player must have no debt. It was explained in [7, 11, 4] that in such a game a player playing according a computable gambling strategy cannot have unlimited gain if the tosses of the coin follow a random zero-one-sequence.

On the other hand, it is quite obvious that, if the zero-one-sequence follows partially a certain computable law, the player may have an unlimited gain. A simple example is a zero-one-sequence which repeats each value twice. Here the player may double his capital every second step just by betting all his remaining capital according to the previous outcome.

I. Althöfer et al. (eds.), Numbers, Information and Complexity, 403-412.

In this paper we investigate the exponent of the increase of the player's capital, λ, under the following assumptions on the game.

1. The player plays a computable gambling strategy, more precisely, he computes his bets from a complete history in a deterministic way.

2. The tosses of the coin follow a zero-one-sequence which belongs to a certain previously fixed set $F \subseteq \{0,1\}^\omega$.

3. The player can bet arbitrary nonnegative amounts not exceeding his capital, in particular, he must not have debts.

It is shown that under these and some additional computability assumptions on the set of admitted zero-one-sequences F there is always a strategy which guarantees the player an exponent λ which depends only upon the size of the constraint F.

Moreover, we show that our result is the best possible in two respects.

1. Regardless which constraint $F \subseteq \{0,1\}^\omega$ we consider, there is always a zero-one-sequence $\xi \in F$ such that $\lambda(\xi)$ cannot be better than the upper bound given by the size of F.

2. Our computability assumption on F guaranteeing the optimal exponent λ is a best one. It cannot be extended to admit larger classes of constraints.

The results of this papers relate several different areas of mathematics and theoretical computer science.

In the first section we give some necessary notation, and we present our notion of game. For these games we derive a description of gambling strategy via computable martingales. In Section 2 we derive an upper bound on the exponent of the increase of the player's capital in terms of Kolmogorov complexity.

The subsequent section introduces an appropriate size measure for sets of zero-one-sequences. It turns out that the Hausdorff dimension, known from fractal geometry, fulfills our requirements of being closely related to Kolmogorov complexity on the one hand and to gambling strategies on the other hand. In the fourth section we discuss the computability requirements which we have to put on our constraints $F \subseteq \{0,1\}^\omega$. Here we state also our main result.

Most of the results presented here are proved in [10]. For the necessary background in computability, random sequences and Kolmogorov complexity we refer the reader to [7], [4] and [1]. For the definition of Hausdorff dimension and their properties see e.g. [2, 3].

NOTATION AND DEFINITIONS

By $\mathbb{N} = \{0,1,2,\ldots\}$ we denote the set of natural numbers. We consider the space $\{0,1\}^\omega$ of infinite zero-one-sequences (ω-words). By $\{0,1\}^*$ we denote the set of finite strings (words) on $\{0,1\}$, including the *empty* word e. For $w \in \{0,1\}^*$ and $b \in \{0,1\}^* \cup \{0,1\}^\omega$ let $w \cdot b$ be their *concatenation*. This concatenation product extends in an obvious way to subsets $W \subseteq \{0,1\}^*$ and $B \subseteq \{0,1\}^* \cup \{0,1\}^\omega$.

Furthermore $|w|$ is the *length* of the word w. By b/n we denote the *length n prefix* of a string $b \in \{0,1\}^*$, $|b| \geq n$, or $b \in \{0,1\}^\omega$, and $\mathbf{A}(b) := \{b/n : n \in \mathbb{N} \wedge n \leq |b|\}$ and $\mathbf{A}(B) := \bigcup_{b \in B} \mathbf{A}(b)$ are the sets of all finite prefixes of $b \in \{0,1\}^* \cup \{0,1\}^\omega$ and $B \subseteq \{0,1\}^* \cup \{0,1\}^\omega$, respectively.

The set of all binary words $\{0,1\}^*$ may be also viewed at as the rooted infinite binary tree, where the empty word e is the root and $w0, w1$ are the successors of the node $w \in \{0,1\}^*$. Then $\{0,1\}^\omega$ is in a natural correspondence with the infinite paths through $\{0,1\}^*$ starting at the root, as any path $\xi \in \{0,1\}^\omega$ is uniquely specified by its finite initial paths $w \in \mathbf{A}(\xi)$.

This much notation suffices to describe our game.

Tree game on the binary tree $\{0,1\}^*$ given a set $F \subseteq \{0,1\}^\omega$ of admitted zero-one-sequences

`Start :` $\quad w := e \quad$ [root node = empty word]

$$\mathcal{V}(e) := 1 \quad \text{[initial capital]} \tag{1}$$

`For` $w := e$ `to` $\xi \in F$ `do`

player bets:	**adversary** chooses
$\mathcal{W}_0(w), \mathcal{W}_1(w) \in [0,1]$ where $\mathcal{W}_0(w) + \mathcal{W}_1(w) \leq 1$	
	$x \in \{0,1\}$ according to $\xi \in F$ and pays $2 \cdot \mathcal{W}_x(w) \cdot \mathcal{V}(w)$

player's capital

$$\mathcal{V}(wx) := \mathcal{V}(w) \cdot (1 + \mathcal{W}_x(w) - \mathcal{W}_{\neg x}(w)) \tag{2}$$

$w := wx$

`Endfor`

We assume that $\mathcal{W}_0 : \{0,1\}^* \to \mathbb{R}$ and $\mathcal{W}_1 : \{0,1\}^* \to \mathbb{R}$ are computable functions. From the Equations (1) and (2) in the above description of our game we can compute in advance the player's capital $\mathcal{V}(w)$ in node w of the binary tree $\{0,1\}^*$ as illustrated in the picture below:

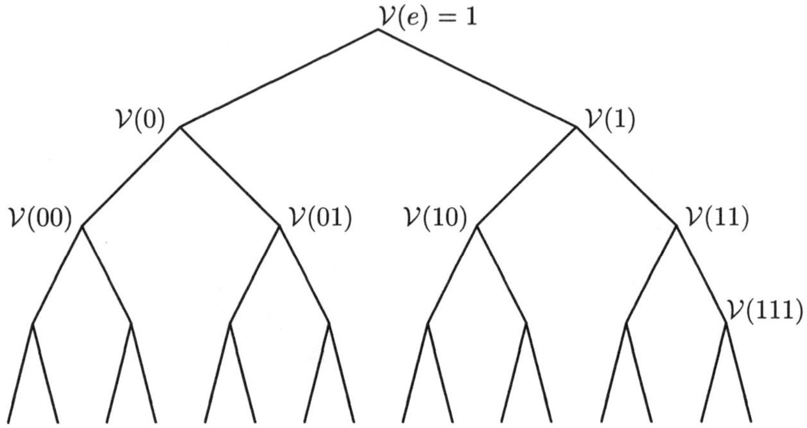

Here one easily observes that the capital function $\mathcal{V}$ has the following property

$$\mathcal{V}(w) = \frac{1}{2} \cdot (\mathcal{V}(w0) + \mathcal{V}(w1)) \ . \tag{3}$$

Conversely, if we have a function $\mathcal{V} : \{0,1\}^* \to \mathbb{R}$ satisfying (1) and (3) then defining

$$\mathcal{W}_x(w) := \begin{cases} \frac{\mathcal{V}(wx)}{\mathcal{V}(w)} & \text{, if } \mathcal{V}(w) > 0 \text{ and} \\ 0 & \text{, otherwise} \end{cases} \tag{4}$$

yields a gambling strategy $(\mathcal{W}_0, \mathcal{W}_1)$ which realizes the capital $\mathcal{V}(w)$ in the node w of the binary tree.

Thus, in the sequel, it suffices to consider (computable) capital functions satisfying (1) and (3). Those functions are also called (*computable*) *martingales* (cf. [7, 11, 4]).

We conclude this section with two examples presenting gambling strategies for given constraints F_1 and F_2.

Example 1. As mentioned above in the introduction let our constrained satisfy $F_1 := \{00, 11\}^\omega$, that is, the adversary repeats its choice once.

A reasonable betting strategy for the player to maximize the growth of his capital would be given by

$$\mathcal{W}_x(w) := \begin{cases} 1, & \text{if } |w| \text{ is odd and } w \in \{0,1\}^* \cdot x \\ 0, & \text{otherwise,} \end{cases}$$

that is, to put every second step all of the capital on letter x if x was previously chosen by the adversary.

One easily calculates that

$$\mathcal{V}_1(w) = \begin{cases} 2^{\lfloor |w|/2 \rfloor}, & \text{if } w \in \mathbf{A}(F_1), \text{ and} \\ 0, & \text{otherwise.} \end{cases}$$

So, asymptotically we have $\log_2 \mathcal{V}_1(w) \approx \frac{|w|}{2}$ for $|w| \to \infty$ and $w \in \mathbf{A}(F_1)$.

Intuitively, it is clear that the player cannot do much better, for at every odd step the adversary might flip a coin to draw his outcome randomly, and, as it is well known (cf. [7, 11, 4]), one cannot win against a random sequence. As we shall prove below, the asymptotic gain of the betting above strategy is optimal. □

The next example is a little bit more involved.

Example 2. Let $F_2 := \{0,1\}^* \cdot 0^\omega$, that is a typical zero-one-sequence in $F2$ has the form

$$\underbrace{x_1 \dots x_m}_{\substack{m \in \mathbb{N} \\ arbitrary}} \cdot \underbrace{0000000\cdots}_{ad\ infinitum} .$$

A reasonable betting strategy for maximizing the player's capital would be to put larger and larger parts of the capital on 0, because the adversary's ultimate behaviour is to draw only zeros. Observe here that, albeit the player is not allowed to make loans, he is allowed to retain arbitrarily small positive amounts.

Thus we might choose

$$\mathcal{W}_x(w) := \begin{cases} 1 - 2^{-(|w|+1)} & , \text{ if } x = 0, \text{ and} \\ 0 & , \text{ otherwise.} \end{cases}$$

If $w \in v \cdot 0^*$ then

$$\begin{aligned} \mathcal{V}_2(w) &\geq \prod_{i=1}^{|v|} 2^{-i} \cdot \prod_{i=|v|+1}^{|w|} 2 \cdot (1 - 2^{-i}) \\ &\geq 2^{|w| - \frac{|v| \cdot (|v|+3)}{2}} \cdot \prod_{i=1}^{\infty} (1 - 2^{-i}) . \end{aligned}$$

Using the fact that $\prod_{i=1}^{\infty}(1-2^{-i}) > 0$ we obtain that $\mathcal{V}_2(w) \geq c_\xi \cdot 2^{|w|}$ as $w \to \xi$ for every $\xi \in F_2$. Here $c_\xi > 0$ is a constant depending on v when $\xi = v \cdot 0^\omega$. □

UPPER BOUNDS BY KOLMOGOROV COMPLEXITY

In this section we derive an upper bound on the exponent of the increase of the player's capital for arbitrary (even non-computable) constraints $F \subseteq \{0,1\}^\omega$. Moreover, we show that, in general, there is no computable gambling strategy which reaches this upper bound.

Before we proceed to the results, we make precise what we mean by the exponent of the increase of the player's capital function $\mathcal{V}$, $\lambda_\mathcal{V}$.

Definition 1. Let $\mathcal{V} : \{0,1\}^* \to [0, \infty)$ be a capital function. Then

$$\lambda_\mathcal{V}(\xi) := \limsup_{n \to \infty} \frac{\log_2 \mathcal{V}(\xi/n)}{n} \tag{5}$$

is the *exponent of the increase* of $\mathcal{V}$ on the zero-one-sequence ξ.

The task of the subsequent sections is, given a constraint $F \subseteq \{0,1\}^\omega$, to maximize the value of $\inf_{\xi \in F} \lambda_\mathcal{V}(\xi)$ for computable $\mathcal{V} : \{0,1\}^* \to [0,\infty)$.

First we give the announced upper bound on $\lambda_\mathcal{V}(\xi)$.

To this end we introduce the *Kolmogorov complexity* of finite and infinite strings.

For a given algorithm (computable partial function) $\mathfrak{A} : \{0,1\}^* \to \{0,1\}^*$ we define the complexity of a word $w \in \{0,1\}^*$ as the length of the shortest program $\pi \in \{0,1\}^*$ for which $\mathfrak{A}$ prints w:

$$K_\mathfrak{A}(w) := \inf\{|\pi| : \mathfrak{A}(\pi) = w\} \ . \tag{6}$$

Then it holds

Theorem 1 (Solomonoff, Kolmogorov) *There is an optimal algorithm $\mathfrak{U}$ such that for all algorithms $\mathfrak{A}$ we have*

$$\forall w \in \{0,1\}^* : \ K_\mathfrak{U}(w) \leq K_\mathfrak{A}(w) + c_{\mathfrak{A},\mathfrak{U}} \ ,$$

for an appropriately chosen constant $c_{\mathfrak{A},\mathfrak{U}}$.

In what follows, when considering the Kolmogorov complexity of a finite string $w \in \{0,1\}^*$, we shall refer to a fixed optimal algorithm $\mathfrak{U}$.

For infinite strings here we shall consider the following notion of Kolmogorov complexity.

Definition 2. The *lower Kolmogorov complexity* of an infinite string $\xi \in \{0,1\}^\omega$ is the value

$$\underline{\kappa}(\xi) := \liminf_{n\to\infty} \frac{K_\mathfrak{U}(\xi/n)}{n} \ .$$

Utilizing Levin's universal semicomputable semimeasure (cf. [12] or [4]) it was shown in [6] that the exponent $\lambda_\mathcal{V}(\xi)$ is bounded from above by $1-\underline{\kappa}(\xi)$ provided the gambler plays according to a computable strategy.

Lemma 2 (Upper bound by Kolmogorov complexity) *Let $\mathcal{V}$ be a computable capital function. Then*

$$\lambda_\mathcal{V}(\xi) \leq 1 - \underline{\kappa}(\xi) \tag{7}$$

for every $\xi \in \{0,1\}^\omega$.

This upper bound, though being helpful as we shall see in the next sections, is inaccessible. From the well-known noncomputability of Levin's universal semicomputable semimeasure one obtains readily the following.

Theorem 3. *There is no computable capital function $\mathcal{V}_{\mathrm{opt}} : \{0,1\}^* \to \mathbb{R}$ such that $\forall \xi \in \{0,1\}^\omega : \lambda_{\mathcal{V}_{\mathrm{opt}}}(\xi) \geq \lambda_\mathcal{V}(\xi)$ for all other computable capital functions $\mathcal{V} : \{0,1\}^* \to \mathbb{R}$.*

As a corollary we obtain further.

Corollary 4. *There is no computable capital function $\mathcal{V} : \{0,1\}^* \to \mathbb{R}$ such that $\lambda_\mathcal{V}(\xi) = 1 - \underline{\kappa}(\xi)$ for all $\xi \in \{0,1\}^\omega$.*

AN APPROPRIATE SIZE MEASURE

So far we did not agree upon the notion of size of a set $F \subseteq \{0,1\}^\omega$. The upper bound on $\lambda_\mathcal{V}$ by Lemma 2, in view of $\inf_{\xi \in F} \lambda_\mathcal{V}(\xi) \leq 1 - \sup_{\xi \in F} \underline{\kappa}(\xi)$, suggests to choose a value like $\sup_{\xi \in F} \underline{\kappa}(\xi)$ as the size of F. This proposal, however, seems to be a bit too artificial.

Take for example a random zero-one-sequence ζ. Those sequences have $\underline{\kappa}(\zeta) = 1$. Thus the size of a singleton $\{\zeta\}$ would be 1 in contrast to the size of the uncountable set F_1 of Example 1 which has a size of $\frac{1}{2}$ only.

Consequently, we follow a different line, mentioning that several papers investigated the relationship between the Kolmogorov complexity of infinite strings and size measures known from information theory and fractal geometry. It turned out in [5], [8] and [10] that the Hausdorff dimension of subsets $F \subset \{0,1\}^\omega$ is closely related to $\sup_{\xi \in F} \underline{\kappa}(\xi)$.

Definition 3. The *Hausdorff dimension* of a set $F \subseteq \{0,1\}^\omega$, $\dim F$, is the smallest real number $\alpha \geq 0$ such that for all $\gamma > \alpha$ it holds

$$\forall \varepsilon > 0 : \exists W \subseteq \{0,1\}^* : F \subseteq W \cdot \{0,1\}^\omega \wedge \sum_{w \in W} \left(2^{-\gamma}\right)^{|w|} < \varepsilon \, .$$

From Definition 3 it is evident that Hausdorff dimension is monotone with respect to set inclusion and that $\dim \{\xi\} = 0$. We observe that Hausdorff dimension fulfills the following stronger property. Let $(F_i)_{i \in \mathbb{N}}$ be a countable family of subsets of $\{0,1\}^\omega$. Then

$$\dim \bigcup_{i \in \mathbb{N}} F_i = \sup_{i \in \mathbb{N}} \dim F_i \, . \tag{8}$$

Eq. (8) implies that $\dim E = 0$ for every at most countable set $E \subseteq \{0,1\}^\omega$.

As a first result we mention a lower bound to Kolmogorov complexity by Hausdorff dimension which states that sets of large dimension contain complex sequences.

Theorem 5 ([5])

$$\dim F \leq \sup\{\underline{\kappa}(\xi) : \xi \in F\}$$

Consider a $\xi \in F$ satisfying $\underline{\kappa}(\xi) \geq \dim F - \varepsilon$. Then Lemma 2 proves $\lambda_\mathcal{V}(\xi) \leq 1 - \underline{\kappa}(\xi) = \dim F - \varepsilon$. Thus we obtain the following worst case behaviour of capital functions.

Lemma 6. *Let $F \subseteq \{0,1\}^\omega$ and let $\mathcal{V} : \{0,1\}^* \to \mathbb{R}$ be a computable capital function. For all $\varepsilon > 0$ there is a $\xi \in F$ such that $\lambda_\mathcal{V}(\xi) \leq 1 - \dim F + \varepsilon$.*

As announced previously, according to Lemma 6 the adversary has always the possibility to limit the exponent of growth of the player's capital function $\lambda_\mathcal{V}$ close to the value $1 - \dim F$ (or even below) when F is his constraint.

Example 1 (continued) Consider again $F_1 := \{00, 11\}^\omega$. One can easily show that $\dim F_1 = \frac{1}{2}$. Thus according to Lemma 6 the asymptotic growth of the capital function $\log_2 \mathcal{V}_1(w) \approx \frac{|w|}{2}$ is optimal. □

WHICH CONSTRAINTS ARE REASONABLE

It follows from the behaviour of computable capital functions that $1 - \dim F$ is a worst case upper bound to the exponent of their growth. In this section we investigate which subsets $F \subseteq \{0,1\}^\omega$ allow for computable betting strategies $(\mathcal{W}_0, \mathcal{W}_1)$ such that the player achieves a guaranteed exponent of the corresponding capital function $\lambda_\mathcal{V}$ which is close to the bound $1 - \dim F$ regardless which infinite sequence $\xi \in F$ the adversary plays.

First we derive an example where the constraint $E \subseteq \{0,1\}^\omega$ is in some sense effectively presented, but nevertheless there is a large gap between $1 - \dim E$ and $\lambda_\mathcal{V}(\zeta)$ for at least one $\zeta \in E$ for all computable capital functions $\mathcal{V}$.

Example 3 ([10], Lemma 6) *There is a countable subset $E \subseteq \{0,1\}^\omega$ such that $\mathbf{A}(E)$ is recursively enumerable*[1] *and contains a random zero-one-sequence ζ. Since $\underline{\kappa}(\zeta) = 1$, as ζ is random, and since $\dim E = 0$, as E is countable, we have $0 = \lambda_\mathcal{V}(\zeta) = 1 - \underline{\kappa}(\zeta) < 1 - \dim E = 1$ for every computable capital function $\mathcal{V}$.* □

Remark. A more subtle consideration of the proof of Lemma 6 of [10] shows that E contains exactly one random zero-one-sequence ζ, and $E \setminus \{\zeta\} \subseteq \{0,1\}^* \cdot 0^\omega$. Thus E might be seen as an effective presentation of the random zero-one-sequence ζ, although infinite random sequences seem to be objects which cannot be presented effectively.

Our Example 3 leads to the conclusion that we have to restrict the range of computability of the constraints.

Definition 4 (Σ_2-definable sets) A subset $F \subseteq \{0,1\}^\omega$ is referred to as *Σ_2-definable* provided there is a computable function $f_F : \mathbb{N} \times \{0,1\}^* \to \{0,1\}$ such that

$$\xi \in F \quad \longleftrightarrow \quad \exists i \in \mathbb{N} : \forall n \in \mathbb{N} : f_F(i, \xi/n) = 0 \ .$$

Remark. The set E of Example 3 can be defined in a similar way: There is a computable function $g_E : \mathbb{N} \times \{0,1\}^* \to \{0,1\}$ such that

$$\xi \in E \longleftrightarrow \forall n \in \mathbb{N} : \exists i \in \mathbb{N} : g_E(i, \xi/n) = 0 \ .$$

Observe, however, that the order of the quantifiers is reversed and, besides that, here the outer quantifier $\forall n$ is related to the sequence ξ.

Now we can derive our main result.

Theorem 7 (Main Theorem) *If $F \subseteq \{0,1\}^\omega$ is Σ_2-definable, then for every $\gamma > \dim F$ there is a computable capital function $\mathcal{V}$ such that*

$$\forall \xi \in F : \lambda_\mathcal{V}(\xi) > 1 - \gamma \ .$$

[1] A subset $W \subseteq \{0,1\}^*$ is called *recursively enumerable* provided $W = \emptyset$ or there is a computable function $f : \mathbb{N} \to \{0,1\}^*$ such that $f(\mathbb{N}) = W$.

If, moreover, $\dim F$ *is a computable real number*[2] *then there is a computable capital function* $\mathcal{V}$ *such that*

$$\forall \xi \in F : \lambda_{\mathcal{V}}(\xi) \geq 1 - \dim F \ .$$

For a proof see [10].

Example 2 (continued) The set $F_2 = \{0,1\}^* \cdot 0^\omega$ is countable. Consequently, $\dim F_2 = 0$. The capital function $\mathcal{V}_2$ introduced above satisfies $\lambda_{\mathcal{V}_2}(\xi) = 1$ whenever $\xi \in F_2$.

It should be mentioned that, in contrast to $\lambda_{\mathcal{V}_1}(\xi) = 0$ for $\xi \in F_1$ where $\mathcal{V}_1$ is the capital function of Example 1, for $\mathcal{V}_2$ we may have even $\lambda_{\mathcal{V}_2}(\xi) = 1$ for some $\xi \notin F_2$ provided the infinitely many ones in ξ are distributed sparsely. □

Finally, combining the results of Theorems 5 and 7 and Lemma 2, we obtain an exact bound on the maximum lower Kolmogorov complexity for Σ_2-definable subsets of $\{0,1\}^\omega$.

Theorem 8 (Exact bound for Σ_2-definable sets)

$$\dim F = \sup\{\underline{\kappa}(\xi) : \xi \in F\}$$

if $F \subseteq \{0,1\}^\omega$ *is* Σ_2*-definable.*

Proof. One inequality is Theorem 5. For the converse inequality, observe that Theorem 7 and Lemma 2 prove that $\gamma > \underline{\kappa}(\xi)$ whenever $\gamma > \dim F$, $\xi \in F$ and $F \subseteq \{0,1\}^\omega$ is Σ_2-definable. Thus, $\dim F \geq \sup_{\xi \in F} \underline{\kappa}(\xi)$. □

Concluding Remark

Our Theorems 7 and 8 in connection with previous results of Ryabko ([5, 6]) and this author ([8, 10]) give evidence that there is a strong coincidence between the concepts of Kolmogorov complexity, gambling strategies and Hausdorff dimension for a class of recursive (computable) sets of infinite zero-one-sequences. The results of the last section show a borderline in the Arithmetical hierarchy[3] up to which this coincidence holds true, and our Example 3 gives evidence that it does not extend much further in the Arithmetical Hierarchy.

References

[1] C. Calude, *Information and Randomness. An Algorithmic Perspective.* Springer-Verlag, Berlin, 1994.

[2] G. A. Edgar, *Measure, Topology, and Fractal Geometry.* Springer, New York, 1990.

[2]A number $\gamma \in \mathbb{R}$ is *computable* provided there is a computable function $f_\gamma : \mathbb{N} \to \mathbb{Q}$ such that $|f_\gamma(n) - \gamma| < 2^{-n}$ for all $n \in \mathbb{N}$.

[3]For the Arithmetical hierarchy of ω-languages see e.g. [9].

[3] K.J. Falconer, *Fractal Geometry*. Wiley, Chichester, 1990.

[4] M. Li and P.M.B. Vitányi, *An Introduction to Kolmogorov Complexity and its Applications*. Springer–Verlag, New York, 1993.

[5] B. Ya. Ryabko, "Noiseless coding of combinatorial sources, Hausdorff dimension, and Kolmogorov complexity", *Problemy Peredachi Informatsii* , 22 1986, No. 3, 16–26 (in Russian; English tranitation: *Problems of Information Transmission* , 22, 1986, No. 3, 170–179).

[6] B. Ya. Ryabko, "An algorithmic approach to prediction problems", *Problemy Peredachi Informatsii* , 29, 1993, No. 2, 96–103 (in Russian).

[7] C.P. Schnorr, *Zufälligkeit und Wahrscheinlichkeit*, Lecture Notes in Math. No. 218, Springer-Verlag, Berlin 1971.

[8] L. Staiger, "Kolmogorov complexity and Hausdorff dimension", *Inform. and Comput.* , 102, 1993, No. 2, 159 – 194.

[9] L. Staiger, "ω-languages", *Handbook of Formal Languages*, (G. Rozenberg and A. Salomaa Eds.), Vol. 3, Springer-Verlag, Berlin 1997, 339 – 387.

[10] L. Staiger, "A tight upper bound on Kolmogorov complexity and uniformly optimal prediction", *Theory of Computing Systems* , 31, 1998, 215 – 229.

[11] M. van Lambalgen, *Random sequences*, Ph. D. Thesis, Univ. of Amsterdam, 1987.

[12] A.K. Zvonkin and L.A. Levin, "Complexity of finite objects and the development of the concepts of information and randomness by means of the theory of algorithms", *Russian Math. Surveys* , 25, 1970, 83 – 124.

ON RANDOM-ACCESS DATA COMPACTION

Frans M.J. Willems, Tjalling J. Tjalkens, and Paul A.J. Volf

Eindhoven Univ. of Technology
Electrical Engineering Department
Eindhoven, The Netherlands

Abstract: Consider a binary I.I.D. sequence that consists of $K = 2^J$ blocks of length T. We are looking for a universal compaction method that allows us to decode a certain block by looking only at certain segments in the codesequence. We have investigated a hierarchical method that encodes the source sequence into a codesequence that consists of 2^{J+1} variable-length segments. For decoding a certain block only $J + 2$ segments need to be accessed. During decoding it is always clear where the next segment that needs to be accessed appears in the codesequence. The cumulative individual redundancy that is achieved by this method, is optimal in the sense that $\frac{1}{2}\log_2 N$ behavior is obtained where $N = 2^J T$. An additional increase of at most one bit per code-segment is possible however.

PROBLEM DESCRIPTION, A BINARY I.I.D. SOURCE

Suppose that the binary sequence $x_1^N = x_1 x_2 \cdots x_N$ of length N consists of K blocks, each with length $T = N/K$ (it is assumed that K divides N), i.e.

$$x_1^N = x_1^T, x_{T+1}^{2T}, \cdots, x_{(K-1)T+1}^{KT}.$$

We want to investigate *universal* compaction methods for such sequences x_1^N that

- achieve optimal redundancy behavior (i.e. $\approx \frac{1}{2}\log_2 N$ bits per parameter), and
- allow expansion of an arbitrary block $x_{(k-1)T+1}^{kT}$ for $k \in \{1, 2, \cdots, K\}$ without having to access "too many" code-segments and code-bits.

To study this problem we first consider a binary *independent and identically distributed* (i.i.d.) source. This i.i.d. source produces a sequence $x_1^N = x_1 x_2 \cdots x_N$

I. Althöfer et al. (eds.), Numbers, Information and Complexity, 413-420.

with components $\in \{0,1\}$ with actual probability $P_a(x_1^N)$. If the source has parameter θ then

$$P_a(1) = 1 - P_a(0) = \theta$$

for some $0 \leq \theta \leq 1$. Now a sequence x_1^N containing a zeros and b ones has probability

$$P_a(x_1^N) = (1-\theta)^a\theta^b.$$

CODES AND REDUNDANCY, KRICHEVSKY-TROFIMOV ESTIMATOR

A *source code* assigns to source sequence x_1^N a binary codeword $c(x_1^N)$ of length $L(x_1^N)$. We consider only *prefix codes* here. In a prefix code no codeword is the prefix of any other codeword. The *individual redundancy* $\rho(x_1^N)$ of a sequence x_1^N is defined as

$$\rho(x_1^N) \triangleq L(x_1^N) - \log_2 \frac{1}{P_a(x_1^N)},$$

i.e. codeword-length minus *ideal* codeword-length. If the actual probabilities $P_a(x_1^N)$ are *not known* we can use instead of $P_a(x_1^N)$ *coding probabilities* $P_c(x_1^N)$ satisfying

$$\begin{aligned} P_c(x_1^N) &> 0 \text{ for all } x_1^N, \text{ and} \\ \sum_{x_1^N} P_c(x_1^N) &= 1. \end{aligned}$$

Now there exists a prefix code (Shannon-Fano code, see Shannon [5]) with codeword-lengths that satisfy

$$L(x_1^N) = \lceil \log_2 \frac{1}{P_c(x_1^N)} \rceil < \log_2 \frac{1}{P_c(x_1^N)} + 1.$$

A good coding probability (see Krichevsky-Trofimov [2]) for a sequence x_1^N that contains a zeroes and b ones is

$$P_e(a,b) \triangleq \int_{\theta=0,1} \frac{1}{\pi\sqrt{(1-\theta)\theta}} \cdot (1-\theta)^a\theta^b d\theta.$$

It can be shown that for all θ and x_1^N with a zeros and b ones (see [6]):

$$\frac{P_c(x_1^N)}{P_a(x_1^N)} = \frac{P_e(a,b)}{(1-\theta)^a\theta^b} \geq \frac{1}{2\sqrt{N}}.$$

Therefore we obbtain for the individual redundancy for all θ and x_1^N with a zeroes and b ones

$$\begin{aligned} \rho(x_1^N) &= L(x_1^N) - \log_2 \frac{1}{P_a(x_1^N)} \\ &= \lceil \log_2 \frac{1}{P_e(a,b)} \rceil - \log_2 \frac{1}{(1-\theta)^a\theta^b} \\ &< \log_2 \frac{(1-\theta)^a\theta^b}{P_e(a,b)} + 1 \leq (\frac{1}{2}\log_2 N + 1) + 1. \end{aligned}$$

STANDARD APPROACH

We can process the entire sequence x_1^N in a standard way (see figure 1) to obtain the Shannon-Fano codeword $c(x_1^N)$. This classical approach:

- achieves optimal redundancy behavior (i.e. $\rho(x_1^N) < \frac{1}{2}\log_2 N + 2$ bits),
- but requires, in principle, the entire codeword $c(x_1^N)$ for decoding a block (see figure1).

SINGLE-BLOCK CODING

To obtain random-accessibility we can encode all the K blocks separately (see figure 2). Let a_k and b_k be the number of zeroes and ones in block k then:

$$\begin{aligned} L(x_{(k-1)T+1}^{kT}) &= \lceil \log_2 \frac{1}{P_e(a_k, b_k)} \rceil \\ &< \log_2 \frac{1}{(1-\theta)^{a_k}\theta^{b_k}} + \frac{1}{2}\log_2 T + 2. \end{aligned}$$

Hence, with $a = \sum_{k=1,K} a_k$, $b = \sum_{k=1,K} b_k$, and $T = N/K$, we get

$$\begin{aligned} L(x_1^N) &= \sum_{k=1,K} L(x_{(k-1)T+1}^{kT}) \\ &< \log_2 \frac{1}{(1-\theta)^a\theta^b} + \frac{K}{2}\log_2 \frac{N}{K} + 2K. \end{aligned}$$

Therefore the single-block coding method:

- has a desirable random-access behavior (it uses only codeword $c(x_{(k-1)T+1}^{kT})$ for decoding block $x_{(k-1)T+1}^{kT}$),

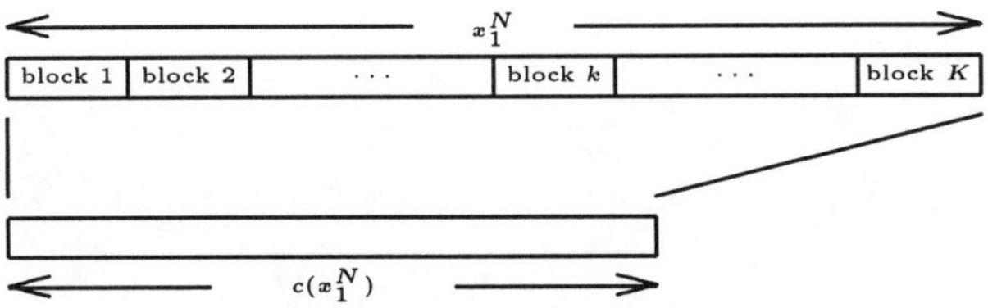

Figure 1 Standard approach.

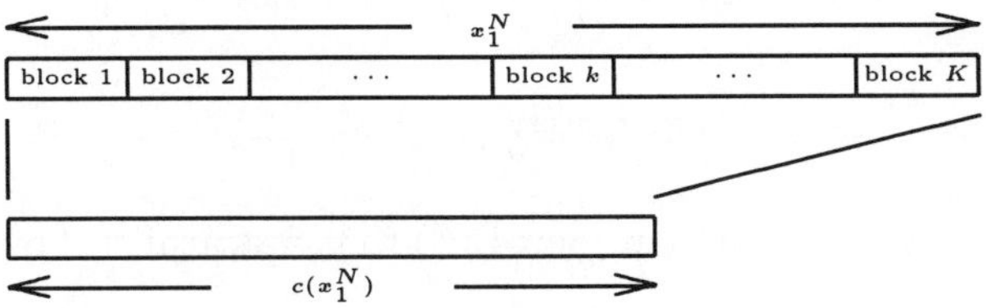

Figure 2 Encoding the blocks separately.

- but does not achieve the optimal redundancy bound (the bound is roughly K times larger than necessary), and
- moreover requires information that tells the decoder where the variable-length code-segments start (roughly $K \log_2 N$ bits).

The question is now: Is there a method that performs better?

ENUMERATIVE APPROACH

Fix the source sequence length N. There are $\binom{N}{b}$ sequences that contain $N-b$ zeroes and b ones. Now, instead of constructing Shannon-Fano codewords for the source sequences x_1^N as before, we form a Shannon-Fano code for all the $b \in \{0, 1, \cdots, N\}$, and then use a fixed-length code to specify which of the sequences with b ones actually occurred. Enumerative methods as described by Schalkwijk [4] and Cover [1] can be used to do this in an efficient way. For the "composition parameter" $b = 0, 1, \cdots, N$ we use the coding probabilities

$$P_c(b) = \binom{N}{b} P_e(N-b, b)$$

to construct the Shannon-Fano code. This yields

$$L(b) = \lceil \log_2 \frac{1}{P_c(b)} \rceil = \lceil \log_2 \frac{1}{\binom{N}{b} P_e(N-b, b)} \rceil.$$

Moreover $L(x_1^N|b)$ code-bits are needed to specify which source sequence with b ones actually occurred where

$$L(x_1^N|b) = \lceil \log_2 \binom{N}{b} \rceil.$$

Lemma: For all u and v

$$\lceil u \rceil + \lceil v \rceil \leq \lceil u + v \rceil + 1.$$

Hence

$$\begin{aligned} & L(b) + L(x_1^N|b) \\ = \; & \lceil \log_2 \frac{1}{\binom{N}{b} P_e(N-b, b)} \rceil + \lceil \log_2 \binom{N}{b} \rceil \\ \leq \; & \lceil \log_2 \frac{1}{P_e(N-b, b)} \rceil + 1, \end{aligned}$$

which is at most 1 bit more than the $L(x_1^N)$ that was achieved by the standard approach. Hence now

$$\rho(x_1^N) < \frac{1}{2} \log_2 N + 3.$$

TWO BLOCKS

We now consider a sequence x_1^N that consists of two blocks. Suppose that this sequence

$$x_1^N = x_1^T x_{T+1}^{2T}$$

contains b ones. Block x_1^T contains b_1 ones, block x_{T+1}^{2T} contains b_2 ones, therefore

$$b_2 = b - b_1.$$

There are

$$\binom{N}{b} = \binom{2T}{b}$$

sequences of length $N = 2T$ with b ones, and

$$\binom{T}{b_1}\binom{T}{b-b_1}$$

sequences of length $N = 2T$ with b_1 ones in the first block and $b_2 = b - b_1$ ones in the second block. Note that

$$\sum_{b_1=\max(0,b-T),\min(T,b)} \binom{T}{b_1}\binom{T}{b-b_1} = \binom{2T}{b}.$$

Now, instead of specifying the source sequence x_1^N given b, we form a Shannon-Fano code for all the possible b_1, and then use a fixed-length code to specify which of the blocks with b_1 ones occurred and then use another fixed-length code to specify which of the blocks with $b_2 = b - b_1$ ones occurred.

For $\max(0, b-T) \leq b_1 \leq \min(T, b)$ we use the coding probabilities

$$P_c(b_1|b) = \frac{\binom{T}{b_1}\binom{T}{b-b_1}}{\binom{2T}{b}}$$

to construct the Shannon-Fano code. This yields

$$L(b_1|b) = \lceil \log_2 \frac{1}{P_c(b_1|b)} \rceil = \lceil \log_2 \frac{\binom{2T}{b}}{\binom{T}{b_1}\binom{T}{b-b_1}} \rceil.$$

Moreover

$$\begin{aligned} L(x_1^T|b_1, b) &= \lceil \log_2 \binom{T}{b_1} \rceil, \\ L(x_{T+1}^{2T}|b_1, b) &= \lceil \log_2 \binom{T}{b-b_1} \rceil. \end{aligned}$$

Now, using the lemma, we get

$$L(x_1^N|b) = L(x_1^T, x_{T+1}^{2T}|b)$$

$$
\begin{aligned}
&= L(b_1|b) + L(x_1^T|b_1,b) + L(x_{T+1}^{2T}|b_1,b) \\
&\leq \lceil \log_2 \frac{\binom{2T}{b}}{\binom{T}{b_1}\binom{T}{b-b_1}} \rceil + \lceil \log_2 \binom{T}{b_1} \rceil + \lceil \log_2 \binom{T}{b-b_1} \rceil \\
&\leq \lceil \log_2 \binom{2T}{b} \rceil + 2.
\end{aligned}
$$

Hence, we need at most two bits more than by specifying $x_1^N = x_1^T, x_{T+1}^{2T}$ in a single fixed-length code.

$K = 2^J$ BLOCKS

Lemma: Let $L(2^j|b)$ denote the number of bits needed to describe 2^j blocks of length T that contain b ones in total, then:

$$
L(2^j|b) \leq \lceil \log_2 \binom{2^j T}{b} \rceil + 2(2^j - 1).
$$

Proof: For $j = 0$, a single block, the statement holds. Suppose that the statement holds for some $j \geq 0$. Then for all $b_1 + b_2 = b$

$$
\begin{aligned}
&L(2^{j+1}|b) \\
&= \lceil \log_2 \frac{\binom{2^{j+1}T}{b}}{\binom{2^j T}{b_1}\binom{2^j T}{b_2}} \rceil + L(2^j|b_1) + L(2^j|b_2) \\
&= \lceil \log_2 \frac{\binom{2^{j+1}T}{b}}{\binom{2^j T}{b_1}\binom{2^j T}{b_2}} \rceil + \lceil \log_2 \binom{2^j T}{b_1} \rceil + 2(2^j - 1) \\
&\qquad + \lceil \log_2 \binom{2^j T}{b_2} \rceil + 2(2^j - 1) \\
&\leq \lceil \log_2 \binom{2^{j+1}T}{b} \rceil + 2(2^{j+1} - 1).
\end{aligned}
$$

Note that we use a Shannon-Fano code to describe b_1 and implicitely $b_2 = b - b_1$.

Suppose that we allocate *exactly* $\lceil \log_2 \binom{2^j T}{b_1} \rceil + 2(2^j - 1)$ bits for describing the first 2^j blocks and then *exactly* $\lceil \log_2 \binom{2^j T}{b_2} \rceil + 2(2^j - 1)$ bits for describing the second 2^j blocks, for $j = 0, J$. In this way we achieve for 2^J blocks that contain b ones a total codewordlength which is *exactly*

$$
\begin{aligned}
L(x_1^N|b) &= \lceil \log_2 \binom{2^J T}{b} \rceil + 2(2^J - 1) \\
&= \lceil \log_2 \binom{N}{b} \rceil + 2(2^J - 1) \text{ bits,}
\end{aligned}
$$

which is $2(2^J - 1)$ bits more than what we achieved with the enumerative approach (i.e 2 bits extra per extra block). In conclusion we obtain for the redundancy

$$\rho(x_1^N) < \frac{1}{2}\log_2 N + 2(2^J - 1) + 3.$$

ACCESS METHOD

To describe the access method we give an example.
Example: Let $J = 2$, i.e. assume that x_1^N consists of $K = 2^J = 4$ blocks. Suppose that we want to decode block 4:

- First (see figure 3) decode b_{1234}. This determines the code for b_{12}.
- Decode b_{12}. Skip the codebits that are allocated for (b_1, block 1 and block 2), i.e. skip ($j = 1$) the next $\lceil \log_2 \binom{2T}{b_{12}} \rceil + 2$ bits. Compute $b_{34} = b_{1234} - b_{12}$. This determines the code for b_3.
- Decode b_3. Skip the codebits that are allocated for block 3, i.e. skip ($j = 0$) the next $\lceil \log_2 \binom{T}{b_3} \rceil$ bits. Compute $b_4 = b_{34} - b_3$. This determines the code for block 4.
- Use the next $\lceil \log_2 \binom{T}{b_4} \rceil$ bits to decode block 4.

HIERARCHY

In figure 4 the hierarchy in the access structure is depicted. The figure shows what code-segments should be read in order to decode a certain block. In general, when x_1^N consists of $K = 2^J$ blocks

$$J + 2 = \log_2 K + 2$$

code-segments have to be read to decode a single block.

CONCLUSIONS, QUESTIONS

The conclusion of this submission is that the proposed scheme achieves

- an acceptable redundancy and
- an acceptable number of accesses.

About the redundancy we can say that it is essentially optimal. The $\frac{1}{2}\log_2 N$ bound is achieved if we ignore the additional two bit increase per extra block. The two bit increase is a very natural consequence of rounding effects related to the code-segments in access-tree structure.

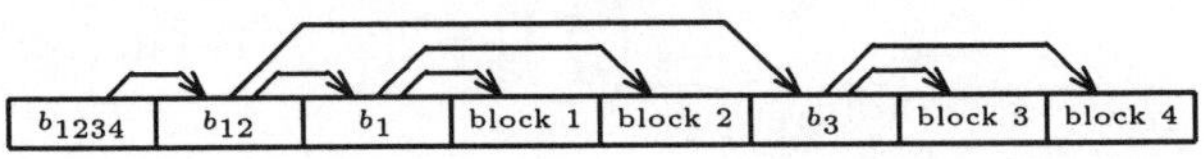

Figure 3 Decoding, jumping from code-block to code-block.

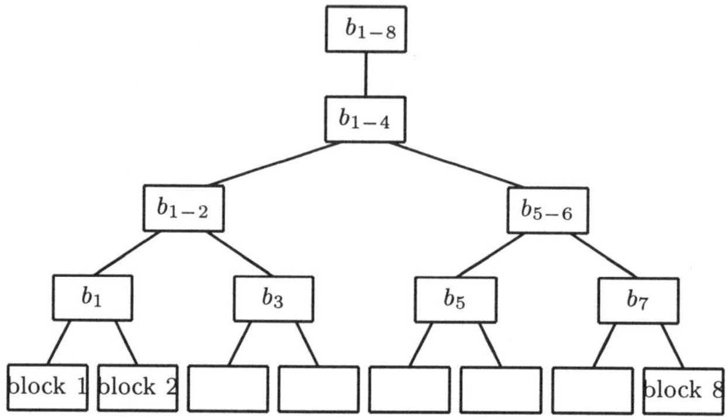

Figure 4 Hierarchy in the access structure.

Whether the number of segment-accesses and the number of codebits that have to be read in order to decode a certain block is the lowest possible, is not clear yet. To solve this problem further research is needed.

References

[1] T.M. Cover, "Enumerative Source Encoding," *IEEE Trans. Inform. Theory,* 19, 1973 , 73-77.

[2] R.E. Krichevsky and V.K. Trofimov, "The Performance of Universal Encoding," *IEEE Trans. Inform. Theory,* 27, 1981, 199-207.

[3] J. Rissanen, "Universal Coding, Information, Prediction, and Estimation," *IEEE Trans. Inform. Theory,* 30, 1984, 629-636.

[4] J.P.M. Schalkwijk, "An Algorithm for Source Coding," *IEEE Trans. Inform. Theory,* 30, 1972, 395-399.

[5] C.E. Shannon, "A Mathematical Theory of Communication," *Bell Sys. Tech. J.,* 27, 1948, 379-424 and 623-657.

[6] F.M.J. Willems, Y.M. Shtarkov and Tj.J. Tjalkens, "The Context Tree Weighting Method: Basic properties," *IEEE Trans. on Inform. Theory,* 41, 1995, 653-664.

UNIVERSAL LOSSLESS CODING OF SOURCES WITH LARGE AND UNBOUNDED ALPHABETS

En-hui Yang and Yunwei Jia

University of Waterloo, Waterloo, Ontario, Canada N2L 3G1

ehyang, yjia@bbcr.uwaterloo.ca

Abstract: A multilevel arithmetic coding algorithm is proposed to encode data sequences with large or unbounded source alphabets. The algorithm first converts the source alphabet into a dynamic tree, and then represents each symbol in the input sequence by its path in the tree and its index in the corresponding leaf. Encoding of the input sequence is then accomplished by encoding the path sequence and the index sequence conditionally. It is shown that the proposed algorithm is universal in the sense that it can achieve asymptotically the entropy rate of any independently and identically distributed integer source with a finite or infinite alphabet, as long as the mean value is finite. The advantages of the proposed algorithm over the traditional adaptive arithmetic coding algorithm are two folds: (1) the proposed algorithm can be used to encode any data sequence no matter whether the corresponding source alphabet is finite or infinite, while the traditional adaptive arithmetic coding algorithm can work only for data sequences with bounded, small alphabets; (2) in the situation in which the traditional adaptive arithmetic coding algorithm can work, the proposed algorithm can reduce coding complexity and improve compression performance. The proposed algorithm is then used to implement the recent Multilevel Pattern Matching(MPM) algorithms. Simulation results show that for a variety of files, the combination of the proposed algorithm with the MPM algorithms results in compression performance better than that afforded by the UNIX Compress algorithm, which is based on the LZ78 algorithm. Other applications of the proposed algorithm are also discussed.

I. Althöfer et al. (eds.), Numbers, Information and Complexity, 421-442.

INTRODUCTION

Consider the following typical data compression system shown in Figure 1. The input data sequence U is first transformed into an integer sequence X, which is then fed to a coder to generate binary codewords.

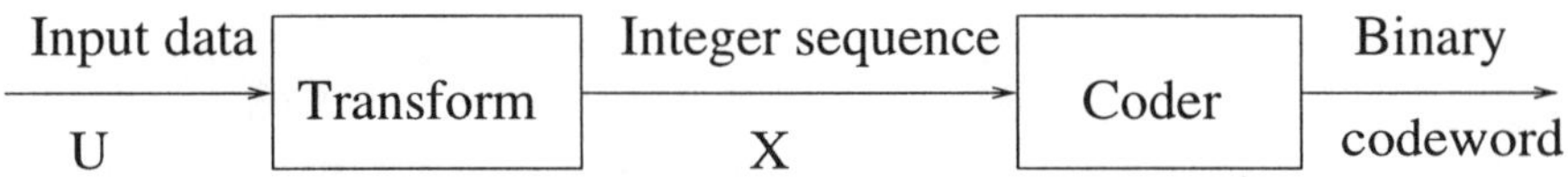

Figure 1 Transform based data compression system

Regardless of what the transform is, in the final step, one often has to efficiently compress the transformed integer sequence with large or even unbounded source alphabets. For example, in run-length coding[5], one has to efficiently encode a sequence of runs of $0'$s and $1'$, which is transformed from the original binary sequence; in entropy constrained scalar and vector quantization[4], one has to efficiently encode a sequence of codeword indices, which is transformed from the original real source; in grammar-based coding[8][12], one has to efficiently compress a sequence of integers with potentially unbounded number of distinct integers. Text compression can also be regarded as compression of integer sequences. If block coding is used in text compression, then integers to be encoded come from a source alphabet which grows exponentially with the block length. For example, if each block to be encoded consists of four 8-bit ASCII codes, the alphabet will be as large as 2^{32}.

When the size of the alphabet from which data sequences are drawn is large enough, however, the problem of universal compression of these data sequences is not as simple as it may look like. Due to the well-known underflow and overflow problems, finite precision implementations of the traditional adaptive arithmetic coding cannot work if the size of the source alphabet exceeds a certain limit. For example, the widely used arithmetic coder by Witten et al.[10] cannot work when the alphabet size is greater than 2^{15}. The improved version of arithmetic coder by Moffat et al.[9] extends the alphabet size to 2^{30} by using low-precision arithmetic, at the expense of compression performance. Another problem associated with the traditional adaptive arithmetic coding is its high coding complexity, which grows linearly with respect to the source alphabet size. On the other hand, although some existing coding schemes can process integer sequences with infinite alphabets, they are not universal in the sense that, for most memoryless sources, their compression rates are strictly above the entropy rates of these sources. For example, Golomb codes [5] are designed for encoding geometric sources with parameter $p = 2^{-l}$ where l is a positive integer. Gallager-Voohis codes[3] are the generalization of Golomb codes to general geometric sources with any parameter p. Both Golomb codes and Gallager-Voohis codes are optimal only in the Huffman coding sense, i.e., each symbol in the input sequence must be assigned a codeword of an integral number of bits long. Their compression rates are usually strictly above the actual entropy rates. Elias codes[2] and their variants[1] can encode any distributed integer sequences, but they can not achieve the entropy rates of these sources either.

In this study, we propose a new practical coding method, called multilevel arithmetic coding (MAC), to encode data sequences with large or even unbounded alphabets. The basic structure of MAC is shown in Figure 2. For any

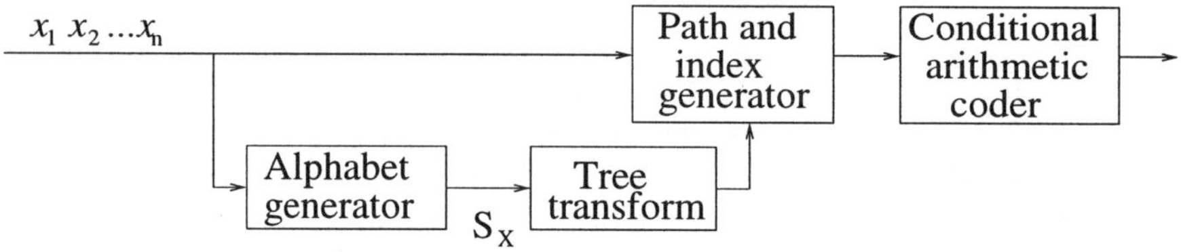

Figure 2 Basic structure of Multilevel Arithmetic Coding

data sequence $X = x_1x_2 \cdots x_n$ to be compressed, let

S_X denote the set that consists of all the distinct symbols appearing in X. In general, as X gets longer and longer, S_X may grow without bound (For some applications, however, no matter how long X is, S_X is always a subset of some fixed finite source alphabet, and hence bounded.). This new method converts the dynamically changing set S_X into a dynamic tree, whose leaves represent small subsets of S_X and, together, form a partition of S_X. For each symbol x_i in the sequence X, let y_i denote the path in the tree from the root to the leaf containing the symbol x_i. Let z_i denote the index of x_i in the corresponding leaf sub-alphabet. The sequence X is then fully represented by the sequences $Y = y_1y_2 \cdots y_n$ and $Z = z_1z_2 \cdots z_n$. From information theory, we have

$$H(X) = H(Y, Z) = H(Y) + H(Z|Y), \tag{1}$$

where $H(X)$, $H(Y, Z)$, and $H(Y)$ are the empirical entropy of the input sequence X, the path and index sequence (Y, Z), and the path sequence Y, respectively, and where $H(Z|Y)$ is the empirical conditional entropy of the index sequence given the path sequence. The above equation implies that to encode X, one may instead encode Y first and then conditionally encode Z given Y. In the proposed algorithm, we take one step further — each symbol in Y is also conditionally encoded given its parent node in the dynamic tree. The resulting coding scheme is indeed a multilevel coding scheme.

Depending on how S_X grows, in the following sections, we distinguish between three different cases and describe our proposed algorithm accordingly. In Section 2, we consider the case that the alphabet is bounded and known to the decoder in advance. In Section 1, we consider the case that the alphabet is unbounded, but both the encoder and decoder know how it grows. In Section 5, we consider the general case that the alphabet may be unbounded and unknown to the decoder.

BOUNDED ALPHABET KNOWN TO THE DECODER

In some situations, the source alphabet is a bounded set of symbols defined before coding, thus known to the decoder. This is the case that will be considered in this section.

Algorithm Description

Let $S = \{1, \cdots, M\}$, $M < \infty$, be a finite source alphabet from which a data sequence $X = x_1 x_2 \cdots x_n$ comes. Assume that S is known to the decoder. Note that $S_X \subset S$. In this case, our proposed algorithm converts S into a binary search tree in advance. It includes the following three steps.

Step 1: Partition the alphabet. Before encoding, the alphabet is first partitioned into sub-alphabets of a smaller, predefined size using a binary search tree (BST) structure, as illustrated in Example 1. A BST is a binary tree with the property that for any two nodes u and v in the tree, the label of u is strictly less(greater, resp.) than the label of v if v is in the right(left, resp.) sub-tree of u. Actually the partition of the alphabet in this case can be performed using a normal binary tree, instead of using BST. We use a BST here because we want to keep our notation consistent throughout the paper.

Example 1: Suppose that the alphabet S consists of 16 symbols numbered from 1 to 16, that is, $S = \{1, \cdots, 16\}$. If each leaf sub-alphabet is defined to contain 4 symbols, the tree structure shown in Figure 3 can be obtained. In Figure 3, the four leaves represent four sub-alphabets S_{X1}, S_{X2}, S_{X3}, and S_{X4}. The $0'$s and $1'$s shown at each branch represent bits in the corresponding path to specify a leaf sub-alphabet.

Step 2: Encode the path. For each symbol in the input sequence, its path in the tree from the root to the corresponding leaf can be represented by a binary sequence $B = b_1 b_2 \cdots b_l$, where l is the number of levels in the tree. Let $\{node_i\}_{i=1}^{l}$ be the sequence of nodes the path traverses. Then the path is encoded by using the conditional arithmetic coding to encode each b_i in B given its parent node $node_i$ in the tree.

Example 1 continued: Suppose the current symbol to be encoded is "9", which is located in S_{X3}. The path for this symbol is specified by 1 0. The first bit, $b_1 = 1$, is encoded conditioning on Node 2. The second bit, $b_2 = 0$, is encoded conditioning on Node 3.

Formally, we represent each branch in the tree by a pair (u, b), where u denotes the node the branch emanates from, and b denotes the label of the branch and takes values in $A = \{0, 1\}$. For each branch (u, b), we associate a count $c(u, b)$ with it. Initially, the count $c(u, b)$ is set to 1. We then encode the path $b_1 b_2 \cdots b_l$ as follows: (1) conditionally encode each b_i given its parent $node_i$ by using the probability $c(node_i, b_i) / \sum_{b \in A} c(node_i, b)$; and (2) increase $c(node_i, b_i)$ by 1.

Step 3: Encode the index. For each symbol in the input sequence, find its index in the leaf sub-alphabet specified by the path. Conditionally encode the index given the path by using the traditional zero order arithmetic coding algorithm which operates on the leaf sub-alphabet.

Example 1 continued: For the symbol "9", its index in the leaf sub-alphabet S_{X3} is 1. Encode this index conditioning on the path, i.e., encode it based on the leaf sub-alphabet S_{X3}. The method to encode the index is the same as that to encode the path, except that now an occurrence count is associated with each leaf sub-alphabet, instead of each node.

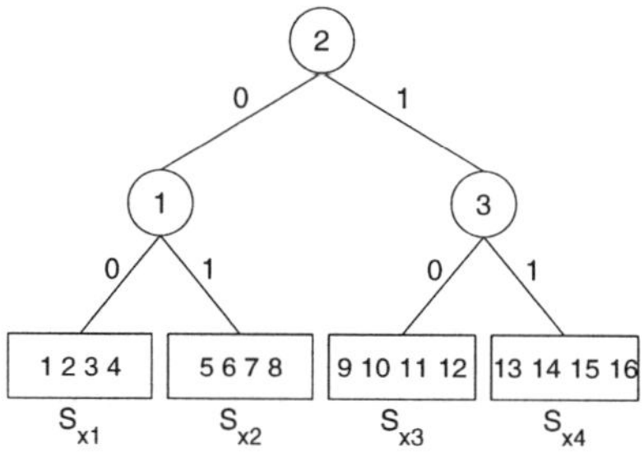

Figure 3 Partition of S into 4 sub-alphabets using a BST

To illustrate the proposed algorithm, let us now look at an example of how to encode a sequence.

Example 2: Suppose that the alphabet and its partition are the same as those in *Example 1* (see Figure 3).

The input sequence is $X = 15\ 2\ 4\ 1\ 16\ 6\ 15\ 3\ 9\ 3$. For the first symbol, "15", its path in the tree is 1 1, and its index in the corresponding leaf is 3. The first bit in the path, 1, is encoded based on Node 2, using the probability $\frac{1}{2}$ (the number of occurrences of 1 divided by the total number of occurrences of 0 and 1 at Node 2 before seeing this symbol. Note that the occurrence counts for 0 and 1 at each node in the tree are initialized to 1 at the beginning of the encoding). The occurrence count of 1 at Node 2 then increases to 2 (the initial one plus the one just occurred). The second bit is encoded based on Node 3, which is specified by the first bit in the path, using the probability $\frac{1}{2}$ (the number of occurrences of 1 divided by the total number of occurrences of 0 and 1 at Node 3 before seeing this symbol). The occurrence count of 1 at Node 3 then increases to 2 (the initial one plus the one just occurred). The index of the symbol "15" is encoded based on S_{X4}, using a probability of $\frac{1}{4}$ (the number of occurrence of index 3 divided by the total number of occurrences of all the symbols in S_{X4} before seeing this symbol). The occurrence count of this index then increases to 2 (the initial one plus the one just occurred). Thus the probability used to encode symbol "15" at this point is $\frac{1}{2} \cdot \frac{1}{2} \cdot \frac{1}{4}$. Similarly, we find the probability to encode the whole sequence is $P_X = \frac{1}{2} \cdot \frac{1}{2} \cdot \frac{1}{4} \cdot \frac{1}{3} \cdot \frac{1}{2} \cdot \frac{1}{4} \cdot \frac{2}{4} \cdot \frac{2}{3} \cdot \frac{1}{5} \cdot \frac{3}{5} \cdot \frac{3}{4} \cdot \frac{1}{6} \cdot \frac{2}{6} \cdot \frac{2}{3} \cdot \frac{1}{5} \cdot \frac{4}{7} \cdot \frac{1}{5} \cdot \frac{1}{4} \cdot \frac{3}{8} \cdot \frac{3}{4} \cdot \frac{2}{6} \cdot \frac{5}{9} \cdot \frac{4}{6} \cdot \frac{1}{7} \cdot \frac{4}{10} \cdot \frac{1}{5} \cdot \frac{1}{4} \cdot \frac{6}{11} \cdot \frac{5}{7} \cdot \frac{2}{8}$.

Algorithm Analysis

Let us first consider a simple case in which $l = 1$. The alphabet S is partitioned into two sub-alphabets S_{X_1} and S_{X_2}, where $S_{X_1} = \{1, \cdots, a\}$, $S_{X_2} = \{a + 1, \cdots, M\}$, and $1 < a < M$. For an input sequence $X = x_1 x_2 \cdots x_n$, let n_1 and n_2 denote the total number of occurrences of symbols from S_{X_1} and S_{X_2} in X, respectively, where $n_1 > 0$, $n_2 > 0$, and $n_1 + n_2 = n$. The following theorem gives an upper bound on the compression rate of the proposed algorithm relative to the traditional arithmetic coding algorithm.

Theorem 1. *For an input sequence $X = x_1 x_2 \cdots x_n$, let R_1 and R_2 denote the compression rates in bits per symbol of the traditional arithmetic coding algorithm and the proposed algorithm using two sub-alphabets as described above, respectively. Then we have*

$$R_2 - R_1 \leq \frac{1}{n} \log(M - 1), \tag{2}$$

where M is the size of the original alphabet S_X, and log *stands for the logarithm relative to base* 2.

Proof: For each symbol x_i in the alphabet S, let f_i denote the number of its occurrences in the input sequence. The zero-order traditional arithmetic coding is based on the following probability

$$P_1 = (M - 1)! \prod_{i=1}^{M} f_i! \; / \; (n + M - 1)!. \tag{3}$$

Assume that the exact arithmetic is used. Then the corresponding compression rate in bits per symbol for the input sequence is

$$R_1 = \frac{1}{n} \log \frac{1}{P_1} = \frac{1}{n} \log \binom{n + M - 1}{M - 1} + \frac{1}{n} \log \frac{n!}{\prod_{i=1}^{M} f_i!}. \tag{4}$$

For the proposed algorithm, the path needs to be encoded first. In this simple case, the path is represented by one bit d_i — d_i is 0 if the symbol comes from the first sub-alphabet S_{X_1}, and 1 if the symbol comes from the second sub-alphabet S_{X_2}. Then $D = d_1 d_2 \cdots d_n$ is the path sequence corresponding to the input sequence. To encode the path sequence using the zero-order arithmetic coding algorithm, the probability used in $P_{path} = n_1! n_2! \, / \, (n + 1)!$. Then, conditioning on the path, the following two probabilities are used to encode symbols from S_{X_1} and S_{X_2} separately:

$$P_{21} = \frac{(a - 1)! \prod_{i=1}^{a} f_i!}{(n_1 + a - 1)!}, \qquad P_{22} = \frac{(M - a - 1)! \prod_{i=a+1}^{M} f_i!}{(n_2 + M - a - 1)!}. \tag{5}$$

Thus the compression rate is give by

$$R_2 = \frac{1}{n}(\log \frac{1}{P_{path}} + \log \frac{1}{P_{21}} + \log \frac{1}{P_{22}})$$

$$= \frac{1}{n} \log \binom{n+1}{1} \binom{n_1 + a - 1}{a-1} \binom{n_2 + M - a - 1}{M - a - 1} + \frac{1}{n} \log \frac{n!}{\prod_{i=1}^{M} f_i!}. \quad (6)$$

Comparing (4) with (6), one can see that their first terms are different, whereas the second terms are the same. The first term in each equation is actually the overhead caused by the initial frequency counts of the corresponding algorithm. The second term is related to the empirical entropy rate of the input sequence. The difference between these two compression rates is thus given by

$$\begin{aligned} R_2 - R_1 &= \frac{1}{n} \log \frac{\binom{n+1}{1}\binom{n_1+a-1}{a-1}\binom{n_2+M-a-1}{M-a-1}}{\binom{n+M-1}{M-1}} \\ &= \frac{1}{n} \log \frac{\binom{n+1}{1}\binom{n_1+a-1}{a-1}\binom{n_2+M-a-1}{M-a-1}}{\frac{n+M-1}{M-1}\binom{n+M-2}{M-2}} \\ &\leq \frac{1}{n} \log \frac{(n+1)(M-1)}{n+M-1} \leq \frac{1}{n} \log(M-1). \end{aligned}$$

This completes the proof of Theorem 1.

Theorem 1 represents the worst case scenario. From Theorem 1, it follows that the compression rate of the proposed algorithm is asymptotically as good as that of the traditional algorithm at least, as $n \to \infty$. In practice, however, the proposed algorithm often outperforms the traditional algorithm, as shown in Section 3. In the following, we give two such practical cases.

Case 1: M is fixed. Almost all the integers in the input sequence come from the first sub-alphabet, that is, $n_2 << n_1$. More strongly, we assume that $\log n_2 << \log n_1$.

Case 2: The alphabet grows proportionally to the length of the input sequence, but most of the symbols in the input sequence come from a small part of the alphabet. That is, $M = \alpha n$ for some $\alpha > 0, a \ll M$, and $n_2 << n_1$, or more strongly, $\log n_2 << \log n_1$.

It can be easily proven by applying the Stirling approximation that in the above two cases, the proposed algorithm can give better compression rates than the traditional algorithm.

In the above discussion, only an one-level tree structure with two sub-alphabets is considered. When the tree has more than one level, each node can be treated in the same way as the simple case discussed above. Thus the arguments given above also hold.

Another important advantage of the MAC algorithm is the reduction of the computational complexity. Suppose that an input sequence has a length of n and an alphabet of size M. In the traditional arithmetic coding algorithm, for each symbol in the input sequence, one has to (1) determine the interval corresponding to this symbol, (2) re-scale the interval while outputting bits, and (3) adjust the related cumulative frequency counts. The time to encode such an input sequence is $O(nM)$. On the other hand, in the MAC algorithm described above, for each symbol in the input sequence, one has to (1) find its path and index, (2) determine the interval corresponding to each bit in the path and that to the index, (3) re-scale the interval while outputting bits, and (4) adjust the related cumulative frequency counts. Comparing with the traditional arithmetic coding algorithm, the MAC algorithm takes extra time in the first two steps. But in step 3, the MAC algorithm saves time because the time used in this step is proportional to the length of the compressed binary sequence. Also in step 4, the MAC algorithm saves a lot of computation time because one just needs to adjust the cumulative frequency counts for those related symbols in the leaf sub-alphabet, whose size is fixed and much smaller than the original alphabet size M. The overall result is the reduction of the computation complexity from $O(nM)$ to $O(n \log M)$. This will be shown in the simulation results in the next section.

Simulation Results

The proposed algorithm was tested on data sequences with the alphabet size of 4096. Table 1 lists the simulation results for one of the test files, "bib". To fit in the algorithm described above, the test file is treated as a sequence of 12-bit integers by converting three consecutive bytes into two 12-bit integers. "bib" is an ASCII file of size 111261 bytes, and thus treated by the coder as a sequence of 74174 12-bit integers. In Table 1, "Number of levels" represents the number of levels in the tree structure. The result of the traditional arithmetic coding (which corresponds to the case of "Number of levels" is 0) is also recorded in the table. The encoding time is based on a Sun Ultra 10 workstation. From

Table 1 Compression results for an integer sequence from an alphabet of $\{0, ..., 4095\}$

Number of levels	0	2	4	6	8	10
Compression rate	11.77	11.59	10.94	10.66	9.88	9.67
Encoding time	12.4	6.3	3.3	3.2	3.3	3.7

Compression rates are expressed in terms of bites per integer. Encoding time is expressed in seconds.

Table 1 one can see that as the number of levels increases, the compression rates given by the MAC algorithm described in this section get better and better, and are generally better than that afforded by the traditional method. By using an 10-level tree structure in the multilevel coding, the compression rate is improved 18% over that of the traditional arithmetic coding for the test file. One can also see that, by using the MAC algorithm to compress the test files, the encoding time can be dramatically reduced compared with that of the traditional arithmetic coding algorithm. For example, by using an 6-level tree structure in the MAC algorithm, the encoding time for "bib" is reduced more than three times over that of the traditional arithmetic coding algorithm. One may also note that after a certain number of levels, the encoding time actually increases slightly with the increase of the number of levels. This is because that the time saved by the MAC algorithm in step 3 and step 4 described before is not enough to compensate for the time used in step 1 and step 2. By using different parameters, the algorithm proposed here can give a good trade-off between compression rates and execution time.

DYNAMIC UNBOUNDED ALPHABET KNOWN TO THE DECODER

In the case discussed in Section 2, there is a prior bound on the alphabet size, so we can reserve necessary memory for the static tree structure in advance. In some applications, however, the source alphabet may increase dynamically and without bound. To facilitate our discussion in this section, we consider the case in which the source alphabet S_X increases dynamically and yet both the encoder and decoder know how the alphabet grows. In the next section, we shall consider the general case in which the decoder does not know how the source alphabet grows.

Algorithm Description

Since the source alphabet grows dynamically, it is now necessary to update the tree structure whenever a new distinct symbol (or integer) is to be encoded. This can be conveniently accomplished by using a dynamically updated BST structure. Initially the coder starts with a BST corresponding to the initial alphabet. If the initial alphabet contains no symbols at all, the coder starts with an empty tree. Suppose that at a certain stage, the BST has l levels and r_1 leaves named S_{X_1} to $S_{X_{r_1}}$ from the left side to the right side of the tree, each one being of size r_2. From the properties of binary trees, we know that (1) $r_1 \leq 2^l$ and (2) there are $r_1 - 1$ node, named Node 1 to Node $r_1 - 1$, in such a tree. The following three rules determine how the BST will be updated when a novel symbol appears in the input stream.

Rule 1: If the last and right-most leaf sub-alphabet, $S_{X_{r_1}}$, is not full when a new symbol appears, add this new symbol into it.

Example 3: Suppose $r_1 = 3, r_2 = 4$, and $l = 2$. The last and right-most leaf sub-alphabet S_{X3} has 2 symbols (not full), and all others are full at the current stage, as shown in Figure 4(a). When a new symbol, the 11th symbol comes, it will be added to S_{X3}, as shown in Figure 4(b). From the root to each leaf, the label on each branch is used to constitute the path to the leaf sub-alphabet.

Rule 2: If the last and right-most leaf sub-alphabet,$S_{X_{r_1}}$, is full when a new symbol appears, and if $r_1 < 2^l$, then insert a new node, Node r_1, and a new leaf, $S_{X_{r_1+1}}$, into the right side of the tree, so that the resulting tree is still a BST. The new symbol is then put into the leaf sub-alphabet $S_{X_{r_1+1}}$.

Example 3 continued: Suppose now the last leaf sub-alphabet, S_{X_3}, holds 4 symbols, as shown in Figure 4(c). When a new symbol, the 13th symbol comes, a new node (Node 3) and a new leaf (S_{X_4}) will be added to the tree structure, as shown in Figure 4(d).

Rule 3: If the last and right-most leaf sub-alphabet, $S_{X_{r_1}}$, is full when a new symbol appears, and if $r_1 = 2^l$, then increase the tree by one level and insert a new node, Node r_1, and a new leaf, $S_{X_{r_1+1}}$, into the right side of the tree, so that the resulting tree is still a BST. The new symbol is then put into the leaf sub-alphabet $S_{X_{r_1+1}}$.

Example 3 continued: Suppose $r_1 = 4, r_2 = 4$, and all the leaf sub-alphabets are full, as shown in Figure 4(e). When a new symbol, the 17th symbol comes, the tree will be increased by one level and a new leaf is added to the tree as well, as shown in Figure 4(f).

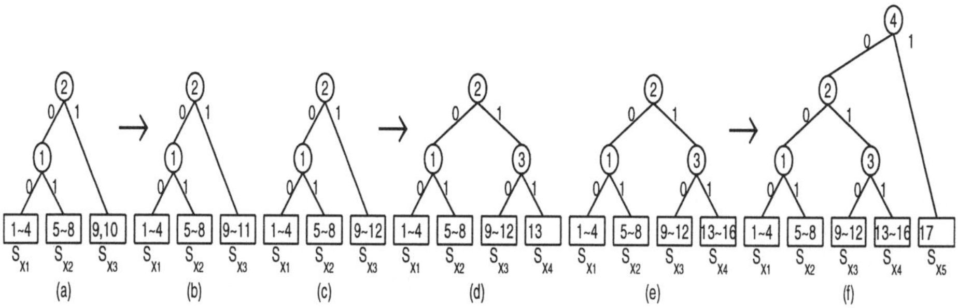

Figure 4 Dynamic BST (Rule 1: (a) and (b); Rule 2: (c) and (d); Rule 3: (e) and (f))

Because both the encoder and decoder know how the alphabet grows, they can update the tree in the same way. So at any point in the encoding/decoding process, they can use the same tree structure to encode/decode the input sequence. Note that in the actual implementation, there is no need to build or store the BST. In fact, there is a simple procedure described below to compute the path, the parent node for each bit in the path, and the index for any symbol in the input sequence.

For any symbol in a BST with l levels , at most l bits are needed to specify its path. For example, in Figure 4(f), symbols in S_{X3} need 3 bits ("0 1 0") to specify their path in the BST, while symbol "17" in S_{X5} just needs 1 bit("1"). That's because the decoder knows that there are totally 5 leaf sub-alphabets in the current BST, and if the path starts with bit "1", the symbol must come from S_{X5}, thus no more bits than the first one are needed to decode the path. Suppose that at the current stage, the BST has l levels and r_1 leaves, each one being predefined to hold at most r_2 symbols. Let $b_1 \cdots b_k, k \leq l$ denote the path for a symbol x, and let $node_i$ denote the parent node in the BST for $b_i, i = 1, \cdots, k$. The following procedure is used to determine the pair $\{node_i, b_i\}_{i=1}^{k}$.

Step 1: Calculate $c = \lfloor \frac{x-1}{r_2} \rfloor$, where $\lfloor \cdot \rfloor$ is the floor function and c is the number (starting from 0) of the leaf containing symbol x.

Step 2: Set $c_1 = c, t_1 = r_1, I_1 = 2^{\lfloor \log(r_1 - 1) \rfloor}$, and $node_1 = I_1$.

Step 3: For $i \geq 1$, compare c_i with I_i.

Step 4: If $c_i < I_i$, then $b_i = 0$. Set $I_{i+1} = \frac{I_i}{2}, node_{i+1} = node_i - I_{i+1}, t_{i+1} = I_i$ and $c_{i+1} = c_i$. Go to step 6.

Step 5: Otherwise, $b_i = 1$. Set $t_{i+1} = t_i - I_i, c_{i+1} = c_i - I_i, I_{i+1} = 2^{\lfloor \log(t_{i+1} - 1) \rfloor}$, and $node_{i+1} = node_i + I_{i+1}$.

Step 6: If $t_{i+1} = 1$, stop. The path and the corresponding node sequence are found. Otherwise, go to Step 3.

After finding the sequence of pairs $\{node_i, b_i\}_{i=1}^{k}$, the encoding of the path is accomplished by encoding each bit b_i in the path based on $node_i$, using the same procedure as that described in Section 2 The next step after encoding of the path is to encode the index of the symbol. For a symbol x, its index can be calculated by $index = x - \lfloor \frac{x-1}{r_2} \rfloor \cdot r_2$. The procedure to encode the index is the same as that described in Section 2.1. Now, let us see an example of how to encode a sequence using the proposed algorithm.

Example 4: Suppose that the initial alphabet contains 8 symbols, and the size of each leaf sub-alphabet is predefined to be 4. The initial tree is shown in Figure 5(a). Suppose the input sequence is $X = 8\ 2\ 9\ 5\ 10\ 5\ 11\ 7\ 12\ 13$. In this example we assume that both the encoder and decoder know how the alphabet grows. In other words, each time when a symbol is coded, both the encoder and the decoder know how to enlarge the current alphabet so that the next symbol to be coded is always in the enlarged alphabet (see Section 3.2 for a practical example). Thus the zero-frequency problem can be avoided effectively. Since the first two symbols, "8" and "2", are from the initial alphabet, they are encoded using the tree shown in Figure 5(a). After these two symbols are encoded, both the encoder and decoder know somehow from some mechanism that they have to enlarge the source alphabet to avoid the zero-frequency problem. That means that at this point, the encoder and decoder know that the next symbol may be new and comes from the enlarged alphabet. By enlarging the alphabet to include the symbol "9", the source alphabet becomes $S_X = \{1, 2, 3, 4, 5, 6, 7, 8, 9\}$. Accordingly, the tree structure is updated from Figure 5(a) to Figure 5(b) in terms of Rule 3. The symbol "9" is then encoded using the tree in Figure 5(b). Note that the frequency counts for the just added branches, and the just added leaf, S_{X3}, are also initialized before encoding this symbol. That is, the number of occurrences of the bit "0" at Node 2 is initialized as 1, so is the bit "1" at Node 2. The number of occurrences of symbol "9", indexed as 1 in S_{X3}, is initialized to 1 as well. Also note that to specify the leaf sub-alphabet S_{X3}, only one bit, "1", is needed in the path because both the encoder and decoder know that there are total three sub-alphabets at the current stage. Thus the probability used to encode the third symbol in the input sequence is $\frac{1}{2} \cdot \frac{1}{1}$. The next symbol in the sequence, "5", is included in the current alphabet, and thus is encoded using the tree shown in Figure 5(b). The fifth symbol, "10", is a new one, so the tree is updated from Figure 5(b) to 5(c) in terms of 1. This symbol is then encoded using the tree in Figure 5(c). Similarly, when the next two new symbols, "11" and "12", appear in the input sequence in positions 7 and 9, the tree is updated from Figure 5(c) to Figure 5(d), and from Figure 5(d) to Figure 5(e), respectively, in terms of Rule 1. When the last symbol, "13", appears, the tree is updated from Figure 5(e) to Figure 5(f), in terms of Rule 2. The product of the probabilities used to encode the input sequence is $P_X = \frac{1}{2} \cdot \frac{1}{4} \cdot \frac{1}{3} \cdot \frac{1}{4} \cdot \frac{1}{2} \cdot \frac{1}{1} \cdot \frac{1}{3} \cdot \frac{2}{4} \cdot \frac{1}{5} \cdot \frac{2}{4} \cdot \frac{1}{3} \cdot \frac{2}{5} \cdot \frac{3}{5} \cdot \frac{2}{6} \cdot \frac{3}{6} \cdot \frac{1}{5} \cdot \frac{3}{7} \cdot \frac{4}{6} \cdot \frac{1}{7} \cdot \frac{4}{8} \cdot \frac{1}{7} \cdot \frac{5}{9} \cdot \frac{1}{2} \cdot \frac{1}{1}$.

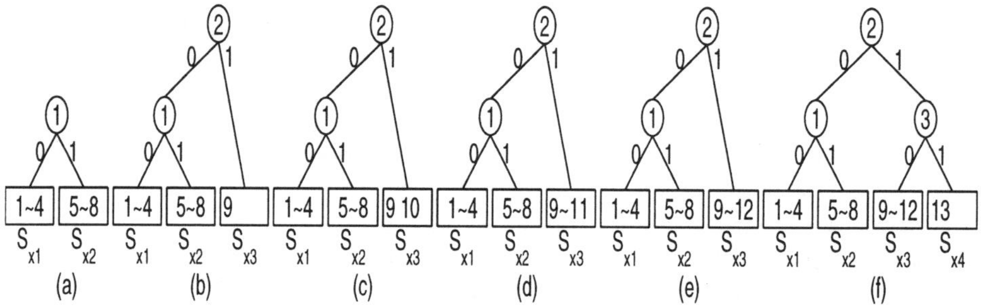

Figure 5 Encode an input sequence with a dynamic alphabet known to the decoder

Applications to the Implementation of the Multilevel Pattern Matching Algorithm

By applying the MAC algorithm to the implementation of the Multilevel Pattern Matching (MPM) algorithm[7], we develop a new data compression algorithm, called the Multilayer Multilevel arithmetic coding (MMAC) algorithm. In this coding method, an input sequence is encoded in a block-by-block manner at a number of layers, $1, 2, \cdots, k$. Each layer i corresponds to an alphabet S_i, which consists of all the "super-symbols" (a block of 2^{k-i} terminal symbols) having appeared so far in the input sequence at layer i. For any $1 \leq i \leq k-1$, S_i will be updated dynamically during the coding process, and initially, S_i consists of only the "escape" symbol, which is used to switch to next layer. The alphabet corresponding to the bottom layer, S_k, consists of all the possible terminal symbols, and will keep the same during the coding process. The algorithm has the following steps:

Step 1: Read a block of 2^{k-1} symbols, $u_1 \cdots u_{2^{k-1}}$.

Step 2: Set the initial layer number to 1: $i = 1$.

Step 3: If the super-symbol $U = u_1 \cdots u_{2^{k-i}}$ has appeared before in the layer i, encode it in this layer using the algorithm described in Section 1.

Step 4: Otherwise, (1) encode "escape" in the layer i using the algorithm described in Section 1; (2) add U into the alphabet S_i; and (3) bisect U into two equal-length parts, $U = U_1U_2$, where $U_1 = u_1 \cdots u_{2^{k-i-1}}$, and $U_2 = u_{2^{k-i-1}+1} \cdots u_{2^{k-i}}$. Increase the layer number i by 1. Feed U_1 and U_2 to Step 3 separately.

Step 5: Go to Step 1 until the end of the input sequence.

With the introduction of the "escape" symbol, when a new symbol appears in the layer $i, i = 1, \cdots, k-1$, it will be encoded in next layers first, and then added to the alphabet S_i. Thus both the encoder and decoder know how the alphabet grows for each layer, and the algorithm described in Section 1 can be applied to each layer directly.

Simulation results

The multi-layer multi-level arithmetic coding algorithm was tested on eight files from Canterbury corpus. They are: "grammar.lsp", "cp.html", "kennedy.xls", "world92.txt", "xargs.1", "sum","ptt5",and "alice29.txt". The average compression rate given by the multi-layer multi-level arithmetic coding algorithm is 3.1004 bits per letter, which is about 29% better than that of the traditional arithmetic coding algorithm (4.3432 bits per letter), and about 5% better than that of UNIX *compress* (3.2573 bits per letter), respectively. In the multilayer multilevel arithmetic coding algorithm, there are two parameters to be selected, the number of layers and the leaf size. Different choices of these two parameters will result in different compression performance. Based on our experiments, for most files, when the number of layers is between 3 to 5, and the leaf size is between 16 to 64, the multilayer multilevel arithmetic coding algorithm can give relatively good compression performance. In the simulation, these two parameter are chosen to be 5 and 16.

UNBOUNDED ALPHABET UNKNOWN TO THE DECODER

In this section, we consider the general case in which the alphabet increases without bound, and the decoder does not know how it grows.

Algorithm Description

To encode a data sequence with an unbounded alphabet unknown to the decoder, we combine Elias coding with the multilevel arithmetic coding algorithm described in Section 1 together. In the following we will take integer sequences as examples to describe the algorithm. But the algorithm works for any symbol sequence since we can always design a one-to-one mapping from symbols to integers. The proposed algorithm works as follows. At the beginning of the coding process, the initial alphabet consists of all the initial symbols plus a special symbol, "escape", which is used to signal to the decoder that the next integer to be encoded is a new distinct integer. The alphabet will be updated dynamically during the coding process, as described below. For each integer x_i in the input sequence $X = x_1 x_2 \cdots x_n$, if it has not appeared before in $x_1 \cdots x_{i-1}$, "escape" is first encoded using the multilevel arithmetic coding algorithm in Section 1. Then, the Elias code of this new symbol is encoded using the zero-order traditional arithmetic coding algorithm on a binary alphabet $\{0, 1\}$, i.e., bit by bit. The tree structure is then updated accordingly as in Section 3.1. If x_i has appeared before, the multilevel arithmetic coding algorithm in Section 3.1 is used to encode its path in the dynamic tree and its index in the corresponding leaf sub-alphabet. In the decoding end, if the decoded symbol is "escape", the decoder will switch to the zero-order traditional arithmetic coding on the binary alphabet to decode the Elias code bit stream, from which the novel symbol can be recovered. This novel symbol is then added to the alphabet and the tree structure is updated accordingly. The following example illustrates the encoding process of a sequence using the above algorithm.

Example 5: Use the same alphabet and the input sequence as those in Example 4, i.e., the initial alphabet contains 8 symbols, $\{1, 2, 3, 4, 5, 6, 7, 8\}$, and the input sequence is $X = 8\ 2\ 9\ 5\ 10\ 5\ 11\ 7\ 12\ 13$. But this time, the decoder does not know how the alphabet grows. Since we need to use an "escape" mechanism to switch from one coding state to another, we need to include the special symbol in the initial alphabet, represented by \$ in Figure 6. By convention, we put it at the first position of the alphabet, as shown in Figure 6(a). The first two symbols in the input sequence are encoded in the same way as in Example 4, except that there are 9 symbols in the initial alphabet and therefore the tree has two levels now. When the encoder encounters the third symbol, "9", the special symbol, "escape", is first encoded using the tree in Figure 6(a). Then the encoder switches to the arithmetic coding of the Elias code of "9" on the binary alphabet. The Elias code of "9" is '00100001'. To encode these eight bits, the probability used by the arithmetic coder is $P = \frac{1}{2} \cdot \frac{2}{3} \cdot \frac{1}{4} \cdot \frac{3}{5} \cdot \frac{4}{6} \cdot \frac{5}{7} \cdot \frac{6}{8} \cdot \frac{2}{9}$. Note that the initial frequency counts for bit "0" and "1" in Elias code bit stream are 1. After the encoding of the Elias code bit stream of "9", this symbol is added to the alphabet, and the tree structure is updated accordingly, as shown in Figure 6(b). Then the encoder switches back to the multilevel arithmetic coding based on the updated tree. The encoder does the similar thing when it encounters new symbols "10","11","12", and "13" for their first time in the input sequence. The tree is updated accordingly, as shown in Figure 6(c) to Figure 6(f). The probability used to encode the whole sequence is the product of the following two probabilities: $P_{Path \cdot Index} = \frac{1}{2} \cdot \frac{1}{1} \cdot \frac{1}{3} \cdot \frac{1}{2} \cdot \frac{1}{4} \cdot \frac{2}{4} \cdot \frac{2}{3} \cdot \frac{1}{5} \cdot \frac{3}{5} \cdot \frac{1}{4} \cdot \frac{1}{4} \cdot \frac{4}{6} \cdot \frac{3}{5} \cdot \frac{2}{6} \cdot \frac{5}{7} \cdot \frac{2}{6} \cdot \frac{2}{5} \cdot \frac{6}{8} \cdot \frac{4}{7} \cdot \frac{3}{7} \cdot \frac{7}{9} \cdot \frac{3}{8} \cdot \frac{1}{6} \cdot \frac{8}{10} \cdot \frac{5}{9} \cdot \frac{4}{8} \cdot \frac{9}{11} \cdot \frac{6}{10} \cdot \frac{5}{9}$, and $P_{Elias} = \frac{1}{2} \cdot \frac{2}{3} \cdot \frac{1}{4} \cdot \frac{3}{5} \cdot \frac{4}{6} \cdot \frac{5}{7} \cdot \frac{6}{8} \cdot \frac{2}{9} \cdot \frac{7}{10} \cdot \frac{8}{11} \cdot \frac{3}{12} \cdot \frac{9}{13} \cdot \frac{10}{14} \cdot \frac{11}{15} \cdot \frac{4}{16} \cdot \frac{12}{17} \cdot \frac{13}{18} \cdot \frac{14}{19} \cdot \frac{5}{20} \cdot \frac{15}{21} \cdot \frac{16}{22} \cdot \frac{17}{23} \cdot \frac{6}{24} \cdot \frac{7}{25} \cdot \frac{18}{26} \cdot \frac{19}{27} \cdot \frac{8}{28} \cdot \frac{20}{29} \cdot \frac{21}{30} \cdot \frac{9}{31} \cdot \frac{22}{32} \cdot \frac{23}{33} \cdot \frac{24}{34} \cdot \frac{25}{35} \cdot \frac{10}{36} \cdot \frac{26}{37} \cdot \frac{27}{38} \cdot \frac{11}{39} \cdot \frac{28}{40} \cdot \frac{12}{41}$, where $P_{Path \cdot Index}$ is for the the encoding of paths and indices and P_{Elias} is for the encoding of Elias codeword bit stream.

Algorithm Analysis

Considering the compression of a positive integer sequence using the algorithm described above, we have the following theorem.

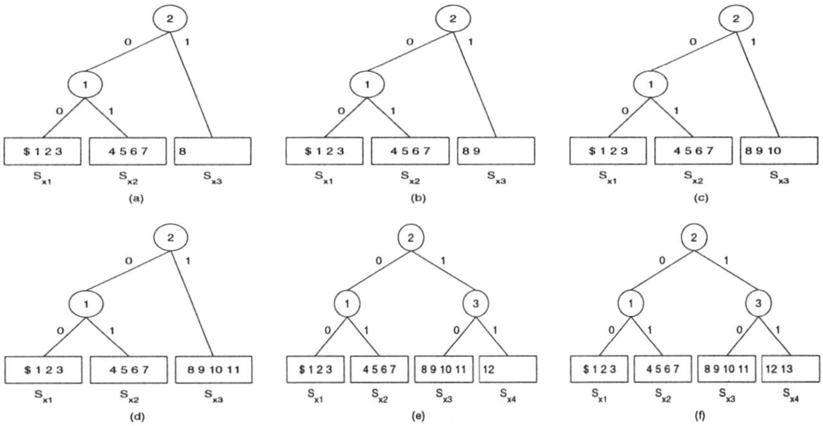

Figure 6 Encode an input sequence with dynamic alphabet unknown to the decoder

Theorem 2. *For any i.i.d. positive integer source* $X = \{x_i\}_{i=1}^{\infty}$ *with a finite mean,* $R(x_1x_2\cdots x_n) \longrightarrow H(X)$ *with probability one as* $n \to \infty$, *where* $R(x_1x_2\cdots x_n)$ *is the compression rate of the sequence* $x_1x_2\cdots x_n$ *using the algorithm described above, and* $H(X)$ *is equal to the entropy rate of* X.

Proof: For an input sequence $x_1x_2\cdots x_n$, let $S = \{y_j\}_{j=1}^{M}$ consist of all the distinct integers appearing in the input sequence. As described in Section 4.1, each symbol y_j, when it appears for the first time in the input sequence, is encoded by Elias code and then the resulting Elias code is encoded again using the zero-order traditional arithmetic coding bit by bit. For the remaining occurrences of y_j, if any, it is encoded by using the multi-level arithmetic coding algorithm described in Section 3.1. Thus, the compression rate resulting from using the proposed algorithm to compress the sequence X has two parts, one contributed from Elias coding (R_E) and the other from multi-level arithmetic coding (R_A). That is

$$R(x_1x_2\cdots x_n) = R_E(x_1x_2\cdots x_n) + R_A(x_1x_2\cdots x_n). \tag{7}$$

To prove Theorem 2, we first show that as $n \to \infty$, $R_E(x_1x_2\cdots x_n)$ goes to 0 with probability one. Let C_j denote the number of bits of Elias code for integer $y_j, j = 1, \cdots, M$, and C denote the total length of all these Elias codewords. Then $C = \sum_{i=1}^{M} C_i$. Because $C_i \leq 1 + \log y_i + 2\log(1+\log y_i)$, we have

$$\begin{aligned} \frac{C}{n} &\leq \frac{1}{n}\sum_{i=1}^{M}[1 + \log y_i + 2\log(1+\log y_i)] \\ &= \frac{M}{n} + \frac{1}{n}\log\prod_{i=1}^{M} y_i + \frac{2}{n}\log\prod_{i=1}^{M}(1+\log y_i). \end{aligned} \tag{8}$$

The first term on the right side of 8, $\frac{M}{n}$, approaches to 0 with probability one as $n \to \infty$, because $E[x_i] < \infty$ from the assumption. For the second term, we have

$$\frac{1}{n} \log \prod_{i=1}^{M} y_i = \frac{\sum_{i=1}^{M} y_i}{n} \cdot \frac{\log \prod_{i=1}^{M} y_i}{\sum_{i=1}^{M} y_i}.$$

If $\sum_{i=1}^{M} y_i$ is finite as $n \to \infty$, then $\frac{1}{n} \sum_{i=1}^{M} y_i \to 0$. Since $[\log \prod_{i=1}^{M} y_i]/\sum_{i=1}^{M} y_i \leq \log e$, $\frac{1}{n} \log \prod_{i=1}^{M} y_i \to 0$ as $n \to \infty$. Otherwise, if $\sum_{i=1}^{M} y_i \to \infty$ as $n \to \infty$, then $\frac{1}{n} \sum_{i=1}^{M} y_i \leq \frac{1}{n} \sum_{i=1}^{n} x_i \to E[X_i] < \infty$ with probability 1 from the strong law of large numbers. Also,

$$\frac{\log \prod_{i=1}^{M} y_i}{\sum_{i=1}^{M} y_i} \leq \frac{\log \left(\frac{\sum_{i=1}^{M} y_i}{M} \right)^M}{\sum_{i=1}^{M} y_i} = \frac{\log \frac{\sum_{i=1}^{M} y_i}{M}}{\frac{1}{M} \sum_{i=1}^{M} y_i} \to 0.$$

Thus, it is also true that $\frac{1}{n} \log \prod_{i=1}^{M} y_i \to 0$ as $n \to \infty$. For the third term on the right side of 8,, as $n \to \infty$, $\frac{2}{n} \log \prod_{i=1}^{M} (1 + \log y_i) \leq \frac{2}{n} \log \prod_{i=1}^{M} (1 + \frac{y_i - 1}{\ln 2}) \to 0$. Thus, from above, we have $\frac{C}{n} \to 0$ with probability one as $n \to \infty$. To encode such a Elias codeword bit stream of length C using the zero-order traditional arithmetic coding algorithm, the probability used is $P_{Elias} = f_0! \cdot f_1! \,/\, (C+1)!$, where f_0 is the total number of occurrences of "0" in the Elias code bit stream and f_1 is that of "1", and $f_0 + f_1 = C$. Thus,

$$R_E(x_1 x_2 \cdots x_n) = \frac{1}{n} \log \frac{1}{P_{Elias}} = \frac{1}{n} \log(C+1) + \frac{1}{n} \log \frac{C!}{f_0! \cdot f_1!}$$

$$\leq \frac{1}{n} \log(C+1) + \frac{C}{n}(-\frac{f_0}{C} \log \frac{f_0}{C} - \frac{f_1}{C} \log \frac{f_1}{C}) \leq \frac{1}{n} \log(C+1) + \frac{C}{n} \to 0$$

with probability one, since $\frac{C}{n} \to 0$ with probability one as $n \to \infty$.

We next show that with probability one,

$$\limsup_{n \to \infty} R_A(x_1 x_2 \cdots x_n) \leq H(X). \tag{9}$$

Let "\$" represent the "escape" symbol, and Z be the sequence obtained from the input sequence $x_1 x_2 \cdots x_n$ by replacing the first appearance of each distinct integer by "\$". In the following discussion, by using the term "a dynamic BST" we mean that the BST is constructed dynamically from the initial BST which contains only one symbol, "\$". By using the term "a static BST" we mean that the BST is constructed to contain $\{y_j\}_{j=1}^{M}$ and "\$" at the beginning of the coding process, and will remain unchanged through the coding process. Note that after the final integer x_n is encoded, the dynamic BST grows to be the same as the static BST. By using the term "a dynamic alphabet" we mean that, starting from the initial alphabet which contains only one symbol, "\$", the alphabet grows dynamically in the way described in Section 5. By using the term "a static alphabet" we mean that the alphabet contains $\{y_j\}_{j=1}^{M}$ and

"\$", and will remain unchanged through the coding process. Using these terms, we describe the following four different coding methods and their compression rates for the sequence Z:

Coding method	Description	Compression rate
A	MAC algorithm using a dynamic BST and a dynamic alphabet	$R_A(x_1x_2\cdots x_n)$
B	MAC algorithm using a static BST and a dynamic alphabet	$R_B(x_1x_2\cdots x_n)$
C	MAC algorithm using a static BST and a static alphabet	$R_C(x_1x_2\cdots x_n)$
D	traditional zero-order arithmetic coding algorithm	$R_D(x_1x_2\cdots x_n)$

The coding method A is the one used in the algorithm described in Section 4.1. We will prove the following results: (1) $R_A(x_1x_2\cdots x_n) \leq R_B(x_1x_2\cdots x_n) \leq R_C(x_1x_2\cdots x_n)$, and (2) $\limsup_{n\to\infty} R_C(x_1x_2\cdots x_n) \leq \limsup_{n\to\infty} R_D(x_1x_2\cdots x_n) \leq H(X)$ with probability one.

$\underline{R_A(x_1x_2\cdots x_n) \leq R_B(x_1x_2\cdots x_n)}$

The only difference between the coding method A and the coding method B is the encoding of the path. Suppose that for a node in the static BST in the method B, the path bit sequence encoded conditionally on this node is $T_{static} = b_1b_2\cdots b_t$. Then, the path bit sequence encoded conditionally on the corresponding node in the dynamic BST in the method A can be expressed as $T_{dynamic} = b_sb_{s+1}\cdots b_t$, where $s \geq 1$. That is, for any node in the dynamic BST, $T_{dynamic}$ is always a suffix of T_{static}. If we assume that the encoding of the path bit sequence on each node is performed backwards[1], i.e., b_t is encoded first, then b_{t-1}, and so on, we can readily see that the probability used to encode $T_{dynamic}$ is no less than that of T_{static}, i.e., $P_{A,path} \geq P_{B,path}$. This implies that, to encode the path bit sequence on each node, the method A needs no more bits than the method B. Since the encoding of the index of each symbol is the same for both coding methods, we conclude that $R_A(x_1x_2\cdots x_n) \leq R_B(x_1x_2\cdots x_n)$.

$\underline{R_B(x_1x_2\cdots x_n) \leq R_C(x_1x_2\cdots x_n)}$

The only difference between the method B and method C is the encoding of the index for each symbol in the sequence Z. Suppose that the current integer x to be encoded is located in the leaf sub-alphabet S_i. For any symbol $\alpha \in S_i$, let $f(\alpha)$ denote the number of occurrences of α in the sequence Z before the current integer x. In the coding method B, the probability used to encode the index of the current integer x is $P_{B,index} = [f(x)+1] \;/\; [\sum_{\alpha\in S_i} f(\alpha) + |S_i|_B]$, where $|S_i|_B$ is the size of the current leaf sub-alphabet S_i in the coding method B. In the coding method C, the probability used to encode the index of the current integer x is $P_{C,index} = [f(x)+1] \;/\; [\sum_{\alpha\in S_i} f(\alpha) + |S_i|_C]$, where $|S_i|_C$ is the

[1] For the traditional arithmetic coding, it is easy to see that both forward encoding and backward encoding yields the same compression rate.

size of the leaf sub-alphabet S_i in the coding method C. Since $|S_i|_B \leq |S_i|_C$, then $P_{B,index} \geq P_{C,index}$. This implies that, to encode the index of the current integer x, the coding method B needs no more bits than the coding method C. Since the encoding of the path for each integer in the sequence Z is the same for both methods, we conclude that $R_B(x_1x_2\cdots x_n) \leq R_C(x_1x_2\cdots x_n)$.

Upper bounding $R_C(x_1x_2\cdots x_n)$ in terms of $R_D(x_1x_2\cdots x_n)$

Suppose that in the static BST in the coding method C, there are l levels, and at each level $i, i = 1, 2, \cdots, l$, there are w_i nodes, numbered as $d_1^{(i)}, d_2^{(i)}, \cdots, d_{w_i}^{(i)}$. Also, let $M_{d_j^{(i)}}, i = 1, 2, \cdots, l, j = 1, 2, \cdots, w_i$, be the number of integers associated with the node $d_j^{(i)}$ in the BST, i.e, the number of distinct integers in all the leaf sub-alphabets of the sub-tree rooted at the node $d_j^{(i)}$. Starting from the bottom level and moving up one level at a time until the top one, we apply Theorem 1 to each node at each level. Then we get

$$n[R_C(x_1x_2\cdots x_n) - R_D(x_1x_2\cdots x_n)] \leq \sum_{i=1}^{l}\sum_{j=1}^{w_i} \log(M_{d_j^{(i)}} - 1)$$

$$< \sum_{i=1}^{l}\sum_{j=1}^{w_i} (M_{d_j^{(i)}} - 2)\log e < lM \log e.$$

Since $l \leq \log M$, we have $R_C(x_1x_2\cdots x_n) - R_D(x_1x_2\cdots x_n) < \frac{\log e}{n} M \log M$. From the assumption that the integer source has a finite mean, we can easily get that $\frac{1}{n} M \log M \to 0$ with probability one. Thus, we conclude that with probability one

$$\limsup_{n\to\infty} R_C(x_1x_2\cdots x_n) \leq \limsup_{n\to\infty} R_D(x_1x_2\cdots x_n).$$

$\limsup_{n\to\infty} R_D(x_1x_2\cdots x_n) \leq H(X)$

For any symbol s which is either a positive integer or the special symbol \$, let $g(s)$ denote the total number of occurrences of s in the sequence Z. It is easy to see that $g(\$) = M$ and $g(s) = 0$ for any positive integer $s \notin \{\$, y_1, y_2, \cdots, y_M\}$. Since the traditional arithmetic coding algorithm assigns the frequency 1 to each symbol $s \in \{\$, y_1, y_2, \cdots, y_M\}$ at the beginning of the encoding process, it is easy to see that

$$R_D(x_1x_2\cdots x_n) = \frac{1}{n}\log\frac{(n+M)!}{M!g(\$)!\prod_{i=1}^{M} g(y_i)!}$$

$$\leq \frac{1}{n}\log\binom{n+M}{M} + \left[-\frac{g(\$)}{n}\log\frac{g(\$)}{n} - \sum_{i=1}^{M}\frac{g(y_i)}{n}\log\frac{g(y_i)}{n}\right]$$

$$\leq \frac{n+M}{n}H\left(\frac{M}{n+M}\right) + \left[-\frac{g(\$)}{n}\log\frac{g(\$)}{n} - \sum_{i=1}^{M}\frac{g(y_i)}{n}\log\frac{g(y_i)}{n}\right] \tag{10}$$

where $H(\frac{M}{n+M})$ denotes the binary entropy function evaluated at $M/(n+M)$. Since M/n goes to 0 with probability one as $n \to \infty$, it follows from (10) that with probability one,

$$\limsup_{n\to\infty} R_D(x_1 x_2 \cdots x_n) \le \limsup_{n\to\infty} \left[-\sum_{j=1}^{\infty} \frac{g(j)}{n} \log \frac{g(j)}{n} \right]$$

$$= \limsup_{n\to\infty} \left[-\sum_{j=1}^{\infty} \frac{g(j)}{n-M} \log \frac{g(j)}{n-M} \right] \tag{11}$$

We are now led to upper bound the summation in (11). Note that $\sum_{j=1}^{\infty} \frac{g(j)}{n-M} = 1$. Consider the Elias coding of positive integers. We can upper bound the entropy of any positive random variable U by the average Elias codeword length of U. For any $J > 0$, we have

$$\begin{aligned}
&-\sum_{j=1}^{\infty} \frac{g(j)}{n-M} \log \frac{g(j)}{n-M} \\
&= -\sum_{j=1}^{J} \frac{g(j)}{n-M} \log \frac{g(j)}{n-M} - \sum_{j=J+1}^{\infty} \frac{g(j)}{n-M} \log \frac{g(j)}{n-M} \\
&\le -\sum_{j=1}^{J} \frac{g(j)}{n-M} \log \frac{g(j)}{n-M} - [\sum_{j=J+1}^{\infty} \frac{g(j)}{n-M}] \log [\sum_{j=J+1}^{\infty} \frac{g(j)}{n-M}] \\
&\quad + \sum_{j=J+1}^{\infty} \frac{g(j)}{n-M} [1 + \log j + 2\log(1+\log j)]
\end{aligned} \tag{12}$$

where the inequality is due to the fact that the Elias codeword length of j is less than $1 + \log j + 2\log(1+\log j)$. Letting $n \to \infty$ in (12) and applying the strong law of large numbers, we get that with probability one,

$$\begin{aligned}
&\limsup_{n\to\infty} -\sum_{j=1}^{\infty} \frac{g(j)}{n-M} \log \frac{g(j)}{n-M} \\
&\le -\sum_{j=1}^{J} p_j \log p_j - (1 - \sum_{j=1}^{J} p_j) \log(1 - \sum_{j=1}^{J} p_j) \\
&\quad + \sum_{j=J+1}^{\infty} p_j [1 + \log j + 2\log(1+\log j)]
\end{aligned} \tag{13}$$

where $p_j = \Pr\{x_1 = j\}$. Since the mean of x_1 is finite, letting $J \to \infty$ yields

$$\limsup_{n\to\infty} -\sum_{j=1}^{\infty} \frac{g(j)}{n-M} \log \frac{g(j)}{n-M} \le -\sum_{j=1}^{\infty} p_j \log p_j = H(X)$$

with probability one, which, together with (11), implies that

$$\limsup_{n\to\infty} R_D(x_1x_2\cdots x_n) \le H(X)$$

with probability one. This completes the proof of (9).

In the above, we have proved that $\limsup_{n\to\infty} R(x_1x_2\cdots x_n) \le H(X)$ with probability one. To complete the proof of Theorem 2, we also have to show that with probability one $\liminf_{n\to\infty} R(x_1x_2\cdots x_n) \ge H(X)$. This is guaranteed by the sample converse theorem of source coding[6]. This completes the proof of Theorem 2.

Simulation Results

Table 2 Compression rates for geometric sources and Poisson sources

	geometric sources ($p = 0.01$)				Poisson sources ($\lambda = 128$)			
Length	10^3	10^4	10^5	10^6	10^3	10^4	10^5	10^6
MAC*	9.74	8.48	8.15	8.10	9.80	7.60	6.02	5.63
Golomb	8.26	8.11	8.11	8.12	8.27	8.26	8.26	8.26
Elias	10.99	10.76	10.77	10.78	11.26	11.26	11.26	11.27
$H_{empirical}$	7.80	8.01	8.07	8.08	5.50	5.52	5.54	5.55

The proposed algorithm was tested on two types of integer sources: a geometric source with a parameter $p = 0.01$ and the Shannon entropy rate 8.0793 bits/integer, and a Poisson source with the mean of 128 and the Shannon entropy rate 5.5462 bits/integer. The lengths of the four test files for each source are 1000, 10000, 100000, and 1000000. The size r_2 of each leaf sub-alphabet was selected to be 256. The parameter r_2 and the integer sequence to be encoded then determine how the corresponding BST is updated sequentially. Tables 2 lists the simulation results. For comparison, the results given by Golomb coding and Elias coding are also included in the table. In the table, the compression rates are expressed in terms of bits per integer, and $H_{empirical}$ represents the empirical entropy of an input sequence.

One can see from Table 2 that, for a geometric source, when the file is small, Golomb coding is better than the proposed coding scheme. This is because Elias code's contribution to the compression rate in the proposed algorithm is not negligible for small files. When the file is large enough, however, we can see that the proposed algorithm outperforms Golomb codes. Also the proposed algorithm outperforms the Elias code, which encodes each integer independently, in all the tested sequences for both distributions. One can also see the clear trend in which the compression rate provided by the proposed algorithm converges to the entropy rate of the source as the length of the input sequence increases.

CONCLUSION

Motivated by overcoming the limitation of the traditional arithmetic coding algorithm which can only handle a small alphabet, we have proposed an algorithm to encode data sequences with bounded or unbounded alphabets in this paper. The algorithm has been investigated under three cases: bounded alphabets known to the decoder, unbounded alphabets known to the decoder, and unbounded alphabets unknown to the decoder. In the first case, a upper bound of the compression rate resulting from applying the multilevel arithmetic coding algorithm to any data sequence is given relative to the traditional arithmetic coding algorithm. In the second case, we apply the multilevel arithmetic coding algorithm to the implementation of MPM algorithm, and simulation results show that the compression performance is better than that given by the Unix *compress*. In the third case, we have proved that for any identically and independently distributed positive integer sequence, the proposed algorithm can asymptotically achieve the entropy rate of the source, and simulation results also demonstrate this. Besides the ability to deal with large or even unbounded source alphabets, the proposed multilevel arithmetic coding algorithm has much lower computation complexity than the traditional arithmetic coding algorithm.

The multilevel arithmetic coding algorithm can be used in many data compression systems as the entropy encoder. It is more suitable and more efficient than the traditional arithmetic coding algorithm, especially when the source alphabet is large or even unbounded. Actually it has been successfully used in grammar-based data compression systems, and very promising results have been obtained[12]. Another application of the multilevel arithmetic algorithm is in block coding for which the large product alphabet makes traditional methods infeasible in practice.

Acknowledgments

This work was supported in part by the Natural Sciences and Engineering Research Council of Canada under Grant RGPIN203035-98 and by the Communications and Information Technology Ontario, Canada.

References

[1] R. Ahlswede, T. S. Han, and K. Kobayashi, "Universal coding of integers and unbounded search trees," *IEEE Trans. Inform. Theory* 43, no.2, 1997, 669–682.

[2] P. Elias, "Universal codeword sets and representations of the integers," *IEEE Trans. Inform. Theory* 21, 1975, 194–203.

[3] R. G. Gallager and D. VanVoorhis, "Optimal Source Codes for Geometrically Distributed Integer Alphabets", *IEEE Trans. on Inform. Theory* 21, 1975, 228–230.

[4] A. Gersho and R. M. Gray, *Vector Quantization and Signal Compression*, Norwell, MA: Kluwer, 1992

[5] S. Golomb, "Run-length encodings," *IEEE Trans. Inform. Theory* 12, 1966, 399–401.

[6] J. C. Kieffer, "Sample converses in source coding theory," *IEEE Trans. Inform. Theory* 37, 1991, 263–268.

[7] J. C. Kieffer, E.-H. Yang, G. Nelson and P. Cosman, "Universal lossless compression via multilevel pattern matching", accepted pending for revisions for publication in *IEEE Trans. Inform. Theory.*

[8] J. C. Kieffer and E.-H. Yang, "Grammar based codes: A new class of universal lossless source codes," *IEEE Trans. Inform. Theory*, revised October 1998.

[9] A. Moffat, R. Neal and I.H. Witten, "Arithmetic coding revisited", *Comm. for ACM* 16, no. 3, 1998, 256–294.

[10] I.H. Witten, R. Neal and J. G. Cleary, "Arithmetic coding for data compression", *Comm. for ACM*, 30, no. 6, 1987, 520–540.

[11] E.-H. Yang and Y. Jia, "Efficient universal compression of integer sequences by using multilevel arithmetic coding", *Proc. of the Sixth Canadian Workshop on Inform. Theory 1999*, Kingston, Ontario.

[12] E.-H. Yang and J. C. Kieffer, "Efficient universal lossless data compression algorithms based on a greedy sequential grammar transform — Part one: without context models", accepted for publication in *IEEE Trans. Inform. Theory.*

METRIC ENTROPY CONDITIONS FOR KERNELS, SCHATTEN CLASSES AND EIGENVALUE PROBLEMS

Bernd Carl

Universität Jena, Fakultät für Mathematik und Informatik, D-07740 Jena, Germany

carl@minet.uni-jena.de

Abstract: In this paper we investigate the problem how the metric entropy of the image $\mathrm{Im}(K)$ of a bounded 'abstract kernel' $K : X \to E'$ mapping an arbitrary set X into the dual E' of a Banach space E reflects the rate of decay of approximation quantities of the induced operator

$$(T_K x)(s) := < x, K(s) > \quad \text{for} \quad x \in E \quad \text{and} \quad s \in X,$$

considered from E into $l_\infty(X)$. In the case of Hilbert spaces, we give sufficient and optimal conditions for the metric entropy of $\mathrm{Im}(K)$ which guarantee that the induced integral operator $T_K : L_2(Y,\nu) \to L_2(X,\mu)$, where $(X,\mu),(Y,\nu)$ are finite measure spaces, belongs to the Schatten classes $S_{q,t}$. In order to illustrate the usefulness of our results we apply them to eigenvalue problems.

INTRODUCTION

Let (X,d) be a metric space and $B(s_o;\varepsilon) := \{s \in X \mid d(s_o,s) \leq \varepsilon\}$ the closed $\varepsilon-$ball with centre s_o. For a bounded set $M \subset X$ let $N(M;\varepsilon)$ be the covering number of M by $\varepsilon-$balls of X which means:

$$N(M;\varepsilon) := \inf \left\{ N | \ \exists s_1,\ldots,s_N \in X \ \text{ such that } \ M \subset \bigcup_{k=1}^{N} B(s_k,\varepsilon) \right\} .$$

We denote the entropy numbers of M by

$$\varepsilon_n(M) := \inf\{\varepsilon \geq 0 \mid N(M;\varepsilon) \leq n\} .$$

It will be convenient to couch the arguments in terms of entropy numbers. For a (bounded linear) operator $T : E \to F$ between Banach spaces E and F the

I. Althöfer et al. (eds.), Numbers, Information and Complexity, 443-451.

nth Gelfand number and the nth Weyl number are defined by

$$c_n(T) := \inf\{||T|_L|| \mid L \subset E \ , \ \text{codim } (L) < n\}$$

and

$$x_n(T) := \sup\{c_n(TA) \mid ||A : l_2 \to E|| \leq 1\} \ ,$$

respectively.

Let us remark that for operators acting between Hilbert spaces these numbers coincide with the well-known singular numbers (cf. [3]). For a bounded function $K : X \to E'$ from an arbitrary set X into the dual E' of a Banach space E we define an operator by

$$(T_K x)(s) := < x, K(s) > \quad \text{for} \quad x \in E \quad \text{and} \quad s \in X \ ,$$

which maps the Banach space E into the Banach space $l_\infty(X)$ of all bounded number families $(\xi_t)_{t\in X}$ with the norm

$$||(\xi_t)|| = \sup_{t\in X} |\xi_t| \ .$$

Occasionally we use this notation also for the space of bounded measurable functions with respect to a measure μ. This situation is insofar universal as the Gelfand and Weyl numbers of a compact operator $S : E \to F$ between Banach spaces E and F are always shared by the Gelfand and Weyl numbers, respectively, of a compact operator $T : E \to C[0,1]$ with values in the space $C[0,1]$ of continuous functions over the interval $[0,1]$ in the sense that

$$c_n(S) = c_n(T) \quad \text{and} \quad x_n(S) = x_n(T) \ .$$

This fact indicates why we study Gelfand and Weyl numbers of operators with values in $l_\infty(X)$.
Moreover, if (X,μ) is a finite measure space we may, by Hölder's inequality, also consider T_K as an operator from E into $L_p(X,\mu)$, $1 \leq p < \infty$, of all $p-$integrable functions with the norm

$$||f||_p = \left(\int\limits_X |f(t)|^p d\mu(t) \right)^{1/p} \ .$$

We will see how the smoothness of the function K expressed in terms of entropy numbers $\varepsilon_n(\text{Im}(K))$ enter the estimates of Weyl numbers $x_n(T_K)$ of the operator T_K, from which we also get estimates of eigenvalues of operators T_K acting on $L_p(X,\mu)$. In particular, we show for an integral operator

$$(T_K f)(s) := \int\limits_Y K(s,t) f(t) d\nu(t) \ ,$$

generated by a Hilbert-Schmidt kernel $K \in L_2(X \times Y,\ \mu \times \nu)$ with the metric entropy condition

$$\min\{\varepsilon_n(\{K(s,\cdot) \in L_2(Y,\nu)|s \in X\})\ ;\ \varepsilon_n(\{K(\cdot,t) \in L_2(X,\mu)|t \in Y\})\} \in l_{r,t}\ ,$$

where $l_{r,t}$ stands for the well-known Lorentz sequence spaces, $0 < r < \infty$, $0 < t \leq \infty$, that T_K belongs to the Schatten classes

$$S_{q,t}\left(L_2(Y,\nu), L_2(X,\mu)\right) := \{S : L_2(Y,\nu) \to L_2(X,\mu) \mid c_n(S) \in l_{q,t}\}\ ,$$

where $\frac{1}{q} = \frac{1}{r} + \frac{1}{2}$. This result refines and generalizes the main theorem in [2].

METRIC ENTROPY CONDITIONS FOR KERNELS, SCHATTEN CLASSES AND EIGENVALUES

We start with the following theorem.

Theorem 2.1 *Let X be an arbitrary set and E a Banach space. For a bounded function $K \in l_\infty(X,E')$ from X into the dual E' of a Banach space E with precompact image* $\mathrm{Im}(K)$ *we define an operator $T_K : E \to l_\infty(X)$ from E into $l_\infty(X)$ by $(T_K x)(s) :=< x, K(s) >$ for $x \in E$ and $s \in X$. Then the Gelfand numbers $c_n(T_K)$ of T_K satisfy the estimate*

$$c_{n+1}(T_K) \leq \varepsilon_n(\mathrm{Im}(K)) \quad \textit{for} \quad n = 1,2,\dots\ .$$

Proof. Without loss of generality we assume $\mathrm{Im}(K)$ to be compact. Let

$$[t] := \{s \in X \mid K(t) = K(s)\} \ \text{ for } \ t \in X\ .$$

On the set

$$\hat{X} := \{[t] \mid t \in X\}$$

we introduce a metric by

$$d([s],[t]) := ||K(s) - K(t)||\ .$$

Hence, we have

$$\varepsilon_n(\hat{X}) = \varepsilon_n(\mathrm{Im}(K))\ ,\ n = 1,2,\dots\ .$$

Define an operator $S : E \to l_\infty(\hat{X})$ by

$$(Sx)([t]) := (T_K x)(t)$$

we get, by definition of the Gelfand numbers,

$$c_n(T_K) = c_n(S)\quad ,\ n = 1,2,\dots\ .$$

From the estimate

$$\begin{aligned} |(Sx)([s]) - (Sx)([t])| &= |(T_K x)(s) - (T_K x)(t)| = \\ |< x, K(s) - K(t) >| &\leq ||x||\ ||K(s) - K(t)|| \\ &\leq ||x||\ d([s],[t])\ , \end{aligned}$$

we infer, for the modulus of continuity $\omega(S;\delta)$,

$$\omega(S;\delta) := \sup_{||x||\leq 1} \sup \left\{ |(Sx)([s]) - (Sx)([t])| : [s],[t] \in \hat{X} \ , d([s],[t]) \leq \delta \right\} ,$$

the estimate

$$\omega(S;\delta) \leq \delta \ .$$

Using the inequality

$$c_{n+1}(S) \leq \omega(S;\varepsilon_n(\hat{X})) \ , \ n = 1,2,\dots \ ,$$

(cf. [49] , [5] and [1], p. 178) we conclude the desired estimate,

$$c_{n+1}(T_K) = c_{n+1}(S) \leq \omega(S;\varepsilon_n(\hat{X})) \leq \varepsilon_n(\hat{X}) = \varepsilon_n(\mathrm{Im}(K)) \ .$$

□

For the proof of the next theorem we need the following well-known lemma (cf. [3] p. 98, 2.7.3).

Lemma 2.2 *Let (X,μ) be a finite measure space, then for the embedding*

$$J : l_\infty(X) \longrightarrow L_p(X,\mu) \ , \ 1 \leq p < \infty \ ,$$

we obtain for the Weyl numbers the estimate

$$x_n(J) \leq \mu^{\frac{1}{p}}(X) n^{-\min\{\frac{1}{2};\frac{1}{p}\}} \quad for \quad n = 1,2,\dots \ .$$

Theorem 2.3 *Let (X,μ) be a finite measure space, E a Banach space and $K \in l_\infty(X,E')$ a bounded function from X into the dual E' of a Banach space E with precompact image* Im(K). *By*

$$(T_K x)(s) := < x, K(s) > \quad for \quad x \in E \quad and \quad s \in X \ ,$$

we define an operator which can be considered as an operator from E into $L_p(X,\mu)$ for $1 \leq p < \infty$. Then for the Weyl numbers of T_K we have

$$x_{2n}(T_K) \leq \mu^{\frac{1}{p}}(X) n^{-\min\{\frac{1}{2};\frac{1}{p}\}} \varepsilon_n(\mathrm{Im}(K)) \quad for \quad n = 1,2,\dots \ .$$

Proof. Using Theorem 2.1 and Lemma 2.2 as well as the following factorization of T_K ,

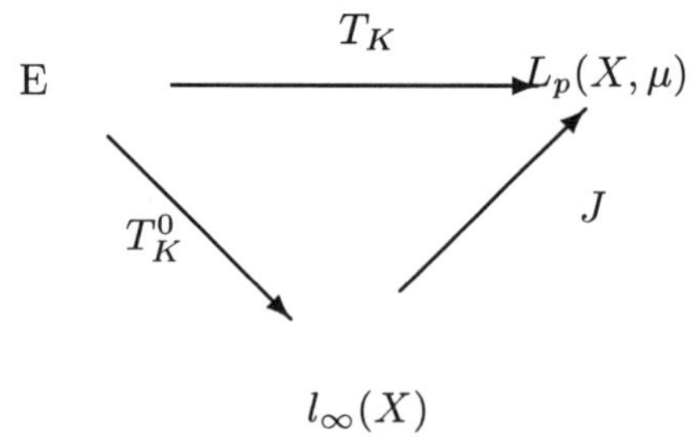

where J is the natural embedding, we get by the multiplicativity of the Weyl numbers and $x_n \le c_n$ the desired assertion,

$$\begin{aligned} x_{2n}(T_K) &\le x_n(J)x_{n+1}(T_K^\circ) \le x_n(J)c_{n+1}(T_K^\circ) \\ &\le \mu^{\frac{1}{p}}(X)n^{-\min\{\frac{1}{2};\frac{1}{p}\}}\varepsilon_n(\mathrm{Im}(K)) \,. \end{aligned}$$

□

The following corollary is a special case of Theorem 2.3.

Corollary 2.4 *Let (X,μ) , (Y,ν) be finite measure spaces and $K \in L_2(X \times Y\ ,\ \mu \times \nu)$ a Hilbert-Schmidt kernel. We put*

$$K_X := \{K(s,\cdot) \mid s \in X\} \subset L_2(Y,\nu)$$

and

$$K_Y := \{K(\cdot,t) \mid t \in Y\} \subset L_2(X,\mu) \,.$$

Then for the Gelfand numbers $c_n(T_K)$ of the induced integral operator

$$T_K : L_2(Y,\nu) \to L_2(X,\mu)\ ,\ (T_K f)(s) := \int_Y K(s,t)f(t)d\nu(t)\ ,\ s \in X\ ,$$

we have

$$c_{2n}(T_K) \le n^{-\frac{1}{2}} \min\left\{\mu^{\frac{1}{2}}(X)\varepsilon_n(K_X)\ ,\ \mu^{\frac{1}{2}}(Y)\varepsilon_n(K_Y)\right\} \quad \textit{for} \quad n = 1,2,\dots \,.$$

Proof. Indeed, the kernel K can be considered as an 'abstract kernel' $K : X \to L_2(Y,\nu)$ with $\varepsilon_n(\mathrm{Im}(K)) = \varepsilon_n(K_X)$. Applying the previous theorem, for $E = L_2(Y,\nu)$ and $p = 2$, we check the estimate

$$\begin{aligned} c_{2n}(T_K) &= x_{2n}(T_K) \le \mu^{\frac{1}{2}}(X)n^{-\frac{1}{2}}\varepsilon_n(\mathrm{Im}(K) \\ &\le \mu^{\frac{1}{2}}(X)n^{-\frac{1}{2}}\varepsilon_n(K_X) \,. \end{aligned}$$

Here we used the fact that the Gelfand and Weyl numbers for operators acting between Hilbert spaces coincide. By duality we infer the following estimate

$$c_{2n}(T_K) = c_{2n}(T_K') \le \mu^{\frac{1}{2}}(Y)n^{-\frac{1}{2}}\varepsilon_n(K_Y) \,.$$

Combining both estimates we conclude the desired assertion,

$$c_{2n}(T_K) \le n^{-\frac{1}{2}} \min\left\{\mu^{\frac{1}{2}}(X)\varepsilon_n(K_X)\ ,\ \mu^{\frac{1}{2}}(Y)\varepsilon_n(K_Y)\right\} \quad \text{for} \quad n = 1,2,\dots \,.$$

□

Using the previous corollary we obtain sufficient and sharp metric entropy conditions for integral operators to be of Schatten classes refining and generalizing the main theorem in [2].

Corollary 2.5 *Let (X,μ) , (Y,ν) be finite measure spaces and $K \in L_2(X \times Y, \mu \times \nu)$ a Hilbert-Schmidt kernel. Then the metric entropy condition of the kernel*

$$min\{\varepsilon_n(K_X), \varepsilon_n(K_Y)\} \in l_{r,t} \ , \ 0 < r < \infty, \ 0 < t \leq \infty \ ,$$

implies that the integral operator

$$(T_K f)(s) = \int_Y K(s,t) f(t) d\nu(t)$$

belongs to the Schatten classes

$$S_{q,t}(L_2(Y,\nu) \ , \ L_2(X,\mu)) \ \ for \ \ \frac{1}{q} = \frac{1}{r} + \frac{1}{2} \ .$$

In particular, if $\min\{\varepsilon_n(K_X) \ , \ \varepsilon_n(K_Y)\} \in l_{2,1}$ *, then T_K is a trace class operator.*

Remark. Since for $0 < r,t < \infty$ the entropy condition $(\varepsilon_n(X)) \in l_{r,t}$ of a metric space is equivalent to

$$\int_0^{\varepsilon_1(X)^t} N(X; \varepsilon^{\frac{1}{t}}) d\varepsilon < \infty$$

we may reformulate the previous corollary in terms of covering numbers. Indeed, if

$$\int_0^{\max\{\varepsilon_1(K_X)^t, \varepsilon_1(K_Y)^t\}} \left(\min\left\{N(K_X; \varepsilon^{\frac{1}{t}}); N(K_Y; \varepsilon^{\frac{1}{t}})\right\}\right)^{\frac{t}{r}} d\varepsilon < \infty \ ,$$

then the integral operator T_K of Corollary 2.5 belongs to the Schatten classes $S_{q,t}$ for $\frac{1}{q} = \frac{1}{r} + \frac{1}{2}$. In the case $t = \infty$ we have the following modification for the metric entropy condition in terms of covering numbers

$$\sup_{\varepsilon > 0} \varepsilon^r \min\{N(K_X; \varepsilon); N(K_Y; \varepsilon)\} \ < \infty \ .$$

Then T_K belongs to $S_{q,\infty}$ with $\frac{1}{q} = \frac{1}{r} + \frac{1}{2}$. Now we are interested in eigenvalues, all Banach spaces under consideration are assumed to be complex. If $T: E \to E$ is a compact or a power compact operator, then $(\lambda_n(T))$ denotes the sequence of all eigenvalues counted according to their algebraic multiplicities and ordered such that $|\lambda_1(T)| \geq |\lambda_2(T)| \geq \ldots \geq 0$. If T has less than n eigenvalues, then we put $\lambda_n(T) = \lambda_{n+1}(T) = \ldots = 0$. We apply Theorem 2.3 to get distributions of eigenvalues of operators satisfying smoothness properties.

Theorem 2.6 *Let (X,μ) be a finite measure space and $K : X \to L_{p'}(X;\mu)$, $1 < p < \infty$, a function with precompact image* $\mathrm{Im}(K)$. *Then by*

$$(T_K f)(s) := < f, K(s) > \quad for \quad f \in L_p(X,\mu), s \in X \ ,$$

we define an operator $T_K : L_p(X,\mu) \to L_p(X,\mu)$ acting on $L_p(X,\mu)$ and for the eigenvalues of T_K we have the estimate

$$|\lambda_{4n-1}(T_K)| \leq e^2 \mu^{\frac{1}{p}}(X) n^{-\min\{\frac{1}{2};\frac{1}{p}\}} \left(\prod_{k=0}^{n-1} \varepsilon_k(\mathrm{Im}(K)) \right)^{\frac{1}{n}}$$

for $n = 1, 2, \ldots$, where we put $\varepsilon_o(\mathrm{Im}(K)) := ||T_K||\ \mu^{-\frac{1}{p}}(X)$.
In particular,

$$(\varepsilon_n(\mathrm{Im}(K))) \in l_{r,t} \quad implies \quad (\lambda_n(T_K)) \in l_{s,t}$$

for $\frac{1}{s} = \frac{1}{r} + \min\left\{\frac{1}{2};\frac{1}{p}\right\}$, $0 < r < \infty$, $0 < t \leq \infty$ ..

Proof. For the proof we use Pietsch's inequality between eigenvalues and Weyl numbers ([3], p. 156)

$$|\lambda_{2n-1}(T_K)| \leq e \left(\prod_{k=1}^{n} x_k(T_K) \right)^{\frac{1}{n}} \quad , \ n = 1, 2, \ldots \ .$$

Because of

$$\left(\prod_{k=1}^{2n} x_k(T_K) \right)^{\frac{1}{2n}} \leq \left(\prod_{k=0}^{n-1} x_{2k}(T_K) \right)^{\frac{1}{n}} \quad ,$$

where we put $x_o(T_K) := x_1(T_K) = ||T_K||$, we obtain by the above eigenvalue inequality,

$$|\lambda_{4n-1}(T_K)| \leq e \left(\prod_{k=0}^{n-1} x_{2k}(T_K) \right)^{\frac{1}{n}} \quad , \ n = 1, 2, \ldots \ .$$

By Theorem 2.3 we have

$$x_{2k}(T_K) \leq \mu^{\frac{1}{p}}(X) k^{-\min\{\frac{1}{2};\frac{1}{p}\}} \varepsilon_k(\mathrm{Im}(K)) \ , \ k = 1, 2, \ldots \ .$$

Hence, we get for $\varepsilon_o(\mathrm{Im}(K)) := x_o(T_K)\ \mu^{-\frac{1}{p}}(X)$,

$$\begin{aligned} |\lambda_{4n-1}(T_K)| &\leq e \left(\prod_{k=0}^{n-1} x_{2k}(T_K) \right)^{\frac{1}{n}} \\ &= e\mu^{\frac{1}{p}}(X)[(n-1)!]^{-\frac{1}{n}\min\{\frac{1}{2};\frac{1}{p}\}} \left(\prod_{k=0}^{n-1} \varepsilon_k(\mathrm{Im}(K)) \right)^{\frac{1}{n}} \\ &\leq e^2 \mu^{\frac{1}{p}}(X) n^{-\min\{\frac{1}{2};\frac{1}{p}\}} \left(\prod_{k=0}^{n-1} \varepsilon_k(\mathrm{Im}(K)) \right)^{\frac{1}{n}} \quad , \end{aligned}$$

because of $[(n-1)!]^{-\frac{1}{n}\min\{\frac{1}{2};\frac{1}{p}\}} \le e\, n^{-\min\{\frac{1}{2};\frac{1}{p}\}}$. Note that this inequality is also true for $n = 1$. Moreover, by ([3], p. 157) we also have the eigenvalue inequality

$$\left(\sum_{k=1}^{n} k^{\frac{t}{r}-1}\, |\lambda_k(T_K)|^t\right)^{\frac{1}{t}} \le c_{r,t} \left(\sum_{k=1}^{n} k^{\frac{t}{r}-1} x_k^t(T_K)\right)^{\frac{1}{t}} .$$

Combining this inequality with the above estimate of Weyl numbers we obtain the remaining assertion of the theorem in terms of Lorentz sequence spaces. □

Finally, we give an application of the previous theorem.

Corollary 2.7 *Let (X,d) be a compact metric space and μ a finite measure on X. If $K : (X \times X, \mu \times \mu) \to \mathbb{C}$ is a measurable Hille-Tamarkin kernel, i. e.*

$$\int_X \left(\int_X |K(s,t)|^{p'} d\mu(t)\right)^{\frac{p}{p'}} d\mu(s) < \infty \quad for \quad 1 < p < \infty ,$$

satisfying the integral Hölder condition in the first variable

$$\left(\int_X |K(s_o,t) - K(s_1,t)|^{p'} d\mu(t)\right)^{\frac{1}{p'}} \le \rho\, d^{\alpha}(s_o,s_1) \ , \ s_o, s_1 \in X ,$$

for $0 < \alpha \le 1$, where $\rho > 0$ is a positive constant, then the eigenvalues of the induced integral operator

$$(T_K f)(s) = \int_X K(s,t) f(t) d\mu(t)) \ , \ s \in X ,$$

acting from $L_p(X,\mu)$ into $L_p(X,\mu)$ satisfy the estimate

$$|\lambda_{4n-1}(T_K)| \le C n^{-\min\{\frac{1}{2};\frac{1}{p}\}} \left(\prod_{k=0}^{n-1} \varepsilon_k^{\alpha}(X)\right)^{\frac{1}{n}} \ , \ n = 1, 2 \dots ,$$

where we put $\varepsilon_0^{\alpha}(X) := ||T_K||$. In particular, if $X = [0,1]^N$ and μ the Lebesgue measure we get

$$|\lambda_n(T_K)| \le C n^{-\min\{\frac{1}{2};\frac{1}{p}\} - \frac{\alpha}{N}} \ , \ n = 1, 2, \dots ,$$

where $C > 0$ is a positive constant not depending on n.

Proof. One can easily show, by Hölder's inequality, that T_K acts in $L_p(X,\mu)$. If we put $K(s) := K(s,\cdot)$, $s \in X$, then we may consider K as a map from X into $L_{p'}(X,\mu)$, $1 < p < \infty$. Because of the integral Hölder condition we have

$$||K(s_o) - K(s_1)||_{p'} \le \rho\, d^{\alpha}(s_o, s_1) \quad \text{for} \quad s_o, s_1 \in X$$

implying that

$$\varepsilon_n(\mathrm{Im}(K)) \le \rho\, \varepsilon_n^{\alpha}(X)\ .$$

Now, the inequality of Theorem 2.6 implies the first assertion of the Corollary. The remaining assertion follows from the just proved estimate by using

$$\varepsilon_n([0,1]^N) \asymp n^{-\frac{1}{N}}\quad ,\ n = 1, 2, \dots\ ,$$

and the monotonicity of the absolute values of eigenvalues. □

Remark. Examples exist which show that the previous results are asymptotically optimal.

References

[1] B. Carl and I. Stephani, "Entropy, Compactness and the Approximation of Operators", Cambridge University Press, 1990.

[2] J. M. Gonzales-Barrios and R. M. Dudley, "Metric entropy conditions for an operator, to be of trace class", *Proc. Amer. Math. Soc.*, 118, 1993, 175-180.

[3] A. Pietsch, *Eigenvalues and s-Numbers*, Leipzig: Geest & Portig K.-G., 1987.

[4] C. Richter, "Entropy, approximation quantities and the asymptotics of the modulus of continuity", *Math. Nachr.*, (to appear).

[5] C. Richter and I. Stephani, "Entropy and the approximation of bounded functions and operators", *Arch. Math.*, 67, 1996, 478-492.

ON SUBSHIFTS AND TOPOLOGICAL MARKOV CHAINS

Wolfgang Krieger

Mathematisches Institut, Universität Heidelberg
Im Neuenheimer Feld 288, 69120 Heidelberg, Germany

INTRODUCTION

Let Σ be a finite alphabet with its discrete topology. On the shift space $\Sigma^{\mathbb{Z}}$ one has the shift S_Σ,

$$S_\Sigma\big((x_i)_{i\in\mathbb{Z}}\big) = (x_{i+1})_{i\in\mathbb{Z}},\ (x_i)_{i\in\mathbb{Z}} \in \Sigma^{\mathbb{Z}}.$$

Subshifts are defined as the closed shift–invariant subsets of the shift spaces $\Sigma^{\mathbb{Z}}$. We recall some notions concerning subshifts, introducing notation and terminology. (An introduction to the theory of subshifts is in [10] and [14]. See also [1].) A word is admissible for a subshift $X \subset \Sigma^{\mathbb{Z}}$ if it appears somewhere on a point $x \in X$. A subshift is uniquely determined by its set of admissible words. A subshift is of finite type if it can be given by a finite set of inadmissible words. We say that a subshift of finite type is irreducible if it has a dense orbit and a dense set of periodic points. For every shift–commuting continuous map ϕ of a subshift $X \subset \Sigma^{\mathbb{Z}}$ into a shift space $\overline{\Sigma}^{\mathbb{Z}}$ there is for some $L \in \mathbb{Z}_+$ a block map Φ that assigns to every admissible word of length $2L+1$ a symbol in $\overline{\Sigma}$, and that determines ϕ by

$$\phi\big((x_i)_{i\in\mathbb{Z}}\big) = \big(\Phi\big((x_j)_{i-L\le j\le i+L}\big)\big)_{i\in\mathbb{Z}},\ (x_i)_{i\in\mathbb{Z}} \in X.$$

We say that ϕ is given by the block map Φ, and we call $[-L, L]$ a coding window. Sofic systems are the subshifts that are the images of subshifts of finite type under continuous shift–commuting maps. An admissible word w of a subshift is said to be synchronizing if for words u, v such that uw and wv are admissible for X, also uwv is admissible for X. A subshift with a dense orbit and a dense set of periodic points that has a synchronizing word we call synchronizing. Sofic systems with a dense orbit and a dense set of periodic points are synchronizing.

I. Althöfer et al. (eds.), Numbers, Information and Complexity, 453-472.

Let Δ be a state space with its discrete topology. (We place no restriction on the cardinality of Δ.) On $\Delta^{\mathbb{Z}}$ one has again the shift S_Δ,

$$S_\Delta\big((\delta_i)_{i\in\mathbb{Z}}\big) = (\delta_{i+1})_{i\in\mathbb{Z}}, \ (\delta_i)_{i\in\mathbb{Z}} \in \Delta^{\mathbb{Z}},$$

and one defines by means of a 0–1–transition matrix $\big(A(\delta,\delta')\big)_{\delta,\delta'\in\Delta}$ a topological Markov chain M_A as the S_Δ–invariant closed set

$$\bigcap_{i\in\mathbb{Z}}\big\{(\delta_i)_{i\in\mathbb{Z}} \in \Delta^{\mathbb{Z}} : A(\delta_i,\delta_{i+1}) = 1\big\}.$$

Let $\overline{\Delta}$ be another state space, and let for some $L \in \mathbb{Z}_+$, Φ be a block map

$$\Phi : \bigcap_{-L\leq j<L} \big\{(\delta_i)_{-L\leq i\leq L} \in \Delta^{[-L,L]} : A(\delta_i,\delta_{i+1}) = 1\big\} \to \overline{\Delta}.$$

Φ determines a continuous shift–commuting map ϕ of the topological Markov chain M_A into $\overline{\Delta}^{\mathbb{Z}}$ by

$$\phi\big((\delta_i)_{i\in\mathbb{Z}}\big) = \big(\Phi\big((\delta_j)_{i-L\leq j\leq i+L}\big)_{i\in\mathbb{Z}}, \ (\delta_i)_{i\in\mathbb{Z}} \in M_A.$$

We say that ϕ is given by Φ and we speak also of a coding window $[-L.L]$. We call a one–to–one shift commuting map ϕ of one topological Markov chain onto another a block conjugacy, if both, ϕ and ϕ^{-1} are given by block maps.

Subshifts of finite type are topologically conjugate to topological Markov chains with a finite state space as can be seen by recoding them using as alphabet the set of their admissible words of length N for sufficiently large N.

The topological Markov chains that we consider in this paper arise from directed graphs G that are labeled with symbols from a finite alphabet Σ. The state space of the Markov chain is here the set of labeled edges of the graph. A transition from edge δ to edge δ' is allowed if the final vertex of δ coincides with the initial vertex of δ'. The topological Markov chain that is associated in this way to the labeled directed graph G we denote by $M(G)$. The points in $M(G)$ are the bi–infinite paths on G.

Every directed graph G with labels taken from a finite alphabet Σ determines a subshift X of $\Sigma^{\mathbb{Z}}$ as the closure of the label sequences of the bi–infinite paths on G. We express this by saying that $M(G)$ projects to X, or that $M(G)$ is an extension of X, or that G determines an extension of X. Denote here the map that assigns to a bi–infinite path on G its label sequence by π_G. A labeling of a directed graph is said to be 1–right resolving if there is for every vertex E of G and every symbol σ in the alphabet at most one edge with initial vertex E and label σ. With view towards a model of a noiseless communication channel proposed by Shannon [16] directed graphs with a 1–right resolving labeling are called Shannon graphs. An edge of a Shannon graph we denote by (σ, E) where E is the initial vertex of the edge and σ is its label. To a finite alphabet Σ there is associated a Shannon graph $G(\Sigma)$ as follows: The vertices of $G(\Sigma)$

are the closed subsets of $\Sigma^{\mathbb{Z}_+}$ and there is an edge in $G(\Sigma)$ with initial vertex E and label σ if

$$\{(x_i)_{i\in\mathbb{Z}_+} \in E : x_0 = \sigma\} \neq \varnothing.$$

The final vertex of this edge is then equal to the set

$$\{(x_{i+1})_{i\in\mathbb{Z}} : x = (x_i)_{i\in\mathbb{N}} \in \Sigma^{\mathbb{N}},\ (\sigma, x) \in E\}.$$

We say that a directed graph is irreducible if it has a path from every vertex to every vertex. The subshifts to which irreducible Shannon graphs project were investigated by Blanchard and Hansel [2] who named these subshifts coded systems. These include the synchronizing subshifts. Blanchard and Hansel used the following definition of a coded system: A subshift $X \subset \Sigma^{\mathbb{Z}}$ is a coded system if there is a set $\mathcal{C}$ of finite words in the alphabet Σ such that X is the closure of the set of $x \in \Sigma^{\mathbb{Z}}$ that carry bi–infinite concatenations of words in $\mathcal{C}$. We denote the coded system that arises in this way from $\mathcal{C}$ by $X(\mathcal{C})$. $\mathcal{C}$ can here be chosen to be a prefix code. In Section 2 we give a characterization of coded systems: A subshift is a coded system if and only if it is the closure of the union of an increasing sequence of irreducible subshifts of finite type.

In Section 3 and 4 we are then concerned with canonical extensions $M(G(X))$ of subshifts X that are given by irreducible Shannon graphs $G(X)$. That an extension is canonical means that the conditions under which the extension exists are invariant under topological conjugacy, and it means that for topologically conjugate subshifts $X \subset \Sigma^{\mathbb{Z}}$ and $\overline{X} \subset \overline{\Sigma}^{\mathbb{Z}}$ with such extensions given by the irreducible Shannon graphs $G(X)$ and $G(\overline{X})$, a topological conjugacy $\phi : X \to \overline{X}$ has a unique lift to a block conjugacy

$$\hat{\phi} : M(G(X)) \to M(G(\overline{X})),$$

that is, $\hat{\phi}$ is the unique block conjugacy of $M(G(X))$ onto $M(G(\overline{X}))$ such that

$$\pi_{G(\overline{X})}\hat{\phi} = \phi\pi_{G(X)}.$$

In other words, the system that consists of the topological Markov chain $M(G(X))$ together with the mapping $\pi_{G(X)}$ is characterized by intrinsic properties, and the interest of a construction of a canonical extension lies in the fact that the invariants of the pair $(M(G(X)), \pi_{G(X)})$ are invariants of X itself.

The construction of a canonical extension that we describe in Section 3 after some preparations in Section 2 uses the forward context. (Compare here e.g. [3], [6], [8], [9], [10], [11], [12], [14], [17].) The subshifts that have this type of canonical extension are called semi–synchronizing. In Section 4 we give another construction of a canonical extension for a class of subshifts that we call a–synchronizing (for asymptotically synchronizing). In both constructions the irreducible Shannon graphs $G(X)$ that determine the extension of the subshift $X \subset \Sigma^{\mathbb{Z}}$ are subgraphs of $G(\Sigma)$ with a vertex set that contains with the initial vertex of an edge in $G(\Sigma)$ also the final vertex of the edge, and $G(X)$ retains all edges of $G(\Sigma)$ that start at one of its vertices.

For the blocks of a subshift $X \subset \Sigma^{\mathbb{Z}}$ we use the notation

$$x_{[i,k]} = (x_j)_{i \le j \le k}, x \in X,$$

and we set

$$X_{[i,k]} = \{x_{[i,k]} : x \in X\}, \ i, k \in \mathbb{Z}, i \le k,$$

with similar notation for one–sided infinite blocks. To facilitate exposition we feel free, if convenient, to identify blocks with the words they carry, without stating this explicitely. Given a block map

$$\Phi : X_{[-L,L]} \to \overline{\Sigma}, \ L \in \mathbb{Z}_+,$$

we set

$$\Phi(a) = \left(\Phi(a_{[j-L,j+L]})\right)_{i+L \le j \le k-L}, a \in X_{[i,k]}, i, k \in \mathbb{Z}, k - i \ge 2L.$$

Γ^+ denotes the set of right–infinite blocks in the forward context of a finite or left–infinite block, e.g.

$$\Gamma^+(x^-) = \{x^+ \in X_{[i,\infty)} : (x^-, x^+) \in X\}, x^- \in X_{(-\infty,i)}, i \in \mathbb{Z}.$$

We also set

$$\omega^-(b) = \left(\bigcup_{n \in \mathbb{N}} \bigcap_{x^+ \in \Gamma^+(b)} \{a \in X_{[i-n,i)} : (a, b, x^+) \in X\} \right)$$
$$\cup \left(\bigcap_{x^+ \in \Gamma^+(b)} \{x^- \in X_{(-\infty,i)} : (x^-, b, x^+) \in X\} \right), \ b \in X_{[i,k]}, i.k \in \mathbb{Z}, i \le k.$$

ω^+ has the time symmetric meaning.

CODED SYSTEMS

We characterize coded systems.

Theorem (2.1). *A subshift $X \subset \Sigma^{\mathbb{Z}}$ is coded if and only if there exists an increasing sequence $X_n \subset \Sigma^{\mathbb{Z}}$, $n \in \mathbb{N}$, of irreducible subshifts of finite type, such that*

$$X = \overline{\bigcup_{n \in \mathbb{N}} X_n}.$$

Proof. We consider first a sequence of irreducible subshifts $X_n \subset \Sigma^{\mathbb{Z}}$ of finite type, $n \in \mathbb{N}$, $X_n \subset X_{n+1}$, and show that

$$X = \overline{\bigcup_{n \in \mathbb{N}} X_n}$$

is coded. For this let $L_n \in \mathbb{N}$, $L_{n+1} \geq L_n$, be such that X_n can be defined by excluding words of length L_n. Let a be a word that appears in a periodic point of X_1 with its length $|a|$ equal to a period of the point. Let $M_n \in \mathbb{N}$, $n \in \mathbb{N}$, be such that

$$M_n|a| > L_n,$$

and set

$$a_n = a^{M_n}, \ n \in \mathbb{N}.$$

Then let $\mathcal{L}_n^o$ be a set of words, such that every word that is admissible for X_n appears as a subword of some word in $\mathcal{L}_n^o$, and such that every word in $\mathcal{L}_n^o$ starts and ends in a_n, $n \in \mathbb{N}$. We claim that

$$X = X\left(\bigcup_{n\in\mathbb{N}} \mathcal{L}_n^o\right).$$

For a proof, denote

$$\mathcal{L}_n = \bigcup_{1\leq m\leq n} \mathcal{L}_m^o,$$

and show that

$$X_n = X(\mathcal{L}_n), \ n \in \mathbb{N}.$$

For this, show by induction that every concatenation of words in $\mathcal{L}_n$ is an admissible word for X_n, $n \in \mathbb{N}$, and that such a concatenation stays admissible for X_n, if it is concatenated further on the left and on the right with words of the form a^k, $k \in \mathbb{N}$. Concerning $\mathcal{L}_1$ one has that every word in $\mathcal{L}_1$ starts and ends in a_1, that the length of a_1 exceeds L_1, and that X_1 is defined by excluding words of length L_1. From this it follows that all subwords of length L_1 of any concatenation of words in $\mathcal{L}_1$, concatenated further on the left and on the right with words of the form a^k, $k \in \mathbb{N}$, is admissible for X_1. For the induction step make similar considerations for the subwords of length L_n of the concatenations in question, $n > 1$.

Conversely, let $X \subset \Sigma^{\mathbb{Z}}$ be a coded system, that contains more than one point. Let $a, b, a \neq b$, be words in a prefix code $\mathcal{C}$ for X, $|a| \leq |b|$. We describe an increasing sequence V_N, $N \in \mathbb{N}$, of finite irreducible Shannon graphs. The irreducible Shannon graph that is obtained as the union of the V_N, $N \in \mathbb{N}$, will project to X. Let w_n, $n \in \mathbb{N}$, be an enumeration of the words that are concatenation of the words in $\mathcal{C}$, and choose $K_n \in \mathbb{N}$, $n \in \mathbb{N}$,

$$K_n < K_{n+1},$$

such that

$$K_n|b| \geq |w_n| + 4|b|, \ n \in \mathbb{N}. \tag{1}$$

Set

$$\begin{aligned} L_n &= |a| + (K_n + 1)|b|, \\ M_n &= |a| + |b| + |w_n|, \ n \in \mathbb{N}. \end{aligned}$$

V_N has vertices

$$\alpha_i, \quad 0 \leq i \leq L_N,$$

together with vertices

$$\beta_{n,j}, \quad 1 \leq j < M_n, \ 1 \leq n \leq N.$$

V_N has no multiple edges and the following transitions are possible in V_N: The transition from α_i to α_{i+1}, $0 \leq i < L_N$, and the transition from $\beta_{n,j}$ to $\beta_{n,j+1}$, $1 \leq j < M_n - 1$, $1 \leq n \leq N$, and also the transition from α_{L_n} to $\beta_{n,1}$, and from β_{n,M_n-1} to α_0, $1 \leq n \leq N$, $N \in \mathbb{N}$. The labeling of these edges is such that one goes from vertex α_0 to vertex α_{L_N} via the vertices α_i, $1 \leq i \leq L_N$, while accepting the word $b \ a \ b^{K_N}$, and from vertex α_{L_n} to vertex α_0 via the vertices $\beta_{n,j}$, $1 \leq j < M_n$ while accepting the word $b \ a \ w_n$, $1 \leq n \leq N$.

We claim that a point

$$(\sigma_i, \gamma_i)_{i \in \mathbb{Z}} \in M(V_N)$$

is uniquely determined by its label sequence $(\sigma_i)_{i \in \mathbb{Z}}$. To confirm this, it is enough to consider the case that the point $(\sigma_i, \gamma_i)_{i \in \mathbb{Z}}$ is periodic. Let then $I \in \mathbb{N}$ be maximal such that in $(\sigma_i)_{i \in \mathbb{Z}}$ there appears a word of length I that is equal to an initial segment of a word of the form b^k, $k \in \mathbb{N}$, and let $i_o \in \mathbb{Z}$ be such that the block

$$(\sigma_i)_{i_o \leq i < i_o + I}$$

carries such an initial segment. Since $\mathcal{C}$ is a prefix code, one sees, using (1), from the labeling of V_N that σ_{i_o} cannot be equal to any of the vertices $\beta_{n,j}$, $1 \leq j < M_n$, $1 \leq n \leq N$, and further, with i^o, $1 < i^o \leq |a| + 1$, minimal such that the word bab starting with the i^o–th symbol has period $|b|$, one sees that

$$\sigma_{i_o} = \alpha_{i^o - 1}. \text{ Q.e.d.}$$

An extension of a subshift $X \subset \Sigma^{\mathbb{Z}}$ can be obtained by using the forward context. For this we define a subgraph $G_{\Gamma^+}(x)$ of $G(\Sigma)$. As vertex set of $G_{\Gamma^+}(x)$ we take the set

$$\{\Gamma^+(x^-) : x^- \in X_{(-\infty,i)}, i \in \mathbb{Z}\}.$$

If the initial vertex of an edge of $G(\Sigma)$ is in this set, then so is its final vertex, and we can take as edges of $G_{\Gamma^+}(x)$ the edges of $G(\Sigma)$ that start in its vertex set.

Theorem (2.2). *Let $X \subset \Sigma^{\mathbb{Z}}$, $\overline{X} \subset \overline{\Sigma}^{\mathbb{Z}}$ be topologically conjugate subshifts. A topological conjugacy $\phi : X \to \overline{X}$ lifts uniquely to a block conjugacy*

$$\hat{\phi} : M(G_{\Gamma^+}(X)) \to M(G_{\Gamma^+}(\overline{X})).$$

One has

$$\hat{\phi}\big((x_i, \Gamma^+(x_{(-\infty,i)}))_{i \in \mathbb{Z}}\big) = \big(\overline{x}_i, \Gamma^+(\overline{x}_{(-\infty,i)})\big)_{i \in \mathbb{Z}}, x \in X, \overline{x} = \phi(x). \qquad (2)$$

Proof. We divide the proof into two parts.

1. In the first part of the proof we show the existence of the lift $\hat{\phi}$. For this let $L \in \mathbb{Z}_+$ be such that ϕ is given by a block map

$$\Phi : X_{[-L,L]} \to \overline{\Sigma}$$

and ϕ^{-1} is given by a block map

$$\overline{\Phi} : \overline{X}_{[-L,L]} \to \Sigma.$$

Define a block map $\hat{\Phi}$ by setting for a block $a \in X_{[-3L,L]}$ and a block

$$(a_i, F_i)_{-3L \le i \le L}$$

of $M\big(G_{\Gamma^+}(X)\big)$,

$$\hat{\Phi}\big((a_i, F_i)_{-3L\le i\le L}\big) = \big(\Phi(a_{[-L.L]}), \overline{F}(a, F_{-L})\big), \tag{3}$$

where

$$\overline{F}(a, F_{-L})\big) = \Phi\{y^+ \in F_{-L} : \Phi(a_{[-3L,-L]}, y^+_{[-L,L]}) = \Phi(a)\}. \tag{4}$$

By $\hat{\Phi}$ there is given a continuous shift commuting map $\hat{\phi}$ of $M\big((G_{\Gamma^+}(X)\big)$ into $M\big(G(\overline{\Sigma})\big)$ such that (2) holds. More generally, let $x \in X$,

$$(x_i, E_i)_{i\in\mathbb{Z}} \in M\big(G_{\Gamma^+}(X)\big),$$

and let

$$y^- \in X_{(-\infty,-4L)}$$

be such that

$$E_{-4L} = \Gamma^+(y^-).$$

Then set

$$\begin{aligned} x' &= (y^-, x_{[-4L,\infty)}), \\ \overline{x}' &\qquad = \phi(x'). \end{aligned}$$

Then

$$\overline{E}(x_{[-3L,L]}, E_{-L}) = \Gamma^+(\overline{x}'_{(-\infty,0)}).$$

It follows that $\hat{\phi}$ is a block homomorphism of $M\big(G_{\Gamma^+}(X)\big)$ into $M\big(G_{\Gamma^+}(\overline{X})\big)$. Define a block map $\hat{\Phi}$ by (3) and (4) interchanging X with $\overline{X}$ and Φ with $\overline{\Phi}$. Then $\hat{\Phi}$ implements a block homomorphism of $M\big(G_{\Gamma^+}(\overline{X})\big)$ onto $M\big(G_{\Gamma^+}(X)\big)$ that is the inverse of $\hat{\phi}$.

2. In the second part of the proof we show uniqueness of the lift. For this let ψ be a block automorphism of $M\big(G_{\Gamma^+}(X)\big)$ that is a lift of the identity. We

prove that ψ itself is the identity. Let $L \in \mathbb{Z}_+$ be such that $[-L, L]$ is a coding window for ψ and its inverse. Let $x \in X$, and

$$(x_i, E_i)_{i \in \mathbb{Z}} \in M\big(G_{\Gamma^+}(X)\big),$$

and set

$$\psi\big((x_i, E_i)_{i \in \mathbb{Z}}\big) = (x_i, \overline{E}_i)_{i \in \mathbb{Z}}.$$

Let $i_o \in \mathbb{Z}$. We show that

$$\overline{E}_{i_o} = E_{i_o}.$$

Let $y^- \in X_{(-\infty, i_o - L)}$ be such that

$$\Gamma^+(y^-) = E_{i_o - L}.$$

Set

$$\begin{aligned} x' &= (y^-, x_{[i_o - L, \infty)}), \\ \psi(x') &= (x'_i, \overline{E}'_i)_{i \in \mathbb{Z}}. \end{aligned}$$

Then

$$\overline{E}_{i_o} = \overline{E}'_{i_o} \subset \Gamma^+(x'_{(-\infty, i_o)}) = E_{i_o}.$$

By symmetry also

$$E_{i_o} \subset \overline{E}_{i_o}. \text{ Q.e.d.}$$

In general, for a coded system X the topological Markov chain $M\big(G_{\Gamma^+}(X)\big)$ does not have an irreducible component that projects to X. We give here an example of a coded system X such that every irreducible component of $M\big(G_{\Gamma^+}(X)\big)$ is a periodic orbit. We use the alphabet

$$\Sigma = \{0, 1, \sigma\}.$$

Let a_ℓ, $\ell \in \mathbb{N}$, be an enumeration of the words in this alphabet, $a_1 = 0$. We define inductively an increasing sequence X_n, $n \in \mathbb{N}$, of irreducible subshifts of $\Sigma^{\mathbb{Z}}$ of finite type. X_1 is the subshift where the symbol σ is excluded unless it appears in the word $0\sigma0\sigma$. Set ℓ_1 equal to 1. Let $n > 1$, and assume that X_m, $1 \le m < n$ and $\ell_m \in \mathbb{N}$, $1 \le m < n$, have been specified. Then let ℓ_n be equal to the minimal index ℓ in

$$\mathbb{N} - \{\ell_m : 1 \le \ell < n\},$$

such that the word a_ℓ is admissible for X_{n-1}, and let X_n be the subshift of $\Sigma^{\mathbb{Z}}$ where the symbol σ is excluded unless it appears in a word of the form

$$a_{\ell_m}\sigma \underbrace{0 \ldots 0}_{m} \sigma, \quad 1 \le m \le n.$$

Set

$$X = \overline{\bigcup_{n \in \mathbb{N}} X_n}.$$

The subshift $X_n, n \in \mathbb{N}$, of finite type being irreducible, X is coded by Theorem (2.1). Here for every admissible word b of X there is an $n \in \mathbb{N}$ such that

$$b = a_{\ell_n}.$$

Indeed, b is admissible for some X_k, $k \in \mathbb{N}$, and if $m \in \mathbb{N}$ is given by

$$b = a_m,$$

then one has by construction that

$$b = a_{\ell_n}$$

for some n, $1 \leq n \leq m + k$. Let now

$$x^- \in X_{(-\infty,0)}, \; E = \Gamma^+(x^-)$$

and let for some $p \in \mathbb{N}$, $a \in X_{[0,p)}$ be such that

$$E = \Gamma^+\big((x^-, a)\big). \tag{5}$$

Set

$$a^{(1)} = a, \; a^{(N)} = (a^{(N-1)}, a), \; N > 1.$$

Every $N \in \mathbb{N}$ determines the $n(N)$, where $a^{(N)}$ carries the word $a_{\ell_{n(N)}}$. From (5) one has that also

$$E = \Gamma^+\big((x^-, a^{(N)})\big), \; N > 1,$$

and it follows that E accepts the words

$$\sigma \underbrace{0 \dots 0}_{n(N)} \sigma, \quad N \in \mathbb{N},$$

and the points $(x^-, a^{(N)})$, $N \in \mathbb{N}$, are now seen to have period p.

SEMI–SYNCHRONIZATION

We want to identify a class of coded systems X such that $M\big(G_{\Gamma^+}(X)\big)$ has an irreducible component that projects to X, and that determines an extension of X that is characterized by intrinsic properties.

Consider a subshift $X \subset \Sigma^{\mathbb{Z}}$. Say that a periodic point u of X is s–presynchronizing (for semi–presynchronizing), if there is an $I \in \mathbb{N}$, such that for all $i \in \mathbb{Z}$,

$$u_{(-\infty,i-I)} \in \omega^-\big(u_{[i-I,i)}\big),$$

or, equivalently, such that for all $i \in \mathbb{Z}$,

$$\Gamma^+(u_{(-\infty,i)}) = \Gamma^+(x_{[i-I,i)}).$$

We denote the set of s–presynchronizing periodic points of X by $P_{sp}(X)$. More generally, say that an $x \in X$ is s–presynchronizing if there are $N_i \in \mathbb{N}$, $i \in \mathbb{Z}$,

$$N_{i+1} - N_i \leq 1,\ i \in \mathbb{Z},$$

$$\lim_{i\to\infty} (i - N_i) = \infty,$$

such that for all $i \in \mathbb{Z}$

$$x_{(-\infty,i-N_i)} \in \omega^-(x_{[i-N_i,i)}),$$

or, equivalently, such that for all $i \in \mathbb{Z}$

$$\Gamma^+(x_{(-\infty,i)}) = \Gamma^+(x_{[i-N_i,i)}).$$

The set of s–presynchronizing points is invariantly associated to a subshift as is seen from the following Lemma.

Lemma (3.1). *Let $X \subset \Sigma^{\mathbb{Z}}$, $\overline{X} \subset \overline{\Sigma}^{\mathbb{Z}}$ be topologically conjugate subshifts, and let $\phi : X \to \overline{X}$ be a topological conjugacy that is given by a 1–block map $\Phi : \Sigma \to \overline{\Sigma}$, with ϕ^{-1} given for some $L \in \mathbb{Z}_+$ by the block map*

$$\overline{\Phi} : \overline{X}_{[-L,L]} \to \Sigma.$$

Let $x \in X$, $\overline{x} = \phi(x)$, and let for some $i \in \mathbb{Z}$, $N \in \mathbb{N}$,

$$\overline{x}_{(i,\infty)} \in \omega^+(\overline{x}_{(i-N,i]}). \tag{1}$$

Then

$$x_{(i,\infty)} \in \omega^+(x_{[i-N-2L,i]}).$$

Proof. Let

$$y^- \in X_{(-\infty,i]},\ y^-_{[i-N-2L,i]} = x_{[i-N-2L,i]},$$

and let

$$\overline{y}^- = \Phi(y^-).$$

Then

$$\overline{y}^-_{[i-N-2L,i]} = \overline{x}_{[i-N-2L,i]}.$$

By (1)

$$(\overline{y}^-, \overline{x}_{(i,\infty)}) \in \overline{X}.$$

By construction

$$(y^-, x_{(i,\infty)}) = \phi^{-1}(\overline{y}^-, \overline{x}_{(i,\infty)}).\ \text{Q.e.d.}$$

We introduce a preorder relation $\lesssim (s)$ into the set $P_{sp}(X)$. For $u, v \in P_{sp}(X)$,

$$u \gtrsim (s) v,$$

will mean that there exists an s–presynchronizing point that is negatively asymptotic to the orbit of u and positively asymptotic to the orbit of v. Denote the equivalence relation on $P_{sp}(X)$ that results from the preorder relation $\lesssim (s)$ by $\approx (s)$, denote the set of $\approx (s)$–equivalence classes by $\Pi_{sp}(X)$, and denote the resulting order relation on $\Pi_{sp}(X)$ by $\leq (s)$. The ordered structure $(\Pi_{sp}(X), \leq (s))$ is by Lemma (3.1) invariantly associated to the subshift X. This structure can also be obtained in a slightly different way. For this call a block $b \in X_{[0,N)}$, $N \in \mathbb{N}$, s–presynchronizing if there is a point $u \in P_{sp}(X)$ such that

$$u_{[0,N)} = b, \ u_{(-\infty,0)} \in \omega^{-}(b).$$

Then introduce a preorder relation $\lesssim (s)$ into the set of s–presynchronizing blocks by writing for s–presynchronizing blocks $b \in X_{[0,N)}, N \in \mathbb{N}$, and $b' \in X_{[0,N')}, N' \in \mathbb{N}$,

$$b' \gtrsim (s) b,$$

if there exists for some $i \in \mathbb{Z}$,

$$i \leq -\max(N' - N, 0)$$

a block $a \in X_{[i,N)}$ such that the block $a_{[i,i+N')}$ carries the word of b' and such that

$$a_{[0,N)} = b, \quad a_{[i,0)} \in \omega^{-}(b).$$

Let $\approx (s)$ be the equivalence relation on the set of s–presynchronizing blocks that results from the preorder relation $\lesssim (s)$, and let $\leq (s)$ be the resulting order relation on the set of $\approx (s)$–equivalence classes of s–presynchronizing blocks. There is then a one–to–one correspondence between the $\approx (s)$–equivalence classes of s–presynchronizing periodic points and the $\approx (s)$–equivalence classes of s–presynchronizing blocks that respects the order relations $\leq (s)$. This correspondence sends the class of a point $u \in P_{sp}(X)$ such that for an $I \in \mathbb{N}$,

$$u_{(-\infty,0)} \in \omega^{-}(u_{[0,I)})$$

into the class of the s–presynchronizing block $u_{[0,I)}$, and it sends the class of an s–presynchronizing block $b \in X_{[0,N)}$, $N \in \mathbb{N}$, into the class of a point $u \in P_{sp}(X)$ such that

$$u_{[0,N)} = b, \quad u_{(-\infty,0)} \in \omega^{-}(b).$$

We denote for $u \in P_{sp}(X)$ by $G_{\Gamma^{+}}(X, u)$ the irreducible component of $G_{\Gamma^{+}}(X)$ that contains the vertices $\Gamma^{+}(u_{(-\infty,i)})$, $i \in \mathbb{Z}$, and we call these irreducible components $G_{\Gamma^{+}}(X, u)$, $u \in P_{sp}(X)$, the s–presynchronizing irreducible components of $G_{\Gamma^{+}}(X)$. The next Lemma says that we have in this way set up a one–to–one correspondence between the $\approx (s)$–equivalence classes of s–presynchronizing periodic points and the s–presynchronizing irreducible components of $G_{\Gamma^+}(X)$ that carries the order relation $\leq (s)$ into the accessability relation.

Lemma (3.2). *For $u, v \in P_{sp}(X)$ there exists a path in $G_{\Gamma^+}(X)$ that connects $G_{\Gamma^+}(X, u)$ to $G_{\Gamma^+}(X, v)$ if and only if $u \gtrsim (s)v$.*

Proof. Let $I \in \mathbb{N}$ be such that

$$\Gamma^+(v_{(-\infty,i)}) = \Gamma^+(v_{[i-I,i)}), \; i \in \mathbb{Z}.$$

If there is a path in $G_{\Gamma^+}(X)$ from $G_{\Gamma^+}(X, u)$ to $G_{\Gamma^+}(X, v)$, then there is an $x \in X$ and a point

$$(x_i, E_i)_{i \in \mathbb{Z}} \in M\big(G_{\Gamma^+}(X)\big)$$

such that for some $J \in \mathbb{N}$, $r \in \mathbb{Z}$,

$$\begin{aligned} (x_i, E_i) &= \big(u_i, \Gamma^+(u_{(-\infty,i)})\big), \; i \leq 0, \\ E_i &= \Gamma^+(v_{(-\infty,i+r)}), \; i \geq J, \\ x_i &= v_{i+r}, \; i > J. \end{aligned}$$

Then

$$\Gamma^+(x_{(-\infty,J+I]}) = \Gamma^+(x_{(J,J+I]}),$$

and the point x is seen to be s–presynchronizing. Q.e.d.

We proceed to show that the structure that is contained in the set of s–presynchronizing irreducible components of $G_{\Gamma^+}(X)$ and its order relation is respected by the lift of a topological conjucacy.

Lemma (3.3). *Let $X \subset \Sigma^{\mathbb{Z}}$, $\overline{X} \in \overline{\Sigma}^{\mathbb{Z}}$ be topologically conjugate subshifts, and let $\phi : X \to \overline{X}$ be a topological conjugacy. Let $u \in P_{sp}(X)$, $\overline{u} = \phi(u)$, and let*

$$(x_i, E_i)_{i \in \mathbb{Z}} \in M\big(G_{\Gamma^+}(X, u)\big).$$

Then

$$\hat{\phi}\big((x_i, E_i)_{i \in \mathbb{Z}}\big) \in M\big(G_{\Gamma^+}(\overline{X}, \overline{u})\big).$$

Proof. Let $L \in \mathbb{Z}_+$ be such that $[-L, L]$ is a coding window for both, ϕ and ϕ^{-1}. Let $i_o \in \mathbb{Z}$. Replace $(x_i, E_i)_{i \in \mathbb{Z}} \in M\big(G_{\Gamma^+}(x, u)\big)$ by an $(x'_i, E'_i)_{i \in \mathbb{Z}} \in M\big(G_{\Gamma^+}(X, u)\big)$ such that for some $r(-), r(+) \in \mathbb{Z}$, and some $j(-), j(+) \in \mathbb{Z}$, $j(-) < i_o - 3L$, $j(+) > i_o + L$,

$$\begin{aligned} (x'_i, E'_i) &= \big(u_{r(-)+i}, \Gamma^+\big(u_{(-\infty,r(-)+i)}\big)\big), \; i \leq j(-), \\ (x'_i, E'_i) &= (x_i, E_i), \; i_o - 3L \leq i \leq i_o + L, \\ (x'_i, E'_i) &= \big(u_{r(+)+i}, \Gamma^+(u_{(-\infty,r(+)+i)})\big), \; i \geq j(+). \end{aligned}$$

Setting

$$\hat{\phi}((x'_i, E'_i)_{i\in\mathbb{Z}}) = (\overline{x}'_i, \overline{E}'_i)_{i\in\mathbb{Z}},$$

one has then from Theorem (2.2) and (3), (4) of Section 2 that

$$\overline{E}_{i_0} = \overline{E}'_{i_0} \in G_{\Gamma^+}(\overline{X}, \overline{u}). \text{ Q.e.d.}$$

Proposition (3.4). *Let $X \subset \Sigma^{\mathbb{Z}}$, $\overline{X} \subset \overline{\Sigma}^{\mathbb{Z}}$ be topologically conjugate subshifts, and let $\phi : X \to \overline{X}$ be a topological conjugacy. Let $u, v \in P_{sp}(X)$, $u \gtrsim (s)v$, $\overline{u} = \phi(u)$, $\overline{v} = \phi(\overline{v})$. Then $\hat{\phi}$ maps a point that is negatively asymptotic to $M(G_{\Gamma^+}(X, u))$ and positively asymptotic to $M(G_{\Gamma^+}(X, v))$ into a point that is negatively asymptotic to $M(G_{\Gamma^+}(\overline{X}, \overline{u}))$ and positively asymptotic to $M(G_{\Gamma^+}(\overline{X}, \overline{v}))$.*

Proof. Use Lemma (3.3) and adapt its proof. Q.e.d.

We are interested in the situation that there is an s–presynchronizing irreducible component of $M(G_{\Gamma^+}(X))$ that projects to X, which will be the case precisely if the corresponding $\approx (s)$–equivalence class of s–presynchronizing periodic points is dense in X. We first note that there can be at most one such s–presynchronizing irreducible component of $M(G_{\Gamma_+}(X))$.

Lemma (3.5). *A dense element of $\Pi_{sp}(X)$ is a minimal element of $(\Pi_{sp}(X), \lesssim (s))$.*

Proof. Let $P \in \Pi_{sp}(X)$ be dense. Let $v \in P_{sp}(X)$, and let $N \in \mathbb{N}$ be such that

$$v_{(-\infty,0)} \in \omega^-(v_{[0,N)}).$$

Since P is dense there is an $u \in P$ such that

$$u_{[0,N)} = v_{[0,N)}.$$

The point

$$(v_{(-\infty,0)}, u_{[0,\infty)})$$

is then s–presynchronizing, and therefore $v \geq (s)u$. Q.e.d.

Subshifts X such that an s–presynchronizing irreducible component of G_{Γ^+} (X) projects to X are called semi–synchronizing. Originally semi–synchronizing subshifts were introduced as the subshifts that have semi–synchronizing blocks [13]. Here a block b is called semi–synchronizing if $\omega^-(b)$ contains a left transitive point. Synchronizing subshifts are semi–synchronizing.

Theorem (3.6). *For a subshift $X \subset \Sigma^{\mathbb{Z}}$ the following are equivalent:*

(a) *X has a semi–synchronizing block.*

(b1) *There exists a unique dense $\approx (s)$–equivalence class of s–presynchronizing periodic points.*

(b2) *There exists a dense $\approx (s)$–equivalence class of s–presynchronizing periodic points.*

(c1) *There exists a unique s–presynchronizing irreducible component of* $M\big(G_{\Gamma^+}(X)\big)$ *that projects to* X.

(c2) *There exists an s–presynchronizing irreducible component of* $M\big(G_{\Gamma^+}(X)\big)$ *that projects to* X.

Proof. Let $X \subset \Sigma^{\mathbb{Z}}$ be a subshift with a semi–synchronizing block

$$b \in X_{[0,N)}, \ \ N \in \mathbb{Z}_+,$$

and let

$$x^- \in \omega^-(b)$$

be left transitive. For an admissible word w of X let $i(w) \in \mathbb{N}$ be such that the block

$$x^-_{[-i(w),-i(w)+N)}$$

carries the same word as b, and such that the word w appears somewhere in the block $x^-_{[-i(w),0)}$. There is then a point $u^{(w)} \in X$ with period $i(w)$ such that

$$u^{(w)}_{[-i(w),0)} = x_{[-i(w),0)}.$$

The periodic points that are constructed in this way are s–presynchronizing and belong to a dense $\approx (s)$–equivalence class.

On the other hand, if there is a dense $\approx (s)$–equivalence class $\mathcal{P}$ of presynchronizing periodic points, then let $u \in \mathcal{P}$, $N \in \mathbb{N}$,

$$u_{(-\infty,0)} \in \omega^-(u_{[0,N)}),$$

and construct for the s–presynchronizing block

$$b = u_{[0,N)}$$

a left transitive point $x^- \in \omega^-(b)$. Such a construction can for instance be described as follows: Let w_k, $k \in \mathbb{N}$, be a list of the admissible words of X, such that every admissible word appears infinitely often on the list. Then let $u^{(k)} \in \mathcal{P}$, $N_k \in \mathbb{N}$, be such that

$$u^{(k)}_{(-\infty,0)} \in \omega^-(u^{(k)}_{[0,N_k)}),$$

and such that the word w_k appears somewhere in the block $u^{(k)}_{[0,N_k)}$, $k \in \mathbb{N}$. The s–presynchronizing blocks $u^{(k)}_{[0,N_k)}$, $k \in \mathbb{N}$, being $\approx (s)$–equivalent, and being $\approx (s)$–equivalent to the s–presynchronizing block $u_{[0,N)}$, there are $i_k \in \mathbb{N}$, $k \in \mathbb{N}$,

$$i_{k+1} - i_k > N_k,$$

and a point $x^- \in \omega^-(b)$ such that

$$x^-_{i_k - N_k + i} = u^{(k)}_i, \ \ 0 \le i < N_k, \tag{2}$$

and

$$x^-_{(-\infty,i_k-N_k)} \in \omega^-(x^-_{[i_k-N_k,i_k)}),\ k \in \mathbb{N}.$$

By (2) and by the choice of the $u^{(k)}$ and N_k, $n \in \mathbb{N}$, x^- is left transitive. Q.e.d.

For a semi-synchronizing subshift X, denote the minimal s-presynchronizing irreducible component of $G_{\Gamma^+}(X)$ by $G_s(X)$. By Proposition (3.4) together with Lemma (3.2) and Lemma (3.5), or alternatively, by Theorem (3.6) together with Lemma (3.1), for topological conjugate subshifts $X \subset \Sigma^{\mathbb{Z}}$, $\overline{X} \subset \overline{\Sigma}^{\mathbb{Z}}$, semi-synchronization of X implies semi-synchronization of $\overline{X}$, a block conjugacy of $M(G_s(X))$ onto $M(G_s(\overline{X}))$, that is a lift of a topological conjucacy $\phi : X \to \overline{X}$, being provided by the restriction of $\hat{\phi}$ to $M(G_s(X))$. Moreover, an adaptation of the second part of the proof of Theorem (2.2) shows that this restriction of $\hat{\phi}$ to $M(G_s(X))$ is the unique lift of ϕ to a block conjucacy of $M(G_s(X))$ onto $M(G_x(X))$. This means that we have obtained in $M(G_s(X))$ a canonical extension of the semi-synchronizing subshift X. The Dyck shifts are prototype examples of semi-synchronizing systems that are not synchronizing.

A-SYNCHRONIZATION

A class of sofic systems that appears to be to a certain extent amenable to analysis are the almost Markov sofic systems ([5], [15]). For instance, by essential use of the bi-resolving canonical extensions that almost Markov sofic systems possess, it was possible to make the first steps towards a classification theory for a certain subclass of these ([4] Section 4-6). One is therefore lead to look for uniformly bi-resolving canonical extensions of more general coded systems. However, it is known that a semi-synchronizing subshift X such that $M(G_s(X))$ is a uniformly left-resolving extension of X, is necessarily synchronizing ([7] Theorem (3.5)). We therefore want to propose here still another notion of synchronization that does yield uniformly bi-resolving canonical extensions beyond the synchronizing case.

For a subshift $X \subset \Sigma^{\mathbb{Z}}$, set

$$\Omega^+(x^-) = \overline{\bigcup_{I \in \mathbb{N}} \omega^+(x^-_{[i-I,i)})},\ x^- \in X_{(-\infty,i)},\ i \in \mathbb{Z}.$$

To obtain an extension of X we define a subgraph $G_{\Omega^+}(X)$ of $G(\Sigma)$. A vertex set of $G_{\Omega^+}(X)$ we take the set

$$\{\Omega^+(x^-) : x^- \in X_{(-\infty,i)},\ i \in \mathbb{Z}\}.$$

If the initial vertex of an edge in $G(\Sigma)$ is in this set, then so is its final vertex, and we can take as edges of $G_{\Omega^+}(X)$ the edges of $G(\Sigma)$ that start in its vertex set.

The proof of the following theorem is patterned after the first part of the proof of Theorem (2.2).

Theorem (4.1). *Let $X \subset \Sigma^{\mathbb{Z}}$, $\overline{X} \subset \overline{\Sigma}^{\mathbb{Z}}$ be topologically conjugate subshifts. A topological conjugacy $\phi : X \to \overline{X}$ lifts to a block conjugacy*

$$\hat{\phi} : M\big(G_{\Omega^+}(X)\big) \to M\big(G_{\Omega^+}(X)\big)$$

such that

$$\hat{\phi}\big((x_i, \Omega^+(x_{-\infty,i}))_{i\in\mathbb{Z}}\big) = \big(\overline{x}_i, \Omega^+(\overline{x}_{(-\infty,i)})\big), \ x \in X, \ \overline{x} = \phi(x). \qquad (1)$$

Proof. Let $L \in \mathbb{Z}_+$ be such that ϕ is given by a block map

$$\Phi : X_{[-L,L]} \to \Sigma,$$

and ϕ^{-1} is given by a block map

$$\overline{\Phi} : \overline{X}_{[-L,L]} \to \Sigma.$$

Define a block map $\hat{\Phi}$ by setting for a block $a \in X_{[-3L,L]}$ and for a block

$$(a_i, F_i)_{-3L\le i\le L}$$

of $M\big(S_{\Omega^+}(X)\big)$

$$\hat{\Phi}\big((a_i, F_i)_{-3L\le i\le L}\big) = \big(\Phi(a_{[-L,L]}), \overline{F}(a, F_{-L})\big), \qquad (2)$$

where

$$\overline{F}(a, F_{-L}) = \Phi\{y^+ \in F_{-L} : \Phi(a_{[-3L,L)}, y^+_{[-L,L]}) = \Phi(a)\}. \qquad (3)$$

By Φ there is then given a continuous shift commuting map of $M\big(G_{\Omega^+}(X)\big)$ into $M\big(G(\overline{\Sigma})\big)$ such that by Lemma (3.1) (1) holds. More generally let $x \in X$,

$$(x_i, E_i)_{i\in\mathbb{Z}} \in M\big(G_{\Omega^+}(X)\big),$$

and let

$$y^- \in X_{(-\infty,-4L)}$$

be such that

$$E_{-4L} = \Omega^+(y^-).$$

Then set

$$x' = (y^-, x_{[-4L,\infty)}),$$
$$\overline{x}' = \phi(x').$$

One has here by Lemma (3.1)

$$\overline{E}(x_{[-3L,L]}, E_{-L}) = \Omega^+(\overline{x}'_{(-\infty,0)}).$$

It follows that $\hat{\phi}$ is a block homomorphism of $M(G_{\Gamma^+}(X))$ into $M(G_{\Omega^+}(\overline{X}))$. Define a block map $\hat{\overline{\Phi}}$ by (2) and (3) interchanging X with $\overline{X}$ and Φ with $\overline{\Phi}$. Then $\hat{\overline{\Phi}}$ implements a block homomorphism of $M(G_{\Omega^+}(\overline{X}))$ onto $M(G_{\Omega^+}(X))$ that is the inverse of $\hat{\phi}$. Q.e.d.

In what follows $\hat{\phi}$ will denote the block conjugacy that is given by a block map $\hat{\Phi}$ that is defined for a topological conjugacy ϕ by (2) and (3). The material that follows now up to the mention of a–synchronization is the analogue of what is in Section 3 before Theorem (3.5). The proofs of Lemma (4.2), Lemma (4.3) and Proposition (4.4) will therefore not be written here.

Consider a subshift $X \subset \Sigma^{\mathbb{Z}}$. Say that a periodic point u of X is a–presynchronizing, if there is an $I \in \mathbb{N}$, such that for all $i \in \mathbb{Z}$,

$$\Omega^+(u_{(-\infty,i)}) = \omega^+(u_{[i-I,i)}).$$

We denote the set of a–presynchronizing periodic points of X by $P_{ap}(x)$. More generally, say that an $x \in X$ is a–presynchronizing if there are $N_i \in \mathbb{N}$, $i \in \mathbb{Z}$,

$$N_{i+1} - N_i \leq 1, \ i \in \mathbb{Z},$$

$$\lim_{i \to \infty} (i - N_i) = \infty,$$

such that for all $i \in \mathbb{Z}$

$$\Omega^+(x_{(-\infty,i)}) = \omega^+(x_{[i-N_i,i)}).$$

By Lemma (3.1) the set of a–presynchronizing points is invariantly associated to a subshift.

We introduce a preorder relation $\lesssim (a)$ into the set $P_{ap}(X)$. For $u, v \in P_{ap}(X)$,

$$u \gtrsim (a)v,$$

will mean that there exists an a–presynchronizing point that is negatively asymptotic to the orbit of u and positively asymptotic to the orbit of v. Denote the equivalence relation on $P_{ap}(X)$ that results from the preorder relation $\gtrsim (a)$ by $\approx (a)$, denote the set of $\approx$ (a)–equivalence classes by $\Pi_{ap}(X)$, and denote the resulting order relation on $\Pi_{ap}(X)$ by $\leq (a)$. The ordered structure $(\Pi_{ap}(X), \leq (a))$ is by Lemma (3.1) invariantly associated to the subshift X. This structure can also be obtained in a slightly different way. For this call a block $b \in X_{[0,N)}$, $N \in \mathbb{N}$, a–presynchronizing if there is a point $u \in P_{ap}(X)$ such that

$$u_{[0,N)} = b, \ \Omega^+(u_{(-\infty,N)}) = \omega^+(b).$$

Then introduce a preorder relation $\lesssim (a)$ into the set of a–presynchronizing blocks by writing for a–presynchromizing blocks $b \in X_{[0,N)}$, $N \in \mathbb{N}$, and $b' \in X_{[0,N')}$, $N' \in \mathbb{N}$,

$$b' \gtrsim (a)b,$$

if there exists for some $i \in \mathbb{Z}$,

$$i \leq -\max(N' - N, 0),$$

a block $a \in X_{[i,N)}$ such that the block $a_{[i,i+N')}$ carries the word of b', and such that

$$a_{[0,N)} = b, \ a_{[i+N',N)} \in \omega^+(a_{[i,i+N)}).$$

Let $\approx (a)$ be the equivalence relation on the set of a–presynchronizing blocks that results from the preorder relation $\lesssim (a)$, and let $\leq (a)$ be the resulting order relation on the set of $\approx (a)$–equivalence classes of a–presynchronizing blocks. There is then a one–to–one correspondence between the $\approx (a)$–equivalence classes of a–presynchronizing period points and $\approx (a)$–equivalence classes of a–presynchronizing blocks that respects the order relations $\leq (a)$. This correspondence sends the class of a point $u \in P_{ap}(X)$ such that for some $I \in \mathbb{N}$

$$\Omega^+(u_{-\infty,i)}) = \omega^+(u_{[i-I,i)}), \ i \in \mathbb{Z},$$

into the class of the a–presynchronizing block $u_{[0,I)}$ and it sends the class of an a–presynchronizing block $b \in X_{[0,N)}$, $N \in \mathbb{N}$, into the class of a point $u \in P_{ap}(X)$ such that

$$u_{[0,N)} = b, \ \Omega^+(u_{(-\infty,N)} = w^+(b).$$

We denote for $u \in P_{ap}(X)$ by $G_{\Omega^+}(X,u)$ the irreducible component of $G_{\Omega^+}(X)$ that contains the vertices $\Omega^+(u_{(-\infty,i)})$, $i \in \mathbb{Z}$. We call the $G_{\Omega^+}(X,u)$, $u \in P_{ap}(X)$ the a–presynchronizing components of $G_{\Omega^+}(X)$.

Lemma (4.2). *Let $u, v \in P_{ap}(x)$. Then there exists a path in $G_{\Omega^+}(X)$ that connects $G_{\Omega^+}(X,u)$ to $G_{\Omega^+}(X,v)$ if and only if $u \gtrsim (a)v$.*

Lemma (4.3). *Let $X \subset \Sigma^{\mathbb{Z}}$, $\overline{X} \in \overline{\Sigma}^{\mathbb{Z}}$ be topologically conjugate subshifts, and let $\phi : X \to \overline{X}$ be a topological conjugacy. Let $u \in P_{ap}(X)$, $\overline{u} = \phi(u)$, and let*

$$(x_i, E_i)_{i \in \mathbb{Z}} \in M\big(G_{\Omega^+}(X,u)\big).$$

Then

$$\hat{\phi}\big((x_i, E_i)_{i \in \mathbb{Z}}\big) \in M\big(G_{\Omega^+}(\overline{X},\overline{u})\big).$$

Proposition (4.4). *Let $X \subset \Sigma^{\mathbb{Z}}$, $\overline{X} \in \overline{\Sigma}^{\mathbb{Z}}$ be topologically conjugate subshifts, and let $\phi : X \to \overline{X}$ be a topologically conjugacy. Let $u, v \in P_{ap}(x)$, $u \gtrsim (a)v$, $\overline{u} = \phi(u)$, $\overline{v} = \phi(v)$. Then $\hat{\phi}$ maps a point that is negatively asymptotic to $M\big(G_{\Omega^+}(X,u)\big)$ and positively asymptotic to $M\big(G_{\Omega^+}(X,v)\big)$ into a point that is negatively asymptotic to $M\big(G_{\Omega^+}(\overline{X},\overline{u})\big)$ and positively asymptotic to $M\big(G_{\Omega^+}(\overline{X},\overline{v})\big)$.*

Imitating the semi–synchronizing case we define now a subshift X to be a–synchronizing if there is a unique dense $\approx (a)$–equivalence class in $\Pi_{as}(X)$.

For an a–synchronizing subshift X denote the unique a–presynchronizing irreducible component of $G(X)$ that corresponds to the dense $\approx$ (a)–equivalence class by $G_a(X)$. $M(G_a(X))$ projects to X. We say that an a–presynchronizing block that corresponds to the dense $\approx$ (a)–equivalence class of a–presynchronizing periodic points of an a–synchronizing subshift is an a–synchronizing block. By Lemma (4.2) and Proposition (4.4), for topologically conjugate subshifts $X \subset \Sigma^{\mathbb{Z}}$, $\overline{X} \subset \overline{\Sigma}^{\mathbb{Z}}$, a–synchronization of X implies a–synchronization of $\overline{X}$, a block conjugacy of $M(G_a(X))$ onto $M(G_a(\overline{X}))$, that is a lift of a topological conjugacy $\phi : X \to \overline{X}$, being provided by the restriction of $\hat{\phi}$ to $M(G_a(X))$. It remains to prove that this restriction is the only lift of ϕ to a block conjugacy of $M(G_a(X))$ onto $M(G_a(\overline{X}))$.

Theorem (4.5). *Let $X \subset \Sigma^{\mathbb{Z}}$, $\overline{X} \subset \overline{\Sigma}^{\mathbb{Z}}$ be topologically conjugate a–synchronizing subshifts. A topological conjugacy $\phi : X \to \overline{X}$ lifts uniquely to a block–conjugacy of $M(G_a(X))$ onto $M(G_a(\overline{X}))$.*

Proof. We prove uniqueness of the lift by showing that for an a–synchronizing subshift X the identity on $M(G_a(X))$ is the only block automorphism that the identity on X can lift to. For this let ψ be a block automorphism of $M(G_a(X))$ that is a lift of the identity on X. We show that ψ is the identity by showing that ψ is the identity on the dense set of left transitive points in $M(G_a(X))$. Let then

$$(x_i, E_i)_{i\in\mathbb{Z}} \in M(G_a(X))$$

be a left transitive point, and set

$$\psi\big((x_i, E_i)_{i\in\mathbb{Z}}\big) = (x_i, \overline{E}_i)_{i\in\mathbb{Z}}.$$

Let $i_o \in \mathbb{Z}$. We have to show that

$$E_{i_o} = \overline{E}_{i_o}. \tag{4}$$

Let $b \in X_{[0,N)}$, $N \in \mathbb{N}$, be an a–synchronizing block of X. Since $(x_i, \overline{E}_i)_{i\in\mathbb{Z}} \in M(G_a(X))$ is left transitive we have indices

$$i \le \bar{i} < i' < i_o$$

such that the blocks $x_{[i-N,i)}$, $x_{[\bar{i}-N,i)}$, $x_{[i'-N,i')}$ carry the word of b, and such that

$$E_i = \overline{E}_{\bar{i}} = E_{i'} = \omega^+(b).$$

It is

$$\overline{E}_{\bar{i}} \subset \omega^+(x_{[i-N,\bar{i})}),$$

and therefore

$$\omega^+(b) \subset \overline{E}_{i'} \subset \omega^+(x_{[i-N,i')}) = \omega^+(b),$$

and we have proved that

$$\overline{E}_{i'} = E_{i'},$$

and (4) follows. Q.e.d.

Synchronizing subshifts are a–synchronizing. The Dyck shifts are prototype examples of a–synchronizing systems X such that $M\big(G_a(X)\big)$ is a 1–left resolving extension of X.

References

[1] M.–P. Béal and D. Perrin, "Symbolic Dynamics and Finite Automata", *Handbook of Formal Languages*, G. Rozenberg and A. Salomaa, Eds., Springer 1997, Vol. 2, 463–506.

[2] F. Blanchard and G. Hansel, "Systèmes codés", *Theoretical Computer Science* 44, 1986, 17–49.

[3] M. Boyle, B. Kitchens, and B. Marcus, "A note on minimal covers for sofic systems", *Proc. Amer. Math. Soc.* 95, 1985, 403–411.

[4] M. Boyle and W. Krieger, "Automorphisms and subsystems of the shift", *J. für die reine und angewandte Mathematik* 437, 1993, 13–28.

[5] M. Boyle, B. Marcus, and P. Trow, *Resolving maps and the Dimension Group for Shifts of Finite Type*, Mem. Amer. Math. Soc. 377, 1987.

[6] I. Csiszár and J. Komlós, "On the equivalence of two models of finite–state noiseless channels from the point at view of the output", *Proceedings of the Colloquium on Information Theory*, A. Renyi and J. Bolyai, eds, Math. Soc. Budapest, 1968, 129–131.

[7] D. Fiebig and U.–R. Fiebig, "Covers for Coded Systems", *Symbolic Dynamics and Its Applications, Contemporary Mathematics* 135, ed. P. Walters, Amer. Math. Soc. 1992, 139-179.

[8] R. Fischer, "Sofic Systems and Graphs", *Monatsh. Math.* 80, 1975, 179–186.

[9] R. Fischer, "Graphs and symbolic dynamics", *Colloq. Math. Soc. János Bólyai 16, Topics in Information Theory*, 1975, 229–244.

[10] B. Kitchens, *Symbolic Dynamics*, Springer 1998.

[11] W. Krieger, "On sofic systems I", *Israel J. Math* 48, 1984, 305–330.

[12] W. Krieger, "On sofic systems II", *Israel J. Math* 60, 1987, 167–176.

[13] W. Krieger, talk given at C.I.R.M., Luminy, July 27, 1987.

[14] D. Lind and B. Marcus, *An Introduction to Symbolic Dynamics and Coding*, Cambridge University Press 1995.

[15] B. Marcus, "Sofic systems and encoding data", *IEEE -IT* 31, 1985, 366–377.

[16] C. Shannon, "A mathematical theory of communication", *Bell System Techn. J.* 27, 1948, 378–432, 623–656.

[17] B. Weiss, "Subshifts of finite type and sofic systems", *Monatsh. Math.* 77, 1973, 462–474.

LARGE DEVIATIONS PROBLEM FOR THE SHAPE OF A RANDOM YOUNG DIAGRAM WITH RESTRICTIONS

Vladimir Blinovsky

Institute for Problems of Information Transmission, RAS,
19 Bolshoi Karetnii, 101447 Moscow, Russia

Abstract: Using the original method from [4], [5] we prove the validity of the local large deviations principle for the shape of a random Young diagram with different constraints on the multiplicity of the rows of equal length.

MAIN RESULT

To the proof of the validity of the process level large deviations principle (LDP) for the trajectories of random walks is devoted a lot of recent work (see for ex. [1], [2], [3]. Let's recall some definitions. Let $\{(\Omega, \mathcal{B}, P_n)\}$ be a sequence of probability spaces, with σ- algebra $\mathcal{B}$ of Borel sets. We say that for this sequence the LDP is true if there exists a lower semicontinuous functional $N : \Omega \to [0, \infty]$, $N \not\equiv 0, \infty$, such that for some sequence $A_n \to \infty$ the following relations are true

$$-\inf_{b\in B^o} N(b) \leq \liminf_{n\to\infty} \frac{\ln P_n(B)}{A_n} \leq \limsup_{n\to\infty} \frac{\ln P_n(B)}{A_n} \leq -\inf_{b\in \bar{B}} N(b) \qquad (1)$$

where B^o is the interior and $\bar{B}$ is the closure of the Borel set B. Roughly speaking the last relations mean that

$$P_n(B) \sim e^{-A_n \inf_{b\in B} N(b)}.$$

In other words we are interested in the rough logarithmic asymptotics of P_n when $n \to \infty$.

Usually, in order to prove the relations (1) which sometimes are called the global LDP, first of all is proved the so-called local LDP, where instead of (1) is proved the validity of the equalities (where we require that Ω is a metric space

I. Althöfer et al. (eds.), Numbers, Information and Complexity, 473-488.

and $\mathcal{B}$ is the $\sigma-$ algebra of Borel sets)

$$\liminf_{\epsilon\to 0}\liminf_{n\to\infty}\frac{P_n(B_\epsilon(y))}{A_n}=\limsup_{\epsilon\to 0}\limsup_{n\to\infty}\frac{\ln P_n(B_\epsilon(y))}{A_n}=-N(y). \quad (2)$$

Here $B_\epsilon(y)=\{x\in\Omega : d(y,x)\leq\epsilon\}$ is the ball of radius ϵ.

Next we consider Young diagrams. Let's recall the definition. A Young diagram of weight n consists of consecutive columns of integer heights and widths. The heights are nonincreasing, the bases are on the same line, and the sum of their areas is equal to n. It is convenient (and we propose this) that the bases of all columns are on the line $x=0$ and the left hand side of the left most column is on the line $y=0$. Next we scale the diagram, dividing the linear sizes of the diagram by $\sqrt{n}$. So the area of the whole diagram becomes unit. We are interested in the shape κ_n of the scaled diagram, which we consider to be continous from the right. κ_n is a piecewise constant, monotonically nonincreasing function. Let $A_n=\sqrt{n}$.

In this paper we consider restricted diagrams. Restrictions consist in the following: the heights (or which is equivalent the lengths) of a Young diagram are not arbitrary but take values from some subset $\mathcal{A}\subset N$ of the natural numbers. On the set of diagrams with given restrictions u we consider the uniform distribution. The case without restriction first was investigated in [5], [6], see also [4]. In that case the total number of diagrams p_n satisfy the Hardy-Ramanujan relation, which in rough logarithmic asymptotics looks as follows:

$$\ln p_n\sim\pi\sqrt{2n/3}.$$

It follows from the well known bijection between Young diagrams and the unordered partitions of n into natural numbers that

$$p_n=\#\{i_1,\ldots,i_n : i_1+2i_2+\ldots+ni_n=n\}.$$

If we consider some restrictions u on the numbers $i_1,\ldots,i_n$, then the number p_n^u of unordered partitions of n with such restrictions depends on n and can be a rather complex function of u. For given u we consider the uniform distribution on κ_n^u, so $P_n(\kappa_n^u)=1/p_n^u$ (more precisely we do not consider p_n^u, but $p^u_{n(1\pm\delta)}=\sum_{n'\in n(1\pm\delta)}p_{n'}^u$, because some restrictions may lead to the situation that $p_n^u=0$ for some values of n,

here $n(1\pm\delta)$ is the set of integers from the range $n(1-\delta),n(1+\delta))$

Next using the original method introduced in [4], [5] for given restrictions u we will find the rough logarithmic asymptotics of the number $\Phi_n=\#\{\kappa_n^u : d(\kappa_n^u,y)<\epsilon\}$. This allows us to state the local LDP in the case of restrictions u. On this way we will obtain the explicit (in general case parametric) formula for the functional (which is called the rate function) N^u, the index u pointing out that we consider the ensemble of Young diagrams under the restrictions u.

The problem of large deviations can be stated in topological spaces. We consider here the L^1- space. Some analysis can be done to obtain the same formula for the rate function N^u in the case of pointwise convergence topology.

In that case using the so called exponential compactness of the shapes of random diagrams one can prove the global LDP. Let's note that considering the L^1- norm we can prove only the local LDP and we cannot prove the validity of the global LDP because in this case the principle of exponential compactness is not valid. The proof of these results in pointwise convergence topology is an exercise from analysis.

The extract of the original method consists of the approximation of the given function y by a piecewise linear function with small enough linear pieces. Then we estimate the number of random functions κ_n^u which are not far from this piecewise linear function. If $(x_1, y(x_1)), \ldots, (x_m, y(x_m))$ is the vertex set of this spline, then the number of shapes κ_n^u which are in $\epsilon-$ neighbourhood of the spline is estimated by the number of κ_n^u such that

$$|\kappa_n^u(x_i) - y(x_i)| < \epsilon.$$

As we will show later, this number can be estimated by the product of possible restrictions of the shapes κ_n^u:

$$\prod_i \#\{\kappa_n^u : |\kappa_n^u(x_i) - y(x_i)| < \epsilon, |\kappa_n^u(x_{i+1}) - y(x_{i+1})| < \epsilon\}.$$

The exponent of the number of κ_n^u for which for given x_i, x_{i+1} the following inequalities are true

$$|\kappa_n^u(x_i) - y(x_i)| < \epsilon, |\kappa_n^u(x_{i+1}) - y(x_{i+1})| < \epsilon,$$

can be estimated using techniques of large deviations for the sum of iid random variables. On this way we will find the number of shapes which are close to the curve $y(x)$. To find the probabilities we need to find the total number p_n^u of Young diagrams with given restrictions u. This can be done in two ways. One is at first to prove the validity of the global LDP for the topology of pointwise convergence and then to find the minimum of the rate function N^u. Then using convexity of N^u one can prove the uniquness $(a.s.)$ of the function $y^{\min}$ on which the minimum of N^u is achieved. Then

$$N^u(y^{\min}) = 0.$$

Another way is to find the asymptotics of the number of solutions of the equations

$$\sum_{j=1}^{n} j i_j = n(1 \pm \delta)$$

under the conditions u on i_j. This number (when $\delta \to 0$) is the number $p^u_{n(1\pm\delta)}$. Note the necessity of the parameter δ here. This parameter allows us to use probabilistic methods for obtaining values p_n^u and N^u. After finding the asymptotics on n we let $\delta \to 0$. This way we avoid difficulties arising from problems concerning the divisibility of n, for example, when the restrictions are such that $i_j = 0$ or a. In this case n should be divisible by a, which cannot be taken into account when using probabilistic methods.

Consider the general class $\mathcal{A}$ of restrictions u which consist of $i_j \in \mathcal{A} \subset N \bigcup 0$. Now we give more precise formulations. First of all we consider functions from $L^1([0,\infty))$ such that $y \geq 0$ $a.s.$ and $\int_0^\infty y dx = 1$. Further we consider only functions y for which there exists a function $\hat{y} \geq 0$, such that $y = \hat{y} a.s.$ and $\hat{y}$ is a monotonically nonincreasing, continuous from the right function. The class of functions with all these properties we denote by $\mathcal{C}$.

Let

$$L_1^u = \inf_{\lambda \in R^1} \left[\int_0^\infty \ln \left(\sum_{i \in \mathcal{A}} e^{i\lambda x} \right) dx - \lambda \right] \tag{3}$$

and

$$L_2^u(z) = \inf_{\lambda \in R^1} \left[\ln \left(\sum_{i \in \mathcal{A}} e^{i\lambda} \right) - z\lambda \right]. \tag{4}$$

We also consider one additional property for $\mathcal{C}$: for the corresponding function $\hat{y}$ and arbitrary $0 \leq x_1 < x_2 < \infty$ the following inequality is valid:

$$L_2^u \left(\frac{\hat{y}(x_1) - \hat{y}(x_2)}{x_2 - x_1} \right) > -\infty. \tag{5}$$

It is easy to see that from (5) and from the expression for L_2^u it follows that if $|\mathcal{A}| < \infty$, then $\hat{y}$ is a continuous function and $\hat{y}'(x) \leq \max a \in \mathcal{A}$ $a.s.$ If $|\mathcal{A}| = \infty$, then (5) is true for the arbitrary function $\hat{y}$. Moreover we exclude the degenerated case, proposing that $0 \in \mathcal{A}$.

Next we formulate our main result.

Theorem 1. *The local LDP is true for κ_n^u with $A_n = \sqrt{n}$ and the rate function N^u, satisfying the following relation*

$$N^u(y) = \begin{cases} L_1^u - \int_0^\infty L_2^u(-\hat{y}'(x))dx, & y \in \mathcal{C}; \\ \infty, & y \notin \mathcal{C}. \end{cases} \tag{6}$$

To derive the relations (2) with the rate function $N^u(y)$ from (6) we shall prove two inequalities.

The first inequality is

$$\limsup_{\epsilon \to 0} \limsup_{n \to \infty} \frac{\ln P_n(B_\epsilon(y))}{\sqrt{n}} \leq -N^u(y) \tag{7}$$

and the second one is

$$\liminf_{\epsilon \to 0} \liminf_{n \to \infty} \frac{\ln P_n(B_\epsilon(y))}{\sqrt{n}} \geq -N^u(y), \tag{8}$$

First of all we note how the expressions (3) and (4) arise. To see this, it is necessary to consider the problem of large deviations of the corresponding sums of independent random variables. L_1^u is the exponent of the number $\#_1^u$ of solutions of the equations

$$\sum_{j=1}^{n} i_j j = n(1 \pm \delta), i_j \in \mathcal{A}, \delta \to 0. \tag{9}$$

The value $L_2^u(z)$ is the exponent of the number $\#_2^u$ of solutions of the equation

$$\sum_{j=1}^{n_1} i_j = n_2(1 \pm \delta), \ \frac{n_2}{n_1} = z(1 \pm \delta), \ i_j \in \mathcal{A}, \ \delta \to 0. \tag{10}$$

Let's note that the number of solutions $\#_1^u$ is the number of unordered partitions of the number $n(1 \pm \delta)$ with multiplicities $i_j \in \mathcal{A}$ and $\#_2^u(z)$ is the number of ordered partitions of the number $n_2(1 \pm \delta)$ into n_1 numbers from $\mathcal{A}$. In this case we suppose that $n_1 z, n_2 \sim \nu\sqrt{n}$ for some $\nu > 0$. We do not reproduce here the detailed proof of the last two facts, but formulate them as statement.

Statement 2. *The following equalities are valid:*

$$\lim_{\delta \to 0} \lim_{n \to \infty} \frac{\ln \#_1^u}{\sqrt{n}} = L_1^u \tag{11}$$

$$\lim_{\delta \to 0} \lim_{n \to \infty} \frac{\ln \#_2^u(z)}{\sqrt{n}} = L_2^u(z). \tag{12}$$

Let's sketch how to prove for example the equality (12) for values $i_j \in \mathcal{A} \bigcap [0, n_2(1 + \delta)]$ with uniform distribution. The probability $p(i_j)$ is equal to $1/\#\{\mathcal{A} \bigcap [0, n_2(1 + \delta)]\}$ and next using Cramer's technique of estimating large deviations of the sum of independent random variables we find the asymptotics of the ratio

$$\frac{\ln \# \left\{ i_1, i_2, \ldots, i_{n_1} : \sum_{j=1}^{n_1} i_j = n_2(1 \pm \delta) \right\}}{\sqrt{n}}, z = \frac{n_2}{n_1}(1 \pm \delta).$$

In other words

$$\lim_{\delta \to 0} \lim_{n \to \infty} \frac{1}{\sqrt{n}} \ln \left(P \left(\sum_{j=1}^{n_1} i_j = n_2(1 \pm \delta), i_j \in \mathcal{A} \bigcap [0, n_2(1 + \delta)] \right) \right.$$
$$\left. \times \#\{\mathcal{A} \bigcap [0, n_2(1 + \delta)]\} \right) = L_2^u(z).$$

Similarly can be proved the equality

$$\lim_{\delta \to 0} \lim_{n \to \infty} \frac{\ln(P(\sum_{j=1}^{n} i_j j = n(1 \pm \delta), i_j \in \mathcal{A} \bigcap [0, n(1 + \delta)]) \#\{\mathcal{A} \bigcap [0, n(1 + \delta)]\}}{\sqrt{n}}$$
$$= L_1^u.$$

The difference in proof here is that the values i_j are taken with coefficients j, i.e. we consider sums of independent, uniformly distributed random variables i_j:

$$p(i_j) = \frac{1}{\#\{\mathcal{A}\bigcap[0, n_2(1+\delta)]\}}.$$

The integral in the expression for L_1^u appears when computing the moment generating function $M(\lambda) = Ee^{x\lambda}$, which is a part of the expression of L_1^u. We substitute the sum $\sum_{j=1}^n \ln \sum_{i\in\mathcal{A}} e^{ij\lambda}$ by the integral $\int_0^\infty \ln(\sum_{i\in\mathcal{A}} e^{i\lambda x})dx$. The validity of such transformations can easily be established by standard considerations from analysis. At last note that the functional $N^u(y)$ is convex. So it has a unique minimum $\tilde{y}$ and from global LDP it follows that $N^u(\tilde{y}) = 0$. The value $\tilde{y}$ can be found by variation of N^u. On this way one can find L_1^u when $L_2^u(z)$ is known. However it needs rather cumbersome calculations. For example it is easy to show that the expressions under inf in (3)(4) are convex and the infimum is achieved in a unique point, which is easy to find setting the first derivatives of this expressions to zero. Then we obtain the parametric form of the L_2^u:

$$L_2^u(h),$$

where $z = z(h)$ can be found setting the coresponding derivative to zero. So it is necessary to variate the functional N^u which is defined parametrically. By similar calculations one can obtain the infimum of the expression for L_1^u.

PROOF OF (7) AND (8)

Now we are going to prove the relations (7), (8). Let's prove at first inequality (7).

If $N(y) = \infty$, then $y \notin \mathcal{C}$, which in turn means that one of the properties which functions from the set $\mathcal{C}$ must possess is not satisfied. It is easy to show (we omit the proof) that in this case for sufficiently small ϵ and for all values of n and functions κ_n^u the relation $\kappa_n^u \notin B_\epsilon(y)$ is valid and hence $P_n(B_\epsilon(y)) = 0$. From here (7) follows in this case.

We exclude the case when $L_1^u = \infty$, which means that does not exists exponential by $\sqrt{n}$ number of partitions of n into numbers with given constrains u.

Suppose now that $N^u(y) < \infty$. The proof of (7) and (8) in some steps uses techniques proposed in [[5]] where they are derived in the case of absence of the restrictions u.

Let $y \in \mathcal{C}$ and $\hat{y}$ be a continuous from the right, monotone function which can be obtained from y by changing y on a set of Lebesgue measure zero. Let's fix $\Delta_1, \Delta_2 > 0$, such that

$$\int_0^{\Delta_1} ydx < \delta, \quad \int_{\Delta_2}^{\infty} ydx < \delta.$$

Consider the decomposition of $\hat{y}$ into the sum of monotone functions,

$$\hat{y} = \hat{y}^1 + \hat{y}^2$$

where $\hat{y}^1$ is absolutely continuous and $\hat{y}^2$ is singular. Without loss of generality let's consider, that $\tilde{y}_2$ is continuous from the right. Our first step in the proof will be to 'localize' the points of discontinuity of the function $\hat{y}^2$ in the interval $[\Delta_1, \Delta_2]$ and on the remaining set, which consists of a finite number of intervals intersecting only in the bounds with the sum of the lengths of these intervals 'almost' equal to $\Delta_2 - \Delta_1$. We estimate the number of curves κ_n^u approximating functions $\hat{y}$ in $L^1([\Delta_1, \Delta_2])$ when $n \to \infty$ and hence when $\Delta_1 \to 0, \Delta_2 \to \infty$ in $L^1([0,\infty))$. Obtaining the upper estimate we take into account some curves which do not belong to κ_n^u. It will be clear that this does not lead to a difference in the exponential estimate.

Let's exclude points of discontinuity $A \subset [\Delta_1, \Delta_2]$ of the function $\hat{y}^2$. As we have already noted, when $|A| < \infty$, then $A = \emptyset$ and the Lebesgue measure $\mu(A) = 0$. In this case we exclude intervals $[c_i, d_i]$ from the next considerations. Because the Lebesgue measure is regular there exists an open set B, such that $A \subset B$ and $\mu(B) < \delta$. The function $\hat{y}_2$ is monotone and determines a measure ν on the interval $[\Delta_1, \Delta_2]$: $\nu([a,b)) = \nu(b) - \nu(a)$, whose support belongs to B, where the set B is the union of at most countably many intervals $:B = \bigcup_{i=1}^{\infty} B_i$. Using the continuity of the measure ν we obtain the existence of some m, such that

$$\nu\left(\bigcup_{i>m} B_i\right) < \delta. \tag{13}$$

Let's add to every interval B_i, $i \leq m$ its bounds and obtain closed intervals $\bar{B}_i$ this way. Then the set $\bigcup_{i=1}^{m} \bar{B}_i$ is the union of finite number of closed intervals, which intersect only on bounds. Let $\{[c_i, d_i]\}$ be the minimal number of such intervals, $\nu(\bigcup_{i=1}^{s}[c_i, d_i]) < \delta$.

The set

$$[\Delta_1, \Delta_2] \setminus \bigcup_{i=1}^{s} [c_i, d_i]$$

consists of a finite number of nonintersecting intervals and c_i, d_i are their bounds. Let's add to every interval its bound. For convenience we denote the new closed intervals by $[a_i, b_i]$, $i = 1, 2, \ldots, p$ where $p = s-1$ or s or $s+1$. Consider the partition of the interval $[a_i, b_i]$ into s_i consecutive subintervals $[e_i^j, f_i^j]$, $j = 1, \ldots, s_i$; $[a_i, b_i] = \bigcup_{j=1}^{s_i}[e_i^j, f_i^j]$. Also we propose that on the ends x of all intervals

$$\hat{y}'(x) \leq C,$$

where C is some constant. It is clear that by slightly 'moving' the ends of the intervals we can establish such a constant C. Now instead of the condition $\kappa_n^u \in B_\epsilon(y)$ we consider the condition

$$|\kappa_n^u(x) - \hat{y}(x)| < \gamma(\epsilon) > 0, \tag{14}$$

for every x, belonging to the set of ends of intervals. Next we show that under the conditions (14) it follows that $\kappa_n^u \in B_\epsilon(y)$ for some $\gamma(\epsilon) \to 0$ when $\epsilon \to 0$.

For $z \geq 0$ the function $L_2^u(z)$ is continuous and

$$L_2^u(z+\xi) - L_2^u(z) \leq L_2^u(\xi), \;\; z, \xi \geq 0. \tag{15}$$

The last inequality follows from the relation for derivations:

$$L_2'(z+\xi) - L_2'(z) = \lambda(z) - \lambda(z+\xi) \leq 0, \;\; z, \xi \geq 0 \tag{16}$$

where $\lambda(z)$ is the solution of the equation

$$\frac{\sum_{i \in \mathcal{A}} i e^{i\lambda}}{\sum_{i \in \mathcal{A}} e^{i\lambda}} = z.$$

The inequality in (16) follows from the relations

$$\begin{aligned} \lambda_z' &= 1/z_\lambda'; \\ z_\lambda' &= \frac{\sum_{i \in \mathcal{A}} i^2 e^{i\lambda}}{\sum_{i \in \mathcal{A}} e^{i\lambda}} - \left(\frac{\sum_{i \in \mathcal{A}} i e^{i\lambda}}{\sum_{i \in \mathcal{A}} e^{i\lambda}} \right)^2 \geq 0. \end{aligned}$$

Suppose that lim sup in (7) is achieved on the sequence of weights $n^1, n^2, \ldots$. For every x which is the end of intervals $\{[e_i^j, f_i^j]\}$ or $\{[c_i, d_i]\}$ there exist not more than $2\gamma(\epsilon)\sqrt{n^{(k)}} + 1$ values of $\kappa_{n^{(k)}}^u(x)$ for which (14) is valid. Next we suppose that the values of $\varphi\sqrt{n}$ for different φ are integers. It will be clear that this proposition does not influence the results and we avoid a lot of routine comments.

Consider the interval $[e_i^j, f_i^j]$ for some i, j and every $2\gamma(\epsilon)\sqrt{n^{(k)}} + 1$ point

$$\left\{ \eta_{ijp} = \left(e_i^j, y(e_i^j \pm \frac{p}{\sqrt{n^{(k)}}}) \right), p = 0, 1, \ldots, 2\gamma(\epsilon)\sqrt{n^{(k)}} \right\}$$

can belong to one or more shapes $\kappa_{n^{(k)}}^u$ such that

$$|\kappa_{n^{(k)}}^u(f_i^j) - y(f_i^j)| < \gamma(\epsilon). \tag{17}$$

At the same time from the above considerations, connected with formula (12) it follows, that for every η_{ijp} the upper bound on the number $\Phi_{ijpn^{(k)}}$ of shapes $\kappa_{n^{(k)}}^u$ such that

$$(e_i^j, \kappa_{n^{(k)}}^u(e_i^j)) = \eta_{ijp} \tag{18}$$

and (17) is true satisfy the relation

$$\lim_{n^{(k)} \to \infty} \frac{\Phi_{ijpn^{(k)}}}{\sqrt{n^{(k)}}} \leq (f_i^j - e_i^j) L_2^u \left(\frac{\hat{y}(e_i^j) - \hat{y}(f_i^j)}{f_i^j - e_i^j} \right) + \chi_{ij}, \tag{19}$$

where $\chi_{ij} \overset{\gamma(\epsilon) \to 0}{\to} 0$ (it is enough to put in (10)

$$\begin{aligned} n_1 &= (f_i^j - e_i^j)\sqrt{n^{(k)}}; \\ n_2 &= (\hat{y}y(e_i^j) - \hat{y}(f_i^j))\sqrt{n^{(k)}}). \end{aligned}$$

Using the same considerations for all p we obtain the asymptotic estimation for the total number $\Phi_{ijn^{(k)}}$ of curves satisfying (17) and (18)

$$\ln \Phi_{ijn^{(k)}} \leq \ln \sum_p \Phi_{ijpn^{(k)}}$$

$$\leq \ln[2\gamma(\epsilon)\sqrt{n^{(k)}}] + \sqrt{n^{(k)}}\Delta x_i^j L_2^u\left(-\frac{\Delta y_i^j}{\Delta x_i^j}\right) + \sqrt{n^{(k)}}\chi_{ij} + o(\sqrt{n^{(k)}}),$$

where $\Delta x_i^j = f_i^j - e_i^j$; $\Delta \hat{y}_i^j = \hat{y}(f_i^j) - \hat{y}(e_i^j)$. Because the restrictions on κ_n^u on different intervals can be considered independently, i.e. the continuation of the curve $\kappa_{n^{(k)}}^u$ on the next right interval with given restrictions on the left does not depend on these restrictions (indeed, there exists such a dependence: it is because the restrictions can lead to diagrams whose weights exeed 1, but as we will show later when $y \in \mathcal{C}$ then these conditions do not essentially change our relations, we simply stop to consider the next restrictions when the area of the diagram reaches 1.) we obtain the estimate on the total number $\Phi_{n^{(k)}}$ of the curves, which satisfy the conditions (14):

$$\begin{aligned}\frac{\Phi_{n^{(k)}}}{\sqrt{n^{(k)}}} &\leq \sum_{i=1}^{p}\sum_{j=1}^{s_i}\left(\frac{\ln(2\gamma(\epsilon)\sqrt{n^{(k)}})}{\sqrt{n^{(k)}}} + \chi_{ij}\right) \\ &+ \sum_{\ell=1}^{s}\left(\frac{\ln(2\gamma(\epsilon)\sqrt{n^{(k)}})}{\sqrt{n^{(k)}}} + \chi_s\right) \\ &+ \sum_{i=1}^{p}\sum_{j=1}^{s_i}\Delta x_i^j L_2^u\left(-\frac{\Delta \hat{y}_i^j}{\Delta x_i^j}\right) \\ &+ \sum_{\ell=1}^{s}\Delta x_\ell L_2^u\left(-\frac{\Delta \hat{y}_\ell}{\Delta x_\ell}\right) + \alpha_{n^{(k)}}(\Delta_1) + \beta_{n^{(k)}}(\Delta_2) + o(1),\end{aligned} \tag{20}$$

where

$$\frac{\Delta \hat{y}_\ell}{\Delta x_\ell} = \frac{\hat{y}(b_\ell) - \hat{y}(a_\ell)}{b_\ell - a_\ell}$$

and $\chi_\ell \overset{\gamma(\epsilon)\to 0}{\to} 0$ is the term arising similar to χ_{ij} but on the interval $[a_\ell, b_\ell]$. Here $\alpha_{n^{(k)}}(\Delta_1)$ and $\beta_{n^{(k)}}(\Delta_2)$ are the estimations of the number of possible restrictions $\kappa_{n^{(k)}}^u$ on the intervals $[0, \Delta_1]$ and $[\Delta_2, \infty)$, respectively. Later we show the validity of the following relations

$$\lim_{n^{(k)}\to\infty} \frac{\alpha_{n^{(k)}}(\Delta_1)}{\sqrt{n^{(k)}}} \overset{\Delta_1\to 0}{\to} 0; \tag{21}$$

$$\lim_{n^{(k)}\to\infty} \frac{\alpha_{n^{(k)}}(\Delta_2)}{\sqrt{n^{(k)}}} \overset{\Delta_2\to\infty}{\to} 0. \tag{22}$$

At first we show, that the contribution from $\sum_\ell$ to the estimation (20) can be made arbitrarily small. Indeed the function L_2^u is convex and from Jensen's inequality it follows

$$\sum_{\ell=1}^{s} \Delta_\ell L_2^u\left(-\frac{\Delta\hat{y}_\ell}{\Delta x_\ell}\right) \le L_2^u\left(-\frac{\sum_\ell \Delta\hat{y}_\ell}{\sum_\ell \Delta x_\ell}\right)\sum_\ell \Delta x_\ell. \tag{23}$$

Because $|\mathcal{A}| = \infty$ (otherwise we do not consider intervals $[c_\ell, d_\ell]$ at all), then $|\sum_\ell \Delta y_\ell| > C_1 > 0$ for some C_1. Moreover

$$\begin{aligned} L_2^u(z) &= \inf_\lambda\left(\ln\left(\sum_{i\in\mathcal{A}} e^{i\lambda}\right) - z\lambda\right) \\ &< \inf_\lambda\left(\ln\left(\sum_{i=0}^{\infty} e^{i\lambda}\right) - z\lambda\right) \\ &= \inf_\lambda\left(-\ln(1-e^\lambda) - z\lambda\right) = (1+z)H\left(\frac{z}{1+z}\right), \end{aligned}$$

where

$$H(\xi) = -\xi\ln\xi - (1-\xi)\ln(1-\xi)$$

is the binary entropy function. Hence

$$\frac{L_2^u(z)}{z} < H\left(\frac{z}{1+z}\right) \overset{z\to\infty}{\longrightarrow} 0. \tag{24}$$

Setting in (24) $z = -\sum_\ell \Delta\hat{y}_\ell / \sum_\ell \Delta x_\ell$ and taking into account that $\sum_\ell \Delta x_\ell = \mu\left(\bigcup_{i>m} B_i\right) < \delta \to 0$ we obtain the relation

$$\frac{L_2^u(z)}{z}\left(-\sum_\ell \Delta\hat{y}_\ell\right) > C\frac{L_2^u(z)}{z} \overset{\sum_\ell \Delta x_\ell \to 0}{\to} 0.$$

From here follows that the sum in the right hand side of (23) can be chosen arbitrarily small. Next using estimate (15) and the decomposition $\hat{y} = \hat{y}^1 + \hat{y}^2$ ($\Delta\hat{y} = \Delta\hat{y}^1 + \Delta\hat{y}^2$) and setting $z = -\Delta\hat{y}^1/\Delta x$, $\xi = -\Delta\hat{y}^2/\Delta x$ we obtain the estimation

$$L_2^u\left(-\frac{\Delta\hat{y}_i^j}{\Delta x_i^j}\right) \le L_2^u\left(-\frac{\Delta\hat{y}_i^{1j}}{\Delta x_i^j}\right) + L_2^u\left(-\frac{\Delta\hat{y}_i^{2j}}{\Delta x_i^j}\right).$$

Let's now estimate the contribution of the second term in the right hand side of the last inequality to the sum over i, j in (20):

$$\sum_{i=1}^{p}\sum_{j=1}^{s_i} \Delta x_i^j L_2^u\left(-\frac{\Delta\hat{y}_i^{2j}}{\Delta x_i^j}\right) \le \left(\sum_{i,j}\Delta x_i^j\right) L_2^u\left(-\frac{\sum_{i,j}\Delta\hat{y}_i^{2j}}{\sum_{i,j}\Delta x_i^j}\right). \tag{25}$$

Here we once more use the convexity of L_2^u and Jensen's inequality. Next

$$\sum_{i,j}\Delta y_i^{2j} < \nu\left(\bigcup_{i,j}[e_i^j, f_i^j]\right) < \delta \to 0$$

and

$$\sum_{i,j} \Delta x_i^j \geq \Delta_2 - \Delta_1 - \delta$$

and so the right hand side of (25) tends to zero.

Next we estimate the contributions to (20) from the terms $\alpha_{n^{(k)}}(\Delta_1)$ and $\beta_{n^{(k)}}(\Delta_2)$. Because $\kappa^u_{n^{(k)}} \in B_\epsilon(y)$, then

$$|\int_0^{\Delta_1} (\kappa^u_{n^{(k)}}(x) - y(x))dx| \leq \int_0^{\Delta_1} |\kappa^u_{n^{(k)}}(x) - y(x)|dx < \epsilon.$$

Hence

$$\int_0^{\Delta_1} \kappa^u_{n^{(k)}} dx < \int_0^{\Delta_1} y dx + \epsilon < \delta + \epsilon.$$

The value $\alpha_{n^{(k)}}(\Delta_1)$ does not exeed the number of diagrams of weight $(\delta+\epsilon)n^{(k)}$:

$$\ln \alpha_{n^{(k)}}(\Delta_1) \sim \pi\sqrt{2(\delta + \epsilon)n^{(k)}/3},$$

hence

$$\frac{\ln \alpha_{n^{(k)}}(\Delta_1)}{\sqrt{n^{(k)}}} < \pi\sqrt{2(\delta + \epsilon)/3}.$$

Similarly

$$\int_{\Delta_2}^{\infty} \kappa^u_{n^{(k)}} dx < \int_{\Delta_2}^{\infty} y dx + \epsilon < \delta + \epsilon$$

and so

$$\frac{\ln \beta_{n^{(k)}}(\Delta_2)}{\sqrt{n^{(k)}}} < \pi\sqrt{(\delta + \epsilon)2/3}.$$

Hence the contributions of $\alpha_{n^{(k)}}$, $\beta_{n^{(k)}}$ to the estimation (20) can be made arbitrarily small.

Taking into account the last considerations we take $\limsup_{\epsilon\to 0}$, $\lim_{n^{(k)}\to\infty}$ from both sides of the estimation (20) and obtain the inequality:

$$\limsup_{\epsilon\to 0} \limsup_{n\to\infty} \frac{\ln \Phi_{n(1\pm\delta)}}{\sqrt{n}} \leq \sum_{i,j} \Delta x_i^j L_2^u \left(\frac{\hat{y}^1(e_i^j) - \hat{y}^1(f_i^j)}{f_i^j - e_i^j} \right) + \xi_o(\delta) + \xi_1(\Delta_1) + \xi_2(\Delta_2), \tag{26}$$

where ξ_o is the contribution of $\sum_\ell$ from (20), ξ_1, ξ_2 are the contributions of α and β correspondingly. As we have already shown these contributions can be made arbitrarily small. Increasing s_j in such a way that $\omega = \max_{i,j} \Delta x_i^j \to 0$ we reduce the right hand side of (26) to the following expression

$$\sum_{i,j} \Delta x_i^j L_2^u \left(\frac{1}{\Delta x_i^j} \int_{f_i^j}^{e_i^j} \hat{y}^{1'}(x)dx \right) = \int_{\bigcup[a_j,b_j]} L(y_c(x))dx,$$

where y_c is the piecewise constant function such that for given partition

$\{[e_i^j, f_i^j]\}$ of the set $\bigcup_j [a_j, b_j]$

$$y_c(x) = \frac{1}{\Delta x_i^j} \int_{f_i^j}^{e_i^j} \hat{y}^{1'}(x)dx, \; x \in [e_i^j, f_i^j)$$

(we omit for a moment $\xi_i, i = 0, 1, 2$). Taking $\lim_{\omega \to 0}$ from both sides of the estimation (26) and using the last equalities we obtain the relations

$$\begin{aligned} \limsup_{\epsilon \to 0} \limsup_{n \to \infty} \frac{\ln \Phi_{n(1\pm\delta)}}{\sqrt{n}} &\leq \limsup_{\omega \to 0} \int_{\bigcup_j [a_j, b_j]} L(y_c(x))dx \qquad (27) \\ \leq \int_{\bigcup_j [a_j, b_j]} \limsup_{\omega \to 0} L(y_c(x)) &= \int_{\bigcup_j [a_j, b_j]} L(\limsup_{\omega \to 0} y_c(x))dx \\ = \int_{\bigcup_j [a_j, b_j]} L(-\hat{y}^{1'}(x))dx &= \int_{\bigcup_j [a_j, b_j]} L(-\hat{y}'(x))dx. \end{aligned}$$

Here in the last inequality we use the Fatou lemma and the first equality follows from the continuity of the function L. The second equality follows from the fact that if $z(x) \in L^1([a, b])$, then for almost all $x_o \in [a, b]$ the following relation is valid

$$\lim_{q \to 0} \frac{1}{|D_q|} \int_{D_q} z(x)dx = z(x_o),$$

where D_q is an arbitrary sequence of closed intervals with nonempty interior such that $\bigcap_q D_q = x$. The last equality in (27) follows from the fact that $\hat{y}^{1'} = \hat{y}'$ *a.s.* Because $\sum_{i,j} (f_i^j - e_i^j) \geq \Delta_2 - \Delta_1 - \delta$ and the value δ can be chosen arbitrarily small, from the absolute continuity of the integral and from the estimation (27) follows the inequality

$$\limsup_{\epsilon \to 0} \limsup_{n \to \infty} \frac{\ln \Phi_{n(1\pm\delta)}}{\sqrt{n}} \leq \int_{[\Delta_1, \Delta_2]} L_2^u(-\hat{y}')dx + \xi_o(\delta) + \xi_1(\Delta_1) + \xi_2(\Delta_2)$$

. Letting $\Delta_1 \to 0, \Delta_2 \to \infty$ we obtain

$$\limsup_{\epsilon \to 0} \limsup_{n \to \infty} \frac{\ln \Phi_{n(1\pm\delta)}}{\sqrt{n}} \leq \int_0^\infty L_2^u(-\hat{y}')dx + \xi_1(\delta). \qquad (28)$$

Now inequality (7) follows from (28) if we set

$$P_n(B_\epsilon(y)) = \frac{\Phi^u_{n(1\pm\delta)}}{p^u_{n(1\pm\delta)}}$$

and use (11) to find values $p^u_{n(1\pm\delta)}$:

$$\lim_{\delta \to 0} \lim_{n \to \infty} \frac{\ln p^u_{n(1\pm\delta)}}{\sqrt{n}} = L_1^u.$$

Hence

$$\lim_{\delta\to 0}\limsup_{\epsilon\to 0}\limsup_{n\to 0}\ln\frac{\Phi_{n(1\pm\delta)}}{p^u_{n(1\pm\delta)}} = -\left(L_1^u - \int_0^\infty L_2^u(-\hat{y}')dx\right) = -N^u(y).$$

This completes the proof of the estimation (7).

Next we prove the estimation (8). Let $\liminf_{n\to\infty}$ in (8) be attained on the sequence $n^{(k)}$. Let's choose Δ_1, Δ_2 as before and consider the partition of the interval $[\Delta_1, \Delta_2]$ into s consecutive intervals $[a_i, b_i]$ of equal length $\Delta = b_i - a_i = (\Delta_2 - \Delta_1)/s$. To obtain an upper bound for $\ln P_n(B_\epsilon(y))$ we consider the contribution from the shapes κ_n^u which do not belong to $B_\epsilon(y)$ and even are not Young diagrams because the area of such diagrams exeeds 1. Now we should restrict our attention only on such shapes κ_n^u which belong to $B_\epsilon(y)$ and are the shapes of diagram of weight $n(1\pm\delta)$. Let's consider only such κ_n^u, which for every $x_o \in \{a_i, b_i; i = 1, 2, \ldots, s\}$ satisfy the relation

$$|\kappa_n^u(x_o) - \hat{y}(x_o)| < \gamma(\epsilon). \tag{29}$$

Obviously we can make such a choice for not too large x_o. Indeed when drawing the diagram adding columns from the right in such a way that the relations (29) are true, it is possible that at last we come to the situation when the diagram already has unit area, but there still exist some intervals $[a_i, b_i]$ on the right which we don't pass yet. Next we show that the sum of the lengths of the remaining intervals will be arbitrary small. If we reach the point $\Delta_3 < \Delta_2$, then the areas under κ_n^u and under $y(x)$ are 'almost' equal. Next for given Δ_1 we choose κ_n^u, such that $\kappa_n(x) = \kappa_n^u(\Delta_1)$, $x \in [0, \Delta_1]$. From

$$|\kappa_n^u(\Delta_1) - \hat{y}(\Delta_1)| < \epsilon_1, \tag{30}$$

follows

$$\int_0^{\Delta_1} \kappa_n^u(x)dx < \Delta_1(\hat{y}(\Delta_1) + \epsilon_1)$$

and

$$\begin{aligned}\int_0^{\Delta_1} |\kappa_n^u(x) - y(x)|dx &\leq \int_0^{\Delta_1} y(x)dx + \int_0^{\Delta_1} \kappa_n^u(x)dx \\ &< \delta + \Delta_1(\hat{y}(\Delta_1) + \epsilon_1)\end{aligned}$$

and $\hat{y}(\Delta_1)\Delta_1 \overset{\Delta_1\to 0}{\to} 0$. If drawing the diagram from the left we reach point Δ_3, it is possible that we stop this drawing. It would be in the case when

$$\int_0^{\Delta_3} \kappa_n^u(x)dx = 1$$

and the last column of height H, multiplied by $\sqrt{n}$ is the multiplicity of the maximal number in an unordered partition of n. Let $H\sqrt{n} \notin \mathcal{A}$. Then the diagram does not satisfy the conditions u and in this case we lift all diagram on $h\sqrt{n}$ in such a way that $(H+h)\sqrt{n} \in \mathcal{A}$ or if the value $H\sqrt{n}$ exeeds all possible multiplicities, we draw the diagram to the left in such a way, that only allowed multiplicities appear in the partition. However, after drawing the last column once more can appear the situation when its height multiplied by $\sqrt{n}$ does not belong to $\mathcal{A}$ and in this case we should lift the whole diagram to provide the necessary multiplicity of the largest element in the partition of the number $n(1 \pm \delta)$. Note, that after such 'additions' the weight of the diagram can increase from n to n' and we should change the scaling from $1/\sqrt{n}$ to $1/\sqrt{n'}$ and $n \sim n'$ and if the possible weights of diagrams are from the range $n(1 \pm \delta')$ $\delta' = \delta/2$ then finaly we obtain diagrams of weight $n(1 \pm \delta)$. Moreover, choosing diagrams satisfying (29) after the above transformations we will obtain shapes κ_n^u which for sufficiently large n also satisfy (29). Our considerations lead to the inequality

$$\int_{\Delta_2}^{\infty} \kappa_n^u(x)dx < \eta,$$

where $\eta \overset{\Delta_2 \to \infty}{\to} 0$ (we omit the easy proof of this fact).

Let's now construct the upper bound on the L^1- distance between y and the shape κ_n^u drawn above. For the pair of monotone, nonincreasing functions z_1, z_2 such that

$$|z_1(x) - z_2(x)| \leq \epsilon_1$$

when $x = a, b;\ a < b$ the following inequality is true

$$\int_a^b |z_1 - z_2| dx \leq (b-a)(z_1(a) - z_1(b) + 2\epsilon_1). \tag{31}$$

Relation (31) follows from the fact that the part of the area bounded by the curves $z_1(x), z_2(x)$ and by the lines $x = a, x = b$ is covered by the rectangle with edges $y = z_1(a) + \epsilon_1,\ y = z_1(b) - \epsilon_1, x = a, x = b$. Let now (19) be true for almost all $x \in \{a_i, b_i\}$, then from (31) we obtain the relations

$$\int_{\Delta_1}^{\Delta_2} (\kappa_n^u(x) - \hat{y}(x))dx = \sum_{i=1}^{s} \int_{a_i}^{b_i} |\kappa_n^u(x) - \hat{y}(x)|dx \tag{32}$$

$$\leq \sum_{i=1}^{s} (b_i - a_i)(\hat{y}(a_i) - \hat{y}(b_i) + 2\epsilon_1) \leq \Delta(\hat{y}(\Delta_1) - \hat{y}(\Delta_2)) + 2\epsilon_1(\Delta_2 - \Delta_1).$$

From (32) it follows that for sufficiently small Δ_1, large Δ_2 and small Δ we can satisfy the following restrictions

$$\int_{\Delta_1}^{\Delta_2} |\kappa_n^u(x) - y(x)|dx < \epsilon. \tag{33}$$

Here it is necessary to note, that when drawing the diagrams from the left, adding the columns to the right sometime we lift the whole diagram to satisfy the restrictions u. It is possible that drawing diagrams in a different way we obtain the same diagram and so the contribution of some diagrams can exeed one. But it is easy to see that for every diagram we obtain a multiplicity less than $O(n)$. And so beginning to draw diagrams of weight $n(1 \pm \delta/2)$ we obtain diagrams of weights from $n(1 \pm \delta)$. Next from (32), (33) and the above considerations we obtain that for sufficiently large n, Δ_2 and small Δ_1, Δ drawing bounds κ_n^u belong to $B_\epsilon(y)$. Now we construct a lower bound on the whole number of such diagrams. As before for the upper bound, the exponent of the number of restrictions κ_n^u on $[a_i, b_i]$ is estimated by the value

$$\sqrt{n}(b_i - a_i) L_2^u \left(-\frac{\hat{y}(b_i) - \hat{y}(a_i)}{b_i - a_i} \right) + o(1) + o(\sqrt{n}),$$

where $o(1) \overset{\gamma(\epsilon) \to 0}{\to} 0$. The contribution of all intervals $[a_i, b_i]$ is estimated by the sum of these values:

$$\sqrt{n} \sum_{i=1}^{s} \Delta x_i L_2^u \left(-\frac{\Delta \hat{y}_i}{\Delta x_i} \right) \tag{34}$$
$$+(o(1) + o(\sqrt{n}) + O(\Delta))(\Delta_2 - \Delta_1).$$

Next it should be taken into account diagrams with differents weights from the interval $n(1 \pm \delta)$. It is clear that this does not change the asymptotics $(n \to \infty,\ \delta \to 0)$ of the estimation. From here using (34) we obtain:

$$\liminf_{n \to \infty} \frac{\ln \Phi_{n(1 \pm \delta)}}{\sqrt{n}} \geq \sum_{i=1}^{s} \Delta x_i L_2^u \left(-\frac{\Delta \hat{y}_i^1}{\Delta x_i} - \frac{\Delta \hat{y}_i^2}{\Delta x_i} \right) + o(1) + o(\delta) \tag{35}$$

$$\geq \sum_{i=1}^{s} \Delta x_i L_2^u \left(-\frac{\Delta \hat{y}_i^1}{\Delta x_i} \right) + o(1) + o(\Delta) = \sum_{i=1}^{s} \Delta x_i L_2^u \left(\frac{\int_{b_i}^{a_i} \hat{y}^{1'}(x) dx}{\Delta x_i} \right)$$

$$+o(1) + o(\delta) \geq \sum_{i=1}^{s} \int_{a_i}^{b_i} L_2^u(-\hat{y}^{1'}(x)) dx + o(1) + o(\delta) = \int_{\Delta_1}^{\Delta_2} L(-\hat{y}'(x)) dx$$

$$+o(1) + 0(\delta).$$

Here the second inequality follows from the following consideration: if $\hat{y}_i^2 \not\equiv 0$ then $L_2^u \left(-\frac{\Delta y}{\Delta x} \right) > -\infty$ for all $\Delta y = y(x_2) - y(x_1), \Delta x = x_2 - x_1,\ x_1, x_2 \in [0, \infty)$ only if $|\mathcal{A}| = \infty$, but in this case L_2^u is a monotone function. Hence there exist two possibilities: $\hat{y}_i^2 = 0$, then the second inequality in (35) is valid or $|\mathcal{A}| = \infty$ and L_2^u is monotone and then the second inequality is also true. In the last inequality we use the convexity of the function L_2^u and Jensen's inequality. Recalling the definition of $N(y)$ and that $P_{n(1 \pm \delta)}^u = \frac{\Phi_{n(1 \pm \delta)}^u}{p_{n(1 \pm \delta)}^u}$ letting $\delta \to 0$ and then $\epsilon \to 0, \Delta_1 \to 0,\ \Delta_2 \to \infty$ we obtain from (35) the estimation (8) in the same way as before when we proved the validity of the upper bound (7).

CONCLUDING REMARKS

From (8) follows the left hand inequality from (1) (if for some b and $\epsilon > 0$, $B_\epsilon(b) \in B^o$, $b \in B^o$, then $P_n(B^o) \geq P_n(B_\epsilon(b))$. The validity of the upper bound from (1) for a compact set $\bar{B}$ can be proved from the local LDP using standard methods. Here it is possible to introduce many examples of choosing the set $\mathcal{A}$. Let's mention two of them.

- A={ 0,1,...}.
- A={ 0,1}

The first example we consider in the paper [5], the rate function in this case is as follows:

$$N^u(y) = \begin{cases} \pi\sqrt{\frac{2}{3}} - \int_0^\infty (1-\hat{y})H\left(\frac{-\hat{y}'}{1-\hat{y}'}\right)dx, & y \in \mathcal{A}; \\ \infty, & y \notin \mathcal{A}. \end{cases}$$

The second example first was considered in paper [6] where it is proved using another method and the rate function in that case is as follows

$$N^u(y) = \begin{cases} \frac{\pi}{\sqrt{3}} - \int_0^\infty H(-\hat{y}')dx, & y \in \mathcal{A}; \\ \infty, & y \notin \mathcal{A}. \end{cases}$$

References

[1] J. Deuschel and D. Stroock, *Large Deviations*, Boston: Academic Press, 1989.

[2] A. Dembo, O. Zeitouni, *Large Deviations Techniques and Applications*, Boston: Jones and Barlett Publishers, 1993.

[3] V. Blinovsky, and R. Dobrushin, "Process level large deviations for a class of piecewise homogeneous random walks", *The Dynkin Festschrift, Markov Processes and their Applications*, Progress in Probability, Boston, Birkhauser, vol.34, 1994, 1-60.

[4] V. Blinovsky, "Large deviations principle for random Young diagram", *Proc. IEEE Symp. on Inf. Theory*, Boston, MIT, 1998.

[5] V. Blinovsky, "Large deviations principle for random Young diagram", *Problems of Information Transmission* 34, No. 1, 1999 (to appear).

[6] A. Dembo, A. Vershik A. and O. Zeitouni, *Large Deviation Principle for Integer Partitions*, manuscript.

BSC: TESTING OF HYPOTHESES WITH INFORMATION CONSTRAINTS*

Marat V. Burnashev

Institute for Problems of Information Transmission, RAS
19 Bolshoi Karetnii, 101447, Moscow, Russia.

Shun-ichi Amari

RIKEN Brain Science Institute, Wako-shi, Hirosawa 2-1, Saitama 351-0198, Japan.

Te Sun Han

Graduate School of Information Systems,
University of Electro–Communications, Chofugaoka 1 - 5 - 1, Chofu, Tokyo 182, Japan.

Abstract: A problem of hypothesis testing on the crossover probability of a BSC is considered. We observe only the channel output and our helper only observes the channel input and can send us some limited amount of information about the input block. What kind of that information allows us to make the best statistical inferences? In particular, what is the minimal information sufficient to get the same results as if we could observe directly all data? Some upper bounds for that minimal amount of information and some related results are obtained.

I. Introduction.

In this paper we consider some particularly interesting cases of the following general problem [1 - 5].

A "statistician" should make certain statistical inferences concerning the system state (for example, estimate some unknown parameter, test some hypotheses, etc.). There are two sets of data (observations): the set $\mathcal{A}$ ("available") and the set $\mathcal{R}$ ("remote").

*The research described in this publication was made possible in part by Grant N 98-01-04108 from the Russian Fund for Fundamental Research and INTAS 94–469.

I. Althöfer et al. (eds.), Numbers, Information and Complexity, 489-500.

The statistician directly observes all data from the set $\mathcal{A}$.

He can not directly observe data from the remote set $\mathcal{R}$, but his "helper" can observe them. Moreover the helper is allowed to send to the statistician some limited amount of information about those data.

The problem is: what kind of information (limited) about those remote data should send the helper in order to allow the statistician to make the best possible statistical inferences (for example, to get the minimum mean–square error for parameter estimation, etc.) ?

There are many practical situations where we meet this kind of problem. For example, in some applications the set $\mathcal{R}$ can be regarded as some "nuisance noise" that has "contaminated" already the data from the set $\mathcal{A}$, and therefore we would like to "remove" (as much as possible) that "contamination" in order to improve our statistical inferences.

We will deal below with discrete–time models and moreover, by "limited amount of information" we will mean that the helper can send information with communication rate not exceeding some prescribed value $R > 0$. Of course, if R is such large that the helper can simply resend to the statistician all data from $\mathcal{R}$ then we come back to a traditional statistical problem (that is not of our interest here). For that reason it is natural to assume that R is small enough in order to avoid such primitive resending.

Nevertheless, even with this assumption there are some cases (sometimes, probably natural) when the optimal solution can be obtained quite easily. Certainly, this will always be the case when data from sets $\mathcal{A}$ and $\mathcal{R}$ both represent independent observations of the same phenomenon.

The situation becomes much more difficult (and more interesting) when there is a sufficiently strong dependence between data from the sets $\mathcal{A}$ and $\mathcal{R}$. We consider mainly the case when neither the statistician, nor the helper can make any good statistical inferences based only on their own data (in other words, there is a very strong dependence between data from $\mathcal{A}$ and $\mathcal{R}$).

The next example illustrates such a case.

Example. Consider the binary symmetrical channel (BSC) [7, 8] with unknown transition probability $0 < p \leq 1/2$ which we will need to estimate or to test some hypotheses about it. The statistician observes only the channel output $\mathcal{A} = (y_1, \dots, y_n)$ and the helper observes only the channel input $\mathcal{R} = (x_1, \dots, x_n)$. We assume also that there is not any prior information about the input block $(x_1, \dots, x_n)$.

It is clear that if the statistician knows nothing about the input block $(x_1, \dots, x_n)$ then he can not draw any reasonable conclusions on the unknown value p.

The fact that the helper may send to the statistician information with rate not exceeding the prescribed value $R > 0$ means that they are allowed to partition the input space $E^n = \{0, 1\}^n$ into $N \leq 2^{Rn}$ arbitrary parts $\{X_1, \dots, X_N\}$ and the helper only informs the statistician about the part X_i to which the input block x^n belongs. It is clear that only the case $N < 2^n$, i.e. $R < 1$, is interesting (otherwise the helper can simply resend the value x^n).

For example, the helper can exactly inform the statistician about the first Rn values $x_1, \ldots, x_{Rn}$ (but then he will send no information about other values x_i). Such a simple partition method of the input space E^n (into cylinder sets $\{X_i\}$) is not generally optimal. From the statistician's point of view the input data $(x_1, \ldots, x_n)$ represent a very severe nuisance parameter.

We can also say that transmission of optimal limited information about block x^n means optimal "compression" of the full information about block x^n. Of course, that optimal "compression" depends on prior information on the transition probability p and the quality criteria used.

Remark. It is clear that the problem will not be changed if the statistician observes the channel input and the helper observes the channel output. We will later use both variants of that problem statement.

In the paper, for the BSC we consider a traditional problem of testing two simple hypotheses concerning the parameter p.

We will point out some partitions $\{X_1, \ldots, X_N\}$ and decision methods that are, probably, asymptotically (when $n \to \infty$) close to optimal ones. Unfortunately, we were not yet able to show that it is not possible to perform better and this remains an open problem.

We limit ourselves here to the BSC (i.e. independent Bernoulli random variables with unknown parameter p) for the following reasons:

1. For a person sufficiently familiar with information theory it is rather clear that in interesting cases some function similar to the *reliability function* of the channel [7,8] should be presented in the solution. From the reliability function point of view the BSC is a very illustrative example (i.e. it contains all essential problems; all other channels are treated using essentially methods developed for BSC; still there are only some lower and upper bounds for the reliability function of the BSC; etc.).

2. All statistical quantities (e.g. Kullback–Leibler information, Fisher information, etc.) have a very simple analytical form and geometrical meaning for the BSC. For that reason in the BSC case all main difficulties of the problem considered will be clearly seen and they will not be additionally complified by questions of more technical type.

We can repeat also a well–known claim: "show us how to deal with the BSC (or Bernoulli distributions) and we will show you how to do the same for a much broader class of channels (distributions)".

Below we write $\log x = \log_2 x$, $\exp_2 x = 2^x$. For any finite set A by $|A|$ its cardinality is denoted. For any function $f(x), x \in A$ by $|f|$ the cardinality of the set $f(A)$ is denoted. In order to distinquish input and output alphabets $E = \{0, 1\}$ we denote them E_{in} and E_{out}, respectively.

II. Testing of two simple hypotheses

1. Statement of the problem and the dual problem

We consider the BSC with some crossover probability p to be tested. We assume that p satisfies one of the two hypotheses: $H_0 : p = p_0$ or $H_1 : p = p_1$, where $0 < p_0 < p_1 \leq 1/2$.

We denote by P and Q the conditional output distributions for H_0 and H_1, respectively. Therefore, the probabilities to get output block $y^n = (y_1, \ldots, y_n)$ provided that the input block was $x^n = (x_1, \ldots, x_n)$ are given, respectively, by

$$P(y^n|x^n) = (1-p_0)^{n-d(x^n,y^n)} p_0^{d(x^n,y^n)}$$

and

$$Q(y^n|x^n) = (1-p_1)^{n-d(x^n,y^n)} p_1^{d(x^n,y^n)} ,$$

where $d(x^n, y^n)$ is the Hamming distance between blocks x^n and y^n (i.e. the minimal number of noncoinciding components on the whole length n).

We are interested in testing those hypotheses in the case that we observe only the channel output and from the helper we only get some limited information about the input block. We consider the minimax statement of the problem.

To be specific, assume that we are allowed to partition the input space E_{in}^n into N parts $\{X_1, \ldots, X_N\}$. After that we observe the channel output $y^n \in E_{\text{out}}^n$ and the helper only informs us to which part X_i belongs the input block x^n. On the basis of observed y^n and the index of X_i we decide in favor of one of the hypotheses H_0 or H_1. In order to avoid overcomplification we only consider nonrandomized decision methods (the problem's essence and the results remain the same). Then the general decision method can be described as follows. For any partition element X_i we choose some set $\mathcal{A}(X_i) \subset E_{\text{out}}^n$ and then depending on the observation y^n make a decision ($\mathcal{A}^c = E_{\text{out}}^n \setminus \mathcal{A}$):

$$y^n \in \mathcal{A}(X_i) \Longrightarrow H_0 ; \quad y^n \in \mathcal{A}^c(X_i) \Longrightarrow H_1 .$$

Define error probabilities of the first kind α_n and the second kind β_n as

$$\alpha_n = \Pr(H_1|H_0) = \max_{i=1,\ldots,N} \max_{x^n \in X_i} P(\mathcal{A}^c(X_i)|x^n) ,$$

$$\beta_n = \Pr(H_0|H_1) = \max_{i=1,\ldots,N} \max_{x^n \in X_i} Q(\mathcal{A}(X_i)|x^n) .$$

Let $\gamma > 0$ be some given constant. We demand that the first kind error probability satisfies the condition

$$\alpha_n = \Pr(H_1|H_0) \leq 2^{-\gamma n} . \tag{1}$$

We are interested in the minimal possible (over all partitions of the input space and all decisions) second kind error probability and we want to minimize (over all partitions of the input set and all decisions) the second kind error probability $\inf \beta_n$.

We consider the asymptotic situation when $n \to \infty$ and $N = 2^{Rn}$, where $0 < R < 1$ is some prescribed constant [1]. Then for the best criteria we denote

$$e(\gamma, R) = \lim_{n \to \infty} \frac{1}{n} \log_2 \frac{1}{\inf \beta_n} > 0 , \tag{2}$$

[1] In order to simplify formulas we don't use integer part sign of value 2^{Rn}

where the infimum is taken over all methods satisfying condition (1).

Our aim is to find (or to get good bounds) for the function $e(\gamma, R)$.

It will be convenient for us to consider also the following dual problem (without helper). Let some constant $0 < r < 1$ be given. We are allowed to choose in advance any set $\mathcal{X} \subset E_{\text{in}}^n$ consisting of $X = 2^{rn}$ input blocks. Let us also know that the input block may only be from the set $\mathcal{X}$.

Now, knowing the set $\mathcal{X}$, we observe the channel output y^n and consider the problem of testing hypotheses H_0 against H_1. For a chosen set $\mathcal{A}$ depending on observation y^n we make the decision:

$$y^n \in \mathcal{A} \Longrightarrow H_0 ; \quad y^n \in \mathcal{A}^c \Longrightarrow H_1 ,$$

and define first kind and second kind error probabilities as

$$\alpha_n = \max_{x^n \in \mathcal{X}} P\left(\mathcal{A}^c | x^n\right) , \quad \beta_n = \max_{x^n \in \mathcal{X}} Q\left(\mathcal{A} | x^n\right) .$$

Let now for the first kind error probability condition (1) be fullfilled. We want to choose a set $\mathcal{X}$ of cardinality $X = 2^{rn}$ and a decision method in order to achieve the minimal possible second kind error probability inf β_n. For this dual problem similarly to (2) we can define the function $e_2(\gamma, R)$. The following result establishes a simple relation between the functions $e(\gamma, R)$ and $e_2(\gamma, R)$.

Proposition 1. *The following relation holds true*

$$e(\gamma, 1-R) = e_2(\gamma, R) ; \qquad 0 \leq R \leq 1 , \quad \gamma > 0 . \tag{3}$$

In order to prove Proposition 1 we will need a simple "covering" lemma (certainly known).

Lemma 1. *Let $\mathcal{X} = \{\mathbf{x}_1, \dots, \mathbf{x}_X\} \subset E^n$ be any set of cardinality X. Then there exist $K = n2^n/X$ "shifts" $\{\mathbf{y}_1, \dots, \mathbf{y}_K\} \subset E^n$ such that the sets $\mathcal{X} + \mathbf{y}_i ; i = 1, \dots, K$, cover the whole space E^n.*

Proof. We choose all K shifts randomly and independently (with returns). Then for any $K > n2^n \ln 2/X$ we have

$$\Pr \{\text{there exists some noncovered point } \mathbf{x} \in E^n\} \leq$$

$$\leq 2^n \Pr \{\text{point } \mathbf{0} \text{ is not covered}\} = 2^n \left(1 - X2^{-n}\right)^K \leq$$

$$\leq \exp\left\{-XK2^{-n} + n \ln 2\right\} < 1 .$$

Therefore among such randomly chosen $K > n2^n \ln 2/X$ shifts there exists a collection, satisfying Lemma 1. □

Proof of Proposition 1. Let the set $\mathcal{X}$ of cardinality $\sim 2^{Rn}$ be the best one for the dual problem, i.e. it gives second kind error probability $\sim 2^{-ne_2(\gamma,R)}$. Due to Lemma 1 the whole input space E_{in}^n can be covered by $N \sim 2^{(1-R)n}$ shifted versions of the set $\mathcal{X}$ (each of them has the same "testing performance").

Reducing some elements of that covering, we can construct a partition of the space E_{in}^n into $\sim 2^{(1-R)n}$ parts. Since we consider the minimax statement of the problem, the "testing performance" of each part will be not be worse than for the original set $\mathcal{X}$, from which follows the inequality

$$e(\gamma, 1-R) \geq e_2(\gamma, R)\,; \qquad 0 \leq R \leq 1\,, \quad \gamma > 0\,.$$

Let us now in the original problem be given some partition $\{X_1, \ldots, X_N\}$ $N \sim 2^{(1-R)n}$, yielding second kind error probability $\sim 2^{-ne(\gamma,1-R)}$. Then there exists some partition element X_i of cardinality $\sim 2^{Rn}$, for which in the dual problem the second kind error probability also does not exceed $2^{-ne(\gamma,1-R)}$ from which follows the opposite inequality

$$e(\gamma, 1-R) \leq e_2(\gamma, R)\,; \qquad 0 \leq R \leq 1\,, \quad \gamma > 0\,.$$

that completes the proof of Proposition 1. □

Therefore due to Proposition 1 it is sufficient to investigate the function $e_2(\gamma, R)$. But first we recall some results for the case that the input block is known.

2. Known input block

Assume first that we know the input block x^n and that we observe the output block y^n. Without loss of generality we may assume that x^n is the all–0 block.

It is clear that for the optimal test the decision set in favor of p_0 is a ball $S(\tau n, \mathbf{0})$ of some radius $\tau(\gamma) \geq p_0 n$ centered at zero. Performing only with values exponential in n for the coefficient $\tau(\gamma)$ we have the condition

$$2^{-\gamma n} \simeq \alpha_n \simeq \binom{n}{\tau n}(1-p_0)^{n-\tau n}p_0^{\tau n} \simeq 2^{nh(\tau)}(1-p_0)^{n-\tau n}p_0^{\tau n}\,,$$

$$h(x) = x\log(1/x) + (1-x)\log(1/(1-x))\,,$$

or

$$\gamma = \tau \log \frac{\tau}{p_0} + (1-\tau)\log\frac{1-\tau}{1-p_0} \equiv D(\tau||p_0)\,. \tag{4}$$

Since we also want to have a small second kind error probability β_n we need to have $p_0 \leq \tau \leq p_1$. The function $D(\tau||p_0)$ is $\cup$–convex in τ and monotonically increasing for $\tau \geq p_0$. Therefore γ should satisfy the condition

$$0 \leq \gamma \leq D(p_1||p_0)\,.$$

For such γ the value $\tau(\gamma)$ is given as the unique root (for $p_0 \leq \tau$) of the equation (4).

For the second kind error probability β_n we have

$$\beta_n \simeq \binom{n}{\tau n}(1-p_1)^{n-\tau n}p_1^{\tau n}$$

or

$$\frac{1}{n}\log\frac{1}{\beta_n} \simeq e(\gamma) = D(\tau||p_1) = \gamma_2 \,, \tag{5}$$

It is convenient to consider $p_0 \le \tau \le p_1$ as a parameter through which both error probabilities can be expressed (see (4) and (5)).

Remark. The function $D(x||y)$ is the divergence for two binomial r.v.'s with parameters x and y, respectively. In other words, it gives the best possible exponential rate for the second kind error probability with fixed first kind error probabilty (so its exponent rate is equal to 0) when testing two simple hypotheses: $H_0 : p = x$ against $H_1 : p = y$.

Examples.

1) Let $\gamma = 0$, then $\tau = p_0$ and $\gamma_2 = D(p_0||p_1)$.

2) Let $\gamma_2 = 0$, then $\tau = p_1$ and $\gamma = D(p_1||p_0)$.

3) Let $\gamma = \gamma_2$, then τ is the unique root of the equation $D(\tau||p_0) = D(\tau||p_1)$, from which follows

$$\tau = \left(\log\frac{1-p_0}{1-p_1}\right) \Big/ \left(\log\frac{p_1(1-p_0)}{p_0(1-p_1)}\right) \tag{6}$$

and

$$\gamma = \gamma_2 = D(\tau||p_0) = D(\tau||p_1) \,.$$

3. Unknown input block and critical rate

As already shown, if we know the input block and $\alpha_n \le 2^{-\gamma n}$ then the best exponent for the second kind error probability $e(\gamma)$ is given by formulas (4)–(5).

If we only know that the input block belongs to some set $\mathcal{X}$ of cardinality $X \sim 2^{rn}$ then for the best chosen such set $\mathcal{X}$ the exponent of the second kind error probability is defined by the function $e_2(\gamma, r)$. It is clear that

$$e_2(\gamma, r) \le e(\gamma)\,; \quad \gamma \ge 0\,, \quad 0 \le r \le 1\,. \tag{7}$$

The function $e_2(\gamma, r)$ is nonincreasing in r and $e_2(\gamma, 0) = e(\gamma)$. Therefore regarding the function $e_2(\gamma, r)$ the following question immediately arises: does there exist an $r > 0$ such that equality in (7) is fullfilled and, if so, what is the maximal such rate $r_{\text{crit}}(\gamma)$? Formally, define $r_{\text{crit}}(\gamma)$ as

$$r_{\text{crit}}(\gamma) = \sup\{r : e_2(\gamma, r) = e(\gamma)\}\,; \quad \gamma \ge 0\,. \tag{8}$$

In other words, what is the maximal cardinality 2^{rn} of the best set $\mathcal{X}$ for which we can achieve the same asymptotical efficiency as for known input block (although we don't know the input block) ?

Similarly we introduce the critical rate $R_{\text{crit}}(\gamma)$ for the original problem

$$R_{\text{crit}}(\gamma) = \inf\{R : e(\gamma, R) = e(\gamma)\}\,; \quad \gamma \ge 0\,. \tag{9}$$

Due to proposition 1 we have

$$R_{\text{crit}}(\gamma) = 1 - r_{\text{crit}}(\gamma)\,; \quad \gamma \geq 0\,. \tag{10}$$

Remark. The value $r_{\text{crit}}(\gamma)$ is similar to the channel capacity C, and the function $e_2(\gamma, r)$ is similar to the reliability function $E(r)$ in information theory [7, 8]. The exact form of the realiability function $E(r)$ is not known till now. Therefore complete investigation of the function $e_2(\gamma, r)$ (for $r > r_{\text{crit}}(\gamma)$) seems to be a rather difficult problem.

III. Estimates for $r_{\text{crit}}(\gamma)$ and $e(\gamma, R)$

1. Lower bound for $r_{\text{crit}}(\gamma)$ (with randomly chosen set $\mathcal{X}$)

As before, let the measure P corresponds to p_0, the measure Q corresponds to p_1 and $0 < p_0 < p_1 \leq 1/2$. We consider all sets $\mathcal{X}$ of cardinality $X \sim 2^{Rn}$ on E_{in}^n. Let also some decision rule be chosen such that the first kind error probability for each set $\mathcal{X}$ does not exceed a given value α_n. Then each $\mathcal{X}$ has its own second kind error probability $\beta_n(\mathcal{X})$. It is clear that there exists some set $\mathcal{X}$ for which the value $\beta_n(\mathcal{X})$ does not exceed the averaged (over all sets $\mathcal{X}$) value $E\beta_n(\mathcal{X})$. Therefore if we are able to calculate (or upperbound) the value $E\beta_n(\mathcal{X})$ then it will give a certain lower bound for $e_2(\gamma, r)$ and $r_{\text{crit}}(\gamma)$.

Such a random choice method (with possible modifications) in information theory represents the most universal tool for obtaining various existence theorems [7, 8].

In order to realize that approach we choose as set $\mathcal{X}$ of cardinality $X \sim 2^{Rn}$ on E_{in}^n randomly and equiprobably X different points $\{x_1, \ldots, x_X\}$ and let y be our observation.

As the acceptance region $\mathcal{A}(\tau)$ in favor of p_0 we use

$$\mathcal{A}(\tau) = \left\{ y : \min_{i=1,\ldots,X} d(y, x_i) \leq \tau n \right\} ,$$

where the value $p_0 \leq \tau \leq p_1$ will be chosen later.

In order to investigate such a test performance, without loss of generality we may assume that the true value of block x is $x_1 = \mathbf{0}$.

If hypothesis p_0 is valid then for the first kind error probability we have

$$\alpha_n \leq P\{w(y) > \tau n | p_0, x_1\} \simeq$$

$$\simeq \binom{n}{\tau n} (1 - p_0)^{(1-\tau)n} p_0^{\tau n} \simeq \exp_2\{-nD(\tau || p_0)\} \, .$$

Let now hypothesis p_1 be valid. If $w(y) \leq \tau n$ then we accept that a decision error takes place. If $w(y) > \tau n$ then we can make a decision error only if in a sphere of radius τn centered at y there is some point x_i. Now for the averaged second kind error probability $E\beta_n$ we have (with $M = |E^n| = 2^n$, V– cardinality of ball of the radius τn in E^n_{in})

$$E\beta_n \leq P\{w(y) \leq \tau n | p_1, x_1\} + 1 - \binom{M-V}{X-1} \Big/ \binom{M-1}{X-1} =$$

$$= \exp_2\{-nD(\tau \| p_1)\} + 1 - \prod_{i=1}^{V-1} \left[1 - \frac{(X-1)}{(M-V-i)}\right] \leq$$

$$\leq \exp_2\{-nD(\tau \| p_1)\} + 1 - \left[1 - \frac{(X-1)}{(M-2V+1)}\right]^{V-1} \leq$$

$$\leq \exp_2\{-nD(\tau \| p_1)\} + \frac{XV}{(M-2V)} \simeq$$

$$\simeq \exp_2\{-nD(\tau \| p_1)\} + \exp_2\{-[1-h(\tau)-r]n\} .$$

Therefore there exists a set $\mathcal{X}$ of cardinality $X \sim 2^{rn}$ for which under the decision rule described the following inequalities are fullfilled

$$\alpha_n \leq \exp_2\{-nD(\tau \| p_0)\} ;$$

$$\beta_n \leq \exp_2\{-n \min\{D(\tau \| p_1), 1-h(\tau)-r\}\} .$$

Therefore for the function $e_2(\gamma, r)$ the following lower bound is valid

$$e_2(\gamma, r) \geq \min\{D(\tau \| p_1), 1-h(\tau)-r\} ; \quad 0 < r < 1 , \tag{11}$$

where the value $p_0 \leq \tau \leq p_1$ for $0 \leq \gamma \leq D(p_1 \| p_0)$ is defined as the unique root of equation $\gamma = D(\tau \| p_0)$.

In particular, if we want to be fulfilled the relation $\beta_n \simeq D(\tau \| p_1)$ (as for $X = 1$) then it is sufficient to have

$$r \leq 1 - h(\tau) - D(\tau \| p_1) .$$

The last result can be formulated in the following form.

Proposition 2. *For critical rate $r_{\text{crit}}(\gamma)$ in the dual problem the following lower bound holds*

$$r_{\text{crit}}(\gamma) \geq 1 - h(\tau) - D(\tau \| p_1) ; \qquad 0 \leq \gamma \leq D(p_1 \| p_0) . \tag{12}$$

where the value $p_0 \leq \tau \leq p_1$ is defined as the unique root of the equation $\gamma = D(\tau \| p_0)$, $0 \leq \gamma \leq D(p_1 \| p_0)$.

Remark. 1) Estimates (11)–(12) remain valid even when we test the composite hypothesis $H_0 : p \leq p_0$ against the simple alternative $H_1 : p = p_1$.

2) Let $\tau = p_1$, i.e. $\gamma = D(p_1 \| p_0)$. Then $D(\tau \| p_1) = 0$ and bound (12) takes the form:

$$r_{\text{crit}}(D(p_1 \| p_0)) \geq 1 - h(p_1) .$$

That bound is defined by a "sphere packing" of the space E_{in}^n by balls of radius $p_1 n$! The reason that knowing only the set $\mathcal{X}$ of cardinality $X \sim 2^{rn}$ we are able to achieve the same performance as if we would know the input block x is the following. For a good set $\mathcal{X}$ (almost all randomly chosen sets $\mathcal{X}$ are such) knowing output block y and set $\mathcal{X}$ it is possible, with small error probability, to identify which of the input blocks $x \in \mathcal{X}$ was really used (under any hypotheses).

2. Case $p_1 = 1/2$

In the special case $p_1 = 1/2$ it is possible to find the function $e_2(\gamma, r)$.

If $p_1 = 1/2$ then for any x^n, y^n we have $Q(y^n|x^n) = 2^{-n}$. Therefore if in the dual problem $\mathcal{A}$ is the set of decision in favor of p_0 then for the second kind error probability we have $\beta_n = |\mathcal{A}|2^{-n}$. Due to simplicity of that expression it is more convenient now to fix the exponent rate of the second kind error probability $0 \le \gamma_2 \le D(p_0||1/2) = 1 - h(p)$ (i.e. $\beta_n \simeq 2^{-\gamma_2 n}$) and to investigate the best exponent rate of the first kind error probability $e_1(\gamma_2, r)$. For given value γ_2 we have for the cardinality of set $\mathcal{A}$

$$|\mathcal{A}| \simeq 2^{(1-\gamma_2)n} .$$

On the other hand, for each input block $x_i \in \mathcal{X}$ the optimal region of decision in favor of p_0 is a ball of some radius τn centerd at x_i. Therefore for the optimal set $\mathcal{X}$ the acception region $\mathcal{A}$ should contain "almost completely" each of those balls. It is clear that in such a set $\mathcal{X}$ all points $\{x_i\}$ should be maximally close to each other (i.e. $\mathcal{X}$ is a ball) and also $\mathcal{A}$ is a ball concentric with it. Therefore we have

$$r_{\text{crit}}(\gamma) = 0, \ \gamma \ge 0 . \tag{13}$$

Let νn and μn be radiuses of balls $\mathcal{X}$ and $\mathcal{A}$, respectively. Since the cardinality $|\mathcal{X}| \simeq 2^{rn}$ then ν and μ are defined from the relations

$$r = h(\nu)\,; \ 1 - \gamma_2 = h(\mu)\,; \quad 0 \le r < 1/2\,, \quad p_0 \le \mu < 1/2\,. \tag{14}$$

We may assume that balls $\mathcal{X}$ and $\mathcal{A}$ are centered at zero. If hypothesis p_0 is valid and input block x has weight νn then due to the law of large numbers the output block y has (with probability close to 1) weight $(p_0 + \nu(1 - 2p_0))n$. Therefore in order to have the first kind error probability α_n small it is necessary that μ satisfies the condition

$$\mu \ge p_0 + \nu(1 - 2p_0)\,. \tag{15}$$

From that condition and (14) follows

$$e_1(\gamma_2, r) = 0\,; \quad r \ge r_0(\gamma_2, p_0) = h\left(\frac{h^{-1}(1-\gamma_2) - p_0}{1 - 2p_0}\right)\,. \tag{16}$$

Now for $r < r_0(\gamma_2, p_0)$ we evaluate the first kind error probability (it will define the function $e_1(\gamma, r)$). The first kind error takes place if the output block has weight greater than μn. We may assume that the input block x has ones on the first νn positions and zeros on the remaining $(1-\nu)n$ positions. Let in denotes the number of errors on the first νn positions and jn denotes the number of errors on the remaining $(1-\nu)n$ positions. Then erroneous decision takes place if $\nu - i + j \geq \mu$. Therefore denoting $z = (1-p_0)/p_0 > 1$ we have

$$\alpha_n \simeq (1-p_0)^n \max \binom{\nu n}{in} \binom{(1-\nu)n}{jn} z^{-(i+j)n} \simeq \tag{17}$$

$$\simeq (1-p_0)^n \exp_2 \left\{ n \max \left[\nu h\left(\frac{i}{\nu}\right) + (1-\nu) h\left(\frac{j}{1-\nu}\right) - (i+j)\log z \right] \right\}.$$

where the maximum is taken over $0 \leq i \leq \nu$; $0 \leq j \leq 1-\nu$; $\nu - i + j \geq \mu$.

It is not difficult to check that at the point where the maximum is attained on the right hand side of (17) equality $\nu - i + j = \mu$ holds (otherwise condition (15) is violated). Therefore from (17) we get

$$e_1(\gamma_2, r) = \log \frac{1}{1-p_0} - \max_{0 \leq i \leq \nu} f(i); \tag{18}$$

$$f(i) = \nu h\left(\frac{i}{\nu}\right) + (1-\nu) h\left(\frac{\mu - \nu + i}{1-\nu}\right) - (2i + \mu - \nu)\log z, \quad z = \frac{1-p_0}{p_0}.$$

It is easy to check that the function $f(i)$ is $\cap$–convex in i and attains its maximum inside the interval $(0, \nu)$. Therefore the optimal value i_0 is the unique root of the equation

$$\log \frac{\nu - i}{i} + \log \frac{1 - \mu - i}{\mu - \nu + i} = 2 \log z,$$

from where denoting $u = (1-2p_0)/p_0^2 = z^2 - 1$ we get

$$i_0 = \frac{\sqrt{[u(\mu-\nu)+1]^2 + 4u\nu(1-\mu)} - u(\mu-\nu) - 1}{2u}, \quad u = \frac{1-2p_0}{p_0^2}. \tag{19}$$

These results can be formulated in the following form.

Proposition 3. *If $p_1 = 1/2$ then $r_{\text{crit}}(\gamma) = 0$, $\gamma \geq 0$, and the best exponent $e_1(\gamma_2, r)$ of the first kind error probability is given by the formula*

$$e_1(\gamma_2, r) = \begin{cases} 0, & r \geq r_0(\gamma_2, p_0); \\ -\log(1-p_0) - f(i_0), & 0 \leq r < r_0(\gamma_2, p_0); \end{cases} \tag{20}$$

where $\mu, \nu, i_0, r_0(\gamma_2, p_0)$ are defined in (14), (16), (15) and (19).

3. A useful counterexample

Unfortunately, we were not able yet to obtain good lower bounds for the critical rate $R_{\text{crit}}(\gamma)$ (or upper bounds for $r_{\text{crit}}(\gamma)$, respectively).

The following counterexample demonstrates some problems arising when one tries to get such results.

We consider the following variant of the dual problem. Let $0 < p_0 < p_1 < 1$ be fixed¿ It is known that input block x^n belongs to some set $\mathcal{X} \subset E_{\text{in}}^n$ of cardinality $X \simeq 2^{rn}$. What is the maximal growth rate $r_{\max}$ of the cardinality of the best set $\mathcal{X}$ such that we can test those hypotheses if we demand only that both error probabilities vanish ? The answer is very simple: $r_{\max} = 1$.

Indeed, we choose as set $\mathcal{X} = \{x_1, \ldots, x_X\}$ all points on a sphere of some radius $rn < n/2$. Then $X \simeq 2^{nh(r)}$. Since the input block has weight rn then due to the law of large numbers the output block with probability close to 1 will have weight $(r + p(1 - 2r))n$. Therefore for large n we will be able to test hypotheses $p = p_0$ and $p = p_1$ with small error probabilities for any $r < 1/2$. It means that $r_{\max} = 1$.

References

[1] R. Ahlswede and I. Csiszár, " Hypothesis testing with communication constraints", *IEEE Trans. Inform. Theory* 32 (4), 1986, 533–542.

[2] Z. Zhang and T. Berger, "Estimation via compressed information", *IEEE Trans. Inform. Theory* 34 (2), 1988, 198–211.

[3] T.S. Han and K. Kobayashi, "Exponential–type error probabilities for multiterminal hypothesis testing", *IEEE Trans. Inform. Theory* 35 (1), 1989, 2–14.

[4] R. Ahlswede and M.V. Burnashev, "On Minimax estimation in the presence of side information about remote data", *The Annals of Statistics* 18 (1), 1990, 141–171.

[5] T.S. Han and S. Amari, "Parameter estimation with multiterminal data compression", *IEEE Trans. Inform. Theory* 41 (6), 1995, 1802–1833.

[6] I.A. Ibragimov and R.Z. Has'minskii, *Statistical Estimation. Asymptotic Theory*, Springer–Verlag, 1981.

[7] R.M. Fano, *Transmission of Information. A Statistical Theory of Communication*, MIT&Wiley, New York–London, 1961.

[8] R.G.Gallager, *Information Theory and Reliable Communication*, Wiley, New York–London–Sydney–Toronto, 1968.

[9] R. Ahlswede and I. Althöfer, "The asymptotic behavior of diameters in the average", *Journal of Combinatorial Theory*, Ser. B 61 (2), 1994, 167–177.

THE AHLSWEDE-DAYKIN THEOREM

Peter C. Fishburn

AT&T Labs–Research, Florham Park, NJ 07932

fish@research.att.com

Lawrence A. Shepp

Rutgers University, Piscataway, NJ 08855

shepp@stat.rutgers.edu

In appreciation to Rudolf Ahlswede

Abstract: In 1978, Rudolf Ahlswede and David Daykin published a theorem which says that a certain inequality on nonnegative real valued functions for pairs of points in a finite distributive lattice extends additively to pairs of lattice subsets. It is an elegant theorem with widespread applications to inequalities for systems of subsets, linear extensions of partially ordered sets, and probabilistic correlation. We review the theorem and its applications, and describe a recent generalization to n-tuples of points and subsets in distributive lattices. Although many implications of the Ahlswede-Daykin theorem follow from the weaker hypotheses of the widely-cited FKG theorem, several important implications are noted to require the stronger hypotheses of the basic theorem of Ahlswede and Daykin.

THE AHLSWEDE-DAYKIN THEOREM

A *lattice* is a partially ordered set $(\Gamma, \prec)$ in which every pair of points $a, b \in \Gamma$ has a unique least upper bound or join

$$a \vee b = \min\{z \in \Gamma : a \preceq z, b \preceq z\}$$

I. Althöfer et al. (eds.), Numbers, Information and Complexity, 501-516.

and a unique greatest lower bound or meet

$$a \wedge b = \max\{z \in \Gamma : z \preceq a, z \preceq b\} .$$

The lattice is *distributive* if

$$a \wedge (b \vee c) = (a \wedge b) \vee (a \wedge c) \quad \text{for all} \quad a, b, c \in \Gamma$$

or, equivalently, if $a \vee (b \wedge c) = (a \vee b) \wedge (a \vee c)$ for all $a, b, c \in \Gamma$. We presume throughout that Γ is finite and recall the useful fact [5, p. 59] that a finite distributive lattice is order-isomorphic for some n to a restriction of $(2^n, \subset)$, the family of subsets of $\{1, 2, \ldots, n\}$ ordered by proper inclusion.

For nonempty $A, B \subseteq \Gamma$, $\vee$ and $\wedge$ are extended to subsets of Γ by

$$\begin{aligned} A \vee B &= \{a \vee b : a \in A, b \in B\} \\ A \wedge B &= \{a \wedge b : a \in A, b \in B\} , \end{aligned}$$

with $A \vee B = \emptyset = A \wedge B$ if A or B is empty. In 1977, Daykin [11] proved that a lattice $(\Gamma, \prec)$ is distributive if and only if

$$|A||B| \leq |A \vee B||A \wedge B| \quad \text{for all} \quad A, B \subseteq \Gamma .$$

This inequality is but one of many implications of a remarkable theorem published the next year by Ahlswede and Daykin [3] that has come to be known as the Ahlswede-Daykin theorem , or the four-functions theorem [6]. For any real-valued function f on Γ, we define the *additive extension* of f, also denoted by f, by

$$f(A) = \sum_{a \in A} f(a) \quad \text{for all} \quad A \subseteq \Gamma .$$

Theorem 1. (Ahlswede-Daykin) *Suppose* $(\Gamma, \prec)$ *is a finite distributive lattice and* $\alpha, \beta, \gamma, \delta : \Gamma \to [0, \infty)$ *satisfy*

$$\alpha(a)\beta(b) \leq \gamma(a \vee b)\delta(a \wedge b) \quad \textit{for all} \quad a, b \in \Gamma .$$

Then

$$\alpha(A)\beta(B) \leq \gamma(A \vee B)\delta(A \wedge B) \quad \textit{for all} \quad A, B \subseteq \Gamma .$$

When $(\Gamma, \prec) = (2^n, \subset)$ with $\vee = \cup$ and $\wedge = \cap$, the hypothesized inequality, $\alpha(a)\beta(b) \leq \gamma(a \vee b)\delta(a \wedge b)$, has the flavor of *log supermodularity* for a probability distribution μ on the ground set 2^n, defined by

$$\mu(a)\mu(b) \leq \mu(a \cup b)\mu(a \cap b) \quad \text{for all} \quad a, b \in 2^n .$$

The hypothesized inequality of Theorem 1 can be viewed as a far-reaching generalization of log supermodularity, which is a key hypothesis of the widely-cited **FKG** theorem of Fortuin, Kasteleyn and Ginibre [18]. The power of the Ahlswede-Daykin theorem lies in its conclusion that the four-functions inequality hypothesized for individual members of Γ is inherited by subsets of Γ under additive extensions.

Proofs of Theorem 1 are included in [3, 6, 16]. The standard approach is to prove the theorem for $(2^n, \subset)$. The general result for $(\Gamma, \prec)$ order-isomorphic to a restriction of $(2^n, \subset)$ then follows by fixing α, β, γ and δ at 0 on the members of 2^n excluded from the isomorphism. The $(2^n, \subset)$ proof shows that the result holds for $n = 1$ and proceeds by induction on n. The overall proof is pleasantly compact — about one page — in view of the theorem's many implications.

Several of those implications, including the FKG theorem, were proved prior to the publication of [3]. We will not dwell on precedence, but instead will indicate how a variety of results follow from Theorem 1 as the root of a tree-like structure. We classify those results into three types.

Type 1 implications follow more or less directly from Theorem 1 by choosing specific forms for α, β, γ and δ. They include Daykin's inequality for distributivity [11], the FKG theorem [18] and Holley's theorem [22], an inequality of Kleitman [27] and Seymour [33], and the Marica-Schönheim inequality [30].

Type 2 implications use direct applications of Theorem 1 or its type 1 implications, but involve other techniques to arrive at their conclusions. The other techniques often include a reformulation of the problem's structure prior to the direct application, and may have one or more steps that require functional extremization or an examination of limit behavior. Examples include the correlational inequalities for linear extensions of Graham, Yao and Yao [21] and Shepp [34], the so-called xyz inequalities of Shepp [35] and Fishburn [13], and universal correlation theorems of Winkler [39] and Brightwell [8].

As we proceed, it will be clear that many implications of the Ahlswede-Daykin theorem follow from the weaker hypotheses of the FKG theorem described in the next section. There are, however, important applications of Theorem 1 which require its stronger hypotheses. Two cases in point occur in the proofs of the strict xyz inequality in [13] and the random permutations theorem [17] mentioned in the next paragraph.

Type 3 implications involve structure for which the hypotheses of Theorem 1 or a type 1 or type 2 implication are false, even under reformulations, but which admit perturbations that allow application of preceding results. The perturbed structure is close to the original, and the disparity between the two can be remedied by methods that lead to the desired conclusion. Our primary example of a type 3 implication is a correlation inequality for match sets of random permutations that was conjectured by Joag-Dev [24] and Prem Goel and proved in Fishburn, Doyle and Shepp [17].

The question of which type characterizes a particular implication is subject to personal judgment and can depend on available proofs, so we acknowledge a degree of latitude in our choices. Nevertheless, we have found the classification useful for an appreciation of the role of the Ahlswede-Daykin theorem, and proceed accordingly.

Section 2 of the paper discusses type 1 implications, section 3 describes type 2 implications, and section 4 outlines our perturbation approach to the match set problem with random permutations. We then conclude with a recent generalization of the Ahlswede-Daykin theorem due to Rinott and Saks [31, 32] and Aharoni and Keich [2].

Prior surveys of much of the material we cover are presented by Graham [19, 20], Winkler [40] and Fishburn [16]. We have borrowed freely from these sources and acknowledge our indebtedness to Ron Graham and Peter Winkler.

TYPE 1 IMPLICATIONS

We assume throughout this section that $(\Gamma, \prec)$ is a finite distributive lattice. Our first implication of Theorem 1 takes $\alpha = \beta = \gamma = \delta = \mu$ with $\mu : \Gamma \to [0, \infty)$. Then *log supermodularity* for μ, i.e.,

$$\mu(a)\mu(b) \leq \mu(a \vee b)\mu(a \wedge b) \quad \text{for all} \quad a, b \in \Gamma ,$$

which becomes the hypothesized inequality of Theorem 1, implies the same form for additive extensions:

$$\mu(A)\mu(B) \leq \mu(A \vee B)\mu(A \wedge B) \quad \text{for all} \quad A, B \subseteq \Gamma .$$

When $\mu \equiv 1$ is added to the hypotheses, log supermodularity is automatic and Theorem 1 yields Daykin's inequality $|A||B| \leq |A \vee B||A \wedge B|$ for all $A, B \subseteq \Gamma$.

Log supermodularity also underlies the following lattice version of the FKG theorem. We say that $f : \Gamma \to \mathrm{R}$ is *nondecreasing* if

$$a \prec b \Rightarrow f(a) \leq f(b), \quad \text{for all} \quad a, b \in \Gamma .$$

Theorem 2. (FKG) *Suppose $\mu : \Gamma \to [0, \infty)$ is log supermodular. Then for all nondecreasing $f, g : \Gamma \to \mathrm{R}$,*

$$\left[\sum_\Gamma \mu(a)f(a)\right]\left[\sum_\Gamma \mu(a)g(a)\right] \leq \left[\sum_\Gamma \mu(a)\right]\left[\sum_\Gamma \mu(a)f(a)g(a)\right] .$$

Proof. It is easily seen that the conclusion is invariant to the addition of a constant c to f and g, so we assume that f and g are positive. Then define α, β, γ and δ for Theorem 1 by $f\mu$, $g\mu$, $fg\mu$ and μ, respectively. For example, $\alpha(a) = f(a)\mu(a)$. The hypotheses of Theorem 2 then imply those of Theorem 1, and the conclusion of Theorem 1 implies that of Theorem 2 when $A = B = \Gamma$. ∎

Several implications of the FKG theorem will be noted later. Other implications and related results are available in Kemperman [26], Joag-Dev, Shepp and Vitale [25], van den Berg and Kesten [38], van den Berg and Fiebig [37], Hwang and Shepp [23], Burton and Franzosa [10], and Bollobás and Brightwell [7].

A probabilistic form of the FKG theorem arises by taking $(\Gamma, \prec) = (2^n, \subset)$ with $\vee = \cup$ and $\wedge = \cap$. Let $\mathcal{B}_n$ denote the Boolean algebra of subsets of 2^n, so each object in $\mathcal{B}_n$ is a set of subsets of $\{1, 2, \ldots, n\}$. We say that $A \in \mathcal{B}_n$ is an *up-set* (order filter) if $(a \in A, a \subset b) \Rightarrow b \in A$, and a *down-set* (order ideal, simplicial complex) if $(a \in A, b \subset a) \Rightarrow b \in A$. Clearly, A is an up-set if and only if its complement $2^n \setminus A$ is a down-set. We normalize $\mu \geq 0$ so that $\sum\{\mu(a) : a \in 2^n\} = 1$, and view its additive extension μ as a probability measure on $\mathcal{B}_n$. The expected value of f with respect to μ is $E(f,\mu) = \sum_{a \in 2^n} \mu(a) f(a)$.

Theorem 3. (FKG) *Suppose μ is a probability measure on $\mathcal{B}_n$ and $\mu(a)\mu(b) \leq \mu(a \cup b)\mu(a \cap b)$ for all $a, b \in 2^n$. Then*

(1) $E(f,\mu)E(g,\mu) \leq E(fg,\mu)$ for all nondecreasing $f, g : 2^n \to \mathrm{R}$;

(2) $\mu(A)\mu(B) \leq \mu(A \vee B)\mu(A \wedge B)$ for all $A, B \in \mathcal{B}_n$,

(3) $\mu(A \cap B) \geq \mu(A)\mu(B)$ for all up-sets $A, B \in \mathcal{B}_n$.

Comments. (1) is tantamount to the inequality of Theorem 2 under normalization. (3) is immediate from (1) by taking $f = 1$ on A, 0 otherwise, and $g = 1$ on B, 0 otherwise. In (2), $A \vee B = \{a \cup b : a \in A, b \in B\}$, which is not generally equal to $A \cup B$. In fact, if A and B are up-sets then $A \vee B = A \cap B$. ∎

An intermediate result between Theorems 1 and 2 was established by Holley [22]. It says that if μ_1, $\mu_2 : \Gamma \to [0, \infty)$ satisfy $\sum_\Gamma \mu_1(a) = \sum_\Gamma \mu_2(a)$ and

$$\mu_1(a)\mu_2(b) \leq \mu_1(a \vee b)\mu_2(a \wedge b) \quad \text{for all} \quad a, b \in \Gamma ,$$

then $\sum_\Gamma \mu_1(a) f(a) \geq \sum_\Gamma \mu_2(a) f(a)$ for every nondecreasing $f : \Gamma \to \mathrm{R}$. The proof by Theorem 1 is similar to the proof of Theorem 2. We add a constant to f to make it positive, define α, β, γ and δ by μ_1, $f\mu_2$, $f\mu_1$ and μ_2, respectively, then use Theorem 1 with $A = B = \Gamma$ to obtain Holley's conclusion. When μ_1 and μ_2 are probability measures on $\mathcal{B}_n$ that satisfy

$$\mu_1(a)\mu_2(b) \leq \mu_1(a \cup b)\mu_2(a \cap b) \quad \text{for all} \quad a, b \in 2^n ,$$

Holley's theorem says that

$$E(f, \mu_1) \geq E(f, \mu_2) \quad \text{for every nondecreasing} \quad f : 2^n \to \mathrm{R} .$$

We mention several further results for $\mathcal{B}_n$.

Theorem 4. ([38, 33]) *Suppose $A, B \in \mathcal{B}_n$. If A is an up-set and B is a down-set, then $2^n|A \cap B| \leq |A||B|$. If both A and B are up-sets or down-sets, then $2^n|A \cap B| \geq |A||B|$.*

Proof. The up-sets conclusion is immediate from Theorem 3(3) on taking $\mu(a) = 2^{-n}$ for each $a \in 2^n$. The other conclusions follow from complementation. ∎

The next theorem involves systems of set differences. Its proof requires a few steps beyond what is immediate from Theorem 1 and could be considered a boundary case between types 1 and 2. For $A, B \in \mathcal{B}_n$, let

$$A - B = \{a \setminus b : a \in A, b \in B\} .$$

Theorem 5. ([30]) *For all $A, B \in \mathcal{B}_n$, $|A - B||B - A| \geq |A||B|$.*

Proof. Let $\mathbf{n} = \{1, 2, \ldots, n\}$. Using Daykin's inequality, we have

$$\begin{aligned}
|A||B| &= |A||\{\mathbf{n} \setminus b : b \in B\}| \\
&\leq |A \vee \{\mathbf{n} \setminus b : b \in B\}||A \wedge \{\mathbf{n} \setminus b : b \in B\}| \\
&= |\{a \cup (\mathbf{n} \setminus b) : a \in A, b \in B\}||\{a \cap (\mathbf{n} \setminus b) : a \in A, b \in B\}| \\
&= |\{\mathbf{n} \setminus (a \cup \mathbf{n} \setminus b) : a \in A, b \in B\}||\{a \setminus b : a \in A, b \in B\}| \\
&= |\{b \setminus a : a \in A, b \in B\}||A - B| \\
&= |B - A||A - B| \,. \quad \blacksquare
\end{aligned}$$

The implication

$$|A - A| \geq |A|$$

of Theorem 5 is known as the *Marica-Schönheim inequality.* Additional facts about the Marica-Schönheim inequality and close relatives are included in Daykin and Lovász [12], Ahlswede and Daykin [4], Aharoni and Holzman [1] and Lengvárszky [29]. Although their proofs go well beyond our type 1 designation, we mention some of their results here before we discuss other type 2 implications in the next section. For the following composite theorem, parts (1) and (4) are proved in [1], (2) is proved in [12], and (3) is proved in [4]. In part (1), we say that A is *weakly separating* [1] if for all distinct i and j in $\{1, 2, \ldots, n\}$, $\{a \in A : i \in a\} = \{a \in A : j \in a\}$ implies that both sets equal A or both are empty. In addition, $\mathcal{B}_s$ denotes the family of sets of subsets of s for $s \in 2^n$.

Theorem 6. *Suppose $A, B \in \mathcal{B}_n$.*

(1) If A is weakly separating, then $|A - A| = A$ if only if there is a partition of $\{1, 2, \ldots, n\}$ into s and t, an up-set S in $\mathcal{B}_s$, and a down-set T in $\mathcal{B}_t$ such that $A = \{a \cup b : a \in S, b \in T\}$.

(2) If $|A| \geq 2$ then there is a bijection $\phi : A \to A$ such that $\phi(a) \neq a$ for all $a \in A$, and $a \setminus \phi(a) \neq b \setminus \phi(b)$ for all $a \neq b$ in A.

(3) If for every $a \in A$, $b \subseteq a$ for some $b \in B$, then $|A - B| \geq |A|$.

(4) If for all a, $a' \in A$, $(a \setminus a') \cap b = \emptyset$ for some $b \in B$, then $|A - B| \geq |A|$.

Part (1) essentially covers all cases of equality for the Marica-Schönheim inequality, and (2) is a strengthened version of the inequality for $|A| > 1$. Part (3) provides a first-order generalization of the Marica-Schönheim inequality, and (4) strengthens (3) by weakening its hypothesis.

Lengvárszky [29] proves that an analogue of the Marica-Schönheim inequality holds for $(\Gamma, \prec)$ when $a - b$ for $a, b \in \Gamma$ is defined in a particular way with $A - B = \{a - b : a \in A, b \in B\}$ for $A, B \subseteq \Gamma$. The paper also considers $|A - A| \geq |A|$ when the lattice is not necessarily distributive.

TYPE 2 IMPLICATIONS FOR LINEAR EXTENSIONS

We assume throughout this section that $(X, \prec)$ is a finite partially ordered set. We do not assume that $(X, \prec)$ is a lattice, let alone a distributive lattice, so implications of the Ahlswede-Daykin and FKG theorems will involve construction of distributive lattices for application of those theorems.

The section focuses on linear extensions of $(X, \prec)$, where $(X, \prec_0)$ is a *linear extension* of $(X, \prec)$ if $<_0$ linearly orders X and $x \prec y \Rightarrow x <_0 y$ for all $x, y \in X$. We say that $x, y \in X$ are *incomparable* in $(X, \prec)$ if $x \neq y$ and neither $x \prec y$ nor $y \prec x$. We let $\mathcal{L}$ denote the set of all linear extensions of $(X, \prec)$ and set $N = |\mathcal{L}|$. We recall [36] that if x and y are incomparable in $(X, \prec)$ then $x <_0 y$ for some linear extension in $\mathcal{L}$, so $\prec = \cap\{<_0 : (X, <_0) \in \mathcal{L}\}$.

A few other notations are used in the section. We let μ denote the *uniform* probability measure on $2^{\mathcal{L}}$, so $\mu(L) = 1/N$ for every $L \in \mathcal{L}$. We take $(x <_0 y) = \{L \in \mathcal{L} : x <_0 y \text{ in } L\}$, the set of linear extensions in which $x <_0 y$. The probability of $(x <_0 y)$ under μ is $\mu(x <_0 y)$, with $\mu(x <_0 y) + \mu(y <_0 x) = 1$ when $x \neq y$. Clearly, $\mu(x <_0 y) = |(x <_0 y)|/N$. Finally, we denote by $\cap_I(a_i <_0 b_i)$ the set of linear extensions of $(X, \prec)$ in which $a_i <_0 b_i$ is true for every $i \in \{1, 2, \ldots, I\}$.

Our first two results for the equally-likely linear extensions model consider two-part partitions of X from different perspectives. Their conclusion, $\mu(A \cap B) \geq \mu(A)\mu(B)$, expresses nonnegative correlation between the defined events A and B: the joint occurrence of A and B is at least as probable as the product of their separate probabilities. When $\mu(B) > 0$, $\mu(A \cap B) \geq \mu(A)\mu(B)$ says that $\mu(A|B) \geq \mu(A)$, or that A is at least as likely to occur when B occurs as it is unconditionally.

Theorem 7. ([21]) *Suppose $\{X_1, X_2\}$ is a nontrivial partition of X and $\prec$ linearly orders X_i for $i = 1, 2$. Let $A = \cap_I(a_i <_0 b_i)$ and $B = \cap_J(c_j <_0 d_j)$ for some I and J with all $a_i, c_j \in X_1$ and all b_i, $d_j \in X_2$. Then $\mu(A \cap B) \geq \mu(A)\mu(B)$.*

Theorem 8. ([34]) *Suppose $(X, \prec)$ is the union of disjoint nonempty partially ordered sets $(X_1, \prec_1)$ and $(X_2, \prec_2)$, with $\prec = \prec_1 \cup \prec_2$. With A and B as in Theorem 7, $\mu(A \cap B) \geq \mu(A)\mu(B)$.*

The intuition behind the theorems is that all elementary events for A and B have the form $(x_1 <_0 x_2)$ for $x_1 \in X_1$ and $x_2 \in X_2$, so realization of one of A and B should enhance the likelihood of the other. We note, however, that this intuition is tenuous because $\mu(A \cap B) \geq \mu(A)\mu(B)$ can be false *except* when $(X, \prec)$ has specialized structure as in the theorems' hypotheses. Examples in Shepp [34] and Graham [20, p. 122] show how the conclusion fails for other structures.

Proofs based on the FKG theorem appear in [28, 34] for Theorem 7 and in [34] for Theorem 8. We sketch the proof of Theorem 7 to illustrate constructions that lead to FKG.

Let $(X_1, \prec) = \{x_1 \prec x_2 \prec \cdots \prec x_m\}$ and $(X_2, \prec) = \{y_1 \prec y_2 \prec \cdots \prec y_n\}$ with $m, n \geq 1$. Let Γ be the set of all strictly increasing m-tuples of integers from $\{1, 2, \ldots, m+n\}$, and for $\alpha = (\alpha_1, \ldots, \alpha_m)$ and $\beta = (\beta_1, \ldots, \beta_m)$ in Γ define a reflexive relation $\leq^*$ on Γ by

$$\alpha \leq^* \beta \quad \text{if} \quad \alpha_i \leq \beta_i \quad \text{for} \quad i = 1, \ldots, m \,.$$

Also define $\alpha \wedge \beta$ and $\alpha \vee \beta$ componentwise by

$$(\alpha \wedge \beta)_i = \min\{\alpha_i, \beta_i\}, \quad (\alpha \vee \beta)_i = \max\{\alpha_i, \beta_i\} .$$

It follows that $(\Gamma, \leq^*)$ is a distributive lattice (reflexive variety).

We next define a log supermodular function v and nondecreasing functions $-f$ and $-g$ on $(\Gamma, \leq^*)$ as follows. Given $\alpha \in \Gamma$, let α^c be the strictly increasing n-tuple of integers in $\{1, 2, \ldots, m+n\} \setminus \{\alpha_1, \ldots, \alpha_m\}$, and let σ_α denote the bijection from X onto $\{1, 2, \ldots, m+n\}$ defined by

$$\sigma_\alpha(x_i) = \alpha_i \quad (i = 1, \ldots, m); \qquad \sigma_\alpha(y_j) = \alpha_j^c \quad (j = 1, \ldots, n) .$$

Also let $(X, \prec_A)$ and $(X, \prec_B)$ denote the ordered sets in which $\prec_A = \{(a_1, b_1), \ldots, (a_I, b_I)\}$ and $\prec_B = \{(c_1, d_1), \ldots, (c_J, d_J)\}$. We then define $v, f, g : \Gamma \to \{0, 1\}$ by

$v(\alpha) = 1 \;\Leftrightarrow\;$ the arrangement of X by increasing values of σ_α is a linear extension of $(X, \prec)$;

$f(\alpha) = 1 \;\Leftrightarrow\;$ the arrangement of X by increasing values of σ_α is a linear extension of $(X, \prec_A)$;

$g(\alpha) = 1 \;\Leftrightarrow\;$ the arrangement of X by increasing values of σ_α is a linear extension of $(X, \prec_B)$.

Once log supermodularity and monotonicity have been verified, we use Theorem 2 to conclude that

$$\sum_\Gamma v(\alpha) \sum_\Gamma f(\alpha) g(\alpha) v(\alpha) \geq \sum_\Gamma f(\alpha) v(\alpha) \sum_\Gamma g(\alpha) v(\alpha) ,$$

where the left-to-right sums are the *numbers of linear extensions* of $(X, \prec)$, of $(X, \prec)$ compatible with $\prec_A$ and $\prec_B$, of $(X, \prec)$ compatible with $\prec_A$, and of $(X, \prec)$ compatible with $\prec_B$. Division by N^2 gives $\mu(A \cap B) \geq \mu(A)\mu(B)$. ∎

Our next two theorems show that some instances of nonnegative (Theorem 9) and positive (Theorem 10) correlation do not require strong hypotheses like those in Theorems 7 and 8.

Theorem 9. *(xyz [35]) For all $x, y, z \in X$,*

$$\mu((x <_0 y) \cap (x <_0 z)) \geq \mu(x <_0 y)\mu(x <_0 z) .$$

Theorem 10. *(xyz [13]) For all mutually incomparable $x, y, z \in X$,*

$$\mu((x <_0 y) \cap (x <_0 z)) > \mu(x <_0 y)\mu(x <_0 z) .$$

Because the nonstrict inequality of Theorem 9 is easily seen to hold when x, y and z are not mutually incomparable, Theorem 10 can be viewed as a strengthening of Theorem 9. We outline a proof of Theorem 9 that uses a limiting argument similar to that used in [34] to prove Theorem 8, and then comment on a substantially different proof for Theorem 10.

Suppose for Theorem 9 that x, y and z are mutually incomparable. Fix an integer $K > |X|$ and let Γ_K be the set of all nondecreasing α from $(X, \prec)$ into $\{1, 2, \ldots, K\}$. Also define $\leq^*$, $\wedge$ and $\vee$ for $\alpha, \beta \in \Gamma_K$ by $\alpha \leq^* \beta$ if $\alpha(x) \geq \beta(x)$, and $\alpha(t) - \alpha(x) \leq \beta(t) - \beta(x)$ for all $t \in X$,

$$\begin{aligned}(\alpha \wedge \beta)(t) &= \min\{\alpha(t) - \alpha(x), \beta(t) - \beta(x)\} + \max\{\alpha(x), \beta(x)\} \\ (\alpha \vee \beta)(t) &= \max\{\alpha(t) - \alpha(x), \beta(t) - \beta(x)\} + \min\{\alpha(x), \beta(x)\} .\end{aligned}$$

Then $(\Gamma_K, \leq^*)$ is a (reflexive) distributive lattice.

Now for $a, b \in X$ let $(a < b)_K = \{\alpha \in \Gamma_K : \alpha(a) \leq \alpha(b)\}$. Then both $(x < y)_K$ and $(x < z)_K$ are up-sets in $(\Gamma_K, \leq^*)$. Indeed, for any $t \neq x$, $(\alpha(x) \leq \alpha(t), \alpha \leq^* \beta) \Rightarrow 0 \leq \alpha(t) - \alpha(x) \leq \beta(t) - \beta(x) \Rightarrow \beta(x) \leq \beta(t)$. This shows that the unusual definition of $\leq^*$ is just right for the up-set calculation. It then follows from Theorem 2 with the uniform measure on Γ_K that

$$\frac{|(x < y)_K \cap (x < z)_K|}{|\Gamma_K|} \geq \frac{|(x < y)_K|}{|\Gamma_K|} \frac{|(x < z)_K|}{|\Gamma_K|} .$$

As $K \to \infty$, the proportion of $\alpha \in \Gamma_K$ that have $\alpha(a) = \alpha(b)$ for $a \neq b$ goes to 0, and it follows by taking limits in the preceding inequality that $\mu((x <_0 y) \cap (x <_0 z)) \geq \mu(x <_0 y)\mu(x <_0 z)$. ∎

Because the limit argument of the preceding proof works only for nonstrict inequality, a different approach is needed for Theorem 10. The following lemma suffices.

Lemma 11. [13] *Suppose x, y and z are mutually incomparable in $(X, \prec)$, and $|X| = n$. Let $N(abc)$ be the number of linear extensions of $\mathcal{L}$ with $a <_0 b <_0 c$ and let*

$$\lambda = \frac{N(yxz)N(zxy)}{[N(xyz) + N(xzy)][N(yzx) + N(zyx)]} .$$

Then $\lambda \leq (n-1)^2/(n+1)^2$ if n is odd, $\lambda \leq (n-2)/(n+2)$ if n is even, and for each $n \geq 3$ some $(X, \prec)$ attains the indicated upper bound on λ.

The bulk of [13] is devoted to the proof of Lemma 11, which features two applications of the Ahlswede-Daykin theorem. The first application uses the preceding embedding technique with $K \to \infty$ and needs only the hypotheses of the FKG theorem. But the second involves an optimization step that requires the stronger hypotheses of Theorem 1 and yields the preceding bounds on λ.

To complete the proof of Theorem 10 let

$$T = \frac{N - N(yxz) - N(zyx)}{N(yzx) + N(zyx)} .$$

Also let $N(ab) = |\{L \in \mathcal{L} : a <_0 b \text{ in } L\}|$. Because $N(xy) = N(zxy) + N(xzy) + N(xyz)$ and $N(xz) = N(yxz) + N(xyz) + N(xzy)$, rearrangement gives

$$\frac{N(xy)N(xz)}{N[N(xyz) + N(xzy)]} = \frac{T + \lambda}{T + 1} .$$

Then $\lambda < 1$ by Lemma 11, so $\mu(x <_0 y)\mu(x <_0 z) < \mu((x <_0 y) \cap (x <_0 z))$. ∎

Fishburn [14, 15] comments further on the strict xyz inequality of Theorem 10. Given $|X| = n$, [14] investigates the maximum value of $(T + \lambda)/(T + 1)$, i.e., of the xyz ratio $\mu(x <_0 y)\mu(x <_0 z)/\mu((x <_0 y) \cap (x <_0 z))$, but does not completely solve the problem. In [15], an application of Theorem 10 is used in a proof that determines all ordered sets $(X, \prec)$ on n points that maximize $\mu(x <_0 y) = N(xy)/N$ when x and y lie in an m-point antichain for fixed m with $n \geq m \geq 2$.

The conclusion of the xyz inequality, which can be rewritten as $N(xyz)N \leq N(xy)N(yz)$, or

$$\mu(x <_0 y <_0 z) \leq \mu(x <_0 y)\mu(y <_0 z) ,$$

is *universal* in the sense that it holds for all ordered sets. It is therefore natural to ask about other universal correlational inequalities. For example, is it always true that

$$\mu(x <_0 y <_0 z <_0 w) \leq \mu(x <_0 y <_0 z)\mu(z <_0 w)?$$

The answer here is "no", as seen by the partially ordered set $(\{x, y, z, w, t\}, \prec)$ in which $\prec$ consists of the chain $y \prec t \prec w$ plus $y \prec z$, $x \prec w$ and $x \prec z$. Then $\mu(x <_0 y <_0 z <_0 w) = 1/4$, whereas $\mu(x <_0 y <_0 z)\mu(z <_0 w) = 15/64 < 1/4$.

The theme of universal inequalities has been pushed to the limit in Winkler [39] and Brightwell [8]. To state their theorems, let $\prec_*$ be an asymmetric binary relation on a set Y. Given an ordered set $(X, \prec)$ with $Y \subseteq X$, let

$$\mu(Y, \prec_*) = \frac{|\{(X, <_0) \in \mathcal{L} : \prec_* \subseteq <_0\}|}{N} .$$

The set of *covering pairs* in $(Y, \prec_*)$ is

$$\Delta(Y, \prec_*) = \{(x, y) \in \prec_* : x \prec_* t \prec_* y \quad \text{for no} \quad t \in Y\} .$$

We say that ordered sets $(Y, \prec_1)$ and $(Y, \prec_2)$ are *compatible* if the transitive closure of $\prec_1 \cup \prec_2$ is irreflexive, i.e., if $\prec_1$ and $\prec_2$ are subsets of a common partial order. In terms of μ as defined here, the xyz inequality of Theorem 9 is

$$\mu(\{x, y, z\}, \{(x, y), (x, z)\}) \geq \mu(\{x, y, z\}, \{(x, y)\})\mu(\{x, y, z\}, \{(x, z)\}) .$$

Theorem 12. ([39]) *Suppose $(Y, \prec_1)$ and $(Y, \prec_2)$ are compatible finite ordered sets. Then*

$$\mu(Y, \prec_1 \cup \prec_2) \geq \mu(Y, \prec_1)\mu(Y, \prec_2)$$

for every finite ordered set $(X, \prec)$ with $Y \subseteq X$ if and only if, for all $x, y, a, b \in Y$,

$$\{(x, y) \in \Delta(Y, \prec_1 \cup \prec_2) \setminus \Delta(Y, \prec_2), (a, b) \in \Delta(Y, \prec_1 \cup \prec_2) \setminus \Delta(Y, \prec_1)\} \Rightarrow (x = a \text{ or } y = b) .$$

Theorem 13. ([8]) *Suppose* $(Y, \prec_1)$ *and* $(Y, \prec_2)$ *are compatible finite ordered sets. Then*

$$\mu(Y, \prec_1 \cup \prec_2) \leq \mu(Y, \prec_1)\mu(Y, \prec_2)$$

for every finite ordered set $(X, \prec)$ *with* $Y \subseteq X$ *if and only if* $\prec_1 \cap \prec_2 = \emptyset$ *and, for all* $x, y, a, b \in Y$,

$$\{(x,y) \in \Delta(Y, \prec_1), (a,b) \in \Delta(Y, \prec_2)\} \Rightarrow (x = b \text{ or } y = a) .$$

The cases of universal nonnegative correlation in Theorem 12 and universal nonpositive correlation in Theorem 13 are extremely limited. The condition of Theorem 12 says that the covering pairs (x, y) and (a, b) must be related as in the xyz hypothesis, i.e., of the form $\{(x, y), (x, z)\}$ or $\{(x, y), (z, y)\}$. The conditions of Theorem 13 seem even more restrictive.

Additional discussion of the universal correlation theme is provided by Brightwell [9].

A TYPE 3 IMPLICATION FOR RANDOM PERMUTATIONS

It is well known that certain instances of the conclusions of Theorems 1 and 2 do not require complete satisfaction of their hypotheses. We illustrate the point with the case of match sets of random permutations from [17].

Let σ be a permutation of $\{1, 2, \ldots, n\}$. The *match set* of σ is its set of fixed points

$$M(\sigma) = \{i \in \{1, 2, \ldots, n\} : \sigma(i) = i\} .$$

We assume that all $n!$ permutations of $\{1, 2, \ldots, n\}$ are equally likely and let $\mu(a)$ for $a \in 2^n$ denote the probability that $M(\sigma) = a$, with $\mu(A) = \sum\{\mu(a) : a \in A\}$ for $A \in \mathcal{B}_n$. Thus, when exactly $T(a)$ permutations σ have match set a, $\mu(a) = T(a)/n!$.

Theorem 14. ([17]) *For all up-sets* $A, B \in \mathcal{B}_n$,

$$\mu(A \cap B) \geq \mu(A)\mu(B) .$$

An easy corollary, similar to the equivalence of (1) and (3) in Theorem 3, says that if f and g are nondecreasing functions from $(2^n, \subset)$ into R, then $E(fg, \mu) \geq E(f, \mu)E(g, \mu)$. However, Theorem 14 is not a direct implication of Theorem 3 because μ is *not* log supermodular. Although $\mu(a)\mu(b) \leq (a \cup b)\mu(a \cap b)$ for most $a, b \in 2^n$, log supermodularity fails when $|a \cup b| = n - 1 > \max\{|a|, |b|\}$. The reason is that no permutation has exactly $n - 1$ fixed points: if $\sigma(i) = i$ for all but one i then $\sigma(i) = i$ for all i. In other words, $\mu(a \cup b) = 0$ when $|a \cup b| = n - 1$.

Despite the breach of log supermodularity, [17] shows how the Ahlswede-Daykin and FKG theorems can be used to prove Theorem 14. We do this by perturbing μ in ways that assign positive probability to $|a| = n - 1$ such that a perturbed μ satisfies the hypotheses of Theorem 1, or satisfies log supermodularity. Given up-sets A and B, the perturbations leave $\mu(A)$, $\mu(B)$ and $\mu(A \cap B)$

unchanged, so the conclusions of Theorems 1 and 3 can be used for these μ values. Unfortunately, our use of perturbations necessitates examination of many special cases, but this may be an unavoidable cost of the perturbation method.

Although our proof of Theorem 14 is very long, a few comments will indicate one way that the Ahlswede-Daykin theorem is involved. With $T(a) = |\{\sigma : M(\sigma) = a\}|$, it is convenient to work with

$$T_i = T(a) \quad \text{when} \quad |a| = n - i \ ,$$

so $T_0 = 1$ (only one permutation has a complete match), $T_1 = 0$ (the breach of log supermodularity), $\sum \binom{n}{i} T_i = n!$, and, by inclusion-exclusion,

$$T_i = i! \sum_{j=0}^{i} (-1)^j / j! \ .$$

The full proof of the theorem assumes that it holds for small n ([24] verifies the result for $n \leq 6$) and considers up-sets A and B that contain every a with $|a| = n - 1$ and do not equal 2^n. The proof divides into two main cases that receive different treatments:

Case 1: $\mu(A \cap B) \geq \mu(A)\mu(B)$ if $A \cup B$ contains a singleton;

Case 2: $\mu(A \cap B) \geq \mu(A)\mu(B)$ if $\min\{|a| : a \in A \cup B\} \geq 2$.

The Case 1 proof assumes that $\{1\} \in A$ and uses the FKG theorem and a matching argument in which $b \in B \setminus A$ with $|b| \leq n-3$ is paired with $b \cup \{1\} \in A \cap B$. The proof for Case 2 uses the Ahlswede-Daykin theorem. Both cases involve perturbations of μ.

In dealing with Case 2, we assume without loss of generality that $A \cap B$ contains all $(n-1)$-sets and work directly with $T(a)$ rather than $\mu(a) = T(a)/n!$. We perturb T to T' on 2^n as follows:

$$T'(a) = \begin{cases} 0 & a = \{1, \ldots, n\} \\ 1/n & |a| = n - 1 \\ T(a) & |a| \leq n - 2 \ . \end{cases}$$

This removes weight 1 from $\{1, \ldots, n\}$ and redistributes it evenly over the $(n - 1)$-sets. To satisfy the hypothesized inequality of Theorem 1, we first define α and β there by

$$\alpha(a) = \begin{cases} 0 & a \notin A \\ T'(a) & a \in A \ , \end{cases} \qquad \beta(b) = \begin{cases} 0 & b \notin B \\ T'(b) & b \in B \ . \end{cases}$$

Because all $(n-1)$-sets are in $A \cap B$, we have $\alpha(A) = \mu(A)n!$ and $\beta(B) = \mu(B)n!$. Next, define γ by

$$\gamma(a) = \begin{cases} 1/(2n) & a = \{1, \ldots, n\} \\ 0 & a \notin A \cap B \\ T'(a) & \text{otherwise} \ . \end{cases}$$

This gives

$$\gamma(A \vee B) = \gamma(A \cap B) = \frac{1}{2n} + \mu(A \cap B)n! \ ,$$

which is slightly greater than $\mu(A \cap B)n!$, so we define $\delta(A \wedge B)$ to be slightly less than $n!$ to make the conclusion of Theorem 1 at A and B agree with $\mu(A \cap B) \geq \mu(A)\mu(B)$. We choose δ constant on sets of fixed cardinality:

$$\delta(a) = \begin{cases} 0 & |a| = n \\ 1/n & |a| = n-1 \\ 1 & |a| = n-2 \\ nT_i & |a| = n-i-1;\ i = 2, \ldots, n-2 \\ 2nT_{n-2} & |a| = 0\ . \end{cases}$$

It follows that, with $\delta_i = \delta(a)$ when $|a| = i$,

$$\delta(A \wedge B) = \sum_{i=0}^{n} \binom{n}{i} \delta_i = 2nT_{n-2} + n \sum_{i=1}^{n-3} \binom{n}{i} T_{n-i-1} + \binom{n}{2} + 1\ .$$

Given α, β, γ and δ, the Case 2 proof now breaks into a number of subcases for the up-sets A and B that depend on n and $k = n - \min\{|a| : a \in A \cap B\}$. All but a finite number of instances of (k, n) satisfy the hypothesized inequality of Theorem 1, and $\mu(A \cap B) \geq \mu(A)\mu(B)$ is obtained from its conclusion. A few instances here use a further perturbation which increases $\delta_0 = 2nT_{n-2}$ but leaves all other parts of α through δ unchanged. The instances of (k, n) that do not satisfy the hypotheses of Theorem 1 use other methods to verify $\mu(A \cap B) \geq \mu(A)\mu(B)$.

A GENERALIZATION

We conclude by describing a generalization of the Ahlswede-Daykin theorem due to Rinott and Saks [31, 32] and, independently, Aharoni and Keich [2]. The generalization applies to n-tuples $\mathbf{a} = (a_1, a_2, \ldots, a_n)$ in Γ^n for $n \geq 2$, and is identical to the Ahlswede-Daykin theorem when $n = 2$. It is too early to say whether a number of interesting applications will arise for $n \geq 3$, but this seems plausible in view of the usefulness of Theorems 1 and 2.

We assume that $(\Gamma, \prec)$ is a finite distributive lattice and take $n \geq 2$. For each $k \in \{1, \ldots, n\}$, let ϕ_k denote the map from Γ^n into Γ defined by

$$\phi_k(\mathbf{a}) = \vee \{\wedge_{i \in S} a_i : \ S \text{ is a } k\text{-set in } \{1, 2, \ldots, n\}\}$$

for all $\mathbf{a} = (a_1, \ldots, a_n) \in \Gamma^n$. For example, when $n = 3$,

$$\begin{aligned} \phi_1(\mathbf{a}) &= a_1 \vee a_2 \vee a_3 \\ \phi_2(\mathbf{a}) &= (a_1 \wedge a_2) \vee (a_1 \wedge a_3) \vee (a_2 \wedge a_3) \\ \phi_3(\mathbf{a}) &= a_1 \wedge a_2 \wedge a_3\ . \end{aligned}$$

With 2^Γ the set of subsets of Γ, we extend ϕ_k to $(2^\Gamma)^n$ by letting

$$\phi_k(\mathbf{A}) = \{\phi_k(\mathbf{a}) : \mathbf{a} \in A_1 \times A_2 \times \cdots \times A_n, \mathbf{a} \in \Gamma^n\}$$

for all $\mathbf{A} = (A_1, \ldots, A_n) \in (2^\Gamma)^n$.

Theorem 15. *Suppose* $(\Gamma, \prec)$ *is a finite distributive lattice,* $n \geq 2$, *and* $f_1, \ldots, f_n, g_1, \ldots, g_n : \Gamma \to [0, \infty)$ *satisfy*

$$\prod_{k=1}^{n} f_k(a_k) \leq \prod_{k=1}^{n} g_k(\phi_k(\mathbf{a})) \quad \textit{for all} \quad \mathbf{a} \in \Gamma^n .$$

Then

$$\prod_{k=1}^{n} f_k(A_k) \leq \prod_{k=1}^{n} g_k(\phi_k(\mathbf{A})) \quad \textit{for all} \quad \mathbf{A} \in (2^\Gamma)^n .$$

The proof in [2] is similar in outline to the proof of Theorem 1 indicated in section 1. It uses $(2^m, \subset)$ in place of $(\Gamma, \prec)$ and proceeds by induction on m after checking the desired result for $m = 0$ and proving it for $m = 1$ with assistance from a result about n-tuples of functions from $\{0, 1\}$ into $[0, \infty)$.

References

[1] R. Aharoni and R. Holzman, "Two and a half remarks on the Marica-Schönheim inequality", *J. London Math. Soc.*, (2), 48, 1993, 385–395.

[2] R. Aharoni and U. Keich, "A generalization of the Ahlswede-Daykin inequality", *Discrete Math.*, 152, 1996,1–12.

[3] R. Ahlswede and D. E. Daykin, "An inequality for the weights of two families of sets, their unions and intersections", *Z. Wahrscheinlichkeitstheorie und Verw. Gebiete*, 43, 1978,183–185.

[4] R. Ahlswede and D. E. Daykin, "Inequalities for a pair of maps $S \times S \to S$ with S a finite set", *Math. Z.*, 165, 1979, 267–289.

[5] G. Birkhoff, *Lattice Theory*, 3rd ed. Providence, RI, Amer. Mathematical Soc., 1967.

[6] B. Bollobás, *Combinatorics*, Cambridge, Cambridge Univ. Press., 1986.

[7] B. Bollobás and G. Brightwell, "Parallel selection with high probability", *SIAM J. Discrete Math.*, 3, 1990, 21–31.

[8] G.R. Brightwell, "Universal correlations in finite posets", *Order*, 2, 1985, 129–144.

[9] G.R. Brightwell, "Some correlation inequalities in finite posets", *Order*, 2, 1986, 387–402.

[10] R. M. Burton Jr. and M. M. Franzosa, "Positive dependence properties of point processes", *Ann. Probab.*, 18, 1990, 359–377.

[11] D.E. Daykin, "A lattice is distributive iff $|A||B| \leq |A \vee B||A \wedge B|$", *Nanta Math.*, 10, 1977, 58–60.

[12] D.E. Daykin and L. Lovász, "The number of values of a Boolean function", *J. London Math. Soc.*, (2) 12, 1976, 225–230.

[13] P.C. Fishburn, "A correlational inequality for linear extensions of a poset", *Order*, 1, 1984, 127–137.

[14] P.C. Fishburn, "Maximizing a correlational ratio for linear extensions of posets", *Order*, 3,1986, 159–167.

[15] P.C. Fishburn, "A note on linear extensions and incomparable pairs", *J. Combin. Theory Ser. A*, 56, 1991, 290–296.

[16] P.C. Fishburn, "Correlation in partially ordered sets", *Discrete Appl. Math.*, 39, 1992, 173–191.

[17] P.C. Fishburn, P.G. Doyle and L.A. Shepp, "The match set of a random permutation has the FKG property", *Ann. Probab.*, 16, 1988, 1194–1214.

[18] C.M. Fortuin, P.N. Kasteleyn and J. Ginibre, "Correlation inequalities for some partially ordered sets" *Comm. Math. Phys.*, 22, 1971, 89–103.

[19] R.L. Graham, "Linear extensions of partial orders and the FKG inequality", *Ordered Sets*, I. Rival, ed., Dordrecht, Reidel., 1982, 213–236.

[20] R.L. Graham, "Applications of the FKG inequality and its relatives", *Proceedings 12th International Symposium on Mathematical Programming.* Berlin, Springer, 1983, 115–131.

[21] R.L. Graham, A.C. Yao and F.F. Yao, "Some monotonicity properties of partial orders", *SIAM J. Algebraic Discrete Methods*, 1, 1980, 251–258.

[22] R. Holley, "Remarks on the FKG inequalities", *Comm. Math. Phys.*, 36, 1974, 227–231.

[23] F.K. Hwang and L.A. Shepp, "Some inequalities concerning random subsets of a set", *IEEE Trans. Information Theory*, 33, 1987, 596–598.

[24] K. Joag-Dev, "Association of matchmakers", mimeo, Department of Statistics, University of Illinois, 1985.

[25] K. Joag-Dev, L.A. Shepp and R.A. Vitale, "Remarks and open problems in the area of the FKG inequality", *IMS Lecture Notes-Monograph Series*, 5, 1984, 121–126.

[26] J.H.B. Kemperman, "On the FKG inequality for measures on a partially ordered space", *Indag. Math.*, 39, 1977, 313–331.

[27] D.J. Kleitman, "Families of non-disjoint sets", *J. Combin. Theory*, 1, 1966, 153–155.

[28] D.J. Kleitman and J. B. Shearer, "Some monotonicity properties of partial orders", *Stud. Appl. Math.*, 65, 1981, 81–83.

[29] Z. Lengvárszky, "The Marica-Schönheim inequality in lattices", *Bull. London Math. Soc.*, 28, 1996, 449–454.

[30] J. Marica and J. Schönheim, "Differences of sets and a problem of Graham", *Canad. Math. Bull.*, 12, 1969, 635–637.

[31] Y. Rinott and M. Saks, "On FKG-type and permanental inequalities", *Proc. 1991 AMS-IMS-SIAM Joint Conf. on Stochastic Inequalities, IMS Lecture Series*, M. Shaked and Y. L. Tong, eds., 1991.

[32] Y. Rinott and M. Saks (n.d.), "Correlation inequalities and a conjecture for permanents", *Combinatorica.*

[33] P.D. Seymour, "On incomparable collections of sets", *Mathematika*, 20, 1973, 208–209.

[34] L.A. Shepp, "The FKG property and some monotonicity properties of partial orders", *SIAM J. Algebraic Discrete Methods*, 1, 1980, 295–299.

[35] L.A. Shepp, "The XYZ conjecture and the FKG inequality", *Ann. Probab.*, 10, 1982, 824–827.

[36] E. Szpilrajn, "Sur l'extension de l'ordre partiel", *Fund. Math.*, 16, 1930, 386–389.

[37] J. van den Berg and U. Fiebig, "On a combinatorial conjecture concerning disjoint occurrences of events", *Ann. Probab.*, 15, 1987, 354–374.

[38] J. van den Berg and H. Kesten, "Inequalities with applications to percolation and reliability", *J. Appl. Probab.*, 22, 1985, 556–569.

[39] P.M. Winkler, "Correlation among partial orders", *SIAM J. Algebraic Discrete Methods*, 4, 1983, 1–7.

[40] P.M. Winkler, "Correlation and order", *Contemp. Math.*, 57, 1986, 151–174.

SOME ASPECTS OF RANDOM SHAPES

Herbert Ziezold

Fachbereich Mathematik/Informatik, Universität Kassel, D-34109 Kassel

ziezold@mathematik.uni-kassel.de

Abstract: Given $x_1, \ldots, x_k$ in $\mathbf{R}^m$, the shape of $\mathbf{x} = (x_1, \ldots, x_k)$ is the equivalence class of $\mathbf{x}$ modulo similarity transformations in $\mathbf{R}^m$. Several metrics on the shape spaces will be introduced. This gives the opportunity to work with mean shapes and to use multivariate statistics, e. g. multidimensional scaling, and nonparametric statistics, e. g. discriminance analysis, for data analysis. Some connections to differential geometry and diffusion processes are also given.

INTRODUCTION

Imagine that we have some object in $\mathbf{R}^m$, $m = 2, 3$ or even $m > 3$. We define k characteristic points of the object, called *landmarks*, for a fixed $k \geq 2$ and measure their coordinates with respect to any Cartesian coordinate system. Thus we get k points $x_i \in \mathbf{R}^m$, $i = 1, 2, \ldots, k$.

By $\mathbf{x}$ we denote the *configuration* $(x_1, \ldots, x_k) \in (\mathbf{R}^m)^k$.

The *shape* $\hat{\mathbf{x}}$ of $\mathbf{x}$ is the equivalence class of $\mathbf{x}$ modulo translations, rotations and scalings, i. e. modulo similarity transformations in $\mathbf{R}^m$. The space of all shapes is denoted by Σ_m^k.

The *size-and-shape* $\tilde{\mathbf{x}}$ of $\mathbf{x}$ is the equivalence class of $\mathbf{x}$ modulo translations and rotations, i. e. modulo Euclidean motions in $\mathbf{R}^m$. The space of all size-and-shapes is denoted by $S\Sigma_m^k$.

SOME HISTORICAL REMARKS

David G. Kendall presented in his fundamental paper [6] in 1984 the topological and probabilistic basics for further research on shape analysis. In [5] he had already given a short report on his current investigations on the subject prompted by archaeological, astronomical, geological and ornithological considerations.

Coming from biometrical problems Fred L. Bookstein was the second researcher influencing the analysis of shapes, e. g. by his Lecture Notes [1] and his 'Orange Book' [2].

I. Althöfer et al. (eds.), Numbers, Information and Complexity, 517-523.

In [15] first definitions and properties of mean size-and-shapes are given together with a strong law of large numbers in metric spaces by which statistical consistency problems can be solved.

The first comprehensive books on the mathematical theory of shapes are [3] and [12]. In these books many applications in biology, medicine, astronomy, archaeology, geography, agriculture and genetics are also presented. They will certainly stimulate much future work on shapes.

In the second of three parts of [14] it is shown how to investigate particles by contours and by set-theoretic methods. A short introduction to the models of Kendall and Bookstein with statistical applications is also given.

PARAMETERIZATIONS OF SHAPES

To be able to use the well known multivariate statistical theory parameterizations of shapes are necessary.

The idea of Bookstein is for dimension $m = 2$:

> Translate, rotate and scale the configuration $\mathbf{x} = (x_1, \dots, x_k) \in (\mathbf{R}^2)^k$ with $x_1 \neq x_2$ such that
>
> $$x_j \to z_j^{\mathrm{B}} = (u_j^{\mathrm{B}}, v_j^{\mathrm{B}})$$
>
> with $z_1^{\mathrm{B}} = (0,0)$ and $z_2^{\mathrm{B}} = (1,0)$.

The real numbers $u_3^{\mathrm{B}}, \dots, u_k^{\mathrm{B}}, v_3^{\mathrm{B}}, \dots, v_k^{\mathrm{B}}$ are called the *Bookstein coordinates* of the shape $\hat{\mathbf{x}}$. For $k = 3$ we get thus a point $(u, v) \in \mathbf{R}^2$ as parameter for the shape of a triangle (x_1, x_2, x_3) with $x_1 \neq x_2$.

Kendall's method runs for the plane as follows:

> Identify $\mathbf{R}^2$ with the complex plane $\mathbf{C}$. Define the Helmert sub-matrix H by the last $k-1$ rows of the Helmert-matrix
>
> $$H^{\mathrm{F}} = \begin{pmatrix} \frac{1}{\sqrt{k}} & \frac{1}{\sqrt{k}} & \frac{1}{\sqrt{k}} & \dots & \frac{1}{\sqrt{k}} \\ -\frac{1}{\sqrt{2}} & \frac{1}{\sqrt{2}} & 0 & \dots & 0 \\ -\frac{1}{\sqrt{6}} & -\frac{1}{\sqrt{6}} & \frac{2}{\sqrt{6}} & \dots & 0 \\ \dots & \dots & \dots & \dots & \dots \\ -\frac{1}{\sqrt{(k-1)(k)}} & -\frac{1}{\sqrt{(k-1)(k)}} & -\frac{1}{\sqrt{(k-1)(k)}} & \dots & \frac{k-1}{\sqrt{(k-1)(k)}} \end{pmatrix}.$$
>
> Define for $\mathbf{x} = (x_1, \dots, x_k) \in \mathbf{C}^k$ with $x_1 \neq x_2$: $\mathbf{z}_0 = (z_2^{(0)}, \dots, z_k^{(0)})^{\mathrm{T}} = H\mathbf{x}$ and $z_j^{\mathrm{K}} = (u_j^{\mathrm{K}}, v_j^{\mathrm{K}}) = \frac{z_j^{(0)}}{z_2^{(0)}}$, $j = 3, \dots, k$.

The real numbers $u_3^{\mathrm{K}}, \dots, u_k^{\mathrm{K}}, v_3^{\mathrm{K}}, \dots, v_k^{\mathrm{K}}$ are called the *Kendall coordinates* of the shape $\hat{\mathbf{x}}$.

If we write

$$\mathbf{z}^B = (u_3^{\mathrm{B}} + iv_3^{\mathrm{B}}, \dots, u_k^{\mathrm{B}} + iv_k^{\mathrm{B}}) \quad \text{and} \quad \mathbf{z}^K = (u_3^{\mathrm{K}} + iv_3^{\mathrm{K}}, \dots, u_k^{\mathrm{K}} + iv_k^{\mathrm{K}}),$$

and if we define H_1 as the lower right $(k-2) \times (k-2)$ partition matrix of the Helmert matrix H^F then one can show that $\mathbf{z}^K = \sqrt{2} H_1 \mathbf{z}^B$.

Having one of these parameterizations or some linear modification of it one can define probability densities on shape spaces. E. g. if one uses the modification $z_1^{\mathrm{B}} = (-1, 0)$ in the above definition of Bookstein coordinates and defines $z = (u, v)$ as the shape parameter of a triangle in the plane and I_m as the m-dimensional unit matrix, one can show the

Proposition 1. *If X_1, X_2, X_3 are independently $N(\mu, \sigma^2 I_m)$-distributed for any $\mu \in \mathbf{R}^2$ and $\sigma > 0$, then the random shape variable Z has the density*

$$f^*(z) = \frac{3}{\pi(3 + |z|^2)^2}, \quad z \in \mathbf{C}.$$

See [12], page 152, for a proof.

A fundamental analysis of normal densities in parameterized shape spaces is done in [4]. See also [3].

METRICS IN SHAPE SPACES

For simplicity we will again only consider configurations in the complex plane $\mathbf{C}$.

Let $\mathbf{x} = (x_1, \ldots, x_k) \in \mathbf{C}^k$ and $\mathbf{y} = (y_1, \ldots, y_k) \in \mathbf{C}^k$ are given. Define $||\mathbf{x}||$ as the Euclidean norm $\sqrt{\sum_{j=1}^k x_j \bar{x}_j}$, S as the unit circle $S^1(1) = \{u \in \mathbf{C} : |u| = 1\}$ and $\mathbf{1} = (1, \ldots, 1) \in \mathbf{C}^k$. Then

$$\delta(\mathbf{x}, \mathbf{y}) = \inf_{u \in S, a \in \mathbf{C}} ||\mathbf{x} - u\mathbf{y} - a\mathbf{1}||$$

gives the 'best fit' of the configurations $\mathbf{x}$ and $\mathbf{y}$ with respect to translations and rotations.

We define

$$d(\tilde{\mathbf{x}}, \tilde{\mathbf{y}}) = \delta(\mathbf{x}, \mathbf{y})$$

as the *size-and-shape distance* between the shapes of $\mathbf{x}$ and $\mathbf{y}$.

Let c be the center $\frac{1}{k} \sum_{j=1}^k x_j$ of $x_1, \ldots, x_k$, let $\mathbf{x}^*$ be the centered k-ad $\mathbf{x} - c\mathbf{1}$ of $\mathbf{x}$ and let $\mathbf{C}_0^k$ be set $\{\mathbf{x} \in \mathbf{C}^k : \mathbf{x}^* \neq \mathbf{0}\}$ of all configurations without those with equal landmarks. We denote by

$$\mathbf{x}^0 = \frac{\mathbf{x}^*}{||\mathbf{x}^*||} \quad \text{for} \quad \mathbf{x} \in \mathbf{C}_0^k$$

the *normed centered* configuration to $\mathbf{x}$ and define the shape distance between $\mathbf{x}$ and $\mathbf{y}$ as

$$D(\hat{\mathbf{x}}, \hat{\mathbf{y}}) = d(\tilde{\mathbf{x}}^0, \tilde{\mathbf{y}}^0) = \delta(\mathbf{x}^0, \mathbf{y}^0).$$

In [3] it is called the *partial Procrustes distance.*

Kendall's Procrustes distance of shapes is defined by

$$\rho(\hat{\mathbf{x}}, \hat{\mathbf{y}}) = \arccos(1 - \frac{1}{2} D(\hat{\mathbf{x}}, \hat{\mathbf{y}})^2).$$

By stereographical projection of the complex plane on the ball $\bar{S}$ with radius $\frac{1}{2}$ touching $\mathbf{C}$ in the origin we get a parameterization of the shapes of non trivial triangles by points on this ball which is trivially isometric to the ball $S^2(\frac{1}{2})$ around the origin of $\mathbf{R}^3$. In [6] the following remarkable theorem is proved.

Theorem 2. *The metric space* (Σ_2^3, ρ) *is isometric to* $(S^2(\frac{1}{2}), d_g)$ *where* d_g *is the geodesic great circle metric on* $S^2(\frac{1}{2})$.

The next result out of [6] supplementing Proposition 1 is not less astonishing:

Theorem 3. *Let* z^S *be the stereographical projection of the shape parameter* z *of a triangle in* $\mathbf{C}$ *on the ball* $\bar{S}$. *If* X_1, X_2, X_3 *are independently* $N(\mu, \sigma^2 I_2)$-*distributed for any* $\mu \in \mathbf{R}^2$ *and* $\sigma > 0$, *then the random shape variable* Z^S *is uniformly distributed on the ball* $\bar{S}$.

These results are serious justifications to use the metric ρ instead of D. But D is more suitable for computing means of shapes and size-and-shapes as it is done e. g. in [3], [9], [13] and [15].

MEAN SHAPES AND ITS APPLICATIONS

As the shape spaces are not linear, the usual definitions of expectations in real spaces are not applicable.

The following definition of means in metric spaces is a straightforward generalization of the fact that the expected value of a real random variable X with existing second moment is the only real value μ which minimizes $\mathrm{E}((X-\mu)^2)$.

Given a random variable X in a metric space $(\mathcal{X}, d)$ an element $\mu \in \mathcal{X}$ is a *Fréchet mean to* X if

$$\mathrm{E}(\mathrm{d}(X,\mu)^2) = \inf_{\alpha \in \mathcal{X}} \mathrm{E}(\mathrm{d}(X,\alpha)^2) \ .$$

We denote by $\mathbb{E}(X)$ the set of Fréchet means to X.

Given elements $z_1, \ldots, z_n$ in a metric space $(\mathcal{X}, d)$ an element $a \in \mathcal{X}$ is a *Fréchet mean to* $z_1, \ldots, z_n$ if

$$\sum_{j=1}^{n} \mathrm{d}(z_j, a)^2 = \inf_{\alpha \in \mathcal{X}} \sum_{j=1}^{n} \mathrm{d}(z_j, \alpha)^2 \ .$$

We denote by $\mathbb{M}(z_1, \ldots, z_n)$ the set of Fréchet means to $z_1, \ldots, z_n$.

Let $\overline{A}$ denote the closure of a set A in a metric space.

The following theorem provides a means to prove statistical consistency of sequences of means of realizations of independent random variables in metric spaces. For the shape spaces (Σ_2^k, ρ) and (Σ_2^k, D) this is done in [10].

Strong law of large numbers. *If* $(\mathcal{X}, d)$ *is a separable metric space and if* $X_1, X_2, \ldots$ *are i.i.d. in* $\mathcal{X}$ *with* $\mathrm{E}(\mathrm{d}(X_1,\alpha)^2) < \infty$, $\alpha \in \mathcal{X}$, *then almost surely*

$$M(\omega) = \cap_{s=1}^{\infty} \overline{(\cup_{r=s}^{\infty} \mathbb{M}(X_1(\omega), \ldots, X_r(\omega)))} \subseteq \mathbb{E}(X_1) \ .$$

Loosely spoken: Every accumulation point of the means is a. s. a Fréchet mean of X_1.

With the help of an algorithm one can compute the *mean size-and-shape* and the *mean shape* to n configurations $\mathbf{x}^{(1)}, \ldots, \mathbf{x}^{(n)}$ in $(\mathrm{S}\Sigma_2^k, \mathrm{d})$ resp. (Σ_2^k, D).

Given two classes of configurations,

$$\mathbf{x}^{(1)}, \ldots, \mathbf{x}^{(r)} \quad \text{and} \quad \mathbf{y}^{(1)}, \ldots, \mathbf{y}^{(s)},$$

one can perform *non-parameterical discriminance analysis* by comparing $\mathrm{d}(\tilde{\mathbf{x}}^{(i)}, \tilde{\mathbf{m}}_{\mathbf{x}})$, $i = 1, \ldots, r$, with $\mathrm{d}(\tilde{\mathbf{y}}^{(i)}, \tilde{\mathbf{m}}_{\mathbf{x}})$, $i = 1, \ldots, s$, where $\tilde{\mathbf{m}}_{\mathbf{x}}$ is the mean size-and-shape to $\mathbf{x}^{(1)}, \ldots, \mathbf{x}^{(r)}$.

For details see [16].

HYPERBOLIC GEOMETRIES FOR THE SIMPLEX SHAPES

We generalize the definition of Bookstein coordinates of triangles in $\mathbf{R}^2$ to non-degenerate simplices in $\mathbf{R}^m$, $m = 2, 3, \ldots$. This means that we now consider the shape spaces Σ_m^{m+1}.

Given $\mathbf{x} = (x_1, \ldots, x_{m+1}) \in (\mathbf{R}^m)^{m+1}$, $m \geq 2$, with $x_i \neq x_j$ for all $i \neq j$ we transform $x_1, \ldots, x_{m+1}$ by translation, rotation, scaling **and reflection** such that

$$x_1 \to (0, 0, \ldots, 0) \in \mathbf{R}^m , \qquad x_2 \to (1, 0, \ldots, 0) \in \mathbf{R}^m$$

$$x_i \to (z_{i,1}, z_{i,2}, \ldots, z_{i,m}) \text{ with } z_{i,j} = 0,\ j \geq i, \text{ and } z_{i,i-1} > 0,\ 3 \leq i \leq m+1 .$$

The thus defined real values $z_{i,j}$, $3 \leq i \leq m+1$, $1 \leq j \leq i-1$, are called the *generalized Bookstein coordinates* of the shape of $\mathbf{x}$. We set

$$\Pi_{\mathbf{x}} = \begin{pmatrix} 1 & z_{31} & z_{41} & \ldots & z_{m+1,1} \\ 0 & z_{32} & z_{42} & \ldots & z_{m+1,2} \\ 0 & 0 & z_{43} & \ldots & z_{m+1,3} \\ \ldots & \ldots & \ldots & \ldots & \ldots \\ 0 & 0 & 0 & \ldots & z_{m+1,m} \end{pmatrix}$$

and define $\mathbf{UT}(m) = \{\Pi_{\mathbf{x}} : \mathbf{x} = (x_1, \ldots, x_{m+1}) \text{ with } x_i \neq x_j \text{ for all } i \neq j\}$. This set is a group with matrix multiplication. With respect to the topology induced by the Euclidean metric in $\mathbf{R}^{\frac{m(m+1)}{2}-1}$ it is even a Lie group.

For the definition of a suitable Riemannian metric on $\mathbf{UT}(m)$ we formally use differentials $\mathrm{d}\mathbf{x}$ etc.. We approximate

$$\Lambda = \Pi_{\mathbf{x}+\mathrm{d}\mathbf{x}} \Pi_{\mathbf{x}}^{-1} = \begin{pmatrix} 1 & \lambda_{12} & \ldots & \Lambda_{1m} \\ 0 & \lambda_{22} & \ldots & \Lambda_{2m} \\ \ldots & \ldots & \ldots & \ldots \\ 0 & 0 & \ldots & \Lambda_{mm} \end{pmatrix}$$

to the first order by $I_m + \mathrm{d}\Lambda$. Let $\lambda_1, \ldots, \lambda_m$ be the eigenvalues of $\Lambda^{\mathrm{T}}\Lambda$ and let $\bar{\lambda}$ be the arithmetic mean $\frac{1}{m}\sum_{i=1}^m \lambda_i$. Then a suitable Riemannian metric is defined by

$$\mathrm{d}s^2 = \frac{1}{m} \sum_{i=1}^{m} (\lambda_i - \bar{\lambda})^2 .$$

This gives, see [12], page 103,

$$\mathrm{d}s^2 = \frac{4}{m^2}\left((m-1)\sum_{i=2}^{m}\mathrm{d}\Lambda_{ii}^2 + \frac{m}{2}\sum_{i<j}\mathrm{d}\Lambda_{ij}^2 - 2\sum_{i<j}\mathrm{d}\Lambda_{ii}\mathrm{d}\Lambda_{jj}\right)$$

In coordinates of $\Pi_{\mathbf{x}}$ we get for $m = 2$ by setting $z_1 = z_{31}$ and $z_2 = z_{32}$:

$$\mathrm{d}s^2 = \frac{\mathrm{d}z_1^2 + \mathrm{d}z_2^2}{z_2^2}\,.$$

This defines the hyperbolic geometry of the Poincaré Plane $\mathbf{HS}^2$.

The much more complicated expression for the differential $\mathrm{d}s^2$ for $m = 3$ is given in [12], page 105 f.

For the more general shape spaces Σ_m^k with $k > m + 1$ the Riemannian structure is analysed in [11].

The following characterization of the above defined hyperbolic geometry in the plane by a diffusion process is proved in [8].

Theorem 4. *Suppose that three landmarks X_1, X_2, X_3 in $\mathbf{R}^2$ move in the following manner:*

$$X_i(t) = G(t)x_i\,,$$

where (x_1, x_2, x_3) is a start configuration and $(G(t))_{t\geq 0}$ is a 'special' Brownian motion on $\mathrm{GL}^+(2,\mathbf{R})$. Let $\sigma(t)$ be the shape of $(X_1(t), X_2(t), X_3(t))$. Then σ is a diffusion process whose intrinsic geometry on the state-space is the hyperbolic geometry of $\mathbf{HS}^2$, making σ into a Brownian motion.

In the proof a computer algebra package for dealing with stochastic differential equations is intensively used.

CONCLUSIONS

It was the purpose of this paper to give the reader a short impression of the mathematical analysis of shapes. In the cited papers and especially in the books [3], [12] and [14] he or she may find much more material to the theory and to many applications.

Note added in proof:

In August 1999 the book [7] has appeared. It contains an algebraic topological and a differential geometrical analysis of the shape spaces Σ_m^k as well as a presentation of results on probability distributions and means in shape spaces.

References

[1] F. L. Bookstein, "The Measurement of Biological Shape and Shape Change", *Lecture Notes in Biomathematics 24*, Springer-Verlag, New York, 1978.

[2] F. L. Bookstein, *Morphometric Tools for Landmark Data: Geometry and Biology*, Cambridge University Press, Cambridge, 1991.

[3] I. L. Dryden and K. V. Mardia, *Statistical Shape Analysis*, Wiley, Chichester, 1998.

[4] C. R. Goodall, "Procrustes methods in the statistical analysis of shape (with discussion)", *Journal of the Royal Statistical Society*, Series B, 53, 1991, 285-339.

[5] D. G. Kendall, "The diffusion of shape", *Advances in Applied Probability,* 9, 1977, 428-430.

[6] D. G. Kendall, "Shape manifolds, Procrustean metrics and complex projective spaces", *Bulletin of the London Mathematical Society* 16, 1984, 81-121.

[7] D. G. Kendall, D. Barden, T. K. Carne, H. Le, *Shape and Shape Theory,* Wiley, Chichester, 1999.

[8] W. S. Kendall, "A diffusion model for Bookstein triangle shape", *Advances in Applied Probability* 30, 1998, 317-334.

[9] J. T. Kent, "New Directions in Shape Analysis", *The Art of Statistical Science*, Wiley, Chichester, 1992, 115-127.

[10] H. Le, "On the consistency of Procrustean mean shapes", *Advances in Applied Probability* 30, 1998, 53-63.

[11] H. Le and D. G. Kendall, "The Riemannian structure of Euclidean shape spaces: a novel environment for statistics", *Annals of Statistics* 21, 1993, 1225-1271.

[12] C. G. Small, *The Statistical Theory of Shape*, Springer-Verlag, New York, 1996.

[13] D. Stoyan and I. S. Molchanov, "Set-valued means of random particles", *Technical Report BS-R9511*, CWI, Amsterdam, 1995.

[14] D. Stoyan and H. Stoyan, *Fractals, Random Shapes and Point Fields*, Wiley, Chichester, 1994. (German edition: Akademie Verlag, Berlin 1992.)

[15] H. Ziezold, "On expected figures and a strong law of large numbers for random elements in quasi-metric spaces", *Transactions of the Seventh Prague Conference on Information Theory, Statistical Decision Functions, Random Processes,* (Prague, 1974), Volumen A. Reidel, Dordrecht, 1977, 591-602.

[16] H. Ziezold, "Mean figures and mean shapes applied to biological figure and shape distributions in the plane", *Biometrical Journal* 36, 1994, 491-510.

DECISION SUPPORT SYSTEMS WITH MULTIPLE CHOICE STRUCTURE

Ingo Althöfer

Friedrich-Schiller-Universität Jena, Fakultät für Mathematik und Informatik,
07740 Jena, Germany

althofer@mipool.uni-jena.de

Abstract: In the "Triple Brain" approach ("3-Hirn" in German) one human and two computers with different programs are involved. Both programs are started and present one solution each. The human is a controller. He inspects the computer solutions and selects one of them. The human is not allowed to outvote the machines.

"Triple Brain" is a "Decision Support System with Multiple Choice Structure": Computer programs (one or several) provide a handful of interesting candidate solutions, and a controller (typically a human) has the final choice among these candidates. This article exhibits and discusses various aspects of Decision Support Systems with Multiple Choice Structure.

Key Words and Phrases: Triple Brain, 3-Hirn, Decision Support System, Multiple Choice, Multiple Choice System, man and machine, k-best algorithm, k-best optimization under side constraints, incremental computing;

INTRODUCTION

Humans are able to think, to feel, and to sense. We can also compute, but not too well. Instead, computers are giants in computing - they crunsh bits and bytes like maniacs. However, they cannot do anything else but computing. By combining the gifts and strengths of man and machine in appropriate ways it is possible to achieve impressive results.

Consider a problem solving situation. In the "Triple Brain" approach ("3-Hirn" in German) one human and two computers with different programs are involved. Both machines are started. In an appropriate moment the human

I. Althöfer et al. (eds.), Numbers, Information and Complexity, 525-540.

stops them and analyses the solutions they propose. Finally he selects one of these computer solutions and realizes it. The human is not allowed to outvote the machines. Using this Triple Brain approach in the game of chess, an amateur player (= this author) together with commercial chess programs was able to play on one level with world class professionals (Lutz, 1996; Althöfer, 1997a,1998a).

The Triple Brain is just one possible way to realize "Decision Support Systems with Multiple Choice Structure" (shortly called "Multiple Choice Systems" in the sequel): One or several computer(s) provide a handful of interesting candidate solutions, and a human controller has the final choice among these candidate solutions. Such systems may be applied successfully in discrete optimization, traffic planning, symbolic computing, computational biology, computer-aided medicine, forecasting (weather, earth quakes, stock markets), and other fields.

This article exhibits and discusses various aspects of "Multiple Choice Systems". It is partly somewhat vague and tentative. Only the future may reveal the full potential of this symbiotic approach with man and machine. In Section 2 we distinguish two types of Multiple Choice Systems: those which use existing programs versus those in which specially developed programs are involved. Section 3 shortly records the success story of Triple Brain in chess. In Section 4 we discuss the issue of k-best optimization under side constraints. Extensions and variants of the Triple Brain Principle are described in Section 5. Finally, Section 6 contains a discussion and some visions.

A short remark to avoid misunderstandings: "Multiple Choice Systems" are not exactly the same as "Multiple Choice Tests" . (The popular understanding of Multiple Choice Tests is that the test person is putting crosses in boxes, thinking only a short moment over each question.) Especially, having the choice between a handful of candidate solutions does not mean that this choice has to be made within a few seconds. Sometimes it can take minutes or even hours or days to select one of two alternatives.

TWO DIFFERENT APPROACHES FOR MULTIPLE CHOICE SYSTEMS: THE USE OF "TRADITIONAL" PROGRAMS VERSUS THE DEVELOPMENT OF SPECIAL SOFTWARE

A "traditional" problem solving program works as follows: The user enters the data and his question, and then the program computes and presents ONE solution. Alternative solutions are not proposed. The user is expected to accept the solution given by the program. Instead, a Multiple Choice System presents a handful of good solutions, and the user has the final choice amongst these alternatives.

(a) Assume a problem class for which several "traditional" programs exist. The user can start two or three or all of them: Each of the programs makes a proposal and the user has the final choice amongst these candidate solutions. If all (or many) programs agree on the same solution the user will have more confidence in this solution as he would have in the suggestion of a single program. In case of different proposals the controller can use his own knowledge (which typically differs from that of the programs) to make the final choice. Technically there are two ways to gather the solutions: either to use one computer, running the programs on this machine one after the other (in such a procedure (repeated) opening and closing of the programs may take a lot of time), or to run the programs simultaneously, using as many computers as there are programs.

Often different programs have rather different user interfaces, and program outputs are typically extensive. These are both no problems, if only a single program is used. Stress arises, if a human works with two (or more) programs simultaneously: Again and again he has to switch between different Input/Output formats. Additionally, the inspection of many details in the program outputs tends to strain a conscientious controller. This "data overflow" becomes a heavy burden, especially if the whole process (with many repeated decision rounds) runs over several hours. "Innocent" readers may not believe in this description, but I faced it again and again during my experiments with chess computers. It is a very hard job to operate two different programs simultaneously.

Sceptics are invited to perform a little experiment: Place two computers side by side on a table and install two different telephone CD programs on them. Take a page of a popular computer magazine with many little advertisements by private persons. Typically in such advertisements no complete address is given, but only a telephone number without a name. It is your task to find out (by the help of the telephone CDs) which persons belong to these telephone numbers. (In real life one would make such a check only for one or very few ads, but this is an experiment.) For a most realistic simulation of a Triple Brain scenario (with many repeated rounds of decision) you should do the following: look for the first number on CD 1, then look for the first number on CD 2, then look for the second number on CD 1, then look for the second number on CD 2, then look for the third number on CD 1, and so on. (So, do not check the whole list by CD 1 first and by CD 2 afterwards!) You will learn at least two things in this experiment: (i) If you do not perform the job artificially slowly there is a good chance that you will be exhausted after the first 50 numbers. (ii) Not only in exceptional cases the (different!) telephone CDs will display different names or one CD will show a name where the other CD has no information.

This telephone example differs from more complicated decision processes as the computers and programs only have to search in a list and do not have to perform many million operations. Nevertheless it gives a good impression of the potential input/output stress in a Triple Brain with traditional programs.

(b) It is nicer to have one program, where the user can force this program to compute not only a single but k alternative solutions. Furthermore it would help when this program presented the alternatives on the monitor in such a way that the user could compare them in a comfortable way. Of course such a program would not make sense in the telephone example mentioned above. However, in chess or in difficult discrete optimization tasks it is important to have a good visualisation of "competing" candidate solutions.

Currently, in most fields there do not exist programs which are able to provide such clear sets of several candidate solutions. It is not an easy exercise to design algorithms for this task. Especially, one has often to deal with the problem that the alternative solutions are only micro mutations of each other and not "real" alternatives. A special case of (b) are programs which first of all compute one solution (their best one) and provide alternatives only on request of the user.

TRIPLE BRAIN IN CHESS: 13 YEARS OF EXPERIMENTS WITH MAN-MACHINE COMBINATIONS

Early in 1985 I prepared for the final exams of my diploma studies. At the same time the concept of Triple Brain rose in my mind, and I began with preliminary experiments in chess. There were mainly four reasons to start these investigations just in the game of chess.

(a) In chess it is well possible to measure the performance of a player (let it be a human, a computer, or some symbiotic system like Triple Brain). These measurements are done worldwide by the one-dimensional "Elo rating numbers" and their national counterparts.

(b) Already in 1985 chess programs were strong players. They were improved from year to year, but even nowadays they are still far away from playing perfect chess. (This is true also for all human players.)

(c) Already in 1985 some ten different programmers designed (independently of each other) commercial chess programs. Although being about equal in strength, these programs had rather different playing styles.

(d) In my youth I was an engaged amateur chess player. Therefore I am familiar with many aspects of the game.

Elo numbers are not explained here in detail (see Elo, 1978 for more information). In the context of this paper it is sufficient to know the following facts:

* The better a player is, the higher is his Elo rating.
* The Elo number of a player is computed from his results against other players who also have Elo numbers.
* Nowadays almost every club player has an Elo number (or a national equivalent, for instance a DWZ = "Deutsche Wertungs-Zahl" (German rating number)).
* Assume two players A and B such that Elo(A) = Elo(B) + 200. Then A should win a (fictive) match over 100 games against B in the average by 75:25. (The expected result does not depend on the absolute Elo numbers of A and B, but only on their difference.) There exists a table in which expected results are listed for all possible Elo differences. (Two concrete examples: Grandmaster Kasparov was in 1998 the world's best player with about 2800 Elo points. The German Grandmaster Arthur Yusupov had (also in 1998) an Elo rating of 2640. The difference is 160 points. So Kasparov should win a match against Yusupov by a score of 70 : 30 or slightly higher. Another example with players who participated also in this Ahlswede-Symposium: The Elo ratings of Ulrich Tamm and Levon Khachatrian are about 2130 and 2170, respectively. So Khachatrian should win a match against Tamm by about 55 : 45.)
* In Sweden there exists a group of computer chess enthusiasts. Since the early 1980's their organisation, SSDF, has played almost 100,000 games between many different chess computers and programs. From the results rating numbers for the programs were computed. These numbers do not predict exactly how good this or that computer performs against human players. But "calibrating games" between computers and humans have shown that the SSDF-ratings are quite comparable to normal Elo numbers. In the rest of this section I do not distinguish between Elo-, SSDF-, and ratings from my own experiments.

Over the years, I performed several chess experiments with Triple Brains. The computers and programs I used (and also the opponents of Triple Brain) became stronger and stronger. In all these matches I was the human controller in Triple Brain. My chess strength (in normal chess without the help of computers) was about Elo 1950 in the year 1980 and decreased slowly as time went by. In 1998 my rating was still something like 1850. In the book "13 Jahre 3-Hirn" ("13 Years with Triple Brain", written in German language) I have described and analysed all my chess experiments with combinations of man and machine.

Chronology, Part I				
Year	Ratings of Computers	Number of Games	Performance of 3-Hirn	Events
1985	1500, 1500	20	1700	private tournament
1987	1800, 1800	20	2050	three tournaments in the region of Bielefeld
1989	2090, 1950	8	2250	match with International Master Dr. Helmut Reefschläger
1992	2260, 2230	22	2500	second match with Reefschläger sparring with the parallel program ZUGZWANG (Uni Paderborn)
1993	2260, 2230	11	2450	Paderborn computer tournament sparring with DEEP THOUGHT (predecessor of IBMs DEEP BLUE) AEGON tournament 93
1994	2260, 2230	7	2550	Clodra mixed tournament
1995	2400, 2330	8	2550	match with International Grand-master (IGM) Christopher Lutz
1996	2370, 2370	6	2390	AEGON tournament 96

Chronology, Part II				
Double-Fritz with Boss				
Year	Ratings of Computers	Number of Games	Performance of 3-Hirn	Events
1996	2350	15	2520	tournament in Apolda match with IGM Timoshchenko
List Triple Brain				
1997	about 2530	8	2720	match in Shuffle Chess with IGM Arthur Yusupov

All results in Part I (except for AEGON 96, the last one) have one pattern in common: The Triple Brain was approximately 200 rating points stronger than the programs which were part of it. So I developed a rule of thumb for myself: "Take two different chess programs of equal strength x and Ingo Althöfer as a controller. Then the resulting Triple Brain will have strength about x + 200". Maybe this rule is only true when the strengths of the programs are not too high in comparision with my own strength, or there was some other flaw in my logic.

The AEGON tournament in April 1996 was a turning point for the Triple Brain experiments. Several things went wrong in this event: Hardware problems, the two programs (Rebel and M-Chess-Pro) did not fit together, and my own expectations were exaggerated (I tried too hard to win the tournament). I was rather disappointed by Triple Brains weak performance and started to think about modifications of the principle.

During the summer in 1996 I developed the concept of "Double-Fritz with Boss". Fritz in its version 4.0 was one of the first chess programs with a 2-best mode. In this mode not only the best but also the second best move (in the opinion of the program) is computed. "Double-Fritz" was my name for Fritz running in this 2-best mode. The Boss was me, having the final choice amongst the two proposals of Fritz. So in contrast to Triple Brain "Double-Fritz with Boss" used only one computer - but the program on this single machine produced two move proposals.

In 1997 I combined the two approaches of Triple Brain and Double-Fritz + Boss: In "List Triple Brain" two different chess programs are involved (on two computers). Each program is running in a k-best mode for some number k equal to or larger than two, and the human controller has the final choice amongst the proposals of the two lists. List Triple Brain (with current hard- and chess software of 1997 and 1998) is tremendously strong. Unfortunately (from my point of view) human top ten masters were not willing to play against this combination. So I was not able to find out if List Triple Brain with me as a controller played stronger than the best human players.

Many observers criticised that in my Triple Brain the controller was not allowed to outvote the computers. For me this renunciation of outvoting had a psychological advantage: When I was not allowed to outvote I was not able to produce terrible blunders. Hence the responsibility for the moves of Triple Brain lay, at least to a large extend, not on my but on the shoulders of the computers (and their programmers). It took me some games to accept the role of only selecting from computer proposals, but after this phase of acclimatisation I always felt comfortable: I did not have to check all possible variations in my own head and could instead concentrate on aspects of the decision process where computers are weak (for instance: finding a move which takes into account the psychological situation of the opponent). On the other hand the missing right to outvote was partially compensated by the controllers right to organize the timing of the computers: I observed their monitors during the computing processes and stopped the programs in moments, when the move proposals seemed to be okay.

K-BEST OPTIMIZATION UNDER SIDE CONSTRAINTS: AVOIDING SOLUTIONS WHICH ARE TOO SIMILAR TO EACH OTHER

Aspects of Multiple Choice Systems may be discussed and analysed in several disciplines: mathematics, computer science (implementation of algorithms, visualisation of candidate solutions, hardware design and adjustments), philosophy (general differences between man and machine in problem solving, comparison of thinking and computing), psychology and medicine (effects of the multiple choice situation on the coordinator: stress, problems of overload and idleness), legal aspects ("who is responsible for severe failures of a Triple Brain?").

Mathematics plays an important role, when algorithms have to be designed which do not produce only a single solution but several alternative candidate solutions. Given a discrete optimization problem (minimize $f : A \to I\!R$), the k-BEST CONCEPT was formulated three decades ago (Hoffman and Pavley, 1959; Bellman and Kalaba, 1960): Find k different solutions $a_1, a_2, \ldots, a_k \in A$, such that there is no other solution $a \in A$ with $f(a) < f(a_i)$ for some $i, 1 \leq i \leq k$. The main goal of this approach, namely to get k "interesting" candidate solutions, is often missed when the k candidates are merely micro mutations of each other instead of real alternatives.

There are several ways to generate "more representative" k-samples of good solutions. In some of these approaches distance functions $d\,(\,.\,,.\,)$ on the set A are used to measure the dissimilarity of solutions (for instance: A is a high-dimensional Hamming space or has some other "natural" metric structure).

(a) Repeated "normal" optimization under successive changes of the optimization problem:

   (i) Assume $m < k$ and that the first m candidate solutions $a_1, a_2, \ldots, a_m$ have already been computed. Then the next subtask may be formulated as to minimize f on the set A_m which is defined by

   $$A_m = \{a \in A| \quad d(a, a_i) \geq d^* \text{ for } i = 1, 2, \ldots, m\}.$$

   Here the distance threshold d^* has to be chosen by the human controller. For $k = 2$ (hence $m = 1$ is the only relevant step) the problem has been solved exemplarily for matroids (WW). All other cases are still open, for instance matroids with $m = 2$.

   (ii) It may be easier not to forbid certain regions of A (by forbidding the d^*-balls around candidate solutions), but instead to modify the objective function such that solutions near previous candidate solutions get worse f-values. For instance, in case of minimum spanning trees all edges e in previous candidate trees may get penalty cost lengths $L(e) + \epsilon$ instead of $L(e)$, for some penalty parameter $\epsilon > 0$. After such modifications of the objective function the "normal" optimization task has to be solved to find the next candidate solution. (In recent years a theory of INCREMENTAL COMPUTING has been growing. The situation under investigation is the following: For some problem instance the minimization task has been solved. Then the instance is slightly changed - and minimization is to be done on this newly created instance, using the computational results found in the original problem.)

(b) In (a) the candidate solutions $a_1, a_2, \ldots, a_k$ do not have symmetric roles. Typically a_1 is better than the other a_i with respect to the original objective function, and so on. A more symmetric approach would ask to minimize the sum $f(a_1) + f(a_2) + \ldots + f(a_k)$ under the side constraint that the a_i are not too similar to each other, by demanding for instance $d(a_i, a_j) \geq d^*$ for all $1 \leq i < j \leq k$. (Minimizing the f-sum would be an optimal criterion, if the human controller made his final choice among the k candidates completely at random.)
(b) seems to be more difficult than (a). Even for matroids the simplest case $k = 2$ is unsolved. Until now no polynomial algorithm has been found. (Althöfer and Wenzel, 1998)

(c) For the problem of finding k "interesting" short paths one may exploit the fact that there exist rather efficient algorithms for finding the k' shortest paths if no side constraints are given. By help of such an algorithm the k' shortest paths are computed for some very large k' ($k' >> k$, for instance $k' = 10,000$). Then, secondary objective functions are used to find k interesting alternatives in the set of these k' paths.

(d) Many discrete optimization problems are difficult, for instance the NP-complete ones. In such cases, often heuristics like local search procedures, genetic algorithms, and greedy constructions are used to find good solutions. In analogy to (c) one may generate k' good candidates by such heuristics for some very large k' and then select k interesting alternatives from these k' candidates by secondary criteria. (A collection of many good solutions may also help to discover, manually or automatically, typical structures in good solutions.)

Research on finding k INTERESTING candidate solutions in discrete optimization problems is still in its infancies. There are many gaps to be filled, both in modelling and in the design of efficient algorithms. In another field, namely "information retrieval" (for instance, searching the World Wide Web with search engines), one of the key questions is that of "recall": How to COVER the space of good solutions by an appropriate number of candidates? So, one is not only interested in real alternatives but in the covering of potentially good solutions. Maybe, optimization people can learn some of the right questions and approaches from the information retrievers.

EXTENSIONS AND VARIANTS OF THE TRIPLE BRAIN PRINCIPLE

The principle of Triple Brain may be extended and modified in several directions. Already in Section 3 we had mentioned the "LIST Triple Brain" where each of the programs does not provide a single solution, but a list of candidate solutions; the human controller has the final choice from these lists. Other ideas are

(a) **Preselection by Majority Rules**
Assume there are more than two different programs, and that the human controller would not have the capacities to inspect all their candidate solutions. Nevertheless all these programs might be run; a protocol might count how often which solution was proposed, and only those two (or three) candidate solutions are presented to the controller which have been proposed most frequently by the programs. In the field of chess I have made rudimentary experiments with such an approach (Althöfer, 1991).

(b) **Successive Fixing**
In many optimization problems the set A, on which a function f has to be minimized, has a high-dimensional product structure. For instance $A = \{0,1\}^n$. In such a situation an iterative procedure with several rounds may be carried out. In the first part of a round good candidate solutions are generated and shown to the human controller. In the second part of the round the controller is allowed to fix some or several coordinates for all the remaining rounds. So the dimension of the optimization problem which was $= n$ originally becomes smaller and smaller from round to round until finally all n coordinates are fixed.

A small example may illustrate the principle: Let $A = \{0,1\}^{10}$. Before round 1 nothing is fixed, so $X(0) = *****\ *****$. In round 1 the controller inspects the candidate solutions and fixes coordinates $2, 8, 9$ to certain values, so for instance $X(1) = *0***\ **10*$. In round 2 candidate solutions respecting $X(1)$ are generated and inspected; coordinates $1, 4, 5, 7$ are fixed ($2, 8, 9$ remain fixed), yielding for instance $X(2) = 00*10\ *110*$. So in round 3 a 3-dimensional problem remains (only coordinates $3, 6, 10$ are free, yet).

Of course, in most optimization problem with such a small dimension like 10 it would be no problem to compute all $2^{10} = 1024$ solutions... Things are different if $A = \{0,1\}^{50}$ or even larger. The size of the example was kept artificially small to make clear the principle.

(c) **Divide and Conquer**
This is a variant of (b), which is useful for instance in routing problems (see Althöfer and Dettborn for an application of a "Divide and Conquer Quadruple Brain" in vehicle routing). A good route from place A to place B has to be determined. Two (or three) different programs are started and make one proposal each. Typically these proposals will not be identical, for instance when A and B are two German towns more than 50 kilometers apart from each other. If the controller has inspected the proposals and decides to route from A to B via some intermediate

place C, he may divide the original problem in two subproblems: finding a good route from A to C, and finding a good route from C to B. In these subproblems he may use the Triple (or Quadruple) Brain approach again.

(d) **Interactive Genetic Algorithms**
In the research group of J. Albert (Computer Science, University of Würzburg) interactive genetic algorithms have been studied. The basic tool is a genetic algorithm in "traditional" form (Goldberg, 1989). However, a human controller is allowed to add a few additional individuals in each generation (= round). By this the user can give new impulses to the population of the genetic algorithm without facing the danger to cause much damage. The additional individuals may be generated by the help of a tool called "phenotype editor" - which again may work in a Triple Brain manner (Schoof, 1998).

(e) **Admissibility Checks**
In experiments with traffic routing it has turned out that even the best commercial programs (in 1998) have lots of errors in their map data (Althöfer and Dettborn). Hence, when programs P1 and P2 are used a necessary condition for selecting the proposal of P1 might be that this route is feasible according to the other program P2, and vice versa. In case of more than two programs a majority criterion might be applied: for instance, only those proposals are acceptable which are feasible in the opinion of at least 70 percent of all the programs.

(f) **Stopping of Repeated Algorithms**
In difficult optimization problems it makes sense to apply some probabilistic heuristic again and again, making several runs with this heuristic. All the time the currently best solution is recorded. The user is allowed to stop when he is satisfied with the best solution found so far. Such a CONTROL BY TIMING was part of my Triple Brain in chess.

(g) **Speedup of Probabilistic Algorithms**
Sometimes an optimization problem is "half difficult" in the following sense concerning execution times : There are several programs which can solve the problem exactly, but these algorithms are not very fast and have some probabilistic structure, resulting in unpredictable run times. Given a problem instance, two or more such algorithms are started on different computers, and the problem is solved when one of them (the fastest one) has finished its computations. This approach makes sense for instance when using the symbolic math programs Maple, Mathematica, MuPAD, and so on for computing orthogonal polynomials. Theoretical investigations of similar scenarios can be found for instance in the paper (Luby, Sinclair, and Zuckerman, 1993).

(h) **Iterated Partial List Reductions**
Sometimes it makes sense to split the reduction process "from k solutions

to one solution" into several steps, for instance first "from k to m" and later "from m to one", where $k > m > 1$. These successive steps of reduction may either be done by different persons (for instance one expert who makes the shortlisting to m promising proposals, and then another expert or a committee who has the final choice) or by the same person at different times (first of all ommitting those candidates which are obviously not "the best"; and the final choice only after the collection of more information). For this second scenario we give an example concerning the translation from one natural language to another (for instance from English to German or vice versa). Commercial programs for this task are not completely useless but also not just perfect (John F. Kennedy: "Ich bin ein Berliner!" → "I am a doughnut!"). Most of these programs have an option where the task is not done fully automatic. Often the translation of a single word is not unique and the correct meaning depends on the context. In such cases the program has a local "Multiple Choice structure" by showing a list of all relevant candidates for this single word, and the user has to make his choice. A refinement (and improvement) of this option might work as follows: The list of all candidate words is shown but the user does not have to decide immediately for his final choice. Instead he can preliminarily reduce the list by omitting only some of the options. The final choice (amongst the remaining candidate words) may be postponed to a later moment when the human has got a better understanding of the whole document. Such a process of repeated list reduction might take even more than two rounds.

(In this special task of computer-supported language translation it may be helpful not only to have the candidate translations for a single word but also the corresponding "back translations" to the initial language. We give an example for this depth-two presentation, using an English-German/German-English dictionary (Weis, 1982):

English	German candidates	Back translations to English
groove	Furche	furrot , rut
	Rille	groove
	Tonspur	sound track
	Gewohnheit	habit

Someone who is fluent in English but not in German will see from the level-two candidates which German candidate word might be the best choice in the context. A referee pointed out that similar ideas have been proposed for instance by Chow and Schwartz, 1989.)

(j) **Deviation Protocols**
A basic part of my Triple Brain concept was that the human is not allowed to outvote the computer(s). A different idea is to allow outvoting but to force the controller to write protocol notes in such situations of

deviation. An example from practice follows. When the Siemens company developed new automatic controlling systems for the German railway company (Deutsche Bahn AG) they included systems of the following type (Kraas, 1997): The computer makes a proposal ("Don't let train X wait for train Y."), and the human controller has the choice either to follow this proposal or to decide differently. However, if the controller rejects the computer option, a little protocol window opens on his computer screen, and he has to type in an explanation for his deviating decision. This protocol note may be discussed later in a group with colleagues, computer programmers, and superiors. An extension of this approach with obligatory explanations for deviations might work as follows: The computer gives its best proposal X_0, k alternatives $X_1, X_2, \ldots, X_k$ and $k+1$ numbers $n_1, n_2, \ldots, n_k, n_{\text{other}}$. If the human controller decides for X_0, he does not have to explain anything. If he opts for alternative $X_i, 1 \leq i \leq k$, he has to write a protocol of length n_i. And if he takes an action that was not in the list of computer proposals he has to explain by a note of length n_{other} (typically $n_{\text{other}} > n_i$ for all i).

DISCUSSION AND VISIONS

It takes only an hour to read this paper. It has taken a month to write it. But it takes much longer to investigate Multiple Choice Systems experimentally in a serious way. Looking at the narrow field of chess, it took me 13 years (with about two months of work per year) to examine aspects of the Triple Brain concept, and even now I am sure not to have understood everything. In most fields outside of chess things are even more complicated especially when it is not so easy to measure and compare the performance of experts, let them be men, machines, or Triple Brains. As an example one can take the diagnosis of heart anomalies by the interpretation of electrocardiogram (ECG) data. Top human cardiologists are supposed to have success rates of about 70 percent, best automatic analysers judge correctly in between 80 and 90 percent (Voss, 1998). What about a Triple Brain consisting of two (sufficiently different) automatic ECG analysers and one top cardiologist as the human controller? Who would be able to quantify the rate of success of such a "team", and how much time (and money) would it take to do this?

In the future concepts like Triple Brain will probably be applied in many fields. Also mathematics will not remain untouched by these ideas for interactive work. There is a huge potential for Multiple Choice Systems in the fields of optimization, symbolic computing, theorem proving, and on a meta level also in the design of brainstorming sessions. Practitioners do not have to wait for theoretical evaluations. Simply start and apply Triple Brains in your own field!

Acknowledgements

Professor Rudolf Ahlswede was my teacher for almost ten years. I joined his research group in 1985 directly after finishing my Diploma Thesis. He was a very tolerant controller, always allowing my brain to work on mathematical and other scientific topics of my own choice. Doctoral dissertation and Habilitation Thesis were fruitful results of his confidence in me. I am gratefully indebted to Rudolf Ahlswede!

An anonymous referee gave many valuable comments and made constructive proposals which improved this paper considerably. Thanks also to her or him!

References

[1] I. Althöfer. "Das Dreihirn - Entscheidungsteilung im Schach", *CSS 6/85*, December 1985, 20-22.

[2] I. Althöfer, "Selective trees and majority systems: two experiments with commercial chess computers", *Proceedings "Advances in Computer Chess 6"* (Editor D. F. Beal), Ellis Horwood, Chichester, 1991, 37-59.

[3] I. Althöfer, "Een experiment met Dreihirn", *Computerschaak 13.5*, October 1993, 186-189.

[4] I. Althöfer, "Doppelfritz mit Chef", *CSS*, 5/96, October 1996, 33-36.

[5] I. Althöfer, "A symbiosis of man and machine beats Grandmaster Timoshchenko", *ICCA-Journal*, 20.1, March 1997, 40-47.

[6] I. Althöfer, "Improving computer performance with a touch of human input" (edited by E. Hallsworth) *Selective Search*, 69, April 1997, 21-25.

[7] I. Althöfer, "List-3-Hirn vs. Grandmaster Yusupov – a report on a very experimental match – Part I: The Games", *ICCA-Journal* 21.1, March 1998, 52-60. "- -Part II: Analysis". *ICCA-Journal*, 21.2, June 1998, 131-134.

[8] I. Althöfer, *13 Jahre 3-Hirn – Meine Schach-Experimente mit Mensch-Maschinen-Kombinationen*, Published by the Author, Jena, 1998, ISBN 3-00-003100-6.

[9] I. Althöfer and T. Dettborn, "Der 3-Hirn-Ansatz in der Routenplanung", *Technical Report*, University of Jena, Institute of Applied Mathematics, April 1998.

[10] I. Althöfer, C. Donninger, U. Lorenz, and V. Rottmann, "On timing, pemanent brain, and human intervention". *Proceedings "Advances in Computer Chess"* (Editors H. J. van den Herik, I. S. Herschberg, and J. W. H. M. Uiterwijk), University of Limburg Press, Maastricht, 1994, 285-296.

[11] I. Althöfer and W. Wenzel. "2-Best solutions under distance constraints: the model and exemplary results for matroids", 1997, to appear in *Advances in Applied Mathematics.*

[12] I. Althöfer and W. Wenzel, "k-Best solutions under distance constraints in valuated Δ–Matroids". 1998, to appear in *Advances in Applied Mathematics.*

[13] R. Bellman and R. Kalaba, "On the k-th best policies", *Journal of the SIAM 8*, 1960, 582-588.

[14] P. J. Brucker and H. W. Hamacher, "k-optimal solution sets for some polynomially solvable scheduling problems", *European Journal of Operations Research*, 41, 1989, 194-202.

[15] Y. L. Chow and R. Schwartz, "The N-best algorithm: an efficient search procedure for finding top N sentence hypotheses". *Proceedings "DARPA Speech & Natural Language Workshop"*, 1989, 199-202.

[16] A. E. Elo, *The Rating of Chess Players, Past and Present*, New York, 1978.

[17] D. Eppstein, *Internet-Bibliography on k-best algorithms*, http://www.ics.uci.edu/~eppstein/bibs/kpath.bib .

[18] D. E. Goldberg, *Genetic Algorithms in Search, Optimization, and Machine Learning*, Addison Wesley, Reading, 1989.

[19] W. Hoffman and R. Pavley, "A method of solution of the Nth best path problem", *Journal of the ACM*, 6, 1959, 506-514.

[20] H.-J. Kraas, Talk at the University of Jena, December 1997.

[21] M. Luby, A. Sinclair, and D. Zuckerman, "Optimal speedup of Las Vegas algorithms", *Information Processing Letters*, 47, 1993, 173-180.

[22] C. Lutz, "Report on the Match 3-Hirn vs. Christopher Lutz", *ICCA-Journal*, 19.2, June 1996, 115-119.

[23] D. McCracken, Man + Computer: a new symbiosis, Communications of the ACM 22, 1979, 587-588.

[24] J. Schoof, " Kooperative Optimierung mit kommunizierenden Algorithmen", Dissertation, University of Würzburg, Faculty of Mathematics and Computer Science,September 1998.

[25] A. M. Turing, "Computing machinery and intelligence", *Mind*, 59, 1950, 433-460.

[26] M. Valvo, "Consulting chess with a computer", *ICCA-Journal*, 13.2, June 1990, 88-98 and *ICCA-Journal*, 13.3, September 1990, 156-162.

[27] A. Voss, Personal communication, 1998.

[28] H. Weigel, "Het Elisto-experiment", *Computerschaak*, 3/85, June 1985, 98-101. (with an introduction by J. Louwman)

[29] H. Weigel, "Best of Four", *MODUL*, 1/88, March 1988, 21-23.

[30] E. Weis (editor), *Pons Kompaktwörterbuch Englisch-Deutsch, Deutsch-Englisch*, Klett, Stuttgart, 1982.

QUANTUM COMPUTERS AND QUANTUM AUTOMATA*

Rūsinš Freivalds

Department of Computer Science, University of Latvia,
Raiņa bulv. 29, Riga, Latvia

Abstract: Quantum computation is a most challenging project involving research both by physicists and computer scientists. The principles of quantum computation differ from the principles of classical computation very much. When quantum computers become available, the public-key cryptography will change radically. It is no exaggeration to assert that building a quantum computer means building a universal code-breaking machine. Quantum finite automata are expected to appear much sooner. They do not generalize deterministic finite automata. Their capabilities are incomparable.

HISTORY

The notion of *quantum* was introduced nearly 100 years ago, namely, in 1900 by Max Karl Ernst Ludwig Planck (b. April 23, 1858 in Kiel, Germany; d. October 4, 1947 in Göttingen, Germany) [20]. He assumed that energy is emanated and absorbed in fixed portions, in quanta. This assumption was so unusual that M. Planck himself considered this assumption only as a useful tool to obtain a certain result. Unfortunately, most of the physicists having made the new physics of the 20th century felt the utmost discomfort of this *drama of ideas*. The new physics produced nice formulas but it was most difficult to understand what these formulas *mean*. They contradicted our common interpretation of the world too much.

In classical physics controversial interpretations have been nothing very much unusual. The discussion on the nature of *light* has brought to us two theories of light: the corpuscular theory where the light is a stream of *photons* and the wave theory where the light is electromagnetic waves. Isaac Newton (b.

*Research supported by Grant No.96.0282 from the Latvian Council of Science

I. Althöfer et al. (eds.), Numbers, Information and Complexity, 541-553.

December 25, 1642 Julian = January 4, 1643 Gregorian, in Lincolnshire, U.K.; d. March 20, 1727 in London, U.K.) supported the corpuscular theory while Christiaan Huygens (b. April 14, 1629 in Hague, now the Netherlands; d. July 8, 1695 in Hague, now the Netherlands) maintained the wave theory. For many decades it seemed that the wave theory has been victorous. Everybody learns in the school that if you take a source of light, a screen and put a wall with a slit in it between the source of light and the screen, then you get a complicated picture on the screen consisting of dark and bright spots. This feature of light is called *difraction*. Since difraction may be observed for waves of a different nature as well (for instance, for waves on a surface of water), this experiment is considered as an invicible argument in support of the wave theory of the light.

Difraction is closely connected with another effect of the wave theory, namely, with *interference* . If you repeat the above-mentioned experiment with a wall with two slits, you get a more complicated picture because the light waves coming from the two slits *interfere*. Interference is an interesting physical phenomenon producing unexpected results. Thomas Young (b. June 13, 1773 in Milverton, U.K.; d. May 10, 1829 in London, U.K.) closed one of the slits in the two slit experiment, and observed that there are some places where the picture becomes not darker but rather brighter. This is illogical! You remove some light but the picture becomes brighter. However physicists explained this result rather easily. The light is waves, and when the waves are in opposites phases, the waves destroy each other.

The development of the new physics went on, and in 1923 Louis Victor Pierre Raymond duc de Broglie (b. August 15, 1892 in Dieppe, France; d. March 19, 1987 in Paris, France) assumed that every particle (for instance, an electron) is a wave as well. And indeed, later many experiments supported this unusual assumption. Particularly, the difraction and interference experiments with electrons were successfully performed.

Quantum mechanics was developed in two different versions. Werner Karl Heisenberg (b. December 5, 1901 in Würzburg, Germany; d. February 1, 1976 in Munich, Germany) developed particle quantum mechanics based on matrices. Erwin Schrödinger (b. August 12, 1887 in Vienna, Austria; d. January 4, 1961 in Alpbach, Austria) developed wave quantum mechanics. Two absolutely different theories for the same object! It was not easy to find out which one was the right one. All the known experiments were not able to distinguish between the two theories.

It was a tremendous surprise when it was established in 1926 that the two theories are equivalent. Every statement provable in one of the theories is provable in the other theory as well. How it is possible? Heisenberg's mechanics deals with particles, i.e. *discrete* objects, while Schrödinger's theory deals with waves, i.e. *continuous* objects. Discrete and continuous have always been considered as opposites.

Establishing this duality was a long story. D.Danin describes in [8] a part of this epizode:

"In the summer of 1925, when the wave mechanics was not yet in existance and the matrix mechanics had just appeared, two theorists from Göttingen went begging to the great David Hilbert, the established head of the Göttingen mathematical school. They asked the world-famous scientist to help them with the matrices. Hilbert listened to them and said something remarkable - each time he had to deal with these square tables they appeared in his calculations as a sort of "a byproduct" in the solutions of the wave equations, "So, if you look for the wave equation which has these matrices you can probably do more with that."

According to the American Edward Condon, the theorists were Max Born and Werner Heisenberg. The episode ended in this way: "They had thought it was a goofy idea and that Hilbert did not know what he was talking about. So he was having a lot of fun pointing out to them later that they could have discovered Schrödinger's mechanics six months earlier if they had paid a little more attention to his words."

One can hardly find a better example demonstrating the blindness of a one-sided approach."

It was not so that a mathematical theorem was proved. There were arguments valid for physicists and even after them there was a need to *understand* why the duality has the place.

Luckily or unluckily, there are many unusual principles in quantum mechanics very much different from the classical physics. *Heisenberg's uncertainty principle* (1927) postulates that no experiment can establish simultaneously the position and the momentum of an electron. This principle was crucially important for the proof of duality between the theories but it was far from trivial to discover the proof.

Any way, it was Max Born (b. December 11, 1882 in Breslau, now Wroclaw, Poland; d. January 5, 1970 in Göttingen, Germany) who produced the explanation. *Schrödinger's psi-waves were the probability waves.*

This explanation satisfied the physicists. This explains why the difraction and interference experiments can be produced with electrons. The position of the discrete particles are decribed by the continuous waves of the probabilities where the electron can be positioned. This implies all the effects of the wave theory.

However a difficulty comes out. This is the two-slit experiment. We know that the probabilities are real numbers between 0 and 1. When adding, these numbers cannot decrease!

The physicists overcame this difficulty by introducing negative probabilities as well. Very soon complex number also were needed to describe the probabilities. For terminological reasons, the physicists call these new complex "probabilities" *the amplitudes* and the relation between the two notions is as follows. While the quantum processes go on and no measurements are performed, you can calculate the amplitudes by formulas reminding the corresponding formulas for probabilities in the classical physics. When you perform a measurement, different outcomes are possible , and the *probability* of each possible outcome is the square of the modulus of the corresponding *amplitude*. Every measurement destroys the object. This is the price for obtaining the information. You cannot make a copy of a particle, i.e. you cannot make another particle to have exactly the same amplitude. Quantum mechanics is very much different from the classical physics.

There is wide-spread belief that quantum physics is very difficult. It is only partly true. The mathematics of quantum physics indeed is not very easy but the real difficulty is of quite different nature. The most difficult part of quantum physics is to *feel* it, to understand what does it all mean. This is a really difficult subject even for the best physicists. No wonder that there were heated discussions on the interpretation of quantum physics. Albert Einstein (b. March 14, 1879 in Ulm, Württemberg, Germany; d. April 18, 1955 in Princeton, New Jersey, U.S.A.) and Niels Henrik David Bohr (b. October 7, 1885 in Copenhagen, Denmark; d. November 18, 1962 in Copenhagen, Denmark) were the most active participants in these discussions. Several interpretations are alive up to this day but the most usually referred interpretation is so-called *Copenhagen interpretation*. However even in the nineties of our century new and new experiments are performed to find out which of the interpretations describes the nature best.

The physicists would prefer to perform genuine experiments for every proposal in these discussions. However in many cases genuine experiments were not possible, and physicists satisfied themselves with thought experiments. One of such thought experiments widely commented even nowadays is due to Erwin Schrödinger.

A photon is directed to a half-silvered mirror. A classically-minded physicist would say that the photon either reflects or goes through the mirror. These are two different possibilities and the experiment is organized so that in one case the transmited component triggers a device that kills a cat placed in a "black box" but in the other case nothing dramatic happens. Hence the classical physicist (or every person not having learned modern physics) would say that after the experiment the cat is *either alive or dead*. Not so for a quantum physicist. A quantum physicist would say that "unless we perform a measurement (i.e. unless we open the black box) the cat is in superposition *alive and dead*".

Of course, such a conclusion was too outrageous even for the physicists. They could allow something extraordinary in the microworld but not for macroscopic objects. A rich ammount of literature exists on the Schrödinger's cat. Try to search Internet with key word "Schrödinger's cat", and you will find many very recent writings as well. Any way, the physicists agree that Schrödinger's cat would be in superposition only for a very short time, and then the quantum noise would destroy the superposition. However for me, this is a good illustration of the essence of *quantum computation*. Just like in the Schrödinger's cat's case, quantum processes allow superposition of several processes (a computer scientist would say, this allows *a massive parallelism*).

This possibility of massive parallelism is very important for Computer Science. It was Nobel prize winner physicist Richard Feynman (b. May 11, 1918,

New York, U.S.A.; d. February 15, 1988 in Los Angeles, U.S.A.) who asked in 1982 what effects can have the principles of quantum mechanics, on computation. Since exact simulation of quantum processes demands exponentional running time, may be there are other computations as well which are performed nowadays by classical computers but might be simulated by quantum processes in much less time.

As for nearly everything genuinely important, this idea came to several person's mind. It went not noticed by Western readers that Yuri Ivanovich Manin (b. February 16, 1937 in Simferopol, now Ukraine; one of the best Soviet mathematicians, then in Moscow University, now in the University of Bonn, Germany) published a small series of two books "The provable and not provable" [17] and "The computable and not computable" [18]. These books and especially their introductions contain authors thoughts on the role of mathematics and the links of mathematics to various areas of science. For instance, the introduction of [18] considers the problem how to describe natural objects (languages, living beings) in precise terms, The introduction contains considerations on the role of DNA in computation (published in 1980!!!). They are immediately followed by the following text (which is presented here in my translation from Russian):

It is possible that to understand these phenomena we lack a mathematical theory of quantum automata. Such objects could show us mathematical models of deterministic processes with absolutely unusual properties. One of the reasons of the much larger capacity of quantum space (compared with the classical space) is the following fact: where in the classical space we have N discrete states, the corresponding quantum space has c^N Planck cells in superposition. In a union of two classical systems with the number of states N_1 and N_2, respectively, the numbers of the states multiply but in the quantum case we get $c^{N_1 \times N_2}$ states.

These imprecise calculations show the much larger potential complexity of the quantum behavior of a system versus its classical simulation. In particular, because of lack of a unique decomposition of the system into elements, the state of the quantum automaton can be considered in many different ways as a composition of different systems of classical automata. (Compare with the following instructive calculation at the end of [22]. "For a quantum-mechanical calculation of the methan molecule we need to perform computation by the sieve method in 10^{42} points. Even if we assume that every point needs 10 elementary operations to be performed, and even if we assume that all the computations are performed at a super-low temperature ($T = 3 \times 10^{-3} K$), then we are to use for the calculation of the methan molecule more energy than is produced on the Earth during a century.")

The first difficulty in implementation of this program is to find a right balance between mathematical and physical principles. The quantum automaton is to be abstract enough: the mathematical model is to use only the most basic quantum principles not restricting the physical implementations. Second, the model of evolution is to be a unitary rotation in a finite-dimensional Hilbert space, and the model of the virtual decomposition into subsystems is to correspond to decomposition of this space into a tensor product. Somewhere in this picture an interaction should be placed which is usually described by Hermitian operators and probabilities"

Well, who ever was the first, but R.Feynman's influence was (and is) so high that rather soon this possibility was explored both theoretically and practically. David Deutsch [9] introduced quantum Turing machines. He made the machine to be a physically realisable model of quantum computers . Quantum Turing machine is a quantum physical counterpart of a probabilistic Turing machine that makes a full use of the quantum superposition principle. D. Deutsch conjectured that it might be more efficient than a classical Turing machine. He also showed the existence of a universal quantum Turing machine. Unfortunately, his universal quantum Turing machine could use exponentionally more time in simulation of a particular quantum Turing machine. This drawback was overcome by Bernstein and Vazirani [6] and Yao [26].

Classical information theory is based on the classical bit as fundamental atom. This classical bit, henceforth called *cbit,* is in one of two classical states *true* and *false* . A *probabilistic* counterpart of the classical bit can be *true* with a probability α and *false* with probability β, where $\alpha + \beta = 1$. A *quantum bit* is very much like to it with the following distinction. For a *quantum bit* α and β are not real but complex numbers with the property $||\alpha||^2 + ||\beta||^2 = 1$.

Every computation done on qbits is performed by means of unitary operators. One of the simplest properties of these operators shows that such a computation is reversible. The result always determines the input uniquely. It may seem to be a very strong limitation for such computations. Luckily, for unlimited quantum algorithms (for instance, for Quantum Turing machines) this is not so. It is possible to embed any irreversible computation in an appropriate environment which makes it reversible. For instance, the computing agent could keep the inputs of previous calculations in successive order. For quantum finite automata the limitation of the automata to be reversible is more sensitive.

Quantum automata might remain a lesser known unusual modification of the standard definitions but two events caused a drastical change. First, P. Shor invented surprising polynomial-time quantum algorithms for computation of discrete logarithms and for factorization of integers. Second, joint research of physicists and computer people have led to a dramatic breakthrough: all the unusual quantum circuits having no classical counterparts (such as quantum bit teleportation) have been physically implemented. Hence universal quantum computers are to come soon. Moreover, since the modern public-key cryptography is based on intractability of discrete logarithms and factorization of integers, building a quantum computer implies building a code-breaking machine.

The above-mentioned features of quantum computers seem unusual and hence one may think that their advent is highly unlikely. On the other hand, in the recent years physicists have performed series of crucial experiments showing that all the basical elements needed for quantum computers can be indeed implemented. A quantum computer with 1 qbit memory has been built in IBM

Almaden Research center and a quantum computer with 4 qbits memory has been built in Los Alamos National Laboratory.

We present results of several authors on complexity of quantum automata. It turns out that for some languages quantum automata have exponentially less size compared with deterministic and even probabilistic automata, while for other languages recognized by deterministic finite automata quantum finite automata do not exist at all.

UNITARY MATRICES

Quantum physics asserts that every transformation of a quantum bit system is *unitary*. This means that the transformation can be performed by a linear operator such that its matrix is unitary. A matrix M is called unitary if

$$MM_T^* = M_T^*M = I,$$

where M_T^* is the conjugate transpose of the matrix M, i.e. the transposition of M and conjugation of its elements, and I is the unit matrix.

The main difficulty in construction of efficient (i.e. small-size) quantum finite automata is to make the needed matrices to be unitary. This is why this Section includes many examples of unitary matrices.

Lemma 1. *For arbitrary real values ϕ, ψ, η, the matrix*

$$\begin{pmatrix} \cos\phi(\cos\eta + i\sin\eta) & \sin\phi(\cos\eta + i\sin\eta) \\ \sin\phi(\cos\psi + i\sin\psi) & -\cos\phi(\cos\psi + i\sin\psi) \end{pmatrix}$$

is unitary.

Corollary 2. *The matrix* $\begin{pmatrix} \frac{1}{\sqrt{2}} & \frac{1}{\sqrt{2}} \\ \frac{1}{\sqrt{2}} & -\frac{1}{\sqrt{2}} \end{pmatrix}$ *is unitary.*

Corollary 3. *The matrix* $\begin{pmatrix} \cos\phi & i\sin\phi \\ i\sin\phi & \cos\phi \end{pmatrix}$ *is unitary.*

This corollary is crucially important for the sequel. It is used to prove that quantum automata (in contrast with deterministic or probabilistic automata) can do the counting modulo arbitrarily large prime numbers using only two states.

The monograph by J. Gruska [13] contains a useful description of all unitary matrices of size 2×2.

Theorem 4. *Every unitary matrix of size 2×2 can be written as follows:*

$$U = e^{i\gamma}\begin{pmatrix} e^{i\alpha} & 0 \\ 0 & e^{-i\alpha} \end{pmatrix}\begin{pmatrix} \cos\phi & i\sin\phi \\ i\sin\phi & \cos\phi \end{pmatrix}\begin{pmatrix} e^{i\alpha} & 0 \\ 0 & e^{-i\alpha} \end{pmatrix}$$

Theorem 5. *Every unitary matrix of size $n \times n$ can be decomposed into a product of n^2 unitary matrices of size $n \times n$ each of which affects only a two-dimensional subspace spanned by two natural basis vectors.*

Definition 6. *We call the matrix*

$$C = \begin{pmatrix} c_{11} & c_{12} & \dots & c_{1\,kn} \\ c_{21} & c_{22} & \dots & c_{2\,kn} \\ \dots & \dots & \dots & \dots \\ c_{kn\,1} & c_{kn\,2} & \dots & c_{kn\,kn} \end{pmatrix}$$

a block-product of the matrices $A = \begin{pmatrix} a_{11} & a_{12} & \dots & a_{1\,k} \\ a_{21} & a_{22} & \dots & a_{2\,k} \\ \dots & \dots & \dots & \dots \\ a_{k\,1} & a_{k\,2} & \dots & a_{k\,k} \end{pmatrix}$

and $B = \begin{pmatrix} b_{11} & b_{12} & \dots & b_{1\,n} \\ b_{21} & b_{22} & \dots & b_{2\,n} \\ \dots & \dots & \dots & \dots \\ b_{n\,1} & b_{n\,2} & \dots & b_{n\,k} \end{pmatrix}$ *if* $C_{(m-1)k+i,(l-1)k+j} = a_{i\,j}b_{m\,l}$.

Lemma 7. *If the matrices A and B are unitary, then their block-product is also a unitary matrix.*

Lemma 8. *For arbitrary prime p, the matrix*

$$\left(\frac{1}{\sqrt{p}}(\cos\frac{2m(p-n)\pi}{p} + i\sin\frac{2m(p-n)\pi}{p})\right) \quad \begin{matrix} m = 0, 1, 2, \dots, p-1 \\ n = 0, 1, 2, \dots, p-1 \end{matrix}$$

is unitary.

Corollary 9. *For arbitrary prime p, there is a unitary matrix C_p of size $p \times p$ such that all the elements c_{ij} of this matrix are of equal modulus $\frac{1}{\sqrt{p}}$.*

Corollary 10. *For arbitrary natural number n, there is a unitary matrix C_n of size $n \times n$ such that all the elements c_{ij} of this matrix are of equal modulus $\frac{1}{\sqrt{n}}$.*

This corollary is used to perform an equiprobable choice among a finite number of possibilities.

QUANTUM FINITE AUTOMATA

We consider 1-way quantum finite automata (QFA) as defined in [15]. Namely, a 1-way QFA is a tuple $M = (Q, \Sigma, \delta, q_0, Q_{acc}, Q_{rej})$ where Q is a finite set of states, Σ is an input alphabet, δ is a transition function, $q_0 \in Q$ is a starting state and $Q_{acc} \subset Q$ and $Q_{rej} \subset Q$ are sets of accepting and rejecting states. The states in Q_{acc} and Q_{rej} are called *halting states* and the states in $Q_{non} = Q - (Q_{acc} \cup Q_{rej})$ are called *non-halting states*. # and \$ are symbols that do not belong to Σ. We use # and \$ as left and right endmarker, respectively. The *working alphabet* of M is $\Gamma = \Sigma \cup \{\#, \$\}$.

A superposition of M is any element of $l_2(Q)$ (the space of mappings from Q to $\mathbb{C}$). For $q \in Q$, $|q\rangle$ denote the unit vector with value 1 at q and 0 elsewhere. All elements of $l_2(Q)$ can be expressed as linear combinations of vectors $|q\rangle$. We will use ψ to denote elements of $l_2(Q)$.

The transition function δ maps $Q \times \Gamma \times Q$ to $\mathbb{C}$. The value $\delta(q_1, a, q_2)$ is an amplitude of $|q_2\rangle$ in the superposition of states to which M goes from $|q_1\rangle$ after reading a. For $a \in \Gamma$, V_a is a linear transformation on $l_2(Q)$ such that

$$V_a(|q_1\rangle) = \sum_{q_2 \in Q} \delta(q_1, a, q_2)|q_2\rangle. \tag{1}$$

We require all V_a to be unitary.

The computation of a QFA starts in the superposition $|q_0\rangle$. Then transformations corresponding to the left endmarker #, the letters of the input word x and the right endmarker \$ are applied. A transformation corresponding to $a \in \Gamma$ consists of two steps.

1. First, V_a is applied. The new superposition ψ' is $V_a(\psi)$ where ψ is the superposition before this step.

2. Then, ψ' is observed with respect to the observable $E_{acc} \otimes E_{rej} \otimes E_{non}$ where $E_{acc} = span\{|q\rangle : q \in Q_{acc}\}$, $E_{rej} = span\{|q\rangle : q \in Q_{rej}\}$, $E_{non} = span\{|q\rangle : q \in Q_{non}\}$. This observation gives $x \in E_i$ with probability equal to the amplitude of the projection of ψ'. After that, the superposition collapses to this projection.

   If we get $\psi' \in E_{acc}$, the input is accepted. If $\psi' \in E_{rej}$, the input is rejected. If $\psi' \in E_{non}$, the next transformation is applied.

We regard these two transformations as reading a letter a. V'_a is the transformation that maps ψ to the non-halting part of $V_a(\psi)$. $V'_a = P_{non}V_a$ where $P_{non}(\psi)$ is a linear tranformation which leaves all non-halting components of the configuration ψ unchanged and maps all accepting and rejecting components to 0. If x is a word consisting of letters $a_1 \ldots a_k$, then V_x denotes $V_{a_k} \ldots V_{a_1}$ and V'_x denotes $V'_{a_k} \ldots V'_{a_1}$.

For a word x, ψ_x is the non-halting part of the QFA's configuration after reading x. It is easy to see that, for any word x and letter a, $\psi_{xa} = V'_a(\psi_x)$.

Indepedently of [15], quantum automata were introduced in [19]. There is one difference between these two definitions. In [15], a QFA is observed after reading each letter (after doing each V_a). In [19], a QFA is observed only after all letters have been read. It is easy to show that any language recognized by a QFA according to the definition of [19] is recognized by a QFA according to [15]. The converse is not true. Any finite language can be recognized in the sense of [15]. However, no finite non-empty language can be recognized in the sense of [19]. Everywhere in this paper, we will use the more general definition of [15].

We are used to the fact that nondeterministic finite automata is a generalization of deterministic finite automata. Likewise, probabilistic, alternating and many other type of finite automata is a generalization of deterministic finite automata. Hence we expect that quantum finite automata can recognize all the regular languages. It turns out not to be the case. The subsequent theorem was proved in [15].

Theorem 11. *The language $L = \{a,b\}^*a$ cannot be recognized by a 1-way quantum finite automaton with bounded error.*

On the other hand, it is easy to see that 1-way QFA can recognize with bounded error only regular languages. Hence the class of languages recognized by 1-way QFAs is a proper subset of regular languages.

The main property of the unitary matrices is reversibility. Hence it is natural to compare the capabilities of 1-way QFA and 1-way reversible finite automata [21].

A 1-way reversible finite automaton (RFA) is a QFA with $\delta(q_1, a, q_2) \in \{0, 1\}$ for all q_1, a, q_2. Alternatively, it can be defined as a deterministic automaton where, for any q_2, a, there is at most one state q_1 such that reading a in q_1 leads to q_2. We use the same definitions of acceptance and rejection. States are partitioned into accepting, rejecting and non-halting states and a word is accepted (rejected) whenever the automaton enters an accepting (rejecting) state. After that, the computation is terminated. Similarly to the quantum case, endmarkers are added to the input word. The starting state is one, accepting (rejecting) states can be multiple. This makes our model different from both [4] (where only one accepting state was allowed) and [21] (where multiple starting states with a non-deterministic choice between them at the beginning were allowed).

It is proved by A. Ambainis and R. Freivalds [1] that some regular languages can be recognized by 1-way QFA with a certain probability of the correct result but not by a higher probability.

Theorem 12. *A language can be recognized by a 1-way QFA with a probability exceeding 7/9 if and only if it can be recognized by a 1-way reversible finite automaton.*

Theorem 13. *The language a^*b^* can be recognized by a 1-way QFA with the probability of correct answer $p = 0.68...$ where p is the root of $p^3 + p = 1$.*

Corollary 14. *There is a language that can be recognized by a 1-QFA with probability 0.68... but not with probability $7/9 + \epsilon$.*

MORE ON UNITARY MATRICES

If we are interested only in the principal capabilities of 1-way QFA but not in the size of the minimal QFA, the following lemma (known by many persons but, probably, never formally published) is useful.

Lemma 15. *For arbitrary 1-way quantum finite automaton A there is an equivalent 1-way quantum finite automaton B such that A and B recognize the same* *language with the same probability of the correct result, and the matrices of the automaton B contain only real numbers (both positive and negative ones).*

For proving lower bounds of parameters of 1-way QFA it is useful to combine Lemma 15 with C. Jordan normal form of matrices and the fundamental property of unitary operators to transform orthonormal vector bases into orthonormal vector bases (see Chapter 9, Section 7 of [12]). This shows that the unitary matrices used in construction of the 1-way QFA can be decomposed into rotations and this geometrical interpretation plays an essential role in all the existing lower bounds.

ADVANTAGES OVER PROBABILISTIC AUTOMATA

We consider a language M_p consisting of words in a single-letter alphabet whose length is divisible to p.

It is easy to see that any deterministic finite automaton recognizing M_p has at least p states. A. Ambainis and R. Freivalds [1] have proved

Theorem 16. *If p is a prime number, then any 1-way probabilistic finite automaton recognizing M_p with probability $\frac{1}{2} + \epsilon$, for a fixed $\epsilon > 0$, has at least p states.*

Theorem 17. *For arbitrary $\epsilon > 0$, there is a 1-way quantum finite automaton with $\lceil 32\log_2 p \rceil$ states recognizing the language M_p.*

Arnolds Ķikusts [14] has improved the number of states for the 1-way QFA used in Theorem 17.

It is not so that quantum automata are always more efficient than deterministic automata. A. Ambainis, A. Nayak, A. Ta-Shma and U. Vazirani [2] proved an exponential lower bound on the size of 1-way quantum finite automata for a family of languages accepted by linear sized deterministic finite automata.

MULTI-TAPE AUTOMATA

In this Section we consider 1-way multi-tape finite automata. They process input information presented on several tapes each of which is read by one 1-way head only. The quantum automata are defined in the natural way, completely in the style used in our Section 3. The results in this Section are taken from the paper [3]. It is proved that 1-way quantum finite multi-tape automata can recognize languages not recognizable by deterministic or probabilistic finite multi-tape automata. In this sense, the results are stronger than those in Section 5.

First, we discuss the following 2-tape language

$$L_1 = \{(\nabla x_1 \nabla x_2 \nabla, \nabla y \nabla) \| x_1 = x_2 = y\},$$

where the words x_1, x_2, y are unary.

R. Freivalds [10] proved

Theorem 18. *The language L_1 can be recognized with arbitrary probability $1-\epsilon$ by a probabilistic 2-tape finite automaton but this language cannot be recognized by a deterministic 2-tape finite automaton.*

The quantum counterpart of this theorem is proved in [3].

Theorem 19. *The language L_1 can be recognized with arbitrary probability $1-\epsilon$ by a quantum 2-tape finite automaton.*

This theorem shows that Theorem 12 fails to have a counterpart for multi-tape automata.

Finally we consider a language which is difficult for a probabilistic recognition:

$$L_2 = \{(x_1 \nabla x_2, y) \| \text{ there is exactly one value } j \text{ such that } x_j = y.\}$$

where the words x_1, x_2, y are binary.

Theorem 20. *The language L_2 cannot be recognized by a 1-way probabilistic 2-tape finite automaton with a bounded error probability.*

Theorem 21. *A quantum finite 2-tape automaton exists which recognizes the language L_2 with a probability $\frac{4}{7}-\epsilon$ for arbitrary positive ϵ.*

References

[1] Andris Ambainis and Rūsiņš Freivalds, "1-way quantum finite automata: strengths, weaknesses and generalizations", *Proc. 39th FOCS,* 1998, $http://xxx.lanl.gov/abs/quant-ph/9802062$

[2] Andris Ambainis, Ashwin Nayak, Amnon Ta-Shma and Umesh Vazirani, "Dense Quantum Coding and a Lower Bound for 1-way Quantum Automata", $http://xxx.lanl.gov/abs/quant-ph/9804043$

[3] Andris Ambainis, Rūsiņš Freivalds and Marek Karpinski, "Multi-tape quantum finite automata", $http://xxx.lanl.gov/abs/quant-ph/9905026$

[4] D. Angluin, "Inference of reversible languages", *Journal of the ACM*, 29, 1982, 741-765.

[5] Paul Benioff, "Quantum mechanical Hamiltonian models of Turing machines", *J. Statistical Physics,* 29, 1982, 515–546.

[6] Ethan Bernstein and Umesh Vazirani, "Quantum complexity theory", *SIAM Journal on Computing,* 26, 1997, 1411–1473.

[7] Daniel Danin, *Inevitability of the strange world,* Molodaya Gvardiya, Moscow, 1962 (in Russian).

[8] Daniel Danin, *Probabilities of the quantum world,* Mir Publishers, Moscow, 1983.

[9] David Deutsch, "Quantum theory, the Church-Turing principle and the universal quantum computer", *Proc. Royal Society London, A400*, 1989, 96–117.

[10] Rūsiņš Freivalds, "Fast probabilistic algorithms", *Lecture Notes in Computer Science,* 74, 1979, 57–69.

[11] Richard Feynman, "Simulating physics with computers", *International Journal of Theoretical Physics,* 21, 6/7, 1982, 467-488.

[12] Felix Gantmacher, *Theory of matrices.* Nauka, Moscow, 1967 (in Russian).

[13] Jozef Gruska, *Quantum Computing.* World Scientific, Singapore, 1999.

[14] Arnolds Ķikusts, "A small 1-way quantum finite automaton", $http://xxx.lanl.gov/abs/quant-ph/9810065$

[15] Attila Kondacs and John Watrous, "On the power of quantum finite state automata", *Proc. 38th FOCS*, 1997, 66–75.

[16] K. de Leeuw, E.F. Moore, C.E. Shannon and N. Shapiro, "Computability by probabilistic machines",*Automata Studies,* C.E. Shannon and J. McCarthy, Eds., Princeton University Press, Princeton, NJ, 1955, 183–212.

[17] Yuri I. Manin, *The provable and not provable,* Sovetskoye Radio, Moscow, 1979, (in Russian).

[18] Yuri I. Manin, *The computable and not computable,* Sovetskoye Radio, Moscow, 1980, (in Russian).

[19] Cristopher Moore, James P. Crutchfield, "Quantum automata and quantum grammars", Manuscript available at $http://xxx.lanl.gov/abs/quant-ph/9707031$

[20] Max Planck, "Über eine Verbesserung der Wien'schen Spectralgleichung", *Verhandlungen der deutschen physikalischen Gesellschaft 2* 1900, S. 202.

[21] Jean-Eric Pin, "On reversible automata", *Lecture Notes in Computer Science,* 583, 401–415.

[22] R. P. Poplavskiy, "Thermodynamical models of information processes", *Uspekhi Fizicheskikh Nauk,* 115, No. 3, 1975, 465–501 (in Russian).

[23] Michael Rabin, "Probabilistic automata", *Information and Control,* 6, 1963, 230–245.

[24] Peter W. Shor, "Algorithms for quantum computation: discrete logarithms and factoring", *Proc. 35th FOCS*, 1994, 124–134.

[25] Daniel R. Simon, "On the power of Quantum Computation", *Proc. 35th FOCS*, 1994, 116–123.

[26] Andrew Chi-Chih Yao, "Quantum circuit complexity", *Proc. 34th FOCS*, 1993, 352–361.

ROUTING IN ALL–OPTICAL NETWORKS: ALGORITHMIC AND GRAPH–THEORETIC PROBLEMS

Luisa Gargano and Ugo Vaccaro

Dipartimento di Informatica ed Applicazioni, Università di Salerno
84081 Baronissi (SA), Italy
{lg,uv}@dia.unisa.it

Abstract: This paper surveys theoretical results for wavelength–routing in all–optical networks and presents several open problems. We focus our attention on graph-theoretical problems and proof techniques.

INTRODUCTION

Optical networks are emerging as key technology in communication networks and are expected to dominate many applications, such as video conferencing, scientific visualisation, real-time medical imaging, high–speed super-computing and distributed computing [19, 20, 34, 37]. The books of Green [19] and McAulay [31] offer a comprehensive overview of the physical theory and applications of this emerging technology.

In WDM (Wavelength Division Multiplexing) optical networks, the bandwidth available in optical fiber is utilised by partitioning it into several channels, each at a different wavelength. Each wavelength can carry a separate stream of data. In general, a WDM network consists of routing nodes interconnected by point–to–point unidirectional optic fiber links. Each link can support a certain number of wavelengths. The routing nodes in the network are capable of routing a wavelength coming in on an input port to one or more output ports, independently of the other wavelengths. The same wavelength on two input ports *cannot* be routed to the same output port. WDM optical networks can be classified into two categories: *switchless* (also called *broadcast–and–select* and *switched.* Each of these in turn can be classified as either *single–hop* (also called *all-optical*) or *multihop* [34]. In switchless networks, the transmission

I. Althöfer et al. (eds.), Numbers, Information and Complexity, 555-578.

from each station is broadcast to all stations in the network. At the receiver, the desired signal is extracted from all the signals. These networks are practically important since the whole network can be constructed out of passive optical components, hence it is reliable and easy to operate. However, switchless networks suffer of severe limitations that make problematic their extension to wide area networks. Indeed it has been proven in [1] that switchless networks require a large number of wavelengths to support even simple traffic patterns. Other drawbacks of switchless networks are discussed in [34]. Therefore, optical switches are required to build large networks.

A switched optical network consists of nodes interconnected by point–to–point optic communication lines. Each of the fiber–optic links supports a given number of wavelengths. The nodes can be terminals, switches, or both. Terminals send and receive signals. Switches direct their input signals to one or more of the output links. Each link is bidirectional and actually consists of a pair of unidirectional links [34].
In this survey we consider switched networks. In this kind of networks, signals for different requests may travel on the same communication link into a node v (on different wavelengths) and then exit v along different links. The only constraint is that no two paths in the network sharing the same optical link have the same wavelength assignment. In switched networks it is possible to "reuse wavelengths" [34], thus obtaining a drastic reduction on the number of required wavelengths with respect to switchless networks [1].

All-optical networks are networks where the information, once transmitted as light, reaches its final destination directly without being converted to electronic form in between. Maintaining the signal in optic form allows to reach high speed in these networks since there is no overhead due to conversions to and from the electronic form. In all-optical networks wavelength translation can be obtained by means of (optical) converters. If there is a converter at a node v, then any path containing v can change its wavelength as it passes through v.

In this survey we will present a theoretical model for communication problems in all–optical networks. We will highlight the most important graph–theoretic and algorithmic problems in the area and present some proofs of known results to illustrate the most effective techniques. Our aim is to illustrate problems and proof–techniques, we refer to the surveys [8, 24] for a more comprehensive list of references.

The graph theoretical model. The optical network will be represented as a graph $G = (V(G), E(G))$, where each undirected edge represents a pair of point–to–point unidirectional optical fiber links connecting a pair of nodes.
A dipath p from x to y in G is the undirected path joining x to y, in which each edge is considered traversed in the direction from x to y. We will use the term edge and link interchangeably, however the term link will always be associated with the direction in which an edge is used, in particular, our algorithms will assign *different* wavelengths to all the signals crossing the same link, i.e., the same edge in the same direction.

We find the above terminology convenient to work with. However, it should be clear to the reader that an equivalent formulation would consider G to be a symmetric directed graph - by replacing each edge of G with the two opposite arcs corresponding to it - and then each dipath would simply become a directed path in the usual sense of the word and conditions on using an edge in one direction then simply translate into conditions on using an arc.

We will also identify wavelengths with colors.

THE DIPATH COLORING PROBLEM

In this section we consider the following problem: Given a graph $G = (V, E)$ and a set $\mathcal{P}$ of dipaths on G, assign a color to each dipath in $\mathcal{P}$ in such a way that two dipaths that share a link must have different color assignments; we will call such a color assignment *valid*. The goal is to use the minimum possible number of colors under the validity constraint.

Given a graph G and a set $\mathcal{P}$ of dipaths on G, the conflict graph of $\mathcal{P}$ in G is the undirected graph with node set $\mathcal{P}$ having an edge between each pair of dipaths in $\mathcal{P}$ that share a link of G. The dipath coloring problem is equivalent to the vertex coloring problem on the corresponding conflict graph.

Example Given the graph G and the set of dipaths $\mathcal{P} = \{p_1 = (a, b, d), p_2 = (c, a, b), p_3 = (d, b), p_4 = (f, d, c, a), p_5 = (f, d, e)\}$ in Figure 1 (a), the conflict graph is given in Figure 1 (b).

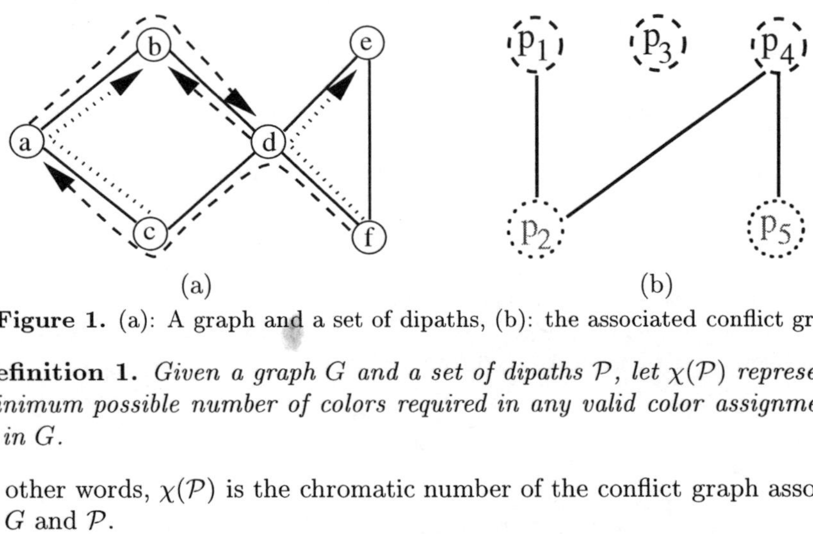

Figure 1. (a): A graph and a set of dipaths, (b): the associated conflict graph.

Definition 1. *Given a graph G and a set of dipaths $\mathcal{P}$, let $\chi(\mathcal{P})$ represent the minimum possible number of colors required in any valid color assignment for $\mathcal{P}$ in G.*

In other words, $\chi(\mathcal{P})$ is the chromatic number of the conflict graph associated to G and $\mathcal{P}$.

Definition 2. *Consider a graph G and a set of dipaths $\mathcal{P}$ on G. For each link e of G let $L(e, \mathcal{P})$ represent the load of the link e, that is, the number of dipaths in $\mathcal{P}$ crossing the link e. The load of $\mathcal{P}$ is defined as $L(\mathcal{P}) = \max_e L(e, \mathcal{P})$ where the maximum is taken over all links of G.*

It is more accurate to denote the above quantities by $\chi_G(\mathcal{P})$, $L_G(e, \mathcal{P})$, and $L_G(\mathcal{P})$ but we will omit the subscript G when it is clear from the context.

Since all dipaths that cross a given link must have different colors, for any set $\mathcal{P}$ it holds

$$\chi(\mathcal{P}) \geq L(\mathcal{P}). \tag{1}$$

Figure 2 gives two examples of sets of dipaths of load 2 that cannot be colored with less than 3 colors. This shows that in (1) the inequality can be strict.

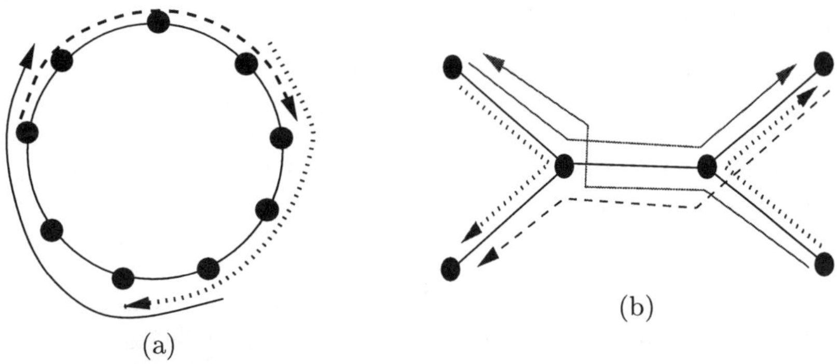

Figure 2 Examples for which the inequality (1) is strict.

We would like to remark that dipath coloring problem just introduced can be seen as a generalization of classical Path Intersection Problem in Graph Theory, see [17, 18, 32] and references therein quoted.

THE ROUTING PROBLEM

Given a graph $G = (V, E)$, a connection request (u, v) from node u to node v asks for choosing a dipath in G from u to v together with a valid color assignment.

Given a set of connection requests $R \subseteq \{(u, v) \mid u, v \in V\}$, in order to route R one has to choose a set of dipaths $\mathcal{P}_R$, one for each $(u, v) \in R$, together with a valid color assignment for $\mathcal{P}_R$.[1] The goal is to minimize the number of used colors. Given a set of requests R, define

$$\chi(R) = \min_{\mathcal{P}_R} \chi(\mathcal{P}_R)$$

and

$$L(R) = \min_{\mathcal{P}_R} L(\mathcal{P}_R)$$

where $\mathcal{P}_R$ ranges over all possible sets of dipaths for the communication requests in R.

[1]We always denote $\mathcal{P}$ and R as sets, however, they can be multisets in case of multiple connection requests for some pair of ordered nodes.

From (1) we get that for any set of requests

$$\chi(R) \geq L(R). \tag{2}$$

Even though $\chi(R)$ and $L(R)$ could be minimized at different points, no such an example is known. Therefore the following question is open.

Open Problem 1 *Is it true that for each R there exists a set $\mathcal{P}_R$ of dipaths for R that satisfies both equalities*

$$\chi(\mathcal{P}_R) = \chi(R) \text{ and } L(\mathcal{P}_R) = L(R)?$$

The routing problem is computationally difficult. Indeed

Theorem 3. *[13] Given a graph G and a set of requests R on G, determining $\chi(R)$ is an NP-complete problem. The problem remains NP-complete even if restricted to some simple graphs like rings and trees.*

Routing on a line

Even though the routing problem is NP-complete in general graphs, it is efficiently solvable when the underlying graph is a line.

Consider a line $L_n = (\{1, 2, \ldots, n\}, \{(i, i+1) \mid i = 1, \ldots, n-1\})$. The routing problem reduces to a dipath coloring problem. In fact, for any request (i, j) there is only one dipath from i to j: the left-to-right dipath $(i, i+1, \ldots, j)$ if $i < j$, the right-to-left dipath $(i, i-1, \ldots, j)$ if $i > j$. Moreover, left-to-right dipaths and right-to-left dipaths use different links and can be coloured independently.

The following theorem follows from the fact that coloring left-to-right (or right-to-left) dipaths on a line is equivalent to color interval graphs [9]. We will give here the simple proof of this result.

Theorem 4. *Let $\mathcal{P}$ be a set of left-to-right (or right-to-left) dipaths on a line,*

$$\chi(\mathcal{P}) = L(\mathcal{P}). \tag{3}$$

Proof: The proof is by induction on n. If $n = 2$ then $L(\mathcal{P}) = |\mathcal{P}| = \chi(\mathcal{P})$. Suppose the equality (3) is true for any set of dipaths on a line on $n - 1$ nodes. Consider a set of dipaths $\mathcal{P}$ on the line with nodes $1, \ldots, n$. Let

$$\begin{aligned} \mathcal{P}_1 &= \{(1, j) \in \mathcal{P} \mid 2 \leq j \leq n\} \\ \mathcal{P}_1' &= \{(2, j) \mid (1, j) \in \mathcal{P}_1,\ 2 < j \leq n\} \\ \mathcal{P}' &= (\mathcal{P} - \mathcal{P}_1) \cup \mathcal{P}_1'. \end{aligned}$$

$\mathcal{P}'$ is a set of dipaths on a line with $n - 1$ nodes, i.e. the nodes $2, \ldots, n$, and $L(\mathcal{P}') \leq L(\mathcal{P})$. By inductive hypothesis we can color $\mathcal{P}'$ with $\chi(\mathcal{P}') = L(\mathcal{P}')$ colors. In order to color the dipaths in $\mathcal{P}$, do the following:

a) Give to the dipath (i, j) with $i > 1$ the same color assigned to it in the coloring of $\mathcal{P}'$;
b) Give to $(1, j)$ the color assigned to $(2, j)$ in the coloring of $\mathcal{P}'$;
c) Color dipaths $(1, 2)$.

Notice that step **c)** needs extra colors only if the load of link (1,2) satisfies $L(\mathcal{P}) = L((1,2), \mathcal{P}) > L(\mathcal{P}')$, Therefore, we obtain a valid color assignment for $\mathcal{P}$ with $L(\mathcal{P}) = \max\{L((1,2), \mathcal{P}), L(\mathcal{P}')\}$ colors.

Routing on a ring

Consider a ring $R_n = (\{1, 2, \ldots, n\}, \{(n, 1)\} \cup \{(i, i+1) \mid i = 1, \ldots, n-1\})$. For any request (i, j) there are two dipaths from i to j: the clockwise and the counterclockwise dipath, see Figure 3. Moreover, clockwise and counterclockwise dipaths use different links of the ring and can be coloured independently.

Counterclockwise dipath from i to j

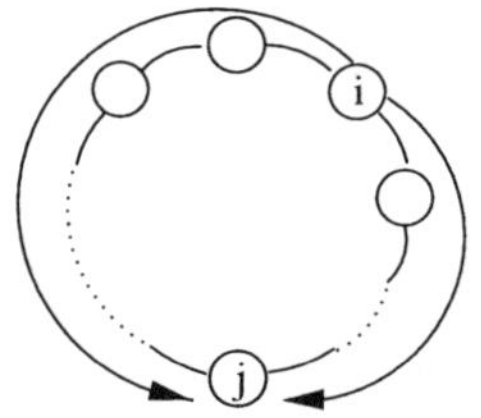

Clockwise dipath from i to j

Figure 3: Routing on a ring.

Theorem 5. *[38] For any set R of requests on a ring there exists an efficient algorithm that determines a set of dipaths $\mathcal{P}_R$ such that $L(\mathcal{P}_R) = L(R)$.*

Theorem 6. *[36] For any set of clockwise (counterclockwise) dipaths $\mathcal{P}$ on a ring*

$$\chi(\mathcal{P}) \leq 2L(\mathcal{P}). \qquad (4)$$

Proof: Split each dipath (i, j) containing the link $(n, 1)$ into two dipaths $(i, 1)$ and $(1, j)$ as shown in Figure 4. Namely, let

$$\mathcal{P}_1 = \{(i, j) \in \mathcal{P} \mid 1 \leq i \leq j \leq n\}$$

be the set of dipaths that do not contain the link $(n, 1)$ and define:

$$\begin{aligned} \mathcal{P}_2 &= \{(i, 1) \mid \text{there is } (i, j) \in \mathcal{P} - \mathcal{P}_1\}, \\ \mathcal{P}_3 &= \{(1, j) \mid \text{there is } (i, j) \in \mathcal{P} - \mathcal{P}_1\}. \end{aligned}$$

The set $\mathcal{P}' = \mathcal{P}_1 \cup \mathcal{P}_2 \cup \mathcal{P}_3$ is a set of dipaths on a line with $n + 1$ nodes (i.e., the nodes $1, 2, \ldots, n, 1'$, where the node $1'$ is a copy of node 1) of load $L(\mathcal{P}') = L(P)$.

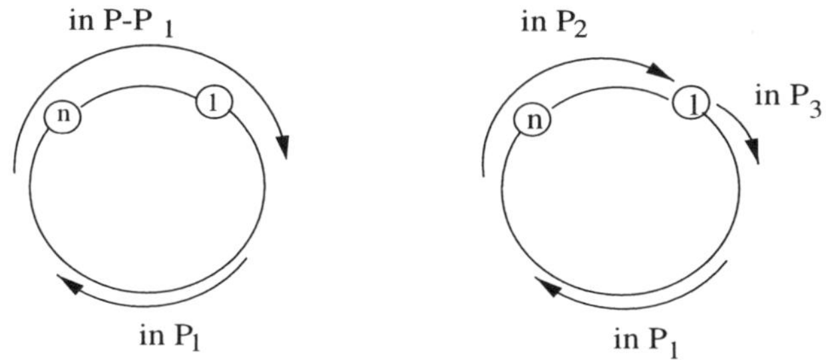

Figure 4: Illustration of dipath splitting on the ring.

We can therefore color $\mathcal{P}'$ with $\chi(\mathcal{P}') = L(\mathcal{P}')$ colors. In order to obtain the desired coloring of the set $\mathcal{P}$, do the following:

a) Assign to each $(i, j) \in \mathcal{P}_1$ the same color as in $\mathcal{P}'$;

b) For each $(i, j) \in \mathcal{P} - \mathcal{P}_1$ if, in the coloring of $\mathcal{P}'$, the color assigned to $(i, 1)$ is equal to the color assigned to $(1, j)$ then assign to (i, j) this color, otherwise if the color assigned to $(i, 1)$ differs from the color assigned to $(1, j)$ then assign to (i, j) an extra color, that is a color that is not used for any other dipath.

The number of extra colors is upper bounded by the load of the link $(n, 1)$ and, therefore, by $L(\mathcal{P})$. Hence,

$$\chi(\mathcal{P}) \leq \chi(\mathcal{P}') + L(\mathcal{P}) \leq 2L(\mathcal{P}).$$

From Theorems 5 and 6, one has the bound $\chi(R) \leq 2L(R)$, for any set R of requests on a ring.

Bounds on $\chi(\mathcal{P})$ have been also found in terms of the clique number $\Theta(\mathcal{P})$ of the conflict graph of $\mathcal{P}$.

Theorem 7. *[23] For any set $\mathcal{P}$ of paths on a ring,*

$$\chi(\mathcal{P}) \leq \frac{3}{2}\Theta(\mathcal{P}). \tag{5}$$

Open Problem 2 *Is it possible to improve the constant factor in the upper bound on $\chi(R)$?*

Routing on a tree

Following are the best known upper and lower bounds for dipath coloring on a tree. We recall that in case of trees the routing problem is NP-complete and is equivalent to the dipath coloring problem.

Theorem 8. *[27] There exists a tree T and a set of dipaths $\mathcal{P}$ on T such that $\chi(\mathcal{P}) \geq \frac{5}{4}L(\mathcal{P})$.*

Theorem 9. *[22] Given a tree T and a set of dipaths $\mathcal{P}$ on T, it is possible to efficiently find a valid color assignment to the dipaths in $\mathcal{P}$ using at most $\frac{5}{3}L(\mathcal{P})$ colors.*

We show now that the problem is solvable exactly for a class of trees. The star $S_n = (\{0,1,\ldots,n\},\{(0,i)|i=1,\ldots,n\})$ is depicted in Figure 5.

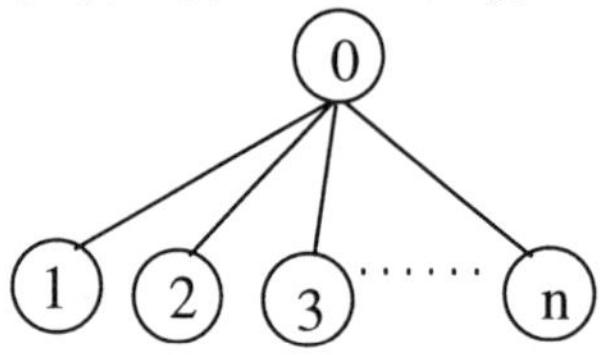

Figure 5. The Star S_n.

Theorem 10. *[15] For any n and set of dipaths $\mathcal{P}$ on S_n*

$$\chi(P) = L(\mathcal{P}).$$

Proof: Given a set of dipaths $\mathcal{P}$ on S_n, define a bipartite multigraph $H = (V(H), E(H))$ where $V(H) = \{a_1,\ldots,a_n\} \cup \{b_1,\ldots,b_n\}$ and $E(H)$ contains the edge (a_i, b_j) if and only if $\mathcal{P}$ contains a dipath from i to j, $1 \le i,j \le n$. The maximum degree Δ_H of a node in H is at most $L(\mathcal{P})$. For an example see Figure 6.

Figure 6. The set of dipaths $\mathcal{P} = \{(0,1),(1,2),(1,3),(2,3),(3,0),(3,1)\}$ on S_3 and the corresponding graph H with $\Delta_H = L(\mathcal{P}) = 2$.

Consider the set of dipaths (i,j), $i,j \ge 1$, of length 2 in $\mathcal{P}$. To color these dipaths is equivalent to edge color H, that is, to assign colors to the edges of H so that no two edges of the same color have a common endpoint. By Hall's theorem [9], the edges of H can be efficiently colored with Δ_H colors.
Once the edge coloring of H has been done, all the dipaths to be still considered are of the type $(i,0)$ or $(0,i)$ and can be now colored using a total of $L(\mathcal{P})$ colors.

A spider is a tree with at most 1 node of degree larger than 2, see Figure 7.

Theorem 11. *[15] If G is a spider then for each set of dipaths $\mathcal{P}$ on G*

$$\chi(\mathcal{P}) = L(\mathcal{P}).$$

Proof: If G is a line or a star then $\chi(\mathcal{P}) = L(\mathcal{P})$. Otherwise G is a spider with at least three legs. In such a case one can first color the dipaths going through the head as in a star. One can then complete the coloring of the dipaths on each leg as in a line.

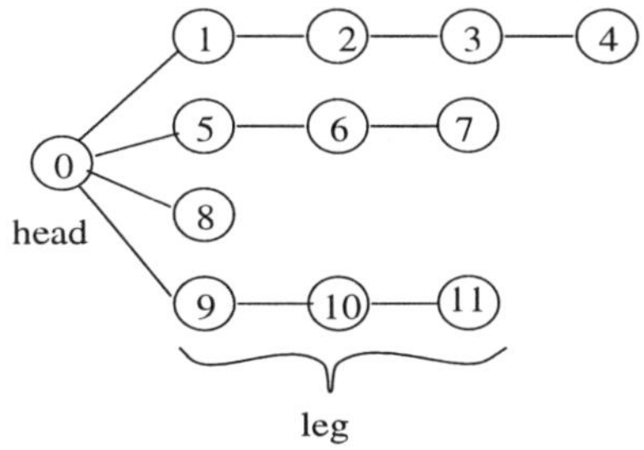

Figure 7: A spider graph

Routing in general graphs

Given a set of requests R on a graph G, let $\mathcal{P}$ be a set of dipaths for R. If ℓ is the maximum length of a dipath in $\mathcal{P}$ then, in the conflict graph of $\mathcal{P}$ each node has degree upper bounded by $(L(\mathcal{P}) - 1)\ell$ and the greedy coloring [9] assures that $\chi(R) \leq \chi(\mathcal{P}) \leq (L(\mathcal{P}) - 1)\ell + 1$.
The following better bound holds in case the maximum length ℓ is large.

Theorem 12. *[1] For any graph $G = (V, E)$ and any set of requests R on G of load $L(R)$,*

$$\chi(R) \leq 3\sqrt{|E|}\, L(R)$$

Proof: Let $\mathcal{P}$ be a set of dipaths for R of load $L(\mathcal{P}) = L(R)$. The number of dipaths of length at least $\sqrt{|E|}$ is at most $2\sqrt{|E|}L(\mathcal{P})$; otherwise there would be a link of load larger than $L(\mathcal{P})$. Give a different color to each of such dipaths. Consider now the remaining dipaths. They have length less than $\sqrt{|E|}$ and conflict with less than $L(\mathcal{P})\sqrt{|E|}$ other dipaths; therefore, the greedy coloring assures that at most $L(\mathcal{P})\sqrt{|E|}$ other colors suffice to color them.

Theorem 13. *[1] There exist a graph G and a set of requests R on G that can be routed with a set of dipaths $\mathcal{P}$ of maximum length ℓ and load $\mathcal{L}(\mathcal{P})$ but $\chi(R) = \Omega(L(\mathcal{P}) \min\{\ell, \sqrt{|E|}\})$.*

The problem of the existence of graphs G for which $\chi(\mathcal{P}) = L(\mathcal{P})$ for any set $\mathcal{P}$ has been considered in [38] and [15].

Theorem 14. *[38] [15] $\chi(\mathcal{P}) = L(\mathcal{P})$ for each set of dipaths $\mathcal{P}$ on G if and only if G is a spider.*

Proof: If G is a spider, the theorem follows from Theorem 11. Suppose now that G is not a spider. If G has a cycle then there exists a set of three dipaths along the edges of this cycle, see Figure 2 (a), of load 2 which cannot be colored with less than three colors. Otherwise G is a tree with at least 2 nodes of degree 3 or more. In such a case we can find a subtree and five dipaths as shown in Figure 2 (b), the load of the dipaths is 2 but at least three colors are needed.

SPECIAL INSTANCES

In this section we will consider some special instances of the routing problem in a graph $G = (V, E)$.

One–to–All Communication: One node v, called the source, has to be connected with each other node in the graph. In this case the set of connection requests is

$$R = \{(v, w) \mid w \in V\}.$$

One–to–Many Communication: One node v, called the source, has to be connected with each other node in a set $W \subseteq V$. In this case the set of connection requests is

$$R = \{(v, w) \mid w \in W\}.$$

All–to–All Communication: Each node v in the network has to be connected with each other node in the graph. In this case the set of connection requests is

$$R = \{(u, v) \mid u, v \in V,\ u \neq v\}.$$

One-to-All Communication

Given a graph $G = (V, E)$, the problem here is to set up $|V| - 1$ dipaths from the source of the One–to–All Communication process to any other node in V. Let $d(v)$ denote the degree of $v \in V$ and $d_{\min}(G) = \min_{v \in V(G)} d(v)$. When v is the source of the process there must exist at least $(|V| - 1)/d(v)$ dipaths out of the $|V| - 1$ dipaths originated at v that share the same edge incident on v. Therefore,

$$\chi(R) \geq \left\lceil \frac{|V| - 1}{d_{\min}(G)} \right\rceil . \tag{6}$$

On the other hand, if G is k–edge–connected, the following upper bound holds.

Theorem 15. *[11] For any k–edge connected graph G on n nodes*

$$\chi(R) \leq \left\lceil \frac{|V| - 1}{k} \right\rceil . \tag{7}$$

Proof Let node v be the source of the process. Partition, in an arbitrary way, the node set $V - \{v\}$ into $s = \lceil (|V| - 1)/k \rceil$ subsets, say $V_1, \ldots, V_s$, of size at most k each. Since G is k–edge–connected, for each $i = 1, \ldots, s$, it is possible to choose k edge–disjoint dipaths to connect v to the k nodes in V_i (see [9], Corollary 3, p. 167); the same color can be assigned to these dipaths. Hence, the information from v to each other node in G can be routed in one round using a total of at most $s = \lceil (|V| - 1)/k \rceil$ colors.

From (6) and (7) we get

Theorem 16. *[11] If G is maximally edge–connected then*

$$\chi(R) = \left\lceil \frac{|V|-1}{d_{\min}(G)} \right\rceil .$$

The above theorem gives the exact value of the number of colors necessary to perform One–to–All Communication in various classes of important networks. By Mader's theorem [29], Theorem 16 gives the exact value of $\chi(R)$ for the wide class of vertex–transitive graphs. In particular, we have

- for the d-dimensional hypercube H_d $\chi_{H_d}(R) = \lceil (2^d-1)/d \rceil$;
- for the $r \times s$ mesh $M_{r,s}$ $\chi_{M_{r,s}}(R) = \lceil (rs-1)/2 \rceil$;
- for the d dimensional torus C_m^d $\chi_{C_m^d}(R) = \lceil (m^d-1)/(2d) \rceil$;
- for any Cayley graph G of degree d $\chi_G(R) = \lceil (|V|-1)/d \rceil$.

For other classes of graphs G for which the edge connectivity is equal to $d_{\min}$ and, therefore, for which $\chi(R) = \left\lceil \frac{n-1}{d_{\min}(G)} \right\rceil$, see the survey paper [10].

As we have already remarked, in case of an arbitrary set of communication requests, the routing problem is NP–complete. In contrast, for the One–to–All communication the computation of $\chi(R)$ can be done in polynomial time also in general graphs by computing at most $\log|V|$ maximum flows on a graph with $O(|V|^2)$ nodes and $O(|V||E|)$ edges. We illustrate now the idea.
Given $G = (V, E)$, let u be the source of the One–to–All Communication process and k be an integer greater than 0. Construct k copies of G: $G_1 = (V_1, E_1), \ldots, G_k = (V_k, E_k)$; for any $v \in V$, let $v_1, \ldots, v_k$ be the copies of v in $G_1 = (V_1, E_1), \ldots, G_k = (V_k, E_k)$, respectively. For any vertex $v \in V - \{u\}$ let $n(v)$ be a new vertex. Define the flow network $G(k) = (V(k), E(k))$ as follows:

$$\begin{aligned} V(k) &= \left(\bigcup_{i=1}^{k} V_i \right) \cup \{s,t\} \cup \left(\bigcup_{\substack{v\in V \\ v\neq u}} n(v) \right) \\ E(k) &= \bigcup_{i=1}^{k} \{(s,u_i)\} \cup \left\{ \bigcup_{i=1}^{k} E_i \right\} \cup \left(\bigcup_{\substack{v\in V \\ v\neq u}} \bigcup_{i=1}^{k} \{(v_i, n(v))\} \right) \cup \left(\bigcup_{\substack{v\in V \\ v\neq u}} \{(n(v),t)\} \right). \end{aligned}$$

Vertex s is the source and vertex t is the sink of the flow network $G(k)$. For any $e \in E(k)$ we set the capacity $c(e)$ of e equal to ∞ if $e = (s, u_i)$, for $i = 1, \ldots, k$, and $c(e) = 1$ otherwise. The flow network $G(k)$ is represented in Figure 8.
The desired algorithm results from the following theorem.

Theorem 17. *[11] There is a flow of value $|V|-1$ in $G(k)$ if and only if $\chi(R) \leq k$.*

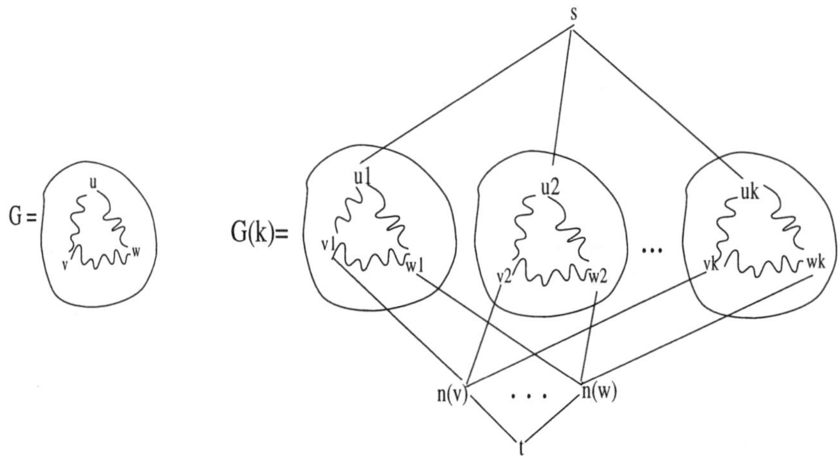

Figure 8. A graph G and the corresponding flow graph $G(k)$

In case of One-to-Many communication, i.e. $R = \{(v, d) \mid d \in D\}$, for some $v \in V$ and $D \subseteq V$, the following similar result holds.

Theorem 18. *[7] For any graph G and any One–to–Many set of requests R on G, there exists an efficient algorithm to compute $\chi(R) = L(R)$.*

It would be interesting to extend the above results to the case in which sole links of the network may fail, see for example [2].

All-to-All (Gossiping)

The set of All–to-All communication requests on a graph $G = (V, E)$ is $R = \{(u, v) \mid u, v \in V,\ u \neq v\}$.

In such a case $L(R)$ coincides with the well studied edge forwarding index of the underlying graph G [21]. Upper bounds on $\chi(R)$ have been obtained for several classes of graphs. In particular, it has been proved that $\chi(R) = L(R)$ for many graphs including trees, rings, hypercubes, and tori [15, 11, 6]. No graph is known for which $\chi(R) > L(R)$ and the following question is open.

Open Problem 3 *Does the equality $\chi(R) = L(R)$ holds for the All-to-All set of requests in any graph G?*

All-to-all on the Ring. The following results was first proved in [11]. We present here a simple proof.

Theorem 19. *The All–to–All communication set of requests on a ring on n nodes satisfies*

$$\chi(R) = L(R) = \left\lceil \frac{n^2 - 1}{8} \right\rceil.$$

Proof: It is known that on a ring C_n on n nodes $L(R) = \left\lceil \frac{n^2-1}{8} \right\rceil$ [21]. We show now that $\chi(R) = L(R)$. We give the proof for a ring on n nodes with n even; a similar proof can be deduced when n is odd. Let $n = 2k$, with $k \geq 1$. The proof is by induction on k.

If $k = 1$ then $\chi(R) = L(R) = 1$. Let $k > 1$. Color the dipaths on ring C_{n-2} on $n-2 = 2(k-1)$ nodes using $\left\lceil \frac{(n-2)^2}{8} \right\rceil$ colors.

In order to obtain a ring C_n on n nodes add to C_{n-2} two opposite nodes x, y. It is easy to see that it is possible to color dipaths connecting x and y from and to each of the old nodes in the ring C_{n-2} using $\lceil \frac{n-2}{2} \rceil$ new colors. We can then use an other new color for the two clockwise (or for the two counterclockwise) dipaths from x to y and from y to x. Moreover, this last color can be utilized for two successive induction steps once using the clockwise direction and the other time using the counterclockwise direction for the dipaths.

The total number of colors used on C_n is

$$\left\lceil \frac{(n-2)^2}{8} \right\rceil + \frac{n-2}{2} + \begin{cases} 0 & \text{if } k \text{ is even;} \\ 1 & \text{if } k \text{ is odd.} \end{cases}$$

Therefore, we get that $\chi(R) \leq \left\lceil \frac{n^2}{8} \right\rceil$. By (2), the theorem holds.

All-to-All on a tree $T = (V, E)$. In order to solve the All–to-All problem on trees, we introduce a generalization of it to weighted trees. A weighted tree is a tree with a weight $w(x)$ assigned to each node $x \in V$. The set of dipaths $\mathcal{P}$ contains $w(x)w(y)$ dipaths from x to y, for all $x, y \in V$, $x \neq y$.

Given $X \subseteq V$ let $w(X) = \sum_{x \in X} w(x)$. Given $e \in E$ let X_1 and X_2 be the node sets of the two trees obtained from T when e is removed; the load of e (in each direction) is then $L(e) = w(X_1)w(X_2)$. Finally, let

$$L(T) = L(\mathcal{P}) = \max_{e \in E} L(e).$$

Theorem 20. *[15] For any weighted tree T there is an efficient algorithm to color all dipaths on T using $L(T)$ colors.*

The rest of this section is devoted to give a sketch of the proof of Theorem 20. There is a natural way to build all trees from a single edge, by adding and splitting leaves. In the following definition we assume that $T = (V, E)$ is a weighted tree of weight $w(T) = w(V) = W$, x is a leaf of T, f is the parent of x, and finally, that w is a positive integer $w < w(x)$.

Definition 21. *The operation Add–Leaf$_w(x, T)$ modifies T as follows:*

- *the weight of x is decreased by w;*
- *a new node y is added with weight w;*
- *the edge (y, x) is added.*

The operation Split-Leaf$_w(x, T)$ modifies T as follows:

- *the weight of x is decreased by w;*
- *a new node y is added with weight w;*
- *the edge (y, f) is added. (Recall that f is the parent of x.)*

We say that an operation Add–Leaf$_w(x, T)$ or Split–Leaf$_w xT$ is legal *if $w + w(x) \leq W$ and $w(x) \leq W/2$.*

We will abbreviate the notation to simply say that we have performed an operation Add-Leaf or Split-Leaf. It is easy to see that if an operation is legal then in the new tree the load cannot be larger than the load of T.

Lemma 22. *[15] Any tree T can be generated starting from a suitable initial star S with $L(S) = L(T)$ by a suitable sequence of legal Add/Split–Leaf operations. Any intermediate T' has $L(T') = L(S) = L(T)$*

Summary of the algorithm for a tree T
Initial step: Determine and color the initial star S
General step: Given a tree T' and its coloring with $L(T')$ colors,
perform the next Add/Split–Leaf operation in the sequence determined by Lemma 22;
let T'' be the resulting tree.
Color T'' using the same $L(T'') = L(T')$ colors as for T'.
Repeat this step until the tree T is obtained.

The correctness of the above algorithm follows from next result.

Lemma 23. *[15] It is possible to color T'' using the same $L(T'') = L(T') = L(S)$ colors.*

WAVELENGTH TRANSLATION

It is possible to reduce the number of colors if we can color different subdipaths of a dipath with different colors, that is, if we can change the color assigned to a dipath in some node. In all-optical networks wavelength translation can be obtained by means of (optical) converters. If there is a converter at a node v, then any dipath containing v can change its wavelength as it passes through v (according to the translation capability of the converter in v).

In networks with converters the notion of wavelength assignment must be generalized: it is now the assignment of a wavelength to each link of the dipath with the restriction that it must be constant on any subdipath that does not go through any converter and that each wavelength change must be allowed by the corresponding converter.

Definition 24. *A converter is a bipartite (undirected) graph $C = (A \cup B, E(C))$ with $|A| = |B| = n$. An input color $a \in A$ can be converted to an output color $b \in B$ iff $(a, b) \in E(C)$. In general a subset $X \subseteq A$ of colors in A can be converted to $Y \subseteq B$ iff there exists a matching between X and Y in C.*

It is known that for each network, if we use a sufficient number of converters in which every input wavelength can be translated into every output wavelength, then we can accommodate any set of requests whose load does not exceed the number of wavelengths [38]. However, the complexity and then the cost of the converters is strictly depending on its degree (intended in the usual sense of the maximum degree of the graph that represents the converter). Networks with limited wavelength conversion will be less costly to implement than networks with full conversion capability. Moreover, in all-optical networks where the conversion is done without transforming the optical signal into electronic form, the conversion efficiency is a strong function of the input and output wavelengths, thus leading to limited conversion capability [33]. Different types of conversion are illustrated in Figure 9. Routing in all-optical networks with converters has been studied in [4, 14, 16, 25, 33, 38].

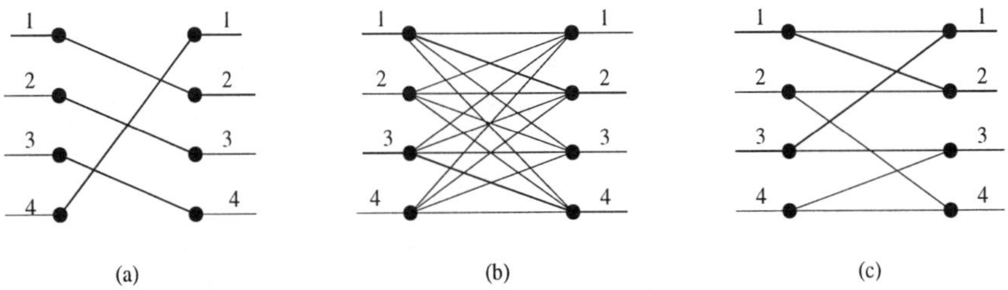

Figure 9: Examples of different types of converters on 4 colors: (a) is a fixed converter, (b) is a full converter, (c) is converter of degree 2

Full Converters

Full converters placement in each node of a graph $G = (V, E)$ allows to color any set of dipaths $\mathcal{P}$ using $L(\mathcal{P})$ colors. A natural question is whether it is possible to reduce the number of nodes with converters by maintaining the property that any set of dipaths $\mathcal{P}$ can be colored with $L(\mathcal{P})$ colors. This brings the following problem.

Minimum Sufficient Set Problem: Find a set of nodes $S \subseteq V$ such that if converters are placed only in the nodes in S then any set of dipaths $\mathcal{P}$ can be colored with $L(\mathcal{P})$ colors The goal is to minimize the size of S.

Theorem 25. *[38] A set S of size 1 is sufficient if G is a ring.*

Proof: Given a set of dipaths $\mathcal{P}$, the coloring is the same as in Theorem 6, noticing that a converter in in node 1 allows to use only $\chi(\mathcal{P}) = L(\mathcal{P})$ colors.

Theorem 26. *There exist graphs for which a set S of size $\Omega(|V|)$ is needed.*

Proof Consider the tree in Figure 10. If there exist two consecutive black nodes in $V - S$ then there exists a subgraph and a set of dipaths $\mathcal{P}$ as in Figure 2 (b) for which $L(\mathcal{P}) = 2 < \chi(P) = 3$.

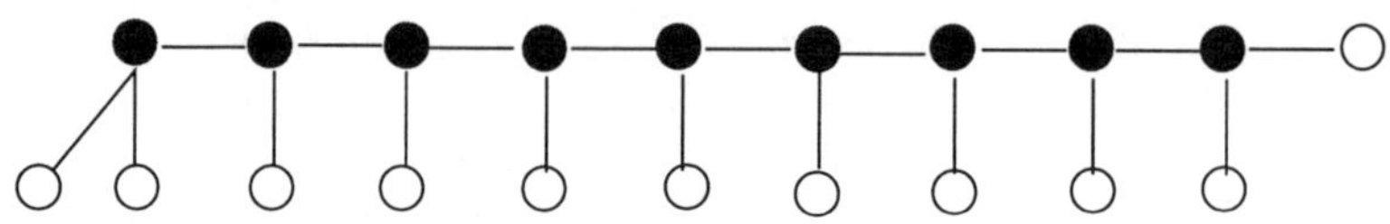

Figure 10: A graph requiring $\Omega(|V|)$ converters.

Theorem 14 can be reformulated as

Theorem 27. *$S = \emptyset$ if and only if G is a spider.*

For a graph $G = (V, E)$ and a set S define $G(S) = (V(S), E(S))$, where $V(S) = V - S \cup \{(s,e)|s \in S, e = (s,v) \in E\}$ and $E(S)$ includes edges in E between nodes in $V - S$, edges $((s,e),v)$ with $e = (s,v)$, and edges $((s,e),(t,e))$ with $s,t \in S$.

Example 5.1 For the tree G in Figure 11 (a) with one converter in the root s, the corresponding graph $G(S)$, with $S = \{s\}$, is given in Figure 11 (b).

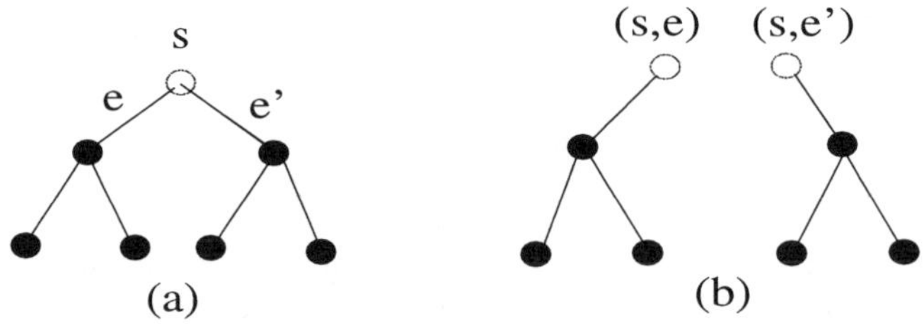

Figure 11: (a): A graph G with $S = \{s\}$, (b): the corresponding graph $G(S)$.

Theorem 28. *[38] S is sufficient for G if and only if each connected component of $G(S)$ is a spider.*

The sufficiency condition does not consider possible reroutings. Given a graph G, suppose that for any set of connection requests R there is some set of dipaths $\mathcal{P}$ for R and a coloring of $\mathcal{P}$ which is valid with respect to S; then S is called *weakly sufficient.*

Theorem 29. *[38] If S is weakly sufficient then it is sufficient.*

In general the Minimum Sufficient Set Problem is computationally difficult.

Theorem 30. *[38] The Minimum Sufficient Set Problem is NP-complete even in planar graphs.*

By reducing the Minimum Sufficient Set Problem to a vertex cover problem Kleinberg and Kumar show the following approximation result.

Theorem 31. *[25] There exists an efficient 2-approximation algorithm for the Minimum Sufficient Set Problem.*

Limited degree converters

We discuss here the following problem: Given a graph G, can we color any set of dipaths $\mathcal{P}$ on G such that $L(\mathcal{P}) \leq \chi$, where χ is the number of available colors, using limited degree converters? The goal is to keep the degree of the converters as small as possible.
We assume in this section that converters are placed in each node of the graph.

Expanders. The effectiveness of a converter C in minimizing the number of colors used in the routing will be shown to be strictly related to the expansion capability of C. In the following we introduce expander graphs and give some property that are used for proving the feasibility of the routing algorithms.

Definition 32. *A bipartite graph $C = (A \cup B, E(C))$ with $|A| = |B| = n$ is called an (α, β) expander if for each subset X of nodes with $|X| \leq \alpha n$ it holds $|N(X)| \geq \beta|X|$, where $N(X)$ denotes the neighbourhood of X in C, that is, the set of nodes which are connected by an edge to a node in X.*

The following Lemma states Tanner inequality [35] for bipartite graphs, see [3] and [12] for a proof.

Lemma 33. *If C is a k–regular bipartite graph on $2n$ nodes*

$$\frac{|N(X)|}{|X|} \geq \frac{k^2}{\lambda^2 + (k^2 - \lambda^2)|X|/n} \tag{8}$$

where λ is the smallest eigenvalue in absolute value of the adjacency matrix of C a part k and $-k$.

A special attention must be given to the class of Ramanujan graphs. These graphs have the property that the absolute value of each eigenvalue of the adjacency matrix, a part k and $-k$, is upper bounded by $2\sqrt{k-1}$, where k is the degree of the nodes. The interest in Ramanujan graphs is due to their good expansion property and to the fact that they are explicitly constructible [28, 30], a compendium can be found in [12]. We will refer to bipartite k–regular Ramanujan graphs simply as Ramanujan graph of degree k.
In case of Ramanujan graphs from Definition 32 and Lemma 33 we get

Corollary 34. *For any $\beta > 1$, Ramanujan graphs of degree k are (α, β) expanders for $\alpha \leq (1 - 4(\beta - 1)(k - 1)/(k - 2)^2)/\beta$.*

It is worth mentioning that better bounds can be obtained if we do not ask for explicit constructions of expanders (converters) but use probabilistic methods. Bassalygo [5] proved that

Theorem 35. *[5] For real numbers $0 < \alpha < 1/\beta < 1$, suppose that an integer k satisfies*

$$k > \frac{H(\alpha) + H(\beta\alpha)}{H(\alpha) - \alpha\beta H(1/\beta)}, \tag{9}$$

where $H(p) = -p\log_2 p - (1-p)\log_2(1-p)$ *is the binary entropy function. Then for any integer* n *there exists a* k*–regular bipartite* (α,β)*–expander on* $2n$ *nodes.*

Moreover, almost all random k–regular bipartite graphs on $2n$ satisfies Theorem 35, see [12].

The following results will be useful in the sequel.

Theorem 36. *[14] Let* $C = (A \cup B, E(C))$ *be an* (α,β) *expander on* $2n$ *nodes, with* $\alpha < 1 < \beta$*. Let* $Y_1,\ldots,Y_h \subset A$ *with* $h \le \beta$ *and* $\sum_{i=1}^{h} |Y_i| \le \lfloor \alpha n \rfloor \beta$*. Then there exists* $B' \subseteq B$ *such that*

i) $|B'| = \sum_{i=1}^{h} |Y_i|$*;*

ii) for each $b \in B'$ *there exists* $a \in \cup_{i=1}^{h} Y_i$ *such that* $(a,b) \in E(C)$*;*

iii) for each $a \in \cup_{i=1}^{h} Y_i$ *it holds* $|\{i \mid 1 \le i \le h,\ a \in Y_i\}| = |\{(a,b) \in E(C) \mid b \in B'\}|$*.*

Proof: Consider $Y_1,\ldots,Y_h \subset A$ with $\sum_{i=1}^{h} |Y_i| \le \lfloor \alpha n \rfloor \beta$ and let

$$Y = \cup_{i=1}^{h} Y_i.$$

Moreover, for each $y \in Y$, let $m(y)$ denote the multiplicity of y

$$m(y) = \{i \ :\ 1 \le i \le h,\ y \in Y_i\}.$$

Notice that

$$1 \le m(y) \le h \le \beta.$$

Consider then the bipartite graph $H = (A' \cup B, E(H))$ where for each $y \in Y$,
- A' contains $m(y)$ different copies of y, say $y^{(1)},\ldots,y^{(m(y))}$.
- $E(H)$ contains the edges (y^i, b), for $i = 1,\ldots,m(y)$, iff $(y,b) \in E(C)$.

Notice that

$$|A'| = \sum_{i=1}^{h} |Y_i|. \tag{10}$$

By definition of H, we have that i), ii) and iii) hold iff H contains a matching which includes all the nodes in A'; therefore, by Hall's theorem (see [9]) it is sufficient to show that for each $Z \subseteq A'$ the neighbourhood $N_H(Z)$ of Z in H satisfies $|N_H(Z)| \ge |Z|$. Consider then $Z \subseteq A'$ and denote by $z(y)$ the number of different copies of $y \in Y$ contained in Z. Notice that

$$0 \le z(y) \le m(y) \le h \le \beta. \tag{11}$$

Moreover, let $Z' = \{y \in Y \mid z(y) > 0\}$. Since C is an (α,β) expander we get

$$|N_H(Z)| = |N_C(Z')| \ge \begin{cases} \beta|Z'| & \text{if } |Z'| \le \lfloor \alpha n \rfloor, \\ \lfloor \alpha n \rfloor \beta & \text{otherwise.} \end{cases}$$

Therefore, using the definition of H, and the relations (10) and (11) we get $|N_H(Z)| \geq |Z|$.

Corollary 37. *Let $C = (A \cup B, E)$ be an $(\alpha, 2)$ expander on $2n$ nodes. For any $X, Y \subset A$ with*

$$|X| + |Y| \leq 2\lfloor \alpha n \rfloor$$

there exists $B' \subseteq B$ with $|B'| = |X| + |Y|$ such that

1) For each $b \in B'$ there exists $a \in X \cup Y$ such that $(a, b) \in E(C)$

2) For each $a \in X \cup Y$ $|\{(a,b) \in E(C) \mid b \in B'\}| = \begin{cases} 2 & \text{if } a \in X \cap Y \\ 1 & \text{otherwise.} \end{cases}$

Lemma 38. *Let $C = (A \cup B, E)$ be an $(\alpha, 2)$ expander on $2n$ nodes of degree $k > 2\sqrt{n}$ as given in Corollary 34. For any $X, Y \subset A$ with $\alpha n < |X| + |Y| \leq n$ it holds $N(X \cup Y) = B$.*

Proof: The proof follows the lines of that of Theorem 36 with the observation that if $k > 2\sqrt{n}$ then for any $X, Y \subset A$ with $|X|+|Y| > \alpha n$ it holds $|X|+|Y| > n - k$. Therefore, each node in B has a neighbor in $X \cup Y$.

Trees. In this section we consider the problem of coloring any set $\mathcal{P}$ of dipaths on a tree T using limited degree converters. Since T is a tree there is only one dipath corresponding to each connection request. If there are more connection requests for the same pair of nodes then some dipaths in $\mathcal{P}$ coincide on all links, however they are considered distinct for the coloring purpose.

We will show the following result.

Theorem 39. *[14] For any $\epsilon > 0$ there exists k s.t. using (explicitly constructible) converters of degree k it is possible to color any set of dipaths $\mathcal{P}$ with $L(\mathcal{P}) \leq \chi(1-\epsilon) - 1$*

We will call *segment* a dipath or any sequence of consecutive links of a dipath in $\mathcal{P}$, formally, given a dipath $p = x_1x_2 \ldots x_\ell$ a segment of p is any dipath $x_ix_{i+1} \ldots x_{j-1}x_j$ with $1 \leq i < j \leq \ell$.

If we fix any node r as the root of T we can then call a link *ascending* (resp. *descending*) (with respect to r) if it is directed toward (resp. away from) the root. In the same way, we say that a segment is ascending (resp. descending) if all its links are ascending (resp. descending). Notice that any dipath $p = x_1x_2 \ldots x_\ell$ that has both ascending and descending links can be partitioned into one ascending segment $x_1x_2 \ldots x_i$ followed by one descending segment $x_i \ldots x_\ell$, for some $1 < i < \ell$; we indicate the ascending segment $x_1x_2 \ldots x_i$ as the *maximal ascending segment* of p (if p has only ascending links then p is ascending and the maximal segment of p is p itself).

Definition 40. *Let T be a tree rooted in r. A set of dipaths is called almost-ascending in T if any descending segment of a dipath in $\mathcal{P}$ has length 1.*

Theorem 41. *[14] Any set of almost-ascending dipaths of load L on a tree T can be efficiently colored (without converters) using exactly L colors.*

Proof: **(Sketch).** If the tree T has diameter 2, that is, T is a star with center r, then by Theorem 10, the result holds. Suppose then that the tree T is not a star. Denote by $\mathcal{P}$ the set of almost–ascending dipaths and consider the conflict graph $G_{\mathcal{P}}$ associated to $\mathcal{P}$. Consider any ascending link (x, y) in T. It is possible to prove that the dipaths containing the link (x, y) form a separating clique in $G_{\mathcal{P}}$. Therefore, the problem can be reduced to coloring almost-ascending sets of dipaths on "smaller" trees.

Given a set of dipaths $\mathcal{P}$ consider the set $\mathcal{P}'$ obtained as follows: $\mathcal{P}'$ contains all the ascending dipaths in $\mathcal{P}$; moreover for each dipath $p = x_1 \ldots, x_\ell$ containing both ascending and descending links, the set $\mathcal{P}'$ contains the almost–ascending dipath $x_1, \ldots, x_i, x_{i+1}$, where $x_1, \ldots, x_i$ is the maximal ascending segment of p. Applying Theorem 41 to $\mathcal{P}'$ we have

Corollary 42. *Given a set of dipaths $\mathcal{P}$ of load L on a tree T, we can color (without converters) all maximal ascending segments of dipaths in $\mathcal{P}$ using at most L colors so that for each node v with children $v_1, \ldots, v_d$, and for each $i = 1, \ldots, d$, we assign different colors to all maximal ascending segments that contain any link (v_j, v), with $i \neq j$, and belong to a dipath in $\mathcal{P}$ which also crosses the descending link (v, v_i).*

The algorithm [for the tree T and set of dipaths $\mathcal{P}$ of load L on T]
Step 1 Root T in any node r.
Step 2 Color the maximal ascending segments applying Corollary 42.
Step 3 In order to complete the coloring we have to assign a color to all the descending segments. This must be done respecting the following
Constraint *If a dipath in $\mathcal{P}$ contains a link e_1 followed by a descending link e_2 then the color assigned to e_1 must be converted into the color assigned to e_2.*
We visit the nodes of T in BFS (Breadth–First Search) manner. For any node v visited we color all the descending segments consisting of a link going from v to a son of v; the coloring is done respecting the above Constraint. In general, suppose we have already considered all the nodes at level $l - 1$ and we are considering a node v at level l. Indicate by f the parent of v and by $v_1, \ldots v_d$ the children of v. We have to assign colors to all the descending segments consisting of the links (v, v_i), for $i = 1, \ldots, d$.
Consider the link (v, v_i) and the dipaths that cross this link.

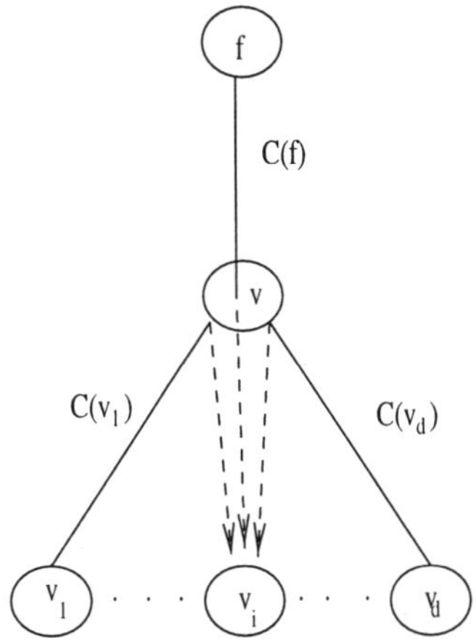

Figure 12: Bold lines denote the segments of dipaths crossing the link (v, v_i) where the colors are already fixed, dashed lines denote the segments where the colors must be assigned.

Consider the set of colors $C(v_j)$, for $j \neq i$, already fixed for the segments using the link (v_j, v) (and belonging to a dipath that also uses the link (v, v_i)). The coloring of the ascending segments done in Step 2 assures that these sets are pairwise disjoint. This implies that we have two sets on which we can apply Corollary 37, namely the set $C = \cup_{j \neq i} C(v_j)$ and $C(f)$, the set of colors already fixed for the segments using the link (f, v) (and belonging to a dipath that also uses the link (v, v_i)). Moreover,

$$|C(f)| + |C| \leq L.$$

By Corollary 34 we obtain that $L \leq n\left(1 - \frac{4(k-1)}{(k-2)^2}\right) - 1$, which gives Theorem 39.

Theorem 39 shows that it is possible to route on a tree any set of dipaths of load arbitrarily close to the number of available colors χ using constant degree converters. We consider now the question of when it is possible to do the same with $L(\mathcal{P}) = \chi$.

If we allow the degree of the converters to depend on the number χ of colors then we can route any set of requests of load up to χ. Namely,

Theorem 43. *[14] For any tree T, if χ colors are available on each link then using (constructible) converters of degree $k > 2\sqrt{\chi}$, it is possible to efficiently assign color to any set of request of load L whenever $L \leq \chi$.*

Proof: The algorithm is the same as given to prove Theorem 39. The only difference is in the converters that, in this case, satisfy Lemma 38.

There exists a class of trees in which constant degree converters allow to route any set of requests having load $L \leq n$, if n colors are available.

We say that a tree is *quasi-binary* if at most one node has degree greater than or equal to 4.

Theorem 44. *[14] Let T be any quasi–binary tree. Then converters of degree 3 allow to efficiently assign colors in a greedy manner to any set of request of load $L \leq n$, where n is the number of colors available on each link.*

Open Problem 4 *Is it possible to use constant degree converters to route any set of dipaths $\mathcal{P}$ of load $L(\mathcal{P}) \leq \chi$ in general trees?*

General Graphs. The following result is the best known for limited degree converters in general graphs.

Theorem 45. *[16] Consider a graph G with χ colors per link. Let the converters be (α, β) expanders. It is possible to color any set of dipaths $\mathcal{P}$ with $L(\mathcal{P}) \leq \delta\chi$, where $\delta = \min\{\alpha(\beta - 1),\ 1 - \alpha\}$.*

Proof: Suppose to have already assigned colors to each dipath in a set $\mathcal{P}'$. We show now that it is possible to assign colors to any dipath p whenever for the set $\mathcal{P} = \mathcal{P}' \cup \{p\}$ it holds $L(\mathcal{P}) < \delta\chi$. Let p consist of the sequence of links $(e_1, e_2, \ldots, e_k)$ for some k. For $i = 1, \ldots, k$ a color c is said busy if on the link e_i it is used by a dipath in $\mathcal{P}'$; for $i = 2, \ldots, k$ a color c is also said busy if on e_{i-1} each color c' that can be converted into c is busy. Otherwise the color c is said idle on e_i. We will show by induction on $i = 1, \ldots, k$ that there are at least $\alpha\chi$ idle colors on e_i. For $i = 1$ this is true since at most $L(\mathcal{P}') \leq \delta\chi \leq (1-\alpha)\chi$ dipaths cross this link. For $i > 1$, suppose there are $\alpha\chi$ idle colors on e_{i-1}. Since the converters are (α, β) expanders, there is a set of $\alpha\beta\chi$ colors on link e_i, each compatible with at least one of the idle colors in e_{i-1}. Since there are at most $\delta\chi \leq (\beta - 1)\alpha\chi$ dipaths crossing the link e_i, there must exist at least $\alpha\chi$ idle colors on e_i. Therefore, it is possible to find a sequence $c_1, \ldots, c_k$ of idle colors on the links $e_1, e_2, \ldots, e_k$ to be assigned to the dipath p.

Open Problem 5 *Is it possible to improve the bound in Theorem 45? Is it possible to improve Theorem 45 for special classes of graphs?*

References

[1] A. Aggarwal, A. Bar-Noy, D. Coppersmith, R. Ramaswami, B. Schieber, M. Sudan, "Efficient Routing and Scheduling Algorithms for Optical Networks", *Proceedings of the 5th Annual ACM-SIAM Symposium on Discrete Algorithms SODA'94*, 1994, 412–423.

[2] R. Ahlswede, L. Gargano, H.S. Haroutunian, L.H. Khachatrian, "Fault–Tolerant Minimum Broadcast Networks", *NETWORKS*, 27, 1996, 293–307.

[3] N. Alon, "Eigenvalues and Expanders", *Combinatorica*, 6, 1986, 83–96.

[4] V. Auletta, I. Caraggiannis, C. Kaklamanis, G. Persiano, "Efficient Wavelength Routing with Low–Degree Converters", *Proceedings of DIMACS Workshop on Optical Networks*, 1998.

[5] L. A. Bassalygo, "Asymptotically Optimal Switching Circuits, *Problems of Information Transmission*, 1981, 206–211.

[6] B. Beauquier, "All-to-All Communication in Some Wavelength-Routed All–Optical Networks", *INRIA Research Report*, 3452.

[7] B. Beauquier, P. Hell, S. Perennes, "Optimal Wavelength-Routed Multicasting", *Discr. Appl. Math* 84, 1998, 15-20.

[8] B. Beauquier, J.-C. Bermond, L. Gargano, P. Hell, S. Perennes, and U. Vaccaro, "Graph Problems Arising from Wavelength–Routing in All–Optical Networks", *2nd Workshop on Optics and Computer Science WOCS*, Geneve, Switzerland, 1997,

[9] C. Berge, *Graphs*, North–Holland.

[10] J.-C. Bermond, N. Homobono, and C. Peyrat, "Large Fault–Tolerant Interconnection Networks", *Graphs and Combinatorics* 5, 1989, 107–123.

[11] J.–C.. Bermond, L. Gargano, S. Perennes, A.A. Rescigno, U. Vaccaro, "Efficient Collective Communication in Optical Networks", *Proceedings of 23th International Colloquium on Automata, Languages, and Programming ICALP 96*, LNCS, 1099, Springer-Verlag, 1996, 574–585.

[12] F.R.K. Chung, *Spectral Graph Theory*, CBMS 92, American Mathematical Society, 1997.

[13] T. Erlebach and K. Jansen. "Scheduling of Virtual Connections in Fast Networks", *Proc. of 4th Workshop on Parallel Systems and Algorithms PASA '96*, 1996, 13–32.

[14] L. Gargano, "Limited Wavelength Conversion in All-Optical Tree Networks", *Proceedings of 25th International Colloquium on Automata, Languages, and Programming ICALP 98*, Aalborg, 1998.

[15] L. Gargano, P. Hell, S. Perennes, "Colouring All Directed Paths in a Symmetric Tree with Applications to WDM Routing", *Proceedings of 24th International Colloquium on Automata, Languages, and Programming ICALP 97*, P. Degano, R. Gorrieri, A. Marchetti–Spaccamela Eds, LNCS, 1256, Springer-Verlag, 1997, 505–515.

[16] O. Gerstel, S. Kutten, R. Ramaswami and G. Sasaki, "Wavelength Conversion in All-Optical Ring Networks", *Proceedings of 6th Annual ACM SIGACT-SIGOPS Symposium on Principles of Distributed Computing PODC'97*, 1997.

[17] M. C. Golumbic and R. E. Jamison. "The Edge Intersection Graphs of Paths in a Tree", *Journ. Comb. Theo.*, Series B, 38, 1985, 8–22.

[18] M. C. Golumbic and R. E. Jamison. "Edge and Vertex Intersection of Paths in a Tree", *Discrete Mathematics*, 55, 1985, 151–159.

[19] P. E. Green, *Fiber-Optic Communication Networks*, Prentice–Hall, 1992.

[20] P. E. Green, "Optical Networking Update", *IEEE J. Selected Areas in Comm.*, vol. 14, 1996, 764–779.

[21] M. C. Heydemann, J-C. Meyer and D. Sotteau, "On Forwarding Indices of Networks", *Discrete Applied Mathematics* 23, 1989, 103–123.

[22] K. Kaklamanis G. Persiano, T. Erlebach, K. Jansen, "Constrained Bipartite Edge Coloring with Applications to Wavelength Routing" *Proceedings of 24th International Colloquium on Automata, Languages, and Programming ICALP 97*, Bologna, Italy, 1997.

[23] I.A. Karapetian, "On Coloring of Arc Graphs", *Dokladi of the Academy of Science of the Armenian SSR*, 70(5), 1980, 306–311.

[24] R. Klasing, "Methods and Problems of Wavelength–Routing in All–Optical Networks", *Proceedings of 23rd International Symposium on Mathematical Foundations of Computer Science MFSC'98*, 1998, LNCS 1450.

[25] J. Kleimberg, E. Kumar, "Wavelength Conversion in Optical Networks", *Proceedings of SODA '99*, 1999.

[26] V. Kumar, "Approximating Circular Arc Colouring and Bandwidth Allocation in All-Optical Ring Networks", *Proceedings of the International Workshop APPROX'98: Approximation Algorithms for Combinatorial Optimization*, Klaus Jansen and Jose Rolim Eds., LNCS 1444, 147-158.

[27] E. Kumar, E. Schwabe, "Improved Access to Optical Bandwidth in Trees", *Proceedings of SODA '97*, 1997.

[28] A. Lubotzky, R. Phillips, P. Sarnak, "Ramanujan Graphs", *Combinatorica* 8, 1988, 261–278.

[29] W. Mader, "Minimale n-fach Kantenzusammenhangende Graphen", *Math. Ann.* 191, 1971, 21-28.

[30] G.A. Margulis, "Explicit Group–Theoretic Constructions of Combinatorial Schemes and their Applications for the Construction of Expanders and Concentrators", *Problemy Peredaci Informacii*, 1988.

[31] A. D. McAulay, *Optical Computer Architectures*, John Wiley, 1991.

[32] C. L. Monma and V. K. Wei. "Intersection Graphs of Paths in a Tree", *Journal of Combinatorial Theory*, Series B, 1986, 141–181.

[33] R. Ramaswami, G.H. Sasaki, "Multiwavelength Optical Networks with Limited Wavelength Conversion" *Proc. of IEEE Infocom 97*, 1997.

[34] R. Ramaswami, "Multi-Wavelength Lightwave Networks for Computer Communication", *IEEE Comm. Magazine* 31, 1993, 78–88.

[35] R.M. Tanner, "Explicit Construction of Concentrators from Generalized n-gons", *SIAM J. Alg. Discr. Meth.*, 5, 1984, 287–293.

[36] A. Tucker. "Coloring a Family of Circular Arcs", *SIAM J. Appl. Math.* 29, No. 3, 1975, 493–502.

[37] R.J. Vetter and D.H.C. Du, "Distributed Computing with High-Speed Optical Networks", *IEEE Computer* 26, 1993, 8–18.

[38] G. Wilfong, P. Winkler, "Ring Routing and Wavelength Translation", *Proceedings of SODA '98*, 1998.

PROVING THE CORRECTNESS OF PROCESSORS WITH DELAYED BRANCH USING DELAYED PC

Silvia M. Mueller, Wolfgang J. Paul, and Daniel Kroening

Abstract:

We show that the programming model of delayed branch is equivalent to what we call *delayed PC*: *all* instruction fetches are delayed by one instruction, not just taken branches. This leads to a very simple new implementation of the delayed branch mechanism. We then prove the correctness of a pipelined machine with delayed PC.

INTRODUCTION

Machine verified correctness proofs for (almost) entire processors have been produced for sequential machines [1], for pipelined machines [2, 3, 4, 5, 6] and for machines with out of order execution [7, 6, 8, 9]. In all non sequential designs cited above either a branch-not-taken strategy is applied or the following actions are performed in a *single cycle*: i) the evaluation of the condition of branch instructions ii) the next PC computation iii) the fetch of the next instruction.

In real machine these three actions are usually performed in two or more cycles in order to reduce cycle time. This does not remain invisible to the programmer: taken *branches* are *delayed* by one or more instructions. The delayed branch semantics is, for example, used in the MIPS [10], the SPARC [11] and the PA-RISC [12] instruction set.

In this paper we show that the programming model of delayed branch is equivalent to what we call *delayed PC*: *all* instruction fetches are delayed by one instruction, not only taken branches. This leads to a very simple new implementation of the delayed branch mechanism. We then prove the correctness of a pipelined machine with delayed PC. Parts of the proof have been verified by machine already.

The paper is organized in the following way: In the next section, we formally define the semantics of a DLX machine [13] with delayed branch and delayed PC, and we show that they are equivalent. We then describe a *sequential* machine DLX_σ with the following features: i) delayed PC, ii) pipelined data

I. Althöfer et al. (eds.), Numbers, Information and Complexity, 579-588.

paths with a 5 stage pipeline, iii) the pipeline stages are clocked in a round robin fashion. In the last section we turn the sequential machine DLX_σ into a pipelined machine DLX_π by only 2 changes: i) the delayed PC of the sequential machine is bypassed, ii) the clocking of the pipeline stages is modified. We then show that the pipelined machine simulates the sequential machine.

DELAYED BRANCH AND DELAYED PC

We consider sequences $I = I_0, I_1, ...$ of DLX instructions started after reset. For registers R and instructions I_i we denote by R_i the value of register R after sequential execution of instruction I_i. By R_{-1} we denote the initial value of R before the execution of I_0. Observe that for sequential machines instruction I_i is fetched from memory address PC_{i-1}.

A (sequential) semantic of delayed branch requires the introduction of state variables which memorize, whether previous instructions were taken branches (or jumps), and memorize the branch target of branches/jumps. We use the variables *bjtaken* and *btarget*. If I_i is a relative branch/jump with immediate constant imm_i or an absolute jump with operand $RS1_{i-1}$, then for machines with delayed branch the branch target is

$$btarget_i = \begin{cases} RS1_{i-1} & \text{for absolute jumps} \\ PC_{i-1} + 4 + imm_i & \text{for relative branch/jumps} \end{cases}$$

Observe that the addition of 4 is an artifact. The variable $bjtaken_i = 1$ indicates that instruction I_i is a jump or a taken branch. The machine is initialized with

$$PC_{-1} = 0 \quad \text{and} \quad bjtaken_{-1} = 0.$$

The delayed branch mechanism is specified by

$$PC_{i+1} = \begin{cases} btarget_i & \text{if} \quad bjtaken_i = 1 \\ PC_i + 4 & \text{otherwise} \end{cases}$$

and by the requirement that delay slots do not contain branch instructions.

The delayed PC mechanism uses a program counter PC' and its delayed version DPC which is used for fetching instructions. They are initialized with

$$DPC_{-1} = 0 \quad \text{and} \quad PC'_{-1} = 4.$$

The computation of the next PC' is completely free of artifacts:

$$PC'_i = \begin{cases} PC'_{i-1} + imm_i & \text{if} \quad bjtaken_i = 1 \ \wedge \ I_i \text{is relative branch/jump} \\ RS1_{i-1} & \text{if} \quad bjtaken_i = 1 \ \wedge \ I_i \text{is absolute branch/jump} \\ PC'_{i-1} + 4 & \text{otherwise} \end{cases}$$

The delayed program counter is simply computed by

$$DPC_{i+1} = PC'_i.$$

The semantics of the jump and link instructions change by the delayed branch mechanism as well. Saving $PC{+}4$ into general purpose register $GPR[31]$ results in a return to the delay slot of the jump and link instruction. Of course, the return should be to the instruction *after* the delay slot. Formally, if I_i is a jump and link instruction, then

$$PC_i = PC_{i-1} + 4$$

because I_i is not in a delay slot, and instruction I_{i+1} fetched from address PC_i is the instruction in the delay slot of I_i. The jump and link instruction I_i should therefore save

$$GPR[31]_i = PC_i + 4 = PC_{i-1} + 8.$$

In the simpler delayed PC mechanism, one simply saves

$$GPR[31]_i = PC'_{i-1} + 4.$$

The equivalence of the two mechanisms is asserted in

Theorem 1. *Suppose a machine with delayed branch and a machine with delayed PC are started with identical memory contents and identical contents of the visible registers, then*

1. $(PC_i, PC_{i+1}) \;=\; (DPC_i, PC'_i)$,
2. *and if I_i is a jump and link instruction, the value $GPR[31]_i$ saved into register 31 during instruction I_i is identical for both machines.*

Proof. The theorem is proven by induction on i. The case $i = -1$ follows from the rules for initializing PC, $bjtaken$, PC', and DPC. Concluding from $i-1$ to i has two parts. The equation

$$DPC_i = PC'_{i-1} = PC_i$$

follows directly from the definition of DPC and the induction hypothesis. The proof of the equation $PC_{i+1} = PC'_i$ has several cases.

If I_i is a branch or jump, instruction I_i is not in a delay slot, and hence $bjtaken_{i-1} = 0$.

If I_i is a taken branch or a relative jump, it then follows for the target address

$$\begin{array}{rcll} PC'_i & = & PC'_{i-1} + imm_i & \\ & = & PC'_{i-2} + 4 + imm_i & \text{because } bjtaken_{i-1} = 0 \\ & = & PC_{i-1} + 4 + imm_i & \text{by the induction hypothesis for } i-2 \\ & = & btarget_i & \end{array}$$

whereas for an absolute jump it follows

$$PC'_i \;=\; RS1_{i-1} \;=\; btarget_i.$$

In both cases, $bjtaken_i = 1$; this implies that

$$PC'_i = btarget_i = PC_{i+1}.$$

In any other case, $bjtaken_i = 0$ and one concludes

$$\begin{aligned} PC'_i &= PC'_{i-1} + 4 & \\ &= PC_i + 4 & \text{by the induction hypothesis} \\ &= PC_{i+1} & \text{by the definition of delayed branch.} \end{aligned}$$

For the second part, suppose I_i is a jump and link instruction. With delayed branch, one then saves $PC_{i-1} + 8$. Because I_i is not in a delay slot, it holds

$$\begin{aligned} PC_{i-1} + 8 &= PC_i + 4 & \\ &= DPC_i + 4 & \text{by induction hypothesis} \\ &= PC'_{i-1} + 4 & \text{by definition of delayed PC.} \end{aligned}$$

This is exactly the value saved in the delayed PC version.

PREPARED SEQUENTIAL MACHINES

The sequential machine DLX_σ is constructed in the following three steps: i) Take a textbook design of a pipelined DLX machine with a classical 5 stage pipeline [14, 13], but without forwarding and interlock. Figure 1 sketches *almost* the data paths of such a machine. Each register, register file or memory is drawn at the end of the stage in which it is written. ii) In stage ID a straightforward circuit $NextPC$ computes the input for PC' which in turn is clocked into DPC (Figure 2). iii) The pipeline stages are updated in a round robin fashion. With proof techniques for sequential machines one shows

Theorem 2. Machine DLX_σ interprets the DLX instruction set with delayed PC semantics.

Theorem 1 implies that machine DLX_σ also interprets the DLX instruction set with delayed branch semantics.

For pipeline stages $k = 0, ..., 4$, nonnegative integers i, and cycles T we denote by

$$I_\sigma(k, T) = i$$

the fact that instruction I_i is in stage k in cycle T. We have

$$I_\sigma(k, T) = i \quad \leftrightarrow \quad T = 5i + k.$$

The content of a register R in cycle T is denoted by R^T.

Suppose execution of instruction I_i is in stage k during cycle T' and the output registers of stage k will be clocked at the end of this cycle. The round robin updating schedule then implies that i) all registers above stage k have already the value they will have after instruction I_i, and ii) all output registers of stages k and below still have the values they had after instruction I_{i-1}. This is asserted in

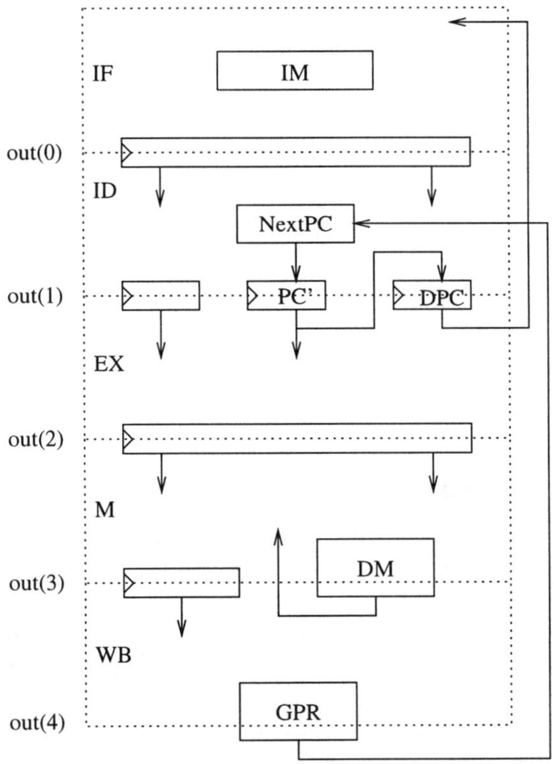

Figure 1 Data flow between the pipeline stages of the DLX_σ design

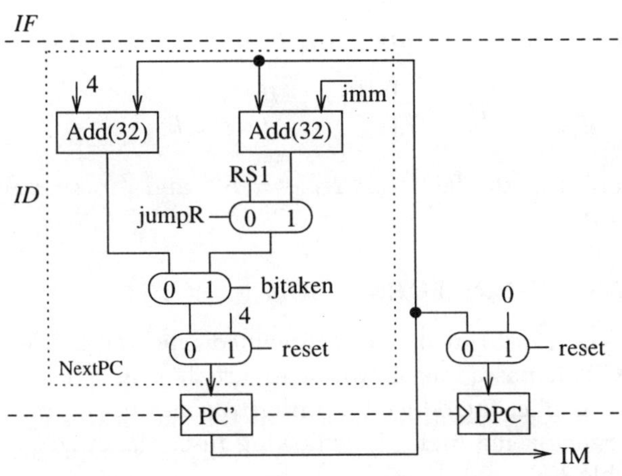

Figure 2 PC environment of the DLX_σ design

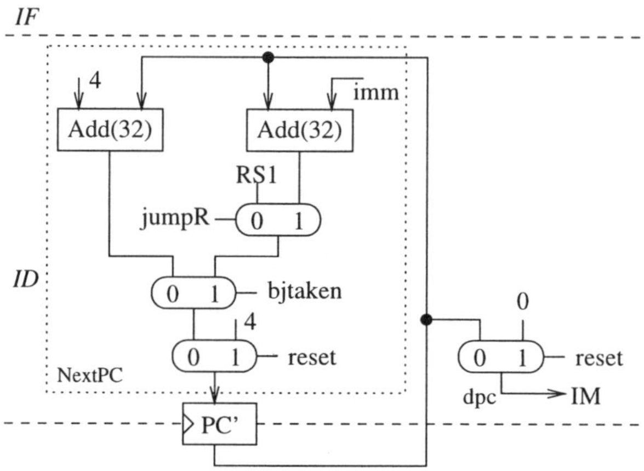

Figure 3 PC environment of the DLX_π design

T	reset	$ue[0]$	$ue[1]$	$ue[2]$	$ue[3]$	$ue[4]$
0	1	1	0	0	0	0
1	0	1	1	0	0	0
2	0	1	1	1	0	0
3	0	1	1	1	1	0
4	0	1	1	1	1	1
...	0	1	1	1	1	1

Table 1 The activation of the update enable signals $ue[4:0]$ after reset. For all i, signal $ue[i]$ enables the update of registers and RAMs in $out(i)$.

Theorem 3. *Let $I_\sigma(k, T') = i$ and let R be an output of stage s. Then*

$$R^{T'} = \begin{cases} R_{i-1} & \text{if} \quad s \geq k \\ R_i & \text{if} \quad s < k \end{cases}$$

A formal proof uses the fact that $R_{i-1} = R^{5i}$ and proceeds for $T = 5i + k$ by induction on k.

PIPELINING AS A TRANSFORMATION

Machine DLX_σ is transformed into a pipelined machine DLX_π in two steps: i) Register DPC is bypassed as shown in Figure 3. This is not surprising; register DPC is an artifact introduced in order to construct a *sequential* machine for a delayed branch semantics. ii) Following reset the stages are updated as indicated in Table 1.

The schedule for this machine is described by the following function I_π:

$$I_\pi(k, T) = i \quad \leftrightarrow \quad T = k + i.$$

stage s	$I_\pi(s,T)$	$I_\pi(s,T-1)$
k-1		i
k	i	i-1

Table 2 Illustration of the scheduling function I_π for the stages $k-1$ and k.

If forwarding and a hardware interlock are added this formula has to be replaced by a more complicated inductive definition [15].

That the pipelined machine simulates the sequential machine is asserted in Theorem 3. In the absence of forwarding and hardware interlock a hypothesis is required about the programs executed.

If we talk about the same register R in the sequential and the pipelined machine, we call one R_σ and the other R_π.

Theorem 4. *Suppose for all i holds that if I_i reads general purpose register R, the instructions I_{i-1}, I_{i-2} and I_{i-3} do not write R. If $I_\pi(k,T) = i$ and R is an output register of stage k, then*

$$R_\pi^{T+1} = R_i.$$

Proof. The proof is done by induction on T. For the cycle $T = 0$, the hypothesis follows from the reset mechanism, e.g.,

$$PC'^{1}_{\pi} \;=\; 4 \;=\; PC'_{-1}.$$

The induction step from $T-1$ to T has 5 cases, one for each stage. They all follow the same pattern. Let R be an *input* register of stage k and let T' be the cycle when instruction I_i is in stage k in machine DLX_σ, i.e., $I_\sigma(k,T') = k$. The technical problem is to argue that

$$R_\sigma^{T'} = R_\pi^{T}.$$

If we can show this for all input registers of stage k then in the corresponding cycles T and T' stage k has in machines DLX_σ and DLX_π the same inputs. Because the stages are identical they produce the same output and the induction step for stage k follows.

The tricky arguments are those dealing with registers R of a stage *below* stage k. We present here only the case $k = 0$ (instruction fetch) and $k = 1$ (decode). For the remaining cases we refer to [15].

Case $k = 0$. In case of stage $k = 0$ (instruction fetch), we have to justify that the delayed PC can be discarded. The input register PC' is an output register of stage 1. We have $I_\pi(0,T) = i$. The scheduling function implies

$$I_\pi(1, T-1) \;=\; I_\pi(0, T-1) - 1 \;=\; i - 2.$$

This is illustrated in Table 2. Using Theorem 3 with stage $s = 1$ we conclude

stage s	$I_\pi(s,T)$	$I_\pi(s,T-1)$
1	i	
2	i-1	
3	i-2	
4	i-3	i-4

Table 3 Illustration of the scheduling function I_π for the stages 1 to 4.

$$\begin{array}{rcll} PC'^T_\pi &=& PC'_{i-2} & \text{by induction hypothesis} \\ &=& DPC_{i-1} & \text{by the construction of delayed PC} \\ &=& DPC^{T'}_\sigma & \text{by Theorem 3} \end{array}$$

Case $k = 1$. The induction step for stage $k = 1$ (reading of the operands) uses the hypothesis about the program. In either design, the decode stage has as inputs some registers $R \in out(0)$ and the register file $GPR \in out(4)$.

For $R \in out(0)$, the scheduling function implies

$$I_\pi(0, T-1) \;=\; I_\pi(1,T) \;=\; I_\sigma(1,T') \;=\; i,$$

as illustrated in Table 2. Using Theorem 3 with stage $s = 0$, we conclude

$$R^T_\pi \;=\; R_i \;=\; R^{T'}.$$

If instruction I_i reads a register $GPR[r]$ only the value $GPR[r]^T$ can be used. The scheduling function implies (Table 3)

$$I_\pi(4, T-1) \;=\; i-4.$$

For $i \geq 4$, we conclude using Theorem 3 with stage $s = 4$ that

$$\begin{array}{rcll} GPR[r]^T_\pi &=& GPR[r]_{i-4} & \text{by induction hypothesis} \\ &=& GPR[r]^{T'}_\sigma . & \end{array}$$

According to the hypothesis of the theorem, instructions $I_{i-3}, \ldots, I_{i-1}$ do not write register $GPR[r]$ and hence

$$GPR[r]_{i-1} \;=\; GPR[r]_{i-4}.$$

$i \leq 3$. The update of the register file GPR is enabled by signal $ue[4]$. The stall engine (Table 1) therefore ensures that the register file is not updated during cycles $t \in \{1,2,3\}$. Thus,

$$GPR_{-1} \;=\; GPR^1_\pi \;=\; \cdots \;=\; GPR^4_\pi .$$

The hypothesis of the theorem implies that instructions I_j with $0 \leq j < 3$ do not write register $GPR[r]$. Hence,

$$GPR[r]_{-1} \;=\; \cdots \;=\; GPR[r]_{i-1}.$$

By Theorem 3 with stage $s = 4$, we conclude

$$GPR[r]^4_\pi \;=\; GPR[r]_{i-1} \;=\; GPR[r]^{T'}_\sigma .$$

CONCLUSION

Using the construction of delayed PC's we have shown the correctness of a pipelined machine with delayed branch.

References

[1] Phillip J. Windley, "Formal modeling and verification of microprocessors", *IEEE Transactions on Computers*, 1995, 44(1), 54–72.

[2] Mark Bickford and Mandayam Srivas, "Verification of a pipelined microprocessor using Clio", *Proceedings of the Mathematical Sciences Institute Workshop on Hardware Specification, Verification and Synthesis: Mathematical Aspects*, Springer, 1990, volume 408 of *LNCS*, 307–332.

[3] James B. Saxe, Stephen J. Garland, John V. Guttag, and James J. Horning, "Using transformations and verification in circuit design", *Technical report*, Digital Systems Research Center, 1991.

[4] Jerry R. Burch and David L. Dill, "Automatic verification of pipelined microprocessor control, *Proc. International Conference on Computer Aided Verification*, 1994.

[5] Jeremy Levitt and Kunle Olukotun, "A scalable formal verification methodology for pipelined microprocessors", *33rd Design Automation Conference (DAC'96)*, Association for Computing Machinery, 1996, 558–563.

[6] Thomas A. Henzinger, Shaz Qadeer, and Sriram K. Rajamani, "You assume, we guarantee: Methodology and case studies", *Proc. 10th International Conference on Computer-aided Verification (CAV)*, 1998.

[7] W. Damm and A. Pnueli, "Verifying out-of-order executions", *Advances in Hardware Design and Verification: IFIP WG 10.5 Internatinal Conference on Correct Hardware Design and Verification Methods (CHARME)*, Chapmann & Hall., 1997, 23–47.

[8] K.L. McMillan, "Verification of an implementation of Tomasulo's algorithm by composition model checking", *Proc. 10th International Conference on Computer Aided Verification*, 1998, 110–121.

[9] A. Shen and X. Shen, "Using term rewriting systems to design and verify processors", *IEEE Micro Special Issue on Modeling and Validation of Microprocessors*, May/June, 1999.

[10] G. Kane and J. Heinrich, "MIPS RISC Architecture", Prentice Hall, 1992.

[11] SPARC International Inc., "The SPARC Architecture Manual", Prentice Hall, 1992.

[12] Hewlett Packard, "PA-RISC 1.1 Architecture Reference Manual", 1994.

[13] J.L. Hennessy and D.A. Patterson, "Computer Architecture: A Quantitative Approach", *Morgan Kaufmann Publishers*, INC., San Mateo, CA, 2nd edition, 1996.

[14] D.A. Patterson and J.L. Hennessy, "The Hardware/Software Interface", *Morgan Kaufmann Publishers*, INC., San Mateo, CA, 1994.

[15] S. M. Mueller and W. J. Paul., "The Complexity of Simple Computer Architectures II", *Lecture notes*, to appear as a book, 1999. Email: {smueller,wjp}@cs.uni-sb.de.

COMMUNICATION COMPLEXITY OF FUNCTIONS ON DIRECT SUMS

Ulrich Tamm

Fakultät Mathematik, Universität Bielefeld, Postfach 100131, 33501 Bielefeld, Germany

tamm@mathematik.uni-bielefeld.de

Abstract: The paper surveys direct sum methods in communication complexity, mostly concentrating on the results obtained by several authors in the research group of Rudolf Ahlswede in Bielefeld. Lower bound techniques are investigated which behave multiplicatively for functions defined on direct sums of sets. Applications, as the exact or asymptotic determination of the communication complexity and the comparison of bounding techniques are discussed.

INTRODUCTION

We survey some results on the communication complexity of *sum-type functions* f_n and *vector-valued functions* f^n, which are defined on the powers $\mathcal{X}^n, \mathcal{Y}^n$ of the sets from the domain of some basic function $f : \mathcal{X} \times \mathcal{Y} \to \mathcal{Z}$. Elements of $\mathcal{X}^n$ and $\mathcal{Y}^n$ are denoted as x^n and y^n, respectively. Hence, e. g., $x^n = (x_1, \ldots, x_n)$ for some $x_1, \ldots, x_n \in \mathcal{X}$. With this notation

$$f^n(x^n, y^n) = \big(f(x_1, y_1), \ldots, f(x_n, y_n)\big), \qquad f_n(x^n, y^n) = \sum_{i=1}^{n} f(x_i, y_i),$$

where it is required that the range $\mathcal{Z}$ is a subset of an additive group G.

The investigations in Bielefeld in this direction trace back to a visit of the scientist celebrated in this volume to Stanford University, where he found his office full of computer printouts. Abbas El Gamal and King Pang told him that the printouts served to get insight into the structure of code pairs (A, B), $A, B \subset \{0,1\}^n$, on which the Hamming distance h_n is constant, i. e. $h_n(a, b) = h_n(a', b')$ for all $a, a' \in A, b, b' \in B$. In order to get rid of all the paper, Ahlswede decided that the problem had to be solved. The result was the joint paper [8] and the following theorem, which he calls the "Four Continents Theorem" (the authors are from Europe, Africa, and Asia and the work was done in America).

I. Althöfer et al. (eds.), Numbers, Information and Complexity, 589-602.

Theorem ([8]) The size $|A \times B|$ of the largest code pair $(A, B), A, B \subset \{0,1\}^n$ on which the binary Hamming distance assumes a constant value is $2^{\lfloor n/2 \rfloor}$.

The theorem was settled by a combinatorial approach, giving the constant distance code pair (for the distance $\lfloor \frac{n}{2} \rfloor$) $A = \{00, 11\}^{\lfloor n/2 \rfloor}, B = \{01, 10\}^{\lfloor n/2 \rfloor}$ and an inductive optimality proof. Later Delsarte and Piret [15] and Hall and van Lint [22] derived the same result via algebraic methods. However, it turned out that the original proof from [8] was extendable to the case that the Hamming distance assumes a specified value δ (i. e. $h_n(a, b) = h_n(a', b') = \delta$ for all $a, a' \in A, b, b' \in B$) on $A \times B$ the size of which has to be maximized.

Such a constant distance code pair was called monochromatic rectangle by Yao [42], who used this concept in order to lower bound the communication complexity $C(f)$ of a function f. The bounds El Gamal and Pang [16] derived on the size of constant distance code pairs with specified value δ enabled them to determine the communication complexity of the binary Hamming distance up to one bit. Ahlswede [1] generalized this result to the Hamming distance over alphabets of size $q = 4, 5$ and also for $q = 3$ (unpublished manuscript).

Theorem ([16], [1]) For $q = 2, 3, 4, 5$ and all positive integers n

$$|C(h_n) - \lceil n \cdot \log q \rceil - \lceil \log(n+1) \rceil| \leq 1 \tag{1}$$

Throughout this paper the logarithm is always taken to the base 2. Unfortunately the method of proof based on the exact determination of the size of largest constant distance code pairs (cf. also [40]) breaks down for alphabet size $q \geq 6$. It is strongly conjectured that (1) holds for all alphabet sizes q and all positive integers n.

The proofs in [16] and [1] are rather involved such that we shall not present them. Let us only mention that research motivated by the study of (1.1) followed several directions. Recently, Haemers [19] has bounded the size of constant distance code pairs by methods from algebraic graph theory. These bounds work quite well for large q. In my thesis [36] (see also [38]), using properties of the Hamming association scheme, (1) was shown to hold for special parameters n, e. g. when $n = p^s - 1$ with p a prime factor of q. Finally, the size of constant distance code pairs for further functions may be determined or at least closely bounded. Ahlswede in [1] also considered the parity of the Hamming distance. For $q = 2$ and $q = 4$ the size of largest constant distance code pairs was exactly determined. Recent progress on this was made in [7].

We shall concentrate, however, on a rather methodical direction of research - the search for possible induction proofs. The Hamming distance has a certain property, it is a sum-type function h_n, with basic function h indicating if $x_i = y_i$ or not. We are interested to conclude from the communication complexity of the basic function f to the communication complexity of the sum-type function f_n and to that of the vector-valued function f^n.

This is not in general possible if one studies lower bounds obtained via largest monochromatic rectangles. However, Ahlswede and Mörs [2] observed that the situation changes if one does not require any more that the function is constant on the rectangle (or code pair) $A \times B$ but that instead the so-called *4–word–property* holds. In this case, we obtain a lower bound $C(f_n) \geq n \cdot C(f)$. This multiplicative behaviour is very useful in the study of the communication complexity of sum-type functions.

Furthermore, one can exploit the structure of the function matrix $M(f_n) = (f_n(x^n, y^n))_{x^n, y^n}$ which can be obtained from the function matrix $M(f)$ of the basic function f in terms of the *Kronecker product.* This is intensively discussed in [36], [4] and for vector-valued funtions in [4] (also for the case of nonidentical functions for each component). Parameters like the rank and the independence number which yield further lower bounds on the communication complexity are multiplicative under the Kronecker product.

A further line of research leading to direct sum methods in communication complexity goes back to Karchmer, Raz, and Wigderson [25], cf. also [28], pp. 42 - 48. Their question was if it is easier to solve communication problems simultaneously than separately. Recall the definition of a vector-valued function $f^n((x_1, \ldots, x_n), (y_1, \ldots, y_n)) = (f(x_1, y_1), \ldots, f(x_n, y_n))$. An obvious upper bound on the communication complexity $C(f^n)$ is obtained by evaluating each component $f(x_i, y_i)$ separately and communicating the result for component i using the optimal protocol for f. Can we do better by considering all components simultaneously? We shall provide a simple example for a function where $C(f) = 2$ but $C(f^n) = \lceil n \cdot \log_2 3 \rceil$. The measure $\lim \sup_{n \to \infty} \frac{1}{n} C(f^n)$ is also called *amortized communication complexity* (see [17] or [31]).

Direct sum methods in communication complexity are useful tools in separating complexity classes. Further applications are the comparison of lower bound techniques and the study of their power (how large can be the gap between the lower bound and the communication complexity). The intuition is that small gaps for the basic function f become large for the vector-valued function f^n.

The paper is organized as follows. After presenting the basic notions on communication complexity and the functions to be studied, the lower bound techniques useful for sum-type and vector-valued functions are introduced in Section 3. The communication complexity of special functions is then studied in Section 4. Applications in Computer Science are discussed in the final section.

BASIC NOTIONS

The notion of communication complexity was introduced by Yao in 1979 [42]. Since then it found many applications in Computer Science, for which we refer to the books by Kushilevitz an Nisan [28] or by Hromkovic [21], see also the survey by Wegener [41] in this volume.

The communication complexity of a function $f : \mathcal{X} \times \mathcal{Y} \to \mathcal{Z}$ (where $\mathcal{X}$, $\mathcal{Y}$, and $\mathcal{Z}$ are finite sets), denoted as $C(f)$, is the number of bits that two persons,

P_1 and P_2, have to exchange in order to compute the function value $f(x,y)$, when initially P_1 only knows $x \in \mathcal{X}$ and P_2 only knows $y \in \mathcal{Y}$.

More specifically, let $\mathcal{Q}$ denote the set of protocols computing f and let $l_P(x,y)$ be the number of bits transmitted for the input (x,y), when the protocol $P \in \mathcal{Q}$ is used. Then the (worst-case) *communication complexity* is

$$C(f) := \min_{P \in \mathcal{Q}} \max_{(x,y) \in \mathcal{X} \times \mathcal{Y}} l_P(x,y).$$

A *protocol* P is a pair of mappings $\phi_1 : \mathcal{X} \times \{0,1\}^* \to \{0,1\}^*$, $\phi_2 : \mathcal{Y} \times \{0,1\}^* \to \{0,1\}^*$. So on input (x,y), the persons starting with P_1 alternatively send binary messages N_1, N_2, N_3, etc., until they both know the result.

There is also a slightly modified model in which communication already stops when one person knows the result. We denote by $C_1(f)$ the communication complexity in this case. Often, Boolean functions are considered, where the difference between the two models is at most the transmission of just one bit for the result. However, the functions we are going to study have a much larger range, such that the gap between $C(f)$ and $C_1(f)$ may be considerable.

Each message depends on the previous messages and on the inputs, hence $N_1 = \phi_1(x)$, $N_2 = \phi_2(y, \phi_1(x))$, $N_3 = \phi_1(x, \phi_1(x)\phi_2(y, \phi_1(x)))$, etc. It is required that the set of messages a person is allowed to send is *prefix-free*, i. e., no possible message is the beginning of another one. This property assures that the other person immediately recognizes the end of the message and can hence start the transmission without delay.

An upper bound on $C(f)$ for any function $f : \mathcal{X} \times \mathcal{Y} \to \mathcal{Z}$ (w. l. o. g. $|\mathcal{X}| \leq |\mathcal{Y}|$) is always obtained from the following *trivial protocol*: P_1 transmits all the bits of his input $x \in \mathcal{X}$. P_2 now is able to compute the function value and returns the result $f(x,y) \in \mathcal{Z}$ (if P_1 must be informed). Hence

$$C_1(f) \leq \lceil \log |\mathcal{X}| \rceil, \qquad C(f) \leq \lceil \log |\mathcal{X}| \rceil + \lceil \log |\mathcal{Z}| \rceil.$$

As mentioned, we shall study the communication complexity of vector-valued and sum-type functions. Let us first present some examples.

i) Let $si : \{0,1\} \times \{0,1\} \to \{0,1\}$ be the logical "and" If we interpret the vectors $x^n, y^n \in \{0,1\}^n$ as representations of two subsets of an n–elementary set ($x_i = 1$ exactly if the i–th element is contained in the subset represented by $x^n = (x_1, \ldots, x_n)$), then the vector–valued function $si^n(x^n, y^n)$ gives the intersection of these two sets, whereas the sum–type function $si_n : \{0,1\}^n \times \{0,1\}^n \to \{0, \ldots, n\} \subset Z$ yields the cardinality of this intersection. The function si_n can be generalized to larger alphabets in two canonical ways: 1) the inner product of x^n and y^n, 2) the sum of the componentwise minima.

ii) The *Hamming distance*, which counts the number of components in which x^n and $y^n \in \{0, \ldots, q-1\}$ differ, is the sum–type function h_n obtained via the basic function $h(x,y) = \begin{cases} 0, & \text{if } x = y \\ 1, & \text{if } x \neq y. \end{cases}$

For the binary Hamming distance the corresponding vector-valued function h^n yields the symmetric difference of the sets represented by x^n and y^n.

We shall also consider the parity of the Hamming distance (the Hamming distance modulo 2). Here we only have to replace the range Z by $Z_2 = Z/2Z$.

iii) A further measure for the distance of two vectors $x^n, y^n \in \{0, \dots, q-1\}^n$ in Coding Theory is the *Lee distance.* Here the basic function is defined by $\ell(x,y) = \min\{|x-y|, q-|x-y|\}$. It can be interpreted as the length of a shortest path on the cycle from x to y. Another distance function is the *taxi metric*, with basic function $t(x,y) = |x-y|$.

LOWER BOUND TECHNIQUES

The aim of this section is to present techniques which allow to conclude from the communication complexity $C(f)$ of the basic function f to the communication complexity of the sum-type function f_n or the vector–valued function f^n. All these bounds are expressed in terms of the *function matrix* $M(f) = \big(f(x,y)\big)_{x\in\mathcal{X},y\in\mathcal{Y}}$ and the *function value matrices* $M_z(f) = (a_{xy})_{x\in\mathcal{X},y\in\mathcal{Y}}$ for all $z \in \mathcal{Z}$ defined by $a_{xy} = \begin{cases} 1 & \text{if } f(x,y) = z \\ 0 & \text{if } f(x,y) \neq z. \end{cases}$

Yao [42] already showed that $C(f) \geq \log D(f)$, where the decomposition number $D(f)$ denotes the minimum size of a partition of $\mathcal{X} \times \mathcal{Y}$ into *monochromatic rectangles*, i. e., products $A \times B$ of pairs $A \subset \mathcal{X}, B \subset \mathcal{Y}$ on which the function is constant. The decomposition number usually is hard to determine, however, further lower bounds can be derived from it. Immediately, we have

$$C_1(f) \geq \left\lceil \log \frac{|\mathcal{X}| \cdot |\mathcal{Y}|}{Lmr\big(M(f)\big)} \right\rceil, \qquad C(f) \geq \left\lceil \log \sum_{z\in\mathcal{Z}} \frac{wt\big(M_z(f)\big)}{lmr\big(M_z(f)\big)} \right\rceil \tag{2}$$

where $Lmr\big(M(f)\big)$ denotes the size of the largest monochromatic rectangle in the function matrix $M(f)$, $lmr\big(M_z(f)\big)$ is the size of the largest monochromatic rectangle on which the function assumes the constant value z, and $wt\big(M_z(f)\big)$ denotes the number of pairs (x,y) with $f(x,y) = 1$.

Yao used this last bound in order to show that for almost all Boolean functions the trivial protocol is optimal up to two bits. Weakening the conditions on the rectangles, further lower bounds are obtained. For instance, it may be no longer required that the function is constant on the rectangle $A \times B$ but that the so called *4-word- property* has to be fulfilled, i. e., for all $a, a' \in A$, $b, b' \in B$

$$f(a,b) - f(a',b) - f(a,b') + f(a',b') = 0$$

Denoting by $Lfw(f)$ the size of the largest rectangle, on which the 4-word-property holds, we obtain

$$C_1(f) \geq \left\lceil \log \frac{|\mathcal{X}| \cdot |\mathcal{Y}|}{Lfw(f)} \right\rceil. \tag{3}$$

A *z-fooling set* $\{(x^{(1)}, y^{(1)}), \ldots, (x^{(N)}, y^{(N)})\}$ for the function value z in $M(f)$ is a set of pairs with $f(x^{(i)}, y^{(i)}) = z$ for all $i = 1, \ldots, N$ such that no two members of the set are in the same monochromatic rectangle. Denoting the size of a z–fooling set (or independent set as in [3]) by $ind(M_z(f))$ and $Ind(f) = \sum_{z \in Z} ind(M_z(f))$ we obtain

$$C_1(f) \geq \lceil \log(\max_{z \in \mathcal{Z}} ind(M_z(f))) \rceil, \qquad C(f) \geq \lceil \log Ind(f) \rceil. \tag{4}$$

where the first bound was derived in [9] and the second one studied in [3].

Mehlhorn and Schmidt [30] observed that $C(f)$ can be lower bounded by the rank of the corresponding function matrices. We shall only use the rank over the reals.

$$C_1(f) \geq \lceil \log \mathrm{rank} M(f) \rceil, \qquad C(f) \geq \lceil \log r(f) \rceil, \text{ where } r(f) = \sum_{z \in \mathcal{Z}} \mathrm{rank} M_z(f) \tag{5}$$

It can be shown that the function f has the same communication complexity as the function g defined by $g(x, y) = c^{f(x,y)}$ for all x, y, when the number c is chosen appropriately ($c \neq 0, |c| \neq 1$). So it is also possible to lower bound $C(f)$ by the rank of $M(g) = \exp(M(f), c) = (c^{f(x,y)})_{x \in \mathcal{X}, y \in \mathcal{Y}}$, the *exponential transform* of the matrix $M(f)$. This yields

$$C_1(f) \geq \lceil \log \mathrm{rank} \exp(M(f), c) \rceil. \tag{6}$$

Central in the following arguments is the observation that the function matrices of the vector-valued and sum-type functions can be expressed in terms of the Kronecker product, defined for two matrices $A = (a_{ij})_{i,j}$ and $B = (b_{kl})_{k,l}$ as $A \otimes B = (a_{ij} \cdot b_{kl})_{i,j,k,l}$. The n-fold Kronecker product of a matrix is denoted as $A^{\otimes n}$. We have (cf. [3], [4], [36], [38])

$$M_{(z_1, \ldots, z_n)}(f^n) = M_{z_1}(f) \otimes M_{z_2}(f) \otimes \ldots \otimes M_{z_n}(f) \tag{7}$$

$$M_z(f_n) = \sum_{\substack{(z_1, \ldots, z_n) \\ z_1 + \ldots + z_n = z}} M_{z_1}(f) \otimes \ldots \otimes M_{z_n}(f) \tag{8}$$

$$\exp\big(M(f_n), c\big) = \big[\exp\big(M(f), c\big)\big]^{\otimes n} \tag{9}$$

It can be shown that the parameters in the bounds (3) - (6) behave multiplicatively. The properties are summarized in the following theorem. For the proofs we refer to the original research papers. Useful for sum-type functions are the 4-word-property, introduced by Ahlswede and Mörs [2] for the Hamming distance and thoroughly analyzed in [6], and the rank of the exponential transform [4]. The rank is multiplicative under the Kronecker product. For vector-valued functions, it is important that the same holds for

$r(f) = \sum_{z \in Z} \text{rank} M_z(f)$. This was derived by Ahlswede and Cai [3] who also found similar results for the independence number $Ind(f)$ and the parameter $\max(Ind(f), r(f))$.

Theorem 1:

$$Lfw(f_n) = n \cdot Lfw(f) \tag{10}$$

$$\text{rank} \exp(M(f_n), c) = (\text{rank}[\exp(M(f), c)])^n \tag{11}$$

$$r(f^n) = r(f)^n \tag{12}$$

$$Ind(f^n) \geq Ind(f)^n \tag{13}$$

$$\max(Ind(f^n), r(f^n)) \geq [\max(Ind(f), r(f))]^n \tag{14}$$

COMMUNICATION COMPLEXITY OF SPECIAL FUNCTIONS

Vector-valued functions

We shall study all vector-valued functions f^n with basic function $f : \{0,1\} \times \{0,1\} \to \{0,1\}$. They fall into four classes: constant functions, projections on one coordinate, the symmetric difference with function matrix $M(h) = \begin{pmatrix} 0 & 1 \\ 1 & 0 \end{pmatrix}$ and its complement, the logical "and" with $M(si) = \begin{pmatrix} 0 & 0 \\ 0 & 1 \end{pmatrix}$ and equivalent functions. In the first case no communication is necessary, projections require n bits of communication.

Theorem 2:

$$C(h^n) = 2n, \qquad C(si^n) = \lceil n \cdot \log 3 \rceil \tag{15}$$

Proof: For the symmetric difference h^n, the trivial protocol requires $\lceil \log 2^n \rceil + \lceil \log 2^n \rceil = 2n$ bits of communication. With the rank lower bound (5) and (12) this can be shown to be optimal, since

$$C(h^n) \geq \log r(h^n) = n \cdot \log r(h) = n \cdot \log(\text{rank} M_0(h) + \text{rank} M_1(h))$$

$$= n \cdot \log(\text{rank} \begin{pmatrix} 1 & 0 \\ 0 & 1 \end{pmatrix} + \text{rank} \begin{pmatrix} 0 & 1 \\ 1 & 0 \end{pmatrix}) = n \cdot \log 4 = 2n$$

For set-intersection the rank lower bound yields

$$C(si^n) \geq \lceil \log r(si^n) \rceil = \lceil n \cdot \log r(si) \rceil = \lceil n \cdot \log(\text{rank} M_0(si) + \text{rank} M_1(si)) \rceil$$

$$= \lceil n \cdot \log(\text{rank} \begin{pmatrix} 1 & 1 \\ 1 & 0 \end{pmatrix} + \text{rank} \begin{pmatrix} 0 & 0 \\ 0 & 1 \end{pmatrix}) \rceil = \lceil n \cdot \log 3 \rceil$$

In order to obtain the same upper bound, we shall modify the trivial protocol, which would require $2n$ bits of transmission. Again, in the first round person P_1 encodes his input $x^n \in \{0,1\}^n$. P_2 then knows both values and hence is able to compute the result $si^n(x^n, y^n)$, which is returned to P_1. However, in knowledge of x^n the set of possible function values is reduced to the set $S(x^n) = \{y^n : y^n \subset x^n\}$. Hence, only $\lceil \log S(x^n) \rceil$ bits have to be reserved for the transmission of $si^n(x^n, y^n)$ such that P_1 can assign longer messages to elements with few subsets. So, in contrast to the trivial protocol, the messages $\{\phi_1(x^n) : x^n \in \{0,1\}^n\}$ are now of variable length. Since the prefix property has to be guaranteed, Kraft's inequality for prefix codes yields a condition, from which the upper bound can be derived. Specifically, we require that to each x^n there corresponds a message $\phi_1(x^n)$ of (variable) length $l(x^n)$ such that for all $x^n \in \{0,1\}^n$ the sum $l(x^n) + \lceil \log S(x^n) \rceil$ takes a fixed value, L say. Kraft's inequality states that a prefix code exists, if $\sum_{x^n} 2^{-l(x^n)} \leq 1$. This is equivalent to $\sum_{x^n} 2^{\lceil \log S(x^n) \rceil} \leq 2^L$. With the choice $L = \lceil \log 3^n \rceil$ Kraft's inequality holds.

Remark: We used the fact that the subsets of an n-elementary set form a lattice. Functions defined on lattices were studied this way in [5], see also [29].

Sum-type functions

I) ONE PERSON HAS TO BE INFORMED ABOUT THE RESULT

As mentioned, the largest 4-word sets and the rank of the exponential transform are suitable lower bounds here. By (10) and (11), these parameters also behave multiplicatively and hence allow to conclude from the communication complexity $C_1(f)$ of the basic function to that of the sum-type function f_n.

The size of largest 4-word-sets have been determined for metrics of sum-type like the Lee distance (see also [11]) and the taxi metric in [6]. If the maximal 4-word set is one row in the function matrix, then the trivial protocol is optimal by bound (3), and by (4) it suffices to study the basic function f, from which the same result for the sum-type function f_n is immediate.

The exponential transform of the function matrix turned out to be very efficient for the analysis of the communication complexity when only one person has to be informed about the result. We summarize several results from [4]

Theorem 3: For the following sum-type functions $f_n\{0,\dots,q-1\} \times \{0,\dots, q-1\} \to Z$, where $q \geq 2$ is a positive integer, holds $C_1(f_n) = \lceil n \cdot \log q \rceil$

i) metrics of sum-type, ii) inner product, iii) sum of the componentwise minima.

Proof: The idea is to choose the parameter c appropriately in $\exp(M(f), c) = (c^{f(x,y)})_{x,y}$. This exponential transform for the basic function f then has full rank q. By (11), $\exp(M(f_n), c)$ has full rank q^n, from which by (16) the lower bound is immediate. The upper bound follows from the trivial protocol.

i) For metrics like Hamming distance, Lee distance, and taxi metric by definition $f(x,y) = 0$ exactly if $x = y$. Letting c tend to 0 the resulting matrix will be the identity matrix, which has full rank q, Now choose $c > 0$ small enough.

ii) The determinant of the exponential transform $\exp(M(f), c)$ of the basic function $f(x,y) = x \cdot y$ of the inner product is the Vandermonde determinant $\prod_{m=2}^{q} \prod_{l=1}^{m-1} (c^{m-1} - c^{m-1-l})$, which is different from 0 for $c \neq 0, -1, 1$.

iii) The determinant of $\exp(M(f), c)$ is $\prod_{m=1}^{q-1} (c-1) \cdot c^{m-1} \neq 0$ for $c \neq 0, 1$.

Remarks: 1) By choosing the parameter $c = e^{\frac{2\pi i}{p}}$ in the exponential transform it is also possible to study sum-type functions modulo the positive integer $p \geq 2$.

2) The inner product demonstrates the power of the approach via the rank of the exponential transform (which has full rank) compared to the rank of the function matrix (which has rank 1) itself.

3) The Kronecker product $M(f) \otimes M(g)$ of the function matrices (not their exponential transforms) of Boolean functions f, g has also been studied, since this is the function matrix of $f \wedge g$ as defined in [28], pp. 42-48.

II) Both persons have to be informed

Determining the communication complexity of sum-type functions when both persons have to be informed is a much harder problem, since the lower bounds are not multiplicative in general. However, we found several inductive rank calculations for the function value matrices $M_z(si_n)$ of the set-intersection function [39], [37], which allow to describe its communication complexity up to one bit.

Theorem 4: $n + \lceil \log(n+1) \rceil - 1 \leq C(si_n) \leq n + \lceil \log(n+1) \rceil$

The upper bound derived from the trivial protocol is assumed for $n = 2^s - 1$. The improvement of the trivial protocol via Kraft's inequality here occasionally allows to save one bit of communication, e. g., for $n = 2^s$ [39] or for $n = 2^s + 1$ and $n = 2^s + 3$ [14] ($s > 2$ is always a positive integer).

We saw in the introduction that also the communication complexity of the binary Hamming distance can be determined up to one bit and that the method of proof via largest monochromatic rectangles extends to small alphabet sizes q but does not yield a general approach. Also, the rank lower bound does not yield a satisfactory result up to now, although we have a lot of information. The function value matrices $M_z(h_n)$ form the Hamming association scheme, intensively studied in Coding Theory, and their eigenvalues are Krawtchouk polynomials - as a set of orthogonal polynomials also an object of intensive research. The problem is to show that Krawtchouk polynomials have only very few integral zeros. There is some hope to prove (1.1) for all q and n this way, since there seems to be some evidence that a Krawtchouk polynomial for $q \geq 3$ can have at most four integral zeros ([26], p. 81).

In [36] we determined the communication complexity of the parities of Hamming and Lee distance. Their function value matrices are simultaneously diagonalizable and hence $r(f)$ is multiplicative. This extends to translation-invariant functions modulo 2.

Theorem 5: Let $f_n : \{0, \ldots, q-1\} \times \{0, \ldots, q-1\} \to Z_2$ be a non-constant sum-type function invariant under translation and let q be an odd prime number. Then for all positive integers n we have $C(f_n) = \lceil n \cdot \log(q) \rceil + 1$.

APPLICATIONS IN COMPUTER SCIENCE

Comparison of Communication Complexity Classes

One reason for studying communication complexity is that here we have a framework that allows to compare different modes of communication. For instance, the equality function which compares two strings of length n ($eq(x^n, y^n)$

$= 1$ if $x^n = y^n$ and 0 otherwise) has deterministic communication complexity n. However, there exist nondeterministic (for $\overline{\text{eq}}$) and randomized protocols which require only $O(\log n)$ bits of communication. So there is an exponential gap between deterministic complexity and nondeterministic or probabilistic complexity. For an intensive analysis of communication complexity classes see [10] and [20]. We point out here that sometimes functions defined on direct sums play an important role. The rank is determined exploiting the block structure of the function matrix, which shows that for these functions the trivial protocol is optimal and hence they have high deterministic communication complexity.

One such function of sum-type is $\sum_{i=1}^{n} si_n(x_i, y_i) \bmod 2$ the inner product (or set intersection) modulo 2. Replacing 0's by 1's and 1's by -1's in the function matrix (this yields the exponential transform with $c = -1$) we obtain a Hadamard matrix of full rank 2^n. So the rank lower bound coincides with the complexity of the trivial protocol. This function is very important in the study of probabilistic protocols, since randomization does not reduce the order of its communication complexity, as shown in [12] or [27]. The same holds for another function related to set intersection, the function indicating if $x^n \subset y^n$, cf. [23] and [35].

The second function we mention in this context is list disjointness defined for $x^n = (x_{(1)}, \ldots, x_{(n)}), y^n = (y_{(1)}, \ldots y_{(n)})$, where $x_{(i)}, y_{(i)} \in \{0,1\}^n$ for all i by

$$ld_n(x^n, y^n) = \begin{cases} 1, & \text{if there is an } i \text{ with } x_{(i)} = y_{(i)} \\ 0, & \text{else} \end{cases}$$

The function matrix $M(ld_n) = \sum_{\vec{j} \in \{0,1\}^n \setminus (1,\ldots,1)} (-1)^{n - \text{wt}(\vec{j})} A_{j_1} \otimes \ldots \otimes A_{j_n}$ is built up of blocks of identity matrices $A_0 = I_n$ and all-one matrices $A_1 = J_n$ of size n. ($\text{wt}(\vec{j})$ denotes the number of 1's in $\vec{j} = (j_1, \ldots, j_n)$). The Kronecker product allows to derive that all eigenvalues are odd, such that $M(ld_n)$ has full rank and ld_n high deterministic communication complexity. Mehlhorn and Schmidt [30] showed that nondeterministic and Las Vegas communication complexity are of smaller order for ld_n. An improved Las Vegas protocol [18] shows that list disjointness (asymptotically) attains the maximum possible quadratic gap between deterministic and Las Vegas complexity.

Amortized Communication Complexity

With Theorem 2 the function si^n can be evaluated much faster considering all n components simultaneously than by componentwise communication of the results for the basic function si, which would cost $2n$ bits. So the amortized communication complexity of the function si is $\frac{1}{n} \lim_{n\to\infty} C(si^n) = \log 3$.

Further with Theorem 2 it is also clear that this is the maximum compression for basic Boolean functions $f : \{0,1\} \times \{0,1\} \to \{0,1\}$. Karchmer, Raz, and Wigderson [25] asked how much better simultaneous computations are compared to the componentwise evaluation of the function f^n for basic functions $f : \{0,1\}^m \times \{0,1\}^m \to \{0,1\}$. They conjectured that the amortized communication complexity $\frac{1}{n}\lim\sup_{n\to\infty} C(f^n)$ cannot differ from $C(f)$ by more than $O(\log m)$ bits. This was further studied in [17] (cf. also [31]), where a partial function is presented with deterministic communication complexity $C(f) = \Theta(\log(m))$ but amortized complexity $O(1)$ and also randomized protocols are studied. Here simultaneous computations can save a lot of communication bits (cf. also [28], pp. 42-48). This contrasts to nondeterministic protocols [24]. The authors of [25] were rather interested in the amortized communication complexity of relations in the study of the circuit depth complexity for the composition of Boolean functions. Related questions had been studied before, e. g., by Paul [33].

Comparison of Lower Bound Techniques

In [3] was considered the function matrix

$$M(f) = \begin{pmatrix} 1 & 1 & 0 & 0 \\ 0 & 0 & 1 & 1 \\ 1 & 0 & 1 & 0 \\ 0 & 1 & 0 & 1 \end{pmatrix} \tag{16}$$

Since the first two rows and also the last two rows sum up to the all-one vector, the matrices $M_0(f)$ and $M_1(f)$ have rank 3, such that $r(f) = 6$. The largest fooling sets in $M_0(f)$ and $M_1(f)$ have size 4 such that $Ind(f) = 8$. Now the vector-valued function f^n has communication complexity $C(f(^n)) = \lceil 3n \rceil$, where the upper bound is obtained from the trivial protocol and the lower bound via the independence number: $C(f^n) \geq \lceil \log Ind(f^n) \rceil = \lceil \log Ind(f)^n \rceil = \lceil n \cdot \log Ind(f) \rceil = \lceil n \cdot 3 \rceil$. On the other hand the rank bound gives only $C(f^n) \geq \lceil n \log 6 \rceil$. So there is a gap of a factor $\frac{3}{\log 6}$ between communication complexity and bound (5). One might ask how large such gaps can be.

For Boolean functions the possible gaps between the communication complexity and the rank lower bound have intensively been studied, because of its relation to a problem in Graph Theory. It was asked by Lovasz and Saks [29] if for every Boolean function $C(f) = (\log \text{rank} M(f))^{O(1)}$. It turns out that this problem is equivalent to to the problem if $\log \chi(\overline{G}) = (\log \text{rank} G)^{O(1)}$, where G is a graph, $\overline{G}$ its complement, and the rank is taken of the adjacency matrix of G. Raz and Spieker [34] gave an example with a non-constant gap between $\log \text{rank} M(f)$ and the communication complexity. A larger gap ($C(f) = \Omega(n)$, $\log \text{rank} M(f) = O(n^{\log_3 2})$) was found for an explicit function by Nisan and Wigderson [32].

Furthermore, in the above example the lower bound obtained by the independence number $Ind(f)$ is better by a factor $\frac{3}{\log 6}$ than the bound obtained via the sum of ranks $r(f)$. Again, there has been more interest in comparing lower bound techniques for Boolean functions (e. g. [9]). Dietzfelbinger, Hromkovic, and Schnitger [13] constructed a sequence of functions via the Kronecker product from a matrix similar to (16) in order to show that the rank bound yields a result worse by a factor $\frac{\log 3}{2}$ than the bound obtained via largest fooling sets. It was further shown that the fooling set method can yield bounds better by at most a factor 2 than the bounds obtained from the rank of the function matrix and that there exist also functions for which the rank bound is better than the fooling set bound.

References

[1] R. Ahlswede, "On code pairs with specified Hamming distances", Combinatorics, Eger, 1987, *Colloquia Math. Soc. J. Bolyai* 52, 1988, 9 - 47.

[2] R. Ahlswede and M. Mörs, "Inequalities for code pairs", *European J. Combinatorics* 9, 1988, 175–188.

[3] R. Ahlswede and N. Cai, "On communication complexity of vector-valued functions", *IEEE Trans. Inform. Theory* 40, no. 6, 1994, 2062 - 2067, also Preprint 91-041, SFB 343, Bielefeld, 1991.

[4] R. Ahlswede and N. Cai, "2-way communication complexity of sum-type functions for one processor to be informed", *Probl. Inform. Transmission* 30, no. 1, 1994, 1 - 10, also Preprint 91-053, SFB 343, Bielefeld 1991.

[5] R. Ahlswede, N. Cai, and U. Tamm, "Communication complexity in Lattices", *Appl. Math. Letters* 6, no. 6, 1993, 53–58.

[6] R. Ahlswede, N. Cai, and Z. Zhang, "A general 4–word–inequality with consequences for 2–way communication complexity", *Advances in Applied Mathematics* 10, 1989, 75–94.

[7] R. Ahlswede and Z. Zhang, "Code pairs with specified parity of the Hamming distances", *Discr. Math.* 188, 1998, 1 - 11.

[8] R. Ahlswede, A. El Gamal, and K. F. Pang, "A two–family extremal problem in Hamming space", *Discr. Math.* 49, 1984, 1–5.

[9] A. V. Aho, J. D. Ullman, and M. Yannakakis, "On notions of information transfer in VLSI circuits", *Proc. ACM STOC*, 1983, 133-139.

[10] L. Babai, P. Frankl, and J. Simon, "Complexity classes in communication complexity theory", *Proc. IEEE FOCS*, 1986, 337-347.

[11] N. Cai, "A bound of sizes of code pairs satisfying the strong 4–words property for Lee distance", *J. System Sci. Math. Sci.* 6, 1986, 129–135.

[12] B. Chor and O. Goldreich, "Unbiased bits from sources of weak randomness and probabilistic communication complexity", *SIAM J. Comp.* 17, no. 2, 1988, 230–261.

[13] M. Dietzfelbinger, J. Hromkovic, and G. Schnitger, "A comparison of two lower bound methods for communication complexity", *Theoret. Comput. Sci.* 168, no. 1, 1996, 39 - 51.

[14] J. Diekmann, "Probabilistische Kommunikationskomplexität", Diploma thesis, Bielefeld, 1997.

[15] P. Delsarte and P. Piret, "An extension of an inequality by Ahlswede, El Gamal and Pang for pairs of binary codes", *Discr. Math.* 55, 1985, 313–315.

[16] A. El Gamal and K. F. Pang, "Communication complexity of computing the Hamming distance", *SIAM J. Comp.* 15, no 4, 1986, 932–947.

[17] T. Feder, E. Kushilevitz, M. Naor, and N. Nisan, "Amortized communication complexity", *SIAM J. Comp.* 24, no. 4, 1995, 736 - 750.

[18] M. Fürer, "The power of randomness for communication complexity", *Proc. ACM STOC*, 1987, 178-181.

[19] W. Haemers, "Disconnected vertex sets and equidistant code pairs", Electron. J. Combin. 4, no. 1, 1997, 10 pp.

[20] B. Halstenberg and R. Reischuk, "Relations between communication complexity classes ", *J. Comput. System Sci.* 41, 1990, 402-429.

[21] J. Hromkovic, *Communication complexity and parallel computing*, Springer, 1997.

[22] J. H. van Lint and J. I. Hall, "Constant distance code pairs", *Proc. Kon. Ned. Akad. v. Wet. (A)* 88, 1985, 41 - 45.

[23] B. Kalyanasundaram and G. Schnitger, "Probabilistic communication complexity of set intersection", *SIAM J. Discr. Math.* 5, 1992, 545-557.

[24] M. Karchmer, E. Kushilevitz, and N. Nisan, "Fractional covers and communication complexity", *SIAM J. Disc. Math.* 8, no. 1, 1995, 76-92.

[25] M. Karchmer, R. Raz, and A. Wigderson, "Super-logarithmic depth lower bounds via direct sum methods in communication complexity", *Proc. 6th IEEE Structure in Complexity Theory*, 1991, 299 - 304

[26] I. Krasikov and S. Litsy, "On integral zeros of Krawtchouk polynomials", *J. Combin. Theory* Ser. A 74, 1996, 71-99.

[27] M. Krause, "Geometric arguments yield better bounds for threshold circuits and distributed computing", PhD thesis, Berlin , 1990, also: *Theoret. Comput. Sci.* 156, no. 1-2, 1996, 99 - 117.

[28] E. Kushilevitz and N. Nisan, *Communication complexity*, Cambridge University Press, 1997.

[29] L. Lovasz and M. Saks, "Communication complexity and combinatorial lattice theory", *J. Comput. System Sci.* 47, 1993, 330-337.

[30] K. Mehlhorn and E. M. Schmidt, "Las Vegas is better than determinism in VLSI and distributed computing", *Proc. ACM STOC*, 1982, 330 - 337.

[31] M. Naor, A. Orlitsky, and P. Shor, "Three results on interactive communication", *IEEE Trans. Inform. Theory* 39, no. 5, 1993, 1608 - 1615.

[32] N. Nisan and A. Wigderson, "On rank vs. communication complexity", *Combinatorica* 15, no. 4, 1995, 557-566.

[33] W. Paul, "Realizing Boolean functions on disjoint sets of variables", *Theoret. Comput. Sci.* 2, 1976, 383-396.

[34] R. Raz and B. Spieker, "On the log rank conjecture in communication complexity", *Combinatorica* 15, no. 4, 1995, 567 - 588.

[35] A. Razborov, "On the distributional complexity of disjointness", *Theoret. Comput. Sci.*, 106, 1992, 385-390

[36] U. Tamm, "Communication complexity of sum-type functions", PhD thesis, Bielefeld, 1991.

[37] U. Tamm, "Still another rank determination of set intersection matrices with an application in communication complexity", *Appl. Math. Letters* 7, 1994, 39 - 44.

[38] U. Tamm, "Communication complexity of sum - type functions invariant under translation", *Inform. and Computation* 116, no. 2, 1995, 162 - 173.

[39] U. Tamm, "Deterministic communication complexity of set intersection", *Discr. Appl. Math.*, 61, 1995, 271 - 283.

[40] J. H. van Lint, "Distance theorems for code pairs", *Combinatorial Mathematics: Proceedings of the Third International Conference, New York, 1985, Ann. New York Acad. Sci.* 555, 1989, 421 - 424.

[41] I. Wegener, "Communication complexity and BDD lower bound techniques", this volume, 1999.

[42] A. C. Yao, "Some complexity questions related to distributive computing", *Proc. ACM STOC*, 1979, 209–213.

ORDERING IN SEQUENCE SPACES: AN OVERVIEW

Peter Vanroose

K.U.Leuven, div. ESAT/PSI, K.Mercierlaan 94,
B–3001 Leuven, Belgium

Peter.Vanroose@esat.kuleuven.ac.be

Abstract: "Creating order" is maybe one of the most important human activities. In its simplest form, ordering is just "sorting", which is a mathematically well understood problem. However, in real life we are often facing practical limitations which inhibit complete sorting. These limitations can be either knowledge (information) restrictions —we don't know the future, we forget the past— or manipulation restrictions —we don't want to carry objects too far—.

A mathematical theory of ordering (with constraints) in sequence spaces was first presented in [7] and [1]. In their setup, an algorithm is sought which "orders" any sequence of length n, i.e., which transforms the sequence $\vec{x}$ into the sequence $\vec{y}$ (of the same length and with the same symbols in it), such that the number of possible resulting sequences $\vec{y}$ is as small as possible. In this sense ordering is a generalization of sorting $\vec{x}$, as this would yield the absolute minimal number of sequences $\vec{y}$.

However, the model imposes extra restrictions on the ordering algorithm: a window of size β moves over the sequence, and the algorithm is only allowed to interchange the symbols within the window; moreover, at any time the algorithm cannot examine the sequence except for π "past" and ϕ "future" symbols.

This simple setup leads to several nice nontrivial mathematical problems, several of which are still unsolved.

INTRODUCTION

This text wants to give a survey overview on the topic of ordering in sequence spaces. Most of the presented results are not new; references to the original source are given where appropriate. Also the proof sketches are in the line of the original proofs, be it that I have tried to present everything in a unified way, using matrix terminology, which also simplified some proofs. This text is not meant to be complete: there are several extensions or variations to the

I. Althöfer et al. (eds.), Numbers, Information and Complexity, 603-613.

basic setup which I will not mention here. The interested reader is referred to the literature. More specifically, I do not consider the following situations: active memory; non-deterministic ordering; the permuting channel; multi-user models; varying number of output symbols; objects of varying length; idle objects; ... See [7] for an overview of model variants.

MOTIVATION

Suppose we want to "order" an arbitrary binary sequence $\vec{x}$ of length n, i.e., we look for a mapping

$$f : \{0,1\}^n \to \{0,1\}^n : \vec{x} \mapsto \vec{y}$$

with the smallest possible range. Unconstrained sorting maps any of the 2^n input sequences to one of the $n+1$ binary sequences where zeroes precede ones, but this at a time complexity cost of at least $n \log n$ and space complexity n.

To obtain output space reduction (cf. lossy compression) with linear time complexity and constant space complexity, *constrained ordering* is needed. For example, the following restrictions could be imposed:

- *at time instant t, only x_t and x_{t+1} may be interchanged;*
- *the decision whether to interchange or not may not depend on y_i for $i < t$ or on x_i for $i > t+2$; it may however depend on $\{x_t, x_{t+1}\}$, on x_{t+2} and on t.*

Only when $x_t \neq x_{t+1}$, some decision has to be taken. This decision determines the value of y_t. So, the ordering algorithm at time instant t can be considered as a mapping $f_t : x_{t+2} \mapsto y_t$, called a *strategy*. There are only 4 possible strategies in this case: f_{00}, f_{01}, f_{10} and f_{11}, where f_{ij} chooses $y_t = i$ if $x_{t+2} = 0$ and $y_t = j$ if $x_{t+2} = 1$.

Amongst the 4^{n-1} possible time-dependent ordering algorithms f —each of which is a sequence of $n-1$ strategies, one for each time instant $t < n$— we want to find that f which minimizes $\tau(f) := \frac{1}{n} \log_2 T(f)$, where $T(f) := \#f(\{0,1\}^n)$ is the number of different n-tuples $\vec{y}$ resulting from f. The number $\tau(f)$ is called the *rate* of f. Finally, we want to find the optimal asymptotic rate $\tau_2(0,2,1) := \lim_{n\to\infty} \inf_{f^{(n)}} \tau(f^{(n)})$, where the infimum is taken over all 4^{n-1} possible algorithms $f^{(n)}$ of length $n-1$.

Note that, as opposed to unconstrained sorting, it makes sense to consider ordering semi-infinite sequences, hence $\tau_2(0,2,1)$ is indeed a useful measure. It turns out that in this particular situation $\tau_2(0,2,1) = \frac{1}{3}\log_2(2+\sqrt{3}) = 0.6333229$. It is straightforward to verify that the periodic algorithm $f = f_{00}f_{01}f_{11}\cdots$ with period 3 achieves this optimal rate. It is not at all evident to prove the optimality of this algorithm!

Clearly, increasing either the knowledge or the manipulation freedom cannot increase the optimal rate. For example, adding y_{t-1} to the knowledge gives optimal rate $\tau_2(1,2,1) = 0.5$, which is even the best possible with the given manipulation constraints. When β symbols $x_t \ldots x_{t+\beta-1}$ can be interchanged at any time instant t, and there are no knowledge constraints, the optimal asymptotic rate is $\frac{1}{\beta}$.

THE BASIC MODEL

The general setup, as introduced by [1], is as follows.

An *ordering machine* of type $(\pi, \beta, \phi, T^+, O^-)_\alpha$ is a device to transform an arbitrary semi-infinite input sequence $\vec{x} = x_0x_1x_2\ldots$ over a source alphabet $\mathcal{A} = \{0, 1, \ldots, \alpha - 1\}$ of size α into a semi-infinite output sequence $\vec{y} = y_0y_1y_2\ldots$ by reordering. (See Figure 1.) It consists of a *look-ahead shift register* of size $\phi - 1$, capable of holding $\phi - 1$ upcoming ("future") input symbols $x_{t+\beta} \cdots x_{t+\beta+\phi-2}$, a *memory box* of size β, capable of holding β unordered symbols from $\mathcal{A}$, and a *look-back shift register* of size π, which holds the last π previous ("past") output symbols $y_{t-1} \cdots y_{t-\pi}$.

$$y_0 \cdots y_{t-\pi-1} \overline{y_{t-\pi} \cdots y_{t-1}}^{\,\pi} \boxed{\overbrace{00\cdots0}^{b_0^{(t)}}\overbrace{11\cdots1}^{b_1^{(t)}}\cdots} \overline{x_{t+\beta} \cdots x_{t+\beta+\phi-2}}^{\,\phi-1} x_{t+\beta+\phi-1} \cdots$$

Figure 1 Situation at time instant t.

The machine can be regarded as moving over the sequence from left to right, or alternatively the sequence moves through the machine from right to left.

The *internal state* of the device at time t consists of the (unordered) contents of the memory box and the (ordered) contents of the two shift registers, and can be represented by the tuple

$$s^{(t)} := (y_{t-\pi}, \ldots, y_{t-1}; b_0^{(t)}, \ldots, b_{\alpha-1}^{(t)}; x_{t+\beta}, \ldots, x_{t+\beta+\phi-2})$$

where $b_i^{(t)}$ $(i \in \mathcal{A})$ represents the number of symbols in the memory box having the value i. Note that $b_i^{(t)} \geq 0$ and $\sum_{i\in\mathcal{A}} b_i^{(t)} = \beta$.

The functioning of the ordering device can be described as follows. At time t, the machine first reads the next upcoming input symbol $x_{t+\beta+\phi-1}$. (So the "knowledge about the future" is indeed ϕ symbols.) Then, depending on this symbol, its internal state $s^{(t)}$, and the time t, the device chooses a symbol y_t from one of the symbols in the memory box. Next, the new symbol $x_{t+\beta+\phi-1}$ is shifted into the look-ahead shift register, the output $x_{t+\beta}$ of the shift register is transferred into the memory box to replace y_t, the output y_t of the memory box is shifted into the look-back shift register, and the output $y_{t-\pi}$ of the shift register is the next output of the entire device.

Denote the collection of all possible internal states of the device by $\mathcal{S}$. The possible *actions* of the device can be described with the aid of a labeled directed graph which has $\mathcal{S}$ as its set of states, and labeled transitions of the form

$$s^{(t)} \xrightarrow{\quad x \;/\; y \quad} s^{(t+1)},$$

with $x = x_{t+\beta+\phi-1}$ and $y = y_{t-\pi}$. A particular ordering algorithm applied to a particular input sequence is a path through this *state transition diagram* which satisfies the x-labels. Its output is the sequence of y-labels.

Clearly $\beta \geq 2$ (because otherwise no ordering can be performed), $\phi \geq 1$ and $\pi \geq 0$. The situation $\phi = 0$ (no future knowledge at all) can also be considered, but does not fit into this graph description of the model. Larger values of ϕ and π mean more knowledge, thus will possibly allow better output space reduction. Larger values of β mean more manipulation freedom, again with a potentially better compression. Full knowledge of past and/or future, written as $\pi = \infty$ and/or $\phi = \infty$, implies knowledge of time, because the start and/or the end of the sequence can be seen. (This is a disputable standpoint, however!)

Two variants to this general setup can be considered. An ordering machine of type $(\pi, \beta, \phi, T^-, O^-)_\alpha$ is a *time-invariant* machine, i.e., the choice for y_t only depends on the state of the machine, not on the time instant. A time-invariant ordering machine can thus be represented by a directed graph which is a subgraph of the generic one, such that exactly one transition with a given x-label leaves a given state. Thus, given a certain starting state and an x-label sequence, there is exactly one path through this graph. Clearly this type of ordering machine has less knowledge.

An ordering machine of type $(\pi, \beta, \phi, T^{-/+}, O^+)_\alpha$ is a machine with an *ordered box*, i.e., the state of the machine is

$$s^{(t)} := (y_{t-\pi}, \ldots, y_{t-1}; x_t, \ldots, x_{t+\beta-1}; x_{t+\beta}, \ldots, x_{t+\beta+\phi-2})$$

Clearly this type of ordering machine has more knowledge than the corresponding types without ordering knowledge. The state transition diagram is similar.

The *asymptotic rate* of an ordering machine f is defined as

$$\tau_\alpha(f) := \lim_{n\to\infty} \frac{1}{n} \log_\alpha \# f(\mathcal{A}^n),$$

where $f(\mathcal{A}^n)$ is the set of all possible output sequences of length n that can be generated by the ordering machine f, and $\log_\alpha$ is the logarithm with base α, further written as just log.

The *optimal asymptotic rate* for the situation $(\pi, \beta, \phi, T^+, O^-)_\alpha$ is

$$\tau_\alpha(\pi, \beta, \phi) := \lim_{n\to\infty} \inf_f \frac{1}{n} \log \# f(\mathcal{A}^n),$$

where the infimum is taken over all possible ordering machines f of type $(\pi, \beta, \phi, T^+, O^-)_\alpha$.

Similarly, $\nu_\alpha(\pi, \beta, \phi)$, $\omega_\alpha(\pi, \beta, \phi)$ and $\gamma_\alpha(\pi, \beta, \phi)$ denote the asymtotic rates for the situations $(\pi, \beta, \phi, T^-, O^-)_\alpha$, $(\pi, \beta, \phi, T^-, O^+)_\alpha$ and $(\pi, \beta, \phi, T^+, O^+)_\alpha$, respectively.

The labeled state transition diagram of a time-invariant ordering machine completely describes its functioning, provided that the initial state is given. A time-varying ordering machine can be completely described by a *sequence* of state transition diagrams, where the t-th diagram in the sequence represents the strategy f_t to be performed at time instant t.

The state transition diagram does not optimally describe the properties of an ordering machine: a certain output sequence could be generated by more than one path through the state diagram, so there is no one-to-one correspondence between paths and output sequences. Which means that the y-labels on branches cannot be discarded. See section 6 for an example of an ordering machine where two different paths produce the same output.

In order to calculate asymptotic rates, it would be nice if there was such a one-to-one correspondence, because the asymptotic growth rate of the number of paths through a state diagram equals the largest eigenvalue $\lambda_{\max}$ of its transition matrix: for a state transition diagram with m states this is the $m \times m$ matrix whose entry (i, j) is the number of transitions from state i to state j. The entries of the n-th power of this matrix are the number of paths of length n between any two states.

Surprisingly, it is possible to find an other state transition diagram where there is indeed a one-to-one mapping between paths and output sequences. This is explained now.

In which ways can a given output sequence $y_0 \dots y_{t-1}$ of a given ordering machine be extended to an output sequence $y_0 \dots y_t$? Let $\mathcal{S}_t$ denote the collection of all states in which the machine can be at time t after generating $y_0 \dots y_{t-1}$. Sets of states of this form will be called *superstates.* In particular, the set $\mathcal{S}$ of all states is the initial superstate $\mathcal{S}_0$ of the machine.

The machine can generate y_t at time t after $y_0 \dots y_{t-1}$ precisely when the time t strategy f_t of the machine can produce output y_t from some state in $\mathcal{S}_t$, under a suitable input $x_{t+\beta+\phi-1}$. Thus, the *superstate transition diagram* has nonempty subsets of $\mathcal{S}$ as its set of states, and labeled transitions of the form

$$\mathcal{S}_t \xrightarrow{\quad y \quad} \mathcal{S}_{t+1},$$

with $y = y_{t-\pi}$, and where S_{t+1} is the set of all states s in $\mathcal{S}$ for which the original (generic) state diagram contains a transition from a state in $\mathcal{S}_t$ to s that produces output y (i.e., with label x/y for some input x).

It was proved in [3] that there is indeed a one-to-one correspondence between walks of length t in the superstate transition diagram starting in superstate $\mathcal{S}$ and output sequences of length t that can be generated by the machine. So the *asymptotic rate* of an ordering machine is $\log \lambda_{\max}$, where $\lambda_{\max}$ is the largest eigenvalue of the transition matrix of the *superstate* transition diagram.

This also means that we may disregard all labels, so the transition matrix of the superstate transition diagram completely describes the ordering machine. For a periodic time-varying ordering machine, this matrix is the product of the composing time-invariant transition matrices, in the correct order. The rate of a period m time-varying ordering machine is $\frac{1}{m} \log \lambda_{\max}$, where $\lambda_{\max}$ is the largest eigenvalue of this product matrix.

KNOWN RESULTS FOR $\alpha = 2$

In the binary case the state transition diagram has $(\beta + 1) \cdot 2^{\pi+\phi-1}$ states for situation O^-. (While for situation O^+ it has $2^{\pi+\beta+\phi-1}$ states.) In each of the $(\beta - 1) \cdot 2^{\pi+\phi-1}$ states with $b_0^{(t)} \neq 0$ and $b_1^{(t)} \neq 0$, there are 4 possibilities for the ordering algorithm: $y = 0$, $y = 1$, $y = x$ or $y = 1 - x$. Hence there are

$$4^{(\beta-1)\cdot 2^{\pi+\phi-1}}$$

different ordering machines of type $(\pi, \beta, \phi, T^-, O^-)_2$ if $\phi \neq 0$. The number of ordering machines of type $(\pi, \beta, \phi, T^+, O^{+/-})_2$ is of course infinite. There are

$$4^{(2^\beta - 2)\cdot 2^{\pi+\phi-1}}$$

different ordering machines of type $(\pi, \beta, \phi, T^-, O^+)_2$ because now 2 out of 2^β states (instead of 2 out of $\beta + 1$) force the output to be either 0 or 1.
The tables below summarize all known results for situations $(\pi, \beta, \phi, T^-, O^-)_2$ and $(\pi, \beta, \phi, T^+, O^-)_2$. As there are three parameters, this should be seen as two threedimensional tables. The parameter ϕ runs from left to right, π runs from top to bottom, and different values of β are found in different sub-tables.

Table 1 Known values of $\nu_2(\pi, 2, \phi)$

π \ ϕ	0	1	2	$\cdots$	∞
0	1	1	0.8791	?*	0.5
1	0.6942	0.6942	?	?	0.5
$\vdots$	$\vdots$	?	?	?	0.5
∞	0.6942	0.5	0.5	$\cdots$	0.5

* $\nu_2(0, 2, 3) = 0.8609$.

Table 2 Known values of $\tau_2(\pi, 2, \phi)$

π \ ϕ	0	1	2	$\cdots$	∞
0	0.6942	0.6333	0.6040	?	0.5
1	0.6942	0.5	0.5	$\cdots$	0.5
$\vdots$	$\vdots$	$\vdots$	$\vdots$	$\ddots$	0.5
∞	0.6942	0.5	0.5	$\cdots$	0.5

Table 3 Known values of $\nu_2(\pi, 3, \phi)$

π \ ϕ	0	1	2	$\cdots$	∞
0	1	1	0.8791	?	0.3333
1	0.5515		?	?	0.3333
$\vdots$	?	?	?	?	$\vdots$
∞	0.5	0.4057	0.3333	$\cdots$	0.3333

Table 4 Known values of $\tau_2(\pi, 3, \phi)$

π \ ϕ	0	1	2	$\cdots$	∞
0	?*	?	?	?	0.3333
1	0.5	0.4057	0.3333	$\cdots$	0.3333
$\vdots$	$\vdots$	$\vdots$	$\vdots$	$\ddots$	$\vdots$
∞	0.5	0.4057	0.3333	$\cdots$	0.3333

* between 0.5515 and 0.5697

Table 5 Known ν_2 values for general β

π \ ϕ	0	1	$\cdots$	$\beta - 1$	$\cdots$	∞
0	1	1	?	?	?	$1/\beta$
1	$\log \psi_\beta$		?	?	?	$1/\beta$
$\vdots$	?	?	?	?	?	$\vdots$
∞	$\approx 2/(\beta + \phi + 1)$			$1/\beta$	$\cdots$	$1/\beta$

$\psi_\beta^\beta = \psi_\beta^{\beta-1} + 1$.

Table 6 Known τ_2 values for general β

π \ ϕ	0	1	$\cdots$	$\beta - 1$	$\cdots$	∞
0	?*	?	?	?	?	$1/\beta$
1	$\approx 2/(\beta + \phi + 1)$			$1/\beta$	$\cdots$	$1/\beta$
$\vdots$	$\vdots$	$\vdots$	$\vdots$	$\vdots$	$\ddots$	$1/\beta$
∞	$\approx 2/(\beta + \phi + 1)$			$1/\beta$	$\cdots$	$1/\beta$

* $\geq \log \psi_\beta$

These results will be derived in the following sections. Most of these were found by [1]. The values for $\tau_2(0, 2, 1)$ and $\tau_2(0, 2, 2)$ follow from a general method introduced in [3]. the expression "$\approx 2/(\beta+\phi+1)$" stands for $\log C(\beta + \phi)$ derived in section 6.

Almost all proofs (except those for $\phi = 0$) make use of the superstate transition matrix of the ordering machine.

BASIC EXTREMAL CASES

No knowledge: $\nu_2(0, \beta, 0) = \nu_2(0, \beta, 1) = 1$.

It suffices to prove that $\nu_2(0, \beta, 1) \geq 1$, because $\nu_2(0, \beta, 0) \geq \nu_2(0, \beta, 1)$, and $\nu_2(\pi, \beta, \phi)$ cannot be larger than 1 since there can be at most 2^n output sequences of length n.

There are $\beta+1$ states $s^{(t)} = (b_0^{(t)}, b_1^{(t)})$. Let row/column i ($i = 0, \ldots, \beta$) of the transition matrix D of an ordering machine correspond to state $(\beta - i, i)$. The rows of D must satisfy the following constraints: (1) first and last row are fixed to $110\ldots0$ and $0\ldots011$ respectively; (2) the sum of row entries is always 2, i.e., there can be two 1's or one 2; (3) the four possible values for the other rows are: $(1,1)$ on and before the diagonal (when both outgoing y-labels are 1), or $(1,1)$ on and after the diagonal (when both y-labels are 0), or $(1,0,1)$ around the diagonal (when x and y labels are opposite), or a 2 on the diagonal (when x and y labels are identical), e.g.:

$$D = \begin{bmatrix} 1 & 1 & 0 & 0 & \cdots & 0 & 0 \\ 1 & 1 & 0 & 0 & \cdots & 0 & 0 \\ 0 & 1 & 0 & 1 & \cdots & 0 & 0 \\ 0 & 0 & 0 & 2 & \cdots & 0 & 0 \\ \vdots & \vdots & \vdots & \vdots & \ddots & \vdots & \vdots \\ 0 & 0 & 0 & 0 & \cdots & 1 & 1 \end{bmatrix} \quad \begin{matrix} (y_t = 0) \\ (y_t = 1) \\ (y_t = 1 - x_{t+\beta}) \\ (y_t = x_{t+\beta}) \\ \\ (y_t = 1) \end{matrix}$$

In this case there is a one-to-one correspondence between paths through the transition diagram and output sequences, hence it is not necessary to consider superstates. All these matrices have an eigenvalue 2, because $S := D - 2I$ is always singular. (All rows of S have weight 0, and there is always overlap of nonzero elements, so there must be a combination of rows that sums to zero.) In all cases $\lambda_{\max} \geq 2$, q.e.d.

One past: $\nu_2(1,\beta,0) = \nu_2(1,\beta,1) = \log\psi_\beta$, where $\psi_\beta^\beta = \psi_\beta^{\beta-1} + 1$.
The optimal strategy is $y_t = y_{t-1}$ (repeat the previous output, if possible). The proof of the optimality is by induction on β, see [1], page 71. This proof was only given for the case $\phi = 0$ but it also holds for $\phi = 1$.

There are 2β superstates $\mathcal{S}_{1,m} := \{(1; i, \beta - i) | i = 0\ldots m\}$ and $\mathcal{S}_{0,m} := \{(0; \beta - i, i) | i = 0 \ldots m\}$ for arbitrary $m \in \{1 \ldots \beta\}$, and the superstate transition matrix is (after merging the pairs of equivalent states $\mathcal{S}_{0,m}$ and $\mathcal{S}_{1,m}$) the $\beta \times \beta$ matrix

$$D = \begin{bmatrix} 0 & 1 & 0 & 0 & \cdots & 0 & 0 \\ 0 & 0 & 1 & 0 & \cdots & 0 & 0 \\ \vdots & \vdots & & & \ddots & \vdots & \vdots \\ 0 & 0 & 0 & 0 & \cdots & 0 & 1 \\ 1 & 0 & 0 & 0 & \cdots & 0 & 1 \end{bmatrix}.$$

Expanding the determinant $|XI - D|$ by its first column, we see that the characteristic equation of this matrix is $X^{\beta-1}(X-1) - 1 = 0$, q.e.d.

Note that this equation uniquely determines ψ_β as it has exactly one positive real root for any value of β. The first few values of $\log\psi_\beta$ are $\log(\sqrt{5}+1) - 1 = 0.694242$, 0.551463, 0.464958, 0.405685, 0.361992, 0.328173, 0.301066, 0.278758, 0.260015.

Full knowledge: $\nu_2(\infty,\beta,\beta-1)=\nu_2(0,\beta,\infty)=\tau_2(1,\beta,\beta-1)=\frac{1}{\beta}$.
Note that knowledge of infinite past or future implies knowledge of time, hence it suffices to determine $\tau_2(1,\beta,\beta-1)$.
Direct part of the proof: produce a blocked output consisting of blocks of β consecutive identical values, as follows:
At time instants $t=k\beta$ (multiples of β), set $y_t=1$ if $b_1^{(t)}+\sum_{i=t}^{t+\beta-2}x_i\geq\beta$, and $y_t=0$ otherwise. At other time instants, set $y_t=y_{t-1}$. This is possible if $\pi\geq 1$, $\phi\geq\beta-1$, and the time instant is known. The output space growth rate is a factor 2 per β symbols, q.e.d. When $\pi=0$, the proof is a little bit more involved; it makes use of the modulo β value of the number of ones in the complete future.
Proof of the converse: when an ordering machine is presented two different β-blocked sequences as input, it cannot produce an identical output sequence for these two.

INFINITE PAST

No future: $\nu_2(\infty,\beta,0)=\omega_2(\infty,\beta,0)=\log C(\beta)\approx 2/(\beta+1)$,
where $C(\beta)$ is the largest root of the equation

$$X^{\beta+1}=X^{\lceil(\beta+1)/2\rceil}+X^{\lfloor(\beta+1)/2\rfloor}.$$

Actually, the number of output sequences of length n equals the number $C(n,\beta)$ defined as the minimal number of leaves of a binary tree of weighted depth n, where the sum of the weights of the two branches leaving from an internal node is at most $\beta+1$. Branch weights must be integers at least 1. Note that for odd values of β, $\log C(\beta)=2/(\beta+1)$, while for even values, $\log C(\beta)$ is only slightly larger than $2/(\beta+1)$: $C(2)=\log(\sqrt{5}+1)-1=0.694242$, $C(4)=0.405685$, $C(6)=0.28776$, $C(8)=0.22318$, $C(10)=0.18234$, $C(12)=0.15416$, $C(14)=0.13354$.

The proof consists of three parts: (1) The minimal number of output sequences is at most $C(n,\beta)$, because the 'minimal' binary tree (with weighted branches) defines an ordering machine: starting at the root, at each internal node, take the left branch and output l_0 zeroes if the left weight is l_0 and there are at least l_0 zeroes in the box; otherwise, take the right branch and output l_1 ones where l_1 is the weight of the right branch. (2) $C(n,\beta)$ is an upper bound for the minimal number of output sequences of an ordering machine for the situation $(\infty,\beta,0,T^-,O^+)$ (with knowledge of order in the box). This is the difficult part of the proof, see [1]. And of course (3) $\omega_2(\infty,\beta,0)\leq\nu_2(\infty,\beta,0)$.

General case: $\nu_2(\infty,\beta,\phi)=\omega_2(\infty,\beta,\phi)=\nu_2(\infty,\beta+\phi,0)=C(\beta+\phi)$ if $\phi\leq\beta-1$, $\nu_2(\infty,\beta,\phi)=1/\beta$ if $\phi\geq\beta-1$,
$\tau_2(\pi,\beta,\phi)=\nu_2(\infty,\beta,\phi)$.
The case $\phi\geq\beta-1$ was already considered before. For the case $\phi\leq\beta-1$, observe that both situations have the same knowledge, but situation

(∞, β, ϕ) has less manipulation freedom. So it suffices to prove that in this situation the algorithm for situation $(\infty, \beta + \phi, 0)$ (described above) can be applied, i.e., that the output object is always present in the box.

RESULTS FOR $\alpha > 2$

The previous sections were solely devoted to the binary alphabet situation. Much less is known about ordering non-binary alphabet sequences. I just mention one result; refer to the literature for more details:

$\nu_\alpha(\infty, 2, 0) = \log \xi_\alpha$ where ξ_α is the largest eigenvalue of the $\alpha \times \alpha$ matrix

$$\begin{bmatrix} 0 & 0 & \cdots & 0 & 1 \\ 0 & 0 & \cdots & 1 & 1 \\ \vdots & \vdots & & \vdots & \vdots \\ 0 & 1 & \cdots & 1 & 1 \\ 1 & 1 & \cdots & 1 & 1 \end{bmatrix}.$$

SOME MORE CASES

The cases considered in the previous three sections are the only infinite parameter families for which the values of ν_α or τ_α have been determined. Some of the other cases have been analysed individually. It is at least remarkable that so few of these values have been determined yet! The value for $\tau_2(0, 2, 1)$ was conjectured in [6], the proof was only given six years later in [3].

- $\nu_2(0, \beta, 2) = 0.879146 = \log \lambda$ where λ satisfies $\lambda^3 = \lambda^2 + \lambda + 1$.
- $\nu_2(0, 2, 3) = 0.860906 = \log \lambda$ where λ satisfies $\lambda^5 = 2\,\lambda^4 - 2$.
- $\tau_2(0, 2, 1) = 0.633323 = \frac{1}{3} \log(2 + \sqrt{3})$.
- $\tau_2(0, 2, 2) = 0.604036 = \frac{1}{6} \log \lambda$ where λ satisfies $\lambda^3 = 12\,\lambda^2 + 4\,\lambda + 1$.
- $\tau_2(0, 3, 0) \leq 0.569663 = \frac{1}{17} \log(10\sqrt{1687} + 412)$.

The value of $\nu_2(\pi, \beta, \phi)$ for specific π, β and ϕ can in principle be determined by exhaustive search: there are only a finite number of possible transition matrices, and for each of these the superstate transition matrix and its largest eigenvalue can be calculated. But this is a tedious work, except for really small values of the parameters.

For the first open case $\nu_2(0, 2, 2)$ there are sixteen state transition diagrams, with 6 states each. One of the optimal algorithms is "$y_t = \max(x_{t+2}, x_{t+3})$ when $b_1^{(t)} = 1$", i.e., always output a 1 if possible, except if both x_{t+2} and x_{t+3} are zero. Note that the state transition matrix has largest eigenvalue 2: the number of paths through the state diagram is much larger than the number of output sequences. E.g., the output "000" can be generated from state $(2, 0; 0)$ by input "010" as well as by "100"; in both cases, the ending state is $\{1, 1; 0\}$.

There are four superstates $\{(2,0;1)\}$, $\{(1,1;0)\}$, $\{(1,1;0),(1,1;1)\}$ and $\{(2,0;1),(1,1;0),(1,1;1)\}$, with superstate transition matrix

$$D = \begin{bmatrix} 0 & 0 & 1 & 0 \\ 1 & 1 & 0 & 0 \\ 0 & 1 & 0 & 1 \\ 0 & 0 & 1 & 1 \end{bmatrix}$$

Note that states $(2,0;0)$, $(0,2;0)$ and $(0,2;1)$ do not occur in any of the four superstates: once the machine has left one of these three states it cannot re-enter them, hence the asymptotic rate does not change when we discard these states. The characteristic equation of this 4×4 matrix is $\lambda^3 = \lambda^2+\lambda+1$, q.e.d.

It turns out that this algorithm "$y_t = \max(x_{t+\beta}, x_{t+\beta+1})$ when $b_1^{(t)} = 1$" is optimal for all values of β, hence $\nu_2(0,\beta,2) = 0.879146$ for arbitrary β.

For time varying ordering machines, an exhaustive search is impossible: the optimal rate in the time-varying case is the infimum over the infinite set of (finite) sequences of superstate transition matrices of $\frac{1}{m}\log\lambda_{\max}$, where $\lambda_{\max}$ is the largest eigenvalue of the product of the m superstate transition matrices in the sequence.

Each particular ordering algorithm thus yields an upper bound on $\tau_2(\pi,\beta,\phi)$. E.g., $\tau_2(0,2,1) \leq 0.6333229$ because the periodic strategy $f_{00}f_{01}f_{11}$ from section 48 with superstates $\{(2,0),(1,1),(0,2)\}$ and $\{(2,0),(1,1)\}$, and superstate transition transition matrix

$$\begin{bmatrix} 2 & 3 \\ 1 & 2 \end{bmatrix}$$

has $\frac{1}{3}\log\lambda_{\max} = \frac{1}{3}\log(2+\sqrt{3}) = 0.6333229$.
$\tau_2(0,3,0) \leq 0.569663$: use the periodic strategy $00m10m10m100m10m1$ where $0/1$ stand for "$y_t = 0/1$ if possible", and m stands for "$y_t =$ the majority vote within the box".

The set of matrix products over which the infimum is to be taken is infinite, but it is a finitely generated multiplicative semigroup. In [3] exactly this observation is used to obtain *lower bounds* on $\tau_2(\pi,\beta,\phi)$. It was even proved that $\tau_2(\pi,\beta,\phi)$ can always be achieved by a periodic strategy. In particular, two τ_2 values were found using this method: $\tau_2(0,2,1) = 0.6333229$, i.e., the above algorithm is optimal; and $\tau_2(0,2,2) = 0.604036$, achieved with the period 6 algorithm $f_{1111}f_{0001}f_{0111}f_{0000}f_{0111}f_{0001}\cdots$ where $f_{v_{00}v_{01}v_{10}v_{11}}$ outputs $y = v_{x_{t+3}x_{t+4}}$ when in state $(1,1;x_{t+2})$.

Again, the calculations can in principle be done for any set of parameters, but become impractical for other than very small values.

And finally here is the superstate transition diagram for the optimal algorithm in the situation $(0,2,3,T^-,O^-)$, viz. $y = \lfloor(x_{\phi+2},x_{\phi+3},x_{\phi+4})/2\rfloor$ when $b_1^{(t)} = 1$ (majority vote amongst the three observed future symbols):

The characteristic equation of the superstate transition matrix is

$$X^4(X+1)^2(X-1)^2(X^2+X+1)(X^2-X+1)(X^5-2\,X^4+2)$$

so the rate of this ordering algorithm is the logarithm of the largest real root of $X^5 - 2\,X^4 + 2$, i.e., $\nu_2(0,2,3) = 0.860906$. This is a new result.

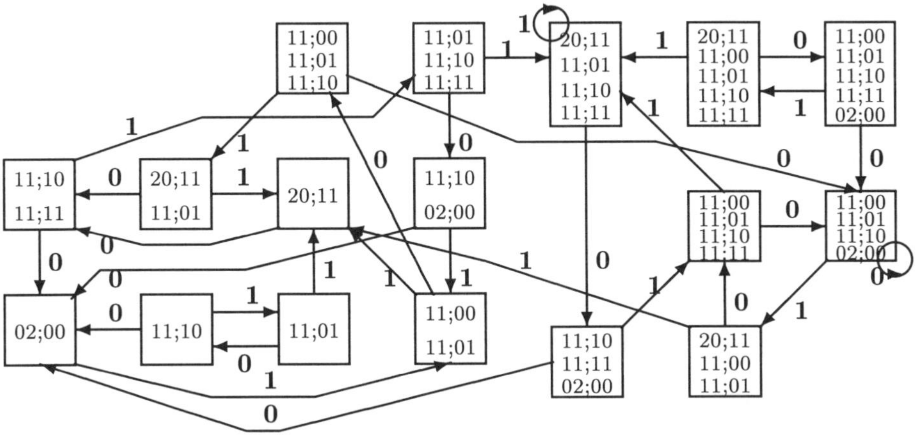

Figure 2 Superstate transition diagram achieving $\nu_2(0,2,3)$.

References

[1] R. Ahlswede, J.-P. Ye and Z. Zhang, "Creating order in sequence spaces with simple machines". *Information and Computation*, 89(1), 1990, 47–94.

[2] R. Ahlswede and Z. Zhang, "Contributions to a theory of ordering for sequence spaces". *Problems of Control and Information theory*, 18(4), 1989, 197–221.

[3] H. D. L. Hollmann and P. Vanroose, "Entropy reduction, ordering in sequence spaces, and semigroups of nonnegative matrices", *Preprint 95–092, SFB 343 "Diskrete Strukturen in der Mathematik"*, Universität Bielefeld, 1995.

[4] U. Tamm, "The influence of memory on creating order". *Preprint 96–031, SFB 343 "Diskrete Strukturen in der Mathematik"*, Universität Bielefeld, 1996.

[5] U. Tamm, "Ballot sequences in creating order". *Preprint, SFB 343 "Diskrete Strukturen in der Mathematik"*, Universität Bielefeld, 1998.

[6] P. Vanroose, *Een ordeningsresultaat voor de situatie* $(0, 2, 1, T^+)$ (in Dutch). PhD supplement, Katholieke Universiteit Leuven, 1989.

[7] Jian-Ping Ye, *Towards a Theory of Ordering in Sequence Spaces.* PhD thesis, Fakultät für Mathematik der Universität Bielefeld, 1988.

COMMUNICATION COMPLEXITY AND BDD LOWER BOUND TECHNIQUES

Ingo Wegener*

LS 2, FB Informatik, Univ. Dortmund
44221 Dortmund, Germany

wegener@ls2.cs.uni-dortmund.de

Abstract: Communication complexity as devised by Yao (1979) has found a lot of applications in the theory of networks, VLSI design, distributed computing, time-space tradeoffs, and in lower bound techniques for the complexity of Boolean functions, in particular for various restricted models of branching programs or binary decision diagrams (BDDs). A survey on lower bound techniques for BDDs based on communication complexity is given and some other BDD lower bound techniques are identified as communication complexity approach based on new variants of communication games.

INTRODUCTION

Information theory deals with all aspects of communication. It contains the theory on the information contents of messages, on the capacity of information channels, and on coding, and it has led to contributions in cryptography. Many of these problems are related to complexity theoretical problems. Yao (1979) has defined a communication game which has turned out to be the core of many computer science problems in different areas like networks, VLSI design, distributed computing, time-space tradeoffs, and complexity of Boolean functions. Communication complexity (see, e.g., Hromkovič (1997) or Kushilevitz and Nisan (1997)) is nowadays a vivid theory.

Boolean functions $f\colon \{0,1\}^n \rightarrow \{0,1\}^m$ play a fundamental role in computer science. Hence, one is interested in the complexity of Boolean functions (see, e.g., Wegener (1987)) with respect to various computation models, in particular, circuits and branching programs . Branching programs have been

*Supported in part by DFG grant We 1066/8-2.

I. Althöfer et al. (eds.), Numbers, Information and Complexity, 615-628.

investigated as model whose size is closely related to the storage space of Turing machine computations. Lower bounds for the general model are hard to obtain and even quadratic lower bounds for explicitly defined functions are still unknown. Therefore, one has considered restricted variants like read-once branching programs. Some types of restricted branching programs have been used since a long time in applications as static representation of Boolean functions but this research community has used the notation binary decision diagram (BDD). Bryant (1986) has observed that only a very special variant of BDDs is really used in applications and he called this variant ordered BDD (OBDD). He observed that a lot of operations which lead to hard problems for general BDDs and even for a lot of restrictions of BDDs can be performed efficiently for OBDDs. As a result, OBDDs are nowadays the state of the art dynamic representation or data structure for Boolean functions. It also has been proved that some generalizations of OBDDs can be handled efficiently. OBDDs and these generalizations are implemented in many CAD tools and have applications in the verification of combinational and sequential circuits, in model checking, logic synthesis, timing analysis, simulation, test pattern generation, graph algorithms, counting problems, and genetic programming (see Bryant (1992), Clarke and Wing (1996), and Wegener (1999)). There is always a tradeoff between the representational power of a BDD variant and the efficiency of the algorithms to perform operations on the BDD variant. In order to estimate the representational power of different BDD variants one needs lower bound techniques working for explicitly defined functions. It has turned out that communication complexity is a major tool to prove such bounds.

In Section 2 the most important communication game investigated in communication complexity is introduced. In Section 3, several BDD variants are defined. In Section 4, it is discussed how lower bounds in communication complexity lead to BDD lower bounds and why upper bounds in communication complexity usually do not support the design of small-size BDDs.

There are some BDD variants whose lower bounds are not based on communication complexity. In Section 5, it is discussed how these bounds can be interpreted as lower bounds for generalized communication games. This establishes a new link between communication complexity and BDD lower bound techniques.

COMMUNICATION COMPLEXITY

The basic communication game is defined as a game between two players Alice and Bob who try to cooperate to evaluate a Boolean function $f\colon \{0,1\}^n \times \{0,1\}^m \rightarrow \{0,1\}$. The input $c = (a, b)$ consists of the part $a \in \{0,1\}^n$ given to Alice and the part $b \in \{0,1\}^m$ given to Bob. Alice and Bob may communicate in order to obtain $f(a, b)$. The communication protocol defines the meaning of messages and is fixed before Alice and Bob obtain the parts of the particular input. It does not matter how difficult it is to follow the protocol. Depending on her input a, Alice computes her first message $m_1(a)$ and sends it to Bob. To be precise, we should prescribe that all messages $m_1(a)$, $a \in \{0,1\}^n$, are prefix-free in order to enable Bob to recognize the end of the first message. Since we only are interested in asymptotic bounds, such details do not matter. Having obtained $m_1(a)$, Bob knows b and $m_1(a)$, computes his message $m_2(b, m_1(a))$, and sends it to Alice. The conversation goes on in an analogous way. Each message only depends on the input given to the player and all messages previously obtained. At some point the protocol declares that the communication is successful which means that some player (or both players) knows $f(a, b)$. The complexity of the protocol on input (a, b) is the number of bits exchanged between Alice and Bob on input (a, b). The complexity of the protocol is the worst case complexity, i.e., the maximal complexity where the maximum is taken over all inputs. This is well-defined, since we have a nonuniform complexity measure for finite Boolean functions. In the asymptotic point of view we investigate sequences of protocols for sequences of Boolean functions. We are used to think of polynomials as small and exponential functions as large. The situation here is different. Alice may send a as her first message and Bob "knows" $f(a, b)$ (knowing means "he can compute"). Hence, protocols of linear length are the worst case and protocols of logarithmic length are called efficient.

Nondeterminism plays a central role in complexity theory and randomization is a key concept for the design of efficient algorithms (see, e.g., Motwani and Raghavan (1995)). A nondeterministic protocol allows that a player chooses the message to be sent out of a list of possible alternatives. The protocol realizes f if for inputs $(a, b) \in f^{-1}(1)$ there is a choice of alternatives leading to the output 1 while for inputs $(a, b) \in f^{-1}(0)$ all choices lead to the output 0. A randomized protocol allows that each player flips coins and the next message depends on the outcome of the coin flips. We distinguish ε-bounded zero-error protocols (for each input the protocol leads to the correct output with a probability of at least $1-\varepsilon$, it may answer "don't know", and it never errs), ε-bounded one-sided error (inputs $(a, b) \in f^{-1}(0)$ are always rejected, and inputs $(a, b) \in f^{-1}(1)$ are accepted with a probability of at least $1 - \varepsilon$), and ε-bounded two-sided error (for each input the output is correct with a probability of at least $1 - \varepsilon$, here we have to assume $\varepsilon < 1/2$ in order to obtain a meaningful model).

BDD MODELS

First, we define the syntax and semantics of general BDDs. We present three equivalent definitions of the semantics, since this simplifies the understanding of the different variants of BDDs.

Syntax of a general BDD

A BDD on the variable set $X_n = \{x_1, \ldots, x_n\}$ consists of a directed acyclic graph $G = (V, E)$ whose inner nodes (non-sink nodes) have outdegree 2 and a labelling of the nodes and edges. The inner nodes get labels from X_n and the sinks get labels from $\{0, 1\}$. For each inner node one of the outgoing edges gets the label 0 and the other one gets the label 1. The size of the BDD is equal to the number of nodes (which approximately is half the number of the edges).

Each node v of a BDD represents a Boolean function $f_v \colon \{0,1\}^n \to \{0,1\}$.

Semantics 1: The computation of $f_v(a)$, $a \in \{0,1\}^n$, starts at v. At nodes labelled by x_i, the outgoing edge labelled by a_i is chosen. Then $f_v(a)$ is equal to the label of the sink finally reached.

Semantics 2: Each input $a \in \{0,1\}^n$ activates all a_i-edges leaving x_i-nodes. Then $f_v(a)$ is equal to the label of the final node on the unique path activated by a and starting at v.

Semantics 3: A sink with label c represents the constant c. Let v be an inner node labelled by x_i whose 0-successor represents f_0 and whose 1-successor represents f_1. Then f_v is defined by Shannon's decomposition rule as $f_v(a) = a_i f_1(a) + \overline{a}_i f_0(a)$.

We are interested in the following BDD variants.

1.) A BDD G is called *read-k-times branching program* (k-BP) if each path of G contains for each i at most k nodes labelled by x_i.

2.) A 1-BP or read-once branching program is also called *free* BDD (FBDD).

3.) An FBDD G is called *ordered* BDD (OBDD) for a variable ordering π (describing a permutation of the variables in X_n) if the labels on each path of G are in the same order as prescribed by π (it is allowed to omit variables).

4.) A BDD G is called k-OBDD for a variable ordering π if it can be partitioned to k layers each fulfilling the ordering requirements given by π (the sequence of variables tested on each path can be partitioned to k consecutive subsequences such that in each subsequence the labels appear in the order prescribed by π).

5.) A BDD G is called k-IBDD (*indexed* BDD) for a vector $\pi = (\pi_1, \ldots, \pi_k)$ of variable orderings if it can be partitioned to k layers where the ith layer fulfills the ordering requirements given by π_i.

A *nondeterministic* BDD may contain binary nondeterministic nodes. Both edges leaving a nondeterministic node are always activated. For a node v we define $f_v(a)$ as 1 if there is a path activated by a and leading from v to a 1-sink. This is the usual OR-nondeterminism. We also may consider AND-nondeterminism and EXOR-nondeterminism defined in the obvious way.

A *randomized* BDD may contain randomized nodes with fan-out 2. Then $f_v(a)$ is a random variable taking values in $\{0, 1, ?\}$ (we also allow ?-sinks with the interpretation "don't know"). Independent coin flips determine for the randomized nodes which of the outgoing edges is activated. The probability that $f_v(a) = 1$ (similarly for 0 and ?) is equal to the probability that the path starting at v and activated by a reaches a 1-sink. Similarly to Section 2, we may define randomized computations of f by BDDs with ε-bounded zero-error, ε-bounded one-sided error, and ε-bounded two-sided error.

In the obvious way we obtain nondeterministic and randomized restricted BDD variants like nondeterministic OBDDs or randomized FBDDs.

We like to work with a BDD variant which allows a small-size representation of the functions we are interested in and which supports the following list of operations which are the most important ones in applications and which can be used as modules for more complicated operations: Let G_f and G_g be BDDs of some type representing f and g, resp., let $a \in \{0,1\}^n$, $c \in \{0,1\}$, $x_i \in X_n$, and let $\otimes$ be a binary Boolean operator.

- evaluation: compute $f(a)$.
- synthesis: compute a BDD of the same type for $h = f \otimes g$.
- satisfiability test: decide whether $f(a) = 1$ for some a.
- equivalence test: decide whether $f(a) = g(a)$ for all a.
- replacement by constants: compute a BDD of the same type for $h = f_{|x_i=c}$.
- replacement by functions: compute a BDD of the same type for $h = f_{|x_i=g}$.
- quantification: compute a BDD of the same type for $(\exists x_i)f = f_{|x_i=0} + f_{|x_i=1}$ or $(\forall x_i)f = f_{|x_i=0} \wedge f_{|x_i=1}$.
- minimization: compute a BDD of the same type representing f with minimal size.

These operations are not independent, e.g., we may perform an equivalence test as EXOR-synthesis followed by a satisfiability test but for some variants we have a more efficient equivalence test. Minimization is of particular importance. Often a function is given by a circuit representation. We start with the inputs and construct a BDD representation of the given type representing the same function as the circuit by a sequence of synthesis operations simulating the gates of the circuit. This may lead to exponential-size representations for simple functions if we are not able to control the size of the representations. The best control is to produce minimal-size representations.

BDDs are like circuits not an adequate representation type. It is obvious that the satisfiability test is NP-complete, the equivalence test is coNP-complete, and the minimization problem is NP-hard (and most probably not contained in NP).

Bryant (1986) has presented efficient algorithms for all operations on π-OBDDs, i.e., OBDDs with a fixed variable ordering π. This was the starting point for all the successful applications mentioned already. But already some simple functions need exponential-size OBDDs. This was the motivation to study the algorithmic behavior of more general BDD variants. The algorithmic use of FBDDs has been proved by Gergov and Meinel (1994) and Sieling and Wegener (1995). Bollig, Sauerhoff, Sieling and Wegener (1998) describe algorithmic properties of k-OBDDs and k-IBDDs while heuristic algorithms for k-IBDDs have been investigated by Jain, Bitner, Abadir, and Fussell

(1997). Many problems have no efficient algorithms for k-BPs and $k \geq 2$. It is quite interesting that also nondeterministic OBDDs can be used in applications. EXOR-OBDDs have been investigated by Gergov and Meinel (1996) and Waack (1997). OR-OBDDs have to be restricted further, since negation may cause an exponential blow-up of the size. Narayan, Jain, Fujita, and Sangiovanni-Vincentelli (1996) have presented promising experiments with partitioned OBDDs with fixed window functions and variable orderings and Bollig and Wegener (1997) have proved theoretical results for this BDD variant. A *partitioned* OBDD (PBDD) with k parts, window functions $w_1, \ldots, w_k$ and variable orderings $\pi_1, \ldots, \pi_k$ representing f consists of k OBDDs $G_1, \ldots, G_k$ where G_i represents $f \wedge w_i$ with variable ordering π_i. The window functions have to fulfill the covering property $w_1 + \ldots + w_k = 1$. This model allows the nondeterministic choice between $G_1, \ldots, G_k$. Randomized BDDs are interesting for complexity theoretical reasons only.

In order to compare the different BDD variants it is not sufficient to compare their algorithmic properties. We also have to compare the representational power of the BDD variants. This is done by simulating one representation by another one and by presenting examples where one variant needs exponential size while another one allows a polynomial-size representation. We are interested in upper and lower bound techniques for BDD variants. In a first step, we are satisfied with an exponential lower bound for some explicitly defined function. In a second step, we like to decide which of the "important" functions, e.g., multiplication, can be represented in polynomial size. Communication complexity is the most powerful technique for proving lower bounds.

LOWER BOUNDS ON BDDS BY COMMUNICATION COMPLEXITY

We distinguish *oblivious* BDD variants from non-oblivious ones. An oblivious BDD can be levelled such that nodes on the same level are labelled by the same variable. Hence, we think of nondeterministic or randomized nodes as lying "between the levels". Edges have to lead from low levels to high levels. This informal definition is not really useful, since each BDD can be considered as oblivious one by defining one level per node (in a topological ordering). The crucial point is that we like to control the number of levels by the maximal depth of the BDD model. Hence, OBDDs, k-OBDDs, k-IBDDs, and their nondeterministic and randomized counterparts are oblivious BDDs while FBDDs and k-BPs are non-oblivious BDDs which are discussed in Section 5.

Let G be a π-OBDD representing $f\colon \{0,1\}^n \times \{0,1\}^m \to \{0,1\}$ and let $A \subseteq X_{n+m}$ be the set of the first n variables according to π and $B = X_{n+m} - A$. We consider the communication game where Alice gets the values of the variables in A. The OBDD G leads to the following one-round communication protocol such that Bob knows $f(a,b)$. Alice follows the path activated by a and her message is the number of the last node w reached by this path. Bob uses his input b and follows the path starting at w and activated by b to find the sink reached by the path activated by (a,b). Hence, $C_1(f)$, the one-round communication complexity of f (given the specific partition of the variables and the property that Alice has to send a message) can be estimated by

$$C_1(f) \leq \lceil \log |G| \rceil \text{ or } |G| > 2^{C_1(f)-1}$$

where $|G|$ denotes the size of G. In order to obtain a lower bound on the OBDD size of a function we have to prove a lower bound for all variable orderings. We are free to choose the number of variables given to Alice. Deterministic one-round communication complexity has a simple interpretation using the communication matrix. This matrix is a $2^n \times 2^m$-matrix which contains $f(a,b)$ as entry in a row a and column b. One-round communication complexity is equal to $\lceil \log(\#\text{rows}) \rceil$ where #rows describes the number of different rows of the matrix. This number is equal to the number of different subfunctions obtained by replacing the variables given to Alice by constants.

One may ask whether we obtain also OBDD upper bounds by considering communication protocols. Communication complexity is not designed for such an approach. The computations performed by the players to compute their next message can be arbitrarily difficult. But BDDs are a nonuniform computation model and if we look for size bounds we are not discussing whether it is easy or difficult to construct a BDD of small size. If we know that $s = C_1(f)$ is small for some partition of the variables between Alice and Bob, then we conclude that an OBDD testing the variables given to Alice before the variables given to Bob has at most 2^s nodes in Bob's part which are reached directly by nodes from Alice's part. In order to obtain an upper bound for the size of the whole π-OBDD representing $f\colon \{0,1\}^n \to \{0,1\}$ we need upper bounds s_i, $1 \leq i \leq n$, on the one-round communication complexity if Alice gets the first i variables with respect to π. It is much easier to describe an OBDD directly and no one has proved upper bounds for OBDDs or other BDD variants using communication complexity.

For further lower bound techniques for $f\colon \{0,1\}^n \to \{0,1\}$ we discuss s-oblivious BDDs of bounded length $l = kn$ (where k is a constant or can depend on n). This model is a generalization of k-OBDDs and k-IBDDs. The sequence $s = (s_1, \ldots, s_l)$ describes the labelling of the levels. Let X_n be partitioned to the set $A(n)$ of variables given to Alice and the set $B(n)$ of variables given to Bob. A layer is a maximal block of consecutive levels owned by the same player. We denote by $ld(G)$ the number of layers of G (layer depth) with respect to the given bipartition of X_n. Alice and Bob agree upon the following communication protocol. The owner of the first layer starts the communication and follows the computation path up to the first node v labelled by a variable of the other player. The player communicates v. Then the other player goes on in the same way until the player reaches a sink and communicates its number. Then both players know the value of $f(a,b)$. The communication takes at most $ld(G)$ rounds and the number of exchanged bits is bounded above by $ld(G)\lceil \log |G| \rceil$. We save one round if it is sufficient that one player knows the value of the function. If the communication complexity of f is denoted by $C(f)$,

$$C(f) \leq ld(G) \lceil \log |G| \rceil \quad \text{or}$$

$$|G| > 2^{C(f)/ld(G)-1}.$$

Since $C(f) \leq n$, we have to ensure that $ld(G)$ is not too large. Moreover, if $ld(G)$ is known to be bounded by r, we know that the number of communication rounds is bounded by r and we can apply lower bounds on the length of protocols for communication games which are restricted to r rounds.

For k-OBDDs and a fixed variable ordering π, we look for lower bounds on $2k$-rounds protocols where $A(n)$ contains for some i the first i variables according to π. If π is not fixed, we may choose some i and have to look for a lower bound which holds for all bipartitions of X_n where $|A(n)| = i$.

The situation for k-IBDDs is much more difficult. If $A(n)$ or $B(n)$ is small, then we cannot expect large lower bounds on the communication complexity. If one player communicates all his or her knowledge, then the other one can compute the value of f. Hence, the communication complexity is bounded above by $\min(|A(n)|, |B(n)|) + 1$. If $A(n)$ and $B(n)$ are not small, then $ld(G)$ cannot be bounded by a small upper bound. The solution is to find subsets of not too few variables such that the number of layers with respect to these variables is small.

In the following we argue with the set $I = \{1, \ldots, n\}$ of indices of the variables. Let (A_0, B_0) be a partition of $I_0 = I$ into sets whose size is at least $n_0 = \lfloor n/2 \rfloor$. If $s = (i_1, \ldots, i_{kn})$ is the index sequence of the levels of a k-IBDD for f, we look for "large" sets $A_k \subseteq A_0$ and $B_k \subseteq B_0$ such that the number of layers with respect to (A_k, B_k) is bounded by $2k$. Then we may apply lower bounds from communication complexity for the bipartition (A_k, B_k) of all variables of a subfunction f^* of f which is obtained by assigning well-chosen constants to all variables x_i, $i \notin A_k \cup B_k$. The sets A_k and B_k can be constructed by the following simple combinatorial approach. Let A_i and B_i be given such that $|A_i|$, $|B_i| \geq n_i$. Then we look at the sequence $(j_1, \ldots, j_n)$ belonging to the variable ordering π_{i+1}. Let r be chosen in such a way that $(j_1, \ldots, j_r)$ contains n_i of the indices in $A_i \cup B_i$. If $(j_1, \ldots, j_r)$ contains at least $\lfloor n_i/2 \rfloor$ elements of A_i, then we define $A_{i+1} = A_i \cap \{j_1, \ldots, j_r\}$ and $B_{i+1} = B_i \cap \{j_{r+1}, \ldots, j_n\}$. Otherwise, $B_{i+1} = B_i \cap \{j_1, \ldots, j_r\}$ and $A_{i+1} = A_i \cap \{j_{r+1}, \ldots, j_n\}$. In both cases $|A_{i+1}|$, $|B_{i+1}| \geq \lfloor n_i/2 \rfloor$. Altogether, $|A_k|$, $|B_k| \geq \lfloor n/2^{k+1} \rfloor$. By construction, it is obvious that the number of layers with respect to A_k and B_k is bounded by $2k$. There are at most two layers in each block for a variable ordering π_i. It may even happen that adjacent layers belong to the same player and can be merged.

For s-oblivious BDDs with at most k levels labelled by the same variable, the situation becomes more difficult. We cannot argue on the blocks which are given for k-IBDDs by the division into k variable orderings. Nevertheless, a similar result as shown above can be obtained by the following fundamental lemma due to Alon and Maass (1988) and proved by arguments borrowed from Ramsey theory.

Let $s = (s_1, \ldots, s_l)$ be a sequence of variables from X_n such that no variable appears more than k times. For each bipartition $X_n = A \cup B$ there exist sets

$A' \subseteq A$ and $B' \subseteq B$ such that $|A'| \geq |A|/2^{2k-1}$, $|B'| \geq |B|/2^{2k-1}$, and the number of layers in s with respect to A' and B' is bounded by $2k+1$.

For s-oblivious BDDs of length kn it is not guaranteed that variables occur at most k times in s. But a simple counting argument proves that at least $\lfloor n/2 \rfloor$ variables occur at most $2k$ times in s. Hence, we can apply the above result for the parameter $2k$ and a subset of at least $\lfloor n/2 \rfloor$ variables.

On balance the result of these investigations is that we obtain lower bounds on the size of k-OBDDs, k-IBDDs, and s-oblivious BDDs of length kn for those functions which have large communication complexity even for subfunctions with a support of approximately $n/2^{2k}$ variables. We only have limited control on the support of the subfunctions but we are free to choose the assignment to the other variables.

This approach, although not always explicitly described in this way, has been used by Jukna (1987), Alon and Maass (1988), Krause (1991), Krause and Waack (1991), and Babai, Nisan, and Szegedy (1992). Nisan and Wigderson (1993) have developed methods to separate $(k-1)$-rounds communication complexity from k-rounds communication complexity, i.e., they have proved for an explicitly defined function that there is a protocol of small length using k rounds while protocols with $k-1$ rounds cannot be short. This result has been used by Bollig, Sauerhoff, Sieling, and Wegener (1998) to prove that some functions representable by polynomial-size k-OBDDs cannot be represented by polynomial-size $(k-1)$-IBDDs. This proof needs a lot of "good assignments" to variables to prepare the application of the lower bound technique from communication complexity. In such a process it is helpful to have a reduction concept which preserves the communication complexity. This concept has been used by many authors, for a clear description see Sauerhoff (1999b).

Let $f: \{0,1\}^n \times \{0,1\}^m \to \{0,1\}$ and $g: \{0,1\}^k \times \{0,1\}^l \to \{0,1\}$. A pair (φ_A, φ_B) of functions $\varphi_A: \{0,1\}^n \to \{0,1\}^k$ and $\varphi_B: \{0,1\}^m \to \{0,1\}^l$ is called rectangular reduction from f to g if $f(a,b) = g(\varphi_A(a), \varphi_B(b))$ for all $(a,b) \in \{0,1\}^n \times \{0,1\}^m$.

Alice can compute $\varphi_A(a)$ and Bob can compute $\varphi_B(b)$. Afterwards they can apply communication protocols for g to evaluate f. Hence, for all types of protocols, the communication complexity of f is bounded above by the communication complexity of g. The notation "rectangular" may be explained as follows. Rectangles of the communication matrix for f are mapped by (φ_A, φ_B) to rectangles of the communication matrix for g.

Our considerations can be generalized to the nondeterministic and randomized case. We refer to Krause and Waack (1991), Krause (1992), and Gergov (1994) for exponential lower bounds for nondeterministic OBDDs and nondeterministic oblivious BDDs. Ablayev (1997), Ablayev and Karpinski (1998), and Sauerhoff (1999b) present exponential lower bounds for randomized OBDDs and Sauerhoff (1999a) has obtained exponential lower bounds for randomized k-OBDDs. All these bounds use communication complexity.

LOWER BOUNDS ON BDDS AND GENERALIZED COMMUNICATION GAMES

In this section we investigate FBDDs and the more general model of k-BPs. There is no obvious way to relate these BDD variants to communication complexity.

The first exponential lower bounds on the size of FBDDs have been proved in 1984 by Žák (1984) and Wegener (1988). Simon and Szegedy (1993) present a general lower bound technique for FBDDs which covers several of the results published before by many authors. Here we discuss another lower bound technique which is influenced by the algorithmic point of view due to Sieling and Wegener (1995). This technique covers indeed all known lower bounds.

The main idea is to generalize the notion of a variable ordering to *graph orderings*. An FBDD is called complete if each variable is tested on each path form the source to a sink. Complete FBDDs G^* describe a graph ordering π_{G^*}. For the input a we obtain the variable ordering $\pi_{G^*}(a)$ which is the ordering of the variables on the path activated by a. Even a polynomial-size FBDD may represent exponentially many variable orderings π_{G^*} describing the variable orderings for all inputs by a complete FBDD G^*. The FBDD G^* also represents a function but this is of no importance. This may be stressed by merging *all* sinks of G^* to a meaningless sink.

For each variable ordering π and each cut index i, we have obtained communication games where Alice gets the first i variables according to π and Bob gets the remaining $n-i$ variables. Now we fix a graph ordering π_{G^*} and a cut line l partitioning the vertex set to an upper part V_A and a lower part V_B, i.e., no edge leads from V_B to V_A. Let $a \in \{0,1\}^n$ be an input. Let $A(a)$ be the set of indices i such that the node on the path $p(a)$ in G^* activated by a and labelled by x_i belongs to V_A. If the input equals a, Alice gets the value of the variables x_i, $i \in A(a)$, and Bob gets the other variables. Again they have the task to evaluate f.

Each FBDD G representing f and respecting the graph ordering π_{G^*} (i.e., the labels on the path activated by a are in the same order as prescribed by π_{G^*} where the FBDD may omit the test of some variables) leads to the following protocol for the generalized one-round communication game. Alice follows the path activated by her partial input and her message contains the number of the first node whose input bit is not known to her. Then Bob follows the rest of the activated path and computes the output. If $C_{1,G^*}(f)$ is the generalized one-round communication complexity of f,

$$C_{1,G^*}(f) \leq \lceil \log |G| \rceil$$

similarly to the special situation of OBDDs and variable orderings. In order to prove lower bounds on the FBDD size of f, we have to consider all graph orderings but for each graph ordering we may choose an appropriate cut line. Then we obtain the following lower bound.

For the input $c \in \{0,1\}^n$ let f_c^* be the subfunction of f obtained by replacing the variables x_i, $i \in A(c)$, by c_i. The generalized one-round complexity of f

in the described situation is equal to $\lceil \log(\#\text{sub}) \rceil$ where #sub describes the number of different subfunctions f_c^*, $c \in \{0,1\}^n$. This leads to lower bounds on the size of FBDDs respecting π_{G^*} and representing f.

Each path p starting at the source and stopping if the first node of V_B is reached describes a partial assignment considered in the lower bound technique. Let f_p be the corresponding subfunction and P be the set of all considered paths p. Many lower bounds are obtained by proving that at most m of the subfunctions f_p are equal. This leads to the lower bound $|P|/m$. This bound is bad if the equivalence classes corresponding to equal subfunctions are of quite different size. Then we can do better by assigning weights $w_p \geq 0$, $p \in P$, to the paths such that the sum of all weights equals 1. If the weight of each equivalence class is bounded above by ε, we need at least $\lceil \varepsilon^{-1} \rceil$ nodes to represent the equivalence classes.

This method covers the lower bound techniques for deterministic FBDDs but it is not possible to obtain lower bounds for nondeterministic or randomized FBDDs. The notion of a graph ordering is no longer useful, since it prescribes which variable has to be tested first. If we start with nondeterministic or randomized nodes, we are allowed to test different variables as first variables on different paths. There are some exponential lower bounds on the nondeterministic FBDD size of explicitly defined functions (e.g., Krause(1988)) but the most general technique which even can be generalized to nondeterministic k-BPs is due to Borodin, Razborov, and Smolensky (1993).

Let G be a nondeterministic FBDD representing f and let $e = (v, w)$ be an edge of G. Let g_e take the value 1 on a iff a activates a path from the source to w via e and let h_e take the value 1 on a iff a activates a path from w to a 1-sink. It follows from the properties of FBDDs that h_e cannot essentially depend on x_i if g_e essentially depends on x_i. It follows from the construction that f is the disjunction of all $g_e \wedge h_e$. We may restrict this disjunction to each subset of all edges such that each path from the source to a 1-sink runs through at least one edge of the chosen subset. It is easily possible to define such a subset of edges where g_e as well as h_e essentially depends on at most $\lceil n/2 \rceil$ variables. We have proved the following statement.

If f can be represented by a nondeterministic FBDD of size s, f is the disjunction of less than $2s$ functions $g_e \wedge h_e$ where the functions g_e and h_e essentially depend on disjoint sets of variables each of size at most $\lceil n/2 \rceil$. (Remember that BDDs of size s have less than $2s$ edges.)

We may interpret this in the following way as nondeterministic communication game. We have three players Alice, Bob, and Carol. The protocol consists of a number t and t partitions (A_i, B_i), $1 \leq i \leq t$, of the set of variables into sets of size at most $\lceil n/2 \rceil$. Carol chooses nondeterministically a number $i \in \{1, \ldots, t\}$ and sends this message to Alice and Bob which implies that Alice sees the part of the input belonging to A_i and Bob sees the part belonging to B_i. They are not allowed to communicate and may output a Boolean value. The input is accepted iff it is accepted by Alice *and* Bob. The communication complexity of this protocol equals $\lceil \log t \rceil$. If f can be represented by a nondeterministic FBDD of size s, the communication game can be solved with a nondeterministic protocol whose length is bounded by $\lceil \log(2s) \rceil$.

Borodin, Razborov, and Smolensky (1993) have introduced the notion (k, a)-rectangle for functions g which can be represented as conjunctions of ka functions g_i each essentially depending on at most $\lceil n/a \rceil$ variables and the additional property that for each variable x_j there are at most k functions g_i essentially depending on x_j. It is easy to see that the functions $g_e \wedge h_e$ considered above are $(1, 2)$-rectangles. Moreover, it is not too difficult to prove that functions representable by nondeterministic k-BPs of size s can be represented as disjunction of $(2s)^{ka-1}$ (k, a)-rectangles. This leads to a communication game where Carol nondeterministically chooses between $(2s)^{ka-1}$ possibilities (message length bounded by $(ka-1)\lceil \log(2s) \rceil$) and every of the other ka players gets access to at most $\lceil n/a \rceil$ variables in such a way that no variable is seen by more than k players. The input is accepted if all other players accept without further communication. This technique has been applied by Borodin, Razborov, and Smolensky (1993) and Jukna (1995). Yao (1983) has presented a technique to obtain lower bounds for randomized algorithms, here k-BPs, by proving lower bounds for deterministic algorithms and random inputs. This technique has been combined by Sauerhoff (1998) with the above technique to obtain lower bounds for randomized k-BPs. Thathachar (1998) has even proved that some explicitly defined functions need exponential size nondeterministic $(k-1)$BPs but can be represented in polynomial size by deterministic k-IBDDs. He also has obtained similar results for the randomized case.

Conclusion

It is shown that even the lower bound techniques for BDD variants which have not been formulated in the framework of communication complexity can be interpreted as methods from communication complexity. This underlines the key role of communication complexity in BDD lower bound techniques.

References

[1] F. Ablayev, "Randomization and nondeterminism are incomparable for ordered read-once branching programs", (The printed title has the misprint "comparable".) *ICALP '97*, LNCS 1256, 1997, 195–202.

[2] F. Ablayev and M. Karpinski, "A lower bound for integer multiplication on randomized ordered read-once branching programs", *ECCC Rep.*, 1998, 98–011.

[3] N. Alon and W. Maass, "Meanders and their applications in lower bound arguments", *Journal of Computer and System Sciences*, 37, 1988, 118–129.

[4] L. Babai, N. Nisan and M. Szegedy, "Multiparty protocols, pseudorandom generators for logspace, and time-space trade-offs", *Journal of Computer and System Sciences*, 45, 1992, 204–232.

[5] B. Bollig, M. Sauerhoff, D. Sieling, and I. Wegener, "Hierarchy theorems for kOBDDs and kIBDDs", *Theoretical Computer Science*, 205, 1992, 45–60.

[6] B. Bollig and I. Wegener, "Complexity theoretical results on partitioned (nondeterministic) binary decision diagrams", *MFCS '97*, LNCS 1295, 1997, 159–168.

[7] A. Borodin, A. Razborov and R. Smolensky, "On lower bounds for read-k-times branching programs", *Computational Complexity*, 3, 1993, 1–18.

[8] R. E. Bryant, "Graph-based algorithms for Boolean function manipulation", *IEEE Trans. on Computers*, 35, 1986, 677–691.

[9] R. E. Bryant, "Symbolic Boolean manipulation with ordered binary decision diagrams", *ACM Computing Surveys*, 24, 1992, 293–318.

[10] E. M. Clarke and J. M. Wing, "Formal methods : State of the art and future directions", *ACM Computing Surveys*, 28, 1996, 626–643.

[11] J. Gergov, "Time-space tradeoffs for integer multiplication on various types of input oblivious sequential machines", *Information Processing Letters*, 51, 1994, 265–269.

[12] J. Gergov and C. Meinel, "Efficient Boolean manipulation with OBDD's can be extended to FBDD's", *IEEE Trans. on Computers*, 43, 1994, 1197–1209.

[13] J. Gergov and C. Meinel, "MOD-2-OBDDs – a data structure that generalizes EXOR-sum-of-products and ordered binary decision diagrams", *Formal Methods in System Design*, 8, 1996, 273–282.

[14] J. Hromkovič, "Communication Complexity and Parallel Computing", 1997, Springer.

[15] J. Jain, J. Bitner, M. Abadir, J. A. Abraham and D. S. Fussell, "Indexed BDDs: Algorithmic advances in techniques to represent and verify Boolean functions", *IEEE Trans. on Computers*, 46, 1997, 1230–1245.

[16] S. Jukna, "Lower bounds on communication complexity", *Math. Logic and Its Applications*, 5, 1987, 22–30.

[17] S. Jukna, "A note on read-k-times branching programs", *RAIRO-Theoretical Informatics and Applications*, 29, 1995, 75–83.

[18] M. Krause, "Exponential lower bounds on the complexity of local and real-time branching programs", *Journal of Information Processing and Cybernetics* (EIK) 24, 1988, 99–110.

[19] M. Krause, "Lower bounds for depth-restricted branching programs", *Information and Computation* 91, 1991, 1–14.

[20] M. Krause, "Separating $\oplus$L from L, NL, co-NL and AL(=P) for oblivious Turing machines of linear access time", *RAIRO Theoretical Informatics and Applications*, 26, 1992, 507–522.

[21] M. Krause and S. Waack, "On oblivious branching programs of linear length", *Information and Computation* 94, 1991, 232–249.

[22] E. Kushilevitz and N. Nisan, "Communication Complexity", Cambridge University Press, 1997.

[23] A. Narayan, J. Jain, M. Fujita and A. Sangiovanni-Vincentelli, "Partitioned ROBDDs – a compact, canonical and efficiently manipulable representation for Boolean functions", *ICCAD'96*, 1996, 547–554.

[24] N. Nisan and A. Wigderson, "Rounds in communication complexity revisited", *SIAM Journal on Computing*, 22, 1993, 211–219.

[25] M. Sauerhoff, "Lower bounds for randomized read-k-times branching programs", *STACS'98*, LNCS 1373, 1998, 105–115.

[26] M. Sauerhoff, *Complexity theoretical results for randomized branching programs*, Ph. D. Thesis, 1999.

[27] M. Sauerhoff, "On the size of randomized OBDDs and read-once branching programs for k-stable functions", *STACS'99*, LNCS 1563, 1998, 488–499.

[28] D. Sieling and I. Wegener, "Graph driven BDDs – a new data structure for Boolean functions", *Theoretical Computer Science*, 141, 1995, 283–310.

[29] J. Simon and M. Szegedy, "A new lower bound theorem for read-only-once branching programs and its applications", *DIMACS Series in Discrete Mathematics and Theoretical Computer Science*, 13, 1993, 183–193.

[30] J. S. Thathachar, "On separating the read-k-times branching program hierarchy", *STOC'98*, 1998, 653–662.

[31] S. Waack, "On the descriptive and algorithmic power of parity ordered binary decision diagrams", *STACS'97*, LNCS 1200, 1997, 201–212.

[32] I. Wegener, *The Complexity of Boolean Functions,* Wiley-Teubner.

[33] I. Wegener, "On the complexity of branching programs and decision trees for clique functions", *Journal of the ACM*, 35, 1988, 461–471.

[34] I. Wegener, "Branching Programs and Binary Decision Diagrams – Theory and Applications", To appear: *SIAM-Monographs in Discrete Mathematics and Applications*, 1999.

[35] A. C. Yao, "Some complexity questions related to distributed computing", *11. STOC*, 1979, 209–213.

[36] A. C. Yao, "Lower bounds by probabilistic arguments", *24. FOCS*, 1983, 420–428.

[37] S. Žák, "An exponential lower bound for one-time-only branching programs", *MFCS'84*, LNCS 176, 1984, 562–566.

REMINISCENCES ABOUT PROFESSOR AHLSWEDE AND A LAST WORD BY THOMAS MANN

Reminiscenses About Professor Ahlswede

Mike Ulrey

The things I remember best about Professor Ahlswede seem to have a common theme: *intense concentration.* My most vivid image of him is with a cigarette dangling from his lips as he works on a problem. Such a scene could be in his office, in a hallway of the math building, or in a restaurant. In those days (late 60's and early 70's), one could still smoke freely in most places. As I recall, he smoked some particularly strong English brand of cigarettes. In any case, he concentrated so strongly on the work at hand, that he hardly took ever time to flick his ashes into an ashtray. Instead the cigarette would continue to burn, the ash growing to an impossible length, the smoke curling upwards, his eyes squinting and watering in self-defense. I was amazed at the length of time he could withstand this self-imposed torture, seemingly without being aware of it. I would secretly make wagers with myself about whether or not he would remove the cigarette before the ashes fell. If the ashes won, they might make a few burn marks in the paper on which he wrote. I wonder if the patterns ever gave him any ideas.

Professor Ahlswede once told me that a good mathematician had to try out at least 100 ideas on the problem at hand. Since I had trouble coming up with 2 or 3, I guess that explains why I'm not in the pure mathematics game any more. I have another vivid image which illustrates this concept. Once we had lunch at the restaurant near the Ohio State University campus. Since it was a warm spring or summer day, we sat at the table on the sidewalk outside the restaurant. As usual, Professor Ahlswede was working on some problem, and since neither of us had any paper, he began to write on the (paper) napkins. After exhausting the supply of napkins, he had to ask the waiter for more.

I. Althöfer et al. (eds.), Numbers, Information and Complexity, 629-632.

When we left, the table and surrounding area were strewn with dozens of math-covered napkins. In my memory, each of the napkins represents an idea, and the collection of them a sort of physical manifestation of the multitude of ideas Professor Ahlswede brought to a problem. I remember noticing the puzzled looks of the waiter and passersby at the curious hieroglyphics on the napkins, and wondered if they thought we were aliens from outer space.

In those days, Professor Ahlswede owned a Chevrolet Camaro. I once asked him why he didn't have a Porsche, and he said that Americans liked them largely because they were foreign and exotic, and besides, a humble Camaro offered a lot of performance for the money. This objectivity took me by surprise and really impressed me, by the way. One day, I was riding with him on the freeways in the Columbus area, probably going about 135-145 km/hr. Not so fast for the autobahn, perhaps, but over the speed limit in Ohio in those days. Not that I cared, mind you – I like fast cars, both Camaros AND Porsches. Anyway, as usual, the flow of ideas could not be stopped, and pretty soon, Professor Ahlswede was using his finger as imaginary chalk, "writing" imaginary mathematics on the inside of the windshield, using it as an imaginary blackboard. Now, as I said before, I like driving fast, but only if one is concentrating on one's driving. It's one thing to drive 135 km/hr while focussing one's eyes 800 meters ahead, quite another if your eyes are focussed 1 meter ahead! I certainly hoped that he returned his attention to the road more often than he did to his cigarette!

Once Professor Ahlswede came to dinner at my parents' house, about 85 km north of Columbus. This time I drove, and we were on the two-lane road leading to my parents' house. This was in a rural area, with small farms and houses spread about a kilometer apart. Professor Ahlswede commented about how the road did not cut through the countryside, but rather was laid down like a ribbon, conforming to every hill and vale, twisting and turning to avoid natural obstacles. It has always been one of my favorrite roads to drive, and I realized that Professor Ahlswede's observation helped explain what made it so enjoyable.

Recently I looked at Professor Ahlswede's Website for the list of papers that he has authored or co-authored. At that time, there were 124 papers (There may already be more now!). Impressive as that is, what impresses me even more is the variety of subjects represented there. Of course, you are all well aware of this, since the mathematical interests of the people at this conference cover such a wide area. Reading through the list of these papers in chronological order reminded me of that road to my parents' house. The trail of papers follows a portion of the mathematical landscape gently turning this way and that as new ideas suggested connections with other areas, driven by a desire to solve unsolved problems, yet always cognizant of what was happening in the world of communication and information transfer.

At the time that Professor Ahlswede visited us, my father had a small airplane (Cessna 172), and he took Professor Ahlswede and me for a ride. Professor Ahlswede sat in the co-pilot (right-hand) seat, which of course has controls which mirror the pilot controls. Let me tell you, he was not shy about putting levers and pressing buttons, much to my father's surprise! Hey, if we got into a sudden drive, my father can pull us out, right? Although this experience was a little disconcerting at the time, the intervening years (and distance from danger) have allowed me to view it as an example of Professor Ahlswede's curiosity and his willingness to experiment with many possiblities to satisfy that curiosity.

In thinking back on these stories, I see another common theme – most of them seem to involve a journey of one sort or another. As evidenced by the crowd gathered here, his life has been a remarkable journey that has touched many lives. I am honored and happy to be part of this celebration.

Thomas Mann: Die schwere Stunde

... Nicht grübeln! Er war zu tief, um grübeln zu dürfen! Nicht ins Chaos hinabsteigen, sich wenigstens nicht dort aufhalten! Sondern aus dem Chaos, welches die Fülle ist, ans Licht emporheben, was fähig und reif ist, Form zu gewinnen. Nicht grübeln: Arbeiten! Begrenzen, ausschalten, gestalten, fertig werden ...

Und es wurde fertig, das Leidenswerk. Es wurde vielleicht nicht gut, aber es wurde fertig. Und als es fertig war, siehe, da war es auch gut. Und aus seiner Seele, aus Musik und Idee, rangen sich neue Werke hervor, klingende und schimmernde Gebilde, die in heiliger Form die unendliche Heimat wunderbar ahnen ließen, wie in der Muschel das Meer saust, dem sie entfischt ist.

LIST OF INVITED LECTURES HELD AT THE SYMPOSIUM "NUMBERS, INFORMATION AND COMPLEXITY" IN BIELEFELD, OCTOBER 8-11, 1998

Birthday Colloquium

A. Sárközy, On divisibility properties of sequences of integers

L. Khachatrian, Correlation inequalities and diametric theorems

G. Dueck, One decade next door to R. Ahlswede

J. Massey, Something new and something blue

45-Minutes Lectures

I. Althöfer, On the design of algorithms for decision support systems

A. Blokhuis, Finite geometry and extremal combinatorics

I. Csiszár, Common randomness capacity

G. Freiman, Structure theory of set addition. Results and problems

R. Freivalds, Quantum computers and quantum automata

G. Frey, Applications of arithmetic geometry to public-key cryptosystems

Z. Füredi, Lotto, footballpool and other covering radius problems

G. Grimmett, Stochastic inequalities and their applications to percolation and disordered systems

W. Haemers, Disconnected vertex sets; a variation on the Lovasz bound for the Shannon capacity of a graph

G. O. H. Katona, The cycle method and its limits

J.-H. Kim, The Ramsey number R(3,t) has asymptotic order of magnitude $t^2 / \log(t)$

J. Körner, Capacity and dimension

A. V. Kostochka, Extremal problems on Δ-systems (a survey)

W. Krieger, Subshifts and topological Markov chains

J. Nesetril, Density vrs colorings

M. Pinkser, Information theoretic methods in filtering

Y. Shtarkov, Some redundancy bounds for sequential estimation of an unknown source model

V. T. Sós, On some extremal set-system problems and information theory

A. Tietäväinen, A method to estimate partial-period correlations

B. Tsybakov, Communications network with self-similar traffic

E. van der Meulen, Some significant results in the theory of multi-user information transfer

J. H. van Lint, The mathematics of the Compact Disc

C. von der Malsburg, The vision code – visual perception from a coding theory point of view

Z.-X. Wan, Some applications of the Anzahl theorems in geometry of classical groups;

I. Wegener, Representations of Boolean functions - complexity, algorithms and applications

J. Ziv, Entropy has many faces

30-Minutes Lectures

H. Aydinian, On d-perfect codes

V. Balakirski, On the structure of a common key constructed by correlated observations and transmission over helping channels

A. Barg, A new upper bound for codes decodable in the list of size 2

C. Bey, Old and new results for the weighted t–intersection problem via AK–methods II

V, Bentkus, Lattice points and the CLT

S. Bezrukov, Some new directions in isoperimetric problems

Y. Bilu, The diophantine equation f(x)=g(y)

V. Blinovskii, Combinatorial approach to process level large deviation problem

M. Burnashev (with T. S. Han and S. I. Amari), On some statistical problems with information constraints

B. Carl, Entropy inequalities and diverse applications

S. Dodunekov (with J. Simonis), Constructions of optimal linear codes

A. Dyachkov (with A. Macula and V. Rykov), New appications and results of superimposed code theory arising from the potentialities of molecular biology

K. Engel, Old and new results for the weighted t–intersection problem via AK–methods I

T. Ericson, Spherical codes

M. Feder, Universal Prediction of Binary Sequences Using Finite Memory

H. Ferreira, On insertion/deletion correction

G. Freiman (with S. Litsyn), Asymptotically exact bounds on the size of high order spectral null codes

L. Gargano, Graph coloring problems arising in optical routing

H.-D. Gronau, On the subword poset

L. Györfi, A large subcode of a Reed-Solomon code is good for asynchronous frequency-hopping

E. Harzheim, On weakly arithmetic progressions

H. Hollmann, A proof of the Welch and Niho conjectures on crosscorrelations of binary m-sequences.

R. Holzman (with R. Aharoni, M. Krivelevich, R. Meshulam), The plank problem from the viewpoint of hypergraph matchings and covers

R. Johannesson, Rudified convolutional encoders

G. Khachatrian, A survey of coding methods for the adder channel

K. Kobayashi, On fix-free codes

J. Koplowitz, Learning with a finite memory

U. Krengel, On the maximal operator for the class of martingales adapted to a given filtration

S. Kurtz, Reducing the space requirement of suffix trees

D. Lazic, Error probability and distance distribution of channel codes

U. Leck, Some new results on Macaulay posets

H. Lefmann, Large approximating independent sets in graphs and hypergraphs

Z. Lonc, Partitions, packings and coverings with antichains and chains

K. Metsch, Covers and partial spreads of finite projective spaces

P. Narayan, Common randomness and secret key capacity

M. Nathanson, Nonabelian additive number theory

W. Paul, Top down design of processors

V. Pless, On self-dual and formally self-dual codes

H.-J. Prömel, Extremal graphs, asymptotic enumeration, and global structure

R. Reischuk, Average case analysis of algorithmic learning

M. Ruszinkó, Intersecting systems

R.-H. Schulz, On check digit systems using antisymmetric mappings

P. Shields, LZ compressible and incompressible sequences

G. Simonyi (with A. Sali), Self-complementary orientations

F. Solove'eva, Switchings and perfect codes

L. Staiger, How much can you win when your adversary is handicapped

T. Tjalkens (with F. Willems), Turnstall codes and arithmetic codes: a geometrical approach

P. Vanroose, Ordering in sequence spaces, an overview

H. Vinck, Code division multiple access for optical communications

F. Willems, Random access data compaction (with T. Tjalkens and P. Volf)

G. Ziegler, Coloring of Hamming graphs, codes, and the 0/1-Borsuk problem

H. Ziezold, Some aspects of shape analysis

The symposium was organized by the Sonderforschungsbereich "Diskrete Strukturen in der Mathematik", University of Bielefeld. Abstracts of the lectures are available in the report

B. Balkenhol and U. Tamm (eds.), *Symposium "Numbers, Information and Complexity" in honour of R. Ahlswede*, Preprint 98-010 Ergänzungsreihe, Sonderforschungsbereich 343 "Diskrete Strukturen in der Mathematik", Bielefeld, Germany, 1998.

BIBLIOGRAPHY OF PUBLICATIONS BY RUDOLF AHLSWEDE

1967

[1] Certain results in coding theory for compound channels, Proc. Colloquium Inf. Th. Debrecen (Hungary), 35–60.

1968

[2] Beiträge zur Shannonschen Informationstheorie im Fall nichtstationärer Kanäle, Z. Wahrscheinlichkeitstheorie und verw. Geb. 10, 1–42.

[3] The weak capacity of averaged channels, Z. Wahrscheinlichkeitstheorie und verw. Geb. 11, 61–73.

1969

[4] Correlated decoding for channels with arbitrarily varying channel probability functions, (with J. Wolfowitz), Information and Control 14, 457–473.

[5] The structure of capacity functions for compound channels, (with J. Wolfowitz), Proc. of the Internat. Symposium on Probability and Information Theory at McMaster University, Canada, April 1968, 12–54.

1970

[6] The capacity of a channel with arbitrarily varying channel probability functions and binary output alphabet, (with J. Wolfowitz), Z. Wahrscheinlichkeitstheorie und verw. Geb. 15, 186–194.

[7] A note on the existence of the weak capacity for channels with arbitrarily varying channel probability functions and its relation to Shannon's zero error capacity, Ann. Math. Stat., Vol. 41, No. 3, 1027–1033.

1971

[8] Channels without synchronization, (with J. Wolfowitz), Advances in Applied Probability, Vol. 3, 383–403.

[9] Group codes do not achieve Shannon's channel capacity for general discrete channels, Ann. Math. Stat., Vol. 42, No. 1, 224–240.

[10] Bounds on algebraic code capacities for noisy channels I, (with J. Gemma), Information and Control, Vol. 19, No. 2, 124–145.

[11] Bounds on algebraic code capacities for noisy channels II, (with J. Gemma), Information and Control, Vol. 19, No. 2, 146–158.

1973

[12] Multi–way communication channels, Proceedings of 2nd International Symposium on Information Theory, Thakadsor, Armenian SSR, Sept. 1971, Akademiai Kiado, Budapest, 23–52.

[13] On two–way communication channels and a problem by Zarankiewicz, Sixth Prague Conf. on Inf. Th., Stat. Dec. Fct's and Rand. Proc., Sept. 1971, Publ. House Chechosl. Academy of Sc., 23–37.

[14] A constructive proof of the coding theorem for discrete memoryless channels in case of complete feedback, Sixth Prague Conf. on Inf. Th., Stat. Dec. Fct's and Rand. Proc., Sept. 1971, Publ. House Czechosl. Academy of Sc., 1–22.

[15] The capacity of a channel with arbitrarily varying additive Gaussian channel probability functions, Sixth Prague Conf. on Inf. Th., Stat. Dec. Fct's and Rand. Proc., Sept. 1971, Publ. House Czechosl. Academy of Sc., 39–50.

[16] Channels with arbitrarily varying channel probability functions in the presence of noiseless feedback, Z. Wahrscheinlichkeitstheorie und verw. Geb. 25, 239–252.

[17] Channel capacities for list codes, J. Appl. Probability, lo, 824–836.

1974

[18] The capacity region of a channel with two senders and two receivers, Ann. Probability, Vol. 2, No. 5, 805–814.

[19] On common information and related characteristics of correlated information sources, (with J. Körner), presented at the 7th Prague Conf. on Inf. Th., Stat. Dec. Fct's and Rand. Proc., included in "Information Theory" by I. Csiszár and J. Körner, Acad. Press, 1981.

1975

[20] Approximation of continuous functions in p–adic analysis, (with R. Bojanic), J. Approximation Theory, Vol. 15, No. 3, 190–205.

[21] Source coding with side information and a converse for degraded broadcast channels, (with J. Körner), IEEE Trans. Inf. Th., Vol. 21, 629–637.

[22] Two contributions to information theory, (with P. Gács), Colloquia Mathematica Societatis János Bolyai, 16. Topics in Information Theory, I. Csiszár and P. Elias Edit., Keszthely, Hungaria, 1975, 17–40.

1976

[23] Bounds on conditional probabilities with applications in multiuser communication, (with P. Gács and J. Körner), Z. Wahrscheinlichkeitstheorie und verw. Geb. 34, 157–177.

[24] Every bad code has a good subcode: a local converse to the coding theorem, (with G. Dueck), Z. Wahrscheinlichkeitstheorie und verw. Geb. 34, 179–182.

[25] Spreading of sets in product spaces and hypercontraction of the Markov operator, (with P. Gács), Ann. Prob., Vol. 4, No. 6, 925–939.

1977

[26] On the connection between the entropies of input and output distributions of discrete memoryless channels, (with J. Körner), Proceedings of the 5th Conference on Probability Theory, Brasov 1974, Editura Academeiei Rep. Soc. Romania, Bucaresti 1977, 13–23.

[27] Contributions to the geometry of Hamming spaces, (with G. Katona), Discrete Mathematics 17, 1–22.

[28] The number of values of combinatorial functions, (with D.E. Daykin), Bull. London Math. Soc., 11, 49–51.

1978

[29] Elimination of correlation in random codes for arbitrarily varying channels, Z. Wahrscheinlichkeitstheorie und verw. Geb. 44, 159–175.

[30] An inequality for the weights of two families of sets, their unions and intersections, (with D.E. Daykin), Z. Wahrscheinlichkeitstheorie und verw. Geb. 43, 183–185.

[31] Graphs with maximal number of adjacent pairs of edges, (with G. Katona), Acta Math. Acad. Sc. Hung. 32, 97–120.

1979

[32] Suchprobleme, (with I. Wegener), Teubner Verlag, Stuttgart, Russian Edition with Appendix by Maljutov 1981 (Book).

[33] Inequalities for a pair of maps $S \times S \to S$ with S a finite set, (with D.E. Daykin), Math. Zeitschrift 165, 267–289.

[34] Integral inequalities for increasing functions, (with D.E. Daykin), Math. Proc. Comb. Phil. Soc., 86, 391–394.

[35] Coloring hypergraphs: A new approach to multi–user source coding I, J. Combinatorics, Information and System Sciences, Vol. 4, No. 1, 76–115.

1980

[36] Coloring hypergraphs: A new approach to multi–user source coding II, J. Combinatorics, Information and System Sciences, Vol. 5, No. 3, 220–268.

[37] Simple hypergraphs with maximal number of adjacent pairs of edges, J. Comb. Theory, Ser. B, Vol. 28, No. 2, 164–167.

[38] A method of coding and its application to arbitrarily varying channels, J. Combinatorics, Information and System Sciences, Vol. 5, No. 1, 10–35.

1981

[39] To get a bit of information may be as hard as to get full information, (with I. Csiszár), IEEE Trans. Inf. Theory, Vol. 27, 398–408.

[40] Solution of Burnashev's problem and a sharpening of Erdös–Ko–Rado, Siam Review, to appear in a book by G. Katona.

1982

[41] Remarks on Shannon's secrecy systems, Probl. of Control and Inf. Theory, Vol. 11, No. 4, 301–318.

[42] Bad Codes are good ciphers, (with G. Dueck), Probl. of Control and Inf. Theory, Vol. 11, No. 5, 337–351.

[43] Good codes can be produced by a few permutations, (with G. Dueck), IEEE Trans. Inf. Theory, IT–28, No. 3, 430–443.

[44] An elementary proof of the strong converse theorem for the multiple–access channel, J. Combinatorics, Information and System Sciences, Vol. 7, No. 3, 216–230.

1983

[45] Note on an extremal problem arising for unreliable networks in parallel computing, (with K.U. Koschnick), Discrete Mathematics 47, 137–152.

[46] On source coding with side information via a multiple–access channel and related problems in multi–user information theory, (with T.S. Han), IEEE Trans. Inf. Theory, Vol. IT–29, No. 3, 396–412.

[47] A two family extremal problem in Hamming space, (with A. El Gamal and K.F. Pang), Discrete mathematics 49, 1–5.

[48] Improvements of Winograd's Result on Computation in the Presence of Noise, IEEE Trans. Inf. Theory, Vol. IT–29, Nov., 11–21.

1985

[49] The rate–distortion region for multiple descriptions without excess rate, IEEE Trans. Inf. Theory, Vol. IT–31, No. 6, 721–726.

1986

[50] Hypothesis testing under communication constraints, (with I. Csiszár), IEEE Trans. Inf. Theory, Vol. IT–32, No. 4, 533–543.

[51] On multiple description and team guessing, IEEE Trans. Inf. Theory, Vol. IT–32, No. 4, 543–549.

[52] Arbitrarily varying channels with states sequence known to the sender, invited paper at a Statistical Research Conference dedicated to the memory of Jack Kiefer and Jacob Wolfowitz, held at Cornell University, July 1983, IEEE Trans. Inf. Theory, Vol. IT–32, No. 5, 621–629.

1987

[53] Optimal coding strategies for certain permuting channels, (with A. Kaspi), IEEE Trans. Inf. Theory, Vol. IT–33, No. 3, 310–314.

[54] Search Problems, (with I. Wegener), English Edition of [32] with Supplement of recent Literature, Wiley–Interscience Series in Discrete Mathematics and Optimization, R.L. Graham, J.K. Leenstra, R.E. Tarjan, edit.

[55] Inequalities for code pairs, (with M. Moers), European J. of Combinatorics 9, 175–181.

[56] Eight problems in information theory
— a complexity problem
— codes as orbits
Contributions to "Open Problems in Communication and Computation", T.M. Cover and B. Gopinath, Editors, Springer Verlag.

[57] On code pairs with specified Hamming distances, Colloquia Mathematica Societatis János Bolyai 52, Combinatorics, Eger (Hungary), 9–47.

1989

[58] Identification via channels, (with G. Dueck), IEEE Trans. Inf. Theory, Vol. 35, No. 1, 15–29.

[59] Identification in the presence of feedback – a discovery of new capacity formulas, (with G. Dueck), IEEE Trans. Inf. Theory, 35, No. 1, 30-39.

[60] Contributions to a theory of ordering for sequence spaces, (with Z. Zhang), Problems of Control and Information Theory, Vol. 18, No. 4, 197–221.

1990

[61] A general 4–words inequality with consequences for 2–way communication complexity, (with N. Cai and Z. Zhang), Advances in Applied Mathematics, Vol. 10, 75–94.

[62] Coding for write–efficient memory, (with Z. Zhang), Information and Computation, Vol. 83, No. 1, 80–97.

[63] Creating order in sequence spaces with simple machines, (with Jian–ping Ye and Z. Zhang), Information and Computation, Vol. 89, No. 1, 47–94.

[64] An identity in combinatorial extremal theory, (with Z. Zhang), Adv. in Math., Vol. 80, No. 2, 137–151.

[65] On minimax estimation in the presence of side information about remote data, (with M.V. Burnashev), Ann. of Stat., Vol. 18, No. 1, 141–171.

[66] Extremal properties of rate–distortion functions, IEEE Trans. Inf. Theory, Vol. 36, No. 1, 166–171.

[67] A recursive bound for the number of complete K–subgraphs of a graph, (with N. Cai and Z. Zhang), "Topics in graph theory and combinatorics" in honour of G. Ringel on the occasion of his 70th birthday, R. Bodendiek, R. Henn (Eds), 37–39.

[68] On cloud–antichains and related configurations, (with Z. Zhang), Discrete Mathematics 85, 225–245.

1991

[69] Reusable memories in the light of the old AV– and new OV–channel theory, (with G. Simonyi), IEEE Trans. Inf. Theory, Vol. 37, No. 4, 1143–1150.

[70] On identification via multi–way channels with feedback, (with B. Verboven), IEEE Trans. Inf. Theory, Vol. 37, No. 5, 1519–1526.

[71] Two proofs of Pinsker's conjecture concerning AV channels, (with N. Cai), IEEE Trans. Inf. Theory, Vol. 37, No. 6, 1647 1649.

1992

[72] Diametric theorems in sequence spaces, (with N. Cai and Z. Zhang), Combinatorica, Vol. 12, No. 1, 1–17.

[73] On set coverings in Cartesian product spaces, *Ergänzungsreihe* SFB 343 "Diskrete Strukturen in der Mathematik", Universität Bielefeld, Nr. 92–005.

[74] Rich colorings with local constraints, (with N. Cai and Z. Zhang), Preprint 89–011, SFB 343 "Diskrete Strukturen in der Mathematik", Universität Bielefeld, J. Combinatorics, Information & System Sciences, Vol. 17, Nos. 3-4, 203–216.

1993

[75] Asymptotically dense nonbinary codes correcting a constant number of localized errors, (with L.A. Bassalygo and M.S. Pinsker), Proc. III International workshop "Algebraic and Combinatorial Coding Theory", June 22–28, 1992, Tyrnovo, Bulgaria, Comptes rendus de l' Académie bulgare des Sciences, Tome 46, No. 1, 35–37.

[76] The maximal error capacity of AV channels for constant list sizes, IEEE Trans. Inf. Theory, Vol. 39, No. 4, 1416–1417.

[77] Nonbinary codes correcting localized errors, (with L.A. Bassalygo and M.S. Pinsker), IEEE Trans. Inf. Theory, Vol. 39, No. 4, 1413–1416.

[78] Common randomness in information theory and cryptography, Part I: Secret sharing, (with I. Csiszár), IEEE Trans. Inf. Theory, Vol. 39, No. 4, 1121–1132.

[79] A generalization of the AZ identity, (with N. Cai), Combinatorica 13 (3), 241–247.

[80] On partitioning the n–cube into sets with mutual distance 1, (with S.L. Bezrukov, A. Blokhuis, K. Metsch, and G.E. Moorhouse), Applied Math. Lett., Vol. 6, No. 4, 17–19.

[81] Communication complexity in lattices, (with N. Cai and U. Tamm), Applied Math. Lett., Vol. 6, No. 6, 53–58.

[82] Rank formulas for certain products of matrices, (with N. Cai), Preprint 92–014, SFB 343 "Diskrete Strukturen in der Mathematik", Universität Bielefeld, Applicable Algebra in Engineering, Communication and Computing, 2, 1–9.

[83] On extremal set partitions in Cartesian product spaces, (with N. Cai), Preprint 92–034, SFB 343 "Diskrete Strukturen in der Mathematik", Universität Bielefeld, Combinatorics, Probability & Computing 2, 211–220.

1994

[84] Note on the optimal structure of recovering set pairs in lattices: the sandglass conjecture, (with G. Simonyi), Preprint 91–082, SFB 343 "Diskrete Strukturen in der Mathematik", Universität Bielefeld, Discrete Math., 128, 389–394.

[85] On extremal sets without coprimes, (with L.H. Khachatrian), Preprint 93–026, SFB 343 "Diskrete Strukturen in der Mathematik", Universität Bielefeld, Acta Arithmetica, LXVI 1, 89–99.

[86] The maximal length of cloud–antichains, (with L.H. Khachatrian), Preprint 91–116, SFB 343 "Diskrete Strukturen in der Mathematik", Universität Bielefeld, Discrete Mathematics, Vol. 131, 9–15.

[87] The asymptotic behaviour of diameters in the average, (with I. Althöfer), Preprint 91–099, SFB 343 "Diskrete Strukturen in der Mathematik", Universität Bielefeld, J. Combin. Theory, Ser. B, Vol. 61, No. 2, 167–177.

[88] 2–way communication complexity of sum–type functions for one processor to be informed, (with N. Cai), Preprint 91–053, SFB 343 "Diskrete Strukturen in der Mathematik", Universität Bielefeld, Problemy Peredachi Informatsii, Vol. 30, No. 1, 3–12.

[89] Messy broadcasting in networks, (with H.S. Haroutunian and L.H. Khachatrian), Preprint 93–075, SFB 343 "Diskrete Strukturen in der Mathematik", Universität Bielefeld, Special volume in honour of J.L. Massey on occasion of his 60th birthday. Communications and Cryptography (Two sides of one tapestry), editors R.E. Blahut, D.J. Costello, U. Maurer, T. Mittelholzer, Kluwer Acad. Publ., 1994, 13–24.

[90] Binary constant weight codes correcting localized errors and defects, (with L.A. Bassalygo and M.S. Pinsker), Preprint 93–025, SFB 343 "Diskrete Strukturen in der Mathematik", Universität Bielefeld, Probl. Peredachi Informatsii, Vol. 30, No. 2, 10–13 (In Russian); Probl. of Inf. Transmission, 102–104.

[91] On sets of words with pairwise common letter in different positions, (with N. Cai), Preprint 91–050, SFB 343 "Diskrete Strukturen in der Mathematik", Universität Bielefeld, Proc. Colloquium on Extremal Problems for Finite Sets, Visograd, Bolyai Soc. Math. Studies, 3, Hungary, 25–38.

[92] On multi–user write–efficient memories, (with Z. Zhang), IEEE Trans. Inf. Theory, Vol. 40, No. 3, 674–686.

[93] On communication complexity of vector–valued functions, (with N. Cai), Preprint 91–041, SFB 343 "Diskrete Strukturen in der Mathematik", Universität Bielefeld, IEEE Trans. Inf. Theory, Vol. 40, No. 6, 2062–2067.

[94] On partitioning and packing products with rectangles, (with N. Cai), Preprint 93–008, SFB 343 "Diskrete Strukturen in der Mathematik", Universität Bielefeld, Combinatorics, Probability & Computing 3, 429–434.

[95] A new direction in extremal theory for graphs, (with N. Cai and Z. Zhang), J. Combinatorics, Information & System Sciences, Vol. 19, No. 3–4, 269–280.

[96] Asymptotically optimal binary codes of polynomial complexity correcting localized errors, (with L.A. Bassalygo and M.S. Pinsker), Preprint 94–055, SFB 343 "Diskrete Strukturen in der Mathematik", Universität Bielefeld, Proc. IV International workshop on Algebraic and Combinatorial Coding Theory, Novgorod, Russia, 1–3.

1995

[97] Localized random and arbitrary errors in the light of AV channel theory, (with L.A. Bassalygo and M.S. Pinsker), Preprint 93–036, SFB 343 "Diskrete Strukturen in der Mathematik", Universität Bielefeld, IEEE Trans. Inf. Theory, Vol. 41, No. 1, 14–25.

[98] Edge isoperimetric theorems for integer point arrays, (with S.L. Bezrukov), Preprint 94–067, SFB 343 "Diskrete Strukturen in der Mathematik", Universität Bielefeld, Applied Math. Letters, Vol. 8, No. 2, 75–80.

[99] New directions in the theory of identification via channels, (with Z. Zhang), Preprint 94–010, SFB 343 "Diskrete Strukturen in der Mathematik, Universität Bielefeld, IEEE Trans. Inf. Theory, Vol. 41, No. 4, 1040–1050.

[100] Towards characterising equality in correlation inequalities, (with L.H. Khachatrian), Preprint 93–027, SFB 343 "Diskrete Strukturen in der Mathematik", Universität Bielefeld, European J. of Combinatorics 16, 315–328.

[101] Maximal sets of numbers not containing $k+1$ pairwise coprime integers, (with L.H. Khachatrian), Preprint 94–080, SFB 343 "Diskrete Strukturen in der Mathematik", Universität Bielefeld, Acta Arithmetica LXX II, 1, 77–100.

[102] Density inequalities for sets of multiples, (with L.H. Khachatrian), Preprint 93–049, SFB 343 "Diskrete Strukturen in der Mathematik", Universität Bielefeld, J. of Number Theory, Vol. 55, No. 2., 170–180.

[103] A splitting property of maximal antichains, (with P.L. Erdös and N. Graham), Preprint 94–048, SFB 343 "Diskrete Strukturen in der Mathematik", Universität Bielefeld, Combinatorica 15 (4), 475–480.

1996

[104] Sets of integers and quasi–integers with pairwise common divisor, (with L. Khachatrian), Acta Arithmetica, LXXIV.2, 141–153.

[105] A counterexample to Aharoni's "Strongly maximal matching" conjecture, (with L.H. Khachatrian), included in "Report on work in progress in combinatorial extremal theory", *Ergänzungsreihe* des SFB 343 "Diskrete Strukturen in der Mathematik", Universität Bielefeld, Nr. 95–004, Discrete Mathematics 149, 289.

[106] Erasure, list, and detection zero–error capacities for low noise and a relation to identification, (with N. Cai and Z. Zhang), Preprint 93–068, SFB 343 "Diskrete Strukturen in der Mathematik", Universität Bielefeld, IEEE Trans. Inf. Theory, Vol. 42, No. 1, 55–62.

[107] Optimal pairs of incomparable clouds in multisets, (with L.H. Khachatrian), Preprint 93–043, SFB 343 "Diskrete Strukturen in der Mathematik", Universität Bielefeld, Graphs and Combinatorics 12, 97–137.

[108] Sets of integers with pairwise common divisor and a factor from a specified set of primes, (with L.H. Khachatrian), Preprint 95–059, SFB 343 "Diskrete Strukturen in der Mathematik", Universität Bielefeld, Acta Arithmetica LXX V 3, 259–276, 1996.

[109] Cross–disjoint pairs of clouds in the interval lattice, (with N. Cai), Preprint 93–038, SFB 343 "Diskrete Strukturen in der Mathematik", Universität Bielefeld, The Mathematics of Paul Erdös, Vol. I; R.L. Graham and J. Nesetril, ed., Algorithms and Combinatorics B, Springer Verlag, 155–164.

[110] Identification under random processes, (with V. Balakirsky), Preprint 95–098, SFB 343 "Diskrete Strukturen in der Mathematik", Universität Bielefeld, Problemy peredachii informatsii (special issue devoted to M.S. Pinsker), vol. 32, no. 1, 144–160, Jan.–March 1996; Problems of Information Transmission, Vol. 32, No. 1, 123–138, 1996.

[111] On common information and related characteristics of correlated information sources, (with J. Körner), *Ergänzungsreihe* des SFB 343 "Diskrete Strukturen in der Mathematik", Universität Bielefeld, Nr. 95–003.

[112] Report on work in progress in combinatorial extremal theory: Shadows, AZ–identity, matching. *Ergänzungsreihe* des SFB 343 "Diskrete Strukturen in der Mathematik", Universität Bielefeld, Nr. 95–004.

[113] Fault–tolerant minimum broadcast networks, (with L. Gargano, H.S. Haroutunian, and L.H. Khachatrian), Preprint 94–032, SFB 343 "Diskrete Strukturen in der Mathematik", Universität Bielefeld, Networks, Vol. 27, No. 4, 1293–1307.

[114] The complete nontrivial–intersection theorem for systems of finite sets, Preprint 95–102, SFB 343 "Diskrete Strukturen in der Mathematik", Universität Bielefeld, J. Combin. Theory, Ser. A, 121–138.

[115] Incomparability and intersection properties of Boolean interval lattices and chain posets, (with N. Cai), Preprint 93–037, SFB 343 "Diskrete Strukturen in der Mathematik", Universität Bielefeld, European J. of Combinatorics 17, 677–687.

[116] Classical results on primitive and recent results on cross–primitive sequences, (with L.J. Khachatrian), Preprint 93–042, SFB 343 "Diskrete Strukturen in der Mathematik", Universität Bielefeld, The Mathematics of P. Erdös, Vol. I; R.L. Graham and J. Nesetril, ed., Algorithms and Combinatorics B, Springer Verlag, 104–116.

[117] Intersecting Systems, (with N. Alon, P.L. Erdös, M. Ruszinko, L.A. Székely), Combinatorics, Probability and Computing 6, 127–137.

[118] Some properties of fix–free codes, (with B. Balkenhol and L.H. Khachatrian), Proceedings First INTAS International Seminar on Coding Theory and Combinatorics, Thahkadzor, Armenia, 20–33, 6–11 October 1996.

[119] Higher level extremal problems, (with N. Cai and Z. Zhang), Preprint 92–031, SFB 343 "Diskrete Strukturen in der Mathematik", Universität Bielefeld, Comb. Inf. & Syst. Sc., Vol. 21, No. 3–4, 185–210.

1997

[120] On interactive communication, (with N. Cai and Z. Zhang), Preprint 93–066, SFB 343 "Diskrete Strukturen in der Mathematik", Universität Bielefeld, IEEE Trans. on Inf. Theory, Vol. 43, No. 1, 22–37.

[121] Identification via compressed data, (with E. Yang and Z. Zhang), Preprint 95–007, SFB 343 "Diskrete Strukturen in der Mathematik", Universität Bielefeld, IEEE Trans. Inf. Theory, Vol. 43, No. 1, 48–70.

[122] The complete intersection theorem for systems of finite sets, (with L.H. Khachatrian), Preprint 95–066, SFB 343 "Diskrete Strukturen in der Mathematik", European J. Combinatorics, 18, 125–136.

[123] Universal coding of integers and unbounded search trees, (with T.S. Han and K. Kobayashi), Preprint 95–001, SFB 343 "Diskrete Strukturen in der Mathematik", Universität Bielefeld, Trans. Inf. Theory, Vol. 43, No. 2, 669–682.

[124] Number theoretic correlation inequalities for Dirichlet densities, (with L.H. Khachatrian), Preprint 93–060, SFB 343 "Diskrete Strukturen in der Mathematik", Universität Bielefeld, J. Number Theory, Vol. 63, No. 1, 34–46.

[125] General edge–isoperimetric inequalities, Part 1: Information theoretical methods, (with Ning Cai), Preprint 94–090, SFB 343 "Diskrete Strukturen in der Mathematik", Universität Bielefeld, European J. of Combinatorics 18, 355–372.

[126] General edge–isoperimetric inequalities, Part 2: A local–global principle for lexicographical solutions, (with Ning Cai), Preprint 94–090, SFB 343 "Diskrete Strukturen in der Mathematik", Universität Bielefeld, European J. of Combinatorics 18, 479–489.

[127] Models of multi–user write–efficient memories and general diametric theorems, (with N. Cai), Preprint 93–019, SFB 343 "Diskrete Strukturen in der Mathematik", Universität Bielefeld, Information and Computation, Vol. 135, No. 1, 37–67.

[128] Shadows and isoperimetry under the sequence–subsequence relation, (with N. Cai), Preprint 95-045, SFB 343 "Diskrete Strukturen in der Mathematik", Universität Bielefeld, Combinatorica 17 (1), 11–29.

[129] Counterexample to the Frankl/Pach conjecture for uniform, dense families, (with L.H. Khachatrian), Preprint 95–114, SFB 343 "Diskrete Strukturen in der Mathematik", Universität Bielefeld, Combinatorica 17 (2), 299–301.

[130] Correlated sources help the transmission over AVC, (with N. Cai), Preprint 95–106, SFB 343 "Diskrete Strukturen in der Mathematik", Universität Bielefeld, IEEE Trans. Inf. Theory, Vol. 135, No. 1, 37–67.

1998

[131] Common randomness in Information Theory and Cryptography, Part II: CR capacity, (with I. Csiszár), Preprint 95–101, SFB 343 "Diskrete Strukturen in der Mathematik", Universität Bielefeld, IEEE Trans. Inf. Theory, Vol. 44, No. 1, 55–62.

[132] The diametric theorem in Hamming spaces – optimal anticodes, (with L.H. Khachatrian) Preprint 96–013, SFB 343 "Diskrete Strukturen in der Mathematik", Universität Bielefeld, Proceedings First INTAS International Seminar on Coding Theory and Combinatorics 1996, Thahkadzor, Armenia, 1-19, 6-11 October 1996; Advances in Applied Mathematics 20, 429-449.

[133] Information and Control: Matching channels, (with N. Cai), Preprint 95–035, SFB 343 "Diskrete Strukturen in der Mathematik", Universität Bielefeld, IEEE Trans. Inf. Theory, Vol. 44, No. 2, 542–563.

[134] Zero–error capacity for models with memory and the enlightened dictator channel, (with N. Cai and Z. Zhang), IEEE Trans. Inf. Theory, Vol. 44, No. 3, 1250–1252.

[135] Code pairs with specified parity of the Hamming distances, (with Z. Zhang), Preprint 96–058, SFB 343 "Diskrete Strukturen in der Mathematik", Universität Bielefeld, Discrete Mathematics 188, 1–11.

[136] Isoperimetric theorems in the binary sequences of finite lengths, (with Ning Cai), submitted to Applied Math. Letters, Vol. 11, No. 5, 121–126.

[137] The intersection theorem for direct products, (with H. Aydinian and L.H. Khachatrian), Preprint 97–051, SFB 343 "Diskrete Strukturen in der Mathematik", Universität Bielefeld, European J. of Combinatorics 19, 649–661.

1999

[138] Construction of uniquely decodable codes for the two–user binary adder channel, (with V.B. Balakirsky), Preprint 97–016, SFB 343 "Diskrete Strukturen in der Mathematik", IEEE Trans. Inf. Theory, Vol 45, No. 1, 326–330.

[139] Arbitrarily varying multiple–access channels, Part I. Ericson's symmetrizability is adequate, Gubner's conjecture is true, (with N. Cai), Preprint 96–068, SFB 343 "Diskrete Strukturen in der Mathematik", Universität Bielefeld, IEEE Trans. Inf. Theory, Vol. 45, No. 2, 742–749.

[140] Arbitrarily varying multiple–access channels, Part II. Correlated sender's side information, correlated messages, and ambiguous transmission, (with N. Cai), Preprint 97–006, SFB 343 "Diskrete Strukturen in der Mathematik", Universität Bielefeld, IEEE Trans. Inf. Theory, Vol. 45, No. 2, 749–756.

[141] A pushing–pulling method: new proofs of intersection theorems, (with L.H. Khachatrian), Preprint 97–043, SFB 343 "Diskrete Strukturen in der Mathematik", Universität Bielefeld, Combinatorica 19(1), 1–15.

[142] A counterexample in rate–distortion theory for correlated sources, (with Ning Cai), Preprint 97–034, SFB 343 "Diskrete Strukturen in der Mathematik", Universität Bielefeld, Applied Math. Letters, 12, No. 7, 1–3.

[143] Identification without randomization, (with Ning Cai), Preprint 98–075, SFB 343 "Diskrete Strukturen in der Mathematik", Universität Bielefeld, IEEE Trans. Inf. Theory 45, No. 7, 2636–2642

[144] On prefix–free and suffix–free sequences of integers, (with L.H. Khachatrian and A. Sárközy), Special volume in honour of R. Ahlswede on occasion of his 60th birthday, editors I. Althöfer, N. Cai, G. Dueck, L. Khachatrian, M. Pinsker, A. Sárközy, I. Wegener, and Z. Zhang, Kluwer Acad. Publ., this volume.

[145] Splitting properties in partially ordered sets and set systems, (with L.H. Khachatrian), Preprint 94–071, SFB 343 "Diskrete Strukturen in der Mathematik", Universität Bielefeld, Special volume in honour of R. Ahlswede on occasion of his 60th birthday, editors I. Althöfer, N. Cai, G. Dueck, L. Khachatrian, M. Pinsker, A. Sárközy, I. Wegener, and Z. Zhang, Kluwer Acad. Publ., this volume.

[146] The AVC with noiseless feedback and maximal error probability: A capacity formula with a trichotomy, (with N. Cai), Preprint 96–064, SFB 343 "Diskrete Strukturen in der Mathematik", Universität Bielefeld, Special volume in honour of R. Ahlswede on occasion of his 60th birthday, editors I. Althöfer, N. Cai, G. Dueck, L. Khachatrian, M. Pinsker, A. Sárközy, I. Wegener, and Z. Zhang, Kluwer Acad. Publ., this volume.

to appear

[147] A counterexample to Kleitman's conjecture concerning an edge–isoperimetric problem, (with Ning Cai), Combinatorics, Probability and Computing ...

[148] On maximal shadows of members in left–compressed sets, (with Zhen Zhang), Preprint 97-026, SFB 343 "Diskrete Strukturen in der Mathematik", Universität Bielefeld, Proceedings of the Rostock Conference, Discrete Applied Math. ...

[149] Network information flow: single source, (with Ning Cai, S.Y. Robert Li, and Raymond W. Yeung), Preprint 98–033, SFB 343 "Diskrete Strukturen in der Mathematik", Universität Bielefeld, IEEE Trans. Inf. Theory ...

[150] On the counting function for primitive sets of integers, (with L.H. Khachatrian and A. Sárközy), Preprint 98–077, SFB 343 "Diskrete Strukturen in der Mathematik", Universität Bielefeld, J. Number Theory ...

[151] On the Hamming bound for nonbinary localized–error–correcting codes, (with L.A. Bassalygo and M.S. Pinsker), Preprint 99-077, SFB 343, Diskrete Strukturen in der Mathematik, Universität Bielefeld, Problemy Per. Informatsii ...

[152] A diametric theorem for edges, (with L.H. Khachatrian), Preprint 97–100, SFB 343 "Diskrete Strukturen in der Mathematik", Universität Bielefeld, J. Comb. Theory. ...

[153] On perfect codes and related concepts, (with H. Aydinian and L.H. Khachatrian), Preprint 98–080, SFB 343 "Diskrete Strukturen in der Mathematik", Universität Bielefeld, Designs, Codes and Cryptography

[154] On the quotient sequence of sequences of integers, (with L.H. Khachatrian and A. Sárkozy), Preprint 98–068, SFB 343 "Diskrete Strukturen in der Mathematik", Universität Bielefeld, Acta Arithmetica ...

submitted

[155] Worst case estimation of permutation invariant functions and identification via compressed data, (with Zhen Zhang), Preprint 97–005, SFB 343 "Diskrete Strukturen in der Mathematik", Universität Bielefeld, submitted to IEEE Trans. Inf. Theory.

[156] General theory of information transfer, Preprint 97–118, SFB 343 "Diskrete Strukturen in der Mathematik", Universität Bielefeld, submitted to IEEE Trans. Inf. Theory.

[157] Quantum data processing, (with Peter Löber), Preprint 99-087, SFB 343 "Diskrete Strukturen in der Mathematik", Universität Bielefeld, submitted to IEEE Trans. Inf. Theory.

[158] Maximal number of constant weight vertices of the unit n–cube contained in a k–dimensional subspace, (with H. Aydinian and L. Khachatrian), submitted to Combinatorica, special issue in honour of P. Erdös.

[159] On primitive sets of squarefree integers, (with L. Khachatrian and A. Särközy), submitted to special volume on number theory in honour of A. Särközy, edited by Periodica Mathematica Hungarica.

[160] Concept of performance parameters for channels, submitted to Problemy Per. Informacii.

SFB 343

Sharp bounds for cloud–antichains of length two, (with L.H. Khachatrian), Preprint 92–012, SFB 343 "Diskrete Strukturen in der Mathematik", Universität Bielefeld, included in [103].

On edge–isoperimetric theorems for uniform hypergraphs, (with N. Cai), Preprint 93–018, SFB 343 "Diskrete Strukturen in der Mathematik", Universität Bielefeld.

A simple proof of the Hook formula by a staircase identity, (with K. Kobayashi), Preprint 95–013, SFB 343 "Diskrete Strukturen in der Mathematik", Universität Bielefeld.

Report on models of write–efficient memories with localized errors and defects, (with M.S. Pinsker), Preprint 97–004 (Ergänzungsreihe), SFB 343 "Diskrete Strukturen in der Mathematik", Universität Bielefeld.